T0290862

Computational Aerodynamics

During the last five decades, computational aerodynamics has emerged as an engineering field that creates new horizons for aerodynamic simulation via sophisticated numerical algorithms. This book is a unique reference on the foundations, methods, and applications of this rapidly developing field. It is designed for a wide audience, from graduate students to experienced researchers and professionals in the aerospace engineering field. The book opens with a presentation of the essential elements of computational aerodynamics, including the relevant mathematical methods of fluid flow and numerical methods for partial differential equations. The introductory chapters are followed by a comprehensive presentation of stability theory, shock capturing schemes, viscous flow, and time integration. The final chapters treat more advanced topics. The book is a rich source of information that is essential for further fundamental research, with applications in aeronautics and a wide range of other fields, such as automotive design, wind turbines, and astrophysics.

Anthony Jameson is widely known as a founder of computational aerodynamics. His work has had a major impact on aircraft design, and he continues to be a prolific researcher in the field. He is currently the Jack E. & Francis Brown Chair in Engineering and Professor of Aerospace Engineering and Ocean Engineering at Texas A & M University. He formerly held chair professorships at Princeton University and Stanford University.

Cambridge Aerospace Series

Editors: Wei Shyy and Vigor Yang

1. J. M. Rolfe and K. J. Staples (eds.): *Flight Simulation*
2. P. Berlin: *The Geostationary Applications Satellite*
3. M. J. T. Smith: *Aircraft Noise*
4. N. X. Vinh: *Flight Mechanics of High-Performance Aircraft*
5. W. A. Mair and D. L. Birdsall: *Aircraft Performance*
6. M. J. Abzug and E. E. Larrabee: *Airplane Stability and Control*
7. M. J. Sidi: *Spacecraft Dynamics and Control*
8. J. D. Anderson: *A History of Aerodynamics*
9. A. M. Cruise, J. A. Bowles, C. V. Goodall, and T. J. Patrick: *Principles of Space Instrument Design*
10. G. A. Khoury (ed.): *Airship Technology, Second Edition*
11. J. P. Fielding: *Introduction to Aircraft Design*
12. J. G. Leishman: *Principles of Helicopter Aerodynamics, Second Edition*
13. J. Katz and A. Plotkin: *Low-Speed Aerodynamics, Second Edition*
14. M. J. Abzug and E. E. Larrabee: *Airplane Stability and Control: A History of the Technologies that Made Aviation Possible, Second Edition*
15. D. H. Hodges and G. A. Pierce: *Introduction to Structural Dynamics and Aeroelasticity, Second Edition*
16. W. Fehse: *Automatic Rendezvous and Docking of Spacecraft*
17. R. D. Flack: *Fundamentals of Jet Propulsion with Applications*
18. E. A. Baskharone: *Principles of Turbomachinery in Air-Breathing Engines*
19. D. D. Knight: *Numerical Methods for High-Speed Flows*
20. C. A. Wagner, T. Hüttl, and P. Sagaut (eds.): *Large-Eddy Simulation for Acoustics*
21. D. D. Joseph, T. Funada, and J. Wang: *Potential Flows of Viscous and Viscoelastic Fluids*
22. W. Shyy, Y. Lian, H. Liu, J. Tang, and D. Viieru: *Aerodynamics of Low Reynolds Number Flyers*
23. J. H. Saleh: *Analyses for Durability and System Design Lifetime*
24. B. K. Donaldson: *Analysis of Aircraft Structures, Second Edition*
25. C. Segal: *The Scramjet Engine: Processes and Characteristics*
26. J. F. Doyle: *Guided Explorations of the Mechanics of Solids and Structures*
27. A. K. Kundu: *Aircraft Design*
28. M. I. Friswell, J. E. T. Penny, S. D. Garvey, and A. W. Lees: *Dynamics of Rotating Machines*
29. B. A. Conway (ed): *Spacecraft Trajectory Optimization*
30. R. J. Adrian and J. Westerweel: *Particle Image Velocimetry*
31. G. A. Flandro, H. M. McMahon, and R. L. Roach: *Basic Aerodynamics*
32. H. Babinsky and J. K. Harvey: *Shock Wave-Boundary-Layer Interactions*
33. C. K. W. Tam: *Computational Aeroacoustics: A Wave Number Approach*
34. A. Filippone: *Advanced Aircraft Flight Performance*
35. I. Chopra and J. Sirohi: *Smart Structures Theory*
36. W. Johnson: *Rotorcraft Aeromechanics vol. 3*
37. W. Shyy, H. Aono, C. K. Kang, and H. Liu: *An Introduction to Flapping Wing Aerodynamics*
38. T. C. Lieuwen and V. Yang: *Gas Turbine Emissions*

39. P. Kabamba and A. Girard: *Fundamentals of Aerospace Navigation and Guidance*
40. R. M. Cummings, W. H. Mason, S. A. Morton, and D. R. McDaniel: *Applied Computational Aerodynamics*
41. P. G. Tucker: *Advanced Computational Fluid and Aerodynamics*
42. Iain D. Boyd and Thomas E. Schwartzentruber: *Nonequilibrium Gas Dynamics and Molecular Simulation*
43. Joseph J. S. Shang and Sergey T. Surzhikov: *Plasma Dynamics for Aerospace Engineering*
44. Bijay K. Sultanian: *Gas Turbines: Internal Flow Systems Modeling*
45. J. A. Kéchichian: *Applied Nonsingular Astrodynamics: Optimal Low-Thrust Orbit Transfer*
46. J. Wang and L. Feng: *Flow Control Techniques and Applications*
47. D. D. Knight: *Energy Disposition for High-Speed Flow Control*
48. Eric Silk: *Introduction to Spacecraft Thermal Design*
49. Antony Jameson: *Computational Aerodynamics*

Computational Aerodynamics

ANTONY JAMESON
Texas A & M University

CAMBRIDGE
UNIVERSITY PRESS

CAMBRIDGE
UNIVERSITY PRESS

University Printing House, Cambridge CB2 8BS, United Kingdom

One Liberty Plaza, 20th Floor, New York, NY 10006, USA

477 Williamstown Road, Port Melbourne, VIC 3207, Australia

314–321, 3rd Floor, Plot 3, Splendor Forum, Jasola District Centre, New Delhi – 110025, India

103 Penang Road, #05–06/07, Visioncrest Commercial, Singapore 238467

Cambridge University Press is part of the University of Cambridge.

It furthers the University's mission by disseminating knowledge in the pursuit of education, learning, and research at the highest international levels of excellence.

www.cambridge.org
Information on this title: www.cambridge.org/9781108837880
DOI: 10.1017/9781108943345

First published 2022

Printed in the United Kingdom by TJ Books Limited, Padstow Cornwall

A catalogue record for this publication is available from the British Library.

ISBN 978-1-108-83788-0 Hardback

The book is dedicated to my wife, Charlotte Ansted-Jameson, without whose persistent encouragement and assistance the book could not have been completed.

Contents

Preface

This book is an attempt both to elucidate the ideas that form the core of the new discipline of computational fluid dynamics (CFD) and to provide a foundation for further development of the discipline and of computational mechanics in general. It is particularly focused on the application of CFD in aeronautical science. Classical aerodynamic theory achieved remarkable results as exemplified in the works of Lamb and Caflisch (1993), Prandtl and Tietjens (1957), and Glauert (1926) but could not treat inherently nonlinear phenomena such as transonic flow. It was the need to predict compressible flow with shock waves that spurred the development of entirely new approaches, such as characteristic-based upwind difference methods, which have spread out to a broader sphere, so that CFD is now ubiquitous in many industries and scientific fields ranging from astrophysics to computational biology.

The text could be used for a graduate course in CFD, but it is also designed to be self-contained and to provide enough information for a research scientist or engineer who might need to engage in further developments. Some of the ideas presented in the book date back to the early development of numerical methods by giants such as Newton, Gauss, and Lagrange. The rapid development of CFD as a new discipline, however, has been accomplished during the last five decades by the combined efforts of many distinguished mathematicians and scientists. The text has also been written with the intent of providing a historical record of some of the key contributions.

The book is laid out as follows. Mathematical models of fluid flows at different levels of fidelity and complexity, ranging from ideal potential flow to the Navier–Stokes equations, are reviewed in Chapter 2, which also presents a detailed analysis for the Jacobian matrices for the equations of gas dynamics. Chapter 3 provides an overview of numerical methods for solving partial differential equations (PDEs), including the three main branches of finite difference, finite element, and finite volume methods. Chapter 4 reviews fundamental stability theory. Chapter 5 presents the basic theory of shock capturing schemes for scalar conservation laws, while Chapter 6 presents the extension of the theory to treat gas dynamics. Chapter 7 presents discretization schemes for flow in complex domains for both structured and unstructured grids. Simulation methods for viscous turbulent flows are reviewed in Chapter 8, including the Reynolds-averaged Navier–Stokes equations (RANS), the Favre-averaged Navier–Stokes equations (FANS), and large eddy simulation (LES). Chapters 9–11 review time integration methods, beginning with an overview in Chapter 9, followed by a more detailed analysis of methods for steady state problems in Chapter 10, and time-accurate methods for unsteady flow simulation in Chapter 11. Chapter 12 reviews concepts of energy stability, including kinetic energy conserving and entropy stable schemes. Chapters 13 and 14 give a brief introduction to high order methods for structured and unstructured meshes.

While computational methods enable engineers to predict the aerodynamics of alternative designs, the question still arises of how to modify a proposed design to improve performance. Chapter 15 addresses the issue of optimum aerodynamic design and describes how this may be achieved via methods based on control theory, which require the solution of a set of adjoint equations in addition to the flow equations.

A review of the mathematical concepts that play an essential role in the formulation of numerical methods for solving PDEs is provided in Appendices A, B, and C, covering vector and function spaces, approximation theory, polynomial interpolation, differentiation, and integration. Potential flow models are reviewed in Appendix D because of their historical importance for early development of CFD, including the formulation of upwind difference methods. Appendix E provides some supplementary notes on the construction of L_∞ stable schemes for conservation laws.

Space limitations preclude the discussion of a number of topics that are the subject of ongoing research. These include the extension of high order methods to treat advection diffusion problems and the Navier–Stokes equations, and also a posteriori error estimation methods, output error estimation methods via adjoint equations, and systematic mesh refinement (AMR) of both mesh spacing and polynomial order (h and p refinement, or hp refinement when both are used in conjunction). As research in these areas unfolds further, a full discussion of these topics could be the subject of another text.

Acknowledgments

My research has benefited greatly from close collaborations with practicing engineers, mathematicians, and scientists, and also with numerous doctoral students and post-doctoral associates, as can be seen from my publications. During the early part of my career, my research was facilitated by the support of NASA and the Office of Naval Research. During the last two decades, it has been facilitated by the continuing support of the National Science Foundation and the Air Force Office of Scientific Research. The attempt to incorporate my class notes into a book was first initiated by Kui Ou. Subsequently Andre Chan took over the task of converting my handwritten notes to LaTeX. I am deeply grateful to Luigi Martinelli for his invaluable help with the many revisions of the original draft into the final text.

1 Introduction and Background

1.1 Introduction

During the last few decades, computational fluid dynamics (CFD) has emerged as an indispensable tool for aerodynamic analysis and design. Recently, however, it has been on a plateau where attached smooth flows can be very accurately predicted, but simulations of separated and turbulent flows continue to be subject to a significant level of uncertainty. In the context of aircraft design, we can now rely on CFD to predict the performance of a commercial aircraft at its design condition for long range cruise but not to predict its behavior at off design conditions such as stall, nor to accurately predict the performance of its high lift system. The very rapid strides currently being made in high performance computing systems are now presenting us with the opportunity to move CFD to an entirely new level within the near future. It will soon be feasible to perform large eddy simulations (LES) in the proper Reynolds number range for hitherto intractable problems of turbulent flows over realistic configurations.

Looming further downstream is the prospect of using direct numerical simulation (DNS) to resolve the full range of turbulent scales in industrial applications. This book aims to expose the kind of thinking that has brought CFD to its present level and will be needed to advance it to the new levels of fidelity that are now within reach.

The main focus of the book is the design and analysis of numerical algorithms for computational aerodynamics. Applications in aeronautical sciences, in particular the need to predict transonic and supersonic flows over complex configurations, have played a key role in driving the emergence of CFD as a distinct discipline during the past 50 years. The importance of transonic flow stems from two reasons. First, it is the most efficient regime for long range flight of jet aircraft. Second, due to its inherent nonlinearity, it has proved intractable to solution by analytical methods. Accordingly, once sufficiently powerful computers became available around 1970–1980, a resort to numerical solution offered the only route forward. It was soon discovered, however, that classical numerical methods were unable to produce acceptable results for flows with shock waves, generally exhibiting spurious oscillations of large amplitude. These difficulties were resolved through the development of a variety of high resolution shock capturing algorithms, which are a major subject of this text. While these techniques had their birth in aeronautical science, they are applicable to systems of nonlinear conservation laws in general, and they have successfully transferred to other branches of computational science, most notably to numerical simulations in astrophysics.

1.2 Focus and Historical Background

1.2.1 Classical Aerodynamics

This chapter surveys some of the principal developments of computational aerodynamics, with a focus on aeronautical applications. It is written with the perspective that computational mathematics is a natural extension of classical methods of applied mathematics, which has enabled the treatment of more complex, in particular nonlinear, mathematical models, and also the calculation of solutions in very complex geometric domains not amenable to classical techniques such as the separation of variables.

This is particularly true for aerodynamics. Efficient flight can be achieved only by establishing highly coherent flows. Consequently there are many important applications where it is not necessary to solve the full Navier–Stokes equations in order to gain an insight into the nature of the flow, and useful predictions can be made with simplified mathematical models. It was already recognized by Prandtl in 1904 (Prandtl 1904, Schlichting & Gersten 1999), essentially contemporaneous with the first successful flights of the Wright brothers, that in flows at the large Reynolds numbers typical of powered flight, viscous effects are important chiefly in thin shear layers adjacent to the surface. While these boundary layers play a critical role in determining whether the flow will separate and how much circulation will be generated around a lifting surface, the equations of inviscid flow are a good approximation in the bulk of the flow field external to the boundary layer. In the absence of separation, a first estimate of the effect of the boundary layer is provided by regarding it as increasing the effective thickness of the body. This procedure can be justified by asymptotic analysis (Van Dyke 1964, Ashley & Landahl 1985).

The classical treatment of the external inviscid flow is based on Kelvin's theorem that in the absence of discontinuities, the circulation around a material loop remains constant. Consequently an initially irrotational flow remains irrotational. This allows us to simplify the equations further by representing the velocity as the gradient of a potential. If the flow is also regarded as incompressible, the governing equation reduces to Laplace's equation. These simplifications provided the basis for the classical airfoil theory of Joukowsky (Glauert 1926) and Prandtl's wing theory (Prandtl & Tietjens 1957, Ashley & Landahl 1985). Supersonic flow over slender bodies at Mach numbers greater than two is also well represented by the linearized equations. Techniques for the solution of linearized flow were perfected in the period 1935–1950, particularly by W.D. Hayes, who derived the supersonic area rule (Hayes 1947).

Classical aerodynamic theory provided engineers with a good insight into the nature of the flow phenomena and a fairly good estimate of the force on simple configurations such as an isolated wing, but could not predict the details of the flow over the complex configuration of a complete aircraft. Consequently, the primary tool for the development of aerodynamic configurations was the wind tunnel. Shapes were tested and modifications selected in the light of pressure and force measurements together with flow visualization techniques. In much the same way that Michelangelo, della Porta, and Fontana could design the dome of St. Peter's through a good physical understanding of stress paths, so could experienced aerodynamicists arrive at efficient shapes through testing guided by good physical insight. Notable examples of the power of this method include the achievement of the Wright brothers in leaving the ground (after first building a wind tunnel) and more recently Whitcomb's discovery of the area rule for transonic flow, followed by his development of aft-loaded supercritical airfoils and winglets (Whitcomb 1956, 1974, 1976). The process was expensive. More than 20,000 hours of wind tunnel testing were expended in the development of some modern designs, such as the Boeing 747.

1.2.2 The Emergence of Computational Aerodynamics and its Application to Transonic Flow

Prior to 1960, computational methods were hardly used in aerodynamic analysis, although they were already widely used for structural analysis. The NACA 6 series of airfoils had been developed during the forties, using hand computation to implement the Theodorsen method for conformal mapping (Theodorsen 1931). The first major success in computational aerodynamics was the introduction of boundary integral methods by Hess and Smith (1962) to calculate potential flow over an arbitrary configuration. Generally known in the aeronautical community as panel methods, these continue to be used to the present day to make initial predictions of low speed aerodynamic characteristics of preliminary designs. It was the compelling need, however, both to predict transonic flow and to gain a better understanding of its properties and character, that was a driving force for the development of computational aerodynamics through the period 1970–1990.

In the case of military aircraft capable of supersonic flight, the high drag associated with high g maneuvers forces them to be performed in the transonic regime. In the case of commercial aircraft, the importance of transonic flow stems from the Breguet range equation. This provides a good first estimate of range as

$$R = \frac{V}{sfc}\frac{L}{D}\log\frac{W_0 + W_f}{W_0}.$$

Here V is the speed, L/D is the lift to drag ratio, sfc is the specific fuel consumption of the engines, W_0 is the landing weight, and W_f is the weight of the fuel burnt. The Breguet equation clearly exposes the multi-disciplinary nature of the design problem. A lightweight structure is needed to minimize W_0. The specific fuel consumption is mainly the province of the engine manufacturers, and in fact the largest advances during the last thirty years have been in engine efficiency. The aerodynamic designer should try to maximize VL/D. This means that the cruising speed should be increased until the onset of drag-rise due to the formation of shock waves. Consequently the best

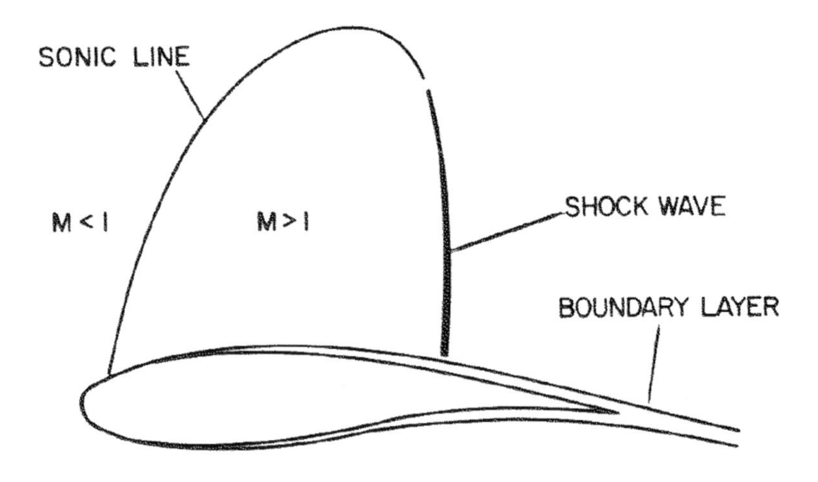

Figure 1.1 Transonic flow past an airfoil.

cruising speed is in the transonic regime. The typical pattern of transonic flow over a wing section is illustrated in Figure 1.1.

Transonic flow had proved essentially intractable to analytic methods. Garabedian and Korn had demonstrated the feasibility of designing airfoils for shock-free flow in the transonic regime numerically by the method of complex characteristics (Bauer, Garabedian, & Korn 1972). Their method was formulated in the hodograph plane, and it required great skill to obtain solutions corresponding to physically realizable shapes. It was also known from Morawetz's theorem (Morawetz 1956) that shock free transonic solutions are isolated points.

A major breakthrough was accomplished by Murman and Cole (1971) with their development of type-dependent differencing in 1970. They obtained stable solutions by simply switching from central differencing in the subsonic zone to upwind differencing in the supersonic zone and using a line-implicit relaxation scheme. Their discovery provided major impetus for the further development of CFD by demonstrating that solutions for steady transonic flows could be computed economically. Figure 1.2, taken from their landmark paper, illustrates the scaled pressure distribution on the surface of a symmetric airfoil. Efforts were soon underway to extend their ideas to more general transonic flows.

Numerical methods to solve transonic potential flow over complex configurations were essentially perfected during the period 1970–1982. The AIAA First Computational Fluid Dynamics Conference, held in Palm Springs in July 1973, signified the emergence of computational fluid dynamics (CFD) as an accepted tool for airplane design and seems to mark the first use of the name CFD. The rotated difference scheme for transonic potential flow, first introduced by the author at this conference,

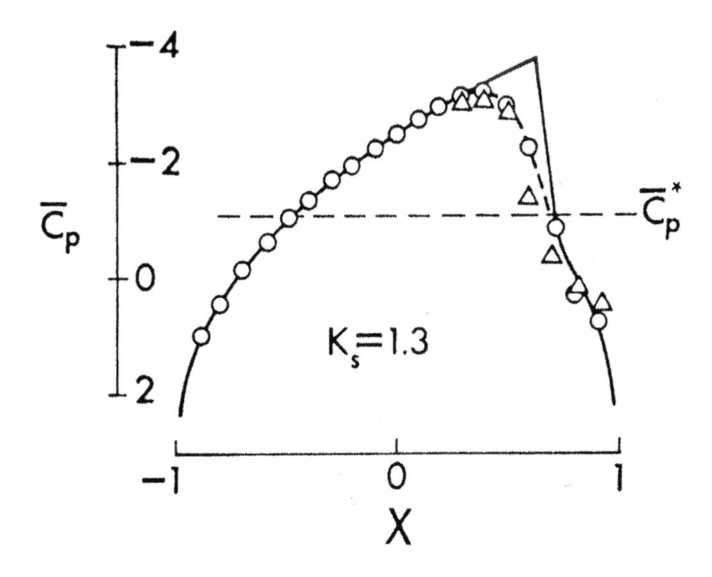

Figure 1.2 Scaled pressure coefficient on surface of a thin, circular-arc airfoil in transonic flow, compared with experimental data; solid line represents computational result.

proved to be a very robust method, and it provided the basis for the computer program FLO22, developed with David Caughey during 1974–1975 to predict transonic flow past swept wings. At the time we were using the CDC 6600, which had been designed by Seymour Cray and was the world's fastest computer at its introduction, but it had only 131,000 words of memory. This forced the calculation to be performed one plane at a time, with multiple transfers from the disk. FLO22 was immediately put into use at McDonnell Douglas. A simplified in-core version of FLO22 is still in use at Boeing today. Figure 1.3, supplied by John Vassberg, shows the result of a recent calculation using FLO22 of transonic flow over the wing of a proposed aircraft to fly in the Martian atmosphere. The result was obtained with 100 iterations on a $192 \times 32 \times 32$ mesh in 7 seconds, using a typical modern workstation. When FLO22 was first introduced at Long Beach, the calculations cost $3,000 for each run. Nevertheless, they found it worthwhile to use it extensively for the aerodynamic design of the C17 military cargo aircraft.

In order to treat complete configurations, it was necessary to develop discretization formulas for arbitrary grids. An approach that proved successful (Jameson & Caughey 1977) is to derive the discretization formulas from the Bateman variational principle that the integral of the pressure over the domain,

$$I = \int_D p d\xi,$$

is stationary (Jameson 1978). The resulting scheme is essentially a finite element scheme using trilinear isoparametric elements. It can be stabilized in the supersonic zone by the introduction of artificial viscosity to produce an upwind bias. The

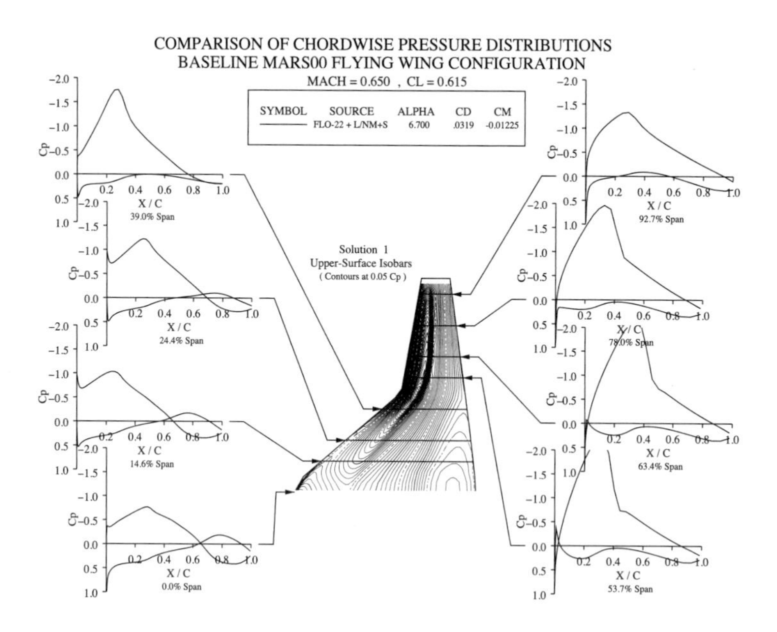

Figure 1.3 Pressure distribution over the wing of a Mars Lander using FLO22.

"hour-glass" instability that results from the use of one point integration scheme is suppressed by the introduction of higher order coupling terms based on mixed derivatives. The flow solvers (FLO27-30) based on this approach were subsequently incorporated in Boeing's A488 software, which was used in the aerodynamic design of Boeing commercial aircraft throughout the 1980s (Rubbert 1998).

In the same period, Perrier was focusing the research efforts at Dassault on the development of finite element methods using triangular and tetrahedral meshes, because he believed that if CFD software was to be really useful for aircraft design, it must be able to treat complete configurations. Although finite element methods were more computationally expensive, and mesh generation continued to present difficulties, finite element methods offered a route toward the achievement of this goal. The Dassault/INRIA group was ultimately successful, and they performed transonic potential flow calculations for complete aircraft such as the Falcon 50 in the early eighties (Bristeau et al. 1985).

1.2.3 The Development of Methods for the Euler and Navier–Stokes Equations

By the 1980s, advances in computer hardware had made it feasible to solve the full Euler equations using software that could be cost-effective in industrial use. The idea of directly discretizing the conservation laws to produce a finite volume scheme had been introduced by MacCormack (MacCormack & Paullay 1972). Most of the early flow solvers tended to exhibit strong pre- or post-shock oscillations. Also, in a workshop held in Stockholm in 1979 (Rizzi & Viviand 1981), it was apparent that none of the existing schemes converged to a steady state. These difficulties were resolved during the following decade.

The Jameson–Schmidt–Turkel scheme (Jameson, Schmidt, & Turkel 1981), which used Runge–Kutta time stepping and a blend of second- and fourth-differences (both to control oscillations and to provide background dissipation), consistently demonstrated convergence to a steady state, with the consequence that it has remained one of the most widely used methods to the present day.

A fairly complete understanding of shock capturing algorithms was achieved, stemming from the ideas of Godunov, Van Leer, Harten, and Roe. The issue of oscillation control and positivity had already been addressed by Godunov (1959) in his pioneering work in the 1950s (translated into English in 1959). He had introduced the concept of representing the flow as piecewise constant in each computational cell and solving a Riemann problem at each interface, thus obtaining a first order accurate solution that avoids nonphysical features such as expansion shocks. When this work was eventually recognized in the West, it became very influential. It was also widely recognized that numerical schemes might benefit from distinguishing the various wave speeds, and this motivated the development of characteristics-based schemes.

The earliest higher order characteristics-based methods used flux vector splitting (Steger & Warming 1981) but suffered from oscillations near discontinuities similar to those of central difference schemes in the absence of numerical dissipation. The Monotone Upwind Scheme for Conservation Laws (MUSCL) of Van Leer (1974) extended the monotonicity-preserving behavior of Godunov's scheme to higher order through the use of limiters. The use of limiters dates back to the flux-corrected transport (FCT) scheme of Boris and Book (1973). A general framework for oscillation control in the solution of nonlinear problems was provided by Harten's concept of Total Variation Diminishing (TVD) schemes. It finally proved possible to give a rigorous justification of the JST scheme (Jameson 1995a, 1995b).

Roe's introduction of the concept of locally linearizing the equations through a mean value Jacobian (Roe 1981) had a major impact. It provided valuable insight into the nature of the wave motions and also enabled the efficient implementation of Godunov-type schemes using approximate Riemann solutions. Roe's flux-difference splitting scheme has the additional benefit that it yields a single-point numerical shock structure for stationary normal shocks. Roe's and other approximate Riemann solutions, such as that due to Osher, have been incorporated in a variety of schemes of Godunov type, including the Essentially Nonoscillatory (ENO) schemes of Harten, Engquist, Osher, and Chakravarthy (1987).

Solution methods for the Reynolds averaged Navier–Stokes (RANS) equations had been pioneered in the seventies by MacCormack and others, but at that time they were extremely expensive. By the 1990s, computer technology had progressed to the point where RANS simulations could be performed with manageable costs, and they began to be fairly widely used by the aircraft industry. The need for robust and reliable methods to predict hypersonic flows, which contain both very strong shock waves and near vacuum regions, gave a further impetus to the development of advanced shock capturing algorithms for compressible viscous flow.

1.3 Overview of the Simulation Process

The essential steps of developing a numerical simulation of a physical problem can be outlined as follows:

1. Formulate a mathematical model of the physical problem that captures the important aspects for the purpose in hand and can provide the desired accuracy. Here it should be noted that models of widely varying complexity and levels of fidelity can be useful. For example, potential flow models based on Laplace's equation can provide reasonably accurate predictions of low speed aerodynamic flows over streamlined shapes at a low computational cost, and this is very useful at an early stage in the design when rapid turnaround is crucial. At the final stage of the design, one would wish to confirm the expected performance using a model with the highest possible fidelity, typically the Reynolds averaged Navier–Stokes equations in actual practice.
2. Analyze the mathematical properties of the model, such as proper formulation of boundary conditions that ensure the existence and convergence of a solution.
3. Formulate a discrete numerical scheme to approximate the mathematical model that has been selected. Analyze the stability, accuracy, and convergence of the scheme. Can we prove, for example, that the error in the numerical approximation decreases as some power of the mesh spacing when the spacing is progressively reduced?
4. Implement the discrete scheme in software that makes efficient use of the available hardware. This is becoming harder with the emergence of parallel systems with multiple levels of parallelism down to multiple threads within each core of multi-core processing chips, which are in turn arranged in parallel clusters. This stage also requires the use of every possible procedure to assure that the software is actually correct.
5. Validate the software by showing that it produces trustworthy results in practice. Here we should distinguish between the questions of whether the software is correct and whether the selected mathematical model adequately represents the physics. To address the first question, we may test whether the results are correct for some limiting situations for which the true answer is known. For example, an arbitrary body has zero drag in inviscid flow. Or is the numerical solution symmetric for flow over a symmetric profile at zero angle of attack? We should also test the convergence of the numerical solution as the grid is refined. Does it exhibit the expected order of accuracy? We may also compare the results with those obtained by other software developed to solve the same problem. Workshops such as the AIAA Drag Prediction Workshops can play a useful role in this process. Finally, once a sufficiently high confidence level has been established for the software, comparisons with experimental data can be used to address the question of whether the mathematical model adequately represents the physical problem of interest.

While mathematical models of fluid flow and their properties are briefly reviewed in the text, it is mainly focused on the third step in the process, the formulation of discrete schemes that yield proper solutions with the desired accuracy and avoid spurious numerical artifacts.

A primary requirement in aeronautical science is the capability to predict steady flows, such as the flow over an airfoil or wing or the flow over a complete aircraft in cruising flight. Accordingly, the issues associated with the prediction of steady solutions are given a substantial emphasis in the text. Three requirements in particular are:

1. The numerical solution should be capable of converging to an exactly steady state within machine zero. This was not the case for any of the early Euler solvers, as reported in the Stockholm workshop of 1979. It is also not the case for some popular high resolution schemes, such as the ENO scheme, which may reach a limit cycle in which the stencil alternates in successive iterations.
2. If the steady state is approached by advancing in time, the final result should be independent of the time step. This is not true of schemes that are formulated using an integrated space and time discretization, such as the Lax–Wendroff scheme.
3. In simulations of inviscid gas dynamics, there should be constant total enthalpy in the steady state. This is not true of many of the standard high resolution shock capturing schemes.

The rate at which the scheme approaches the steady state is also crucial. This is measured both by the number of iterations required to reach a steady state within some tolerance and also the cost of each iteration. The most widely used methods in current use are actually based on advancing the unsteady flow equations in time with varyingly drastic modifications to accelerate convergence to a steady state. Multigrid schemes have proved particularly effective, and these will be examined in detail.

It turns out that there is a close interplay between steady state methods and implicit time stepping schemes, because the nonlinear equations that need to be solved in each implicit time step have a form that closely resembles the steady state problem. Accordingly, many of the techniques that have proved useful for steady state problems can be directly carried over to the formulation of implicit schemes for unsteady problems. For these reasons, steady state simulation techniques are the subject of careful analysis in this text.

2 Mathematical Models of Fluid Flow

2.1 Overview

The Navier–Stokes equations state the laws of conservation of mass, momentum, and energy for the flow of a gas in thermodynamic equilibrium. Using the Cartesian tensor notation, let x_i be the coordinates; p, ρ, T, E, and H the pressure, density, temperature, and total energy and enthalpy; and u_i the velocity components, where for three-dimensional flow i ranges from 1 to 3. Also, we use the summation convention that a repeated index implies a sum over that index. Each conservation equation has the differential form

$$\frac{\partial w}{\partial t} + \frac{\partial f_j}{\partial x_j} = 0, \tag{2.1}$$

which can be derived from the integral form

$$\frac{\partial}{\partial t} \int_D w dV + \int_B f_j n_j dS = 0,$$

for a control volume D with boundary B, where dV and dS are the volume and area elements, and n_j are the components of the surface normal. The integral form remains valid in the presence of discontinuities such as shock waves and is the basis of the finite volume method, which is discussed in detail in subsequent chapters. For the mass equation,

$$w = \rho, \qquad f_j = \rho u_j. \tag{2.2}$$

For the i momentum equation,

$$w_i = \rho u_i, \quad f_{ij} = \rho u_i u_j + p\delta_{ij} - \tau_{ij}, \tag{2.3}$$

where τ_{ij} is the viscous stress tensor, which for a Newtonian fluid is proportional to the rate of strain tensor and the bulk dilatation, and δ_{ij} is the Kronecker delta

$$\delta_{ij} = \begin{cases} 1 & \text{if} \quad i = j \\ 0 & \text{if} \quad i \neq j \end{cases}.$$

If μ and λ are the coefficients of viscosity and bulk viscosity, then

$$\tau_{ij} = \mu \left(\frac{\partial u_i}{\partial x_j} + \frac{\partial u_j}{\partial x_i} \right) + \lambda \delta_{ij} \left(\frac{\partial u_k}{\partial x_k} \right). \tag{2.4}$$

Typically $\lambda = -2\mu/3$. For the energy equation,

$$w = \rho E, \quad f_j = \rho H u_j - \tau_{jk} u_k - \kappa \frac{\partial T}{\partial x_j}, \tag{2.5}$$

where κ is the coefficient of heat conduction, and H is the total enthalpy

$$H = E + \frac{p}{\rho}.$$

In the case of a perfect gas, the pressure is related to the density and energy by the equation of state

$$p = (\gamma - 1)\rho \left(E - \frac{1}{2} q^2 \right),$$

where

$$q^2 = u_i u_i,$$

and γ is the ratio of specific heats. The coefficient of thermal conductivity and the temperature satisfy the relations

$$k = \frac{c_p \mu}{Pr}, \quad T = \frac{p}{R\rho},$$

where c_p is the specific heat at constant pressure, R is the gas constant, and Pr is the Prandtl number. Also, the speed of sound c is given by the ratio

$$c^2 = \frac{\gamma p}{\rho},$$

and a key dimensionless parameter governing the effects of compressibility is the Mach number

$$M = \frac{q}{c},$$

where q is the magnitude of the velocity.

If the flow is inviscid, the boundary condition that must be satisfied at a solid wall is

$$\boldsymbol{u} \cdot \boldsymbol{n} = u_i n_i = 0,$$

where n denotes the normal to the surface. Viscous flows must satisfy the "no-slip" condition

$$\boldsymbol{u} = 0.$$

Viscous solutions also require a boundary condition for the energy equation. The usual practice in pure aerodynamic simulations is either to specify the isothermal condition

$$T = T_0,$$

or to specify the adiabatic condition

$$\frac{\partial T}{\partial n} = 0,$$

corresponding to zero heat transfer. The calculation of heat transfer requires an appropriate coupling to a model of the structure.

For an external flow, the flow variables should approach free-stream values,

$$p = p_\infty, \quad \rho = \rho_\infty, \quad T = T_\infty, \quad u = u_\infty,$$

far upstream at the inflow boundary. If any entropy is generated, the density downstream at the outflow boundary cannot recover to ρ_∞ if the pressure recovers to p_∞. In fact, if trailing vortices persist downstream, the pressure does not recover to p_∞. In general it is necessary to examine the incoming and outgoing waves at the outer boundaries of the flow domain. Boundary values should then only be imposed for quantities transported by the incoming waves. In a subsonic flow, there are four incoming waves at the inflow boundary and one escaping acoustic wave. Correspondingly four quantities should be specified. At the outflow boundary, there are four outgoing waves, so one quantity should be specified. One way to do this is to introduce Riemann invariants corresponding to a one-dimensional flow normal to the boundary, as will be discussed in Section 2.8. In a supersonic flow, all quantities should be fixed at the inflow boundary, while they should all be extrapolated at the outflow boundary. The proper specification of inflow and outflow boundary conditions is particularly important in the calculation of internal flows; otherwise spurious wave reflections may severely corrupt the solution.

An indication of the relative magnitudes of the inertial and viscous terms is given by the Reynolds number

$$Re = \frac{\rho U L}{\mu},$$

where U is a characteristic velocity and L a representative length. The viscosity of air is very small, and typical Reynolds numbers for the flow past a component of an aircraft such as a wing are of the order of 10^7 or more, depending on the size and speed of the aircraft. In this situation, the viscous effects are essentially confined to thin boundary layers covering the surface. Boundary layers may nevertheless have a global impact on the flow by causing separation. Unfortunately, unless they are controlled by active means, such as suction through a porous surface, boundary layers are unstable and generally become turbulent.

Using dimensional analysis, Kolmogorov's theory of turbulence (Kolmogorov 1941a) estimates the length scales of the smallest persisting eddies to be of order $\frac{1}{Re^{3/4}}$ in comparison with the macroscopic length scale of the flow. Accordingly, the mesh requirements for the full simulation of all scales of turbulence can be estimated as growing proportionally to $Re^{9/4}$ and are clearly beyond the reach of current computers. Turbulent flows may be simulated by the Reynolds-averaged Navier–Stokes (RANS) equations, in which statistical averages are taken of rapidly fluctuating components. Denoting fluctuating parts by primes and averaging by an overbar, this leads to the appearance of Reynolds stress terms of the form $\overline{\rho u_i' u_j'}$, which cannot be determined from the mean values of the velocity and density. Estimates of these additional terms must be provided by a turbulence model. The simplest turbulence models augment the molecular viscosity by an eddy viscosity that crudely represents the effects of turbulent mixing and is estimated with some characteristic length scale such as the boundary layer thickness. A rather more elaborate class of models introduces two additional equations for the turbulent kinetic energy and the rate of dissipation. Existing turbulence models are adequate for particular classes of flow for which empirical correlations are available, but they are generally not capable of reliably predicting more complex phenomena, such as shock wave–boundary layer interaction. The current status of turbulence modeling is reviewed by Wilcox (2006), Haase et al. (1997), Leschziner (2003), and Durbin and Pettersson Reif (2011). The treatment of viscous turbulent flows will be discussed in more detail in Chapter 8.

Outside the boundary layer, excellent predictions can be made by treating the flow as inviscid. Setting $\sigma_{ij} = 0$ and eliminating heat conduction from equations (2.2, 2.3, and 2.5) yields the Euler equations for inviscid flow. These are a very useful model for predicting flows over aircraft. According to Kelvin's theorem, a smooth inviscid flow that is initially irrotational remains irrotational. This allows one to introduce a velocity potential ϕ such that $u_i = \partial\phi/\partial x_i$. The Euler equations for a steady flow now reduce to

$$\frac{\partial}{\partial x_i}\left(\rho\frac{\partial\phi}{\partial x_i}\right) = 0. \tag{2.6}$$

In a steady inviscid flow, it follows from the energy equation (2.5) and the continuity equation (2.2) that the total enthalpy is constant:

$$H = \frac{c^2}{\gamma - 1} + \frac{1}{2}u_i u_i = H_\infty, \tag{2.7}$$

where the subscript ∞ is used to denote the value in the far-field. According to Crocco's theorem, vorticity in a steady flow is associated with entropy production through the relation

$$\boldsymbol{u} \times \boldsymbol{\omega} + T\nabla\mathcal{S} = \nabla H = 0,$$

where $\boldsymbol{u}$ and $\boldsymbol{\omega}$ are the velocity and vorticity vectors, T is the temperature, and $\mathcal{S}$ is the entropy. Thus, the introduction of a velocity potential is consistent with the assumption of isentropic flow.

Substituting the isentropic relationship p/ρ^γ = constant and the formula for the speed of sound, equation (2.7) can be solved for the density as

$$\frac{\rho}{\rho_\infty} = \left[1 + \frac{\gamma - 1}{2}{M_\infty}^2\left(1 - \frac{u_i u_i}{{u_\infty}^2}\right)\right]^{\frac{1}{\gamma-1}}.$$

It can be seen from this equation that

$$\frac{\partial\rho}{\partial u_i} = -\frac{\rho u_i}{c^2},$$

and correspondingly in isentropic flow,

$$\frac{\partial p}{\partial u_i} = \frac{dp}{d\rho}\frac{\partial \rho}{\partial u_i} = -\rho u_i. \tag{2.8}$$

Substituting $\frac{\partial \rho}{\partial x_j} = \frac{\partial \rho}{\partial u_i}\frac{\partial u_i}{\partial x_j}$, the potential flow equation (2.6) can be expanded in quasi-linear form as

$$c^2\frac{\partial^2 \phi}{\partial {x_i}^2} - u_i u_j \frac{\partial^2 \phi}{\partial x_i \partial x_j} = 0. \tag{2.9}$$

If the flow is locally aligned, say with the x_1 axis, equation (2.9) reads as

$$(1 - M^2)\frac{\partial^2 \phi}{\partial {x_1}^2} + \frac{\partial^2 \phi}{\partial {x_2}^2} + \frac{\partial^2 \phi}{\partial {x_3}^2} = 0, \tag{2.10}$$

where M is the Mach number u_1/c. The change from an elliptic to a hyperbolic partial differential equation as the flow becomes supersonic is evident.

The potential flow equation (2.9) also corresponds to the Bateman variational principle that the integral over the domain of the pressure,

$$I = \int_D p d\xi,$$

is stationary. Here $d\xi$ denotes the volume element. Using the relation (2.8), a variation δp results in a variation

$$\delta I = \int_D \frac{\partial p}{\partial u_i}\delta u_i d\xi = -\int \rho u_i \frac{\partial}{\partial x_i}\delta\phi d\xi,$$

or, on integrating by parts with appropriate boundary conditions,

$$\delta I = \int_D \frac{\partial}{\partial x_i}\left(\rho\frac{\partial \phi}{\partial x_i}\right)\delta\phi d\xi.$$

Then $\delta I = 0$ for an arbitrary variation $\delta\phi$ if equation (2.6) holds.

The equations of inviscid supersonic flow admit discontinuous solutions – both shock waves, which satisfy the Rankine–Hugoniot jump conditions (Liepmann & Roshko 1957), and contact discontinuities. Only compression shock waves are admissible, corresponding to the production of entropy. Expansion shock waves cannot occur because they would correspond to a decrease in entropy. Because shock waves generate entropy, they cannot be exactly modeled by the potential flow equations. The amount of entropy generated is proportional to $(M-1)^3$, where M is the Mach number upstream of the shock. Accordingly, weak solutions admitting isentropic jumps that conserve mass but not momentum are a good approximation to shock waves, as long as the shock waves are quite weak (with a Mach number <1.3 for the normal velocity component upstream of the shock wave). Stronger shock waves tend to separate the flow, with the result that the inviscid approximation is no longer adequate. Thus this model is well balanced, and it has proved extremely useful for the prediction of the cruising performance of transport aircraft. An estimate of the pressure drag arising from shock waves is obtained because of the momentum deficit through an isentropic jump.

If one assumes small disturbances about a free stream in the x_i direction and a Mach number close to unity, equation (2.10) can be reduced to the transonic small disturbance equation, in which M^2 is estimated as

$$M_\infty^2[1 - (\gamma + 1)\partial\phi/\partial x_1].$$

This is the simplest nonlinear model of compressible flow.

The final level of approximation is to linearize equation (2.10) by replacing M^2 by its free stream value M_∞^2. In the subsonic case, the resulting Prandtl–Glauert equation can be reduced to Laplace's equation by scaling the x_i coordinate by $(1 - M_\infty{}^2)^{1/2}$. Irrotational incompressible flow satisfies Laplace's equation, as can be seen by setting ρ = constant in equation (2.6). With limits on available computing power and the cost of the calculations, one has to make a trade-off between the complexity of the mathematical model and the complexity of the geometric configuration to be treated.

The computational requirements for aerodynamic simulation are a function of the number of operations required per mesh point, the number of cycles or time steps needed to reach a solution, and the number of mesh points needed to resolve the important features of the flow. Algorithms for the three-dimensional transonic potential flow equation require about 500 floating-point operations per mesh point per cycle. The number of operations required for an Euler simulation is in the range of 1,000–5,000 per time step, depending on the complexity of the algorithm. The number of mesh intervals required to provide an accurate representation of a two-dimensional inviscid transonic flow is of the order of 160 wrapping around the profile and 32 normal to the airfoil. Correspondingly, about 200,000 mesh cells are sufficient to provide adequate resolution of three-dimensional inviscid transonic flow past a swept wing, and this number needs to be increased to provide a good simulation of a more complex configuration, such as a complete aircraft. The requirements for viscous simulations by means of turbulence models are much more severe. Good resolution of a turbulent boundary layer needs about 32 intervals inside the boundary layer, with the result that a typical mesh for a two-dimensional Navier–Stokes calculation contains 512 intervals wrapping around the profile and 64 intervals in the normal direction. A corresponding mesh for a swept wing would have, say, $512 \times 64 \times 256 \approx$ 8,388,608 cells, leading to a calculation at the outer limits of current computing capabilities. Figure 2.1 gives an indication of the boundaries of the complexity of problems that can be treated with different levels of computing power. The vertical axis indicates the geometric complexity and the horizontal axis the equation complexity.

High-performance computing hardware and software continue to evolve at a rapid pace, and the onset of systems with exaflop performance will enable large eddy simulations (LES) and direct numerical simulations (DNS) of complex separated and turbulent flows, with applications to the design of real engineering systems such as turbomachinery. LES and DNS will be further discussed in Chapter 8.

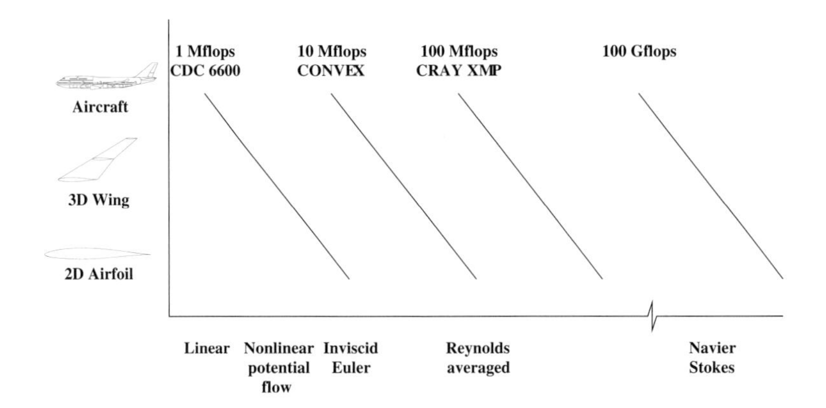

Figure 2.1 Complexity of the problems that can be treated with different classes of computer (1 flop = 1 floating-point operation per second; 1 Mflop = 10^6 flops; 1 Gflop = 10^9 flops).

2.2 Coordinate Transformations

In order to calculate solutions for flows in complex geometric domains, it is often useful to introduce body-fitted coordinates through global or, as in the case of isoparametric elements, local transformations. With the body now coinciding with a coordinate surface, it is much easier to enforce the boundary conditions accurately. Suppose that the mapping to computational coordinates (ξ_1, ξ_2, ξ_3) is defined by the transformation matrices

$$K_{ij} = \frac{\partial x_i}{\partial \xi_j}, \quad K_{ij}^{-1} = \frac{\partial \xi_i}{\partial x_j}, \quad J = \det(K), \tag{2.11}$$

where for the sake of simplicity it is assumed that K and J are independent of time.

Also define

$$S = JK^{-1}.$$

The elements of S are the cofactors of K, and in a finite volume discretization, they are just the face areas of the computational cells projected in the x_1, x_2, and x_3 directions. Using the Levi–Civita permutation symbol

$$\epsilon_{ijk} = \begin{cases} +1 & \text{if} \quad (i,j,k) \text{ is } (1,2,3),(2,3,1), \text{ or } (3,1,2) \\ -1 & \text{if} \quad (i,j,k) \text{ is } (3,2,1),(1,3,2), \text{ or } (2,1,3) \\ 0 & \text{if} \quad i=j \quad \text{or} \quad j=k \quad \text{or} \quad k=i \end{cases},$$

we can express the elements of S as

$$S_{ij} = \frac{1}{2}\epsilon_{jpq}\epsilon_{irs}\frac{\partial x_p}{\partial \xi_r}\frac{\partial x_q}{\partial \xi_s}. \tag{2.12}$$

Moreover,

$$\frac{\partial}{\partial \xi_i} S_{ij} = \frac{1}{2}\epsilon_{jpq}\epsilon_{irs}\left(\frac{\partial^2 x_p}{\partial \xi_r \partial \xi_i}\frac{\partial x_q}{\partial \xi_s} + \frac{\partial x_p}{\partial \xi_r}\frac{\partial^2 x_q}{\partial \xi_s \partial \xi_i}\right).$$
$$= 0.$$

Using the chain rule, equation (2.1) now yields

$$\frac{\partial \boldsymbol{w}}{\partial t} + \frac{\partial \boldsymbol{f}_j}{\partial \xi_i}\frac{\partial \xi_i}{\partial x_j} = 0.$$

It follows on multiplying by J that

$$\frac{\partial (J\boldsymbol{w})}{\partial t} + S_{ij}\frac{\partial \boldsymbol{f}_j}{\partial \xi_i} = 0,$$

and using equation (2.12),

$$S_{ij}\frac{\partial \boldsymbol{f}_j}{\partial \xi_i} = \frac{\partial S_{ij}\boldsymbol{f}_j}{\partial \xi_i}.$$

Thus we obtain the transformed conservation laws

$$\frac{\partial (J\boldsymbol{w})}{\partial t} + \frac{\partial \boldsymbol{F}_i}{\partial \xi_i} = 0, \tag{2.13}$$

where

$$\boldsymbol{F}_i = S_{i,j}\boldsymbol{f}_j(\boldsymbol{w}). \tag{2.14}$$

Defining scaled contravariant velocity components as

$$U_i = S_{ij}u_j,$$

the flux formulas may be expanded as

$$\boldsymbol{F}_i = \begin{bmatrix} \rho U_i \\ \rho U_i u_1 + S_{i1}p \\ \rho U_i u_2 + S_{i2}p \\ \rho U_i u_3 + S_{i3}p \\ \rho U_i H \end{bmatrix}. \tag{2.15}$$

If we choose a coordinate system so that the boundary is at $\xi_l = 0$, the wall boundary condition for inviscid flow is now

$$U_l = 0.$$

2.3 Analysis of the Equations of gas Dynamics: The Jacobian Matrices

The Euler equations for the three-dimensional flow of an inviscid gas are obtained by eliminating the viscous stress and the heat conduction term from the Navier–Stokes equations. Following the notation of Section 2.1, the conservation laws for mass, momentum, and energy can be written in combined form as

$$\frac{\partial \boldsymbol{w}}{\partial t} + \frac{\partial}{\partial x_i} \boldsymbol{f}_i(\boldsymbol{w}) = 0, \tag{2.16}$$

where the state and flux vectors are

$$\boldsymbol{w} = \rho \begin{bmatrix} 1 \\ u_1 \\ u_2 \\ u_3 \\ E \end{bmatrix}, \quad \boldsymbol{f}_i = \rho u_i \begin{bmatrix} 1 \\ u_1 \\ u_2 \\ u_3 \\ H \end{bmatrix} + p \begin{bmatrix} 0 \\ \delta_{i1} \\ \delta_{i2} \\ \delta_{i3} \\ 0 \end{bmatrix}.$$

Also,

$$p = (\gamma - 1)\rho \left(E - \frac{u^2}{2} \right), \quad H = E + \frac{p}{\rho} = \frac{c^2}{\gamma - 1} + \frac{u^2}{2},$$

where u is the speed, and c is the speed of sound:

$$u^2 = u_i u_i, \quad c^2 = \frac{\gamma p}{\rho}.$$

Let m_i and e denote the momentum components and total energy,

$$m_i = \rho u_i, \quad e = \rho E = \frac{p}{\gamma - 1} + \frac{m_i m_i}{2\rho}.$$

Then $\boldsymbol{w}$ and $\boldsymbol{f}$ can be expressed as

$$\boldsymbol{w} = \begin{bmatrix} \rho \\ m_1 \\ m_2 \\ m_3 \\ e \end{bmatrix}, \quad \boldsymbol{f}_i = u_i \begin{bmatrix} \rho \\ m_1 \\ m_2 \\ m_3 \\ e \end{bmatrix} + p \begin{bmatrix} 0 \\ \delta_{i1} \\ \delta_{i2} \\ \delta_{i3} \\ u_i \end{bmatrix}. \tag{2.17}$$

In a finite volume scheme, the flux needs to be calculated across the interface between each pair of cells. Denoting the face normal components by n_i, the flux is

$$\boldsymbol{f} = n_i \boldsymbol{f}_i.$$

This can be expressed in terms of the conservative variables $\boldsymbol{w}$ as

$$\boldsymbol{f} = u_n \begin{bmatrix} \rho \\ m_1 \\ m_2 \\ m_3 \\ e \end{bmatrix} + p \begin{bmatrix} 0 \\ n_1 \\ n_2 \\ n_3 \\ u_n \end{bmatrix}, \tag{2.18}$$

where u_n is the normal velocity

$$u_n = n_i u_i = \frac{n_i m_i}{\rho}.$$

Also

$$p = (\gamma - 1)\left(e - \frac{m_i m_i}{2\rho}\right).$$

In smooth regions of the flow, the equations can also be written in quasilinear form as

$$\frac{\partial \boldsymbol{w}}{\partial t} + A_i \frac{\partial \boldsymbol{w}}{\partial x_i} = 0,$$

where A_i are the Jacobian matrices

$$A_i = \frac{\partial \boldsymbol{f}_i}{\partial \boldsymbol{w}}.$$

The composite Jacobian matrix at a face with normal vector $\boldsymbol{n}$ is

$$A = A_i n_i = \frac{\partial \boldsymbol{f}}{\partial \boldsymbol{w}}.$$

All the entries in $\boldsymbol{f}_i$ and $\boldsymbol{f}$ are homogeneous of degree 1 in the conservative variables $\boldsymbol{w}$. It follows that $\boldsymbol{f}_i$ and $\boldsymbol{f}$ satisfy the identities

$$\boldsymbol{f}_i = A_i \boldsymbol{w}, \quad \boldsymbol{f} = A\boldsymbol{w}.$$

This is a consequence of the fact that if a quantity q can be expressed in terms of the components of a vector $\boldsymbol{w}$ as

$$q = \prod_j w_j^{\alpha_j}, \text{ where } \sum_j \alpha_j = \alpha,$$

then it is called a homogeneous function of degree α, and

$$\begin{aligned} \frac{\partial q}{\partial w_i} w_i &= \sum_i \alpha_i \prod_j w_j^{\alpha_j} \\ &= \sum_i \alpha_i q \\ &= \alpha q. \end{aligned}$$

In order to evaluate A, note that $\frac{\partial u_n}{\partial w}$ and $\frac{\partial p}{\partial w}$ are row vectors:

$$\frac{\partial u_n}{\partial \boldsymbol{w}} = \frac{1}{\rho}\left[-u_n, n_1, n_2, n_3, 0\right]$$

and

$$\frac{\partial p}{\partial \boldsymbol{w}} = (\gamma - 1)\left[\frac{u^2}{2}, -u_1, -u_2, -u_3, 1\right].$$

Accordingly,

$$\frac{\partial \boldsymbol{f}}{\partial \boldsymbol{w}} = \frac{\partial}{\partial \boldsymbol{w}}(u_n \boldsymbol{w}) + \begin{bmatrix} 0 \\ n_1 \\ n_2 \\ n_3 \\ u_n \end{bmatrix} \left[\frac{\partial p}{\partial \boldsymbol{w}} \right] + \begin{bmatrix} 0 \\ 0 \\ 0 \\ 0 \\ p \end{bmatrix} \left[\frac{\partial u_n}{\partial \boldsymbol{w}} \right].$$

Then the Jacobian matrix can be assembled as the sum of a diagonal matrix and two outer products of rank 1,

$$\begin{aligned} A &= u_n I + \rho \begin{bmatrix} 1 \\ u_1 \\ u_2 \\ u_3 \\ H \end{bmatrix} \left[\frac{\partial u_n}{\partial \boldsymbol{w}} \right] + \begin{bmatrix} 0 \\ n_1 \\ n_2 \\ n_3 \\ u_n \end{bmatrix} \left[\frac{\partial p}{\partial \boldsymbol{w}} \right] \\ &= u_n I + \begin{bmatrix} 1 \\ u_1 \\ u_2 \\ u_3 \\ H \end{bmatrix} [-u_n, n_1, n_2, n_3, 0] \\ &\quad + (\gamma - 1) \begin{bmatrix} 0 \\ n_1 \\ n_2 \\ n_3 \\ u_n \end{bmatrix} \left[\tfrac{u^2}{2}, -u_1, -u_2, -u_3, 1 \right]. \end{aligned}$$

Note that every entry in the Jacobian matrix can be expressed in terms of the velocity components u_i and the speed of sound c since

$$H = \frac{c^2}{\gamma - 1} + \frac{u^2}{2}.$$

It may also be directly verified that

$$\boldsymbol{f} = A\boldsymbol{w},$$

because u_n is homogeneous of degree 0, with the consequence that $\frac{\partial u_n}{\partial \boldsymbol{w}}$ is orthogonal to $\boldsymbol{w}$

$$\frac{\partial u_n}{\partial \boldsymbol{w}} \boldsymbol{w} = 0,$$

while p is homogenous of degree 1 so that

$$\frac{\partial p}{\partial \boldsymbol{w}} \boldsymbol{w} = p.$$

Thus

$$Aw= u_n w + \begin{bmatrix} 1 \\ n_1 \\ n_2 \\ n_3 \\ u_n \end{bmatrix} \left[\frac{\partial p}{\partial w} \right] w = f.$$

The special structure of the Jacobian matrix enables the direct identification of its eigenvalues and eigenvectors. Any vector in the three-dimensional subspace orthogonal to the vectors $\frac{\partial u_n}{\partial w}$ and $\frac{\partial p}{\partial w}$ is an eigenvector corresponding to the eigenvalue u_n, which is thus a triple eigenvalue. It is easy to verify that the vectors

$$r_0 = \begin{bmatrix} 1 \\ u_1 \\ u_2 \\ u_3 \\ \frac{u^2}{2} \end{bmatrix} \quad r_1 = \begin{bmatrix} 0 \\ 0 \\ n_3 \\ -n_2 \\ u_2 n_3 - u_3 n_2 \end{bmatrix}$$

$$r_2 = \begin{bmatrix} 0 \\ -n_3 \\ 0 \\ n_1 \\ u_3 n_1 - u_1 n_3 \end{bmatrix} \quad r_3 = \begin{bmatrix} 0 \\ n_2 \\ -n_1 \\ 0 \\ u_1 n_2 - u_2 n_1 \end{bmatrix}$$

are orthogonal to both $\frac{\partial u_n}{\partial w}$ and $\frac{\partial p}{\partial w}$. However, r_1, r_2, and r_3 are not independent since

$$\sum_{k=1}^{3} n_k r_k = 0.$$

Three independent eigenvectors can be obtained as

$$v_1 = n_1 r_0 + c r_1, \; v_2 = n_2 r_0 + c r_2, \; ; v_3 = n_3 r_0 + c r_3,$$

where c is the speed of sound.

In order to verify this, note that the middle three elements of r_k, $k = 1,2,3$, are equal to $n \times i_k$, where i_k is the unit vector in the kth coordinate direction. Also, the last element of v_k is equal to $i_k \cdot (u \times n)$. If the vectors v_k are not independent, they must satisfy a relation of the form

$$v = \sum_{k=1}^{3} \alpha_k v_k = 0$$

for some non-zero vector α. The first element of v is

$$\sum_{k=1}^{3} \alpha_k n_k = \alpha \cdot n.$$

For the next three elements of $\boldsymbol{v}$ to be zero,

$$\sum_{k=1}^{3} \alpha_k \left(\boldsymbol{n} \times \boldsymbol{i}_k\right) = \boldsymbol{n} \times \boldsymbol{\alpha} = 0,$$

which is only possible if $\boldsymbol{\alpha}$ is parallel to $\boldsymbol{n}$, so that $\boldsymbol{\alpha} \cdot \boldsymbol{n} \neq 0$.

In order to identify the remaining eigenvectors, denote the column vectors in A as

$$\boldsymbol{a}_1 = \begin{bmatrix} 1 \\ u_1 \\ u_2 \\ u_3 \\ H \end{bmatrix} \quad \boldsymbol{a}_2 = \begin{bmatrix} 0 \\ n_1 \\ n_2 \\ n_3 \\ u_n \end{bmatrix}.$$

Then

$$\rho \frac{\partial u_n}{\partial \boldsymbol{w}} \boldsymbol{a}_1 = 0, \quad \rho \frac{\partial u_n}{\partial \boldsymbol{w}} \boldsymbol{a}_2 = 1,$$

$$\frac{\partial p}{\partial \boldsymbol{w}} \boldsymbol{a}_1 = c^2, \quad \frac{\partial p}{\partial \boldsymbol{w}} \boldsymbol{a}_2 = 0.$$

Now consider a vector of the form

$$\boldsymbol{r} = \boldsymbol{a}_1 + \beta \boldsymbol{a}_2,$$

where β is a scalar factor. Then

$$\begin{aligned} A\boldsymbol{r} &= u_n(\boldsymbol{a}_1 + \beta \boldsymbol{a}_2) + \beta \boldsymbol{a}_1 + c^2 \boldsymbol{a}_2 \\ &= \lambda(\boldsymbol{a}_1 + \beta \boldsymbol{a}_2) = \lambda \boldsymbol{r}, \end{aligned}$$

if

$$u_n + \beta = \lambda, \quad \beta u_n + c^2 = \lambda \beta,$$

which is the case if

$$\beta^2 = c^2.$$

Thus the vectors

$$\boldsymbol{v}_4 = \boldsymbol{a}_1 + c\boldsymbol{a}_2, \quad \boldsymbol{v}_5 = \boldsymbol{a}_1 - c\boldsymbol{a}_2$$

are the eigenvectors corresponding to the eigenvalues $u_n + c$ and $u_n - c$. Written in full,

$$\boldsymbol{v}_4 = \begin{bmatrix} 1 \\ u_1 + n_1 c \\ u_2 + n_2 c \\ u_3 + n_3 c \\ H + u_n c \end{bmatrix}, \quad \boldsymbol{v}_5 = \begin{bmatrix} 1 \\ u_1 - n_1 c \\ u_2 - n_2 c \\ u_3 - n_3 c \\ H - u_n c \end{bmatrix}.$$

2.4 Two-Dimensional Flow

The equations of two-dimensional flow have the simpler form

$$\frac{\partial \boldsymbol{w}}{\partial t} + \frac{\partial \boldsymbol{f}_1(\boldsymbol{w})}{\partial x_1} + \frac{\partial \boldsymbol{f}_2(\boldsymbol{w})}{\partial x_2} = 0,$$

where the state and flux vectors are

$$\boldsymbol{w} = \begin{bmatrix} \rho \\ m_1 \\ m_2 \\ e \end{bmatrix}, \quad \boldsymbol{f}_1 = u_1 \boldsymbol{w} + p \begin{bmatrix} 0 \\ 1 \\ 0 \\ u_1 \end{bmatrix}, \quad \boldsymbol{f}_2 = u_2 \boldsymbol{w} + p \begin{bmatrix} 0 \\ 0 \\ 1 \\ u_2 \end{bmatrix}.$$

Now the flux vector normal to an edge with normal components n_1 and n_2 is

$$\boldsymbol{f} = n_1 \boldsymbol{f}_1 + n_2 \boldsymbol{f}_2.$$

The corresponding Jacobian matrix is

$$\begin{aligned} A = \frac{\partial \boldsymbol{f}}{\partial \boldsymbol{w}} &= n_1 \frac{\partial \boldsymbol{f}_1}{\partial \boldsymbol{w}} + n_2 \frac{\partial \boldsymbol{f}_2}{\partial \boldsymbol{w}} \\ &= u_n I + \rho \begin{bmatrix} 1 \\ u_1 \\ u_2 \\ H \end{bmatrix} \frac{\partial u_n}{\partial \boldsymbol{w}} + \begin{bmatrix} 0 \\ n_1 \\ n_2 \\ u_n \end{bmatrix} \frac{\partial p}{\partial \boldsymbol{w}}. \end{aligned}$$

The eigenvalues of $\boldsymbol{A}$ are

$$u_n, \ u_n, \ u_n + c, u_n - c$$

with corresponding eigenvectors

$$v_1 = \begin{bmatrix} 1 \\ u_1 \\ u_2 \\ \frac{u^2}{2} \end{bmatrix} \quad v_2 = \begin{bmatrix} 0 \\ -cn_2 \\ cn_1 \\ c(u_2 n_1 - u_1 n_2) \end{bmatrix}$$

$$v_3 = \begin{bmatrix} 1 \\ u_1 + n_1 c \\ u_2 + n_2 c \\ H + u_n c \end{bmatrix} \quad v_4 = \begin{bmatrix} 1 \\ u_1 - n_1 c \\ u_2 - n_2 c \\ H - u_n c \end{bmatrix}.$$

2.5 One-Dimensional Flow

The equations of one-dimensional flow are further simplified to

$$\frac{\partial \boldsymbol{w}}{\partial t} + \frac{\partial}{\partial x} \boldsymbol{f}(\boldsymbol{w}) = 0,$$

where the state vector is

$$w = \begin{bmatrix} \rho \\ \rho u \\ \rho E \end{bmatrix} = \begin{bmatrix} \rho \\ m \\ e \end{bmatrix},$$

and the flux vector is

$$f = \begin{bmatrix} \rho u \\ \rho u^2 + p \\ \rho u H \end{bmatrix} = u w + \begin{bmatrix} 0 \\ p \\ up \end{bmatrix}.$$

Now the Jacobian matrix is

$$\begin{aligned} A = \frac{\partial f}{\partial w} &= uI + \rho \begin{bmatrix} 1 \\ u \\ H \end{bmatrix} \frac{\partial u}{\partial w} + \begin{bmatrix} 0 \\ 1 \\ u \end{bmatrix} \frac{\partial p}{\partial w} \\ &= uI + \begin{bmatrix} 1 \\ u \\ H \end{bmatrix} [-u, 1, 0] + (\gamma - 1) \begin{bmatrix} 0 \\ 1 \\ u \end{bmatrix} \left[\frac{u^2}{2}, -u, 1 \right]. \end{aligned}$$

The eigenvalues of A are

$$u, u + c, u - c,$$

with the corresponding eigenvectors

$$v_1 = \begin{bmatrix} 1 \\ u \\ \frac{u^2}{2} \end{bmatrix} \quad v_2 = \begin{bmatrix} 1 \\ u + c \\ H + uc \end{bmatrix} \quad v_3 = \begin{bmatrix} 1 \\ u - c \\ H - uc \end{bmatrix}.$$

Then

$$A = V \Lambda V^{-1},$$

where Λ is a diagonal matrix containing the eigenvalues of A. Also, the rows ${\boldsymbol{r}^{(1)}}^T$, ${\boldsymbol{r}^{(2)}}^T$, and ${\boldsymbol{r}^{(3)}}^T$ of V^{-1} are left eigenvectors of A satisfying

$${\boldsymbol{r}^{(k)}}^T A = \Lambda^{(k)} {\boldsymbol{r}^{(k)}}^T, \quad k = 1, 2, 3.$$

Setting

$$\bar{\gamma} = \frac{\gamma - 1}{c^2},$$

it may be verified that

$$V^{-1} = \begin{bmatrix} 1-\bar{\gamma}\frac{u^2}{2} & \bar{\gamma}u & -\bar{\gamma} \\ \frac{1}{2}\left(\bar{\gamma}\frac{u^2}{2}-\frac{u}{c}\right) & -\frac{1}{2}\left(\bar{\gamma}u-\frac{1}{c}\right) & \frac{\bar{\gamma}}{2} \\ \frac{1}{2}\left(\bar{\gamma}\frac{u^2}{2}+\frac{u}{c}\right) & -\frac{1}{2}\left(\bar{\gamma}u+\frac{1}{c}\right) & \frac{\bar{\gamma}}{2} \end{bmatrix}$$

$$= \begin{bmatrix} 1 & 0 & 0 \\ -\frac{u}{2c} & \frac{1}{2c} & 0 \\ \frac{u}{2c} & -\frac{1}{2c} & 0 \end{bmatrix} - \bar{\gamma}\begin{bmatrix} 1 \\ -\frac{1}{2} \\ -\frac{1}{2} \end{bmatrix}\begin{bmatrix} \frac{u^2}{2} & -u & 1 \end{bmatrix}.$$

2.6 Transformation to Alternative Sets of Variables: Primitive Form

The quasi-linear equations of gas dynamics can be simplified in various ways by transformations to alternative sets of variables.

Consider the equations of three-dimensional flow,

$$\frac{\partial \boldsymbol{w}}{\partial t} + A_i \frac{\partial \boldsymbol{w}}{\partial x_i} = 0,$$

where

$$A_i = \frac{\partial \boldsymbol{f}_i}{\partial \boldsymbol{w}}.$$

Under a transformation to new set of variables $\tilde{\boldsymbol{w}}$,

$$\tilde{M}\frac{\partial \tilde{\boldsymbol{w}}}{\partial t} + A_i \tilde{M}\frac{\partial \tilde{\boldsymbol{w}}}{\partial x_i} = 0,$$

where

$$\tilde{M} = \frac{\partial \boldsymbol{w}}{\partial \tilde{\boldsymbol{w}}}.$$

Multiplying by $\frac{\partial \tilde{\boldsymbol{w}}}{\partial \boldsymbol{w}} = M^{-1}$,

$$\frac{\partial \tilde{\boldsymbol{w}}}{\partial t} + \tilde{A}_i \frac{\partial \tilde{\boldsymbol{w}}}{\partial x_i} = 0,$$

where

$$\tilde{A}_i = \tilde{M}^{-1} A_i \tilde{M}, \quad A_i = \tilde{M}\tilde{A}_i\tilde{M}^{-1}.$$

In the case of the primitive variables,

$$\tilde{\boldsymbol{w}} = \begin{bmatrix} \rho \\ u_1 \\ u_2 \\ u_3 \\ p \end{bmatrix},$$

we find that

$$\tilde{M} = \begin{bmatrix} 1 & 0 & 0 & 0 & 0 \\ u_1 & \rho & 0 & 0 & 0 \\ u_2 & 0 & \rho & 0 & 0 \\ u_3 & 0 & 0 & \rho & 0 \\ \frac{u^2}{2} & \rho u_1 & \rho u_2 & \rho u_3 & \frac{1}{\gamma-1} \end{bmatrix}$$

$$\tilde{M}^{-1} = \begin{bmatrix} 1 & 0 & 0 & 0 & 0 \\ -\frac{u_1}{\rho} & \frac{1}{\rho} & 0 & 0 & 0 \\ -\frac{u_2}{\rho} & 0 & \frac{1}{\rho} & 0 & 0 \\ -\frac{u_3}{\rho} & 0 & 0 & \frac{1}{\rho} & 0 \\ (\gamma-1)\frac{u^2}{2} & -(\gamma-1)u_1 & -(\gamma-1)u_2 & -(\gamma-1)u_3 & \gamma-1 \end{bmatrix}$$

and

$$A_1 = \begin{bmatrix} u_1 & \rho & 0 & 0 & 0 \\ 0 & u_1 & 0 & 0 & \frac{1}{\rho} \\ 0 & 0 & u_1 & 0 & 0 \\ 0 & 0 & 0 & u_1 & 0 \\ 0 & \rho c^2 & 0 & 0 & u_1 \end{bmatrix}$$

$$A_2 = \begin{bmatrix} u_2 & 0 & \rho & 0 & 0 \\ 0 & u_2 & 0 & 0 & 0 \\ 0 & 0 & u_2 & 0 & \frac{1}{\rho} \\ 0 & 0 & 0 & u_2 & 0 \\ 0 & 0 & \rho c^2 & 0 & u_2 \end{bmatrix}$$

$$A_3 = \begin{bmatrix} u_3 & 0 & 0 & \rho & 0 \\ 0 & u_3 & 0 & 0 & 0 \\ 0 & 0 & u_3 & 0 & 0 \\ 0 & 0 & 0 & u_3 & \frac{1}{\rho} \\ 0 & 0 & 0 & \rho c^2 & u_3 \end{bmatrix}.$$

2.7 Symmetric Form

Consider the equations of one-dimensional flow in primitive variables:

$$\frac{\partial \rho}{\partial t} + u\frac{\partial \rho}{\partial x} + \rho\frac{\partial u}{\partial x} = 0$$

$$\frac{\partial u}{\partial t} + u\frac{\partial u}{\partial x} + \frac{1}{\rho}\frac{\partial p}{\partial x} = 0$$

$$\frac{\partial p}{\partial t} + \rho c^2\frac{\partial u}{\partial x} + u\frac{\partial p}{\partial x} = 0.$$

Then subtracting the first equation multiplied by c^2 from the third equation, we find that

$$\frac{\partial p}{\partial t} - c^2\frac{\partial \rho}{\partial t} + u\left(\frac{\partial p}{\partial x} - c^2\frac{\partial \rho}{\partial x}\right) = 0.$$

This is equivalent to a statement that the entropy

$$S = \log\left(\frac{p}{\rho^\gamma}\right) = \log p - \gamma \log \rho$$

is constant since

$$dS = \frac{dp}{p} - \gamma\frac{d\rho}{\rho} = \frac{1}{p}(dp - c^2 d\rho).$$

If the entropy is constant, then

$$dp = c^2 d\rho.$$

With this substitution the first equation becomes

$$\frac{1}{c^2}\frac{\partial p}{\partial t} + \frac{u}{c^2}\frac{\partial p}{\partial x} + \rho\frac{\partial u}{\partial x} = 0,$$

and now the first two equations can be rescaled as

$$\begin{aligned} \frac{1}{\rho c}\frac{\partial p}{\partial t} + \frac{u}{\rho c}\frac{\partial p}{\partial x} + c\frac{\partial u}{\partial x} &= 0 \\ \frac{\partial u}{\partial t} + \frac{c}{\rho c}\frac{\partial p}{\partial x} + u\frac{\partial u}{\partial x} &= 0. \end{aligned}$$

Thus if we write the equations in terms of the differential variables

$$d\bar{\boldsymbol{w}} = \begin{bmatrix} \frac{dp}{\rho c} \\ du \\ dp - c^2 d\rho \end{bmatrix},$$

we obtain the symmetric form

$$\frac{\partial \bar{\boldsymbol{w}}}{\partial t} + \bar{A}\frac{\partial \bar{\boldsymbol{w}}}{\partial x} = 0,$$

where

$$\bar{A} = \begin{bmatrix} u & c & 0 \\ c & u & 0 \\ 0 & 0 & u \end{bmatrix}.$$

These transformations can be generalized to the equations of three-dimensional flow. A convenient scaling is to set

$$d\bar{w} = \begin{bmatrix} \frac{dp}{c^2} \\ \frac{\rho}{c} du_1 \\ \frac{\rho}{c} du_2 \\ \frac{\rho}{c} du_3 \\ \frac{dp}{c^2} - d\rho \end{bmatrix}. \tag{2.19}$$

This eliminates the density from the transformation matrices $\bar{M} = \frac{\partial w}{\partial \bar{w}}$ and $\bar{M}^{-1} = \frac{\partial \bar{w}}{\partial w}$. The equations now take the form

$$\frac{\partial \bar{w}}{\partial t} + \bar{A}_i \frac{\partial \bar{w}}{\partial x_i} = 0, \tag{2.20}$$

where the transformed Jacobian matrices

$$\bar{A}_i = \bar{M}^{-1} A_i \bar{M} \tag{2.21}$$

are simultaneously symmetrized as

$$\bar{\bar{A}}_1 = \begin{bmatrix} u_1 & c & 0 & 0 & 0 \\ c & u_1 & 0 & 0 & 0 \\ 0 & 0 & u_1 & 0 & 0 \\ 0 & 0 & 0 & u_1 & 0 \\ 0 & 0 & 0 & 0 & u_1 \end{bmatrix} \tag{2.22}$$

$$\bar{\bar{A}}_2 = \begin{bmatrix} u_2 & 0 & c & 0 & 0 \\ 0 & u_2 & 0 & 0 & 0 \\ c & 0 & u_2 & 0 & 0 \\ 0 & 0 & 0 & u_2 & 0 \\ 0 & 0 & 0 & 0 & u_2 \end{bmatrix} \tag{2.23}$$

$$\bar{\bar{A}}_3 = \begin{bmatrix} u_3 & 0 & 0 & c & 0 \\ 0 & u_3 & 0 & 0 & 0 \\ 0 & 0 & u_3 & 0 & 0 \\ c & 0 & 0 & u_3 & 0 \\ 0 & 0 & 0 & 0 & u_3 \end{bmatrix}, \tag{2.24}$$

while

$$\bar{M} = \frac{\partial w}{\partial \bar{w}} = \begin{bmatrix} 1 & 0 & 0 & 0 & -1 \\ u_1 & c & 0 & 0 & -u_1 \\ u_2 & 0 & c & 0 & -u_2 \\ u_3 & 0 & 0 & c & -u_3 \\ H & cu_1 & cu_2 & cu_3 & -\frac{u^2}{2} \end{bmatrix}$$

and

$$\bar{M}^{-1} = \frac{\partial \bar{w}}{\partial w} = \begin{bmatrix} \bar{\gamma}\frac{u^2}{2} & -\bar{\gamma}u_1 & -\bar{\gamma}u_2 & -\bar{\gamma}u_3 & \bar{\gamma} \\ -\frac{u_1}{c} & \frac{1}{c} & 0 & 0 & 0 \\ -\frac{u_2}{c} & 0 & \frac{1}{c} & 0 & 0 \\ -\frac{u_3}{c} & 0 & 0 & \frac{1}{c} & 0 \\ \bar{\gamma}(u^2 - H) & -\bar{\gamma}u_1 & -\bar{\gamma}u_2 & -\bar{\gamma}u_3 & \bar{\gamma} \end{bmatrix},$$

where

$$\bar{\gamma} = \frac{\gamma - 1}{c^2}.$$

The combined Jacobian matrix

$$A = n_i A_i$$

can now be decomposed as

$$A = n_i \bar{M} \bar{A}_i \bar{M}^{-1}.$$

Corresponding to the fact that $\bar{A}$ is symmetric, one can find a set of orthogonal eigenvectors, which may be normalized to unit length. Then one can express

$$\bar{A} = \bar{V} \Lambda \bar{V}^{-1},$$

where the diagonal matrix Λ contains the eigenvalues $u_n, u_n, u_n, u_n + c$, and $u_n - c$ as its elements. The matrix $\bar{V}$ containing the corresponding eigenvectors as its columns is

$$\bar{V} = \begin{bmatrix} \frac{1}{\sqrt{2}} & -\frac{1}{\sqrt{2}} & 0 & 0 & 0 \\ \frac{n_1}{\sqrt{2}} & \frac{n_1}{\sqrt{2}} & 0 & -n_3 & n_2 \\ \frac{n_2}{\sqrt{2}} & \frac{n_2}{\sqrt{2}} & n_3 & 0 & -n_1 \\ \frac{n_3}{\sqrt{2}} & \frac{n_3}{\sqrt{2}} & -n_2 & n_1 & 0 \\ 0 & 0 & 1 & n_2 & n_3 \end{bmatrix},$$

and $\bar{V}^{-1} = \bar{V}^T$. The Jacobian matrix can now be expressed as

$$A = V \Lambda V^{-1}, \tag{2.25}$$

where

$$V = \bar{M}\bar{V}, \quad V^{-1} = \bar{V}^T \bar{M}^{-1}.$$

This decomposition is often useful.

Since

$$c^2 = \frac{dp}{d\rho} = \gamma \frac{p}{\rho},$$

it follows that in isentropic flow,

$$2c\,dc = \frac{\gamma}{\rho}dp - \frac{\gamma p}{\rho^2}d\rho = \frac{\gamma}{\rho}dp - \frac{c^2}{\rho}d\rho = \frac{\gamma - 1}{\rho}dp.$$

Thus,

$$\frac{dp}{\rho c} = \frac{2}{\gamma - 1}dc,$$

so the equations can also be expressed in the same form, (2.20)–(2.24), for the variables

$$\bar{w} = \begin{bmatrix} \frac{2c}{\gamma-1} \\ u_1 \\ u_2 \\ u_3 \\ S \end{bmatrix}$$

with the transformation matrices

$$\bar{M} = \frac{dw}{d\bar{w}} = \begin{bmatrix} \frac{\rho}{c} & 0 & 0 & 0 & -\frac{p}{c^2} \\ \frac{\rho u_1}{c} & \rho & 0 & 0 & -\frac{p u_1}{c^2} \\ \frac{\rho u_2}{c} & 0 & \rho & 0 & -\frac{p u_2}{c^2} \\ \frac{\rho u_3}{c} & 0 & 0 & \rho & -\frac{p u_3}{c^2} \\ \frac{\rho H}{c} & \rho u_1 & \rho u_2 & \rho u_3 & -\rho\frac{p}{c}\frac{u^2}{2} \end{bmatrix}$$

and

$$\bar{M}^{-1} = \frac{d\bar{w}}{dw} = \begin{bmatrix} \bar{\gamma}\frac{u^2}{2} & -\bar{\gamma}u_1 & -\bar{\gamma}u_2 & -\bar{\gamma}u_3 & \bar{\gamma} \\ -\frac{u_1}{\rho} & \frac{1}{\rho} & 0 & 0 & 0 \\ -\frac{u_2}{\rho} & 0 & \frac{1}{\rho} & 0 & 0 \\ -\frac{u_3}{\rho} & 0 & 0 & \frac{1}{\rho} & 0 \\ \frac{\bar{\gamma}}{p}(u^2 - H) & -\frac{\bar{\gamma}u_1}{p} & -\frac{\bar{\gamma}u_2}{p} & -\frac{\bar{\gamma}u_3}{p} & \frac{\bar{\gamma}}{p} \end{bmatrix},$$

where

$$\bar{\gamma} = \frac{\gamma - 1}{\rho c}.$$

2.8 Riemann Invariants

In the case of one-dimensional isentropic flow, these equations reduce to

$$\frac{2}{\gamma - 1}\frac{\partial c}{\partial t} + \frac{2u}{\gamma - 1}\frac{\partial c}{\partial x} + c\frac{\partial u}{\partial x} = 0$$

$$\frac{\partial u}{\partial t} + \frac{2c}{\gamma - 1}\frac{\partial c}{\partial x} + u\frac{\partial u}{\partial x} = 0$$

$$\frac{\partial S}{\partial t} + u\frac{\partial S}{\partial x} = 0.$$

Now the first two equations can be added and subtracted to yield

$$\frac{\partial R^+}{\partial t} + (u + c)\frac{\partial R^+}{\partial x} = 0$$

and

$$\frac{\partial R^-}{\partial t} + (u - c)\frac{\partial R^-}{\partial x} = 0,$$

where R^+ and R^- are the Riemann invariants

$$R^+ = u + \frac{2c}{\gamma - 1}, R^- = u - \frac{2c}{\gamma - 1},$$

which remain constant as they are transported at the wave speeds $u + c$ and $u - c$. The Riemann invariants prove to be useful in the formulation of far field boundary conditions designed to minimize wave reflection.

2.9 Symmetric Hyperbolic Form

Equation (2.20) can be written in terms of the conservative variables as

$$\frac{\partial \bar{w}}{\partial w}\frac{\partial w}{\partial t} + \bar{A}_i\frac{\partial \bar{w}}{\partial w}\frac{\partial w}{\partial x_i} = 0.$$

Here $\frac{\partial \bar{w}}{\partial w} = \bar{M}^{-1}$. Now multiplying by $\bar{M}^{T^{-1}}$, the equation is reduced to the symmetric hyperbolic form

$$Q\frac{\partial w}{\partial t} + \hat{A}_i\frac{\partial w}{\partial x_i} = 0,$$

where Q is symmetric and positive definite, and the matrices $\hat{A}_i$ are symmetric:

$$Q = (MM^T)^{-1}, \quad \hat{A}_i = \bar{M}^{T^{-1}}\bar{A}_i\bar{M}^{-1}.$$

This form could alternatively be derived by multiplying the conservative form (2.3) of the equations by Q. Then

$$\begin{aligned} Q\frac{\partial f_i}{\partial x_i} &= \bar{M}^{T^{-1}}\bar{M}^{-1}\frac{\partial f}{\partial x_i} \\ &= \bar{M}^{T^{-1}}\bar{M}^{-1}A_i\frac{\partial w}{\partial x_i} \end{aligned}$$

$$= \bar{M}^{T^{-1}} \bar{M}^{-1} M \bar{A}_i \bar{M}^{-1} \frac{\partial \boldsymbol{w}}{\partial x_i}$$

$$= \bar{M}^{T^{-1}} \bar{A}_i \bar{M}^{-1} \frac{\partial \boldsymbol{w}}{\partial x_i}.$$

A symmetric hyperbolic form can actually be obtained for any choice of the dependent variables. For example, (2.20) could be written in terms of the primitive variables as

$$N \frac{\partial \tilde{\boldsymbol{w}}}{\partial t} + \bar{A}_i N \frac{\partial \tilde{\boldsymbol{w}}}{\partial x_i} = 0,$$

where

$$N = \frac{\partial \bar{\boldsymbol{w}}}{\partial \tilde{\boldsymbol{w}}} = \begin{bmatrix} 0 & 0 & 0 & 0 & \frac{1}{c^2} \\ 0 & \frac{\rho}{c} & 0 & 0 & 0 \\ 0 & 0 & \frac{\rho}{c} & 0 & 0 \\ 0 & 0 & 0 & \frac{\rho}{c} & 0 \\ -1 & 0 & 0 & 0 & \frac{1}{c^2} \end{bmatrix}.$$

Then multiplying by N^T, we obtain the symmetric hyperbolic form

$$N^T N \frac{\partial \tilde{\boldsymbol{w}}}{\partial t} + N^T \bar{A}_i N \frac{\partial \tilde{\boldsymbol{w}}}{\partial x_i} = 0.$$

2.10 Entropy Variables

A particular choice of variables that symmetrizes the equations can be derived from functions of the entropy

$$S = \log\left(\frac{p}{\rho^\gamma}\right) = \log p - \gamma \log \rho.$$

The last equation of (2.20) is equivalent to the statement that

$$\rho \frac{\partial S}{\partial t} + \rho u_i \frac{\partial S}{\partial x_i} = 0,$$

which can be combined with the mass conservation equation multiplied by S,

$$S \frac{\partial \rho}{\partial t} + S \frac{\partial}{\partial x_i}(\rho u_i) = 0,$$

to yield the entropy conservation law

$$\frac{\partial(\rho S)}{\partial t} + \frac{\partial(\rho u_i S)}{\partial x_i} = 0. \tag{2.26}$$

This is a special case of a generalized entropy function defined as follows. Given a system of conservation laws with the general form

$$\frac{\partial \boldsymbol{w}}{\partial t} + \frac{\partial}{\partial x_i} \boldsymbol{f}_i(\boldsymbol{w}) = 0, \tag{2.27}$$

suppose that we can find a scalar function $U(\boldsymbol{w})$ such that

$$\frac{\partial U}{\partial \boldsymbol{w}}\frac{\partial \boldsymbol{f}_i}{\partial \boldsymbol{w}} = \frac{\partial G_i}{\partial \boldsymbol{w}}, \tag{2.28}$$

and $U(\boldsymbol{w})$ is a convex function of $\boldsymbol{w}$. Then $U(\boldsymbol{w})$ is an entropy function with an entropy flux $G_i(\boldsymbol{w})$ since multiplying (2.27) by $\frac{\partial U}{\partial \boldsymbol{w}}$, we obtain

$$\frac{\partial U}{\partial \boldsymbol{w}}\frac{\partial \boldsymbol{w}}{\partial t} + \frac{\partial U}{\partial \boldsymbol{w}}\frac{\partial \boldsymbol{f}_i}{\partial \boldsymbol{w}}\frac{\partial \boldsymbol{w}}{\partial x_i} = 0, \tag{2.29}$$

and using (2.28), this is equivalent to

$$\frac{\partial U(\boldsymbol{w})}{\partial t} + \frac{\partial G_i}{\partial \boldsymbol{w}}\frac{\partial \boldsymbol{w}}{\partial x_i} = 0,$$

which is in turn equivalent to the generalized entropy conservation law

$$\frac{\partial U(\boldsymbol{w})}{\partial t} + \frac{\partial}{\partial x_i}G_i(\boldsymbol{w}) = 0. \tag{2.30}$$

Now, introduce dependent variables

$$\boldsymbol{v}^T = \frac{\partial U}{\partial \boldsymbol{w}}.$$

Then the equations are symmetrized. This can be shown as follows. Define the scalar functions

$$Q(\boldsymbol{v}) = \boldsymbol{v}^T\boldsymbol{w} - U(\boldsymbol{w}) \tag{2.31}$$

and

$$R_i(\boldsymbol{v}) = \boldsymbol{v}^T\boldsymbol{f}_i - G_i(\boldsymbol{w}). \tag{2.32}$$

Then

$$\frac{\partial Q}{\partial \boldsymbol{v}} = \boldsymbol{w}^T + \boldsymbol{v}^T\frac{\partial \boldsymbol{w}}{\partial \boldsymbol{v}} - \frac{\partial U}{\partial \boldsymbol{w}}\frac{\partial \boldsymbol{w}}{\partial \boldsymbol{v}} = \boldsymbol{w}^T$$

and

$$\frac{\partial R_i}{\partial \boldsymbol{v}} = \boldsymbol{f}_i^T + \boldsymbol{v}^T\frac{\partial \boldsymbol{f}_i}{\partial \boldsymbol{w}}\frac{\partial \boldsymbol{w}}{\partial \boldsymbol{v}} - \frac{\partial G_i}{\partial \boldsymbol{w}}\frac{\partial \boldsymbol{w}}{\partial \boldsymbol{v}} = \boldsymbol{f}_i^T.$$

Hence $\frac{\partial \boldsymbol{w}}{\partial \boldsymbol{v}}$ is symmetric,

$$\frac{\partial w_j}{\partial v_k} = \frac{\partial^2 Q}{\partial v_j \partial v_k} = \frac{\partial w_k}{\partial v_j},$$

and $\frac{\partial \boldsymbol{f}_i}{\partial v}$ is symmetric,

$$\frac{\partial f_{ij}}{\partial v_k} = \frac{\partial^2 R_i}{\partial v_j \partial v_k} = \frac{\partial f_{ik}}{\partial v_j},$$

and correspondingly

$$\frac{\partial \boldsymbol{f}_i}{\partial \boldsymbol{w}} \frac{\partial \boldsymbol{w}}{\partial \boldsymbol{v}} = A_i \frac{\partial \boldsymbol{w}}{\partial \boldsymbol{v}}$$

is symmetric. Thus, the conservation law (2.27) is converted to the symmetric form

$$\frac{\partial \boldsymbol{w}}{\partial \boldsymbol{v}} \frac{\partial \boldsymbol{v}}{\partial t} + A_i \frac{\partial \boldsymbol{w}}{\partial \boldsymbol{v}} \frac{\partial \boldsymbol{v}}{\partial x_i} = 0.$$

Multiplying the conservation law (2.27) by $\frac{\partial \boldsymbol{w}}{\partial \boldsymbol{v}}$ from the left produces the alternative symmetric form

$$\frac{\partial \boldsymbol{v}}{\partial \boldsymbol{w}} \frac{\partial \boldsymbol{w}}{\partial t} + \frac{\partial \boldsymbol{v}}{\partial \boldsymbol{w}} A_i \frac{\partial \boldsymbol{w}}{\partial x_i} = 0$$

since

$$\frac{\partial \boldsymbol{v}}{\partial \boldsymbol{w}} A_i = \frac{\partial \boldsymbol{v}}{\partial \boldsymbol{w}} \left(A_i \frac{\partial \boldsymbol{w}}{\partial \boldsymbol{v}} \right) \frac{\partial \boldsymbol{v}}{\partial \boldsymbol{w}}.$$

Moreover, multiplying (2.27) by $\boldsymbol{v}^T$ recovers (2.29) and the corresponding entropy conservation law (2.30).

Conversely, if $U(\boldsymbol{w})$ satisfies the conservation law (2.30), then

$$\frac{\partial U}{\partial \boldsymbol{w}} \frac{\partial \boldsymbol{w}}{\partial t} = -\frac{\partial G_i}{\partial \boldsymbol{w}} \frac{\partial \boldsymbol{w}}{\partial x_i},$$

but

$$\frac{\partial U}{\partial \boldsymbol{w}} \frac{\partial \boldsymbol{w}}{\partial t} = -\frac{\partial U}{\partial \boldsymbol{w}} \frac{\partial \boldsymbol{f}_i}{\partial \boldsymbol{w}} \frac{\partial \boldsymbol{w}}{\partial x_i}.$$

These can both hold for all $\frac{\partial \boldsymbol{w}}{\partial x_i}$ only if

$$\frac{\partial U}{\partial \boldsymbol{w}} \frac{\partial \boldsymbol{f}_i}{\partial \boldsymbol{w}} = \frac{\partial G_i}{\partial \boldsymbol{w}},$$

so $U(\boldsymbol{w})$ is an entropy function if it is a convex function of $\boldsymbol{w}$.

The property of convexity enables $U(\boldsymbol{w})$ to be treated as a generalized energy since U becomes unbounded if $\boldsymbol{w}$ becomes unbounded. It can be shown that $-\rho S$ is a convex function of $\boldsymbol{w}$. Accordingly it follows from the entropy equation (2.26) that we can take

$$U(\boldsymbol{w}) = -\frac{\rho S}{\gamma - 1} = \frac{\gamma}{\gamma - 1} \rho \log \rho - \frac{\rho}{\gamma - 1} \log p$$

and

$$G_i(\boldsymbol{w}) = -\frac{\rho u_i S}{\gamma - 1}$$

as a generalized entropy function and flux, where the scaling factor $\frac{1}{\gamma-1}$ is introduced to simplify the resulting expressions.

Using the formula

$$\frac{\partial p}{\partial \boldsymbol{w}} = (\gamma - 1)\left[\frac{u^2}{2}, -u_1, -u_2, -u_3, 1\right],$$

we find that

$$\boldsymbol{v}^T = \frac{\partial U}{\partial \boldsymbol{w}} = \left[\frac{\gamma - S}{\gamma - 1} - \frac{\rho}{p}\frac{u^2}{2}, \frac{\rho u_1}{p}, \frac{\rho u_2}{p}, \frac{\rho u_3}{p}, \frac{-\rho}{p}\right],$$

where

$$u^2 = u_i u_i.$$

The reverse transformation is

$$\boldsymbol{w}^T = p\left[-v_5, v_2, v_3, v_4, \left(1 - \frac{1}{2}(v_2^2 + v_3^2 + v_4^2)\right) v_5\right].$$

Now the scalar functions $Q(v)$ and $R_i(v)$ defined by (2.31) and (2.32) reduce to

$$Q(v) = \rho, R_i(v) = m_i = \rho u_i.$$

Accordingly,

$$\boldsymbol{w}^T = \frac{\partial p}{\partial \boldsymbol{v}} = \frac{\partial p}{\partial \boldsymbol{w}}\frac{\partial \boldsymbol{w}}{\partial \boldsymbol{v}}$$

and

$$\boldsymbol{f}_i^T = \frac{\partial m_i}{\partial \boldsymbol{v}} = \frac{\partial m_i}{\partial \boldsymbol{w}}\frac{\partial \boldsymbol{w}}{\partial \boldsymbol{v}},$$

where

$$\frac{\partial \rho}{\partial \boldsymbol{w}} = [1\ 0\ 0\ 0\ 0]$$

and

$$\frac{\partial m_i}{\partial \boldsymbol{w}} = [0\ \delta_{i1}\ \delta_{i2}\ \delta_{i3}\ 0].$$

Thus the first four rows of $\frac{\partial \boldsymbol{w}}{\partial v}$ consist of $\boldsymbol{w}^T, \boldsymbol{f}_1^T, \boldsymbol{f}_2^T$, and $\boldsymbol{f}_3^T$, and correspondingly, since $\frac{\partial \boldsymbol{w}}{\partial v}$ is symmetric, its first four columns are $\boldsymbol{w}, \boldsymbol{f}_1, \boldsymbol{f}_2$, and $\boldsymbol{f}_3$. The only remaining element is $\frac{\partial \boldsymbol{w}_5}{\partial v_5}$. A direct calculation reveals that

$$\frac{\partial \boldsymbol{w}_5}{\partial v_5} = \rho E^2 + \left(E + \frac{u^2}{2}\right) p.$$

Thus

$$\frac{\partial w}{\partial v} = \begin{bmatrix} \rho & \rho u_1 & \rho u_2 & \rho u_3 & \rho E \\ \rho u_1 & \rho u_1^2 + p & \rho u_2 u_1 & \rho u_3 u_1 & \rho u_1 E + u_1 p \\ \rho u_2 & \rho u_1 u_2 & \rho u_2^2 + p & \rho u_3 u_2 & \rho u_2 E + u_2 p \\ \rho u_3 & \rho u_1 u_3 & \rho u_2 u_3 & \rho u_3^2 + p & \rho u_3 E + u_3 p \\ \rho E & \rho u_1 E + u_1 p & \rho u_2 E + u_2 p & \rho u_3 E + u_3 p & \rho E^2 + \left(E + \frac{u^2}{2}\right) p \end{bmatrix}$$

$$= \frac{1}{\rho} w w^T + p \begin{bmatrix} 0 & 0 & 0 & 0 & 0 \\ 0 & 1 & 0 & 0 & u_1 \\ 0 & 0 & 1 & 0 & u_2 \\ 0 & 0 & 0 & 1 & u_3 \\ 0 & u_1 & u_2 & u_3 & E + \frac{u^2}{2} \end{bmatrix}.$$

2.11 The Shock Jump Conditions for Gas Dynamics

The general jump conditions for a moving shock wave are

$$f_R - f_L = u_S (w_R - w_L), \tag{2.33}$$

where w and $f(w)$ are the state and flux vectors, the subscripts L and R denote the conditions on the left and right side of the shock wave, and u_S is the shock speed. Expanding (2.33) separately for mass, momentum, and energy:

$$\rho_L (u_L - u_S) = \rho_R (u_R - u_S) = m \tag{2.34}$$

$$\rho_L u_L (u_L - u_S) + p_L = \rho_R u_R (u_R - u_S) + p_R \tag{2.35}$$

$$\rho_L \left(i_L + \frac{u_L^2}{2} \right) (u_L - u_S) + u_L p_L = \rho_R \left(i_R + \frac{u_R^2}{2} \right) (u_R - u_S) + u_R p_R, \tag{2.36}$$

where i is the specific internal energy and m is the mass flow through the shock wave.

For a perfect gas,

$$i = \frac{p}{(\gamma - 1)\rho}.$$

Equations (2.34) and (2.35) are called the mechanical conditions because they are independent of the thermodynamic properties and hold for any compressible fluid. They can be rearranged to give a variety of useful relations. Substituting (2.34) in (2.35),

$$m u_L + p_L = m u_R + p_R$$

or

$$m(u_L - u_S) + p_L = m(u_R - u_S) + p_R,$$

and hence conservation of momentum can be expressed as

$$\rho_L(u_L - u_S)^2 + p_L = \rho_R(u_R - u_S)^2 + p_R = P, \tag{2.37}$$

where P is the momentum flux.

In addition, we can write (2.35) as

$$p_R - p_L = m(u_L - u_R) = m\big(u_L - u_S - (u_R - u_S)\big) \tag{2.38}$$

or

$$p_R - p_L = m^2\left(\frac{1}{\rho_L} - \frac{1}{\rho_R}\right),$$

so

$$m^2 = \frac{p_R - p_L}{\frac{1}{\rho_L} - \frac{1}{\rho_R}} = \frac{p_R - p_L}{\rho_R - \rho_L}\rho_R\rho_L. \tag{2.39}$$

Consequently,

$$(u_L - u_S)^2 = \frac{p_R - p_L}{\frac{1}{\rho_L} - \frac{1}{\rho_R}} = \frac{p_R - p_L}{\rho_R - \rho_L}\frac{\rho_R}{\rho_L} \tag{2.40}$$

and

$$(u_R - u_S)^2 = \frac{p_R - p_L}{\frac{1}{\rho_L} - \frac{1}{\rho_R}} = \frac{p_R - p_L}{\rho_R - \rho_L}\frac{\rho_L}{\rho_R}. \tag{2.41}$$

Also,

$$\frac{p_R - p_L}{\rho_L} = (u_L - u_S)(u_L - u_S)\big(u_L - u_S - (u_R - u_S)\big)$$

and

$$\frac{p_R - p_L}{\rho_R} = (u_R - u_S)(u_L - u_S)\big(u_L - u_S - (u_R - u_S)\big),$$

so

$$\left(\frac{1}{\rho_L} + \frac{1}{\rho_R}\right)(p_R - p_L) = (u_L - u_S)^2 - (u_R - u_S)^2 \tag{2.42}$$

and

$$\left(\frac{1}{\rho_L} - \frac{1}{\rho_R}\right)(p_R - p_L) = (u_L - u_R)^2$$

or

$$u_L - u_R = \sqrt{\left(\frac{1}{\rho_L} - \frac{1}{\rho_R}\right)(p_R - p_L)}.$$

Moreover, (2.38) can be expanded as

$$p_R - p_L = \rho_R(u_R - u_S)(u_L - u_S) - \rho_L(u_L - u_S)(u_R - u_S),$$

so

$$(u_R - u_S)(u_L - u_S) = \frac{p_R - p_L}{\rho_R - \rho_L}. \tag{2.43}$$

Combining the mass and energy equations (2.34) and (2.36), we find that

$$\begin{aligned} m\left(i_L + \frac{u_L^2}{2}\right) - m\left(i_R + \frac{u_R^2}{2}\right) &= u_R p_R - {}_L p_L \\ &= (u_R - u_S)p_R - (u_L - u_S)p_L + u_S(p_R - p_L). \end{aligned}$$

Hence, substituting from (2.38),

$$m\left(i_L + \frac{u_L^2}{2}\right) - m\left(i_R + \frac{u_R^2}{2}\right) = (u_R - u_S)p_R - (u_L - u_S)p_L + u_S m(u_L - u_R),$$

so

$$\begin{aligned} m\left(i_L + \frac{u_L^2}{2} - u_L u_S + \frac{u_S^2}{2}\right) &+ (u_L - u_S)p_L \\ = m\left(i_R + \frac{u_R^2}{2} - u_R u_S + \frac{u_S^2}{2}\right) &+ (u_R - u_S)p_R \end{aligned}$$

or dividing by m,

$$i_L + \frac{p_L}{\rho_L} + \frac{1}{2}(u_L - u_S)^2 = i_R + \frac{p_R}{\rho_R} + \frac{1}{2}(u_R - u_S)^2.$$

Thus the energy equation reduces to

$$h_L + \frac{1}{2}(u_L - u_S)^2 = h_R + \frac{1}{2}(u_R - u_S)^2, \tag{2.44}$$

where h is the specific enthalpy $i + \frac{p}{\rho}$.

For a perfect gas,

$$h = \frac{c^2}{\gamma - 1},$$

and in a steady flow,

$$\frac{c^2}{\gamma - 1} + \frac{u^2}{2} = \frac{1}{2}\frac{\gamma + 1}{\gamma - 1}c^{*2},$$

where c^* is the speed of sound at the point where the flow has exactly sonic speed $u = c$. Now the energy equation (2.44) takes the form

$$\frac{c_L^2}{\gamma - 1} + \frac{1}{2}(u_L - u_S)^2 = \frac{c_R^2}{\gamma - 1} + \frac{1}{2}(u_R - u_S)^2 = \frac{1}{2}\frac{\gamma + 1}{\gamma - 1}c^{*2}.$$

Also the momentum equation (2.37) can be expressed as

$$\frac{\gamma+1}{\gamma-1}\rho + p_L = 2\frac{\gamma+1}{\gamma-1}\left(\frac{1}{2}(u_L - u_S)^2 + \frac{c_L^2}{\gamma-1}\right)\rho_L = \rho_L {c^*}^2$$

and

$$\frac{\gamma+1}{\gamma-1}\rho + p_R = 2\frac{\gamma+1}{\gamma-1}\left(\frac{1}{2}(u_R - u_S)^2 + \frac{c_R^2}{\gamma-1}\right)\rho_R = \rho_R {c^*}^2$$

with the consequence that

$$\frac{p_R - p_L}{\rho_R - \rho_L} = {c^*}^2,$$

and (2.43) reduces to Prandtl's relation

$$(u_R - u_S)(u_L - u_S) = {c^*}^2.$$

It is also possible to eliminate the velocities from the jump relation (2.44) for the energy by combining it with (2.42) to obtain

$$\begin{aligned}\frac{1}{2}\left(\frac{1}{\rho_L} + \frac{1}{\rho_R}\right)(p_R - p_L) &= \frac{1}{2}(u_L - u_S)^2 - \frac{1}{2}(u_R - u_S)^2 \\ &= h_L - h_R = i_L + \frac{p_L}{\rho_L} - \left(i_R + \frac{p_R}{\rho_R}\right),\end{aligned}$$

so

$$i_R - i_L = \frac{1}{2}(p_R + p_L)\left(\frac{1}{\rho_L} - \frac{1}{\rho_R}\right). \tag{2.45}$$

This relationship, known as the Hugoniot equation, connects the left and right thermodynamic states of any shock wave.

For a perfect gas, the Hugoniot equation takes the form

$$\frac{1}{\gamma-1}\left(\frac{p_R}{\rho_R} - \frac{p_L}{\rho_L}\right) = \frac{1}{2}(p_R + p_L)\left(\frac{1}{\rho_L} - \frac{1}{\rho_R}\right),$$

which reduces to

$$\frac{p_R}{p_L} = \frac{(\gamma+1)\rho_R - (\gamma-1)\rho_L}{(\gamma+1)\rho_L - (\gamma-1)\rho_R} \tag{2.46}$$

or

$$\frac{\rho_R}{\rho_L} = \frac{(\gamma+1)p_R - (\gamma-1)p_L}{(\gamma+1)p_L - (\gamma-1)p_R}. \tag{2.47}$$

It may now be verified that the entropy

$$S = \log\left(\frac{p}{\rho^\gamma}\right) = \log p - \gamma \log \rho$$

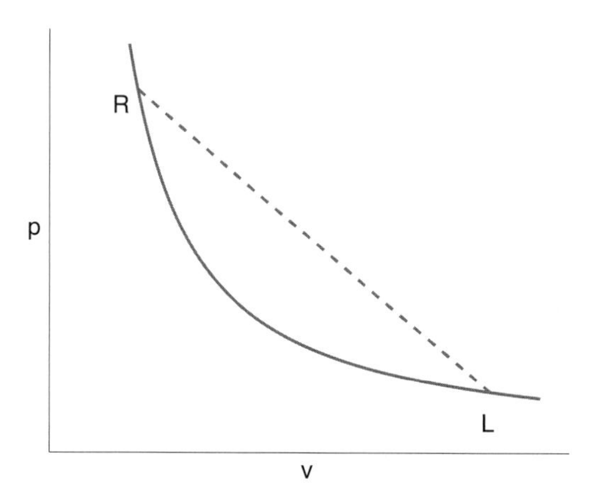

Figure 2.2 Hugoniot curve.

increases in a transition from p_L and ρ_L to p_R and ρ_R. Thus, only a compression shock is physically possible. In the case of an extreme shock where the incoming Mach number approaches infinity and $p_L \to 0$, the density ratio $\frac{\rho_R}{\rho_L}$ approaches the limiting value $\frac{\gamma+1}{\gamma-1}$.

The relations (2.46) and (2.47) are often expressed in terms of the specific volume $v = \frac{1}{\rho}$. For a given left state p_L and v_L, the curve connecting possible right states p_R and v_R is a hyperbola extending to the left, as illustrated in Figure 2.2.

To find the right state given the left state, note that (2.40) can be expressed as

$$\frac{p_R - p_L}{v_R - v_L} = -\frac{(u_L - u_S)^2}{v_L^2},$$

which defines a straight line through the left state that intersects the Hugoniot curve at the right state.

The Hugoniot curve can also be extended to the right of p_L and v_L to give states from which a transition to p_L and v_L through a shock wave is possible. The right Hugoniot curve from p_L and v_L would pass through p_R and v_R. However, it does not coincide with the left Hugoniot curve from p_L and v_L.

According to (2.39),

$$m^2 = \rho_L \frac{p_R - p_L}{1 - \frac{\rho_L}{\rho_R}} = \rho_R \frac{p_R - p_L}{\frac{\rho_R}{\rho_L} - 1}.$$

This can be combined with (2.45) to obtain alternative expressions for the mass flow m. From (2.45),

$$1 - \frac{\rho_L}{\rho_R} = \frac{2(p_R - p_L)}{(\gamma + 1)p_R + (\gamma - 1)p_L}$$

and

$$\frac{\rho_R}{\rho_L} = \frac{2(p_R - p_L)}{(\gamma+1)p_L + (\gamma-1)p_R}.$$

Thus

$$m^2 = 2\rho_L\big((\gamma+1)p_R + (\gamma-1)p_L\big) = 2\rho_R\big((\gamma+1)p_L + (\gamma-1)p_R\big),$$

and according to (2.38), the velocity jump can be expressed as

$$u_L - u_R = \frac{p_R - p_L}{m}.$$

3 Numerical Methods for the Solution of Partial Differential Equations

3.1 Introduction

The numerical methods that have been widely used for the solution of partial differential equations (PDEs), both in fluid dynamics and in other disciplines, fall into three main branches: finite difference methods, finite element methods and finite volume methods. These are reviewed in the next sections. A fourth branch, spectral methods, enable very accurate simulations provided that the geometric domain is simple. They have proved useful, for example, for problems such as the study of the decay of isotropic turbulence in a rectangular domain with periodic boundary conditions. Spectral methods may be viewed as a variant of finite element methods, in which the numerical solution is represented by an expansion in global rather than local basis functions.

3.2 Finite Difference Methods

3.2.1 Finite Difference Approximations to Derivatives

Finite difference methods were used in the earliest attempts to solve PDEs by numerical methods, dating back to the work of Richardson at the beginning of the twentieth century (Richardson 1911). They represent the simplest possible approach to the discretization of a PDE. One simply substitutes finite difference approximations for each derivative that appears in the equations.

To approximate the first derivative $f'(x)$ of a function $f(x)$, let f_j denote the value $f(x_j)$ at the point $x_j = jh$ on a uniform mesh with interval h. Then we can represent $f'(x_j)$ by the forward, backward, or central difference approximations

$$
\begin{aligned}
D_x^+ f_j &= \frac{f_{j+1} - f_j}{h} \\
D_x^- f_j &= \frac{f_j - f_{j-1}}{h} \\
D_x f_j &= \frac{f_{j+1} - f_{j-1}}{2h}.
\end{aligned}
$$

Expanding $f(x_j + h)$ in a truncated Taylor series,

$$f(x_j + h) = f(x_j) + hf'(x_j) + \frac{h^2}{2} f''(x_j + \theta h),$$

where $0 \leq \theta \leq 1$, it can be seen that

$$D_x^+ f_j = f'(x_j) + \frac{h}{2} f''(x_j + \theta h),$$

and accordingly the error

$$\left| D_x^+ f_j - f'(x_j) \right| \leq M \frac{h}{2},$$

where $f''(x)$ satisfies the bound $|f''(x)| \leq M$. Similarly

$$\left| D_x^- f_j - f'(x_j) \right| = \mathcal{O}(h).$$

In order to obtain an error estimate for $D_x f$, we expand the Taylor series for $f(x_j + h)$ and $f(x_j - h)$ to an extra term as

$$f(x_j + h) = f(x_j) + hf'(x_j) + \frac{h^2}{2!} f''(x_j) + \frac{h^3}{3!} f'''(x_j + \theta_1 h)$$

and

$$f(x_j - h) = f(x_j) + hf'(x_j) - \frac{h^2}{2!} f''(x_j) - \frac{h^3}{3!} f'''(x_j + \theta_2 h),$$

where $0 \leq \theta_1 \leq 1, 0 \leq \theta_2 \leq 1$. Now

$$D_x f_j - f'(x_j) = \frac{h^2}{12} \left(f'''(x_j + \theta_1 h) + f'''(x_j - \theta_2 h) \right).$$

Assuming that $f'''(x)$ is smooth, the right hand side is equal to $\frac{h^2}{6} f'''(x_j + \theta h)$, where $-1 \leq \theta \leq 1$ by the mean value theorem. Accordingly,

$$\left| D_x f_j - f'(x_j) \right| \leq M \frac{h^2}{6},$$

where $|f'''(x)| \leq M$.

Note that if $f(x)$ is a parabola, $f'''(x) = 0$, so that $D_x f$ is exact. Thus the central difference approximation is equivalent to evaluating the first derivative of a parabola that fits the data at the three points x_{j-1}, x_j, and x_{j+1}. Correspondingly, finite difference approximations with an order of accuracy $O(h^n)$ may be obtained by fitting the data on a stencil of $n + 1$ points by a Lagrange interpolation polynomial $p_n(x)$ of degree n and setting

$$D_x f_j = p_n'(x_j),$$

as is shown in Appendix C, Sections 2 and 3. This procedure may be used to obtain both centered and offset or one-sided formulas. For example, it yields the second order accurate one sided approximations

$$D_x^+ f(x_j) = \frac{-3f_j + 4f_{j+1} - f_{j+2}}{2h} = f'(x_j) + \mathcal{O}(h^2)$$

and

$$D_x^- f(x_j) = \frac{3f_j - 4f_{j-1} + f_{j-2}}{2h} = f'(x_j) + \mathcal{O}(h^2),$$

which can be used at boundary points. The same procedure may also be used to derive higher order finite difference approximations on non uniform grids.

Second and higher order derivatives can be approximated by repeated application of the differencing formula. For example, $f''(x_j)$ can be approximated as

$$\frac{D_x f_{j+1} - D_x f_{j-1}}{2h} = \frac{f_{j+2} - 2f_j + f_{j-2}}{4h^2}.$$

However, this spreads out the approximation over an unnecessarily wide five-point stencil, with the consequence that an odd-even mode $f(x_j) = (-1)^j$ is not recognized and is actually perceived as a function with a second derivative equal to zero. Discretizations that use non-compact formulas of this kind are prone to numerical instabilities.

A better approximation is obtained by setting

$$D_x f\left(x_{j+\frac{1}{2}}\right) = \frac{f_{j+1} - f_j}{h}, \quad D_x f\left(x_{j-\frac{1}{2}}\right) = \frac{f_j - f_{j-1}}{h}$$

and

$$\begin{aligned} D_{xx} f(x_j) &= \frac{1}{h}\left(D_x f\left(x_{j+\frac{1}{2}}\right) - D_x f\left(x_{j-\frac{1}{2}}\right)\right) \\ &= \frac{1}{h^2}(f_{j+1} - 2f_j + f_{j-1}). \end{aligned}$$

This formula can also be interpreted as

$$D_{xx} f(x_j) = D_x^+ D_x^- f(x_j) = D_x^- D_x^+ f(x_j).$$

Expansion of the terms in a truncated Taylor series now shows that

$$D_{xx} f(x_j) = f''(x_j) + \frac{h^2}{12} f^{(4)}(x_j + \theta h),$$

where $-1 \leq \theta \leq 1$, or

$$\left|D_{xx} f(x_j) - f''(x_j)\right| \leq M \frac{h^2}{12}, \tag{3.1}$$

where $|f^{(4)}(x)| \leq M$.

The same approximation to $f''(x_j)$ can be obtained by differentiating the parabola

$$p_2(x) = f_j + (f_{j+1} - f_{j-1})\frac{x}{2h} + (f_{j+1} - 2f_j + f_{j-1})\frac{x^2}{2h},$$

which interpolates the data at x_{j-1}, x_j, and x_{j+1}.

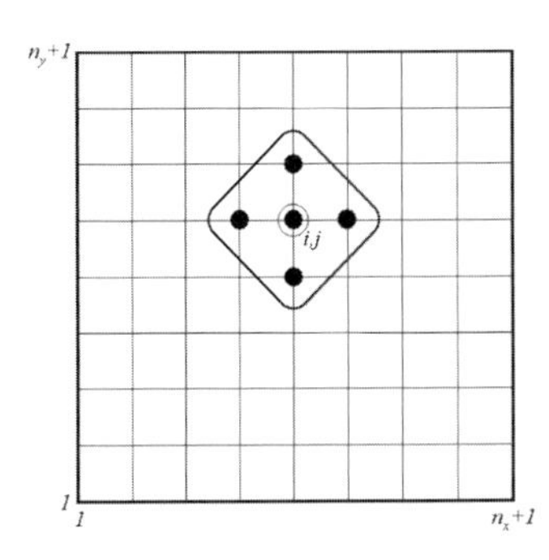

Figure 3.1 Grid and five-point stencil.

3.2.2 Finite Difference Approximation for Laplace's Equation

As an example of the application of the finite difference method to the solution of a PDE, consider the numerical solution of Laplace's equation

$$\frac{\partial^2 u}{\partial x^2} + \frac{\partial^2 u}{\partial y^2} = 0$$

on a rectangular domain with Dirichlet boundary conditions. Using a Cartesian mesh with intervals Δx and Δy in the x and y directions, let $v_{i,j}$ denote the approximate solution at $x = i\Delta x$, $y = j\Delta y$. Then the PDE can be represented at every interior point by the Laplace difference equation

$$\frac{v_{i+1,j} - 2v_{i,j} + v_{i-1,j}}{\Delta x^2} + \frac{v_{i,j+1} - 2v_{i,j} + v_{i,j-1}}{\Delta y^2} = 0 \tag{3.2}$$

on the five-point stencil consisting of i,j and the four nearest neighbors, as illustrated in Figure 3.1. According to the foregoing error analysis summarized in (3.1), this is a second order accurate discretization of the PDE. If the mesh has n_x and n_y intervals in the x and y directions, this yields $(n_x - 1)(n_y - 1)$ equations for the unknowns $v_{i,j}$ at the interior mesh points where the boundary values are specified by the Dirichlet boundary conditions. Then the numerical solution can be obtained by solving the linear system of equations.

3.2.3 Finite Difference Approximation of a Conservation Law

As another example, consider the scalar conservation law

$$\frac{\partial u}{\partial t} + \frac{\partial}{\partial x} f(u) = 0.$$

A semi-discrete central difference approximation on a uniform mesh with interval Δx is

$$\frac{dv_j}{dt} + \frac{f_{j+1} - f_{j-1}}{2\Delta x} = 0,$$

where v_j denotes the numerical solution at $x_j = j\Delta x$ and $f_j = f(v_j)$, and the time dependent solution is obtained by advancing the coupled set of ordinary differential equations (ODEs) in time. The numerical solution may or may not be stable, remaining bounded or growing without bound as the number of time steps is increased, depending on both the time discretization scheme used to solve the ODEs and the space discretization scheme.

In the case of the linear advection equation

$$\frac{\partial u}{\partial t} + a\frac{\partial u}{\partial x} = 0,$$

which represents wave motion at a speed a, a fully discrete scheme can be obtained by combining a central difference spatial discretization and a forward Euler time discretization. Denoting the numerical solution at $t = n\Delta t$ and $x = j\Delta x$ by v_j^n, this can be written as

$$v_j^{n+1} = v_j^n - \frac{\lambda}{2}(v_{j+1}^n - v_{j-1}^n),$$

where the parameter $\lambda = \frac{a\Delta t}{\Delta x}$ is the fraction of the mesh width covered by propagation of the wave during one time step. This scheme proves to be unstable, as will be analyzed in the next chapter. If $a > 0$, corresponding to right traveling waves, the true solution depends on data along the backward characteristics to the left. This motivates the use of an upwind spatial discretization

$$v_j^{n+1} = v_j^n - \lambda(v_j^n - v_{j-1}^n).$$

If $0 < \lambda < 1$, this scheme is stable, since v_j^{n+1} is a convex combination of v_j^n and v_{j-1}^n,

$$v_j^{n+1} = (1-\lambda)v_j^n + \lambda v_{j-1}^n,$$

where the coefficients of v_j^n and v_{j-1}^n are non-negative. Consequently,

$$\begin{aligned}|v_j^{n+1}| &\leq (1-\lambda)|v_j^n| + \lambda|v_{j-1}^n| \\ &\leq \max_j |v_j^n|.\end{aligned}$$

It should be noted, however, that the central difference scheme is stable when it is combined with a higher order Runge–Kutta time discretization scheme, although it may exhibit oscillatory behavior in the neighborhood of a discontinuity in the solution. The systematic analysis of these issues is the topic of the following chapters.

3.2.4 Error Analysis for the Discrete Poisson Equation

In this section, we derive error bounds for discrete solutions of Poisson's equation. Suppose that u satisfies Poisson's equation in a square domain $0 \leq x \leq 1, 0 \leq y \leq 1$, with Dirichlet boundary conditions on the boundary $\mathcal{B}$

$$Lu = u_{xx} + u_{yy} = f \quad \text{in } \mathcal{D} \tag{3.3}$$

$$u = u_b \quad \text{on } \mathcal{B}.$$

For the sake of simplicity, we calculate the discrete solution $u_h = v$ on a Cartesian mesh with equal intervals $\Delta x = \Delta y = h$, where we use the notation v to suppress the subscript h when it is not needed. Then, v satisfies the net equation

$$L_h v = f \quad \text{in } \mathcal{D}_h \tag{3.4}$$

$$v = u_b \quad \text{in } \mathcal{B}_h, \tag{3.5}$$

where $\mathcal{D}_h$ consists of the interior mesh points, $\mathcal{B}_h$ consists of the boundary points, $f_{i,j}$ is the value of f at the mesh point i,j, and

$$L_h v_{i,j} = \frac{1}{h^2}(v_{i+1,j} + v_{i-1,j} + v_{i,j+1} + v_{i,j-1} - 4v_{i,j}). \tag{3.6}$$

The local truncation error is defined as

$$\tau_h = L_h u - f,$$

where u is the exact solution, or equivalently since u satisfies (3.3) as

$$\tau_h = L_h u - Lu.$$

According to the error estimate (3.1) for the second difference approximations to $\frac{\partial^2 u}{\partial x^2}$ and $\frac{\partial^2 u}{\partial y^2}$,

$$\tau_h \leq \frac{h^2}{12}(M_x + M_y), \tag{3.7}$$

where $\left|\frac{\partial^4 u}{\partial x^4}\right| \leq M_x$, $\left|\frac{\partial^4 u}{\partial y^4}\right| \leq M_y$. Now, we derive a global error bound for the solution error $v - u$ using arguments based on the maximum principle.

First, we note that if $f = 0$, v cannot have an extremum at any interior point in $\mathcal{D}_h$ because according to (3.4) and (3.6), it cannot have a value greater or less than the values of all four of its neighbors. Accordingly, $\max_{\mathcal{D}_h} |v| \leq \max_{\mathcal{B}_h} |u_B|$, and we may conclude that the Laplace discretization is stable in the sense that the discrete solution remains bounded as $h \to 0$.

To derive a bound on the error $v - u$, we need a more refined statement of the maximum principle, as follows. Let w be any net function defined on $\mathcal{D}_h + \mathcal{B}_h$ that satisfies

$$L_h w \geq 0 \quad \text{in } \mathcal{D}_h.$$

Then

$$\max_{\mathcal{D}_h} w \leq \max_{\mathcal{B}_h} w.$$

Alternatively, if

$$L_h w \leq 0 \quad \text{in } \mathcal{D}_h,$$

then

$$\min_{\mathcal{D}_h} w \geq \min_{\mathcal{B}_h} w.$$

In the first case, it follows from (3.5) that at every interior point, $w_{i,j} \leq$ the maximum value of its neighbors, with the consequence that w cannot have a maximum in $\mathcal{D}_h$. In the second case, $w_{i,j} \geq$ the minimum value of its neighbors, and hence w cannot have a minimum in $\mathcal{D}_h$.

Next, we derive a bound on the maximum value of any net function w. Let

$$N = \max_{\mathcal{D}_h}(L_h w),$$

and consider the function

$$N\phi + w,$$

where

$$\phi = \frac{x^2}{2}.$$

Then,

$$L_h\phi = 1 \quad \text{in } \mathcal{D}_h, \qquad \max_{\mathcal{B}_h} \phi = \frac{1}{2}.$$

Consequently,

$$L_h(N\phi + w) \geq 0 \quad \text{in } \mathcal{D}_h,$$

and by the maximum principle,

$$N\phi + w \leq \max_{\mathcal{B}_h}(w) + \frac{1}{2}N.$$

Since $\phi \geq 0$, it follows that

$$w \leq \max_{\mathcal{B}_h}(w) + \frac{1}{2}N.$$

Applying the same argument to $N\phi - w$,

$$-w \leq \max_{\mathcal{B}_h}(-w) + \frac{1}{2}N.$$

Thus,

$$\left|w\right| \leq \max_{\mathcal{B}_h} \left|w\right| + \frac{1}{2} \max_{\mathcal{D}_h}(L_h w). \tag{3.8}$$

It follows from (3.4) that the solution v satisfies the bound

$$|v| \le \max_{\mathcal{B}_h} |v| + \frac{1}{2} \max_{\mathcal{D}_h} |f|.$$

Note that this proof could be carried out with alternative choices of ϕ that satisfy $L_h\phi = 1$, such as $\frac{1}{2}y^2$ or $\frac{1}{4}(x^2+y^2)$. Note also that the maximum principle holds for any set of discrete equations which is diagonally dominant.

Now, consider $e_h = v - u$. Since u and v satisfy (3.3) and (3.4), it follows from the definition (3.7) that

$$L_h e_h = -\tau_h \quad \text{in } \mathcal{D}_h. \tag{3.9}$$

Also, since $v = u = u_b$ on the boundary,

$$e_h = 0 \quad \text{in } \mathcal{B}_h.$$

Hence, by the estimate (3.8) it follows that

$$\begin{aligned} |e_h| &\le \frac{1}{2} \max_{\mathcal{D}_h} |\tau_h| \\ &\le \frac{h^2}{24}(M_x + M_y). \end{aligned}$$

Note also that the discrete Poisson equation (3.9), which defines e_h, has the same form as the discrete Poisson equation (3.4), which defines $u_h = v$, with $-\tau_h$ substituted for f. Now if we retain additional terms in the Taylor series for u,

$$\tau_h = \frac{h^2}{12}\left(\frac{\partial^4 u}{\partial x^4} + \frac{\partial^4 u}{\partial y^4}\right) + \frac{h^4}{180}\left(\frac{\partial^6 u}{\partial x^6} + \frac{\partial^6 u}{\partial y^6}\right) + \mathcal{O}(h^6).$$

Thus, we can regard e_h as a discrete approximation to a function $h^2 e_2$, which satisfies the Poisson equation

$$Le_2 = -\tau_2,$$

where

$$\tau_2 = \frac{1}{12}\left(\frac{\partial^4 u}{\partial x^4} + \frac{\partial^4 u}{\partial y^4}\right).$$

Now,

$$L_h(u_h - (u + h^2 e_2)) = -h^4 \tau_4 + \mathcal{O}(h^6),$$

where

$$\tau_4 = \frac{1}{180}\left(\frac{\partial^6 u}{\partial x^6} + \frac{\partial^6 u}{\partial y^6}\right) + \frac{1}{12}\left(\frac{\partial^4 e_2}{\partial x^4} + \frac{\partial^4 e_2}{\partial y^4}\right).$$

Hence, $u_h - (u + h^2 e_2)$ is an approximation to $h^4 e_4$ where e_4 satisfies

$$Le_4 = -\tau_4.$$

In this way, we can recursively define an error expansion

$$u_h = u + h^2 e_2 + h^4 e_4 + \cdots ,$$

where each error coefficient u_k satisfies a Poisson equation, provided that u has a sufficient number of bounded derivatives. Then, $u_h - h^2 e_2$ is a fourth order accurate approximation to u, and so on.

One could correct u_h by a second order accurate discrete approximation e_{2h} to e_2 by solving the discrete Poisson equation

$$L_h e_{2h} = \tau_{2h},$$

where τ_{2h} is calculated by substituting finite difference approximations to $\frac{\partial^4 u}{\partial x^4}$ and $\frac{\partial^4 u}{\partial y^4}$. A much less expensive way to obtain a fourth order accurate approximation is to use Richardson extrapolation. Let u_{2h} be a solution on a mesh with an interval $2h$, consisting of every second point in each coordinate direction. Then,

$$u_{2h} = u + 4h^2 e_2 + 16h^4 e_4 + \cdots ,$$

and since the error coefficients e_k are independent of h, we can take a linear combination of u_h and u_{2h} such that the first error term is canceled. In this manner, we can obtain the improved solution

$$\frac{4}{3} u_h - \frac{1}{3} u_{2h} = u + \mathcal{O}(h^4)$$

at the mesh points of the coarser mesh.

3.2.5 Iterative Solution of the Laplace Difference Equation

The solution of the difference equations for an elliptic problem requires the inversion of a large sparse matrix. While fast methods are available for the discrete Laplacian operator, in the case of more general elliptic operators, the operation count of direct inversion can become prohibitive for solutions on fine meshes. This issue, which is accentuated in three-dimensional problems, has motivated the development of iterative methods for the solution of the difference equations. These methods, commonly called relaxation methods, were pioneered in the period 1930–1945 by Southwell (1946), before the advent of electronic computers.

We shall consider here the iterative solution of the Laplace difference equation (3.2) on a mesh with equal intervals $\Delta x = \Delta y = h$, as a representative example. Let $v^n_{i,j}$ donate the approximate solution at the mesh point i,j after n iterations. The two classic approaches are the Jacobi and Gauss–Seidel methods. In both methods, we sweep over the mesh, updating the solution at each mesh point in turn.

In the Jacobi method, we solve the equations at each point using values from the previous iteration at the neighboring points to obtain a provisional value $\tilde{v}_{i,j}$,

$$\tilde{v}_{i,j} = \frac{1}{4}\left(v^n_{i+1,j} + v^n_{i-1,j} + v^n_{i,j+1} + v^n_{i,j-1}\right), \tag{3.10}$$

and then we update the solution at that point by the formula

$$v_{i,j}^{n+1} = v_{i,j}^n + r\left(\tilde{v}_{i,j} - v_{i,j}^n\right), \tag{3.11}$$

where r is a relaxation factor, which may be chosen to optimize the rate of convergence. In the Gauss–Seidel method, we solve the equations at each point using the latest available values of the solution at the neighboring points as they are generated during the sweep. If we march row by row from left to right, the provisional value is

$$\tilde{v}_{i,j} = \frac{1}{4}\left(v_{i+1,j}^n + v_{i-1,j}^{n+1} + v_{i,j+1}^n + v_{i,j-1}^{n+1}\right), \tag{3.12}$$

and then $v_{i,j}^{n+1}$ is updated as in the Jacobi method by the formula (3.11). It is well known that both the Jacobi and Gauss–Seidel methods converge for diagonally dominant systems of equations. The diagonal coefficient of the Laplace difference equations, however, is exactly equal to the sum of the coefficients of the neighboring points in the stencil, complicating the analysis of the methods. The performance of alternative iterative methods has actually been exhaustively studied in classic works of Young, Varga, and Wachspress (Wachspress 1963, Varga 2000, Young 2003).

As one might expect, the Gauss–Seidel method converges faster than the Jacobi method, in fact much faster with an optimum choice of the relaxation factor. On the other hand, the Jacobi method allows all the mesh points to be updated simultaneously if a parallel computer with a sufficient number of processors is available, whereas the Gauss–Seidel method with successive ordering, as defined in (3.12), requires sequential processing. In the early days of relaxation methods, it was a common practice to search for the mesh point with the largest residual error and update that point next. This proved inconvenient when electronic computers were introduced, because it prevented the use of simple loops to implement the updating process. A major breakthrough was achieved by Young when he proved that the Gauss–Seidel method converges with a fixed ordering (Young 2003). Subsequently, it was shown that the same asymptotic rate of convergence can be obtained with different orderings, including the checkerboard ordering with alternate red and black points, which allows simultaneous computation of the updates at points of the same color (Varga 2000).

We can regard each step in the iterative process as corresponding to a time level in a time dependent process. This allows the iterative process to be viewed as a discretization of a time dependent problem with well known properties. It turns out that this approach can provide valuable insights into the behavior of the iterative schemes with comparatively simple analysis, and we use it in the following paragraphs to analyze both the Jacobi and Gauss–Seidel methods. Consider first the Jacobi method, as it is defined by (3.10) and (3.11). We can rewrite (3.10) as

$$\tilde{v}_{i,j} - v_{i,j}^n = \frac{h^2}{4} L_h v_{i,j},$$

where L_h is the discrete Laplacian operator, and hence according to (3.11),

$$v_{i,j}^{n+1} = v_{i,j}^n + r\frac{h^2}{4} L_h v_{i,j}. \tag{3.13}$$

Now, if we set

$$\Delta t = r\frac{h^2}{4},$$

we obtain the standard forward Euler finite difference approximation to the heat equation

$$\frac{\partial u}{\partial t} = \frac{\partial^2 u}{\partial x^2} + \frac{\partial^2 u}{\partial y^2}.$$

Moreover, we can write (3.13) as

$$v_{i,j}^{n+1} = (1-4\beta)v_{i,j}^n + \beta(v_{i+1,j}^n + v_{i-1,j}^n + v_{i,j+1}^n + v_{i,j-1}^n),$$

where $\beta = \frac{\Delta t}{h^2}$. If $r \leq 1$ so that $\Delta t \leq \frac{h^2}{4}$ and $\beta \leq \frac{1}{4}$, all the coefficients on the right are non-negative. It follows that

$$\begin{aligned}\left|v_{i,j}^{n+1}\right| &\leq (1-4\beta)\left|v_{i,j}^n\right| + \beta\left(\left|v_{i+1,j}^n\right| + \left|v_{i-1,j}^n\right| + \left|v_{i,j+1}^n\right| + \left|v_{i,j-1}^n\right|\right) \\ &\leq \max_{i,j}\left|v_{i,j}^n\right|.\end{aligned}$$

Thus, the discrete solution is stable in the maximum norm. Since the discrete scheme is consistent with the heat equation, it now follows from the Lax equivalence theorem, which will be proved in Chapter 4, that it converges to the true solution of the heat equation as the mesh interval $h \to 0$. And since solutions of the heat equation decay, we conclude that the Jacobi method must be convergent as the mesh is refined, provided that $r \leq 1$. Moreover, since this condition implies that $\Delta t \leq \frac{h^2}{4}$, we can also infer that the number of iterations needed for convergence is proportional to $\frac{1}{h^2}$. These conclusions are confirmed by more rigorous algebraic analysis.

We next consider the Gauss–Seidel method with row by row successive ordering, as defined by (3.12) and (3.11). We can solve (3.11) for $\tilde{v}_{i,j}$ as

$$\tilde{v}_{i,j} = \frac{1}{r}v_{i,j}^{n+1} + \left(1 - \frac{1}{r}\right)v_{i,j}^n.$$

Substituting this in (3.12) and dividing by h^2, we obtain

$$\begin{aligned}&\frac{1}{h^2}\left(v_{i+1,j}^n - \frac{2}{r}v_{i,j}^{n+1} - 2\left(1-\frac{1}{r}\right)v_{i,j}^n + v_{i-1,j}^{n+1}\right) + \\ &\quad + \frac{1}{h^2}\left(v_{i,j+1}^n - \frac{2}{r}v_{i,j}^{n+1} - 2\left(1-\frac{1}{r}\right)v_{i,j}^n + v_{i,j-1}^{n+1}\right) = 0.\end{aligned} \tag{3.14}$$

The first term can be rearranged as

$$\begin{aligned}&\frac{1}{h^2}\left(v_{i+1,j}^n - 2v_{i,j}^n + v_{i-1,j}^n\right) - \frac{1}{h^2}\left(v_{i,j}^{n+1} - v_{i,j}^n - v_{i-1,j}^{n+1} + v_{i-1,j}^n\right) + \\ &\qquad - \frac{1}{h^2}\left(\frac{2}{r} - 1\right)\left(v_{i,j}^{n+1} - v_{i,j}^n\right).\end{aligned}$$

Now with a time step $\Delta t = h$, set

$$\left(\frac{2}{r} - 1\right) = \alpha h, \quad r = \frac{2}{1 + \alpha h}. \tag{3.15}$$

Then if $v_{i,j}$ is a net function corresponding to a smooth function $u(x, y)$, the first term is an approximation to

$$u_{xx} - u_{xt} - \alpha u_t.$$

Similarly, the second term is an approximation to

$$u_{yy} - u_{yt} - \alpha u_t.$$

Thus, we can regard (3.14) as an approximation to a wave equation of the form

$$u_{xx} - u_{xt} + u_{yy} - u_{yt} - 2\alpha u_t = 0. \tag{3.16}$$

Under the transformation

$$X = x, \quad Y = y, \quad T = t + \frac{1}{2}x + \frac{1}{2}y,$$

the chain rule yields

$$\begin{aligned}
u_x &= u_X X_x + u_Y Y_x + u_T T_x = u_X + \frac{1}{2}u_T \\
u_t &= u_X X_t + u_Y Y_t + u_T T_t = u_T \\
u_{xx} &= u_{XX} + u_{XT} + \frac{1}{4}u_{TT} \\
u_{xt} &= u_{XT} + \frac{1}{2}u_{TT},
\end{aligned}$$

with similar expressions for u_{yy} and u_{yt}. Thus, (3.16) is transformed to a damped wave equation in the diagonal form

$$\frac{1}{2}u_{TT} + 2\alpha u_T = u_{XX} + u_{YY}, \tag{3.17}$$

where the damping coefficient α is positive if $r < 2$. With an optimum choice of α, this gives a much faster rate of convergence than the heat equation. In fact, because the time step is proportional to h, we can expect convergence with a number of iterations proportional to $\frac{1}{h}$.

We shall now show that the red-black checkerboard ordering leads to the same equation, and accordingly, it can be expected to yield the same asymptotic rate of convergence. We regard the red and black points as belonging to time levels n and $n + \frac{1}{2}$ separated by half a time step. Now, the Gauss–Seidel equations become

$$\tilde{v}_{i,j} = \frac{1}{4}\left(v_{i+1,j}^{n+\frac{1}{2}} + v_{i-1,j}^{n+\frac{1}{2}} + v_{i,j+1}^{n+\frac{1}{2}} + v_{i,j-1}^{n+\frac{1}{2}}\right)$$

and

$$v_{i,j}^{n+1} = v_{i,j}^{n} + r(\tilde{v}_{i,j} - v_{i,j}^{n}) = (1-r)v_{i,j}^{n} + r\tilde{v}_{i,j}.$$

Then with αh defined as before by (3.15),

$$v_{i,j}^{n+1} = \frac{\alpha h - 1}{\alpha h + 1} v_{i,j}^{n} + \frac{2}{\alpha h + 1}\tilde{v}_{i,j}.$$

Now multiplying by $1 + \alpha h$,

$$(1+\alpha h)v_{i,j}^{n+1} - 2v_{i,j}^{n+\frac{1}{2}} + (1-\alpha h)v_{i,j}^{n} = 2\left(\tilde{v}_{i,j} - v_{i,j}^{n+\frac{1}{2}}\right)$$

$$= \frac{1}{2}\left(v_{i+1,j}^{n+\frac{1}{2}} - 2v_{i,j}^{n+\frac{1}{2}} + v_{i-1,j}^{n+\frac{1}{2}} + v_{i,j+1}^{n+\frac{1}{2}} - 2v_{i,j}^{n+\frac{1}{2}} + v_{i,j-1}^{n+\frac{1}{2}}\right).$$

Dividing by h^2 and setting $\Delta t = h$, the left side is an approximation to $\frac{1}{4}u_{tt} + \alpha u_t$, while the right side is an approximation to $\frac{1}{2}(u_{xx} + u_{yy})$. Thus, we recover the damped wave (3.17) without the need for a coordinate transformation.

Code fragment for the solution of Laplace's equation on a Cartesian mesh by Gauss–Seidel iteration

```
do step istep = 1,nstep

   do j = 2,ny
   do i = 2,nx

      upr = 0.25*(u(i+1,j)+u(i-1,j)+u(i,j+1)+u(i,j+1))
      u(i,j) = u(i,j) + r*(upr-u(i,j))

   end do
   end do

end do step
```

3.2.6 Multigrid Method

A further dramatic improvement in the rate of convergence can be achieved by the introduction of multiple grids, where corrections are calculated on coarser grids based on the residual errors on fine grids, in such a way that the coarser grids drive the solution on the finest grid toward the true discrete solution. Such a procedure was first formally proposed for the Laplace difference equations by Fedorenko (1964). It was subsequently proved by Bakhvalov (1966) that converged solutions could be obtained with a number of iterations independent of the number of unknowns. These proofs were extended to general elliptic systems by Nicolaides (1975) and Hackbusch (1978). A major advance was achieved by Brandt (1977), who showed that elliptic problems could actually be solved in about 10 iterations, and also introduced the full approximation scheme (FAS) for nonlinear problems.

Consider a linear elliptic problem

$$Lu = f.$$

Suppose that it is discretized on a mesh M_h with interval h as

$$L_h v_h = f_h. \tag{3.18}$$

This could, for example, be the Poisson difference equation on a two-dimensional Cartesian grid with equal intervals $\Delta x = \Delta y = h$. Suppose now that we are solving (3.18) by an iterative process in which v_h^n is the result after n iterations. Then, the exact solution would be obtained by adding a correction δv_h such that

$$L_h \left(v_h^n + \delta v_h\right) = f_h,$$

or

$$L_h \delta v_h + R_h = 0, \tag{3.19}$$

where the residual is

$$R_h = L_h v_h^n - f_h.$$

We can now introduce a coarser mesh M_{2h} with an interval $2h$ by deleting alternate mesh points in each coordinate direction, as illustrated in Figure 3.2. The basic idea of the multigrid process is to transfer the calculation of the correction to the coarser grid by the equation

$$L_{2h} \delta v_{2h} + I_{2h}^h R_h = 0. \tag{3.20}$$

Here, L_{2h} is a discretization of the operator L on the coarser grid, while the operator I_{2h}^h aggregates the fine grid residuals to a corresponding weighted average on the coarse grid. In the case of the Cartesian mesh under consideration, a natural choice for the weights in I_{2h}^h is illustrated in Figure 3.3.

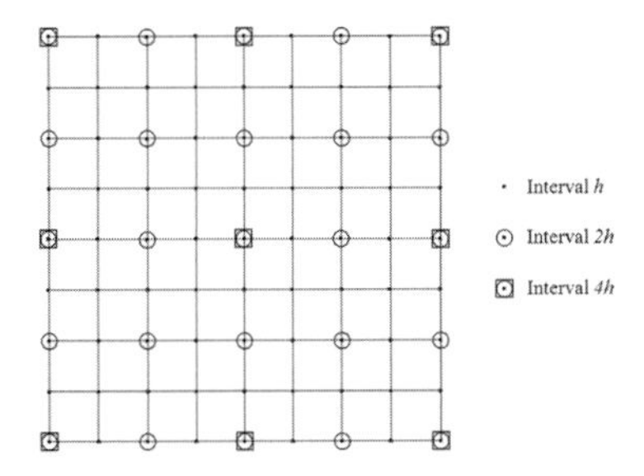

Figure 3.2 Nested grids with intervals h, $2h$, and $4h$.

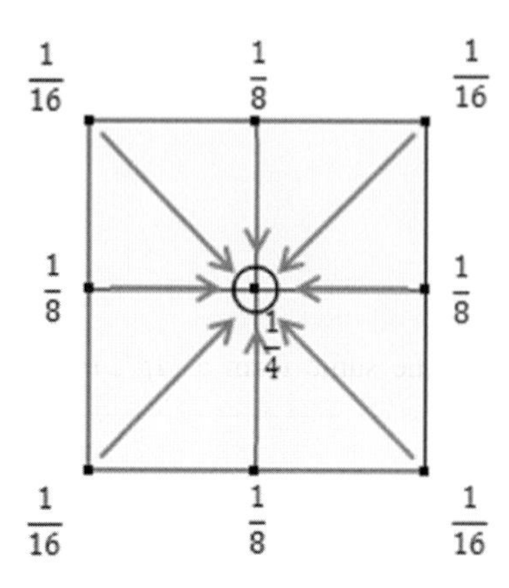

Figure 3.3 Aggregation stencil for the residuals.

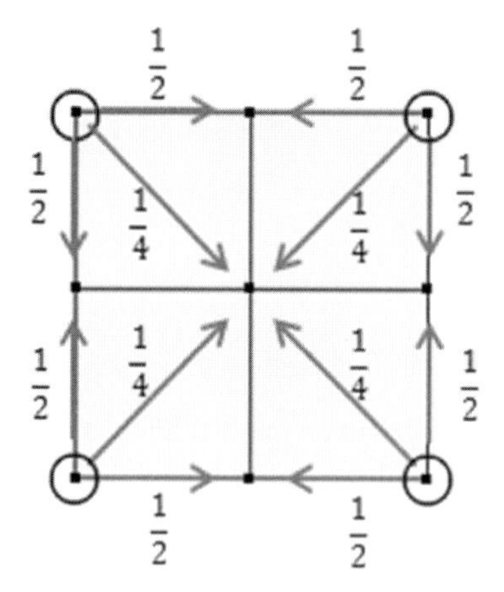

Figure 3.4 Linear interpolation from the coarse to the fine grid.

We can now interpolate the coarse grid correction to the fine grid to update the solution by the formula

$$v_h^{new} = v_h + I_h^{2h} \delta v_{2h},$$

where I_h^{2h} is an interpolation operator. For this purpose, we may simply use linear interpolation from the nearest coarse grid points, as illustrated in Figure 3.4.

In principle, we could calculate δv_{2h} exactly by solving the corresponding system of linear equations. But we can also apply the same iterative scheme to the coarse grid equation (3.20). For this purpose, we transfer the solution v_h to give an initial value $v_{2h}^{(0)}$ on the coarse mesh M_{2h}. In the case of a Cartesian mesh, one may simply set $v_{2h}^{(0)}$ equal to the value v_h at the coincident fine mesh point. Now, after some number of iterations, we can again double the mesh interval to produce a yet coarser mesh M_{4h}, and transfer the correction equation to this mesh as

$$L_{4h} \delta v_{4h} + I_{4h}^{2h} R_{2h}.$$

Here, R_{2h} is the residual on M_{2h}, which should be recalculated using the updated coarse mesh solution values, and I_{4h}^{2h} is an aggregation operator of the same form as I_h^{2h}. Now, we can update the solution on M_{2h} by interpolating the correction from M_{4h},

$$v_{2h}^{new} = v_{2h} + I_{2h}^{4h}\delta v_{2h},$$

where I_{2h}^{4h} is an interpolation operator of the same form as I_h^{2h}, typically a linear interpolation operator. Now, however, the correction to be transferred to the fine grid is

$$\delta v_{2h} = v_{2h}^{new} - v_{2h}^{(0)},$$

where $v_{2h}^{(0)}$ is the initial value originally transferred from the fine grid.

This process may be extended recursively to successively coarser grids until the coarsest mesh is reduced to two intervals in some coordinate direction. For example, we might solve the Poisson difference equation on the following sequence of grids with seven levels:

$$\begin{gathered} 128 \times 128 \\ 64 \times 64 \\ 32 \times 32 \\ 16 \times 16 \\ 8 \times 8 \\ 4 \times 4 \\ 2 \times 2. \end{gathered}$$

With Dirichlet boundary conditions, there is only one unknown on the coarsest grid, with the consequence that it can immediately be determined exactly.

In the case of a relaxation method such as the Gauss–Seidel method with successive ordering, an update at any mesh point does not influence the solution at its neighbor to the left until the following iteration. Thus, on a mesh with n intervals, n iterations will be required to transfer information from the far end to the near end of the mesh. It is evident that the number of iterations required for convergence must be some multiple of the number of mesh intervals in any coordinate direction. With a multigrid scheme, however, this information can be transferred across the mesh in a number of iterations proportional to the number of intervals on the coarsest mesh. This is the fundamental reason why multigrid schemes can yield very rapid convergence.

Full approximation scheme (FAS)

In the case of a nonlinear equation, say

$$L(v) = f,$$

discretized as

$$L_h(v_h) = f_h, \tag{3.21}$$

the addition of a correction δv_h would give

$$L_h(v_h^n + \delta v_h) = L_h(v_h^n) + N_h \delta v_h + \mathcal{O}\|\delta v_h\|^2,$$

where N_h is the Jacobian matrix $\frac{\partial L_h}{\partial v_h}$. Then, the correction equation (3.19) would take the form

$$N_h(\delta v_h) + R_h = 0,$$

where

$$R_h = L_h(v_h^n) - f_h.$$

In order to avoid the need to explicitly introduce the Jacobian matrix, Brandt (1977) proposed the following "full approximation scheme" (FAS). Considering again the linear equation (3.18) suppose that the solution is transferred to the coarse mesh M_{2h} with the initial value $v_{2h}^{(0)}$, and f_{2h} represents the corresponding source term. The initial coarse grid residual is

$$R_{2h}^{(0)} = L_{2h} v_{2h}^{(0)} - f_{2h}.$$

Now, we can add $L_{2h} v_{2h}^{(0)} - f_{2h} - R_{2h}^{(0)}$ to the correction equation (3.20) to obtain

$$L_{2h} v_{2h} - f_{2h} + I_{2h}^h R_h - R_{2h}^{(0)} = 0.$$

Here, we solve for the updated coarse grid solution v_{2h} instead of the correction δv_{2h}, using an equation of the same form as the fine grid equation with the additional source term

$$P_{2h} = I_{2h}^h R_h - R_{2h}^{(0)}. \tag{3.22}$$

With this procedure, the correction to be transferred to the fine mesh is based on the difference $v_{2h} - v_{2h}^{(0)}$, so the fine grid update takes the form

$$v_h^{new} = v_h + I_{2h}^h (v_{2h} - v_{2h}^{(0)}).$$

The full approximation scheme can easily be adapted to nonlinear problems. In order to solve the nonlinear equation (3.21), one simply replaces the coarse grid correction equation (3.2.6) by the corresponding nonlinear correction equation

$$L_{2h}(v_{2h}) - f_{2h} + P_{2h} = 0,$$

where the source term P_{2h} is defined by (3.22). It should be noted that in evaluating P_{2h}, the fine grid residuals R_h should be calculated with the latest available solution values after the completion of the iterations on the fine grid.

Implementation of the full approximation scheme requires three transfer operators. First, we need to transfer the solution to the coarse grid,

$$v_{2h}^{(0)} = Q_{2h}^h(v_h),$$

using a solution transfer operator Q_{2h}^h. In the case of a Cartesian mesh with coincident coarse and fine mesh points, we may simply inject the fine mesh values to the coarse

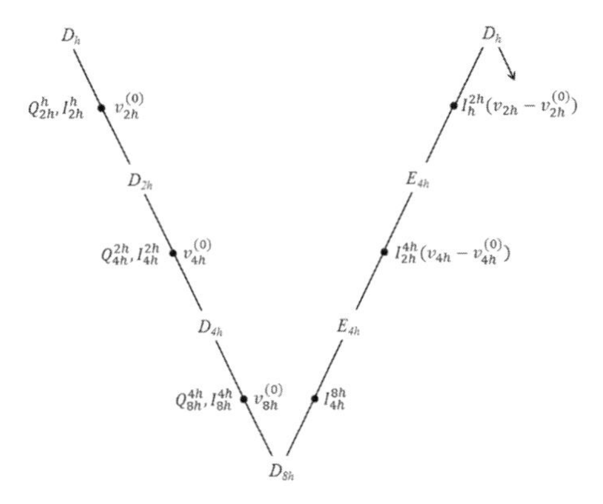

Figure 3.5 Mutigrid V cycle. D denotes three updates of the solution at any level, followed by a recalculation of the residuals. E denotes one update of the solution.

mesh. Then, we need a residual transfer operator I^h_{2h} to transfer the fine mesh residuals to the coarse mesh. Finally, we need an interpolation operator I^{2h}_h to transfer the coarse mesh corrections to the fine mesh. These transfer operations are often called "restriction" (of the solution from the fine to the coarse mesh) and "prolongation" (of the solution from the coarse to the fine mesh).

Early works on multigrid methods suggested the use of adaptive cycles, in which a sufficient number of iterations would be used to reduce the residual by a fixed fraction at any grid level. Such procedures are subject to the risk, particularly in the case of nonlinear problems, that the desired threshold might never be reached at some coarse grid level, with the result that they fail to return to the fine grid. In practice, it has been found that fixed cycles are very effective. Figure 3.5 illustrates a V-cycle in which the solution is updated three times on each level on the way down, and once on the way up. On the way down, it is also necessary to recalculate the residuals with the updated solution values.

With a fixed cycle, it may be observed that the computational cost of an update is divided by four in two-dimensional calculations, or eight in three-dimensional calculations, each time the mesh interval is doubled. Thus, the computational cost of the entire cycle relative to an equal number of updates on the fine mesh is of the order of

$$1 + \frac{1}{4} + \frac{1}{16} + \frac{1}{64} + \cdots \leq \frac{4}{3}$$

in two dimensions and

$$1 + \frac{1}{8} + \frac{1}{64} + \frac{1}{512} + \cdots \leq \frac{8}{7}$$

in three dimensions, plus the relatively small overhead of the mesh transfer operations.

Modal analysis

An important issue with multigrid methods is that a high frequency mode on the fine grid may be "aliased" to a lower frequency mode on the coarser grid, as illustrated in

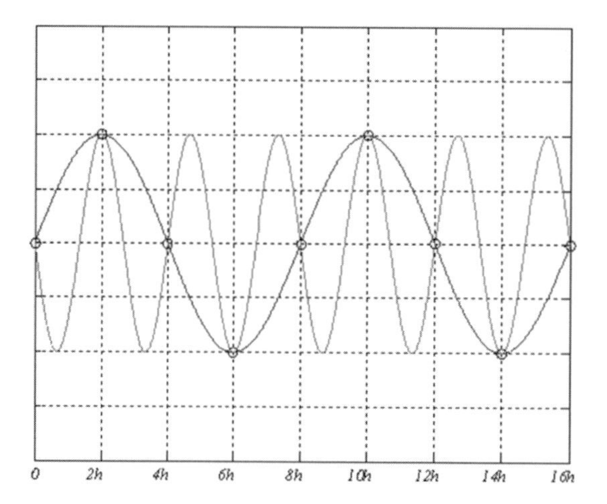

Figure 3.6 Aliasing. $\sin \frac{\pi x}{4h}$ and $\sin \frac{3\pi x}{4h}$ are indistinguishable on a grid with interval $2h$.

Figure 3.6. Consequently, an inappropriate correction may be calculated on the coarse grid. In order to prevent this, enough iterations should be performed on each grid level to sufficiently attenuate the high frequency modes. This can be analyzed using modal analysis, as presented in the following paragraphs.

Consider the one-dimensional problem

$$u_{xx} = f.$$

With a mesh interval h, this may be discretized as

$$\frac{v_{j+1} - 2v_j + v_{j-1}}{h^2} = f_j. \tag{3.23}$$

Denote the solution after n iterations by v_j^n. In a Gauss–Seidel process, we set

$$v_{j+1}^n - 2\tilde{v}_j + v_{j-1}^{n+1} = h^2 f_j \tag{3.24}$$

and

$$v_j^{n+1} = v_j^n + r(\tilde{v}_j - v_j^n), \tag{3.25}$$

where r is the relaxation factor. According to (3.25),

$$\tilde{v}_j = \left(1 - \frac{1}{r}\right) v_j^n + \frac{1}{r} v_j^{n+1},$$

and hence, we can eliminate $\tilde{v}_j$ from (3.24) to obtain

$$v_{j+1}^n - 2\left(1 - \frac{1}{r}\right) v_j^n - \frac{2}{r} v_j^{n+1} + v_{j-1}^{n+1} = h^2 f_j. \tag{3.26}$$

We can rewrite (3.26) in terms of the error $e_j^n = v_j^n - v_j$, where v_j is the exact solution. Thus, subtracting (3.23) from (3.26), the Gauss–Seidel scheme may be written as

$$e_{j+1}^n - 2\left(1 - \frac{1}{r}\right)e_j^n - \frac{2}{r}e_j^{n+1} + e_{j-1}^{n+1} = 0. \tag{3.27}$$

Consider now a Fourier mode of the error, which we assume to have the form

$$e_j^n = g^n e^{i\omega x_j},$$

where g is growth factor of the amplitude during a single iteration, and ω is the frequency. Substituting this mode in (3.27) and dividing through by $g^n e^{i\omega x_j}$,

$$e^{i\xi} - 2\left(1 - \frac{1}{r}\right) - \frac{2}{r}g + ge^{-i\xi} = 0,$$

where ξ is the mesh wave number ωh. Thus, the growth factor is

$$g(\xi) = -\frac{2\left(1 - \frac{1}{r}\right) - e^{i\xi}}{\frac{2}{r} - e^{-i\xi}}.$$

We want to make sure that higher frequency modes with wave numbers in the range $\frac{\pi}{2} \leq \xi \leq \pi$ are well damped.

If we use a pure Gauss–Seidel process with $r = 1$,

$$g(\xi) = \frac{e^{i\xi}}{2 - e^{-i\xi}},$$

$$g\left(\frac{\pi}{2}\right) = \frac{i}{2+i}, \quad \left|g\left(\frac{\pi}{2}\right)\right| = \frac{1}{\sqrt{5}},$$

$$g(\pi) = -\frac{1}{3}.$$

More generally, if we take $r \neq 1$,

$$g(\pi) = \frac{2-3r}{2+r},$$

and the choice $r = \frac{2}{3}$ gives $g(\pi) = 0$. These results suggest that if we use multigrid with a Gauss–Seidel scheme, the high frequency modes will be well damped if we use a relaxation factor near 1.

It should be noted that each time the mesh is coarsened, the frequency band in the range $\frac{\pi}{2} \leq \xi \leq \pi$ is halved, with the consequence that the lower frequency bands on the fine mesh appear as the high frequency band of a mesh at some coarser level. Thus, the multigrid scheme may be viewed as a process in which error modes in each frequency band are smoothed at the different grid levels.

3.3 Finite Element Methods

3.3.1 Historical Background

Finite element methods originated in structural analysis based on the idea of subdividing a complex structure into small sub-elements, each with a known relationship between displacements and externally applied forces, and then calculating the global solution by solving a set of discrete equations in which the local forces and displacements are balanced. The earliest description of such an approach is generally attributed to Turner et al. (1956). It was subsequently realized that finite element methods for general elliptic problems could be derived from variational principles, and this provided a route to rigorous formulations with accompanying error estimates.

The variational formulation was pioneered by Clough (1960, 2004), Argyris (1954a, 1954b, 1954c, 1960), and Zienkiewicz (1995, 2004). It had actually been anticipated in a paper by Courant (1943). Subsequently an elaborate mathematical theory of finite element methods has been developed by mathematicians, particularly those belonging to or associated with the French school (Ciarlet, Raviart, Brezzi, Oden, among others). Unfortunately this theory is often presented in language that is not readily understood by practitioners who actually have to calculate solutions to real problems in structural analysis, fluid dynamics, or other disciplines.

3.3.2 General Formulation

In contrast to finite difference methods, which approximate the differential operators appearing in a PDE, finite element methods approximate the solution. The approximate trial solution is represented as an expansion in a set of basis functions, which is then inserted in the exact differential equation. Suppose that we wish to solve an equation of the form

$$Lu = f. \tag{3.28}$$

Let ϕ_j be a set of independent basis functions. Then we insert a trial solution

$$u_h = \sum_{j=1}^{n} u_j \phi_j \tag{3.29}$$

into (3.28) to obtain

$$Lu_h - f = r_h,$$

where r_h is the residual error.

The coefficients u_j are now chosen to make the residual error as small as possible. In general there are not enough degrees of freedom to reduce the error to zero, so we try to minimize the error by choosing u_j such that the residual is orthogonal to each of n independent test functions ψ_i, yielding the weighted residual equations

$$(r_h, \psi_i) = (Lu_h - f, \psi_i) = 0, \quad i = 1, \ldots, n, \tag{3.30}$$

where (u, v) is an inner product in an appropriate space. Inserting the trial solution, we obtain a linear system of equations for the n unknowns u_j. Important properties of the method, such as accuracy, computational cost, and complexity of the formulation, all depend on the choice of basis and test functions.

The Galerkin method uses the basis functions ϕ_j as the test functions, while Petrove–Galerkin methods use distinct basis and test functions. If we choose basis functions that are zero outside a region of local support, then the weighted residual equations (3.30) will be sparse, with each element coupled only to some nearby neighbors. This is the essence of the finite element method, as it enables the equations for problems in complex domains to be assembled from local equations with a simple repetitive form.

An alternative approach is to expand the solution in global basis functions, such as Fourier series for problems with periodic boundary conditions. This is the approach taken in spectral methods. These can provide very accurate approximations but are generally only feasible for solutions in very simple geometric domains.

3.3.3 Linear Elements

The simplest basis functions with compact support are piecewise linear functions that have the value unity at a single node and zero at every other node. In a one-dimensional problem, this leads to "hat" functions as illustrated in Figure 3.7. Introducing nodes $x_j, j = 1, \ldots, n$ from the left to the right boundary of the domain,

$$\phi_j(x) = \begin{cases} 0, & x < x_{j-1}, \\ \dfrac{x - x_{j-1}}{x_j - x_{j-1}}, & x_{j-1} \leq x \leq x_j, \\ \dfrac{x - x_j}{x_{j+1} - x_j}, & x_j \leq x \leq x_{j+1}, \\ 0, & x > x_{j+1}. \end{cases}$$

The trial solutions now belong to the space of piecewise linear functions. The detailed formulation of the finite element method is most easily clarified by considering some specific applications.

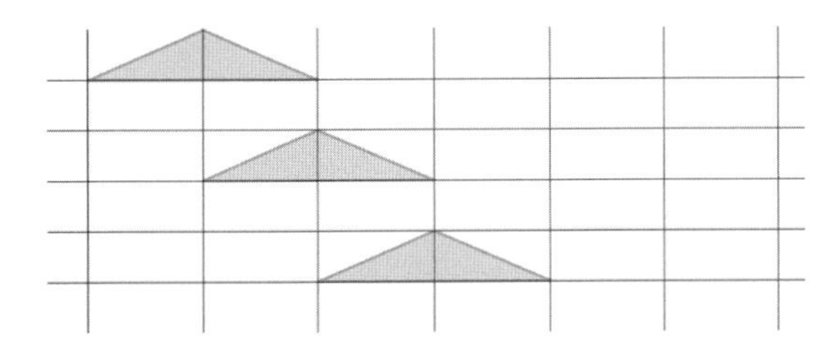

Figure 3.7 Linear basis functions.

Galerkin method for the one-dimensional Poisson equation

Consider the one-dimensional Poisson equation

$$u_{xx} = f \tag{3.31}$$

on a domain from a to b with Dirichlet boundary conditions. This could represent, for example, steady heat conduction with a constant conductivity κ. The distribution of the temperature T is then governed by

$$\frac{d}{dx}\kappa\frac{dT}{dx} = q(x),$$

where q is the rate of heat input. Then we recover (3.31) with $u = \kappa T$, $f = q$.

Using the superscript prime to denote differentiation with respect to x, the Galerkin method leads to the residual equation

$$\int_a^b u''\phi_i dx = \int_a^b f\phi_i dx.$$

With a piecewise linear trial solution, the second derivative u_h'' is not defined at the nodes. However, we can circumvent this difficulty by integrating the residual equation by parts to obtain

$$-\int_a^b u_h'\phi_i' dx = \int_a^b f\phi_i dx.$$

The use of integration by parts in this manner to reduce the order of the highest derivative of the trial solution is a key ingredient of the success of the finite element method.

Now, inserting the expansion (3.29) into the residual equation, we obtain the linear system of equations

$$\sum_{j=1}^{n} S_{ij}u_j = -b_i,$$

where

$$S_{ij} = \int_a^b \phi_i'\phi_j' dx \tag{3.32}$$

$$b_i = \int_a^b f\phi_i dx, \tag{3.33}$$

or in matrix vector form,

$$\mathbf{S}\,\mathbf{u} = -\mathbf{b}, \tag{3.34}$$

where $\mathbf{u}$ is the solution vector

$$\mathbf{u}^T = [u_1, u_2, \ldots, u_n].$$

It is evident that the matrix S is symmetric: $S_{ij} = S_{ji}$. In deference to the origins of the finite element method in structural analysis, S is commonly called the stiffness matrix.

Each test function ϕ_j has a non-zero derivative only in the elements $[x_{j-1}, x_j]$ and $[x_j, x_{j+1}]$. Accordingly the entries S_{ij} are zero if $j < i - 1$ or $j > i + 1$, and **S** is tridiagonal. For a uniform mesh with interval h, the derivatives ϕ_j' are piecewise constant with the values

$$\phi_j'(x) = \begin{cases} 0, & x < x_{j-1} \\ \dfrac{1}{h}, & x_{j-1} \le x \le x_j \\ -\dfrac{1}{h}, & x_j \le x \le x_{j+1} \\ 0, & x > x_{j+1}. \end{cases}$$

Multiplying by the mesh width h to perform the integrations, it can be seen that the stiffness matrix has the form

$$S = \frac{1}{h}\begin{bmatrix} 2 & -1 & & & & & \\ -1 & 2 & -1 & & & & \\ & -1 & 2 & -1 & & & \\ & & \ddots & \ddots & \ddots & & \\ & & & -1 & 2 & -1 & \\ & & & & -1 & 2 & -1 \\ & & & & & -1 & 2 \end{bmatrix}.$$

Here the boundary values u_1 and u_n are absorbed in the right hand side vector **b**. In order to evaluate the right hand side at interior points, we can represent f by a piecewise linear approximation of the same kind as the expansion of the trial solution:

$$f_h = \sum_{j=1}^{n} f_j \phi_j(x),$$

where $f_j = f(x_j)$. Substituting this in (3.33), the right hand side values can be expressed as

$$b_i = \sum_{j=1}^{n} M_{ij} f_j,$$

where

$$M_{ij} = \int_a^b \phi_i \phi_j dx. \tag{3.35}$$

Like the stiffness matrix S_j, the matrix M is symmetric and tridiagonal. Following again the traditions of structural analysis, M is commonly called the mass matrix.

For a uniform mesh, M has the form

$$M = h\begin{bmatrix} \frac{2}{3} & \frac{1}{6} & & & & & \\ \frac{1}{6} & \frac{2}{3} & \frac{1}{6} & & & & \\ & \frac{1}{6} & \frac{2}{3} & \frac{1}{6} & & & \\ & & \ddots & \ddots & \ddots & & \\ & & & \frac{1}{6} & \frac{2}{3} & \frac{1}{6} & \\ & & & & \frac{1}{6} & \frac{2}{3} & \frac{1}{6} \\ & & & & & \frac{1}{6} & \frac{2}{3} \end{bmatrix}.$$

The equations at each interior point now reduce to

$$\frac{u_{j-1} - 2u_j + u_{j+1}}{h^2} = \frac{1}{6} f_{j-1} + \frac{2}{3} f_j + \frac{1}{6} f_{j+1}.$$

The left hand side is identical to the finite difference approximation, but the right side differs in replacing f_j by a weighted average of the neighboring values.

In the case of the Poisson equation with Neumann boundary conditions

$$u'(a) = 0, \quad u'(b) = 0,$$

it is evident that the solution for u is non-unique by the possible addition of a constant. This implies a compatibility condition on the source term f. In fact, integrating (3.31) by parts, we find that

$$\int_a^b f\,dx = u'(b) - u'(a) = 0. \tag{3.36}$$

The physical meaning of this condition is simply that a steady solution of the heat conduction equation is not possible unless the total net heat flux is zero. Similar issues arise with the discrete system. With the boundary values u_1 and u_n treated as unknowns, the mass and stiffness matrices on a uniform mesh now take the form

$$M = h \begin{bmatrix} \frac{1}{3} & \frac{1}{6} & & & & & \\ \frac{1}{6} & \frac{2}{3} & \frac{1}{6} & & & & \\ & \frac{1}{6} & \frac{2}{3} & \frac{1}{6} & & & \\ & & \ddots & \ddots & \ddots & & \\ & & & \frac{1}{6} & \frac{2}{3} & \frac{1}{6} & \\ & & & & \frac{1}{6} & \frac{2}{3} & \frac{1}{6} \\ & & & & & \frac{1}{6} & \frac{1}{3} \end{bmatrix}$$

and

$$S = \frac{1}{h} \begin{bmatrix} 1 & -1 & & & & & \\ -1 & 2 & -1 & & & & \\ & -1 & 2 & -1 & & & \\ & & \ddots & \ddots & \ddots & & \\ & & & -1 & 2 & -1 & \\ & & & & -1 & 2 & -1 \\ & & & & & -1 & 1 \end{bmatrix}.$$

It is easily seen that the constant vector

$$\mathbf{e}^T = [1, 1, \ldots, 1]$$

is an eigenvector of S corresponding to a zero eigenvalue

$$S\,e = 0.$$

Multiplying (3.34) on the left by $\mathbf{e}^T$:

$$\mathbf{e}^T\mathbf{b} = -\mathbf{e}^T S\mathbf{u} = 0.$$

It follows that the data f_j must satisfy the discrete compatibility condition

$$\sum_{i=1}^{n}\sum_{j=1}^{n} M_{ij} f_j = \frac{1}{2} f_1 + f_2 + f_3 + \cdots + \frac{1}{2} f_n = 0.$$

This is equivalent to approximating the integral compatibility condition (3.36) by trapezoidal integration, which is exact for the piecewise linear representation of the source term f.

The Galerkin method for the diffusion equation

In order to illustrate the Galerkin method for time dependent problems, consider now the one-dimensional diffusion equation

$$\frac{\partial u}{\partial t} = \sigma \frac{\partial^2 u}{\partial x^2}$$

on a domain from a to b with the Neumann boundary conditions

$$\left.\frac{\partial u}{\partial x}\right|_a = 0, \quad \left.\frac{\partial u}{\partial x}\right|_b = 0.$$

This could represent unsteady heat conduction with no heat source and a constant conductivity σ. Substituting the trial solution u_h and integrating by parts as before, we now obtain the residual equation

$$\int_a^b \frac{\partial u_h}{\partial t} \phi_i dx = \sigma \phi_i \left.\frac{\partial u_h}{\partial x}\right|_a^b - \sigma \int \frac{\partial u_h}{\partial x} \frac{\partial \phi_i}{\partial x} dx = 0,$$

where the boundary term vanishes on applying the Neumann boundary conditions. Substituting the expansion (3.29) for u_h, the equations reduce to

$$\sum_{j=1}^{n} M_{ij} \frac{du_j}{dt} = -\sigma S_{ij} u_j, \tag{3.37}$$

where M and S are the mass and stiffness matrices defined by the same equations (3.35) and (3.32) as before, or in matrix vector form,

$$M \frac{d\mathbf{u}}{dt} = -\sigma \, S \, \mathbf{u}. \tag{3.38}$$

Thus the time derivatives of the solution vector $\mathbf{u}$ are coupled by the mass matrix, leading to an implicit scheme, and in order to advance the solution by standard time integration techniques, it is necessary to solve the linear system (3.37) at every time step.

In comparison with a finite difference method, this is a significant additional computational cost, particularly for solutions in multi-dimensional domains, for which the number of degrees of freedom may be very large. The mass matrix does play a role, however, in maintaining accuracy on non-uniform meshes, by compensating for the fact that the discretization terms corresponding to ϕ_{xx} in the neighborhood of a node are not symmetric about that node.

The Galerkin method for linear advection

As a third example, consider the linear advection equation

$$\frac{\partial u}{\partial t} + a \frac{\partial u}{\partial x} = 0$$

in the domain from 0 to 1. The exact solution is constant along the characteristics

$$x - at = \xi$$

describing wave motion at the speed a. The wave propagation is to the right if $a > 0$, as we shall assume. Then u should be specified at the left boundary,

$$u(0,t) = g(t),$$

but the value at the right boundary cannot be specified because it is determined by the evolving solution. Replacing u by the trial solution u_h in (3.38), we obtain the residual equation

$$\int_0^1 \phi_i \left(\frac{\partial u_h}{\partial t} + a \frac{\partial u_h}{\partial x} \right) dx = 0.$$

Then, substituting the expansion (3.29), we obtain the equations

$$\sum_{j=1}^{n} M_{ij} \frac{du_j}{dt} + a \sum_{j=1}^{n} S_{ij} u_j = 0,$$

where the entries of the mass matrix are the same as before,

$$M_{ij} = \int_0^1 \phi_i \phi_j dx,$$

but the entries of the stiffness matrix are now

$$S_{ij} = \int_0^1 \phi_i \phi_j' dx.$$

With u_1 specified at the left boundary and the right boundary value u_n treated as an unknown, the stiffness matrix on a uniform mesh with interval h is

$$S = h \begin{bmatrix} 0 & \frac{1}{2} & & & & & \\ -\frac{1}{2} & 0 & \frac{1}{2} & & & & \\ & -\frac{1}{2} & 0 & \frac{1}{2} & & & \\ & & \ddots & \ddots & \ddots & & \\ & & & -\frac{1}{2} & 0 & \frac{1}{2} & \\ & & & & -\frac{1}{2} & 0 & \frac{1}{2} \\ & & & & & -\frac{1}{2} & \frac{1}{2} \end{bmatrix}.$$

It can be seen that the entries are skew symmetric with the exception of the diagonal entry in the last row.

At every interior point, the equations reduce to

$$\frac{1}{6}\frac{du_{j-1}}{dt} + \frac{2}{3}\frac{du_j}{dt} + \frac{1}{6}\frac{du_{j+1}}{dt} = a\frac{u_{j+1} - u_{j-1}}{2h}. \tag{3.39}$$

The interior discretization is exactly the same as the discretization that would be obtained by representing u_h as a cubic spline and setting $\frac{\partial u_h}{\partial x}$ equal to the derivative of the spline. It is also the same as the fourth order compact difference scheme, which will be described in the chapter on high order methods.

The interior equation (3.39) is a central difference approximation that takes no account of the fact that the true solution depends only on data to the left, along the backward characteristics. This highlights a disadvantage of the Galerkin method for convection-dominated problems, which are typical in fluid dynamics: it provides no mechanism for biasing the discretization in an upwind direction corresponding to the physics of wave propagation. On the other hand, upwind discretizations can be introduced very easily in finite difference and finite volume schemes, and they have actually proved very beneficial.

3.3.4 Variational Foundations of the Finite Element Method

In order to clarify the variational foundations of the finite element method, we examine in this section the one-dimensional elliptic equation

$$Lu = f \tag{3.40}$$

on a domain from a to b, where

$$Lu = -\frac{d}{dx}p\frac{du}{dx} + qu$$

with $p(x) \geq p_{\min} > 0$, $q(x) \geq 0$, and Dirichlet or Neumann boundary conditions

$$u(a) = u_L \text{ or } u'(a) = 0$$
$$u(b) = u_R \text{ or } u'(b) = 0$$

either at both ends, or one at each end. This is the most general second order self-adjoint linear operator. We seek a solution $u(x)$ in a space $\mathcal{S}$ of twice-differentiable functions that satisfy any Dirichlet boundary conditions that may be specified.

The solution of this problem minimizes the functional

$$I(u) = \frac{1}{2}\int_a^b (pu'^2 + qu^2 - 2fu)\,dx.$$

Suppose that $u(x)$ is modified by the addition of $\delta u(x) = \epsilon v(x)$, where ϵ is small. The corresponding variation of I is

$$\begin{aligned}\delta I &= I(u + \delta u) - I(u)\\ &= \epsilon V_1 + \epsilon^2 V_2,\end{aligned}$$

where

$$V_1 = \int_a^b (pu'v' + quv - fv)\, dx$$

and

$$V_2 = \frac{1}{2}\int_a^b (pv'^2 + qv^2)\, dx \geq 0.$$

Integrating the first term in V_1 by parts,

$$V_1 = pu'\,\delta u\Big|_a^b + \int_a^b \left(-(pu')' + qu - f\right) v\, dx.$$

Here the boundary terms vanish with either Dirichlet or Neumann boundary conditions, so that

$$V_1 = (Lu - f, v).$$

If u satisfies (3.40) together with the boundary conditions, $V_1 = 0$ for all v consistent with the Dirichlet boundary conditions, and consequently $I(u)$ is a minimum. Note that in order to satisfy (3.40), u must belong to the space of twice-differentiable functions. On the other hand, the variational statement that $V_1 = 0$ for all admissible v leads to the weak form

$$\int_a^b (pu'v' + quv - fv)\, dx = 0$$

for all test functions v that belong to the same space as the solution.

It is evident that this can be satisfied by taking u and v from an enlarged space of once-differentiable functions. This enables the use of a trial space of piecewise linear functions for the discrete solution. Introducing the generalized inner product

$$a(u,v) = \int_a^b (pu'v' + quv)\, dx,$$

the weak form can be written as

$$a(u,v) = (f,v) \tag{3.41}$$

for all admissible test functions v. Moreover, the cost functional can be written as

$$I(u) = \frac{1}{2}a(u,u) - (f,u).$$

Here, the term $\frac{1}{2}a(u,u)$ may be regarded as a generalized energy. Also, $a(u,u)$ is a legitimate norm of u provided that it is coercive, or positive definite, in the sense that

$$a(u,u) \geq c\|u\|^2$$

for a constant $c > 0$. This is the reason for requiring $p(x) \geq p_{\min} > 0$.

Consider now the minimization of $I(u_h)$ where u_h belongs to the subspace $\mathcal{S}_h$ spanned by the basis functions $\phi_i(x)$, $i = 1, \ldots, n$. If we modify u_h by a variation $\delta u_h = \epsilon v_h$, where v_h belongs to the same subspace, the resulting variation of I is

$$\delta I = \epsilon V_1 + \mathcal{O}(\epsilon^2),$$

where

$$V_1 = \int_a^b (pu_h' v_h' + qu_h v_h - f v_h)\, dx,$$

and as before, $I(u_h)$ is minimized if $V_1 = 0$ for all admissible v_h. Since v_h can be expressed as a linear combination of the basis functions, this will be the case if

$$\int_a^b (pu_h' \phi_i' + qu_h \phi_i - f \phi_i)\, dx = 0, \quad i = 1, \ldots, n.$$

These are just the Galerkin equations. Thus the Galerkin method finds the trial solution u_h that minimizes $I(u_h)$ over the space $\mathcal{S}_h$.

We can write the discrete variational equations as

$$a(u_h, v_h) = (f, v_h) \tag{3.42}$$

for all v_h in $\mathcal{S}_h$. In practice, if u_h is represented by the expansion

$$u_h = \sum_{j=1}^{n} u_j \phi_j(x),$$

we substitute the basis functions ϕ_i, $i = 1, \ldots, n$ for v_h to obtain the linear system

$$\sum_{j=1}^{n} S_{ij} u_j = b_i, \quad i = 1, \ldots, n,$$

where S is a generalized stiffness matrix with entries

$$S_{ij} = a(\phi_i, \phi_j)$$

and

$$b_i = (f, \phi_i).$$

As another example, consider the solution of Laplace's equation

$$\nabla^2 u = 0 \quad \text{in } \mathcal{D}$$

with Dirichlet boundary conditions

$$u = u_b \tag{3.43}$$

on the boundary $\mathcal{B}$. Then we seek solutions in the space of twice-differentiable functions satisfying (3.43). The variational form is

$$a(u, v) = 0$$

for all admissible test functions v, where

$$a(u, v) = \int_{\mathcal{D}} \nabla u \cdot \nabla v \, dV,$$

and the cost functional is

$$I(u) = \frac{1}{2} \int q^2 dV,$$

where

$$\mathbf{q} = \nabla u.$$

If u is the velocity potential of an ideal fluid flow, it can be seen that $I(u)$ is simply the kinetic energy. This is minimized by the true solution u, and correspondingly the discrete solution u_h minimizes the kinetic energy over the space $\mathcal{S}_h$, which may now be enlarged to the space of once-differentiable functions.

Error estimates based on the variational form

An error estimate for the solution error $u - u_h$ can be derived by the following argument. First, since the trial space $\mathcal{S}_h$ is a subspace of the solution space $\mathcal{S}$, u satisfies

$$a(u, v_h) = (f, v_h)$$

for all v_h in $\mathcal{S}_h$, and hence it follows from (3.41) and (3.42) that

$$a(u - u_h, v_h) = 0. \tag{3.44}$$

In other words, the error is orthogonal to all test functions v_h in $\mathcal{S}_h$. But this implies that u_h also minimizes $a(u - u_h, u - u_h)$ over $\mathcal{S}_h$ since

$$a(u - u_h - \epsilon v_h, u - u_h + \epsilon v_h) = a(u - u_h, u - u_h) - 2\epsilon a(u - u_h, v_h) + \epsilon^2 a(v_h, v_h),$$

where the middle term vanishes according to the orthogonality condition (3.44). Thus u_h minimizes the energy $\frac{1}{2}a(u-u_h, u-u_h)$ of the error, and provided that $a(u, u)$ meets the requirements for a legitimate norm, we can regard this as a norm of the error.

Also, we can substitute u_h for v_h in (3.44) to obtain

$$a(u - u_h, u_h) = 0.$$

Hence

$$a(u, u_h) = a(u_h, u_h)$$

and

$$\begin{aligned} a(u - u_h, u - u_h) &= a(u, u) - 2a(u, u_h) + a(u_h, u_h) \\ &= a(u, u) - a(u_h, u_h). \end{aligned}$$

Thus the energy of the error is equal to the error in the energy, and it can also be seen that if the error is not zero, the discrete energy $a(u_h, u_h)$ always underestimates the true energy $a(u, u)$.

We can now obtain an error estimate for $e_h = u - u_h$ by comparing it with the error of a piecewise linear interpolate u_I to the true solution u. Now assume that $u''(x)$ is continuous and $|u''(x)| \leq M$. Then on an interval of width h from x_{j-1} to x_j, say,

$$\max |u(x) - u_I(x)| \leq M\frac{h^2}{8}$$

and

$$\max |u'(x) - u'_I(x)| \leq Mh.$$

This can be proved as follows. Let $\Delta(x) = u(x) - u_I(x)$. Since $\Delta = 0$ at each end of the interval, Δ reaches a maximum at some point $\hat{x}$ in the interval, where $\Delta'(\hat{x}) = 0$. Moreover,

$$\Delta'(x) = \int_{\hat{x}}^{x} \Delta''(r)\, dr = \int_{\hat{x}}^{x} u''(r)\, dr$$

since $u''_I = 0$. Hence

$$|\Delta'(x)| \leq Mh.$$

Also, suppose that $\tilde{x}$ is the nearer of the mesh points x_{j-1} and x_j to $\hat{x}$. Then

$$\Delta(\tilde{x}) = \Delta(\hat{x}) + (\tilde{x} - \hat{x})\Delta'(\hat{x}) + \frac{(\tilde{x} - \hat{x})^2}{2}\Delta''(w)$$

where w lies between $\hat{x}$ and $\tilde{x}$. Here $\Delta(\tilde{x}) = 0$, $\Delta'(\hat{x}) = 0$, $\Delta''(w) = u''(w)$, and $|\tilde{x} - \hat{x}| \leq \frac{h}{2}$. Hence

$$|\Delta(\hat{x})| = \max |\Delta(x)| \leq M\frac{h^2}{8}.$$

It now follows that

$$a(u - u_I, u - u_I) \leq CM^2h^2$$

for some constant C. But since u_h minimizes $a(e_h, e_h)$ over the space $\mathcal{S}_h$, and u_I is in $\mathcal{S}_h$, it also follows that

$$a(e_h, e_h) \leq a(u - u_I, u - u_I) \leq CM^2h^2.$$

3.3.5 Multi-dimensional Finite Element Schemes with Linear Elements

Finite element methods are very easy to define for triangular and tetrahedral meshes using piecewise linear trial solutions. Then we introduce basis functions ϕ_j that have the value unity at the jth node and zero at all other nodes, as illustrated in Figure 3.8 for a triangular mesh. We can visualize the basis function of each node as a tent surrounding the node. Consider now Laplace's equation in a triangulated domain $\mathcal{D}$ with Dirichlet boundary conditions on the boundary $\mathcal{B}$,

$$\begin{aligned} u_{xx} + u_{yy} &= 0 \quad \text{in} \quad \mathcal{D} \\ u \quad &\text{specified on} \quad \mathcal{B}. \end{aligned} \tag{3.45}$$

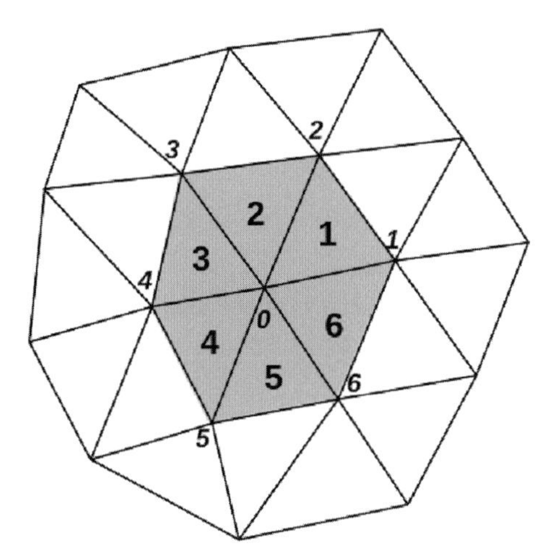

Figure 3.8 Test function and stencil for a node in a triangular mesh.

The corresponding weak form is obtained by multiplying (3.45) by a test function $\psi(x, y)$ and integrating by parts to obtain

$$\int_{\mathcal{B}} \psi \nabla u_h \cdot \mathbf{n} \, dl - \int_{\mathcal{D}} \nabla u_h \cdot \nabla \psi \, d\mathcal{S} = 0, \tag{3.46}$$

where $\mathbf{n}$ is the unit normal to the boundary. The trial solution is

$$u_h = \sum_{j=1}^{n} u_j \phi_j(x, y).$$

Following the Galerkin method, we require u_h to satisfy the weak form (3.46) for each test function $\psi = \phi_j$. With Dirichlet boundary conditions, we only need to solve for u_j at the interior nodes, for which $\phi_j = 0$ on $\mathcal{B}$ with the consequence that the boundary integral vanishes. Referring to Figure 3.8, consider the equation for the node labeled zero. $\nabla\phi_0$ is nonzero only in the neighboring triangles, which are shaded in the figure. Moreover, both ∇u_h and $\nabla\phi_0$ are constant in each of these triangles. Thus, the residual equation for node 0 reduces to

$$r_0 = \sum_{\text{neighbors}} \mathcal{S}_k \nabla u_{hk} \cdot \nabla \phi_{0_k} = 0, \tag{3.47}$$

where $\mathcal{S}_k$ is the area of the kth triangle, and ∇u_{hk} and $\nabla\phi_{0_k}$ are the gradients of u_h and ϕ_0 in that triangle.

The gradient of any piecewise linear function u_h on a triangular mesh can be evaluated very easily using Gauss' theorem

$$\int_{\mathcal{D}} \frac{\partial u_h}{\partial x} \, d\mathcal{S} = \oint_{\mathcal{B}} u_h \, dy.$$

Since u_h is linear and $\frac{\partial u_h}{\partial x}$ is constant in each triangle, $\frac{\partial u_h}{\partial x}$ can be evaluated exactly by trapezoidal integration. Suppose that the vertices are numbered 1–3 in a counterclockwise sense. Then,

$$\frac{\partial u_h}{\partial x} = \frac{1}{2\mathcal{S}} \sum_{\text{edges}} (u_{k+1} + u_k)(y_{k+1} - y_k),$$

where vertex 1 is also counted as vertex 4, and $\mathcal{S}$ is the area of the triangle. Noting that the contribution of u_2, say, is $u_2(y_3 - y_1)$, this formula can be written as

$$\frac{\partial u_h}{\partial x} = \frac{1}{2\mathcal{S}}(u_1(y_2 - y_3) + u_2(y_3 - y_1) + u_3(y_1 - y_2)). \tag{3.48}$$

Here each nodal value is multiplied by the increment in y along the opposite edge. Similarly,

$$\frac{\partial u_h}{\partial y} = -\frac{1}{2\mathcal{S}}(u_1(x_2 - x_3) + u_2(x_3 - x_1) + u_3(x_1 - x_2)). \tag{3.49}$$

We can also evaluate the area in the same way as

$$\mathcal{S} = \oint_{\mathcal{B}} x\, dy$$

or in discrete form

$$\mathcal{S} = \frac{1}{2}(x_1(y_2 - y_3) + x_2(y_3 - y_1) + x_3(y_1 - y_2)). \tag{3.50}$$

Direct assembly of the contributions to the residual equation at each node would require a data structure that includes lists of the triangles surrounding each node. Moreover, the number of these may vary from one node to the next. These complications can be avoided by an indirect assembly procedure in which the contributions of each triangle are accumulated to the residual equation of each of its three vertices in a loop over the triangles. This procedure can be implemented very simply with the following data structure.

First, we require a list of the nodes with their Cartesian coordinates:

x(n), n = 1, . . . , nnode
y(n), n = 1, . . . , nnode.

We need to distinguish the boundary nodes at which the Dirichlet boundary condition is specified. This is most conveniently accomplished by ordering the nodes so that the boundary nodes are placed at the beginning of the list, n = 1, . . . , nbnd. Next, we require a list of the cells containing the three vertices of each cell:

ndc(1,n), n = 1, . . . , ncell
ndc(2,n), n = 1, . . . , ncell
ndc(3,n), n = 1, . . . , ncell.

The vertices should be consistently numbered in the counter-clockwise sense to ensure that the cell areas are calculated to be positive. Finally, we need lists of the solution values, nodal residuals and nodal areas:

$$\begin{aligned} &\mathrm{u(n)}, &&\mathrm{n} = 1, \dots, \mathrm{nnode} \\ &\mathrm{res(n)}, &&\mathrm{n} = 1, \dots, \mathrm{nnode} \\ &\mathrm{s(n)}, &&\mathrm{n} = 1, \dots, \mathrm{nnode}. \end{aligned}$$

The solution values should be set to the correct boundary values at the boundary nodes, $\mathrm{n} = 1, \dots, \mathrm{nbnd}$, and to an initial guess, which is typically zero at the interior nodes $\mathrm{n} = \mathrm{nbnd} + 1, \dots, \mathrm{nnode}$. The nodal areas are the sums of the areas of the triangles surrounding each node, which can be pre-calculated by the formula (3.50).

The solution requires the inversion of a sparse matrix with a dimension equal to the number of interior nodes. This grows as the mesh is refined and may be very large. We can avoid a direct matrix inversion by simply advancing the heat equation

$$\frac{\partial u}{\partial t} + \nabla^2 u = 0$$

in time until it reaches a steady state. Since we are not interested in time accuracy, there is no need to include a mass matrix, which would result from the term $\int \frac{\partial u}{\partial t} \phi_0 ds$. Instead, we can use a lumped form

$$\mathcal{S}_0 \frac{du_{h0}}{dt} = r_0,$$

where $\mathcal{S}_0$ is the sum of the areas of the triangles surrounding node 0. This procedure is very similar to solving the equations iteratively by the Jacobi method.

We need to specify a time step and a number of steps sufficient to approach the steady state to the desired level of accuracy. We can then solve Laplace's equation with the simple computer program displayed below. Within each time step, the residuals are first set to zero. Then in the main loop over the cells, the cell area and derivatives $\frac{\partial u_h}{\partial x}$ and $\frac{\partial u_h}{\partial y}$ are calculated by the formulas (3.50), (3.48), and (3.49). The residual equation (3.47) requires a contribution $\nabla u_h \cdot \nabla \phi_0$ multiplied by the cell area from each of the surrounding cells. Hence, a given cell must send to each vertex in turn the inner product of ∇u_h with the gradient of the basis function centered at that vertex. According to the formulas (3.49) and (3.50), the derivatives with respect to x and y of the test function centered at node 1 are simply

$$\phi_x = \frac{y_2 - y_3}{2\mathcal{S}}, \quad \phi_y = -\frac{x_2 - x_3}{2\mathcal{S}},$$

with corresponding formulas for nodes 2 and 3. The required inner products are sent to the three vertices by the last three statements of the loop. Since these must be multiplied by the cell area, only one division by the cell area is needed, so the formulas u_x and u_y are simplified by not dividing by the area. Also the minus signs for $\frac{\partial u_h}{\partial y}$ and ϕ_y cancel, and only one division by 2 is needed overall. After completion of the loop over the cells, the solution is updated at the interior nodes only in the next loop running from nbnd + 1 to nnode. This completes the time step.

This program contains only 19 statements or lines of code. It is not very efficient. The time step has to be specified with a small enough value to satisfy the stability limit for the heat equation, which will be discussed in Chapter 4. Nevertheless, it illustrates the way in which the finite element method can be simplified by the introduction of advantageous data structures.

Code fragment for the solution of Laplace's equation on a triangular mesh

The solution values u(n) have been pre-assigned to the correct values in the boundary nodes, n = 1, ..., nbnd, and zero at every interior node, and the polygonal areas s(n) associated with each node have been pre-calculated.

```
% Outer time step loop
   timestep_loop: do istep = 1,nstep

% Set the nodal residuals to zero at the beginning of the time step
      do n = 1,nnode

         res(n)  = 0

      end do

% Loop over the cells to calculate the contributions of each cell
% to the nodal residuals
      do n = 1,ncell

           n1  = ndc(1,n)
           n2  = ndc(2,n)
           n3  = ndc(3,n)

         area  = x(n1)*(y(n2)-y(n3))  + x(n2)*(y(n3)-y(n1))  + x(n3)*(y(n1)-y(n2))
           ux  = u(n1)*(y(n2)-y(n3))  + u(n2)*(y(n3)-y(n1))  + u(n3)*(y(n1)-y(n2))
           uy  = u(n1)*(x(n2)-x(n3))  + u(n2)*(x(n3)-x(n1))  + u(n3)*(x(n1)-x(n2))

       res(n1)  = res(n1)  + 0.5*((y(n2)-y(n3))*ux +  (x(n2)-x(n3))*uy)/area
       res(n2)  = res(n2)  + 0.5*((y(n3)-y(n1))*ux +  (x(n3)-x(n1))*uy)/area
       res(n3)  = res(n3)  + 0.5*((y(n1)-y(n2))*ux +  (x(n1)-x(n2))*uy)/area

      end do

% Loop to update the solution at the interior nodes
      do n = nbnd+1,nnode

         u(n)  = u(n)  - dt*res(n)/s(n)

      end do

   end do timestep_loop
```

3.3.6 Further Analysis of the Discrete Laplacian

If we consider the stencil illustrated in Figure 3.8, evaluation of formulas (3.47), (3.48), and (3.49) reduces the equation for node 0 to the form

$$r_0 = \sum s_{k0}(u_k - u_0),$$

where s_{k0} are the entries of the stiffness matrix between node k and 0. If we consider edge 20, say, the total contribution of u_2, applying the formulas (3.47) and (3.48) to the triangles 012 and 023 with areas $\mathcal{S}_1$ and $\mathcal{S}_2$ respectively, is

$$\begin{aligned} s_{20} = {} & \frac{1}{4\mathcal{S}_1}((y_0 - y_1)(y_2 - y_1) + (x_0 - x_1)(x_2 - x_1)) \\ & + \frac{1}{4\mathcal{S}_2}((y_0 - y_3)(y_2 - y_3) + (x_0 - x_3)(x_2 - x_3)). \end{aligned}$$

Let l_{k0} be the length of the edge $k0$. Then, we find that

$$\begin{aligned} s_{20} &= \frac{l_{01}l_{21}\cos\theta_{012}}{2l_{01}l_{21}\sin\theta_{012}} + \frac{l_{03}l_{23}\cos\theta_{032}}{2l_{03}l_{23}\sin\theta_{032}} \\ &= \frac{1}{2}(\cot\theta_{012} + \cot\theta_{032}), \end{aligned}$$

where θ_{012} and θ_{032} are the angles between edges l_{01} and l_{21} and l_{03} and l_{23}, as illustrated in Figure 3.9.

In the case of the Poisson equation

$$u_{xx} + u_{yy} = f,$$

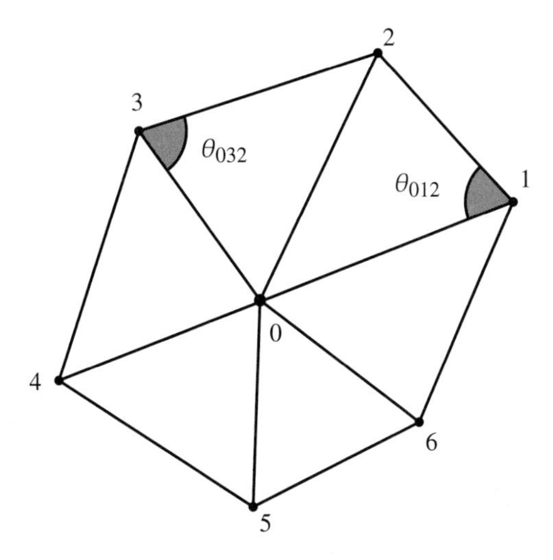

Figure 3.9 Stiffness matrix.

we must also evaluate

$$\int f\phi_0 dS,$$

assuming f is represented by a piecewise linear approximation. Carrying out the integrations, we find that the equation for node 0 is

$$\sum_{k=1}^{n} s_{k0}(u_k - u_0) = M_{00} f_0 + M_{0k} f_k,$$

where n is the number of neighbors, and the coefficients

$$M_{00} = \frac{1}{6}\sum_{k=1}^{n} s_k, \quad M_{0k} = \frac{s_k + s_{k+1}}{12}$$

are entries of the stiffness matrix.

Figure 3.10 illustrates the stiffness and mass coefficients that result from three alternative local triangulations of a Cartesian mesh with $\Delta x = \Delta y = h$. On dividing by h^2, all three recover the standard finite difference approximation to $\nabla^2 u$. On the right hand side, on the other hand, the mass coefficients sum to unity only in case (b). In cases (a) and (c), they sum to 4/3 and 2/3 respectively, with the consequence that the discretizations are locally consistent with

$$\nabla^2 u = \frac{4f}{3}$$

and

$$\nabla^2 u = \frac{2f}{3},$$

respectively. This highlights the fact that with linear elements, the Galerkin method yields discretizations that are not necessarily locally consistent, although they will be consistent on the average, leading to global second order accuracy.

In order to illustrate this further, Figure 3.11 shows the stiffness coefficients for a local triangulation with four edges meeting at node 0 in an asymmetric configuration. On dividing by h^2, we now find that the stiffness coefficients are consistent with an approximation to

$$Lu = u_{xx} + 3u_{yy}.$$

Given a local discretization of the form

$$L_h u_h = \sum_{k=1}^{n} s_{k0}(u_k - u_0),$$

we can check its consistency by a Taylor series expansion about the node 0:

$$L_h u_h = \sum_{k=1}^{n} s_{k0}\left(\Delta x_{k0}\frac{\partial u}{\partial x} + \Delta y_{k0}\frac{\partial u}{\partial y} + \frac{\Delta x_{k0}^2}{2}\frac{\partial^2 u}{\partial x^2}\right.$$
$$\left. + \Delta x_{k0}\Delta y_{k0}\frac{\partial^2 u}{\partial x \partial y} + \frac{\Delta y_{k0}^2}{2}\frac{\partial^2 u}{\partial y^2} + \cdots\right),$$

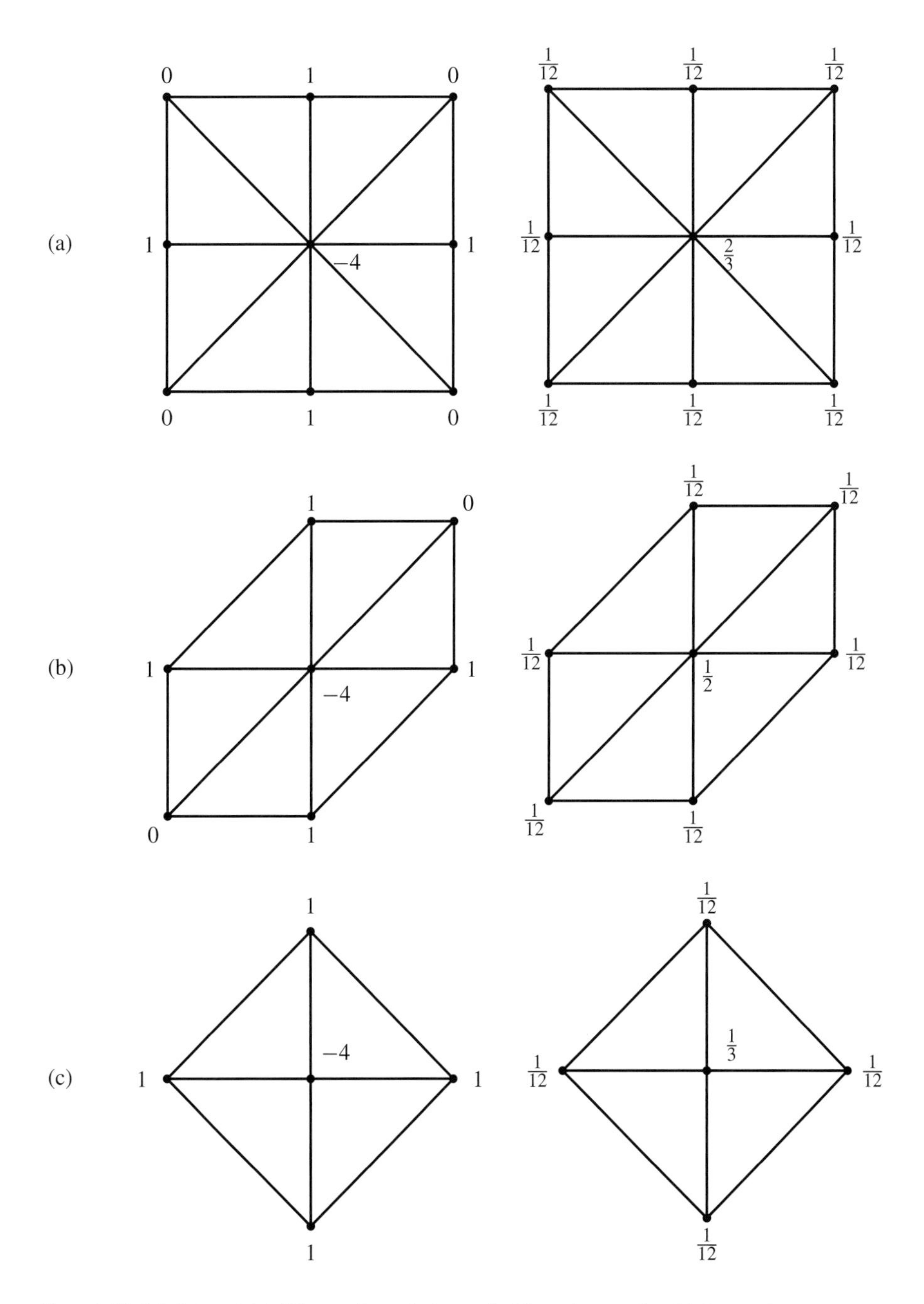

Figure 3.10 (a),(b), and (c): Three triangulations of a Cartesian mesh with interval $\Delta x = \Delta y = h$. Left: Stiffness coefficients. Right: Mass coefficients divided by h^2.

where Δx_{k0} and Δy_{k0} are the displacements between nodes k and 0 in the x and y coordinate directions. For this to be consistent with the Laplacian, the coefficients of $\frac{\partial u}{\partial x}$, $\frac{\partial u}{\partial y}$, and $\frac{\partial^2 u}{\partial x \partial y}$ must be zero, and the coefficients of $\frac{\partial^2 u}{\partial x^2}$ and $\frac{\partial^2 u}{\partial y^2}$ must be equal.

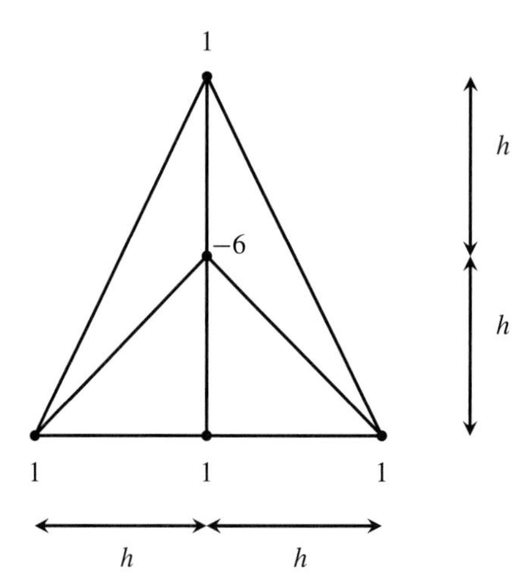

Figure 3.11 Stiffness coefficients for a local triangulation.

Hence, we require

$$\sum_{k=1}^{n} s_{k0}\Delta x_{k0} = 0,$$

$$\sum_{k=1}^{n} s_{k0}\Delta y_{k0} = 0,$$

$$\sum_{k=1}^{n} s_{k0}\Delta x_{k0}\Delta y_{k0} = 0,$$

$$\sum_{k=1}^{n} s_{k0}\Delta x_{k0}^2 = \sum_{k=1}^{n} s_{k0}\Delta y_{k0}^2.$$

In the case of a triangulation in which node 0 has no more than four neighbors, it will only be possible to satisfy these constraints if the nodes are aligned as in a Cartesian mesh. The absence of local consistency does not prevent finite element solutions with linear elements from attaining second order accuracy, but it may lead to locally noisy solutions, depending on the triangulation. Certainly, the quality of the solution is strongly dependent on the quality of the mesh.

3.3.7 Isoparametric Bilinear and Trilinear Elements

The simplest finite element method for a mesh with quadrilateral cells uses trial solutions with unique nodal values and a bilinear form inside each cell,

$$u_h = a_0 + a_1 x + a_2 y + a_3 xy,$$

where the local coefficients are chosen so that u_h matches the nodal values. The resulting trial solutions are continuous across the edges of a Cartesian mesh because either x or y is constant along every edge. Consequently, the trial solutions in the cells

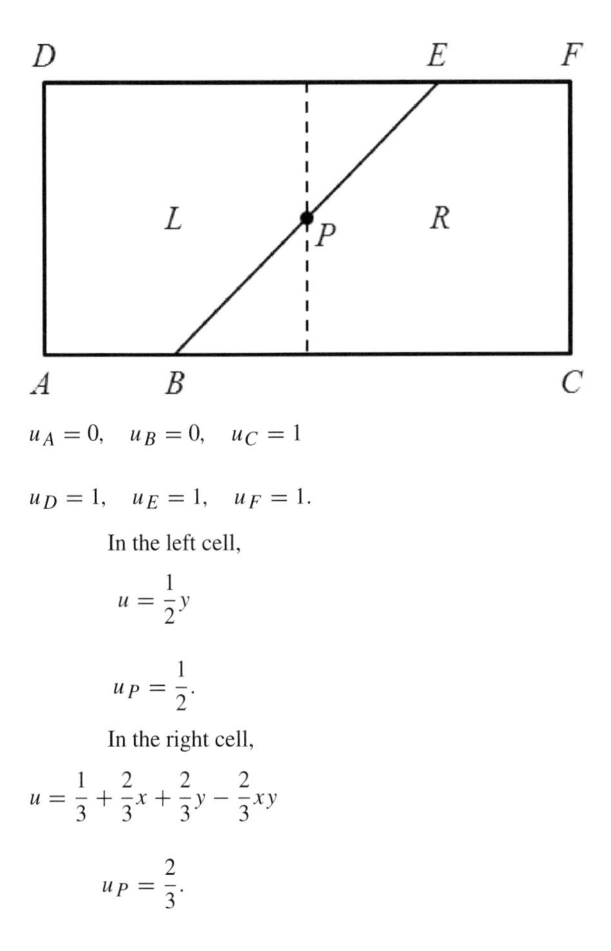

Figure 3.12 Nonrectangular elements.

on either side of an edge vary linearly along the edge and must coincide, since they have the same value at each end of the edge. Unfortunately, this is not true for nonrectangular elements, as can be seen from the counterexample shown in Figure 3.12 of two cells separated by a diagonal edge.

Requiring the use of Cartesian meshes would be far too restrictive. In fact, it would obviate the apparent advantage of finite element over finite difference methods in the treatment of complex geometrical shapes. The need to avoid such a restriction leads us to the introduction of isoparametric bilinear and trilinear elements. These use local bilinear and trilinear formulas of the same type as the formulas used to represent the trial solution to map each element to a square or a cube. This is the reason for calling these elements isoparametric. Now the solution is implicitly defined by the equations for the solution and the local position within each element.

The two-dimensional isoparametric bilinear element uses a local bilinear mapping of the physical domain and a square reference element $-\frac{1}{2} \leq \xi \leq \frac{1}{2}$, $-\frac{1}{2} \leq \eta \leq \frac{1}{2}$, as illustrated in Figure 3.13. Labeling the corners from 1 to 4, we set

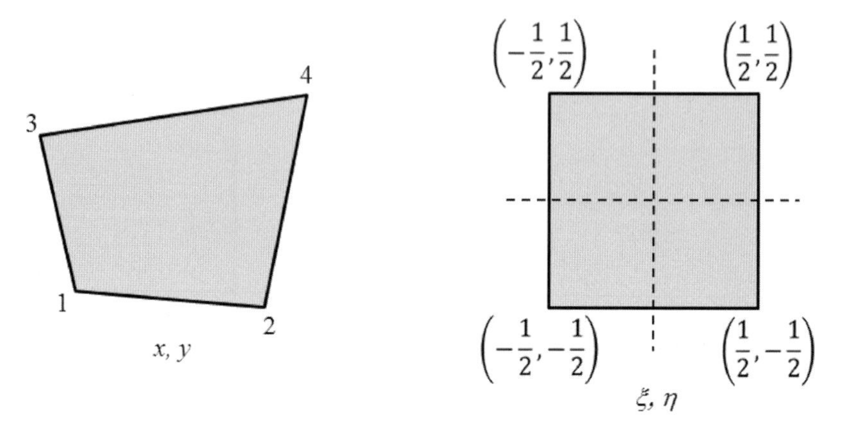

Figure 3.13 Isoparametric bilinear elements.

$$
\begin{aligned}
x = &\left(\frac{1}{2}-\xi\right)\left(\frac{1}{2}-\eta\right)x_1+\left(\frac{1}{2}+\xi\right)\left(\frac{1}{2}-\eta\right)x_2 \\
&+\left(\frac{1}{2}-\xi\right)\left(\frac{1}{2}+\eta\right)x_3+\left(\frac{1}{2}+\xi\right)\left(\frac{1}{2}+\eta\right)x_4
\end{aligned} \tag{3.51}
$$

and

$$
\begin{aligned}
y = &\left(\frac{1}{2}-\xi\right)\left(\frac{1}{2}-\eta\right)y_1+\left(\frac{1}{2}+\xi\right)\left(\frac{1}{2}-\eta\right)y_2 \\
&+\left(\frac{1}{2}-\xi\right)\left(\frac{1}{2}+\eta\right)y_3+\left(\frac{1}{2}+\xi\right)\left(\frac{1}{2}+\eta\right)y_4.
\end{aligned} \tag{3.52}
$$

Now, we define the local trial solution by a similar formula as

$$
\begin{aligned}
u_h = &\left(\frac{1}{2}-\xi\right)\left(\frac{1}{2}-\eta\right)u_1+\left(\frac{1}{2}+\xi\right)\left(\frac{1}{2}-\eta\right)u_2 \\
&+\left(\frac{1}{2}-\xi\right)\left(\frac{1}{2}+\eta\right)u_3+\left(\frac{1}{2}+\xi\right)\left(\frac{1}{2}+\eta\right)u_4.
\end{aligned} \tag{3.53}
$$

Along an edge $\xi =$ constant, u_h, x, and y are all linear in η, and hence they match the values from the solution in the neighboring element on the other side of the edge. The same is true for an edge $\eta =$ constant, and thus the trial solution is continuous across every edge. Now within each cell, we can evaluate the derivatives by the chain rule:

$$
\begin{aligned}
\frac{\partial u_h}{\partial x} &= \frac{\partial u_h}{\partial \xi}\xi_x + \frac{\partial u_h}{\partial \eta}\eta_x \\
\frac{\partial u_h}{\partial y} &= \frac{\partial u_h}{\partial \xi}\xi_y + \frac{\partial u_h}{\partial \eta}\eta_y,
\end{aligned}
$$

where the mapping derivative matrix

$$
K = \begin{bmatrix} x_\xi & x_\eta \\ y_\xi & y_\eta \end{bmatrix}
$$

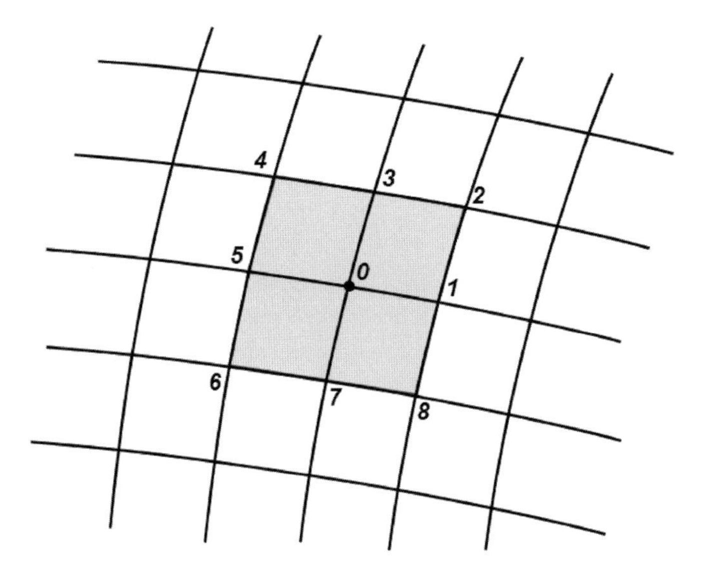

Figure 3.14 Stencil for isoparametric bilinear element centered at node 0.

is determined by differentiating (3.51) and (3.52). Then

$$\begin{bmatrix} \xi_x & \xi_y \\ \eta_x & \eta_y \end{bmatrix} = K^{-1} = \frac{1}{J} \begin{bmatrix} y_\eta & -x_\eta \\ -y_\xi & x_\xi \end{bmatrix},$$

where

$$J = \det(K) = x_\xi y_\eta - x_\eta y_\xi. \tag{3.54}$$

Consider an interior node as illustrated in Figure 3.14, where it is labeled zero. The basis function ϕ_0 centered at this node is nonzero only in the four surrounding quadrilaterals, in which we need to form the integral

$$\int_{\mathcal{D}} \nabla u_h \cdot \nabla \phi_0 dS,$$

where the domain $\mathcal{D}$ consists of the four surrounding cells. In the local coordinate system, this is transformed to

$$\int \int_{\mathcal{D}} J \nabla u_h \cdot \nabla \phi_0 d\xi d\eta, \tag{3.55}$$

where J is the determinant of the mapping matrix in each cell defined by (3.54). Finally, the integral (3.55) is evaluated numerically. The simplest procedure is to use single point integration for each cell, with these formulas evaluated at the cell center $\zeta = 0$, $\eta = 0$. In this case the derivatives of the mapping functions and trial solution defined by (3.51)–(3.53) reduce to the simple formulas

$$x_\xi = \frac{1}{2}(x_4 - x_3 + x_2 - x_1), \qquad x_\eta = \frac{1}{2}(x_4 - x_2 + x_3 - x_1)$$

$$y_\xi = \frac{1}{2}(y_4 - y_3 + y_2 - y_1), \qquad y_\eta = \frac{1}{2}(y_4 - y_2 + y_3 - y_1)$$

$$\frac{\partial u_h}{\partial \xi} = \frac{1}{2}(u_4 - u_3 + u_2 - u_1), \quad \frac{\partial u_h}{\partial \eta} = \frac{1}{2}(u_4 - u_2 + u_3 - u_1).$$

Single point integration is consistent with second order accuracy. However, if it is used for Laplace's equation with bilinear elements on a Cartesian mesh with equal intervals in each coordinate direction, as illustrated in Figure 3.15, it leads to a discrete Laplacian with a rotated stencil. If the nodes are colored red and black in a checkerboard pattern, denoted by squares and circles in the figure, it can be seen that the two sets of nodes are completely decoupled, resulting in effect in two independent solutions, each determined by the boundary values of the nodes with the same color (shape). In more complicated problems, this can provoke an instability known as the hour glass instability. Jameson and Caughey (1977, Jameson 1978) showed that this may be prevented by adding terms proportional to $h^2 \frac{\partial^2}{\partial x \partial y} \frac{\partial^2}{\partial x \partial y} u_h$ to recouple the discrete solution at the two sets of nodes. Otherwise, the instability may be prevented by using higher order Gauss integration with four quadrature points in each cell, with correspondingly increased computational expense.

Three-dimensional isoparametric elements can be defined in a similar manner by trilinear mappings to a cubic reference element extending from $-\frac{1}{2}$ to $\frac{1}{2}$ in each coordinate direction:

$$x = 8 \sum_{i=1}^{8} \left(\frac{1}{4} + \xi_i \xi\right) \left(\frac{1}{4} + \eta_i \eta\right) \left(\frac{1}{4} + \zeta_i \zeta\right) x_i$$

$$y = 8 \sum_{i=1}^{8} \left(\frac{1}{4} + \xi_i \xi\right) \left(\frac{1}{4} + \eta_i \eta\right) \left(\frac{1}{4} + \zeta_i \zeta\right) y_i$$

$$u_h = 8 \sum_{i=1}^{8} \left(\frac{1}{4} + \xi_i \xi\right) \left(\frac{1}{4} + \eta_i \eta\right) \left(\frac{1}{4} + \zeta_i \zeta\right) u_i,$$

as illustrated in Figure 3.16. Now, the mapping matrix K has entries

$$K_{ij} = \frac{\partial x_i}{\partial \xi},$$

which can be evaluated directly. Then,

$$\frac{\partial u_h}{\partial x_j} = \sum_{i=1}^{3} \frac{\partial u_h}{\partial \xi_i} \frac{\partial \xi_i}{\partial x_j},$$

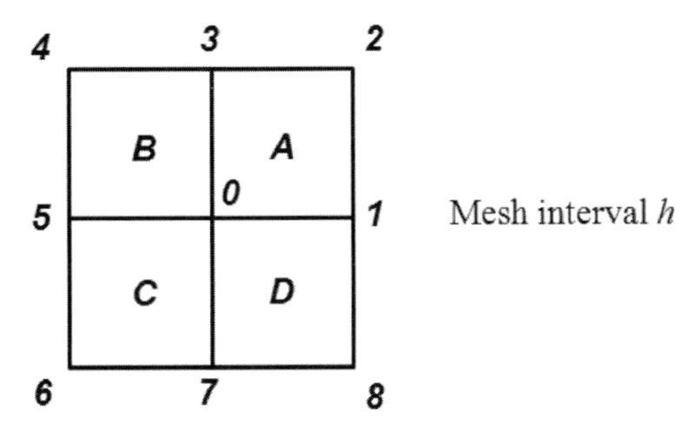

Test function $\phi_0 = 1, \phi_k = 0, k \neq 1.$

Derivatives in cell A:

$$\frac{\partial u_h}{\partial x} = \frac{u_2 - u_3 + u_1 - u_0}{2h}, \quad \phi_x = -\frac{1}{2h}$$
$$\frac{\partial u_h}{\partial y} = \frac{u_2 - u_1 + u_3 - u_0}{2h}, \quad \phi_y = -\frac{1}{2h}$$

Derivatives in cell B:

$$\frac{\partial u_h}{\partial x} = \frac{u_3 - u_4 + u_0 - u_5}{2h}, \quad \phi_x = \frac{1}{2h}$$
$$\frac{\partial u_h}{\partial y} = \frac{u_3 - u_0 + u_4 - u_5}{2h}, \quad \phi_y = -\frac{1}{2h}$$

Derivatives in cell C:

$$\frac{\partial u_h}{\partial x} = \frac{u_0 - u_5 + u_7 - u_6}{2h}, \quad \phi_x = \frac{1}{2h}$$
$$\frac{\partial u_h}{\partial y} = \frac{u_0 - u_7 + u_5 - u_6}{2h}, \quad \phi_y = \frac{1}{2h}$$

Derivatives in cell D:

$$\frac{\partial u_h}{\partial x} = \frac{u_1 - u_0 + u_8 - u_7}{2h}, \quad \phi_x = -\frac{1}{2h}$$
$$\frac{\partial u_h}{\partial y} = \frac{u_1 - u_8 + u_0 - u_7}{2h}, \quad \phi_y = \frac{1}{2h}$$

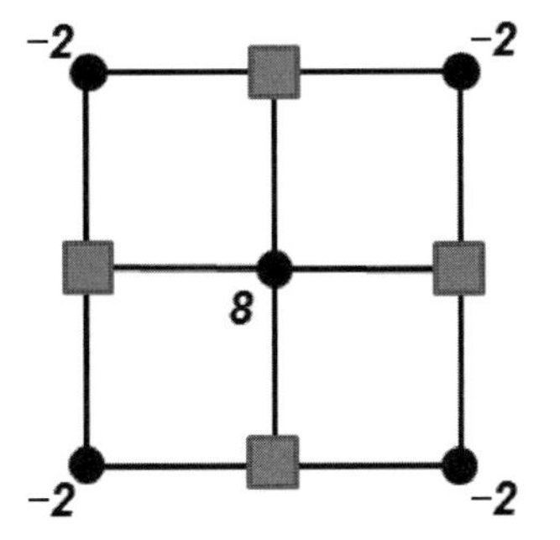

Discrete Laplacian

$$r_0 = \sum_{\text{cells}} h^2 \left(\frac{\partial u_h}{\partial x}\phi_x + \frac{\partial u_h}{\partial y}\phi_y \right) = -2(u_2 + u_4 + u_6 + u_8 - 4u_0)$$

Figure 3.15 Rotated Laplacian with bilinear elements and single point integration.

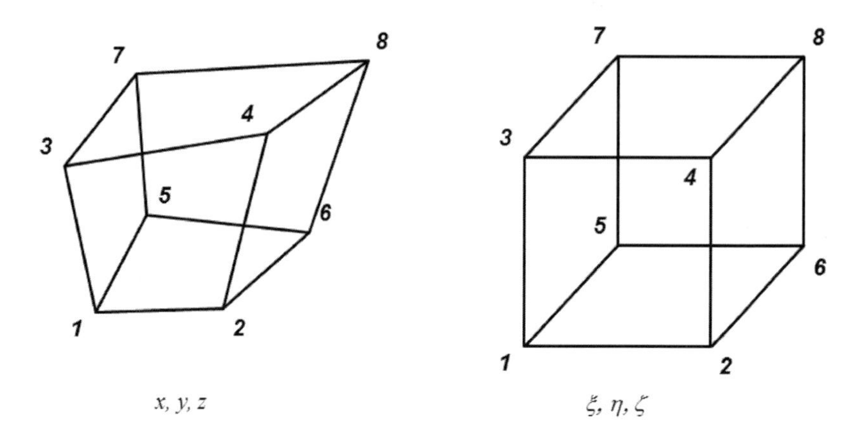

Figure 3.16 Three-dimensional isoparametric element.

where $\frac{\partial \xi_i}{\partial x_j}$ are the entries of K^{-1}, while an integral of a quantity f over a cell is transformed as

$$\int fV = \int\int Jf d\xi d\eta d\zeta,$$

where $J = \det(K)$.

3.4 Finite Volume Methods

Finite volume methods directly approximate the integral form of the conservation laws – in the case of fluid dynamics, conservation of mass, momentum, and energy. The domain is divided into a large number of small discrete control volumes, and the conservation laws are approximated for each control volume. This is a conceptually simple and intuitively appealing way to discretize any time-dependent system of conservation laws. In the aerospace community it seems to have been first adopted by MacCormack, although he did not immediately publish his ideas (MacCormack & Paullay 1972). Finite volume methods were also independently developed for the simulation of complex turbulent flows in a variety of industrial applications. These developments have been described, for example, by Patankar and Spalding (1972).

In order to illustrate the formulation of a finite volume method, consider the two-dimensional time-dependent conservation law

$$\frac{\partial \boldsymbol{w}}{\partial t} + \frac{\partial}{\partial x}\boldsymbol{f}(\boldsymbol{w}) + \frac{\partial}{\partial y}\boldsymbol{g}(\boldsymbol{w}) = 0.$$

This could, for example, be the Euler equations for inviscid compressible flow. This equation can be expressed in integral form for a domain $\mathcal{D}$ and boundary $\mathcal{B}$ as

$$\frac{d}{dt}\int_{\mathcal{D}} \boldsymbol{w} dS + \oint_{\mathcal{B}} (\boldsymbol{f} dy - \boldsymbol{g} dx) = 0, \tag{3.56}$$

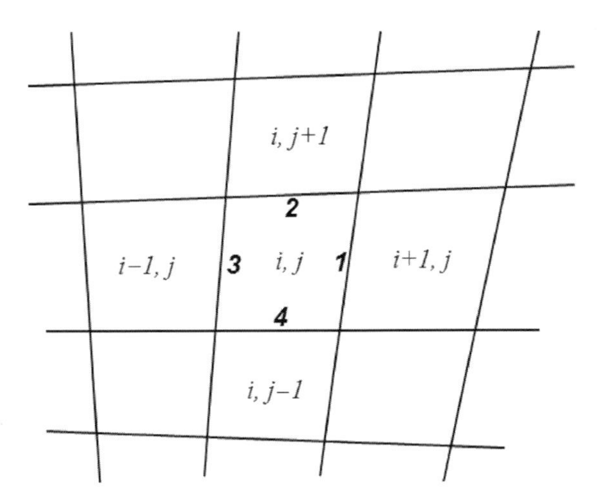

Figure 3.17 Quadrilateral control volumes.

where $\mathcal{S}$ is the area. Suppose now that the domain is subdivided into small quadrilateral control volumes as illustrated in Figure 3.17. Assuming for the moment a regular pattern, we can label the cells with double subscripts i,j and obtain a semi-discrete scheme by applying (3.56) to each control volume. This leads to the semi-discrete system of ordinary differential equations

$$\frac{d}{dt}\left(S_{i,j}\boldsymbol{w}_{i,j}\right) + \boldsymbol{Q}_{i,j} = 0, \tag{3.57}$$

where $S_{i,j}$ is the cell area, and $\boldsymbol{Q}_{i,j}$ is the net flux out of the cell. This may be evaluated as

$$\boldsymbol{Q}_{i,j} = \sum_{k-1}^{4}(\Delta y_k \boldsymbol{f}_k - \Delta x_k \boldsymbol{g}_k), \tag{3.58}$$

where $\boldsymbol{f}_k$ and $\boldsymbol{g}_k$ denote estimates of the flux vectors f and g on the kth side, Δx_k and Δy_k are the increments of x and y along the side with appropriate signs corresponding to an outward normal, and the sum is taken over the four sides of the cell. Interpreting $\boldsymbol{w}_{i,j}$ as the average value of the state vector $\boldsymbol{w}$ in the cell, (3.57) would be an exact statement if the values of $\boldsymbol{f}$ and $\boldsymbol{g}$ were exactly known along each side.

In fact we have to recover estimates of f_k and g_k from the average values of w in the neighboring cells separated by the edge. The simplest estimate is just an arithmetic average,

$$\boldsymbol{f}_1 = \frac{1}{2}(\boldsymbol{f}_{i+1,j} + \boldsymbol{f}_{i,j})$$

on side 1, for example, where $\boldsymbol{f}_{i,j} = \boldsymbol{f}(\boldsymbol{w}_{i,j})$. An alternative is to set

$$\boldsymbol{f}_1 = \boldsymbol{f}(\boldsymbol{w}_1),$$

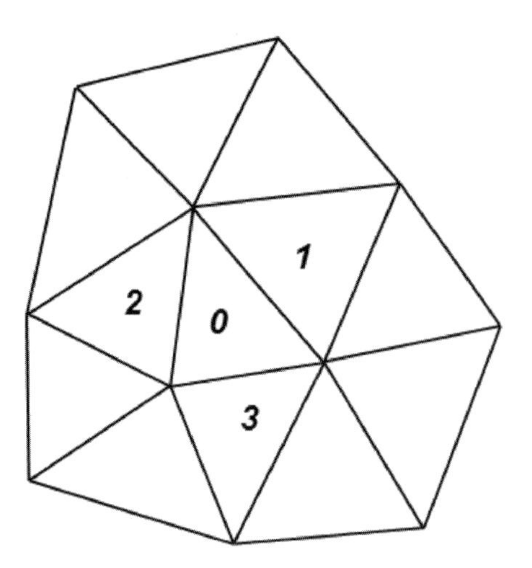

Figure 3.18 Triangular control volumes.

where

$$w_1 = \frac{1}{2}(w_{i+1,j} + w_j).$$

In the case that the domain is subdivided by a Cartesian grid, either of these approximations reduce to central difference schemes, and accordingly we can expect to obtain second order accuracy; but on an irregular grid the attainable accuracy will be degraded.

The scheme (3.57) with the flux (3.58) actually reduces on a Cartesian grid with mesh intervals Δx and Δy to

$$\frac{dw_{i,j}}{dt} + \frac{f_{i+1,j} - f_{i-1,j}}{2\Delta x} + \frac{g_{i,j+1} - g_{i,j-1}}{2\Delta y} = 0,$$

after dividing through by the cell area $\Delta x \Delta y$. This leads to a decoupling between the odd and even numbered cells, allowing undesirable oscillatory modes. Schemes that are satisfactory in practice can be obtained by using upwind biased estimates of the flux vectors, as will be discussed in the following section, or by adding appropriate dissipative terms. These should have a magnitude proportional to some power of the mesh width, in order to ensure that the discretization is consistent with the true equation in the limit as the mesh width is reduced to zero.

The same discretization procedure can easily be applied to triangular meshes or unstructured quadrilateral meshes, or meshes with mixed cells. Figure 3.18 illustrates a triangular mesh where, for the sake of convenience, the cell of interest is labeled with the subscript 0 and its neighbors with the subscripts 1–3. The finite volume discretization may now be written as:

$$S_0 \frac{dw_0}{dt} + \sum_{k=1}^{3} (f_{0k} \Delta y_{0k} - g_{0k} \Delta x_{0k}) = 0,$$

where Δx_{0k} and Δy_{0k} are the increments in x and y along the edge $0k$ separating cells 0 and k, and f_{0k} and g_{0k} are estimates of the flux vectors along that edge.

In order to apply a similar procedure to an unstructured quadrilateral mesh, it is necessary to introduce a cell numbering system and data structure which identifies the neighbors of each cell. In the case of a mesh with mixed triangular and quadrilateral cells and possibly other polyhedral cells, the number of neighbors varies from cell to cell. Since each edge separates exactly two cells, it turns out that the calculation of the flux balances can be simplified by looping over the edges. After the flux across each edge has been calculated, the accumulated flux balances of the two neighboring cells are updated by adding it to the balance in the cell receiving the flux and subtracting it from the balance in the other cell.

Finite volume methods have the following advantages.

1. Like finite element methods, they can easily be applied to complex geometric domains, provided that a suitable mesh generating procedure is available. In the case of a mesh of mixed polyhedral cells, they can actually be formulated more easily than finite element schemes.
2. The integral form of the equations is valid in the presence of discontinuities, whereas the differential form is not. Thus finite volume methods are well suited to the treatment of flows with shock waves or contact discontinuities.
3. They facilitate the construction of upwind biased schemes, which enable the direction of wave propagation to be distinguished in convection-dominated problems, as will be discussed in Chapters 5–7.
4. They automatically produce zero residuals for a uniform flow, because then the equations reduces to the sum of the projected side lengths or face areas over a closed control volume. This is not necessarily the case with finite difference schemes using a mapping to curvilinear coordinates.

Accordingly, it is not surprising that most commercial CFD software is based on finite volume methods, and so are the majority of CFD programs that have been developed in the aerospace community. They do have the disadvantage that it is quite hard to construct a finite volume scheme with better than second order accuracy. Two requirements must be met in order to attain higher order accuracy. First, it is necessary to implement higher order quadrature schemes to calculate the integrated fluxes across each edge or face. Second, it is necessary to implement procedures to recover higher order estimates of the flux vectors at the quadrature points from the known average values of the state in a stencil of neighboring cells.

3.4.1 Nodal and Cell-Vertex Schemes

While the original formulation of finite volume methods is based on the concept of updating cell average values, one can associate point values of the solution at the cell centroids. Since the difference between the average and the centroidal value is of order h^2 for a mesh interval of order h, we do not need to distinguish these values for second order accurate discretizations. This facilitates the formulation of alternative

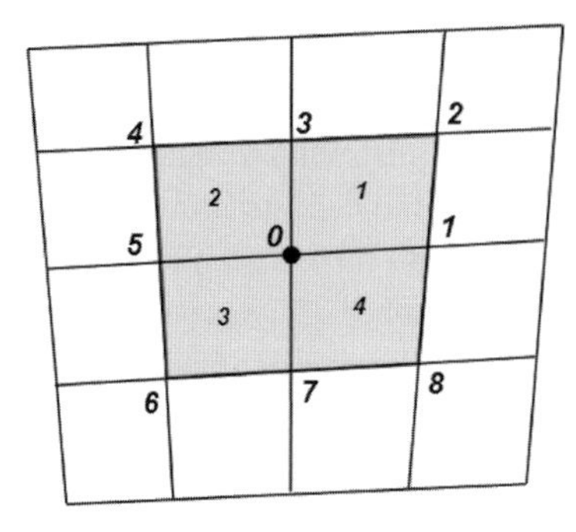

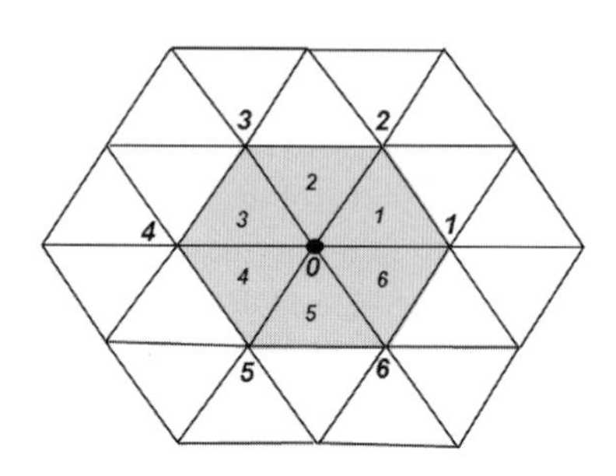

Figure 3.19 Cell vertex schemes on quadrilateral meshes (a) and triangular meshes (b). Shaded area is control volume.

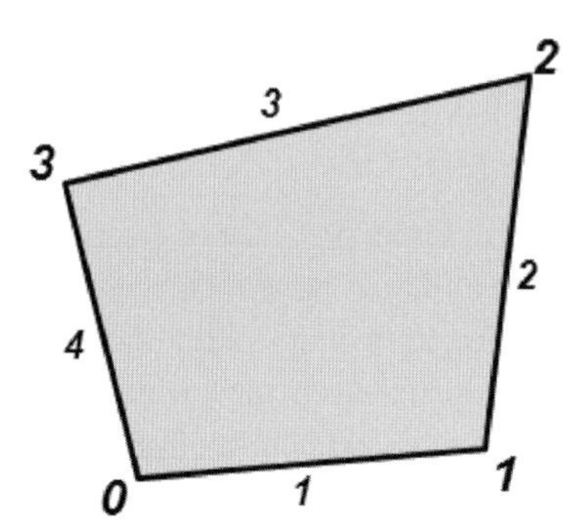

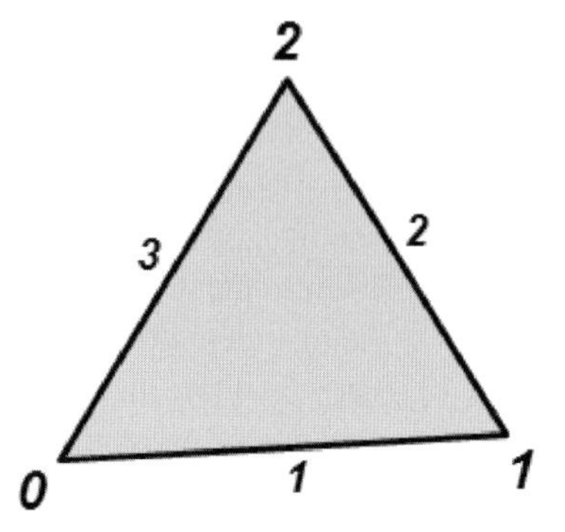

Figure 3.20 Flux balance for a cell. (a) Quadrilateral cell 0, 1, 2, 3; (b) Triangular cell 0, 1, 2.

finite volume schemes in which the solution is stored at the mesh nodes, as illustrated in Figure 3.19 for quadrilateral and triangular meshes in a two-dimensional domain. We can approximate the flux balance for each cell by trapezoidal integration:

$$Q = \sum_{\text{sides}} \bar{f}\Delta y - \bar{g}\Delta x,$$

where Δx an Δy are the increments in x and y along each side, and $\bar{f}$ and $\bar{g}$ are the arithmetic averages of $f(w)$ and $g(w)$ between the vertices joined by the side, as illustrated in Figure 3.20. Then, we can update the solution at each node by forming a central volume, which is the union of the cells surrounding that node. Labeling the center node by the subscript zero as illustrated in Figures 3.19 and 3.20, the discrete conservation law takes the form

$$\left(\sum_{\text{cells}} S_k\right)\frac{dw_0}{dt} + \sum_{\text{cells}} Q_k = 0.$$

These discretizations are closely related to the corresponding finite element discretizations that would be obtained with bilinear isoparametric elements on a quadrilateral mesh or linear elements on a triangular mesh. They do not easily provide

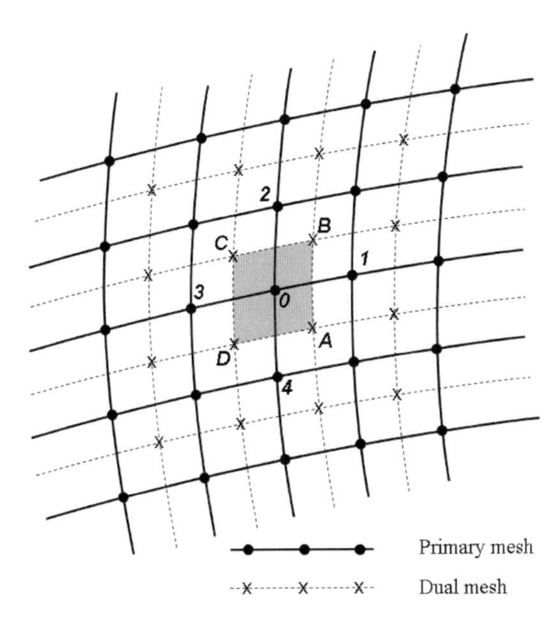

Figure 3.21 Discretization with dual control volumes.

for upwind fluxes but can be stabilized by the addition of artificial diffusive terms (Jameson 1993), as will be discussed in Chapter 7. An alternative nodal finite volume scheme that allows for upwind fluxes can be formulated by introducing a dual mesh connecting the cell centers, as illustrated in Figure 3.21 for a quadrilateral mesh.

3.5 Some Classical Schemes

We examine here some classical schemes for the solution of time dependent problems, ranging from linear advection up to systems of conservation laws. We consider first the Lax–Wendroff scheme (Lax & Wendroff 1960). Suppose that the linear advection equation

$$\frac{\partial u}{\partial t} + a\frac{\partial u}{\partial x} = 0 \tag{3.59}$$

is discretized on a uniform mesh with space and time intervals Δx and Δt. The true solution at time $t + \Delta t$ can be expanded as a Taylor series:

$$u(t + \Delta t) = u(t) + \Delta t\frac{\partial u}{\partial t} + \frac{\Delta t^2}{2}\frac{\partial^2 u}{\partial t^2} + \mathcal{O}(\Delta t^3).$$

Also

$$\frac{\partial u}{\partial t} = -a\frac{\partial u}{\partial x}$$

and

$$\frac{\partial^2 u}{\partial t^2} = -\frac{\partial}{\partial t}\left(a\frac{\partial u}{\partial x}\right)$$
$$= -a\frac{\partial}{\partial x}\frac{\partial u}{\partial t}$$
$$= a^2\frac{\partial^2 u}{\partial x^2}.$$

In order to obtain a second order accurate discretization, we now advance the numerical solution $v(x,t)$ to $v(x,t+\Delta t)$ by taking the first three terms of the Taylor series and approximating the time derivatives by finite difference approximations to the corresponding spatial derivatives. Thus, we replace $\frac{\partial v}{\partial t}$ and $\frac{\partial^2 v}{\partial t^2}$ by finite difference approximations to $-a\frac{\partial v}{\partial x}$ and $a^2\frac{\partial^2 v}{\partial x^2}$. This leads to the Lax–Wendroff scheme

$$v_j^{n+1} = v_j^n - \frac{\lambda}{2}(v_{j+1}^n - v_{j-1}^n) + \frac{\lambda^2}{2}(v_{j+1}^n - 2v_j^n + v_{j-1}^n), \tag{3.60}$$

where $\lambda = \frac{a\Delta t}{\Delta x}$.

It will be proved in the next chapter that the Lax–Wendroff scheme is stable in a Euclidean norm, provided that the time step satisfies the restriction $|\lambda| \leq 1$. In the case that $\lambda = 1$, the scheme is exact, perfectly representing propagation along the characteristics

$$v_j^{n+1} = v_{j-1}^n.$$

In order to extend the Lax–Wendroff scheme to the nonlinear equation

$$\frac{\partial u}{\partial t} + \frac{\partial}{\partial x}f(u) = 0, \tag{3.61}$$

we again expand $u(t+\Delta t)$ as a Taylor series. Now

$$\frac{\partial u}{\partial t} = -\frac{\partial f}{\partial x}$$

and

$$\frac{\partial^2 u}{\partial t^2} = -\frac{\partial}{\partial t}\frac{\partial f}{\partial x}$$
$$= -\frac{\partial}{\partial x}\left(\frac{\partial f}{\partial u}\frac{\partial u}{\partial t}\right)$$
$$= \frac{\partial}{\partial x}\left(a(u)\frac{\partial f}{\partial x}\right),$$

where $a(u) = \frac{\partial f}{\partial u}$ is the wave speed. This leads to the discretization

$$v_j^{n+1} = v_j^n - \frac{\Delta t}{\Delta x}(f_{j+1} - f_{j-1}) + \frac{\Delta t^2}{\Delta x^2}\left(a_{j+\frac{1}{2}}(f_{j+1} - f_j) - a_{j-\frac{1}{2}}(f_j - f_{j-1})\right),$$

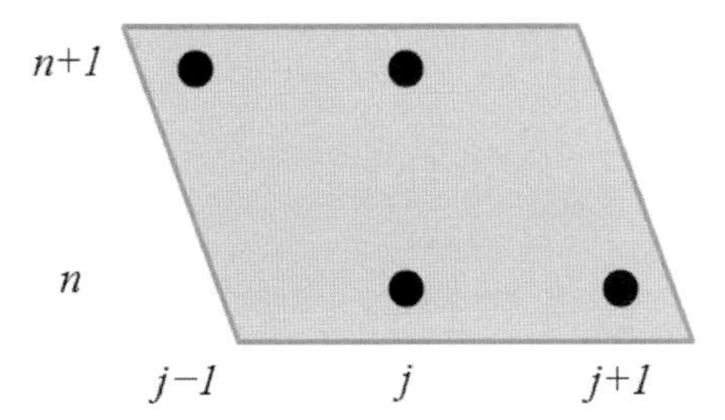

Figure 3.22 Skewed stencil of the MacCormack scheme.

where $f_j = f(v_j)$, and $a_{j+\frac{1}{2}}$ is an estimate of $a(u)$ at the midpoint of the interval from x_j to x_{j+1}. For example, we can evaluate $a_{j+\frac{1}{2}}$ as $\frac{\partial f}{\partial u}$ for $u = \frac{1}{2}(v_{j+1} + v_j)$.

The same method can be used to derive a corresponding discretization for a nonlinear system of equations of the form

$$\frac{\partial \boldsymbol{w}}{\partial t} + \frac{\partial}{\partial x}\boldsymbol{f}(\boldsymbol{w}) = 0.$$

Now,

$$\begin{aligned}\frac{\partial^2 \boldsymbol{w}}{\partial t^2} &= -\frac{\partial}{\partial t}\frac{\partial \boldsymbol{f}}{\partial x} \\ &= -\frac{\partial}{\partial x}\left(\frac{\partial \boldsymbol{f}}{\partial w}\frac{\partial \boldsymbol{w}}{\partial t}\right) \\ &= \frac{\partial}{\partial x}\left(A(\boldsymbol{w})\frac{\partial \boldsymbol{f}}{\partial x}\right),\end{aligned}$$

where $A(\boldsymbol{w})$ is the Jacobian matrix $\frac{\partial \boldsymbol{f}}{\partial \boldsymbol{w}}$. Accordingly, we obtain an equation of the same form as (3.5), with $a_{j+\frac{1}{2}}$ replaced by an estimate $A_{j+\frac{1}{2}}$ of $\frac{\partial \boldsymbol{f}}{\partial \boldsymbol{w}}$.

In order to avoid the need to evaluate the Jacobian matrix, various two-step schemes have been devised in which provisional intermediate values of the solution are obtained in the first step, corresponding to advancing by $\frac{1}{2}\Delta t$ or Δt, and in the second step the spatial derivatives are evaluated using these provisional values. The most famous of these schemes was proposed by MacCormack in a celebrated paper (MacCormack 1969). The MacCormack scheme uses a skewed stencil as illustrated in Figure 3.22. The first step of the MacCormack scheme predicts the solution of the nonlinear equation (3.61) by the forward difference scheme

$$\tilde{v}_j = v_j^n - \frac{\Delta t}{\Delta x}(f_{j+1}^n - f_j^n),$$

where $f_j^n = f(v_j^n)$. The second step takes the average of the forward difference at time level n and the backward difference of the provisional solution at time level $n+1$:

$$v_j^{n+1} = v_j^n - \frac{1}{2}\frac{\Delta t}{\Delta x}(f_{j+1}^n - f_j^n) - \frac{1}{2}\frac{\Delta t}{\Delta x}(\tilde{f}_j - \tilde{f}_{j-1}),$$

where $\tilde{f}_j = f(\tilde{v}_j)$. The second step may be simplified by substituting the equation of the first step to obtain

$$v_j^{n+1} = \frac{1}{2}(v_j^n + \tilde{v}_j) - \frac{1}{2}\frac{\Delta t}{\Delta x}(\tilde{f}_j - \tilde{f}_{j-1}).$$

The scheme may also be formulated with backward differences in the first step and forward difference in the second step. In either case, the formulas reduce identically to the Lax–Wendroff scheme for the case of linear advection, $f(u) = au$. The scheme is formally second order accurate, while it also minimizes the computational effort to advance a time step.

The Lax–Wendroff and MacCormack schemes are examples of integrated space-time discretization schemes. If they are used to obtain a steady state solution by evolution in time toward a steady state, the steady state solution will generally depend on the time step in the case of multi-dimensional problems. The alternative approach of first using a spatial discretization to define a semi-discrete scheme and then using a time discretization to solve the resulting ODEs enables steady state solutions that are independent of the time step, and it also provides more flexibility in the design of accurate and robust discretization procedures for complex nonlinear problems.

Although the Lax–Wendroff and MacCormack schemes are stable in the Euclidean norm for linear problems, they actually produce oscillatory solutions when they are used to treat discontinuous solutions, such as the propagation of a discontinuity by the linear advection equation, as well as the treatment of nonlinear conservation laws that admit solutions with shock waves. This has led to the development of a variety of non-oscillatory shock capturing schemes, both for scalar conservation laws and systems of conservation laws, which are supported by now well developed theoretical analysis based on error measures that are more appropriate than the Euclidean norm for the treatment of problems with discontinuous solutions. The next chapter presents the classical theory of stability pioneered by Lax. The following chapters develop the theory of shock capturing schemes, first for scalar nonlinear conservation laws and then for systems of conservation laws that generally have solutions with multiple waves that may be traveling in opposite directions.

4 Fundamental Stability Theory

4.1 Introduction

This chapter examines the stability of difference schemes for initial value problems defined by ordinary or partial differential equations. Three simple examples are examined first: an ordinary differential equation, the linear advection equation, which is the prototype for hyperbolic equations, and the diffusion equations. These serve to illustrate that in all three cases the numerical scheme can become unstable if the time step is too large, or in the case of a hyperbolic equation, if the difference scheme does not contain the proper region of dependence. The use of implicit schemes can remove the limit on the time step, at the expense of greater solution complexity.

Next, the general definitions of consistency, convergence, and stability are introduced in terms of an arbitrary norm, leading to the Lax equivalence theorem that consistent and stable schemes must converge to the true solution in the limit as the mesh interval and time step are reduced toward zero. Then stability in the Euclidean norm is examined, and the von Neumann stability test is introduced as a convenient way to deduce the stability of any linear scheme.

Nonlinear conservation laws generally admit solutions containing discontinuities, such as shock waves in a fluid flow. This motivates the need for difference schemes in conservation form. Moreover, the linear stability theory is no longer adequate, and schemes that pass a von Neumann test can easily admit highly oscillatory solutions in the neighborhood of shock waves. Thus alternative measures of stability are needed, leading to the introduction of concepts such as total variation diminishing (TVD) and local extremum diminishing (LED) schemes. Generally it proves necessary to use upwind-biased schemes to meet these criteria, either by direct construction or through the introduction of a controlled amount of artificial diffusion.

4.2 Stability and Accuracy of Simple Difference Schemes for Ordinary Differential Equations

Consider as a model a simple ordinary differential equation

$$\frac{du}{dt} = \alpha u, \quad t > 0$$

with a given initial condition $u(0)$ where α may be complex. The solution is

$$u = u(0)e^{\alpha t},$$

which will decay if $Re(\alpha) < 0$. The stability region of the true equation is thus the left half plane. To construct a discrete approximation, we introduce a discrete time step Δt and denote the discrete solution at $t = n\Delta t$ by v_n, so that v_n corresponds to $u(n\Delta t)$. The use of v rather than u to represent the discrete solutions serves to remind us that the discrete and exact solutions are not the same.

Now representing the derivative $\frac{du}{dt}$ by a simple difference formula $\frac{v^{n+1}-v^n}{\Delta t}$ leads to the forward Euler scheme

$$v^{n+1} = v^n + \alpha \Delta t v^n = \lambda v^n, \tag{4.1}$$

where

$$\lambda = 1 + \alpha \Delta t.$$

We set $v^0 = u(0)$, $v^n = v(n\Delta t)$, and it follows by repeated substitution that

$$v(n\Delta t) = \lambda^n v^0 = \lambda^n u(0).$$

If α is real and negative, the true solution decays exponentially. In this case, however, if Δt is increased to the point that $\alpha \Delta t < -2$, then $\lambda < -1$, and the discrete solution will oscillate with increasing amplitude. This is the simplest example of numerical instability. In fact to ensure that the discrete solution decreases monotonically, we need $\lambda \geq 0$, or $|\alpha \Delta t| \leq 1$.

Consider now a system of linear ordinary differential equations of the form

$$\frac{du}{dt} = Au, \quad t > 0 \tag{4.2}$$

with $u(0)$ given, where u is now an n-dimensional vector and A an $n \times n$ matrix. Suppose that A has a complete set of right eigenvectors satisfying

$$Av^{(k)} = \lambda^{(k)} v^{(k)}, \quad k = 1, n.$$

Let V be the right eigenvector matrix with $v^{(k)}$ as the columns. Now

$$A = V \Lambda V^{-1},$$

where Λ is a diagonal matrix with the eigenvalues as its entries, and if we multiply (4.2) by Λ^{-1} and set $\hat{u} = V^{-1}u$, it is reduced to

$$\frac{d\hat{u}}{dt} = \Lambda \hat{u},$$

where now each component equation is independent. In general the eigenvalues may be complex, and the system is stable if $Re(\lambda^{(k)}) \leq 0, k = 1, n$. This leads to the concept of the stability region of the difference equation (4.1), which is defined to be the region in the complex plane containing values of $\alpha \Delta t$ for which the scheme is

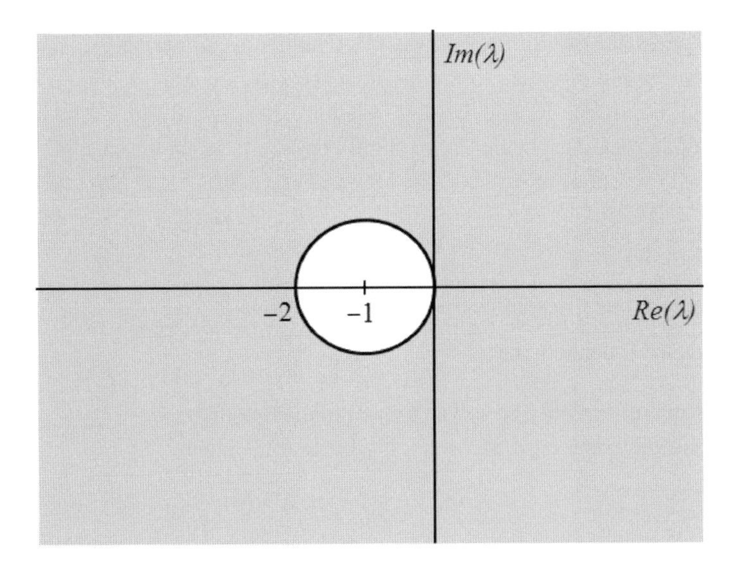

Figure 4.1 Stability region of the forward Euler time stepping scheme.

stable. In this case these are values of $\alpha \Delta t$ such that λ lies inside or on the circle $e^{i\theta}$. If $\lambda = e^{i\theta}$ then

$$\alpha \Delta t = e^{i\theta} - 1.$$

Thus the stability region of the forward Euler scheme (4.1) is a unit circle centered at the point -1, as illustrated in Figure 4.1.

Now in many real systems, there may be large variations in the magnitude of the eigenvalues, with fast and slow modes defined by large and small eigenvalues respectively. The integration of systems of this kind, generally called stiff systems, presents the difficulty that the stability limit for Δt is set by the largest eigenvalue, but the rate of decay is governed by the smallest eigenvalue. For simplicity consider the case where all the eigenvalues are real and negative. Then the stability limit is

$$\Delta t < \frac{2}{|\lambda_{\max}|},$$

where $\lambda_{\max}$ is the largest (negative) eigenvalue, and the rate of decay is $e^{\lambda_{\min} t}$, where $\lambda_{\min}$ is the smallest (negative) eigenvalue. To track this will require a number of time steps of the order of $\left|\frac{\lambda_{\max}}{\lambda_{\min}}\right|$, which may be very large.

This motivates the introduction of the implicit backward Euler scheme

$$\frac{v^{n+1} - v^n}{\Delta t} = \alpha v^{n+1}, \tag{4.3}$$

where the right hand side is evaluated at the end of the time step. In the case of a system, this becomes

$$v^{n+1} = v^n + \Delta t A v^{n+1}, \tag{4.4}$$

which requires the solution of a system of equations to advance a time step. The description "implicit" refers to the fact that the solution v_{n+1} at the end of the time step is only implicitly defined by (4.4), whereas the forward Euler scheme (4.1) is an example of an explicit scheme that explicitly defines v_{n+1}. In the case of the nonlinear system

$$\frac{du}{dt} = f(u),$$

(4.4) is replaced by the nonlinear system of equations

$$v^{n+1} = v^n + \Delta t f(v^{n+1}),$$

which in general can only be solved by an iterative method.

In the simple linear case, the implicit scheme (4.3) yields

$$v(n\Delta t) = \lambda^n v^0 = \lambda^n u(0),$$

where now

$$\lambda = \frac{1}{1 - \alpha \Delta t}.$$

In order to determine the boundary of the stability region, we set $\lambda = e^{i\theta}$ as before. This yields

$$\alpha \Delta t = 1 - e^{-i\theta}.$$

This is a unit circle to the right of the imaginary axis, as shown in Figure 4.2, but now the scheme is stable in the domain outside the circle. Accordingly, suppose that α lies in the left half plane

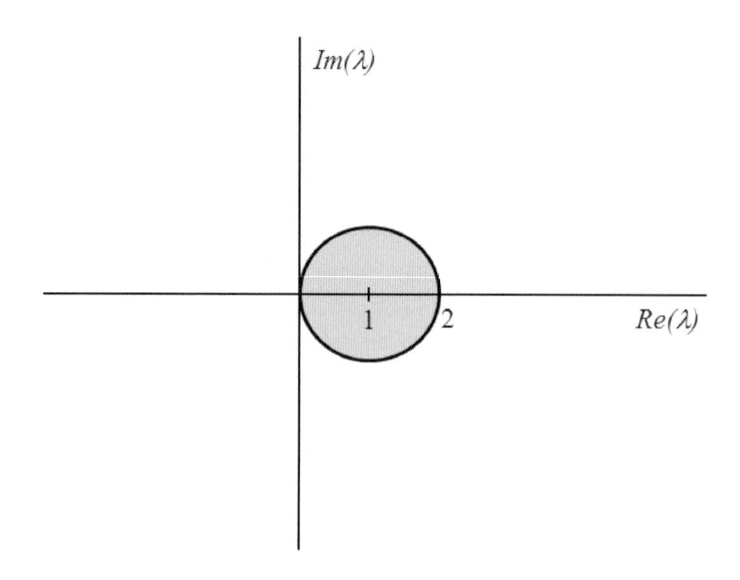

Figure 4.2 Stability region of the backward Euler time stepping scheme.

$$\alpha = -\beta + i\gamma$$

with $\beta > 0$. Then

$$\lambda = \frac{1}{1 + \beta\Delta t + i\gamma\Delta t}.$$

Then $|\lambda| \leq 1$ when $Re(\alpha) \leq 0$, so the stability region contains the entire left half plane $Re(\alpha)\Delta t \leq 0$. Difference schemes that are stable for all values of $\alpha\Delta t$ in the left half plane are called A-stable and are generally suitable for the integration of stiff systems. This is the simplest example of such a scheme.

So far we have not considered the accuracy of the discretizations (4.1) and (4.3). For this simple model it is easy to make a direct comparison of the discrete and true solutions. For simplicity set $u(0) = 1$. Then the true solution is

$$u(n\Delta t) = e^{\alpha n \Delta t} = 1 + n\alpha\Delta t + \frac{n^2}{2}(\alpha\Delta t)^2 + \frac{n^3}{3}(\alpha\Delta t)^3 + \cdots$$

For the explicit scheme (4.1),

$$v(n\Delta t) = \lambda^n = 1 + n\alpha\Delta t + \frac{n(n+1)}{2!}(\alpha\Delta t)^2 + \frac{n(n+1)(n+2)}{3!}(\alpha\Delta t)^3 + \cdots$$

giving an error

$$v(n\Delta t) - u(n\Delta t) = \frac{n}{2}(\alpha\Delta t)^2 + \frac{3n^2 + 2n}{6}(\alpha\Delta t)^3 + \cdots$$

To integrate for a fixed time t, one has $\Delta t = \frac{t}{n}$, so that as $\Delta t \to 0$ the error approaches

$$\frac{\alpha^2}{2} t\Delta t = O(\Delta t),$$

and the scheme is seen to be first order accurate. The same conclusion holds for the implicit scheme (4.3).

We can improve the accuracy by centering the evaluation of αv at the midpoint of the time step with the scheme

$$\frac{v^{n+1} - v^n}{\Delta t} = \frac{\alpha}{2}(v^{n+1} + v^n).$$

This scheme is called the trapezoidal rule as it corresponds to using trapezoidal integration for the formula

$$u(t + \Delta t) = u(t) + \int_t^{t+\Delta t} \alpha u(s) ds.$$

Now

$$v(n\Delta t) = \lambda^n v^0 = \lambda^n u(0),$$

where

$$\lambda = \frac{1 + \frac{1}{2}\alpha\Delta t}{1 - \frac{1}{2}\alpha\Delta t}.$$

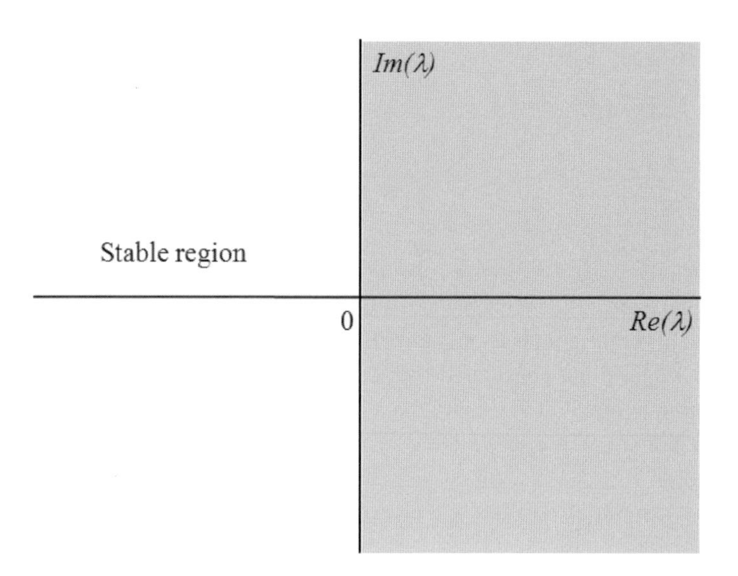

Figure 4.3 Stability region of the implicit trapezoidal time stepping scheme.

It is easy to verify that this scheme is A-stable. In fact $|\lambda| = 1$ on the imaginary axis, and accordingly the stability region is exactly the left half plane, as illustrated in Figure 4.3. A comparison of the power series expansion verifies that the scheme is second order accurate, with an error after a fixed time t of order Δt^2 in the limit as $\Delta t \to 0$, and the number of time steps $n \to \infty$.

4.3 The Linear Advection Equation

The simplest example of an initial value problem for a hyperbolic equation is provided by the linear advection equation

$$\frac{\partial u}{\partial t} + a\frac{\partial u}{\partial x} = 0, \quad t > 0, \quad -\infty \leq x \leq \infty, \tag{4.5}$$

which defines the solution $u(x,t)$ for a given initial condition

$$u(x,0) = f(x).$$

For convenience assume that a is positive. Then this equation describes a simple wave motion in which the solution is transported to the right at a speed a. Along a line $x - at = \xi$, as illustrated in Figure 4.4, we have

$$\frac{du}{dt} = \frac{d}{dt}u(at + \xi, t) = a\frac{\partial u}{\partial x} + \frac{\partial u}{\partial t} = 0.$$

Thus u is constant along this line, which is a characteristic – that is a line along which $\frac{\partial}{\partial t} = a\frac{\partial}{\partial x}$ is an internal operator – giving no information about derivatives normal to the line. This means that

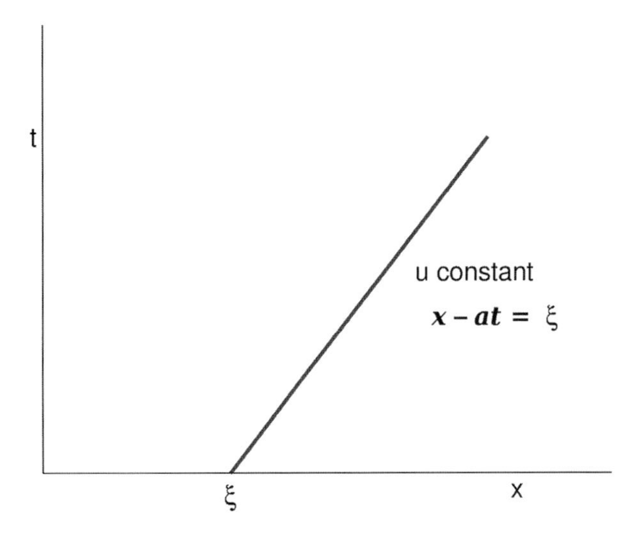

Figure 4.4 Characteristic line.

1. u cannot be arbitrarily specified along such a line, because u is constant according to the equation
2. data given along the line is not sufficient to continue the solution into a larger region.

The general solution of (4.5) has the form

$$u(x,t) = f(\xi) = f(x - at). \tag{4.6}$$

This represents the solution uniquely in terms of the initial values. Conversely every u of the form (4.6) is a solution of (4.5) provided it is differentiable, as can be verified by substituting $f(x - at)$ for $u(x,t)$. Note that $u(x,t)$ depends only on previous values along the characteristic passing through the point (x,t) and hence only on $f(x)$ at a single point on the axis.

Consider now a difference approximation with interval $\Delta x, \Delta t$ of the form

$$\frac{v(x,t+\Delta t) - v(x,t)}{\Delta t} + a\frac{v(x+\Delta x,t) - v(x,t)}{\Delta x} = 0.$$

This is consistent with (4.5) in the limit as Δx and $\Delta t \to 0$. Set

$$v_j^n = v(n\Delta t, j\Delta x), \quad f_j = f(j\Delta x), \quad \lambda = a\frac{\Delta t}{\Delta x},$$

where the superscript n and subscript j denote the location on a space-time mesh. Then

$$v_j^{n+1} = v_j^n - \lambda(v_{j+1}^n - v_j^n). \tag{4.7}$$

It is convenient to introduce the identity operator I and right shift operator E such that

$$Iv_j^n = v_j^n, \quad Ev_j^n = v_{j+1}^n.$$

Then we find that

$$\begin{aligned} v_j^n &= (I + \lambda I - \lambda E)v_j^{n-1} \\ &= (I + \lambda I - \lambda E)^n v_j^0 \\ &= \sum_{m=0}^{n} \binom{n}{m} (1+\lambda)^m I(-\lambda E)^{n-m} f_j \\ &= \sum_{m=0}^{n} \binom{n}{m} (1+\lambda)^m (-\lambda)^{n-m} f_{j+n-m}. \end{aligned}$$

The domain of dependence of $v(x,t)$ is thus $x, x+\Delta x, \ldots x+n\Delta x$, whereas the true solution $u(x,t)$ actually depends on the initial data at the point $x - at$. Therefore in general the solution of the difference equations will not converge to the true solution in the limit as Δx and $\Delta t \to 0$.

The scheme is in fact unstable. If errors $\pm\epsilon$ are added to f_j at consecutive odd and even points, the resulting eror in v_j^n is

$$\epsilon \sum_{m=0}^{n} \binom{n}{m} (1+\lambda)^n \lambda^{n-m} = (1+2\lambda)^n \epsilon.$$

Thus for a fixed λ, the resulting error in v grows exponentially with n and is unbounded for a fixed time t when the mesh is refined so that Δx and $\Delta t \to 0$ while $n \to \infty$. The necessity for the domain of dependence of the difference scheme to contain the true domain of dependence was first recognized by Courant, Friedrichs, and Lewy (Math Ann, 100, 1923, p22) and is known as the CFL condition.

We can satisfy the CFL condition by using backward differencing in the space direction:

$$\frac{v(x,t+\Delta t) - v(x,t)}{\Delta t} + a\frac{v(x,t) - v(x-\Delta x,t)}{\Delta x} = 0$$

or

$$v_j^{n+1} = v_j^n - \lambda(v_j^n - v_{j-1}^n). \tag{4.8}$$

This is the simplest example of an upwind scheme in which the space discretization is taken backward in the direction of the wave motion. Now

$$v_j^n = (I - \lambda I + \lambda E^{-1})v_j^0 = \sum_{m=0}^{n} \binom{n}{m} (1-\lambda)^m \lambda^{n-m} f_{j-(n-m)}.$$

The domain of dependence of $v(x,t)$ is the wedge highlighted in Figure 4.5, which covers the axis between x and $x - n\Delta x = x - \frac{at}{x}$. Provided that $0 \leq \lambda \leq 1$, the point $x - at$ lies inside this interval, and the CFL condition is satisfied. The ratio λ, generally

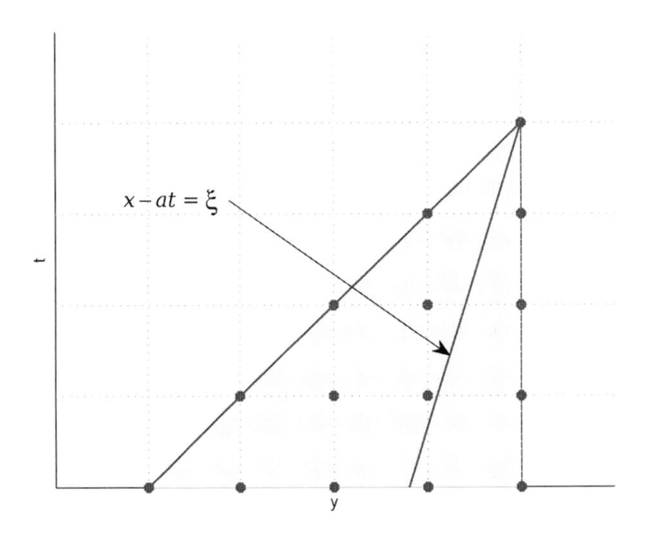

Figure 4.5 Domain of dependence of the discrete scheme and the true solution.

known as the CFL number, measures the number of mesh intervals Δx traveled during one time step by a front moving at the wavespeed a. If $\lambda > 1$, or equally $\Delta t > \frac{\Delta x}{a}$, the backward characteristic from (x,t) lies to the left of the wedge defining the numerical domain of dependence. Thus the CFL condition places a restriction on the admissible time step.

In order to verify the stability of the upwind scheme (4.8) when the CFL condition is satisfied, it is convenient to measure the maximum norm $\|v\| = \max_j |v_j|$ of the solution. According to (4.8),

$$v_j^{n+1} = (1-\lambda)v_j^n + \lambda v_{j-1}^n,$$

where now both the coefficients λ and $1-\lambda$ lie between zero and 1. Then

$$|v_j^{n+1}| \leq (1-\lambda)|v_j^n| + \lambda|v_{j-1}^n| \leq (1-\lambda)\|v^n\| + \lambda\|v^n\| = \|v^n\|_\infty.$$

Thus

$$\|v^{n+1}\|_\infty = \max_j |v_j^{n+1}| \leq \|v^n\|,$$

and the solution is bounded in the maximum norm.

Note that in general, however, schemes that satisfy the CFL condition may still be unstable. A well know example is the scheme

$$v_j^{n+1} = v_j^n + \frac{\lambda}{2}(v_{j+1}^n - v_{j-1}^n),$$

which combines the forward Euler time stepping scheme with central differencing in space. It is shown in Section 4.6 that the CFL condition is necessary but not sufficient for stability.

4.4 Initial Boundary Value Problems

Additional complications arise when hyperbolic equations have to be solved in a finite space domain, as it becomes necessary to examine whether waves are traveling in or out of the boundaries. Consider the linear advection equation in a finite domain:

$$\frac{\partial u}{\partial t} + a\frac{\partial u}{\partial x} = 0, \quad t \geq 0, \quad 0 \leq x \leq 1.$$

The case when the wave speed a is positive is illustrated in Figure 4.4. Now the initial data

$$u(x,0) = f(x), \quad 0 \leq x \leq 1$$

does not define the solution in the wedge along the characteristic passing through $x = 0$. In order to define a solution in this region, boundary data

$$u(0,t) = g(t), \quad t > 0$$

must be defined at the left boundary. The solution is continuous across the characteristic $x = at$ only if f and g satisfy the compatibility condition

$$f(0) = g(0).$$

At the right hand boundary, $x = 1$; on the other hand, the initial data $f(x,0)$ and left boundary data $g(t)$ determine the solution because the values $u(1,t)$ can be traced along the backward characteristics, which intersect either the initial data or the left boundary. Therefore no boundary condition should be specified at $x = 1$. If the wave speed is negative, the role of the boundaries is reversed. In general a boundary condition is required at an inflow boundary, which is crossed by incoming waves, and no boundary condition should be specified at an outflow boundary, which is crossed by outgoing waves. In the case of a system of hyperbolic equations, such as the Euler equations of fluid flow, the solution generally contains waves traveling in opposite directions. Then the number of conditions that should be specified at a boundary should be equal to the number of incoming waves.

4.5 The Diffusion Equation

As a third example, consider the diffusion equation

$$\frac{\partial u}{\partial t} = \sigma\frac{\partial^2 u}{\partial x^2}, \quad t > 0, \quad -\infty \leq x \leq \infty$$

with the initial data

$$u(x,0) = f(x).$$

This is the simplest example of a parabolic equation. It describes, for example, the diffusion of temperature along a bar due to heat conduction.

The use of forward difference in time and central difference in space leads to the standard explicit difference scheme

$$\frac{v(x,t+\Delta t)-v(x,t)}{\Delta t}=\sigma\frac{v(x+\Delta x,t)-2v(x,t)+v(x-\Delta x,t)}{\Delta x^2}.$$

Using the same notation as before, with superscripts to denote time intervals and subscripts to denote space intervals, this may be written as

$$u_j^{n+1}=u_j^n+\beta(u_{j+1}^n-2u_j^n+u_{j-1}^n), \tag{4.9}$$

where

$$\beta=\sigma\frac{\Delta t}{\Delta x^2}.$$

Again it turns out that too large a time step leads to instability. Suppose that errors $\pm\epsilon$ are added to v_j^n at consecutive odd and even points. The resulting error in v_j^{n+1} is

$$\epsilon_j^{n+1}=(1-4\beta)\epsilon. \tag{4.10}$$

If $\beta>\frac{1}{2}$ this will result in growing oscillations of opposite sign at successive time steps. If $0\le\beta\le\frac{1}{2}$, on the other hand, we can establish stability in the maximum norm. Rewriting (4.9) as

$$v_1^{n+1}=\beta v_{j-1}^n+(1-2\beta)v_j^n+\beta v_{j+1}^n,$$

it follows that

$$\begin{aligned}|v_j^{n+1}|&\le\beta|v_{j-1}^n|+(1-2\beta)|v_j^n|+\beta|v_{j+1}^n|\\&\le\beta\|v^n\|_\infty+(1-2\beta)\|v^n\|_\infty+\beta\|v^n\|_\infty=\|v^n\|_\infty.\end{aligned}$$

Thus

$$\|v^{n+1}\|_\infty=\max_j|v_j^{n+1}|\le\|v^n\|_\infty.$$

It can be seen from (4.10) that negative values of β will also lead to growing oscillations. This corresponds to the instability of the true solution when the diffusion coefficient is negative. In fact we should expect any consistent difference scheme to produce growing solutions when the true solution is unstable, because otherwise the discrete solution could not properly approximate the true solution.

The stability condition $\beta<\frac{1}{2}$, or

$$\Delta t<\frac{\Delta x^2}{2\sigma},$$

is a severe restriction on the time step when the mesh interval Δx is reduced toward zero. Again we can remove this restriction by adopting an implicit scheme. A second order accurate implicit scheme can be derived by using the trapezoidal rule to integrate the formula

$$u(t + \Delta t) = u(t) + \int_t^{t+\Delta t} u(s)ds.$$

This leads to the Crank–Nicolson scheme

$$v_j^{n+1} = v_j^n + \frac{\beta}{2}(v_{j+1}^{n+1} - 2v_j^{n+1} + v_{j-1}^{n+1}) + \frac{\beta}{2}(v_j^n - 2v_j^n + v_{j-1}^n).$$

This can be shown to be stable for arbitrarily large time steps by the von Neumann test that will be introduced later in this chapter.

4.6 Review of Stability Theory

It is apparent from the foregoing examples that we need some systematic procedures for assessing the stability and accuracy of difference schemes. This section surveys the general theory originally developed by Lax and Richtmyer (1956). For more details the reader is referred to the book by Richtmyer and Morton (1994). The theory introduces formal definitions of the notions of consistency, convergence, and stability in terms of general norms. The theory as it is presented is confined to linear initial value problems.

The first step is to define well posed initial value problems. The solution u will be regarded as an element of a function space with a norm $\|u\|$, as discussed in Appendix A. The exact definition of the norm is immaterial, provided that it satisfies the axioms of a norm. Suppose that the evolution of u in time is described by the equation

$$\frac{d}{dt}u(t) = Au(t) \tag{4.11}$$

with initial data

$$u(0) = u_0.$$

Here A is a differential operator in space if the governing equation is a partial differential equation. For simplicity A is assumed to be independent of time. We define a genuine solution as $u(t)$ such that

$$\left\| \frac{u(t + \Delta t) - u(t)}{\Delta t} - Au(t) \right\| \to 0 \quad \text{as} \quad \Delta t \to 0, 0 \leq t \leq T. \tag{4.12}$$

The transformation from the initial data to the solution at time t can be expressed as

$$u(t) = E(t)u_0,$$

where $E(t)$ is the evolution operator of the true solution. E has the semi-group property that

$$E(s + t) = E(s)E(t) \quad \text{for} \quad s \geq 0,\ t \geq 0$$

and

$$E(0) = I,$$

where I is the identity operator.

An initial value problem is called properly posed if the domain of $E(t)$ is in the space in which u_n is defined and the operators $E(t)$ are uniformly bounded, that is,

$$\|E(t)u_0\| \leq K\|u_0\|, \quad 0 \leq t \leq T$$

for all u_0. This means that for two different initial conditions u_1 and u_2, the difference

$$\|E(t)(u_1 - u_2)\| \leq K\|u_1 - u_2\|,$$

with the consequence that the solution depends continuously on the initial data.

We consider difference approximations with the form

$$B_1 v^{n+1} = B_0 v^n$$

with initial data

$$v^0 = u_0,$$

where the superscript denotes the time level $t = n\Delta t$. The operator B_1 is included on the left to allow for implicit schemes. For an explicit scheme, B_1 will be the identity operator I. We suppose that (4.12) can be solved to yield

$$v^{n+1} = C(\Delta t)v^n,$$

where the discrete evolution operator

$$C = B_1^{-1} B_0$$

will depend on the time step. The framework is now in place for the formal definitions of consistency, convergence, and stability. We consider the behavior of the discrete solution in the limit as the time step and mesh width both approach zero. Essentially consistency means that in this limit, the difference approximation (4.12) reduces to the true differential equation (4.11), while convergence means that discrete solution approaches the true solution, and stability means that the discrete solutions remain bounded. To make these notions precise, we need a measure of the difference between the true and discrete solutions. The values of the discrete solution are only calculated at the mesh points, but we may regard them as defining a function in the same space as the true solution by assuming the value at the intermediate points to be defined by an interpolation rule. Alternatively we may regard the difference equation that connects values of $u(x)$ with neighboring values $v(x + k\Delta x)$ for some range of fk as holding for all x, thus providing a definition of $v(x)$ for all x, although in practice we only calculate the values of v on a finite mesh. In either case it is legitimate to measure the difference between the discrete and the true solution by the norm in the space containing the true solution.

In the discrete approximation, $\frac{du}{dt}$ is approximated by

$$\frac{v^{n+1} - v^n}{\Delta t} = \frac{C(\Delta t) - I}{\Delta t} v^n.$$

This should approach Au^n as Δt and $\Delta x \to 0$. Accordingly the formal definition of consistency is that for all genuine solutions $u(t)$ of (4.11),

$$\left\| \left\{ \frac{C(\Delta t) - I}{\Delta t} - A \right\} u(t) \right\| \to 0 \quad \text{as} \quad \Delta t \to 0,\ 0 \le t \le T.$$

Since this is applied only to genuine solutions, it follows from (4.12) that we can substitute $\frac{u(t+\Delta t)-u(t)}{\Delta t}$ for $Au(t)$. The quantity

$$d = \frac{u(t+\Delta t) - C(\Delta t)u(t)}{\Delta t} = \frac{(E(\Delta t) - C(\Delta t))u(t)}{\Delta t} \tag{4.13}$$

is called the discretization or truncation error. We can now alternatively express the consistency condition as

$$\|d\| \to 0 \quad \text{as} \quad \Delta t \to 0, \quad 0 \le t \le T \tag{4.14}$$

whenever $u(t)$ is a genuine solution of (4.11). By (4.14) we mean that for any $\epsilon > 0$, however small, we can find some $\sigma > 0$ such that

$$\|(E(\Delta t) - C(\Delta t))u(t)\| < \epsilon \Delta t, \quad 0 \le t \le T \tag{4.15}$$

whenever $0 < \Delta t < \sigma$.

We describe the scheme as pth order accurate if the norm of the discretization error

$$\|d\| = \mathcal{O}(\Delta t^p), \quad \text{as} \quad \Delta t \to 0. \tag{4.16}$$

For example, consider the upwind difference scheme (4.8) for the advection equation. The discretization error is

$$d(x,t) = \frac{u(x,t+\Delta t) - u(x,t)}{\Delta t} + a\frac{u(x,t) - u(x-\Delta x,t)}{\Delta x},$$

which may be expanded in a truncated Taylor series

$$\frac{\partial u}{\partial t} + \frac{\Delta t}{2}\frac{\partial^2 u}{\partial t^2} + a\left(\frac{\partial u}{\partial x} + \frac{\Delta x}{2}\frac{\partial^2 u}{\partial x^2}\right),$$

where $\frac{\partial^2 u}{\partial t^2}$ and $\frac{\partial^2 u}{\partial x^2}$ are evaluated at $t + \theta_1, \Delta t$, and $x - \theta_2 \Delta x$ respectively with $0 \le \theta_1 \le 1, 0 \le \theta_2 \le 1$. Since u satisfies

$$\frac{\partial u}{\partial t} + a\frac{\partial u}{\partial x} = 0,$$

it follows that

$$d(x,t) = \frac{\Delta t}{2}\left(\frac{\partial^2 u}{\partial t^2} + a\frac{\Delta x}{\Delta t}\frac{\partial^2 u}{\partial x^2}\right).$$

Thus $d(x,t) = O(\Delta t)$ for a fixed ratio $\frac{\Delta t}{\Delta x}$, and correspondingly $\|d\| = O(\Delta t)$, so the scheme is first order accurate.

Convergence is defined as the condition that as we reduce the time step, and correspondingly refine the space interval, the difference between the true solution

$E(t)u_0$ and the discrete solution $C(\Delta t)^n u_0$ at a fixed time t approaches zero. Thus we require that

$$\|C(\Delta t_j)^{n_j} u_0 - E(t)u_0\| \to 0$$

for any sequence of step sizes Δt_j and number of steps n_j such that $n_j \Delta t_j \to t$, $\Delta t_j \to 0$, $n_j \to \infty$.

Stability is defined by the condition that the discrete solution at any time $t \leq T$ remains bounded as the time step $\Delta t \to 0$ and the number of time steps $n \to \infty$. Thus we require that the operators $C(\Delta t)^n$ are uniformly bounded, and the scheme is said to be stable if for some $\tau > 0$,

$$\|C(\Delta t)^n\| \leq K\|u\|$$

for all u on the solution space whenever

$$\begin{aligned} 0 &\leq \Delta t \leq \tau \\ 0 &\leq n\Delta t \leq T. \end{aligned}$$

It should be noted that in the stability condition (4.16), the operator $C(\Delta t)$ is modified as Δt is refined. Thus this concept must be distinguished from that of a convergent matrix B say, which satisfies the condition $\|B^n\| \to 0$ as $n \to \infty$, where B is fixed. By expressing B in terms of the Jordan form, it may be verified that a matrix is convergent if its spectral radius ρ, defined as the magnitude of its largest eigenvalue, satisfies the condition $\rho < 1$. Consider again, for example, the upwind scheme (4.8) for the advection equation. Taking the case of an initial boundary value problem on a finite domain, as discussed in Section 4.4, denote the solution at time level n by the vector v^n, with element $v_j^n, j = 1, n$. The scheme can be written in vector-matrix notation as

$$v^{n+1} = Bv^n,$$

where the matrix B is a bidiagonal matrix of the form

$$B = \begin{bmatrix} 1-\lambda & 0 & 0 & \cdots \\ \lambda & 1-\lambda & 0 & \cdots \\ 0 & \lambda & 1-\lambda & \cdots \\ 0 & 0 & \cdots & \cdots \end{bmatrix},$$

where λ is the CFL number,

$$\lambda = a\frac{\Delta t}{\Delta x}.$$

Since B is triangular, its eigenvalues μ_k are its diagonal entries,

$$\mu_j = 1 - \lambda, \quad j = 1, m,$$

where m is the dimension of v. Thus

$$\rho = |1 - \lambda| < 1 \quad \text{if} \quad 0 < \lambda < 2.$$

This means that for a fixed mesh with a fixed boundary value at the inflow boundary

$$v_0^n = c,$$

the solution will eventually converge to the constant solution

$$v_j^n = c,$$

as $n \to \infty$, provided that the arithmetic can be performed exactly, even if $\lambda > 1$, violating the CFL condition. In fact because of the multiplicity of the single eigenvalue $1 - \lambda$, the solution exhibits initial growth to increasingly large value as n is increased whenever $\lambda > 1$, and in practice this will generally result in an arithmetic overflow in the computations before the solution begins to decay. We have seen in Section 4.3 that the solution is bounded in the maximum norm when $0 \le \lambda \le 1$. Using the maximum norm, the induced matrix norm of B is

$$\begin{aligned} \|B\| = \max \sum_j |b_{ij}| &= 1, \quad \text{if} \quad 0 \le \lambda \le 1 \\ &= 2\lambda - 1, \quad \text{if} \quad \lambda > 1. \end{aligned}$$

Thus if $0 < \lambda < 1$,

$$\|B^n\| \le \|B\|^n = 1,$$

whereas if $\lambda > 1$ there is no bound on $\|B^n\|$, which may be very large for some finite n even if $\|B^n\| \to 0$ as $n \to \infty$. These excursions become unbounded as the mesh width $\Delta x \to 0$ and $n \to \infty$, conforming the instability of the scheme when CFL condition is violated.

It follows from the Lax equivalence theorem, which is proved next, that any consistent scheme that violates the CFL condition must be unstable.

THEOREM 4.6.1 (The Lax equivalence theorem) *For a properly posed initial value problem and a consistent finite different approximation, stability is necessary and sufficient for convergence.*

Proof

(1) If the scheme is convergent it must be stable. Otherwise we could find initial data u_0 and a sequence of the steps Δt_j and number of steps n_j such that $\|C(\Delta t_j)^{n_j} u_0\| \to \infty$ with $\tau = n_j \Delta t_j \le T$, and which thus does not converge to $E(t)u_0$ since $\|E(t)u_0\|$ is bounded.

(2) The error

$$v - u = (C(\Delta t)^n - E(n\Delta t))u_0$$

can be expanded as follows. The final error is the sum of the errors at each step amplified by the operator $C(\Delta t)$ for each of the remaining time steps. Thus

$$\begin{aligned} v - u = {} & C(\Delta t)^{n-1}(C(\Delta t) - E(\Delta t))u_0 \\ & + C(\Delta t)^{n-2}(C(\Delta t) - E(\Delta t))E(\Delta t)u_0 \end{aligned}$$

$$+ C(\Delta t)^{n-3}(C(\Delta t) - E(\Delta t))E(2\Delta t)u_0$$
$$\ldots$$
$$+ C(\Delta t)(C(\Delta t) - E(\Delta t))E((n-2)\Delta t)u_0$$
$$+ (C(\Delta t) - E(\Delta t))E((n-1)\Delta t)u_0,$$

where it can be seen that all the interior terms cancel each other. Now by the consistency condition (4.15), we can choose $\sigma > 0$ such that for any $\epsilon > 0$,

$$\|(C(\Delta t) - E(\Delta t))E(k\Delta t)u_0\| \leq \epsilon \Delta t, \quad 0 \leq k \leq n-1$$

whenever $0 < \Delta t < \sigma$. Also, stability means that $C(\Delta t)^{n-k+1}$ are uniformly bounded, so that

$$\|C(\Delta t)^{n-k+1}(C(\Delta t) - E(\Delta t))E(k\Delta t)u_0\| \leq K\epsilon\Delta t, \quad 0 \leq k \leq n$$

for some K independent of u_0. Then by the triangle inequality,

$$\|v - u\| \leq Kn\epsilon\Delta t,$$

where $n\Delta t \to t$ as $\Delta t \to 0, n \to \infty$, and we can make ϵ as small as we please by choosing the bound σ or Δt small enough. In fact if the scheme is pth order accurate, according to (4.16),

$$\|(C(\Delta t) - E(\Delta t))E(k\Delta t)u_0\| = O(\Delta t^{p+1}),$$

whence for a fixed time t

$$\|v - u\| = O(n\Delta t^{p+1}) = O(\Delta t^p)$$

justifying the definition (4.16). □

A corresponding theorem that consistency and stability, if carefully defined, guarantee convergence was proved by Dahlquist (1956) for both linear and nonlinear ordinary differential equations in 1956.

An immediate consequence of the Lax equivalence theorem is that the CFL condition is necessary for stability. Suppose that a consistent scheme violates the CFL condition. Then we can find initial data such that the discrete solution does not converge to the true solution as the mesh is refined. In the case, for example, of the upwind scheme (4.8) for the advection equation with $\lambda > 1$, we have only to take initial data $f(x) = 0$ in the region of dependence of the difference scheme and nonzero in an interval containing the backward characteristic from the point (x, t). Then the discrete solution $v(x, t)$ will remain zero as the mesh is refined with a fixed ration $\frac{\Delta t}{\Delta x}$, while the true solution is nonzero. But according to the equivalence theorem, the discrete solution would converge to the true solution if the difference scheme were stable. We conclude, therefore, that it must be unstable. It should be noted, however, that the converse proposition is not true: not all schemes that satisfy the CFL condition are stable.

4.7 Stability in the Euclidean Norm: Von Neumann Test

In this section we restrict our attention to linear initial value problems with constant coefficients. Then it is possible to use Fourier transforms to derive a very simple test for stability, the von Neumann test. To illustrate the procedure, consider the simple advection equation

$$\frac{\partial u}{\partial t} + a\frac{\partial u}{\partial x} = 0, \quad t \geq 0, \quad -\infty \leq x \leq \infty \tag{4.17}$$

with initial data

$$u(x,0) = f(x).$$

The Fourier transform of u is

$$\hat{u}(\omega,t) = \frac{1}{\sqrt{2\pi}}\int_{-\infty}^{\infty} e^{-i\omega x} u(x,t)dx,$$

and then u can be recovered by the reverse transform

$$u(x,t) = \frac{1}{\sqrt{2\pi}}\int_{-\infty}^{\infty} e^{i\omega x}\hat{u}(\omega,t)d\omega. \tag{4.18}$$

Furthermore, u and $\hat{u}$ satisfy Parseval's relation

$$\|u\|_2 = \|\hat{u}\|_2,$$

where the Euclidean norms $\|u\|_2$ and $\|\hat{u}\|_2$ are defined as

$$\|u\|_2 = \left(\int_{-\infty}^{\infty} u(x)^2 dx\right)^{\frac{1}{2}}, \quad \|\hat{u}\|_2 = \left(\int_{-\infty}^{\infty} |\hat{u}(\omega)|^2 d\omega\right)^{\frac{1}{2}}.$$

Integrating by parts, the Fourier transform of $\frac{\partial u}{\partial x}$ is found to be

$$\frac{1}{\sqrt{2\pi}}\int_{-\infty}^{\infty} e^{-i\omega x}\frac{\partial u}{\partial x}dx = \frac{i\omega}{\sqrt{2\pi}}\int_{-\infty}^{\infty} e^{-i\omega x} u dx = i\omega\hat{u}.$$

Thus, taking the Fourier transform of (4.17), we find that $\hat{u}(\omega,t)$ satisfies the ordinary differential equation

$$\frac{d\hat{u}}{dt} = -ia\omega\hat{u},$$

where

$$\hat{u}(\omega,0) = \hat{f}(\omega).$$

This has the solution

$$\hat{u}(\omega,t) = e^{-ia\omega t}\hat{f}(\omega), \tag{4.19}$$

and the solution for $u(x,t)$ can then be recovered by the inversion formula (4.18). Suppose now that (4.17) is approximated by a difference formula of the form

$$v(x,t+\Delta t) = v(x,t) + \Delta t \sum_k a_k v(x + k\Delta x, t), \tag{4.20}$$

where the sum is over some finite stencil and approximates $a\frac{\partial v}{\partial x}$. Then we can assume that (4.20) holds for all x, although in practice it would only be solved on a fixed mesh. Thus we can also take the Fourier transform of (4.20), noting that

$$\frac{1}{\sqrt{2\pi}}\int_{-\infty}^{\infty} e^{-iwx} v(x + k\Delta x, t)dx = \frac{1}{\sqrt{2\pi}}\int_{-\infty}^{\infty} e^{-i\omega(x-k\Delta x)} v(x,t)dx = e^{ik\omega\Delta x}\hat{v}.$$

Introducing the mesh wave number $\xi = \omega\Delta x$, the result can be written as

$$\hat{v}(\omega, t+\Delta t) = g\hat{v}(\omega,t), \tag{4.21}$$

where

$$g = 1 + \Delta t \sum_k a_k e^{ik\xi}. \tag{4.22}$$

This is the amplification factor for one time step. For a fixed ratio $\frac{\Delta t}{\Delta x}$, we may regard g as a function $g(\omega, \Delta t)$ of ω and Δt.

Now the consistency condition states that the discretization error d of a pth order accurate scheme satisfies the estimate

$$\|d\| = O(\Delta t^p)\ as\ \Delta t \to 0.$$

According to (4.13),

$$d = \frac{E(\Delta t) - C(\Delta t)}{\Delta t} u(t),$$

where $E(\Delta t)$ and $C(\Delta t)$ are the true and discrete evolution operators for a time step. It follows from (4.19) and (4.21) that

$$\hat{d}(\omega) = \frac{e^{-ia\omega\Delta t} - g(\omega,\Delta t)}{\Delta t}\hat{v}(\omega).$$

Since by Parseval's formula $\|\hat{d}\|_2 = \|d\|_2$, it follows that

$$\|\hat{d}\|_2 = O(\Delta t^p)$$

for any $\hat{u}$ in the solution space, and thus

$$|e^{-ia\omega\Delta t} - g(\omega,\Delta t)| = O(\Delta t^{p+1})$$

for a pth order accurate scheme. This provides another way to evaluate the order of accuracy. It follows, moreover, from (4.22) that after n time steps,

$$\hat{v}(\omega, n\Delta t) = g(\omega,\Delta t)^n \hat{f}(\omega).$$

Then by Parseval's formula

$$\|v(n\Delta t)\|_2 = \|\hat{v}(n\Delta t)\|_2 = \left[\int_{-\infty}^{\infty} |g(\omega,\Delta t)|^{2n}|\hat{f}(\omega)|^2 d\omega\right]^{\frac{1}{2}}.$$

Clearly there can be no growth in the Euclidean norm if

$$|g(\omega,\Delta t)| \leq 1.$$

In fact the solution of an equation with a source term such as

$$\frac{\partial u}{\partial t} = \sigma\frac{\partial^2 u}{\partial x^2} + \alpha u$$

may exhibit growth. The stability condition

$$\|v(n\Delta t)\|_2 = \|C(\Delta t)^n v(0)\|_2 \leq K\|v(0)\|_2, \quad 0 \leq t \leq T \tag{4.23}$$

is then equivalent to

$$\|\hat{v}(n\Delta t)\|_2 \leq K\|\hat{v}(0)\|_2 \tag{4.24}$$

or for any $\hat{v}(0)$,

$$\int_{-\infty}^{\infty} |g(\omega,\Delta t)|^{2n}|\hat{v}(0)|^2 d\omega \leq K^2\int_{-\infty}^{\infty} |\hat{v}(0)|^2 d\omega. \tag{4.25}$$

This implies

$$|g(\omega,\Delta t)|^n < K,$$

or for a fixed time t, setting $n = \frac{t}{\Delta t}$,

$$|g(\omega,\Delta t)| < K^{\frac{\Delta t}{t}}.$$

Now $K^{\frac{\Delta t}{t}}$ is bounded by $1 + C\Delta t$ in the interval $0 \leq \Delta t \leq \tau$ for some C, so this means that for stability,

$$|g(\omega,\Delta t)| \leq 1 + O(\Delta t) \quad \text{as} \quad \Delta t \to 0. \tag{4.26}$$

This is the von Neumann condition. For the case of a single dependent variable, it is both necessary and sufficient, because if (4.26) is satisfied, then (4.25), (4.24), and (4.23) are also satisfied.

If one has a system of equations with n dependent variables, we find on taking the Fourier transform of the difference equations that

$$\hat{v}(\omega,t+\Delta t) = G(\omega,\Delta t)\hat{v}(\omega,t),$$

where the amplification factor G is now an $n \times n$ matrix. The stability condition becomes

$$\|G(\omega,\Delta t)\| \leq 1 + O(\Delta t) \quad \text{as} \quad \Delta t \to 0.$$

Suppose $G(\omega, \Delta t)$ has an eigenvalue λ with a magnitude $|\lambda| > 1$. Then if the corresponding eigenvector was taken as the initial data $\hat{v}(0)$, the solution would grow. Thus if ρ is the spectral radius of G,

$$\rho = \max_k |\lambda^{(k)}| \quad \text{for} \quad k = 1, \dots, n,$$

the condition

$$\rho(G) \leq 1 + O(\Delta t)$$

is necessary for stability. It is not in general sufficient, however, because $\|G\|$ may exceed unity even though $\rho(G) < 1$.

Instead of assuming the difference equations to hold for all x and taking the infinite Fourier transform, one may apply a discrete Fourier transform to the mesh equations to derive the same result. In this case we define $\hat{v}(\omega)$ for a mesh function v_j as

$$\hat{v}(\omega) = \Delta x \sum_{j=-\infty}^{\infty} v_j e^{-i\omega x_j}.$$

Then v_j can be recovered by the reverse transform

$$v_j = \frac{1}{2\pi} \int_{-\frac{\pi}{\Delta x}}^{\frac{\pi}{\Delta x}} \hat{v}(\omega) e^{i\omega x_j} dw.$$

This is equivalent to considering v_j to be the jth component of the complex Fourier series for the function $\hat{v}(\omega)$, which is periodic over the interval $(-\frac{\pi}{\Delta x}, \frac{\pi}{\Delta x})$, while $x_j = j\Delta x$ is regarded as its frequency. Parseval's theorem now takes the form

$$\Delta x \sum_{j=-\infty}^{\infty} v_j^2 = \frac{1}{2\pi} \int_{-\frac{\pi}{\Delta x}}^{\frac{\pi}{\Delta x}} |\hat{v}(\omega)|^2 d\omega.$$

Now multiply the difference equation

$$v_j^{n+1} = v_j^n + \Delta t \sum_k a_k v_{j+k} \tag{4.27}$$

by $e^{-i\omega x_j}$ and take the sum from $j = -\infty$ to ∞. Thus, since

$$\sum_{j=-\infty}^{\infty} a_k v_{j+k}^n e^{-i\omega x_j} = \sum_{j=-\infty}^{\infty} a_k v_j^n e^{-i\omega x_{j-k}} = \sum_{j=-\infty}^{\infty} a_k v_j^n e^{-i\omega x_j} e^{ik\omega\Delta x} = a_k e^{i\omega k\Delta x} \hat{v},$$

we find that

$$\hat{v}^{n+1}(\omega) = g\hat{v}^n(\omega),$$

where, setting $\omega\Delta x = \xi$, we recover (4.22) for the amplification factor

$$g = 1 + \Delta t \sum_k a_k e^{ik\xi}.$$

In fact to carry out a von Neumann test, one simply needs to substitute a trial solution of the form

$$v_j^n = \hat{v}e^{i\omega x_j}$$

into the difference equation (4.27). Then one finds immediately that

$$v_j^{n+1} = g\hat{v}e^{i\omega x_j},$$

where g is given by (4.22), and if $|g| > 1$, this particular solution will grow.

The von Neumann test provides a quick and easy way to check the stability of simple difference equations for a single dependent variable with constant coefficients. It is less easy to check the stability for a system because of the difficulty in general of determining a bound on $\|G\|$. When the equations have variable coefficients and are nonlinear, they may nevertheless behave locally like a linear system with constant coefficients as the mesh is refined, because the variations in both the solution and the coefficients become very small between neighboring mesh points as $\Delta x \to 0$. Thus it is often useful to apply a von Neumann test to a locally linearized problem with locally frozen coefficients. If the scheme is unstable for this simplified problem, it is unlikely to be stable for the true problem.

4.8 Examples of the Von Neumann Test

4.8.1 Schemes for Linear Advection

To illustrate the use of the von Neumann test, consider the linear advection equation

$$\frac{\partial u}{\partial t} + a\frac{\partial u}{\partial x} = 0$$

with initial data

$$u(x,0) = f(x).$$

Suppose that this is approximated by a forward Euler scheme in time, and central differences in space,

$$v_j^{n+1} = v_j^n - \frac{\lambda}{2}(v_{j+1}^n - v_{j-1}^n),$$

where

$$\lambda = a\frac{\Delta t}{\Delta x}.$$

Substituting a Fourier mode $v_j^n = \hat{v}e^{i\omega x_j}$, we find that

$$v_j^{n+1} = \hat{v}e^{i\omega x_j} + \frac{\lambda}{2}(\hat{v}e^{i\omega(x_j+\Delta x)} - \hat{v}e^{i\omega(x_j-\Delta x)}) = gv_j^n,$$

where, setting $w\Delta x = \xi$,

$$g = 1 - \frac{\lambda}{2}(e^{i\xi} + e^{-i\xi}) = 1 - \lambda i \sin \xi.$$

Clearly $|g| > 1$ for all ξ and λ. Thus the scheme is unstable, despite the fact that it satisfies the CFL condition if $-1 \leq \lambda \leq 1$.

The oldest stabilization is due to Lax and Friedrichs who proposed the scheme

$$v_j^{n+1} = \frac{1}{2}(v_{j+1}^n + v_{j-1}^n) - \frac{\lambda}{2}(v_{j+1}^n - v_{j-1}^n),$$

in which v_j^n is replaced by the average of its neighbors. Now we find that

$$g = \frac{1}{2}(e^{i\xi} + e^{-i\xi}) - \frac{\lambda}{2}(e^{i\xi} - e^{-i\xi}) = \cos \xi - \lambda i \sin \xi$$

and

$$|g|^2 = \cos^2 \xi + \lambda^2 \sin^2 \xi \leq 1 \text{ if } |\lambda| \leq 1.$$

This scheme is stable but very inaccurate because in effect one has added $\frac{1}{2}(v_{j+1}^n - 2v_j^n + v_{j-1}^n)$ to the right hand side, which approximates an artificial diffusive term $\frac{1}{2}\Delta x \frac{\partial^2 v}{\partial x^2}$.

An alternative is to use the upwind scheme

$$v_j^{n+1} = v_j^n - \lambda(v_j^n - v_{j-1}^n). \tag{4.28}$$

Now we find that

$$g = 1 - \lambda(1 - e^{-i\xi}) = 1 - \lambda + \lambda e^{-i\xi}.$$

Thus g lies on a circle of radius λ centered at $1 - \lambda$, as sketched in Figure 4.6. Clearly g lies in the unit disk if $0 \leq \lambda \leq 1$, confirming stability when the CFL condition is satisfied. This is in agreement with the previous result obtained in Section 4.3 that the scheme is stable in the maximum norm when the CFL condition is satisfied. This scheme is also first order accurate. Equation (4.28) can be written as

$$v_j^{n+1} = v_j^n - \frac{\lambda}{2}(v_{j+1}^n - v_{j-1}^n) - \frac{\lambda}{2}(v_{j+1}^n - 2v_j^n + v_{j-1}^n).$$

The third term on the right approximates the artificial diffusive term $\frac{1}{2}\lambda\Delta x \frac{\partial^2 v}{\partial x^2}$, so the scheme is less diffusive than the Lax–Friedrichs scheme when $|\lambda| < 1$. When one considers a nonlinear equation, the wave speed $a = a(u)$ may change sign. The direction of the space differencing then has to be switched as the wave speed changes sign, and this may lead to improper solutions as will be discussed later. The application of upwind differencing to systems of equations with waves traveling in opposite directions also becomes quite complex.

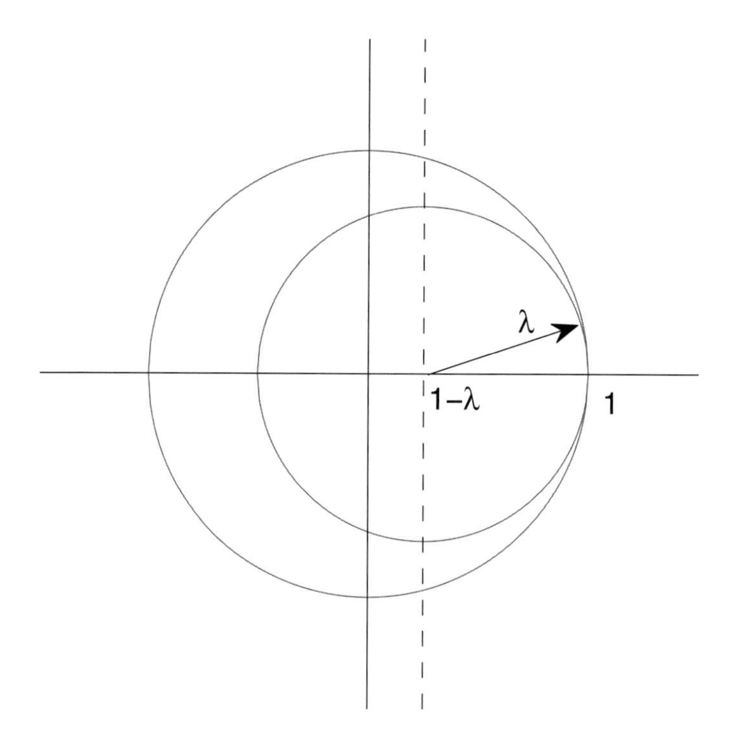

Figure 4.6 Locus of amplification factor for the upwind scheme.

Let us now consider the Lax–Wendroff scheme described in Section 3.5. For the linear advection equation, the scheme can be written as

$$v_j^{n+1} = v_j^n - \frac{\lambda}{2}(v_{j+1}^n - v_{j-1}^n) + \frac{\lambda^2}{2}(v_{j+1}^n - 2v_j^n + v_{j-1}^n).$$

Making the substitution $v_j^n = \hat{v}e^{i\omega x_j}$, we now find that

$$v_j^{n+1} = g v_j^n,$$

where

$$\begin{aligned} g &= 1 - \lambda i \sin \xi - \lambda^2(1 - \cos \xi) \\ &= 1 - 2\lambda^2 \sin^2 \frac{\xi}{2} - 2\lambda i \sin \frac{\xi}{2} \cos \frac{\xi}{2}. \end{aligned}$$

Now,

$$\begin{aligned} |g| &= 1 - 4\lambda^2 \sin^2 \frac{\xi}{2} + 4\lambda^4 \sin^4 \frac{\xi}{2} + 4\lambda^2 \sin^2 \frac{\xi}{2} \cos^2 \frac{\xi}{2} \\ &= 1 - 4(\lambda^2 - \lambda^4) \sin^4 \frac{\xi}{2}, \end{aligned}$$

and hence $|g| \leq 1$ when $|\lambda| \leq 1$ since in that case $\lambda^4 \leq \lambda^2$.

Finally, let us consider the implicit central difference scheme

$$v_j^{n+1} = v_j^n - \frac{\lambda}{4}(v_{j+1}^{n+1} - v_{j-1}^{n+1}) - \frac{\lambda}{4}(v_{j+1}^n - v_{j-1}^n).$$

Making the substitution

$$v_j^n = g^n \hat{v} e^{i\omega x_j},$$

we now find that

$$g\left\{1 + \frac{\lambda}{2}(e^{i\omega\Delta x} - e^{-i\omega\Delta x})\right\} = 1 - \frac{\lambda}{2}(e^{i\omega\Delta x} - e^{-i\omega\Delta x})$$

or

$$g = \frac{1 - \frac{1}{2}\lambda i \sin\xi}{1 + \frac{1}{2}\lambda i \sin\xi}.$$

Thus

$$|g| = 1$$

for all ξ, λ. The scheme is thus unconditionally stable for an arbitrarily large time step Δt. This scheme suffers drawbacks, however, when it is applied to more complex problems. First, the cost of solving the implicit equations can become large either in multi-dimensional or nonlinear problems. Second, it proves highly oscillatory in nonlinear problems with shock waves.

The use of the Fourier technique is not limited to schemes defined between two time levels. As a simple example of a multi-level scheme, suppose that central differencing is used in both time and space. This produces the "leapfrog" scheme

$$\frac{v(x,t+\Delta t) - v(x,t-\Delta t)}{2\Delta t} + a\frac{v(x+\Delta x,t) - v(x-\Delta x,t)}{2\Delta x} = 0$$

or

$$v_j^{n+1} = v_j^{n+1} - \lambda(v_{j+1}^n - v_{j-1}^n).$$

The substitution

$$v_j^n = g^n \hat{v} e^{iwx_j}$$

now yields

$$g = \frac{1}{g} - \lambda(e^{i\omega\Delta x} - e^{-i\omega\Delta x}).$$

Thus g satisfies the quadratic equation

$$g^2 + 2\lambda i \sin\xi g - 1 = 0$$

with roots

$$g = -\lambda i \sin \xi \pm \sqrt{1 - \lambda^2 \sin^2 \xi}.$$

If $|\lambda| \leq 1$, we find that for both roots

$$|g|^2 = 1.$$

If $|\lambda| > 1$, then for $\xi = \frac{\pi}{2}$ one root is $-i(\lambda + \sqrt{\lambda^2 - 1})$, which lies outside the unit disk. Thus the scheme is stable whenever $|\lambda| \leq 1$, which is preciously the limit imposed by the CFL condition.

It can be verified by Taylor series expansion that the leapfrog scheme is second order accurate. This is a consequence of centering both the time and space differences on the same point. Leapfrog is attractive because its computational complexity is less than that of any other second order accurate scheme. Unfortunately it is unstable for the diffusion equation, or if any diffusive term is added to the advection equation to produce an advection-diffusion problem. It also shares the disadvantages with all multi-level schemes that some other scheme must be used at the first time step to generate v_j^1.

4.8.2 Schemes for the Diffusion Equation

The von Neumann test is equally useful in the analysis of schemes for the diffusion equation

$$\frac{\partial u}{\partial t} = \sigma \frac{\partial^2 u}{\partial x^2}$$

with initial data

$$u(x, 0) = f(x).$$

Consider first the standard explicit scheme

$$v_j^{n+1} = v_j^n + \beta(v_{j+1}^n - 2v_j^n + v_{j-1}^n),$$

where

$$\beta = \sigma \frac{\Delta t}{\Delta x^2}.$$

Now the substitution $v_j^n = \hat{v} e^{i w x_j}$ yields

$$v_j^{n+1} = \left\{ 1 + \beta(e^{i\omega\Delta x} - 2 + e^{-i\omega\Delta x}) \right\} \hat{v} e^{i\omega x_j}$$

or

$$v_j^{n+1} = g v_j^n,$$

where, setting $\omega \Delta x = \xi$,

$$g = 1 - 2\beta(1 - \cos \xi) = 1 - 4\beta \sin^2 \frac{\xi}{2}.$$

Clearly $g < 1$ if $\beta > 0$, but if $\beta > \frac{1}{2}$ it is possible that $g < -1$, $|g| > 1$, so the stability condition is

$$0 \leq \beta \leq \frac{1}{2},$$

which is satisfied for positive σ if $\Delta t \leq \frac{\Delta x^2}{2\sigma}$. This stringent condition is the same as the result that was obtained in Section 4.5 for stability in the maximum norm. Actually the scheme is unsatisfactory when $\beta = \frac{1}{2}$. Then

$$v_j^{n+1} = \frac{1}{2}(v_{j+1}^n - v_{j-1}^n).$$

Consider initial data consisting of a spike at the origin, $f_0 = 1$, $f_j = 0$ when $j \neq 0$. The initial steps are sketched in Figure 4.7. The initial spike separates into two spikes, which in turn split into three. This has little resemblance to the true solution. However, as Δx is refined with fixed initial data, in accordance with the theory, the initial spike near the origin would be spread out over increasing numbers of mesh points, allowing proper convergence.

Moving now to the implicit Crank Nicolson scheme

$$v_j^{n+1} = v_j^n + \frac{\beta}{2}(v_{j+1}^{n+1} - 2v_j^n + v_{j-1}^n) + \frac{\beta}{2}(v_j^n - 2v_j^n + v_{j-1}^n),$$

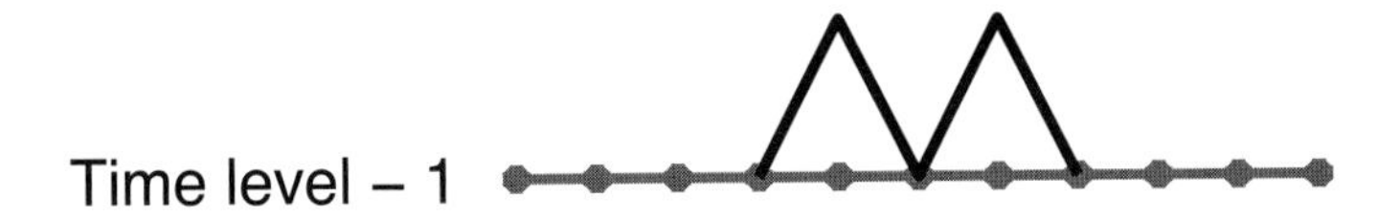

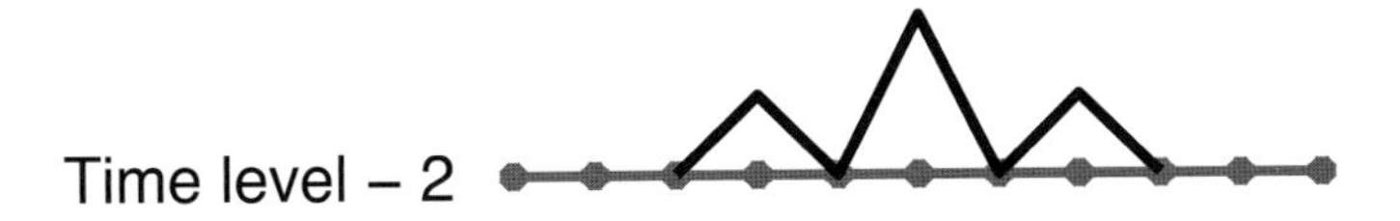

Figure 4.7 Evolution of diffusion equation with maximum stable time step.

the substitution

$$v_j^n = g^n e^{i\omega x_j}$$

yields the amplification factor

$$g = \frac{1 - \beta(1 - \cos\xi)}{1 + \beta(1 - \cos\xi)} = \frac{1 - 2\beta\sin^2\frac{\xi}{2}}{1 + 2\beta\sin^2\frac{\xi}{2}}.$$

Thus the scheme is seen to be unconditionally stable.

4.9 Stability Theory for Initial Boundary Value Problems

A stability theory for linear boundary value problems based on modal trial solutions has been worked out by Gustaffson, Kreiss, and Sundstrom (Kreiss 1968, Gustafsson, Kreiss, & Sundström 1972, Gustafsson 1975). For a full exposition of the theory, the reader is referred to the textbook by Gustaffson Bertil and Joseph (2013). The application of the theory to problems in fluid mechanics has been reviewed by Oliger and Sundstrom (1978). Procedures have also been developed for the construction of boundary conditions designed to allow outgoing waves to pass through the outer boundary without reflection (Engquist & Majda 1977, Bayliss & Turkel 1982). The availability of this body of theory provides a solid foundation for the development of programs to treat practical aerodynamic problems.

4.10 Nonlinear Conservation Laws and Discontinuous Solutions

Solutions of nonlinear conservation laws are not necessarily continuous. They may contain both shock waves and contact discontinuities, and also expansion fans. This leads to additional considerations in the formulation of discretization schemes, which will now be discussed.

As a first example, we consider the inviscid Burgers' equation,

$$\frac{\partial u}{\partial t} + \frac{\partial}{\partial x}\left(\frac{u^2}{2}\right) = 0, \quad u(x,0) = u_0(x).$$

This can be written in quasilinear form as

$$\frac{\partial u}{\partial t} + u\frac{\partial u}{\partial x} = 0$$

and has characteristics

$$x - ut = \xi.$$

On such a line,

$$\frac{du}{dt} = \frac{d}{dt}u(ut+\xi,t)$$
$$= \left(u + t\frac{du}{dt}\right)\frac{\partial u}{\partial x} + \frac{\partial u}{\partial t},$$

whence

$$\left(1 - t\frac{\partial u}{\partial x}\right)\frac{du}{dt} = u\frac{\partial u}{\partial x} + \frac{\partial u}{\partial t} = 0.$$

Thus, u is constant along characteristics, so we can construct solutions by propagating the initial data along the characteristics. If

$$u_0(x) = 0, \quad x \leq 0$$
$$= 1, \quad x > 0,$$

the solution will therefore be

$$u = 0, \quad x \leq 0$$
$$= 1, \quad x > 0.$$

In the intermediate interval, we can suppose that u varies linearly from 0 to 1 and that these values are propagated at the corresponding speeds, giving

$$u = \frac{x}{t}, \quad 0 \leq t.$$

This is an expansion fan for which

$$u_t = -\frac{x}{t^2}, \quad uu_x = \frac{x}{t^2},$$

and hence the equation is satisfied. The solution is illustrated in Figure 4.8.

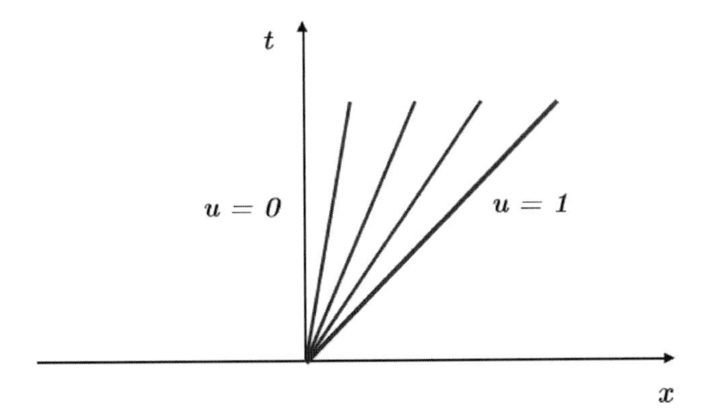

Figure 4.8 Expansion fan.

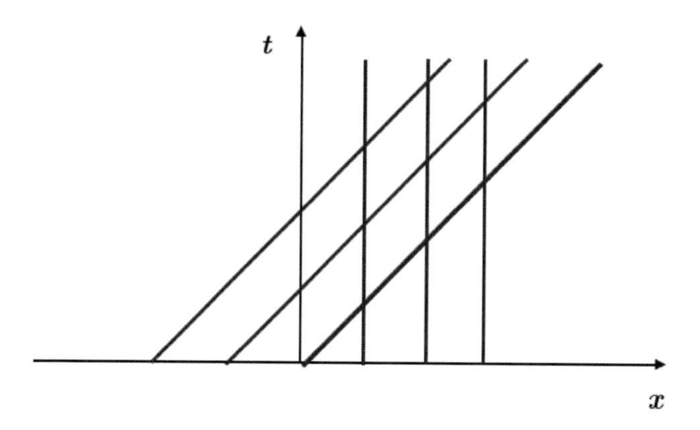

Figure 4.9 Nonunique solution with crossed-over characteristics.

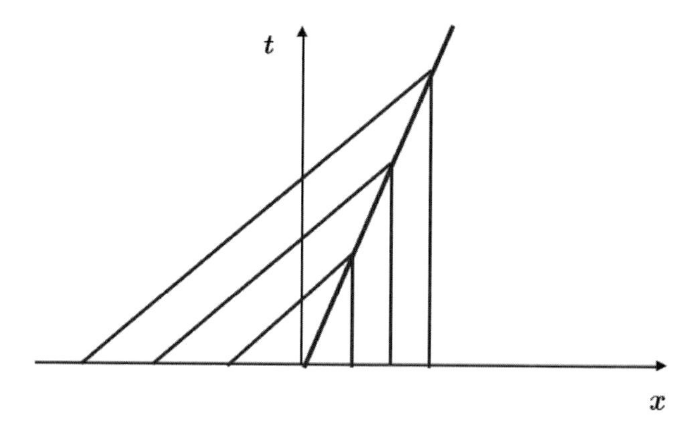

Figure 4.10 Solution with left and right states separated by a discontinuity.

Suppose, on the other hand, that

$$\begin{aligned} u_0(x) &= 1, \quad x \leq 0 \\ &= 0, \quad x > 0. \end{aligned}$$

Then above the line $x = t$, the characteristics from the two segments of the initial data cross each other, so that the requirement that u is constant along characteristics leads to a conflict, as illustrated in Figure 4.9 To resolve this, one must allow for separate solutions on each side of a boundary across which u is discontinuous, and u_t and u_x are not defined, as illustrated in Figure 4.10. Further analysis is then required to determine the location of the discontinuity and the jump conditions across it.

4.10.1 Weak Form of a Conservation Law

The derivatives appearing in the differential form of a conservation law are not defined at discontinuities. This difficulty can be circumvented by introducing the weak form, in which the differential equation is multiplied by a smooth test function and integrated by parts over space and time to transfer the derivative from the solution to the test function.

Consider the general nonlinear scalar conservation law,

$$\begin{aligned}\frac{\partial u}{\partial t} + \frac{\partial}{\partial x} f(u) &= 0 \\ u(x,0) &= u_0(x).\end{aligned} \tag{4.29}$$

This has the form of a divergence and represents a conservation law for a vector with components u, f. Multiply by any smooth function $w(x,t)$, which vanishes for large x, t, and integrate over x and t to obtain

$$0 = \int_0^\infty \int_{-\infty}^\infty \left(\frac{\partial u}{\partial t} + \frac{\partial f}{\partial x} \right) w \, dx dt, \tag{4.30}$$

or

$$0 = \int_0^\infty \int_{-\infty}^\infty \left(u \frac{\partial w}{\partial t} + f \frac{\partial w}{\partial x} \right) dx dt + \int_{-\infty}^\infty u_0 w \, dx. \tag{4.31}$$

Then if u satisfies (4.31) for all such w, it is considered to be a "weak solution" of (4.29). If u is differentiable, one can integrate (4.31) by parts to recover (4.30), and since (4.30) holds for all w, (4.29) follows. If there is a line C in the x, t plane across which u is discontinuous, it is necessary to integrate separately over the regions either side of C. Let subscripts L and R denote values on either side of the boundary C as illustrated in Figure 4.11. Then,

$$\iint_L u \frac{\partial w}{\partial t} dx dt = - \int_{-\infty}^{x_C} u_0 w \, dx - \int_C u w \, dx - \iint_L w \frac{\partial u}{\partial t} dx dt,$$

and

$$\iint_R f \frac{\partial w}{\partial t} dx dt = - \int_C f w \, dt - \iint_R w \frac{\partial f}{\partial x} dx dt.$$

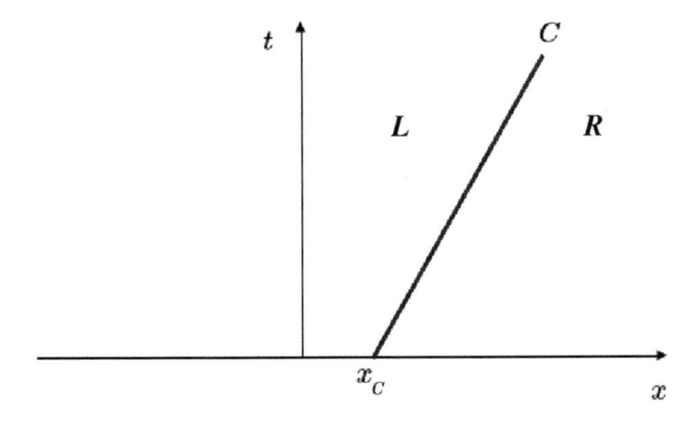

Figure 4.11 Path of discontinuity in $x - t$ plane.

Thus, (4.31) reduces to

$$0 = -\iint_L w\left(\frac{\partial u}{\partial t} + \frac{\partial f}{\partial x}\right) dxdt - \iint_R w\left(\frac{\partial u}{\partial t} + \frac{\partial f}{\partial x}\right) dxdt + \int_C w\,(S[u] - [f])\,dt,$$

where

$$[u] = u_R - u_L, \quad [f] = f_R - f_L, \quad S = \frac{dx}{dt}.$$

Now we can choose a test function w that differs from zero only in an arbitrarily narrow strip containing C. It follows that the discontinuous solution must satisfy the jump condition

$$[f] = S[u]. \tag{4.32}$$

Note that if the equation is written in quasilinear form as

$$u_t + a(u)u_x = 0,$$

then on multiplying w and integrating by parts, the double integral becomes

$$\int_0^{\infty}\int_{-\infty}^{\infty} u\left(\frac{\partial w}{\partial t} + \frac{\partial}{\partial x}a(u)w\right) dxdt,$$

and this still does not permit discontinuous solutions to be defined, because it requires the existence of $\frac{\partial}{\partial x}a(u)$. Thus, it is necessary to write the equation in conservation form in order to define weak solutions.

4.10.2 Shock Waves

Consider again the example of the inviscid Burgers' equation (4.10), for which $f(u) = \frac{u^2}{2}$, and

$$[f] = \frac{1}{2}\left(u_R^2 - u_L^2\right).$$

Thus, we obtain the jump condition that a discontinuity propagates at a speed

$$S = \frac{1}{2}(u_L + u_R),$$

where u_L and u_R are the values to the left and right of the discontinuity yielding the solution illustrated in Figure 4.10. Note that weak solutions satisfying this relationship are not necessarily unique.

In the case that

$$\begin{aligned} u_0(x) &= 0, \quad x \leq 0, \\ &= 1, \quad x > 0, \end{aligned}$$

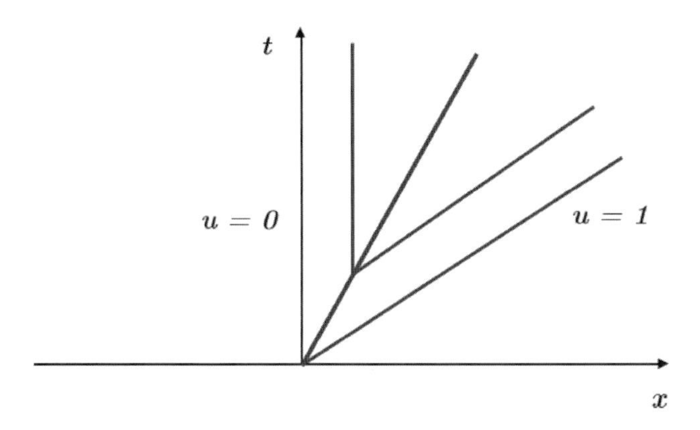

Figure 4.12 Improper expansion jump.

one finds that

$$
\begin{aligned}
u &= 0, \quad x \le \frac{t}{2}, \\
&= 1, \quad x > \frac{t}{2}
\end{aligned}
$$

satisfies the differential equation and the jump conditions. This solution, illustrated in Figure 4.12, is an alternative to the expansion fan. To restore uniqueness, we need the additional condition that a discontinuity will only be permitted if the characteristics on both sides converge on the discontinuity, in this case that

$$u_L > S > u_R,$$

where S is the speed of the discontinuity. Such a jump is called a shock. The condition that the characteristics must converge on the jump is called an entropy condition, because in the case of fluid dynamics, it corresponds to the condition that entropy cannot decrease and hence that discontinuous expansions are impossible.

Finally if

$$
\begin{aligned}
u_0(x) &= u_L, \quad x \le 0, \\
&= u_R, \quad x > 0,
\end{aligned}
$$

the entropy-satisfying solution is a shock moving with a speed $S = \frac{1}{2}(u_R + u_L)$ if $u_L > u_R$, and an expansion fan

$$
\begin{aligned}
u &= \frac{x}{t} \quad \text{for} \quad u_L t \le x \le u_R t, \\
u &= u_L, \quad x < u_L t, \\
u &= u_R, \quad x > u_R t
\end{aligned}
$$

if $u_L < u_R$.

4.10.3 Difference Schemes in Conservation Form

Discrete solutions that satisfy the proper jump conditions can be obtained if the discrete solution is written in a conservation form that precisely mimics the conservation law. Let

$$D_x f = \frac{1}{\Delta x}\left\{g\left(x+\frac{1}{2}\Delta x\right) - g\left(x-\frac{1}{2}\Delta x\right)\right\},$$

where at the mesh point $j\Delta x$, the function g depends on the mesh function v at some number of neighboring mesh points, for example

$$g\left(x+\frac{1}{2}\Delta x\right) = g(v_{j-1}, v_j, v_{j+1}, v_{j+2})$$

$$g\left(x-\frac{1}{2}\Delta x\right) = g(v_{j-2}, v_{j-1}, v_j, v_{j+1}),$$

and g is a function that satisfies the consistency condition

$$g(u,u,u,\ldots) = f(u).$$

Set

$$v(x,t+\Delta t) = v(x,t) - \frac{\Delta t}{\Delta x}\left\{g\left(x+\frac{1}{2}\Delta x\right) - g\left(x-\frac{1}{2}\Delta x\right)\right\}. \tag{4.33}$$

Now, we prove a theorem due to Lax and Wendroff.

THEOREM 4.10.1 (Lax and Wendroff) *If the discrete solution converges in the limit as Δt and $\Delta x \to 0$, it satisfies the weak form (4.31) and consequently the jump condition (4.32).*

Proof For this purpose, we multiply (4.33) by a test function $w(x,t)$, sum over the time steps from 0 to ∞, and integrate over x from $-\infty$ to ∞. Now, using superscripts to denote the time level,

$$w^{n+1}(v^{n+1} - v^n) + w^n(v^n - v^{n-1}) \cdots + w^1(v^1 - v^0) = w^{n+1}v^{n+1} - (w^{n+1} - w^n)v^n$$
$$-(w^n - w^{n-1})v^{n-1} \ldots - (w^1 - w^0)v^0 - w^0 v^0.$$

Thus, assuming that w^n vanishes for large t, we obtain

$$-\sum_0^\infty \int_{-\infty}^{\infty} \frac{w(x,t+\Delta t) - w(x,t)}{\Delta t} v(x,t)\,dx\,\Delta t - \int_{-\infty}^{\infty} w(x,0)u_0(x)\,dx$$
$$-\sum_0^\infty \int_{-\infty}^{\infty} \frac{w\left(x+\frac{1}{2}\Delta x\right) - w\left(x-\frac{1}{2}\Delta x\right)}{\Delta x} g\,dx\,\Delta t = 0.$$

In the limit as Δt and $\Delta x \to 0$, this reduces to the equation for the weak solution, (4.31). The theorem of Lax and Wendroff follows. □

5 Shock Capturing Schemes I: Scalar Conservation Laws

5.1 Introduction

A major achievement in the early development of computational fluid dynamics (CFD) was the formulation of non-oscillatory shock capturing schemes. The first such scheme was introduced by Godunov in his pioneering work first published in 1959 (Godunov 1959). Godunov also showed that non-oscillatory schemes with a fixed form are limited to first order accuracy. This is not sufficient for adequate engineering simulations. Consequently there were widespread efforts to develop "high resolution" schemes, which circumvented Godunov's theorem by blending a second or higher order accurate scheme in smooth regions of the flow with a first order accurate non-oscillatory scheme in the neighborhood of discontinuities. This is typically accomplished by the introduction of logic that detects local extrema and limits their formation or growth.

Here we first discuss the formulation of non-oscillatory schemes for scalar conservation laws in one or more space dimensions and illustrate the construction of schemes that yield second order accuracy in the bulk of the flow but are locally limited to first order accuracy at extrema. In the following chapters, we discuss the formulation of finite volume schemes for systems of equations, such as the Euler equations of gas dynamics, and analyze the construction of interface flux formulas with favorable properties, such as sharp resolution of discontinuities and assurance of positivity of the pressure and density. The combination of these two ingredients leads to a variety of schemes that have proved successful in practice.

5.2 The Need for Oscillation Control

5.2.1 Odd-Even De-coupling

Consider the linear advection equation for a right running wave:

$$\frac{\partial u}{\partial t} + a\frac{\partial u}{\partial x} = 0, \quad a > 0.$$

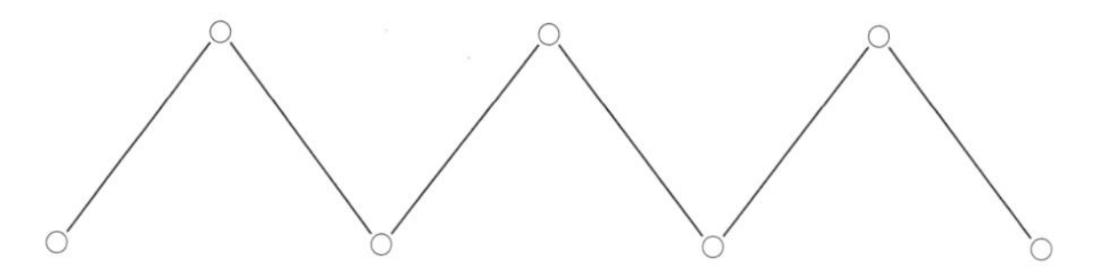

Figure 5.1 Odd-even mode.

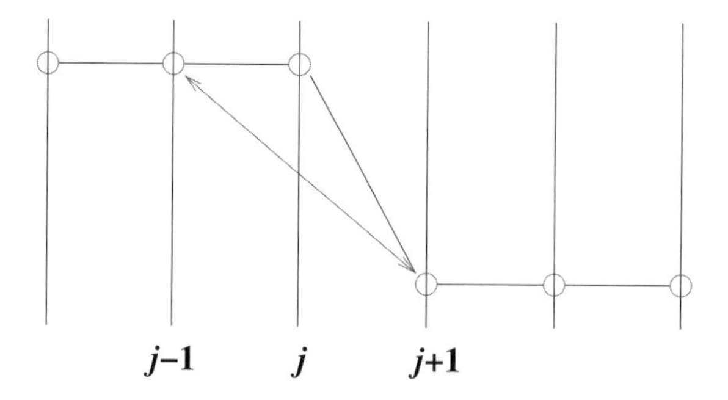

Figure 5.2 Propagation of right running wave.

Representing the discrete solution at the mesh point x_j by v_j, a semi-discrete scheme with central differences is

$$\frac{dv_j}{dt} + \frac{a}{2\Delta x}\left(v_{j+1} - v_{j-1}\right) = 0.$$

Then an odd-even mode

$$v_j = (-1)^j,$$

as illustrated in Figure 5.1, gives

$$\frac{dv_j}{dt} = 0.$$

Thus an odd-even mode is a stationary solution, and it is apparent that odd-even decoupling needs to be removed by some means, such as the addition of artificial viscosity or upwinding.

5.2.2 Propagation of a Step Discontinuity

Consider the propagation of a step as a right running wave by the central difference scheme as illustrated in Figure 5.2. Now the discrete derivative

$$D_x v_j = \frac{v_{j+1} - v_{j-1}}{2\Delta x} < 0,$$

and hence with $a > 0$,

$$\frac{dv_j}{dt} > 0,$$

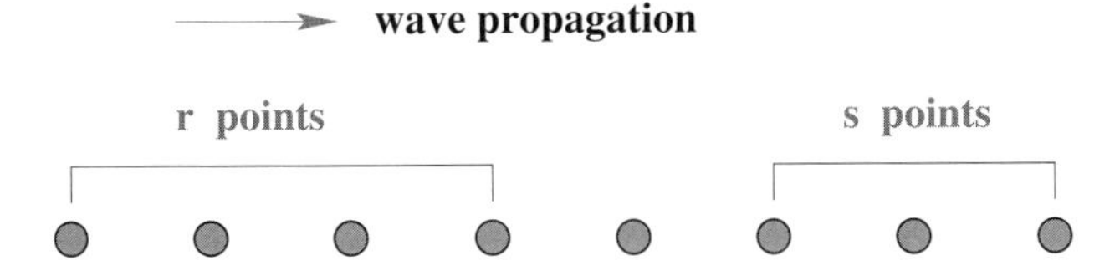

Figure 5.3 Iserles' barrier theorem.

giving an overshoot. On the other hand, the upwind scheme

$$D_x^- v_j = \frac{v_j - v_{j-1}}{\Delta x}$$

correctly yields

$$\frac{dv_j}{dt} = 0.$$

5.3 Iserles' Barrier Theorem

Both the need to suppress odd-even modes and the need to prevent overshoots in the propagation of a step discontinuity motivate the use of an upwind scheme. However, purely upwind schemes are also subject to limitations as a consequence of Iserles' barrier theorem (Iserles 1982).

Consider the approximation of the linear advection equation by a semi-discrete scheme with r upwind points and s downwind points, as illustrated in Figure 5.3. The theorem states that the maximum order of accuracy of a stable scheme is

$$\min(r + s, 2r, 2s + 2). \tag{5.1}$$

This is a generalization of an earlier result of Engquist and Osher that the maximum order of accuracy of a stable upwind semi-discrete scheme is two ($s = 0$ in formula (5.1)). It may also be compared to Dahlquist's result that A-stable linear multistep schemes for ODEs are at most second order accurate. One may conclude from Iserles' theorem that upwind-biased schemes may be preferred over purely upwind schemes as a route to attaining higher order accuracy.

5.4 Stability in the L_∞ Norm

Consider the nonlinear conservation law for one dependent variable with diffusion:

$$\frac{\partial u}{\partial t} + \frac{\partial}{\partial x} f(u) = \frac{\partial}{\partial x}\left(\mu(u)\frac{\partial u}{\partial x}\right), \quad \mu \geq 0. \tag{5.2}$$

With zero diffusion, (5.2) is equivalent in smooth regions to

$$\frac{\partial u}{\partial t} + a(u)\frac{\partial u}{\partial x} = 0,$$

where the wave speed is

$$a(u) = \frac{\partial f}{\partial u}.$$

The solution is constant along characteristics

$$x - a(u)t = \xi,$$

so extrema remain unchanged as they propagate unless the characteristics converge to form a shock wave – a process that does not increase extrema. With a positive diffusion coefficient, the right hand side of (5.2) is negative at a maximum and positive at a minimum. Thus in a true solution of (5.2), extrema do not increase in absolute value. It follows that L_∞ stability is an appropriate criterion for discrete schemes, consistent with the properties of true solutions of the nonlinear conservation law (5.2).

Consider the general discrete scheme

$$v_i^{n+1} = \sum_j c_{ij} v_j^n, \tag{5.3}$$

where the solution at time level $n+1$ depends on the solution over an arbitrarily large stencil of points at time level n. A Taylor series expansion of (5.3) about the point x_i at time t yields

$$\begin{aligned} v(x_i) &+ \Delta t \frac{\partial v}{\partial t}(x_i) + \frac{\Delta t^2}{2}\frac{\partial^2 v}{\partial t^2}(x_i) \\ &= \sum_j c_{ij}\left[v(x_i) + (x_j - x_i)\frac{\partial v(x_i)}{\partial x} + \frac{(x_j - x_i)^2}{2}\frac{\partial^2 v(x_i)}{\partial x^2} + \cdots \right]. \end{aligned}$$

For consistency with any equation with no source term, as is the case for (5.2),

$$\sum_j c_{ij} = 1. \tag{5.4}$$

This condition can also be inferred from the observation that if v_j^n is constant everywhere, there should be no change in the solution. Also, it follows from (5.3) that

$$|v_i^{n+1}| \leq \sum_j |c_{ij}||v_j^n| \leq \sum_j |c_{ij}|\; \|v^n\|_\infty,$$

and hence

$$\|v^{n+1}\|_\infty \leq \max_i \sum_j |c_{ij}|\|v^n\|_\infty,$$

where equality is realized if

$$v_j^n = \operatorname{sgn}(c_{ij})$$

for the row for which $\sum_j |c_{ij}|$ has its maximum value. Thus for L_∞ stability,

$$\max_i \sum_j |c_{ij}| \leq 1. \tag{5.5}$$

If one writes (5.3) in matrix vector form,

$$v^{n+1} = Cv^n,$$

this may be recognized as the condition that the induced matrix norm

$$\|C\|_\infty = \sup_v \frac{\|Cv\|_\infty}{\|v\|_\infty} \leq 1.$$

Now (5.4) and (5.5) together imply

$$c_{ij} \geq 0 \tag{5.6}$$

because if any $c_{ij} < 0$, then on taking absolute values in (5.4), the sum would have a value > 1. Therefore the discrete scheme (5.3) is L_∞ stable if and only if the coefficients c_{ij} are non-negative. Note that this also implies that given initial data that is everywhere non-negative, the discrete scheme (5.3) has the "positivity" property that it will preserve a non-negative solution at every time step.

Using the consistency condition (5.4), the discrete scheme can be written as

$$\begin{aligned} v_i^{n+1} &= \left(\sum_j c_{ij}\right) v_i^n + \sum_{j\neq i} c_{ij}(v_j^n - v_i^n) \\ &= v_i^n + \Delta t \sum_{j\neq i} a_{ij}(v_j^n - v_i^n), \end{aligned} \tag{5.7}$$

where

$$c_{ij} = \Delta t(a_{ij}).$$

This displays the scheme as a forward Euler time stepping scheme. Moreover, comparing (5.7) with (5.3),

$$c_{ii} = 1 - \Delta t \sum_{j\neq i} a_{ij}.$$

Thus (5.6) can only be satisfied if

$$a_{ij} \geq 0, \quad j \neq i,$$

and the time step also satisfies the constraint

$$\Delta t \leq \frac{1}{\sum_{j\neq i} a_{ij}}. \tag{5.8}$$

This is a generalization of the Courant–Friedrichs–Lewy (CFL) condition for schemes with an arbitrary stencil.

In the case of linear advection,

$$\frac{\partial u}{\partial t} + a\frac{\partial u}{\partial x} = 0,$$

the simplest L_∞ stable scheme is the upwind scheme

$$\begin{aligned} v_j^{n+1} &= v_j^n - \lambda(v_j^n - v_{j-1}^n) \\ &= (1-\lambda)v_j^n + \lambda v_{j-1}^n, \end{aligned} \tag{5.9}$$

where λ is the CFL number $\frac{a\Delta t}{\Delta x}$, and $0 \le \lambda \le 1$. Here v^{n+1} is a convex combination of v_j^n and v_{j-1}^n as was observed in Sections 3.2.3 and 4.3. When $\lambda = 1$, this scheme exactly represents propagation along the characteristics.

It should also be noted that the discussion in this section is not limited to the one-dimensional problem and applies equally to the case of multi-dimensional equations discretized on arbitrary unstructured grids. Supplementary notes are provided in Appendix E on the construction of both explicit and implicit schemes that are stable in the L_1, L_2 and L_∞ norms.

5.5 Local Extremum Diminishing (LED) Schemes

L_∞ stability does not exclude the possibility that a monotonically decreasing profile (Figure 5.4(a)) could develop into an oscillatory profile (Figure 5.4(b)). This motivates the stricter requirement that local extrema cannot grow in the numerical solution; nor can new local extrema be created. Such a scheme will be called local extremum diminishing (LED).

Consider now the discrete scheme

$$v_i^{n+1} = \sum_j c_{ij} v_j^n, \tag{5.10}$$

where the coefficients $c_{ij} \neq 0$ only for the nearest neighbors to the mesh point i, in one or more space dimensions. The argument of the previous section can now be repeated, where the summations are now limited to the nearest neighbors of the mesh point i, denoted by the set N_i. For consistency with an equation with no source term,

$$\sum_{j \in N_i} c_{ij} = 1,$$

while for no increase in $|v_i|$,

$$\sum_{j \in N_i} |c_{ij}| \le 1.$$

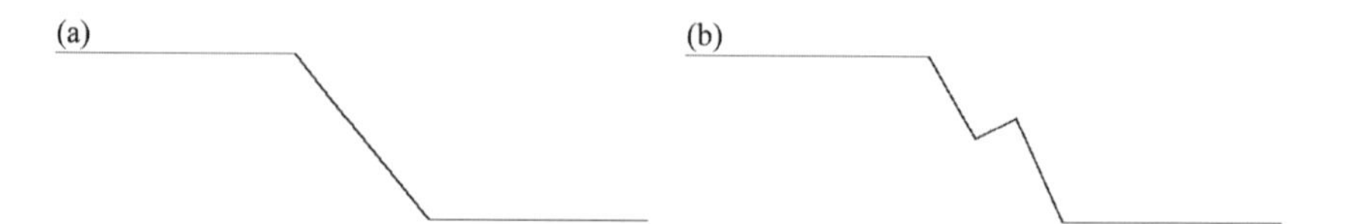

Figure 5.4 Local oscillation.

Thus the scheme is LED if for all i

$$\begin{aligned} c_{ij} &\geq 0, \\ c_{ij} &= 0, \quad \text{if } j \text{ is not a neighbor of } i. \end{aligned} \tag{5.11}$$

5.6 Total Variation Diminishing (TVD) Schemes

A useful measure of the oscillation of a one-dimensional function $u(x)$ is its total variation

$$TV(u) = \int_a^b \left| \frac{\partial u}{\partial x} \right| dx.$$

Correspondingly, the total variation of a discrete solution, say $v_j, j = 0, 1, \ldots, n$, is defined as

$$TV(v) = \sum_{j=1}^{n} |v_{j+1} - v_j|.$$

It was proposed by Harten (1983) that a good recipe for the construction of non-oscillatory shock capturing schemes is to require that the total variation of the discrete solution cannot increase. Such schemes are called total variation diminishing (TVD). This criterion has been widely used and also extended to the concept of total variation bounded (TVB) schemes, which permit only a bounded increase in the total variation.

Let $x_j, j = 1, 2, \ldots, n-1$, be the interior extrema of $u(x)$, and augment these with the endpoints $x_0 = a$ and $x_n = b$. Referring to Figure 5.5, it can be seen that over an interval in which $u(x)$ is increasing, say x_{j-1} to x_j, the contribution to $TV(u)$ is $u(x_j) - u(x_{j-1})$, while over the following interval in which $u(x)$ is decreasing, the contribution to $TV(u)$ is $u(x_j) - u(x_{j+1})$. Thus,

$$\begin{aligned} TV(u) = 2 &\sum(\text{interior maxima}) \\ - 2 &\sum(\text{interior minima}) \\ \pm &\,(\text{the end values}). \end{aligned}$$

Similarly in the discrete case,

$$\begin{aligned} TV(v) = 2 &\sum(\text{interior maxima}) \\ - 2 &\sum(\text{interior minima}) \\ \pm &\, v_0 \pm v_n. \end{aligned}$$

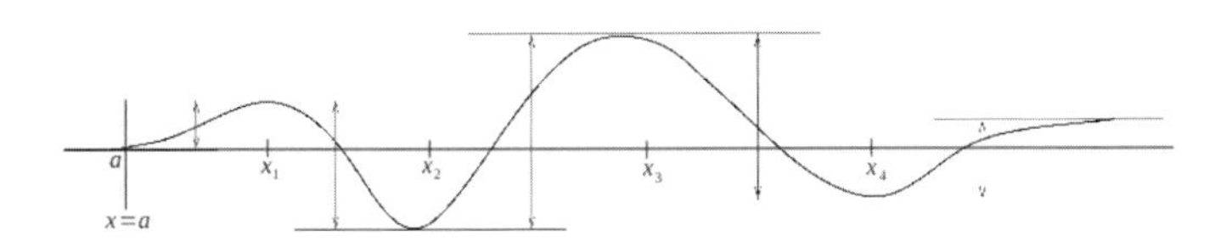

Figure 5.5 Equivalent LED and TVD schemes (one-dimensional case).

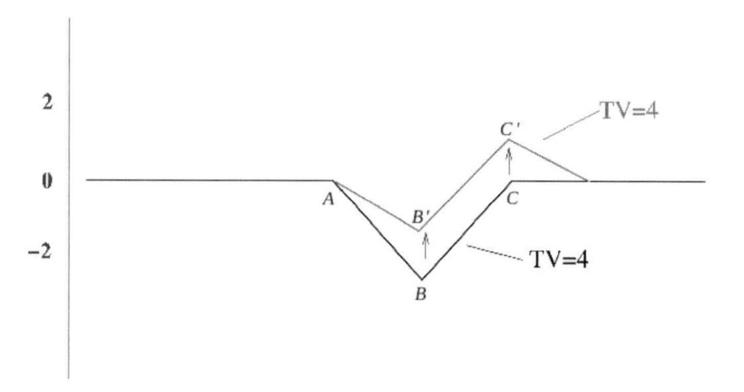

Figure 5.6 Distinction between LED and TVD schemes.

If the initial data is, say, monotonically decreasing, the total variation is

$$TV(v^0) = v_0^0 - v_n^0.$$

If the end values are fixed, the introduction of any interior extrema will produce an increase in the total variation. Thus a TVD scheme preserves the monotonicity of initial data that is monotonic. Also if the end values are fixed, the discrete scheme will be TVD if interior maxima cannot increase, interior minima cannot decrease, and no new extrema are introduced. Accordingly LED schemes are TVD. The converse is not necessarily true. This is illustrated in Figure 5.6, in which the upper and lower profiles have the same total variation with a value equal to 4. Thus a TVD scheme would allow a shift from the lower to the upper profile, which would not be permitted by an LED scheme because it would incur the formation of a new maximum at C'. In this sense the LED criterion is more stringent than the TVD criterion while still consistent with the properties of the nonlinear conservation law (5.2).

Harten derived conditions for the coefficients of a three-point TVD scheme in one dimension that are essentially equivalent to the positivity condition (5.11) for an LED scheme. The conditions for a multipoint TVD scheme derived by Jameson and Lax (1986) would be hard to realize in practice. The LED criterion is directly applicable to multi-dimensional discretizations on both structured and unstructured meshes, whereas the use of $\int_D \|\nabla u\| \, dS$ as a measure of total variation can lead to anomalous results on a triangular mesh using any of the standard norms (Jameson & Lax 1986). For the foregoing reasons, the analysis in the following sections is based on the LED principle.

5.7 Semi-discrete L_∞ Stable and LED Schemes

It is often convenient to separate the formulation of the time stepping and space discretization schemes by first using a semi-discretization to reduce the continuous equation to a set of ordinary differential equations with the general form

$$\frac{dv_i}{dt} = \sum_j a_{ij} v_j, \tag{5.12}$$

or in matrix vector notation

$$\frac{dv}{dt} = Av.$$

Assuming that the discrete values $v_i(t)$ correspond to a sufficiently smooth function $v(x,t)$ at $x = j\Delta x$, a Taylor series expansion yields

$$\frac{dv_i}{dt} = \sum_j a_{ij} \left[v(x_i) + (x_j - x_i)\frac{\partial v(x_i)}{\partial x} + \left(\frac{x_j - x_i}{2}\right)^2 \frac{\partial^2 v(x_i)}{\partial x^2} + \cdots \right].$$

Accordingly, the semi-discrete (5.12) is consistent with a differential equation with no source term only if

$$\sum_j a_{ij} = 0.$$

As in the case of the fully discrete scheme, this condition may also be inferred from the observation that if v_j is constant everywhere, the solution should remain unchanged. Then (5.12) can be written as

$$\frac{dv_i}{dt} = \left(\sum_j a_{ij}\right) v_i + \sum_{j \neq i} a_{ij}(v_j - v_i)$$

or

$$\frac{dv_i}{dt} = \sum_{j \neq i} a_{ij}(v_j - v_i).$$

Suppose that

$$a_{ij} \geq 0, \quad j \neq i. \tag{5.13}$$

Then if v_i is a maximum,

$$v_j - v_i \leq 0$$

and

$$\frac{dv_i}{dt} \leq 0. \tag{5.14}$$

Similarly if v_i is a minimum,

$$v_j - v_i \geq 0$$

and

$$\frac{dv_i}{dt} \geq 0. \tag{5.15}$$

Now $\|v\|_\infty$ can increase only if the maximum increases or the minimum decreases. But

$$\left|\frac{dv_i}{dt}\right| \leq \|A\|_\infty \|v\|_\infty;$$

so if $|v_i| < \|v\|_\infty$, there is a time interval $\epsilon > 0$ during which it cannot become an extremum, while if $|v_i| = \|v\|_\infty$, it follows from (5.14) or (5.15) that

$$\frac{d}{dt}|v_i| \leq 0.$$

Thus condition (5.13) is sufficient to ensure that $\|v\|_\infty$ does not increase.

Suppose that condition (5.13) is not satisfied. Then if $v_i = 1$, and if for $j \neq i$, $v_j = 1$ if $a_{ij} \geq 0$ and $v_j = 0$ if $a_{ij} < 0$, one obtains

$$\frac{dv_i}{dt} > 0,$$

and $\|v\|_\infty$ will increase. Accordingly, the semi-discrete scheme (5.12) is L_∞ stable if and only if the positivity condition (5.13) is satisfied.

As in the case of the discrete scheme, the semi-discrete scheme will be LED if it is limited to a compact stencil of nearest neighbors with non-negative coefficients $a_{ij} \geq 0$, $a_{ij} = 0$ if j is not a neighbor of i because then the same argument can be repeated while examining the behavior of v_i with respect only to its neighbors.

Note that any semi-discrete scheme with non-negative coefficients can be converted into a corresponding discrete L_∞ stable or LED scheme

$$v_i^{n+1} = v_i^n + \Delta t \sum_{j \neq i} a_{ij}(v_j^n - v_i^n),$$

where the time step must satisfy the restriction (5.8) to ensure the coefficient of v_i^n is non-negative. Accordingly, we shall often restrict our discussion in the following sections to the semi-discrete case.

5.8 Growth of the L_∞ Norm with a Source Term

Consider the general semi-discrete scheme

$$\frac{dv_i}{dt} = \sum_j a_{ij} v_j,$$

where

$$\sum_j a_{ij} = \alpha.$$

It now follows from a Taylor series expansion, as in Section 5.7, that the semi-discrete is constant with the differential equation

$$\frac{du}{dt} = Lu + \alpha u,$$

where L is a source-free differential operator. Subtracting $\left(\sum_j a_{ij} - \alpha\right) v_i$, the scheme can be written with no loss of generality as

$$\frac{dv_i}{dt} = \alpha v_i + \sum_{j \neq i} a_{ij}(v_j - v_i).$$

Set

$$v_i = \omega_i e^{\alpha t}.$$

Then

$$\frac{dv_i}{dt} = \left(\frac{d\omega_i}{dt} + \alpha \omega_i\right) e^{\alpha t}.$$

Consequently

$$\frac{d\omega_i}{dt} = \sum_{j \neq i} a_{ij}(\omega_j - \omega_i).$$

Therefore, if $a_{ij} \geq 0$, $\|\omega\|_\infty$ does not increase, and the growth of $\|v\|_\infty$ is bounded by $e^{\alpha t}$.

5.9 Accuracy Limitation on L_∞ Stable and LED Schemes

Suppose that the linear advection equation

$$\frac{\partial u}{\partial t} + a\frac{\partial u}{\partial x} = 0$$

is approximated on a uniform grid by a fixed scheme with the same coefficients at every mesh point:

$$v_j^{n+1} = \sum_{k=-\infty}^{\infty} c_k v_{j+k}^n.$$

A Taylor series expansion now yields

$$\begin{aligned}
v(x_j) &+ \Delta t \frac{\partial v}{\partial t}(x_j) + \frac{\Delta t^2}{2}\frac{\partial^2 v}{\partial t^2}(x_j) \\
&= \sum_{k=-\infty}^{\infty} c_k \left[v(x_j) + k\Delta x \frac{\partial v(x_j)}{\partial x} + \frac{k^2 \Delta x^2}{2}\frac{\partial^2 v(x_j)}{\partial x^2} + \cdots \right].
\end{aligned}$$

In order to realize second order accuracy, the coefficients c_k must satisfy the conditions

$$\sum_k c_k = 1 \tag{5.16}$$

$$\sum_k k c_k = -\lambda$$
$$\sum_k k^2 c_k = \lambda^2,$$

where

$$\lambda = a\frac{\Delta t}{\Delta x}.$$

Since $c_k \geq 0$, we can set

$$\alpha_k = \sqrt{c_k}, \quad \beta_k = k\sqrt{c_k} = k\alpha_k. \tag{5.17}$$

Then

$$\sum_k \alpha_k^2 = 1$$
$$\sum_k \alpha_k \beta_k = -\lambda$$
$$\sum_k \beta_k^2 = \lambda^2,$$

and consequently

$$\left(\sum_k \alpha_k \beta_k\right)^2 = \left(\sum_k \alpha_k^2\right)\left(\sum_k \beta_k^2\right).$$

But by the Cauchy–Schwarz inequality,

$$\left(\sum_k \alpha_k \beta_k\right)^2 \leq \left(\sum_k \alpha_k^2\right)\left(\sum_k \beta_k^2\right),$$

with equality only if the vectors α and β are aligned or for some scale r, $\beta_k = r\alpha_k$ for all k. According to the definition (5.17) of α_k and β_k, it follows that there can only be one nonzero c_k. Moreover it follows from (5.16) that this must have the value unity. Taking

$$c_k = 0, \quad k \neq -1$$
$$c_{-1} = 1,$$

we recover the standard upwind scheme

$$v_j^{n+1} = v_j^n - \lambda(v_j^n - v_{j-1}^n)$$

with a CFL number $\lambda = 1$, for which the scheme corresponds to exact propagation along characteristics. A choice such as $c_{k-2} = 1$ corresponds to the upwind scheme applied over double intervals with $\lambda = 1$. It is evident that no solution is possible for time steps such that $\lambda \neq 1$.

We conclude that it is not possible for a discrete scheme with fixed non-negative coefficients at every mesh point to yield better than first order accuracy. In order to overcome this barrier, it is necessary to consider schemes in which the discretization is locally adapted to the solution. Typically these schemes enforce positivity or the LED principle only in the neighborhood of extrema, which can be detected by a change of sign in the slope measured by $\Delta v_{j+\frac{1}{2}} = v_{j+1} - v_j$. The earliest examples of schemes of this type are Boris and Book's flux corrected transport (FCT) scheme (Boris & Book 1973) and Van Leer's monotone upstream conservative limited (MUSCL) scheme (Van Leer 1974).

5.10 Artificial Diffusion and Upwinding

Taking as the simplest possible example the linear advection equation

$$\frac{\partial u}{\partial t} + a\frac{\partial u}{\partial x} = 0, \tag{5.18}$$

which represents right running waves of the form

$$u(x,t) = f(x - at),$$

when $a > 0$, an upwind semi-discretization is

$$\begin{aligned} \frac{dv_j}{dt} + \frac{a}{\Delta x}(v_j - v_{j-1}) = 0, &\qquad a > 0, \\ \frac{dv_j}{dt} + \frac{a}{\Delta x}(v_{j+1} - v_j) = 0, &\qquad a < 0. \end{aligned}$$

This can be equivalently written as

$$\frac{dv_j}{dt} + \frac{1}{2}\frac{a}{\Delta x}(v_{j+1} - v_{j-1}) - \frac{1}{2}\frac{|a|}{\Delta x}(v_{j+1} - 2v_j + v_{j-1}) = 0,$$

in which $\frac{\partial u}{\partial x}$ is approximated by a second order accurate central difference formula modified by an approximation to $-\frac{1}{2}|a|\Delta x\frac{\partial^2 u}{\partial x^2}$. Thus upwinding is equivalent to the addition of artificial diffusion with a coefficient proportional to $\frac{1}{2}|a|\Delta x$, the mesh width multiplied by half the wave speed.

We may consider a class of schemes with artificial diffusion of the form

$$\frac{dv_j}{dt} + \frac{a}{2\Delta x}(v_{j+1} - v_{j-1}) - \frac{\alpha|a|}{2\Delta x}(v_{j+1} - 2v_j + v_{j-1}) = 0,$$

where α is a parameter to be chosen. This can be written in conservative form as

$$\Delta x\frac{dv_j}{dt} + h_{j+\frac{1}{2}} - h_{j-\frac{1}{2}} = 0, \tag{5.19}$$

where the numerical flux is

$$h_{j+\frac{1}{2}} = \frac{1}{2}a(v_{j+1} + v_j) - \frac{1}{2}\alpha|a|(v_{j+1} - v_j).$$

Then

$$h_{j+\frac{1}{2}} = av_j - \frac{1}{2}(\alpha|a| - a)(v_{j+1} - v_j)$$

and

$$h_{j-\frac{1}{2}} = av_j - \frac{1}{2}(\alpha|a| + a)(v_j - v_{j-1}),$$

so

$$\Delta x \frac{dv_j}{dt} = \frac{1}{2}(\alpha - 1)|a|(v_{j+1} - v_j) + \frac{1}{2}(\alpha|a| + a)(v_{j-1} - v_j),$$

and the scheme is LED if $\alpha \geq 1$.

We can also consider the energy stability of this scheme. Suppose (5.18) holds in the interval $0 \leq x \leq L$, with $a > 0$, and it is solved with $u(0,t)$ specified and $u(L,t)$ free, which are the proper boundary conditions for a right running wave. Multiplying by u and integrating from 0 to L,

$$\begin{aligned}
\int_0^L u\frac{\partial u}{\partial t}dx &= \frac{d}{dt}\int_0^L \frac{u^2}{2}dx \\
&= -a\int_0^L u\frac{\partial u}{\partial x}dx \\
&= \frac{1}{2}a\big(u(0)^2 - u(L)^2\big).
\end{aligned}$$

If $u(0,t)$ is fixed equal to zero, the energy decays if $u(L) \neq 0$.

Suppose the semi-discretization is over n equal intervals from $x_0 = 0$ to $x_n = L$. Let v_0 be fixed corresponding to the inflow boundary condition. The outflow boundary value v_n should depend on the solution. To complete the semi-discrete scheme at x_n, we extrapolate the value at an extra mesh point x_{n+1}, $v_{n+1} = v_n$. Now, multiplying (5.19) by v_j and summing by parts, we find that

$$\begin{aligned}
\frac{dE}{dt} &= \Delta x \sum_{j=0}^{n} v_j \frac{dv_j}{dt} \\
&= -\sum_{j=1}^{n} v_j\left(h_{j+\frac{1}{2}} - h_{j-\frac{1}{2}}\right) \\
&= h_{\frac{1}{2}}v_1 - \sum_{j=1}^{n-1} h_{j+\frac{1}{2}}(v_{j+1} - v_j) - h_{n+\frac{1}{2}}v_n.
\end{aligned}$$

In the case $\alpha = 0$, now

$$\frac{dE}{dt} = \frac{1}{2}av_0v_1 - \frac{1}{2}av_n^2,$$

and when $v_0 = 0$, the discrete energy decays if $v_n \neq 0$, as in the case of the true solution.

If $\alpha > 0$, $v_0 = 0$, and $v_{n+1} = v_n$:

$$\frac{dE}{dt} = -\frac{1}{2}av_n^2 - \frac{1}{2}\alpha a v_1^2 - \frac{1}{2}\alpha a \sum_{j=1}^{n-1}(v_{j+1} - v_j)^2.$$

Accordingly, there is a negative contribution to $\frac{dE}{dt}$ from every interior interface. Thus finally the semi-discrete scheme is energy stable if $\alpha \geq 0$ and LED if $\alpha \geq 1$.

5.11 The Lax–Friedrichs and Lax–Wendroff Schemes

As was mentioned in Section 4.8, one of the first proposals for a discretization of the linear advection equation was the scheme of Lax and Friedrichs in which they replace v_j^n on the right by the arithmetic average of the neighbors to obtain

$$v_j^{n+1} = \frac{1}{2}(v_{j+1}^n + v_{j-1}^n) - \frac{1}{2}\lambda(v_{j+1}^n - v_{j-1}^n),$$

or

$$v_j^{n+1} = \frac{1}{2}(1 - \lambda)v_{j+1}^n + \frac{1}{2}(1 - \lambda)v_{j-1}^n.$$

This scheme satisfies the CFL condition if $|\lambda| \leq 1$, and in that case, the coefficients of v_{j+1}^n and v_{j-1}^n are non-negative, so the scheme is LED. It can be written in viscosity form as

$$v_j^{n+1} = v_j^n - \frac{1}{2}\lambda(v_{j+1}^n - v_{j-1}^n) + \frac{1}{2}(v_{j+1}^n - 2v_j^n - v_{j-1}^n),$$

or as

$$v_j^{n+1} = v_j^n - \frac{\Delta t}{\Delta x}(h_{j+\frac{1}{2}} - h_{j-\frac{1}{2}}),$$

where

$$h_{j+\frac{1}{2}} = \frac{1}{2}a(v_{j+1} + v_j) - \alpha_{j+\frac{1}{2}}(v_{j+1} - v_j),$$

with the viscosity coefficient of fixed magnitude

$$\alpha_{j+\frac{1}{2}} = \frac{1}{2}\frac{\Delta x}{\Delta t},$$

independent of the wave speed a.

For the scalar conservation law

$$\frac{\partial u}{\partial t} + \frac{\partial}{\partial x}f(u) = 0,$$

the Lax–Friedrichs scheme is

$$v_j^{n+1} = \frac{1}{2}(v_{j+1}^n + v_{j-1}^n) - \frac{1}{2}\frac{\Delta t}{\Delta x}(f_{j+1} - f_{j-1}),$$

or

$$v_j^{n+1} = v_j^n - \frac{\Delta t}{\Delta x}\left(h_{j+\frac{1}{2}} - h_{j-\frac{1}{2}}\right),$$

with the numerical flux

$$h_{j+\frac{1}{2}} = \frac{1}{2}(f_{j+1} + f_j) - \frac{1}{2}\frac{\Delta x}{\Delta t}(v_{j+1} - v_j).$$

This is very diffusive, but it does maintain a positive coefficient of diffusion over the entire domain and converges to the correct vanishing viscosity solution as the mesh is refined.

In order to satisfy the CFL condition over the entire solution domain $\mathbb{D}$, the time step must satisfy the restriction

$$\frac{\Delta t}{\Delta x}\max_{\mathbb{D}}\left|f'(v)\right| \leq 1.$$

Accordingly, if the solution is advanced by the maximum permissible time step,

$$\frac{\Delta x}{\Delta t} = \max_{\mathbb{D}} f'(v),$$

and we may express the Lax–Friedrichs flux as

$$h_{j+\frac{1}{2}} = \frac{1}{2}(f_{j+1} + f_j) - \frac{1}{2}\max_{\mathbb{D}}\left|f'(v)\right|(v_{j+1} - v_j). \tag{5.20}$$

The Lax–Wendroff scheme for the linear advection equation,

$$\begin{aligned} v_j^{n+1} &= v_j^n - \frac{1}{2}\lambda(v_{j+1}^n - v_{j-1}^n) + \frac{1}{2}\lambda^2(v_{j+1}^n - 2v_j^n + v_{j-1}^n) \\ &= \frac{1}{2}(\lambda^2 - \lambda)v_{j+1}^n + (1 - \lambda^2)v_j^n + \frac{1}{2}(\lambda^2 + \lambda)v_{j-1}^n, \end{aligned}$$

satisfies the CFL condition if $|\lambda| \leq 1$. If $|\lambda| < 1$, however, the coefficient of either v_{j+1}^n or v_{j-1}^n is negative, so it is not L_∞ stable or LED. In conservation form, the scheme is

$$v_j^{n+1} = v_j^n - \frac{\Delta t}{\Delta x}\left(h_{j+\frac{1}{2}} - h_{j-\frac{1}{2}}\right) = 0,$$

where

$$h_{j+\frac{1}{2}} = \frac{1}{2}a(v_{j+1}^n + v_j^n) - \frac{1}{2}\lambda a(v_{j+1}^n - v_j^n)$$

with a viscosity coefficient $\frac{1}{2}\lambda a = \frac{1}{2}|\lambda||a|$, which is just enough for stability in the Euclidean norm but not enough for L_∞ stability.

5.12 LED Schemes for Nonlinear Conservation Laws

We now consider scalar nonlinear conservation laws of the form

$$\frac{\partial u}{\partial t} + \frac{\partial}{\partial x} f(u) = 0$$

in the interval $0 \leq x \leq L$, where the boundary values $u(0)$ and $u(L)$ are specified at an inflow boundary and free at an outflow boundary. We shall restrict our discussion to cases where $f(u)$ is a convex function with $f''(u) > 0$. We can write the conservation law in integral form as

$$\frac{d}{dt} \int_0^L u dx + f(u(L)) - f(u(0)) = 0. \tag{5.21}$$

Suppose that the domain is divided into a uniform grid of cells of width Δx_j, where the j-th cell extends from $x_j - \frac{1}{2}\Delta x$ to $x_j + \frac{1}{2}\Delta x$. Applying (5.21) separately to each cell, we obtain the semi-discrete finite volume scheme

$$\Delta x \frac{dv_j}{dt} + h_{j+\frac{1}{2}} - h_{j-\frac{1}{2}} = 0,$$

where v_j represents the average value of u in cell j, and $h_{j+\frac{1}{2}}$ is the numerical flux across the interface separating cells j and $j+1$. Introducing artificial diffusion, we define the numerical fluxes

$$h_{j+\frac{1}{2}} = \frac{1}{2}(f_{j+1} + f_j) - \alpha_{j+\frac{1}{2}}(v_{j+1} - v_j),$$

where

$$f_j = f(v_j),$$

and $\alpha_{j+\frac{1}{2}}$ is the coefficient of artificial diffusion. Define also a numerical estimate of the wave speed $a(u) = f'(u)$ as

$$a_{j+\frac{1}{2}} = \begin{cases} \dfrac{f_{j+1} - f_j}{v_{j+1} - v_j}, & v_{j+1} \neq v_j \\ \left.\dfrac{\partial f}{\partial u}\right|_{v_j}, & v_{j+1} = v_j. \end{cases} \tag{5.22}$$

Now the numerical fluxes can be expressed as

$$\begin{aligned} h_{j+\frac{1}{2}} &= f_j + \frac{1}{2}\left(f_{j+1} - f_j\right) - \alpha_{j+\frac{1}{2}}\left(v_{j+1} - v_j\right) \\ &= f_j - \left(\alpha_{j+\frac{1}{2}} - \frac{1}{2} a_{j+\frac{1}{2}}\right)\left(v_{j+1} - v_j\right) \end{aligned}$$

and

$$h_{j-\frac{1}{2}} = f_j - \frac{1}{2}(f_j - f_{j-1}) - \alpha_{j-\frac{1}{2}}(v_j - v_{j-1})$$
$$= f_j - \left(\alpha_{j-\frac{1}{2}} + \frac{1}{2}a_{j-\frac{1}{2}}\right)(v_j - v_{j-1}).$$

Thus the semi-discrete scheme reduces to

$$\Delta x \frac{dv_j}{dt} = \left(\alpha_{j+\frac{1}{2}} - \frac{1}{2}a_{j+\frac{1}{2}}\right)(v_{j+1} - v_j) + \left(\alpha_{j-\frac{1}{2}} + \frac{1}{2}a_{j-\frac{1}{2}}\right)(v_{j-1} - v_j).$$

Accordingly it is LED if for all j

$$\alpha_{j+\frac{1}{2}} \geq \frac{1}{2}|a_{j+\frac{1}{2}}|.$$

5.13 The First Order Upwind Scheme

The least diffusive LED scheme is obtained by setting

$$\alpha_{j+\frac{1}{2}} = \frac{1}{2}|a_{j+\frac{1}{2}}|$$

to produce the diffusive flux

$$d_{j+\frac{1}{2}} = \frac{1}{2}|a_{j+\frac{1}{2}}|\Delta v_{j+\frac{1}{2}},$$

where

$$\Delta v_{j+\frac{1}{2}} = v_{j+1} - v_j.$$

This is the pure first order upwind scheme, since if $a_{j+\frac{1}{2}} > 0$,

$$d_{j+\frac{1}{2}} = \frac{1}{2}\frac{f_{j+1} - f_j}{v_{j+1} - v_j}(v_{j+1} - v_j) = \frac{1}{2}(f_{j+1} - f_j),$$

and if $a_{j+\frac{1}{2}} < 0$,

$$d_{j+\frac{1}{2}} = -\frac{1}{2}(f_{j-1} - f_j).$$

Thus,

$$h_{j+\frac{1}{2}} = \begin{cases} f_j & \text{if} \quad a_{j+\frac{1}{2}} > 0 \\ \frac{1}{2}(f_{j+1} + f_j) & \text{if} \quad a_{j+\frac{1}{2}} = 0 \\ f_{j+1} & \text{if} \quad a_{j+\frac{1}{2}} < 0. \end{cases} \tag{5.23}$$

Thus, the upwind scheme is the least diffusive first order accurate LED scheme.

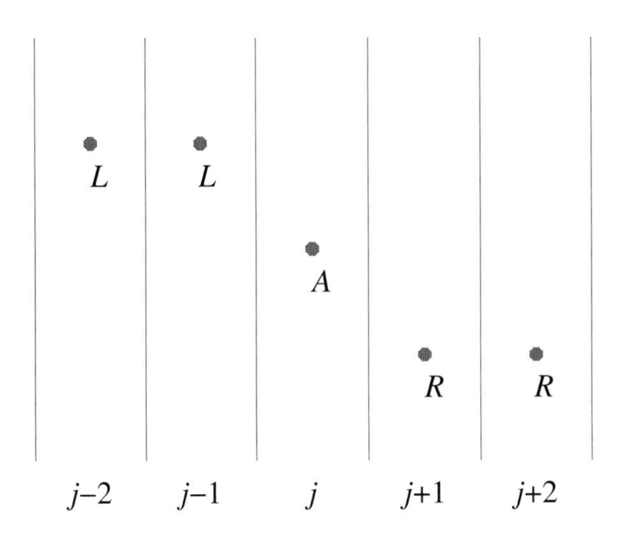

Figure 5.7 Shock structure of the upwind scheme.

5.14 Shock Structure of the Upwind Scheme

The first order upwind scheme supports a numerical structure of a stationary shock with a single interior point, as illustrated in Figure 5.7, where the upstream and downstream values are denoted by the subscripts L and R, and the transitional value located at cell j is denoted by the subscript A. The jump condition for a shock moving at a speed S is

$$f_R - f_L = S(u_R - u_L).$$

For a stationary shock, $f_R = f_L$, and since the characteristics converge on a shock, the true wave speed changes sign. This is also the case for the numerical wave speed. Assuming $f(u)$ is a convex function of u, as illustrated in Figure 5.8,

$$a_{j-\frac{1}{2}} = \frac{f_A - f_L}{v_A - v_L} > 0, \quad a_{j+\frac{1}{2}} = \frac{f_R - f_A}{v_R - v_A} < 0.$$

Specifically in the case of Burgers' equation,

$$f(u) = \frac{u^2}{2}, \quad v_R = -v_L, \quad a_{j-\frac{1}{2}} = \frac{1}{2}(v_A + v_L) > 0, \quad a_{j+\frac{1}{2}} = \frac{1}{2}(v_R + v_A) < 0.$$

Accordingly the numerical fluxes are

$$h_{j-\frac{1}{2}} = f_{j-1} = f_L, \quad h_{j+\frac{1}{2}} = f_{j+1} = f_R,$$

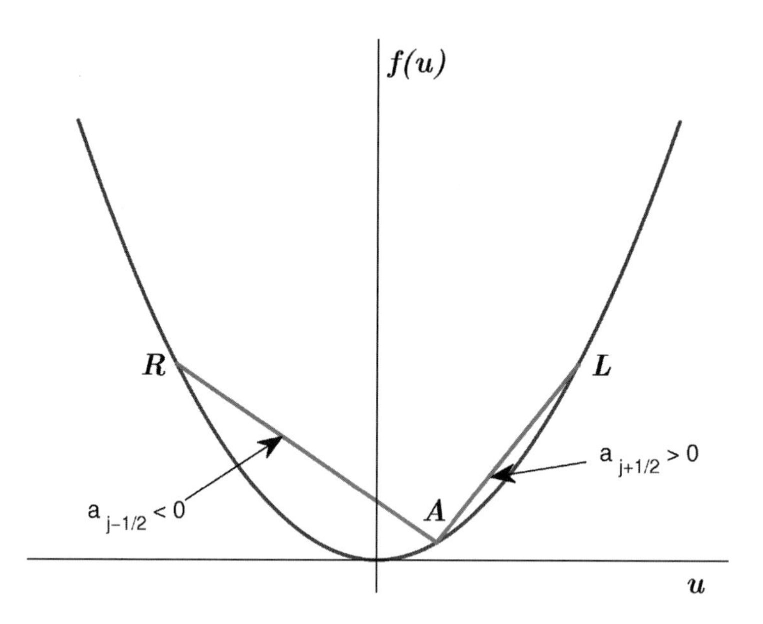

Figure 5.8 Convex flux.

while the interface fluxes between cells to the left of j are equal to f_L, and those between cells to the right of j are equal to f_R. Hence

$$\frac{d}{dt}v_{j-1} = \frac{f_L - f_L}{\Delta x} = 0$$

$$\frac{d}{dt}v_j = \frac{f_R - f_L}{\Delta x} = 0$$

$$\frac{d}{dt}v_{j+1} = \frac{f_R - f_R}{\Delta x} = 0.$$

With an appropriate choice of the numerical flux, a single point numerical structure can also be obtained for a stationary shock in gas dynamics, as will be discussed in Chapter 6.

5.15 Upwinding and Conservation

In early formulations of upwind schemes, the scheme itself was switched according to the sign of the wave speed $a_j = \frac{\partial f_j}{\partial v_j}$.

$$\frac{dv_j}{dt} + \frac{f_j - f_{j-1}}{\Delta x} = 0, \quad a_j > 0$$

$$\frac{dv_j}{dt} + \frac{f_{j+1} - f_j}{\Delta x} = 0, \quad a_j < 0.$$

Suppose that there is a transition between j and $j+1$

$$a_k > 0, \quad k \leq j$$

$$a_k < 0, \quad k \geq j+1.$$

Now

$$\frac{dv_j}{dt} + \frac{f_j - f_{j-1}}{\Delta x} = 0$$

corresponding to a numerical flux

$$h_{j+\frac{1}{2}} = f_j,$$

while

$$\frac{dv_{j+1}}{dt} + \frac{f_{j+2} - f_{j+1}}{\Delta x} = 0$$

corresponding to

$$h_{j+\frac{1}{2}} = f_{j+1}.$$

Accordingly, the scheme is not conservative, since the interior fluxes do not cancel when the solution values are summed over j. In fact, with f_0 and f_n fixed since both boundaries are inflow boundaries,

$$\Delta x \sum_{j=1}^{n-1} \frac{dv_j}{dt} = -f_1 + f_0 - f_2 + f_1 \cdots - f_j + f_{j-1} - f_{j+2} + f_{j+1} \cdots - f_n + f_{n-1}$$

$$= f_0 - f_n - f_j + f_{j+1},$$

whereas the true solution satisfies

$$\frac{d}{dt} \int_{x_0}^{x_n} u dx = - \int_{x_0}^{x_n} \frac{\partial f}{\partial x} dx = f(u(x_0)) - f(u(x_n)).$$

By upwinding the flux rather than the scheme, we ensure that the interior fluxes are cancelled in a telescopic sum, with the result that the correct conservation law is preserved by the numerical scheme.

5.16 Shortcomings of the Upwind Scheme

Aside from being only first order accurate, the pure upwind scheme (5.23) has the serious shortcoming that it can admit discontinuous expansions, which violate the entropy condition. This is easily apparent for the inviscid Burgers' equation. Suppose that

$$v_j^n = \begin{cases} -1, & 1 \leq j \leq J, \\ 1, & J < j \leq N, \end{cases}$$

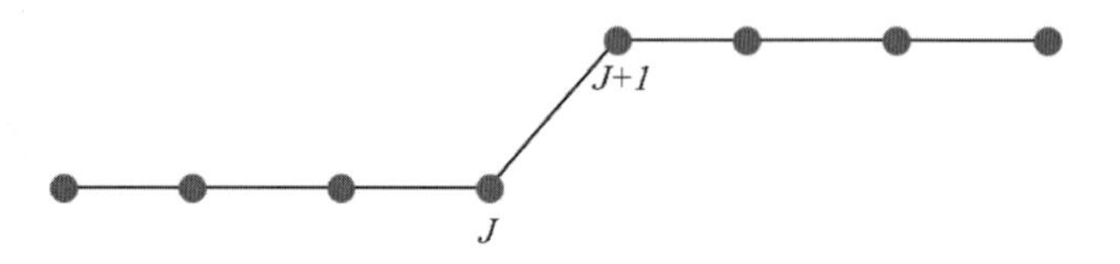

Figure 5.9 Stationary expansion shock.

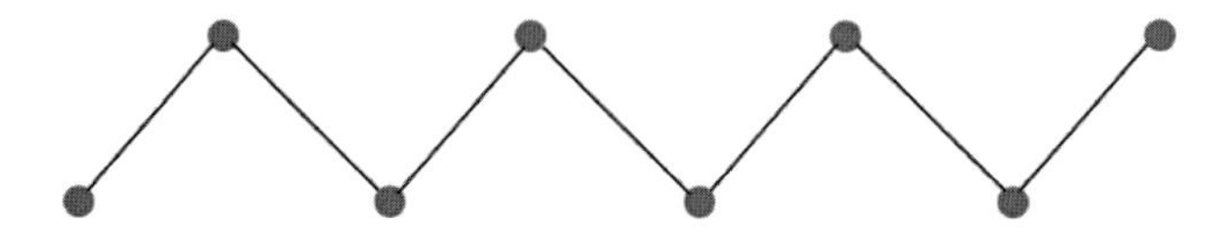

Figure 5.10 Oscillatory stationary solution.

as illustrated in Figure 5.9. Then,

$$a_{J+\frac{1}{2}} = \frac{1}{2}(v^n_{J+1} + v^n_J) = 0,$$

and

$$h_{J+\frac{1}{2}} = \frac{1}{4}({v^n_{J+1}}^2 + {v^n_J}^2) = \frac{1}{2},$$

while

$$h_{j+\frac{1}{2}} = \begin{cases} \frac{1}{2}{v^n_j}^2 = \frac{1}{2}, & j < J, \\ \frac{1}{2}{v^n_{j+1}}^2 = \frac{1}{2}, & j > J. \end{cases}$$

Thus,

$$h_{j+\frac{1}{2}} - h_{j-\frac{1}{2}} = 0$$

at every interior point, and the discrete solution is a stationary expansion shock, whereas the true solution is an expansion fan.

In fact, the situation is even worse than this. Any combination of values $v^n_j = \pm 1$ with $v_1 = -1$ and $v_N = 1$ leads to the constant value

$$h_{j+\frac{1}{2}} = \frac{1}{2}$$

at every interface and is consequently a stationary discrete solution, such as the one illustrated in Figure 5.10. This has motivated the search for alternative formulations that exclude discontinuous expansions in the discrete solution.

5.16.1 The Entropy Fix

Consider the viscosity form of the upwind scheme

$$h_{j+\frac{1}{2}} = \frac{1}{2}(f_{j+1} + f_j) - \alpha_{j+\frac{1}{2}}(v_{j+1} - v_j),$$

where

$$\alpha_{j+\frac{1}{2}} = \frac{1}{2}|a_{j+\frac{1}{2}}|,$$

and $a_{j+\frac{1}{2}}$ is the discrete approximation (5.22) to the wave speed $f'(u)$. It can be seen that the viscosity coefficient is a discontinuous function of u at the point s where $f'(s) = 0$, whereas the true solution is continuous once an expansion fan has been initiated about the point where $u = s$. Referring to Figure 5.9, it can also be seen that in order to initiate the development of an expansion fan, we need $\frac{dv_J}{dt} > 0$ and $\frac{dv_{J+1}}{dt} < 0$. This will be the case if $h_{J+\frac{1}{2}} < h_{J-\frac{1}{2}}$ and $h_{J+\frac{1}{2}} < h_{J+\frac{3}{2}}$. We can achieve this by requiring the viscosity coefficient to be positive, $\alpha_{j+\frac{1}{2}} > 0$, at interfaces where $a_{j+\frac{1}{2}} = 0$. These considerations suggest replacing the pure absolute value $|a_{j+\frac{1}{2}}|$ by rounding it out above zero to produce a smooth viscosity coefficient that is everywhere positive. Thus, for example, we may set

$$\alpha_{j+\frac{1}{2}} = \frac{1}{2}\tilde{a}_{j+\frac{1}{2}},$$

where $\tilde{a}$ is defined for a threshold ϵ as

$$\tilde{a} = \begin{cases} |a| & \text{if } |a| > \epsilon, \\ \dfrac{1}{2}\left(\epsilon + \dfrac{a^2}{\epsilon}\right) & \text{if } |a| \leq \epsilon. \end{cases}$$

This modification of the wave speed, which was first introduced by Harten, is generally known as the "entropy fix."

In upwind schemes for the Euler equations, it is prudent to apply the entropy fix to all the wave speeds. Unfortunately, there is no precise rule for choosing the threshold ϵ. It should be large enough to prevent discontinuous expansions but as small as possible in order to avoid a significant loss of accuracy. Thus, there is still a need for schemes that are provably convergent to entropy satisfying-solutions. Some examples are presented in the next sections. They belong to classes of schemes known as monotone schemes and E-schemes.

5.16.2 The Local Lax–Freidrichs Scheme

A more drastic way to increase the artificial viscosity is to choose the coefficient $\alpha_{j+\frac{1}{2}}$ as

$$\alpha_{j+\frac{1}{2}} = \max_{v \in I} \left|f'(v)\right|,$$

where I is the interval between v_j and v_{j+1}, without regard to which is larger. For a convex flux with $f'(v) > 0$, this is equivalent to setting

$$\alpha_{j+\frac{1}{2}} = \max\left(\left|f'(v_j)\right|, \left|f'(v_{j+1})\right|\right).$$

This is frequently called the local Lax–Friedrichs (LLF) scheme, because the viscosity coefficient is proportional to the maximum value of $f'(v)$ over the interval between v_j and v_{j+1} instead of the whole domain, as in (5.20).

5.16.3 The Engquist–Osher (EO) Upwind Scheme

An alternative construction of a first order accurate upwind scheme was proposed by Engquist & Osher (1981). Assuming that the flux function $f(u)$ is convex, with a minimum at u^* where $f'(u^*) = 0$, it can be split as

$$f(u) = f^+(u) + f^-(u),$$

where

$$f^+(u) = \begin{cases} f(u), & u \geq u^*, \\ f(u^*), & u \leq u^*, \end{cases}$$

and

$$f^-(u) = \begin{cases} f(u), & u \leq u^*, \\ f(u^*), & u \geq u^*. \end{cases}$$

In the case of Burger's equation:

$$f(u) = \frac{u^2}{2},$$

$$f^+(u) = \begin{cases} \frac{u^2}{2}, & u \geq 0, \\ 0, & u \leq 0, \end{cases}$$

$$f^-(u) = \begin{cases} \frac{u^2}{2}, & u \leq 0, \\ 0, & u \geq 0. \end{cases}$$

Now the numerical flux is defined as

$$h_{j+\frac{1}{2}} = f^+(v_j) + f^-(v_{j+1}).$$

In zones where the wave speed has a constant sign, this is identical to the upwind scheme defined in Section 5.13, but it is different in transition zones where the wave speed changes sign.

Consider, for example, the case of Burger's equation with $v_j < 0$, $v_{j+1} > 0$. The Engquist–Osher flux is

$$h^{EO}_{j+\frac{1}{2}} = f_j^+ + f_{j+1}^- = 0,$$

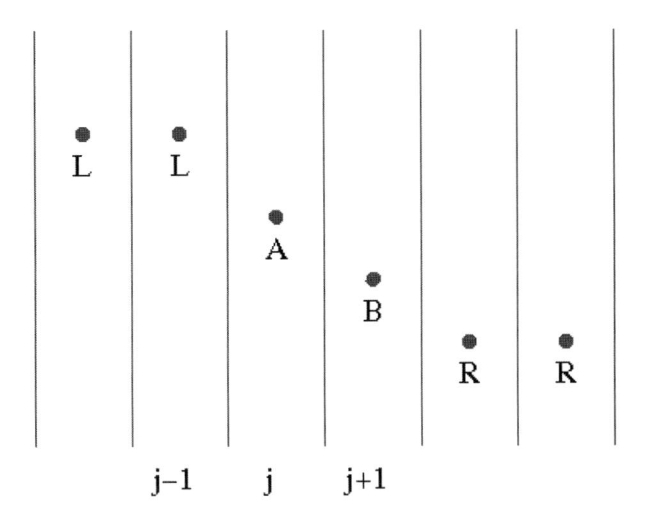

Figure 5.11 Shock structure of the Engquist–Osher scheme.

while the upwind flux is

$$h^U_{j+\frac{1}{2}} = \begin{cases} f_j & \text{if } v_{j+1} + v_j > 0, \\ \frac{1}{2}(f_j + f_{j+1}) & \text{if } v_{j+1} + v_j = 0, \\ f_{j+1} & \text{if } v_{j+1} + v_j < 0. \end{cases}$$

This modification rules out a steady state solution with a discontinuous expansion such as that illustrated in Figure 5.9. In the case of a shock, there may be a transition from left state v_j with $a(v_j) > 0$ to a right state v_{j+1} with $a(v_{j+1}) < 0$, so that

$$h^{EO}_{j+\frac{1}{2}} = f^+_j + f^-_{j+1} = f_j + f_{j+1}.$$

Consider again Burger's equation where we now examine the numerical structure of a stationary shock with left and right states $v_L > 0$ and $v_R < 0$. It now turns out that a structure with two interior points is possible, as illustrated in Figure 5.11. Here

$$v_L \geq v_A \geq 0, \quad 0 \geq v_B \geq v_R,$$

and

$$h_{j-\frac{1}{2}} = f_L,$$

$$h_{j+\frac{1}{2}} = f^+(v_A) + f^-(v_B) = f_A + f_B,$$

$$h_{j+\frac{3}{2}} = f_R.$$

Thus every point is in equilibrium if

$$f_A + f_B = f_L = f_R,$$

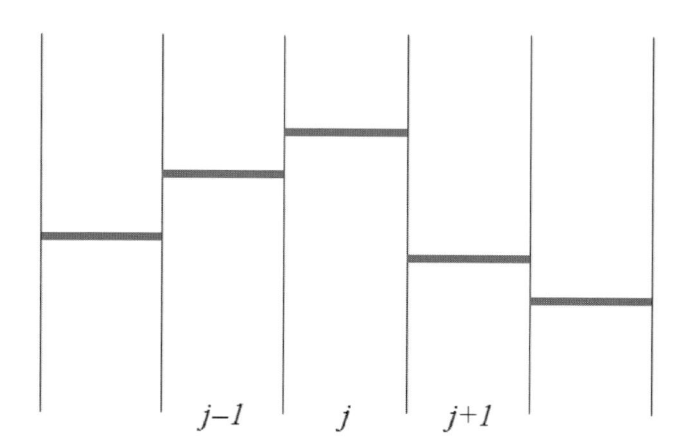

Figure 5.12 Piecewise constant solution.

or

$$v_A^2 + v_B^2 = v_L^2 = v_R^2.$$

Hence, given v_A,

$$v_B = -\sqrt{v_L^2 - v_A^2}.$$

5.16.4 The Godunov Scheme

In 1959, Godunov proposed a scheme to solve the Euler equations of gas dynamics (Godunov 1959), which at first remained little known outside Russia but, following the work of Van Leer (1973, 1974, 1977a, 1977b, 1979), eventually became very influential in the development of CFD. While Godunov originally proposed his scheme for the Euler equations, it may also be applied to conservation laws in general. His idea was to regard the solution as piecewise constant in each cell, with a value equal to the average value in the cell, as sketched in Figure 5.12. Then the interface flux $h_{j+\frac{1}{2}}$ is taken to be the exact solution of a Riemann problem defined as the evolution of the state from an initial condition with constant values to the left and right of the interface. The solution to the problem is self similar, depending only on the variable $\xi = \frac{x}{t}$, where $x = 0$ at the interface. Hence, the interface fluxes are constant for a time interval small enough that the waves from different interfaces do not interact with each other.

The possible solutions for a scalar conservation law are displayed in Figure 5.13.

1. right running shock wave,
2. right running expansion fan,
3. an expansion fan containing $f'(v) = 0$,
4. left running expansion fan,
5. left running shock wave.

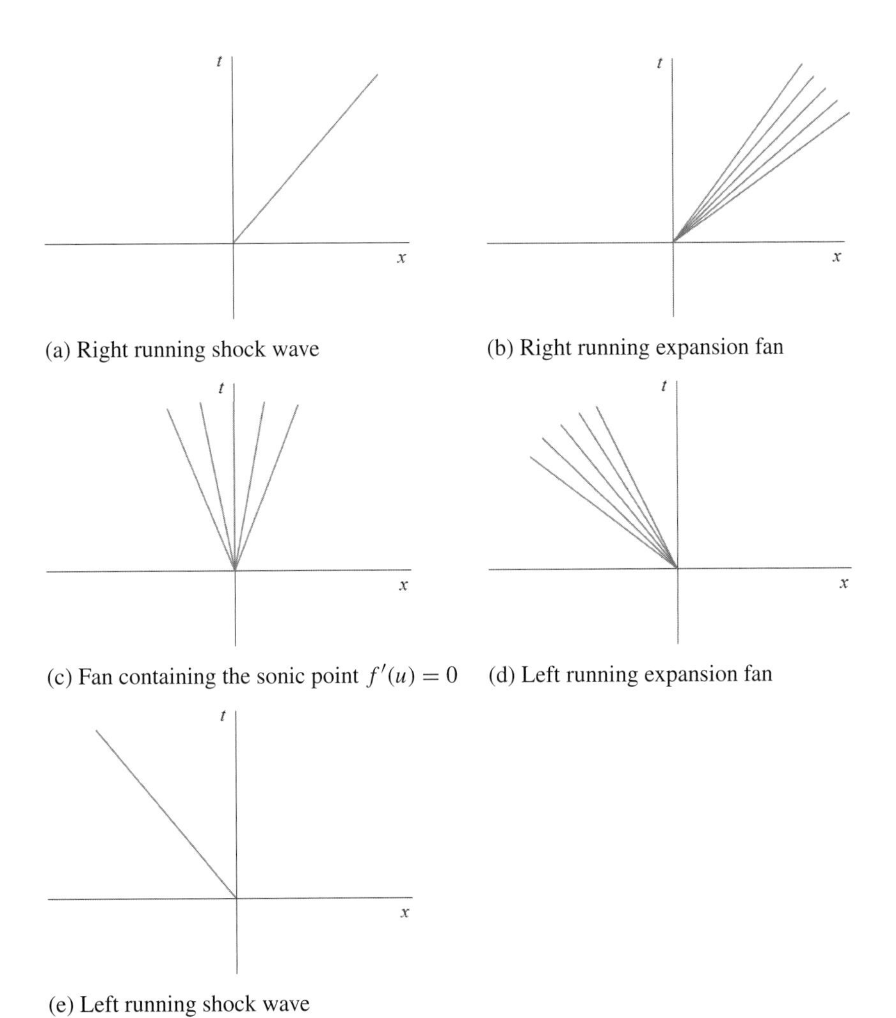

(a) Right running shock wave

(b) Right running expansion fan

(c) Fan containing the sonic point $f'(u) = 0$

(d) Left running expansion fan

(e) Left running shock wave

Figure 5.13 The Riemann problem for a scalar conservation law.

In cases (a) and (b), $h_{j+\frac{1}{2}}$ is constant, equal to the left state f_j, while in cases (d) and (e), $h_{j+\frac{1}{2}}$ is equal to the right state f_{j+1}. In all these cases, the Godunov flux is equal to the upwind flux. It differs only in the case of an expansion fan containing the speed $f'(v) = 0$, where the Godunov flux is equal to zero. As in the case of the Engquist–Osher scheme, this ensures that initial data containing a discontinuous expansion will evolve into an expansion fan. The range of possibilities in the case of gas dynamics is considerably more complicated, as will be discussed in the next chapter.

5.16.5 Monotone Schemes

The local Lax–Friedrichs (LLF), Engquist–Osher (EO) and Godunov schemes are examples of monotone schemes, which may be characterized as follows. Suppose that the discrete scheme for a time step is expressed as

$$v_j^{n+1} = G(v_{j-q}^n, \dots, v_{j+p}^n), \tag{5.24}$$

where in general G is a nonlinear function of the values of the discrete solution at time level n over a stencil of $p + q + 1$ points. The scheme is said to be monotone if

$$\frac{\partial G}{\partial v} \geq 0 \text{ for all arguments,}$$

with the consequence that

$$\frac{\partial v_j^{n+1}}{\partial v_k^n} \geq 0 \text{ for all } k.$$

This requirement is more stringent than the requirement for positive coefficients in the scheme (5.3), which was previously discussed. Suppose that the update (5.24) is consistent with a differential equation with no source term. Then, if the data at time level n is constant, all derivatives are zero, so $\frac{\partial u}{\partial t} = 0$, and the solution must remain unchanged. Hence, G must satisfy the consistency condition

$$G(v, v, \dots, v) = v,$$

leading to

$$v_{j+1}^n = v_j^n$$

if v_j^n is constant.

According to the definition, a monotone scheme satisfies

$$G(u)_j \leq G(v)_j \text{ for all } j,$$

if

$$u_j \leq v_j \text{ for all } j.$$

It also satisfies the maximum principle

$$\min_{k \in S} v_k \leq G(v)_j \leq \max_{k \in S} v_k, \tag{5.25}$$

where S is the stencil for the mesh point j. To prove this, take

$$w_i = \begin{cases} \max\limits_{k \in S} v_k & \text{if } i \in S, \\ v_i & \text{otherwise.} \end{cases}$$

Then, $v_j \leq w_j$ for all j, and hence,

$$G(v)_j \leq G(w)_j \text{ for all } j,$$

but $G(w)_j = v_k$ since w_i = constant in S. Similarly, if we take

$$w_i = \begin{cases} \min\limits_{k \in S} v_k & \text{if } i \in S, \\ v_i & \text{otherwise,} \end{cases}$$

it follows that $v_j \geq w_j$ for all j, and

$$G(v)_j \geq G(w)_j = v_k.$$

A three-point monotone scheme can be constructed as

$$v_j^{n+1} = v_j^n - \frac{\Delta t}{\Delta x}\left(h(v_j^n, v_{j+1}^n) - h(v_{j-1}^n, v_j^n)\right),$$

where we require the numerical flux $h(v_L, v_R)$ to satisfy the conditions

$$\frac{\partial h}{\partial v_L} \geq 0, \quad \frac{\partial h}{\partial v_R} \leq 0.$$

Then,

$$\frac{\partial v_j^{n+1}}{\partial v_{j-1}^n} = \frac{\Delta t}{\Delta x}\frac{\partial h}{\partial v_L}\bigg|_{v_{j-1}^n} \geq 0,$$

$$\frac{\partial v_j^{n+1}}{\partial v_{j+1}^n} = -\frac{\Delta t}{\Delta x}\frac{\partial h}{\partial v_R}\bigg|_{v_{j+1}^n} \geq 0,$$

and

$$\frac{\partial v_j^{n+1}}{\partial v_j^n} = 1 - \frac{\Delta t}{\Delta x}\left(\frac{\partial h}{\partial v_L}\bigg|_{v_j^n} - \frac{\partial h}{\partial v_R}\bigg|_{v_j^n}\right) \geq 0$$

if the time step satisfies the restriction

$$\frac{\Delta t}{\Delta x}\left(\frac{\partial h}{\partial v_L}\bigg|_{v_j^n} - \frac{\partial h}{\partial v_R}\bigg|_{v_j^n}\right) \leq 1.$$

L_∞ stability follows from (5.25). Let $v_{\min}^n$ and $v_{\max}^n$ be the minimum and maximum values of v_j^n over all j. Then for all j,

$$v_{\min}^n \leq \min_{k\in S} v_k^n \leq v_j^{n+1} \leq \max_{k\in S} v_k^n \leq v_{\max}^n$$

and

$$\|v^{n+1}\|_\infty = \max_j \left|v_j^{n+1}\right| \leq \max\left(|v_{\min}^n|, |v_{\max}^n|\right) \leq \|v^n\|_\infty.$$

Similarly, if a monotone scheme has a three-point stencil, it will be LED.

It has been proved by Harten et al. (1976), and also by Crandall and Majda (1980), that monotone schemes converge to entropy-satisfying solutions. However, monotone schemes are at most first order accurate, as was originally proved by Godunov.

It may be verified that the upwind scheme (5.23) is not monotone by considering the case of an expansion $v_R > v_L$ containing the point u^* where $f'(u^*) = 0$, as displayed in Figure 5.14. In the case that $f(v_R) < f(v_L)$,

$$\frac{f(v_R) - f(v_L)}{v_R - v_L} < 0,$$

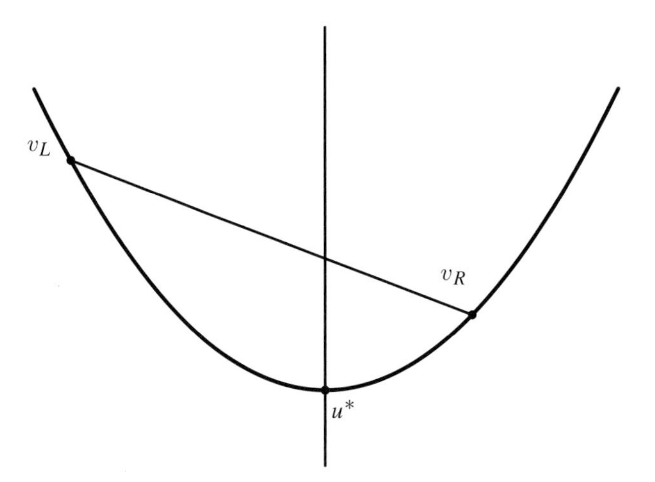

Figure 5.14 Nonmonotonicity of the upwind scheme.

so the upwind flux is

$$h(v_L, v_R) = f(v_R),$$

and since $v_R > u^*$, $\frac{\partial h}{\partial v_R} > 0$, in contradiction to the requirement.

The flux for the local Lax–Friedrichs (LLF) scheme may be written as

$$h(v_L, v_R) = \frac{1}{2}(f_R + f_L) - \frac{1}{2}a(v_R - v_L),$$

where $a = \max f'(u)$ for u in the interval between v_L and v_R. The LLF scheme is monotone if the flux function is strictly convex. In that case, $f''(u) > 0$ for all u, so that $a = \max(|f'(v_R)|, |f'(v_L)|)$. In the case, for example, that $|f'(v_R)| > |f'(v_L)|$ and $v_R > 0$,

$$\frac{\partial h}{\partial v_L} = \frac{1}{2}(f'(v_L) + |f'(v_R)|) \geq 0,$$

$$\frac{\partial h}{\partial v_R} = \frac{1}{2}(f'(v_R) - |f'(v_R)| - v_R f''(v_R)) \leq 0,$$

while the other possibilities can be checked in a similar manner. The Engquist–Osher (EO) flux may be written as

$$h(v_L, v_R) = f^+(v_L) + f^-(v_R)$$

and is evidently monotone by its construction.

5.16.6 E-schemes

The concept of E-schemes was introduced by Osher (1984), who proved that these schemes converge to entropy-satisfying solutions. A scheme is said to be an E-scheme

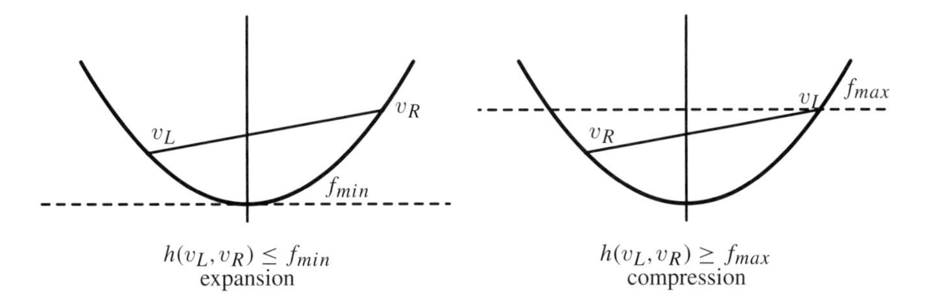

Figure 5.15 E-scheme in regions of expansion and compression.

if its flux satisfies the following conditions. In a region of expansion for which $v_R > v_L$,

$$h(v_L, v_R) \leq f_{\min}$$

in the interval $v_L \leq u \leq v_R$, and in a region of compression for which $v_L > v_R$,

$$h(v_R, v_L) \geq f_{\max}$$

in the interval $v_R \leq u \leq v_L$, where for a convex flux, $f_{\max} = f(v_R)$ or $f(v_L)$. Some possibilities are displayed in Figure 5.15.

We can prove that monotone schemes are E-schemes if the flux function is convex by the following argument. In a region of expansion with $v_R > v_L$, the critical case is when $f(u)$ has a minimum at u^* in the interval between v_L and v_R. For an E-scheme, we require

$$h(v_L, v_R) \leq f_{\min} = f(u^*).$$

For consistency,

$$f(u^*) = h(u^*, u^*).$$

Now, since $\dfrac{\partial h}{\partial v_L} \geq 0$, $\dfrac{\partial h}{\partial v_R} \leq 0$, $v_L < u^* < v_R$,

$$h(v_L, v_R) \leq h(u^*, u^*).$$

In a region of compression for which $v_L > v_R$, we need

$$h(v_L, v_R) \geq f_{\max} = f(s),$$

where $s =$ either v_L or v_R, since $f(u)$ is convex. Again, for consistency, $f(s) = h(s, s)$. Then, if $s = v_L$,

$$h(v_L, v_R) \geq h(s, s) = h(v_L, v_L)$$

because $\frac{\partial h}{\partial v_R} \leq 0$ and $v_L > v_R$, and if $s = v_R$,

$$h(v_L, v_R) \geq h(s, s) = h(v_R, v_R)$$

because $\frac{\partial h}{\partial v_R} \geq 0$ and $v_L > v_R$.

5.17 The Jameson–Schmidt–Turkel Scheme

According to Section 5.9, an LED scheme with fixed coefficients is limited to first order accuracy. We now examine ways of constructing switched higher order schemes that revert to first order accuracy only in the vicinity of extrema. One of the simplest such formulations is the Jameson-Schmidt-Turkel (JST) scheme (Jameson et al. 1981, Jameson 1995a, 2017b).

This scheme blends low and high order diffusion using a switch that eliminates the high order diffusion at extrema. Suppose that the conservation law,

$$\frac{\partial u}{\partial t} + \frac{\partial}{\partial x} f(u) = 0,$$

is approximated by the semi-discrete finite volume scheme

$$\Delta x \frac{dv_j}{dt} + h_{j+\frac{1}{2}} - h_{j-\frac{1}{2}} = 0.$$

In the JST scheme, the numerical flux is

$$h_{j+\frac{1}{2}} = \frac{1}{2}(f_{j+1} + f_j) - d_{j+\frac{1}{2}},$$

where the diffusive flux has the form

$$d_{j+\frac{1}{2}} = \epsilon^{(2)}_{j+\frac{1}{2}} \Delta v_{j+\frac{1}{2}} - \epsilon^{(4)}_{j+\frac{1}{2}} \left(\Delta v_{j+\frac{3}{2}} - 2\Delta v_{j+\frac{1}{2}} + \Delta v_{j-\frac{1}{2}} \right),$$

with

$$\Delta v_{j+\frac{1}{2}} = v_{j+1} - v_j.$$

Let $a_{j+\frac{1}{2}}$ be the numerically estimated wave speed

$$a_{j+\frac{1}{2}} = \begin{cases} \dfrac{f_{j+1} - f_j}{v_{j+1} - v_j}, & v_{j+1} \neq v_j, \\ \dfrac{\partial f}{\partial u}\Big|_{u=v_j}, & v_{j+1} = v_j, \end{cases}$$

as before. Then we have the following.

THEOREM 5.17.1 *The JST scheme is LED if whenever v_j or v_{j+1} is an extremum,*

$$\epsilon^{(2)}_{j+\frac{1}{2}} \geq \frac{1}{2}|a_{j+\frac{1}{2}}|, \quad \epsilon^{(4)}_{j+\frac{1}{2}} = 0. \tag{5.26}$$

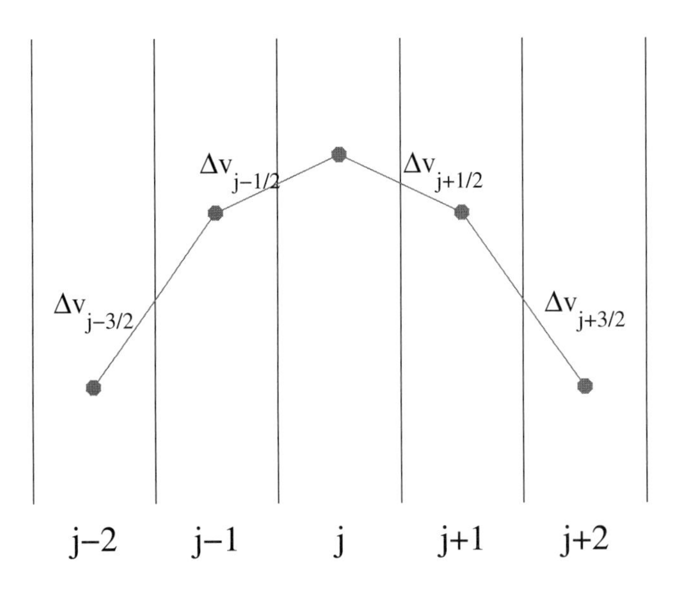

Figure 5.16 JST scheme at an extremum. $\Delta v_{j+\frac{3}{2}}$ and $\Delta v_{j-\frac{1}{2}}$ have opposite signs. Also $\Delta v_{j+\frac{1}{2}}$ and $\Delta v_{j-\frac{3}{2}}$ have opposite signs. Thus, $\epsilon^{(4)}_{j+\frac{1}{2}} = 0$ and $\epsilon^{(4)}_{j-\frac{1}{2}} = 0$.

Proof Suppose v_j is an extremum. Then the second condition ensures that

$$\epsilon^{(4)}_{j+\frac{1}{2}} = 0, \quad \epsilon^{(4)}_{j-\frac{1}{2}} = 0.$$

Hence the scheme reduces to the three-point scheme

$$\Delta x \frac{dv_j}{dt} = \left(\epsilon^{(2)}_{j+\frac{1}{2}} - \frac{1}{2} a_{j+\frac{1}{2}}\right) \Delta v_{j+\frac{1}{2}} - \left(\epsilon^{(2)}_{j-\frac{1}{2}} + \frac{1}{2} a_{j-\frac{1}{2}}\right) \Delta v_{j-\frac{1}{2}},$$

and according to the first condition, the coefficients of both $v_{j+1} - v_j$ and $v_{j-1} - v_j$ are non-negative, satisfying the requirements for an LED scheme. Figure 5.16 illustrates the JST scheme at a local maximum. □

In order to construct coefficients $\epsilon^{(2)}_{j+\frac{1}{2}}$ and $\epsilon^{(4)}_{j+\frac{1}{2}}$ satisfying conditions (5.26), define the function

$$R(u, v) = \left| \frac{u - v}{|u| + |v|} \right|^q,$$

where $q \geq 1$. Then, if u and v have opposite signs,

$$R(u, v) = 1.$$

Now set

$$\epsilon^{(2)}_{j+\frac{1}{2}} = \alpha_{j+\frac{1}{2}} Q_{j+\frac{1}{2}}$$

and

$$\epsilon^{(4)}_{j+\frac{1}{2}} = \beta_{j+\frac{1}{2}}\left(1 - Q_{j+\frac{1}{2}}\right),$$

where

$$Q_{j+\frac{1}{2}} = R\left(\Delta v_{j+\frac{3}{2}}, \Delta v_{j-\frac{1}{2}}\right).$$

Since $\Delta v_{j+\frac{3}{2}}$ and $\Delta v_{j-\frac{1}{2}}$ have opposite signs if either v_j or v_{j+1} is an extremum, the scheme will be LED if

$$\alpha_{j+\frac{1}{2}} \geq \frac{1}{2}\left|a_{j+\frac{1}{2}}\right|.$$

Typically

$$\beta_{j+\frac{1}{2}} = K\left|a_{j+\frac{1}{2}}\right|,$$

where in the case of steady state calculations, K can be tuned to maximize the rate of convergence to a steady state.

5.18 Essentially Local Extremum Diminishing (ELED) Schemes

The JST scheme as presented in Section 5.17 has the disadvantage that it reverts to first order accuracy at smooth extrema. In order to circumvent this loss of accuracy, we can relax slightly the requirements of an LED scheme by introducing the concept of an essentially local extremum diminishing (ELED) scheme, defined as a scheme that becomes LED in the limit as the mesh width $\Delta x \to 0$.

The JST scheme can be converted to an ELED scheme by redefining the switching function as

$$R(u, v) = \left|\frac{u - v}{\max((|u| + |v|), \epsilon\Delta x^r)}\right|^q,$$

where a threshold has been introduced in the denominator. The coefficient ϵ should have the physical dimensions of u divided by a length scale to the power r. It is shown in Jameson (1995a) that the scheme is second order accurate and ELED if

$$q \geq 2, \quad r = \frac{3}{2}.$$

5.19 Symmetric Limited Positive (SLIP) Schemes

We now discuss schemes that obtain second order accuracy by subtracting an anti-diffusive term based on neighboring values to cancel the first order diffusive term but limit the anti diffusion in the vicinity of extrema in order to preserve the LED property. This idea was first introduced in the flux-corrected transport (FCT) scheme proposed by Boris and Book (1973). Their formulation used two stages. The first stage consisted of a first order accurate positive scheme. The second stage added negative diffusion to cancel the first order error, subject to a limit on its magnitude at any location where it would cause the appearance of a new extremum in the solution. In this section we show how the same result can be accomplished with a single stage scheme.

We can subtract the anti-diffusion symmetrically or from the upwind side, leading to schemes that will be classified as symmetric limited positive (SLIP) and upstream limited positive (USLIP) schemes, respectively. A symmetric scheme of this type was proposed by Jameson (1984). A short time later, Helen Yee proposed a very similar scheme (Yee 1985), which has been popularly known as a symmetric TVD scheme.

The formulation here uses the concept of limited averages (Jameson 1993, 1995a). A limited average $L(u,v)$ of u and v is defined as an average with the following properties:

- P1: $L(u,v) = L(v,u)$
- P2: $L(\alpha u, \alpha v) = \alpha L(u,v)$
- P3: $L(u,u) = u$
- P4: $L(u,v) = 0$ if u and v have opposite signs; otherwise $L(u,v)$ has the sign of u and v or the same sign as whichever is nonzero if $u = 0$ or $v = 0$.

The first three properties are satisfied by the arithmetic average. It is the fourth property that distinguishes the limited average.

Suppose now that the conservation law

$$\frac{\partial u}{\partial t} + \frac{\partial}{\partial x} f(u) = 0$$

is approximated by the semi-discrete scheme

$$\Delta x \frac{dv_j}{dt} + h_{j+\frac{1}{2}} - h_{j-\frac{1}{2}} = 0$$

as before. Define the numerical flux as

$$h_{j+\frac{1}{2}} = \frac{1}{2}(f_{j+1} + f_j) - d_{j+\frac{1}{2}},$$

where $d_{j+\frac{1}{2}}$ is the numerical diffusive flux. Let

$$\Delta v_{j+\frac{1}{2}} = v_{j+1} - v_j.$$

By subtracting $\frac{1}{2}(\Delta v_{j+\frac{3}{2}} + \Delta v_{j-\frac{1}{2}})$ from $\Delta v_{j+\frac{1}{2}}$, we could produce a diffusive term

$$d_{j+\frac{1}{2}} = \alpha_{j+\frac{1}{2}} \left(\Delta v_{j+\frac{1}{2}} - \frac{1}{2} \left(\Delta v_{j+\frac{3}{2}} + \Delta v_{j-\frac{1}{2}} \right) \right)$$

that approximates $\frac{1}{2} \alpha \Delta x^3 \frac{\partial^3 u}{\partial x^3}$. However this would lead to a scheme that does not satisfy the LED conditions (5.11) in Section 5.5.

In order to circumvent this, we replace the arithmetic average of $\Delta v_{j+\frac{3}{2}}$ and $\Delta v_{j-\frac{1}{2}}$ by their limited average and define

$$d_{j+\frac{1}{2}} = \alpha_{j+\frac{1}{2}} \left(\Delta v_{j+\frac{1}{2}} - L\left(\Delta v_{j+\frac{3}{2}}, \Delta v_{j-\frac{1}{2}} \right) \right).$$

Hence $\alpha_{j+\frac{1}{2}} L(\Delta v_{j+\frac{3}{2}}, \Delta v_{j-\frac{1}{2}})$ is a limited anti-diffusive term, and it will be verified below that the scheme is LED if $\alpha_{j+\frac{1}{2}} \geq \frac{1}{2}|a_{j+\frac{1}{2}}|$.

We first give some examples of limited averages. Define

$$S(u,v) = \frac{1}{2}\left(\text{sgn}\,(u) + \text{sgn}\,(v)\right)$$

so that

$$S(u,v) = \begin{cases} 1 & \text{when } u > 0 \text{ and } v > 0 \\ 0 & \text{when } u \text{ and } v \text{ have opposite sign} \\ -1 & \text{when } u < 0 \text{ and } v < 0. \end{cases}$$

Some well known limited averages are

- Min mod: $L(u,v) = S(u,v)\min(|u|,|v|)$
- Van Leer: $L(u,v) = S(u,v) 2\frac{|u||v|}{|u|+|v|}$
- Superbee: $L(u,v) = S(u,v)\max\{\min(2|u|,|v|), \min(|u|,2|v|)\}$.

Limited averages can be characterized in the following manner. Define

$$\phi(r) = L(1,r) = L(r,1)$$

so that by P2, setting $\alpha = \frac{1}{u}$,

$$L\left(1, \frac{v}{u}\right) = \frac{1}{u} L(u,v).$$

Hence,

$$L(u,v) = \phi\left(\frac{v}{u}\right) u,$$

and similarly,

$$L(u,v) = \phi\left(\frac{u}{v}\right) v.$$

It follows on setting $u = 1, v = r$ that

$$\phi(r) = r\phi\left(\frac{1}{r}\right).$$

Also, by P4,

$$\phi(r) = 0, \quad r < 0,$$

$$\phi(r) \geq 0, \quad r \geq 0.$$

Using these properties we can now prove that the SLIP scheme is LED if

$$\alpha_{j+\frac{1}{2}} \geq \frac{1}{2}\left|a_{j+\frac{1}{2}}\right|$$

for all j, as follows. Define

$$r^{+} = \frac{\Delta v_{j+\frac{3}{2}}}{\Delta v_{j-\frac{1}{2}}}, \quad r^{-} = \frac{\Delta v_{j-\frac{3}{2}}}{\Delta v_{j+\frac{1}{2}}}.$$

Then,

$$L\left(\Delta v_{j+\frac{3}{2}}, \Delta v_{j-\frac{1}{2}}\right) = \phi(r^{+})\Delta v_{j-\frac{1}{2}}$$

and

$$L\left(\Delta v_{j+\frac{1}{2}}, \Delta v_{j-\frac{3}{2}}\right) = \phi(r^{-})\Delta v_{j+\frac{1}{2}}.$$

Hence, the semi-discrete scheme is reduced to

$$\begin{aligned}
\Delta x \frac{dv_j}{dt} &= -\frac{1}{2}a_{j+\frac{1}{2}}\Delta v_{j+\frac{1}{2}} - \frac{1}{2}a_{j-\frac{1}{2}}\Delta v_{j-\frac{1}{2}} \\
&\quad + \alpha_{j+\frac{1}{2}}\left(\Delta v_{j+\frac{1}{2}} - \phi(r^{+})\Delta v_{j-\frac{1}{2}}\right) - \alpha_{j-\frac{1}{2}}\left(\Delta v_{j-\frac{1}{2}} - \phi(r^{-})\Delta v_{j+\frac{1}{2}}\right) \\
&= \left\{\alpha_{j+\frac{1}{2}} - \frac{1}{2}a_{j+\frac{1}{2}} + \alpha_{j-\frac{1}{2}}\phi(r^{-})\right\}\Delta v_{j+\frac{1}{2}} \\
&\quad - \left\{\alpha_{j-\frac{1}{2}} + \frac{1}{2}a_{j-\frac{1}{2}} + \alpha_{j+\frac{1}{2}}\phi(r^{+})\right\}\Delta v_{j-\frac{1}{2}}.
\end{aligned}$$

Since $\phi(r^{+}) \geq 0$ and $\phi(r^{-}) \geq 0$, the coefficient of $\Delta v_{j+\frac{1}{2}}$ is non-negative and the coefficient of $\Delta v_{j-\frac{1}{2}}$ is non-positive under the stated condition. The limiters enable the SLIP scheme to be effectively represented as a three-point scheme.

Figures 5.17 and 5.18 illustrate the behavior of the SLIP scheme for an odd-even mode and a shock wave. For an odd-even mode, $\Delta v_{j+\frac{3}{2}}$ and $\Delta v_{j-\frac{1}{2}}$ have the same sign,

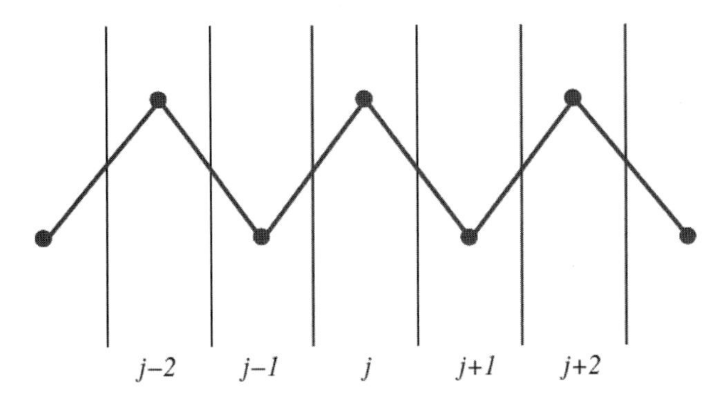

Figure 5.17 SLIP scheme for odd-even mode. $\Delta v_{j-\frac{1}{2}}$ and $\Delta v_{j+\frac{3}{2}}$ have the same sign leaving diffusive terms proportional to fourth difference in place.

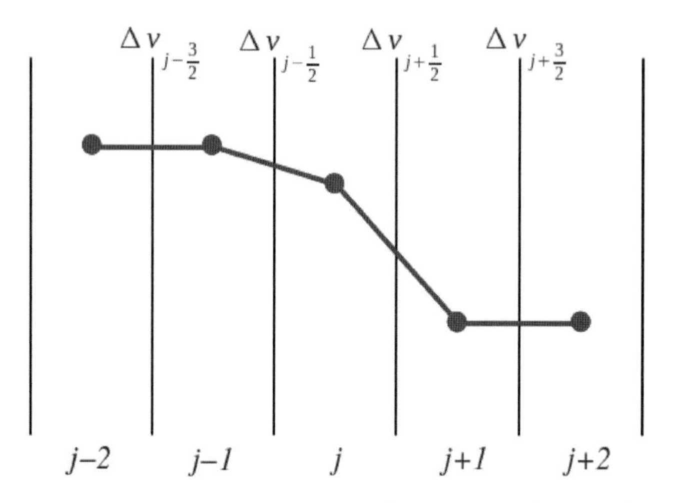

Figure 5.18 SLIP scheme at a shock. $L\left(\Delta v_{j+\frac{1}{2}}, \Delta v_{j-\frac{3}{2}}\right) = 0$, $L\left(\Delta v_{j+\frac{3}{2}}, \Delta v_{j-\frac{1}{2}}\right) = 0$, yielding a three-point scheme at cell j.

opposite to that of $v_{j+\frac{1}{2}}$, and $v_{j+\frac{1}{2}}$ is attenuated as it would be with simple diffusion using fourth differences. With $\alpha_{j+\frac{1}{2}} = \frac{1}{2}|a_{j+\frac{1}{2}}|$, the scheme allows a stationary shock with one interior point because

$$L\left(\Delta v_{j+\frac{3}{2}}, \Delta v_{j-\frac{1}{2}}\right) = 0, \quad L\left(\Delta v_{j+\frac{1}{2}}, \Delta v_{j-\frac{3}{2}}\right) = 0,$$

with the consequence that

$$h_{j+\frac{1}{2}} = f_{j+1} = f_R\,, \quad h_{j-\frac{1}{2}} = f_{j-\frac{1}{2}} = f_L.$$

A general class of limiters satisfying conditions P1–P4 can be constructed as the arithmetic average multiplied by a switch:

$$L(u, v) = \frac{1}{2}D(u, v)(u + v),$$

where $0 \leq D(u, v) \leq 1$ and $D(u, v) = 0$ if u and v have opposite signs This is realized by the formula

$$D(u, v) = 1 - \left|\frac{u - v}{u + v}\right|^q,$$

where q is a positive integer. This definition contains some of the previously defined limiters as follows:

$$q = 1 \text{ gives minmod}$$

$$q = 2 \text{ gives van Leer limiters since } \frac{1}{2}\left(1 - \left|\frac{u - v}{u + v}\right|^2\right)(u + v) = \frac{2uv}{u + v}.$$

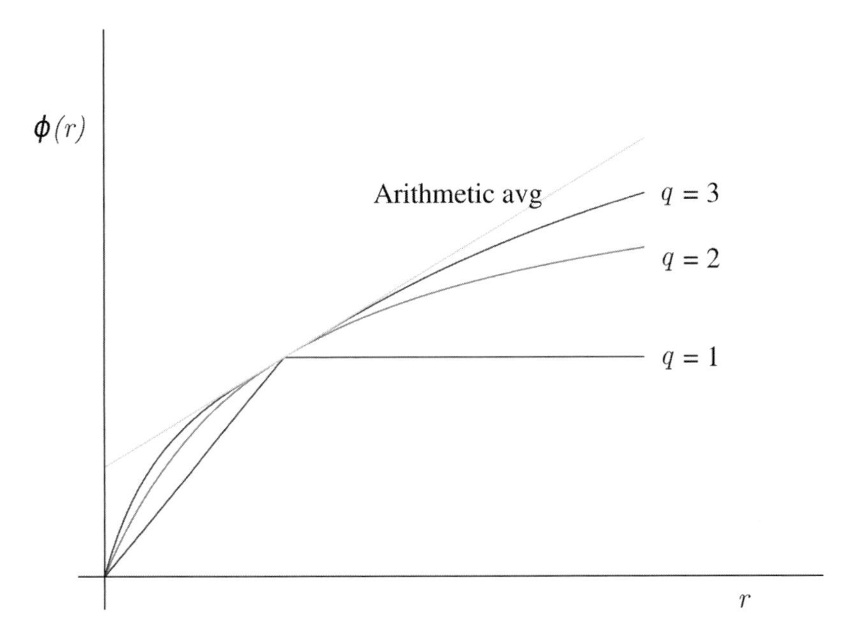

Figure 5.19 Switch ϕ as a function of r and q.

As $q \to \infty$, $L(u, v)$ approaches a limit set by the arithmetic mean if u and v have the same sign and zero if they have opposite signs. The corresponding switching function

$$\phi(r) = L(1, r) = \frac{1}{2}(1 + r)\left(1 - \left|\frac{1 - r}{1 + r}\right|^q\right) \tag{5.27}$$

is illustrated in Figure 5.19. Note that $\phi(r) \to q$ as $r \to \infty$.

With this class of limiters, the SLIP scheme actually recovers a variant of the JST scheme since

$$D(u, v) = 1 - R(u, v),$$

where $R(u, v)$ is the switch used in the JST scheme. Set

$$Q_{j+\frac{1}{2}} = R\left(\Delta v_{j+\frac{3}{2}}, \Delta v_{j-\frac{1}{2}}\right).$$

Then the SLIP scheme can be written as

$$\begin{aligned} d_{j+\frac{1}{2}} &= \alpha_{j+\frac{1}{2}} \left\{\Delta v_{j+\frac{1}{2}} - \frac{1}{2}\left(1 - Q_{j+\frac{1}{2}}\right)\left(\Delta v_{j+\frac{3}{2}} + \Delta v_{j-\frac{1}{2}}\right)\right\} \\ &= \alpha_{j+\frac{1}{2}} Q_{j+\frac{1}{2}} \Delta v_{j+\frac{1}{2}} - \frac{1}{2}\alpha_{j+\frac{1}{2}}\left(1 - Q_{j+\frac{1}{2}}\right)\left(\Delta v_{j+\frac{3}{2}} - 2\Delta v_{j+\frac{1}{2}} + \Delta v_{j-\frac{1}{2}}\right). \end{aligned}$$

This is the JST scheme with $K = \frac{1}{2}$.

5.20 Upstream Limited Positive (USLIP) Schemes

By adding the anti-diffusive correction purely from the upstream side, one may derive a family of upstream limited positive (USLIP) schemes. Corresponding to the SLIP scheme, a USLIP scheme is obtained by setting

$$d_{j+\frac{1}{2}} = \alpha_{j+\frac{1}{2}} \left\{ \Delta v_{j+\frac{1}{2}} - L\Big(\Delta v_{j+\frac{1}{2}}, \Delta v_{j-\frac{1}{2}}\Big) \right\}$$

if $a_{j+\frac{1}{2}} > 0$ and

$$d_{j+\frac{1}{2}} = \alpha_{j+\frac{1}{2}} \left\{ \Delta v_{j+\frac{1}{2}} - L\Big(\Delta v_{j+\frac{1}{2}}, \Delta v_{j+\frac{3}{2}}\Big) \right\}$$

if $a_{j+\frac{1}{2}} < 0$. Now an analysis similar to that in Section 18 reveals that if $a_{j+\frac{1}{2}} > 0$ and $a_{j-\frac{1}{2}} > 0$ while $\alpha_{j+\frac{1}{2}} = \frac{1}{2} a_{j+\frac{1}{2}}$, the scheme reduces to

$$\Delta x \frac{dv_j}{dt} = -\left\{ \frac{1}{2}\phi(r^+) a_{j+\frac{1}{2}} + \left(1 - \frac{1}{2}\phi(r^-)\right) a_{j-\frac{1}{2}} \right\} \Delta v_{j-\frac{1}{2}},$$

where

$$r^+ = \frac{\Delta v_{j+\frac{1}{2}}}{\Delta v_{j-\frac{1}{2}}}, \quad r^- = \frac{\Delta v_{j-\frac{3}{2}}}{\Delta v_{j-\frac{1}{2}}}.$$

Thus the coefficient of $v_{j-1} - v_j$ is nonnegative, and the scheme is LED if $\phi(r)$ satisfies the additional constraint $\phi(r) \leq 2$.

5.21 Reconstruction

An alternative approach to constructing a high resolution scheme is reconstruction. This was first introduced by van Leer in the Monotone Upstream-centered Schemes for Conservation Laws (MUSCL) (van Leer 1974). It can be described as follows. Suppose the interface flux $h_{j+\frac{1}{2}}$ of a first order LED scheme is constructed as a function of v_j and v_{j+1}. Instead we evaluate $h_{j+\frac{1}{2}}$ from values v_L and v_R, which represent estimates of the solution at the left and right side of the interface that take account of the gradient of the solution in cells j and $j+1$. However, in order to prevent the formation of a new extremum or the growth of an existing extremum, these values are limited based on a comparison with the gradients in neighboring cells.

In order to illustrate the process, consider the case of linear advection

$$\frac{\partial u}{\partial t} + a \frac{\partial u}{\partial x} = 0$$

for a right running wave, $a > 0$. Suppose this is approximated by the semi-discrete scheme,

$$\Delta x \frac{dv_j}{dt} + h_{j+\frac{1}{2}} - h_{j-\frac{1}{2}} = 0.$$

Using an upwind flux in a first order scheme,

$$h_{j+\frac{1}{2}} = av_j, \quad h_{j-\frac{1}{2}} = av_{j-1}$$

and

$$\Delta x \frac{dv_j}{dt} = -a(v_j - v_{j-1}),$$

which satisfies the conditions for an LED scheme. Suppose that we have an estimate v'_j for the slope in cell j. Then in order to improve the accuracy, we can set

$$h_{j+\frac{1}{2}} = av_L,$$

where

$$v_L = v_j + \frac{1}{2}\Delta x v'_j.$$

Note that since this is a finite volume scheme, v_j represents the average value of v in cell j. But with a linear variation of v in the cell, this is also the point value at cell center. With higher order reconstruction using polynomials of degree > 1 to represent the solution, it becomes necessary to distinguish between the point value at the cell center and the average value in the cell, since these are no longer the same. Consistent with the use of an upwind flux, we can estimate v'_j from the upwind side as

$$v'_j = \frac{v_j - v_{j-1}}{\Delta x} = \frac{\Delta v_{j-\frac{1}{2}}}{\Delta x},$$

so that

$$h_{j+\frac{1}{2}} = a\left(v_j + \frac{1}{2}\Delta v_{j-\frac{1}{2}}\right)$$

$$h_{j-\frac{1}{2}} = a\left(v_{j-1} + \frac{1}{2}\Delta v_{j-\frac{3}{2}}\right)$$

and

$$\Delta x \frac{dv_j}{dt} = -\frac{1}{2}a(3v_j - 4v_{j-1} + v_{j-2}) = -2a(v_j - v_{j-1}) + \frac{1}{2}a(v_j - v_{j-2}).$$

This is a second order accurate estimate of $\frac{\partial v}{\partial x}$, but it violates the conditions for an LED scheme because the coefficient of $v_{j-2} - v_j$ is negative. In order to remedy this, we can estimate the slope as

$$v'_j = \frac{1}{\Delta x} L\left(\Delta v_{j+\frac{1}{2}}, \Delta v_{j-\frac{1}{2}}\right),$$

where L is a limited average satisfying properties P1–P4 in Section 5.19. Now define

$$r^+ = \frac{\Delta v_{j+\frac{1}{2}}}{\Delta v_{j-\frac{1}{2}}}, r^- = \frac{\Delta v_{j-\frac{3}{2}}}{\Delta v_{j-\frac{1}{2}}}.$$

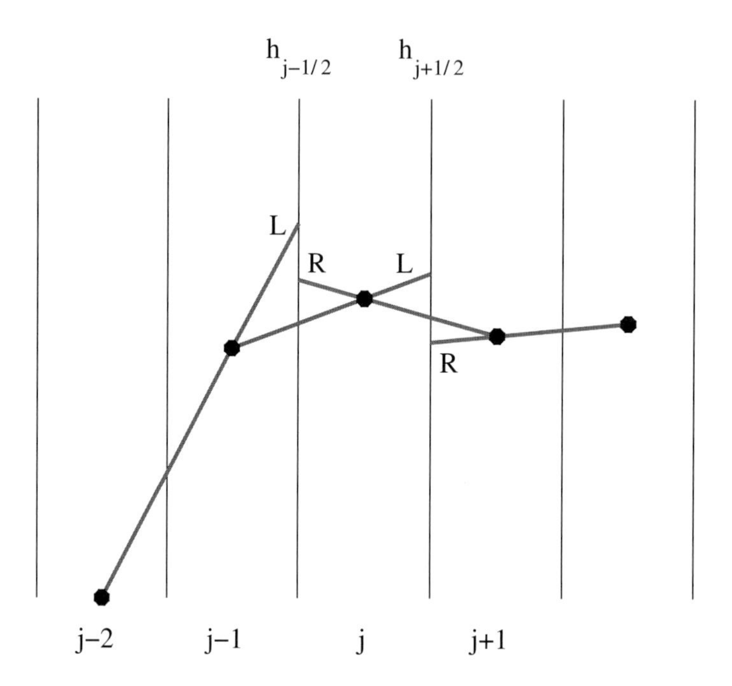

Figure 5.20 Reconstruction for linear advection.

Then, using the notation of Section 5.19,

$$h_{j+\frac{1}{2}} = a\left(v_j + \frac{1}{2}\phi(r^+)(v_j - v_{j-1})\right)$$

$$h_{j-\frac{1}{2}} = a\left(v_{j-1} + \frac{1}{2}\phi(r^-)(v_j - v_{j-1})\right)$$

and

$$\Delta x \frac{dv_j}{dt} = -a\left(1 + \frac{1}{2}\phi(r^+) - \frac{1}{2}\phi(r^-)\right)(v_j - v_{j-1}).$$

Now to ensure that the coefficient of $v_{j-1} - v_j$ is non-negative, we must also require that

$$\phi(r^+) - \phi(r^-) \geq -2. \tag{5.28}$$

If $\phi(r^+) \geq 0$ with $\phi(r) = 0$ when $r \leq 0$, this is satisfied if $\phi(r)$ satisfies the additional requirement that

$$\phi(r) \leq 2, \quad r > 0.$$

It was pointed out by Spekreijse (1987) that actually (5.28) is satisfied if for all r

$$\phi(r) \geq -\alpha, \quad \phi(r) \leq 2 - \alpha,$$

where $0 \leq \alpha \leq 2$. Thus the class of admissible limiters can be expanded to include limiters that are negative for some range of r.

For second order accuracy, v'_j should approach $\frac{\Delta v_{j-\frac{1}{2}}}{\Delta x}$ as $\Delta v_{j+\frac{1}{2}} \to \Delta v_{j-\frac{1}{2}}$, thus recovering the second order accurate upwind scheme. This implies

$$\phi(1) = 1,$$

consistent with property P3. These constraints are satisfied by the MinMod, van Leer, and Superbee limiters, with the latter reaching the limits $\phi(r) \leq 2r$ in the interval $0 < r \leq \frac{1}{2}$, and $\phi(r) \leq 2$ in the interval $2 \leq r \leq \infty$. Notice also that when v_j is a maximum, as illustrated in Figure 5.20, it is $v_{L,j-\frac{1}{2}}$ that becomes too large and causes $\frac{dv_j}{dt}$ to become positive in the absence of limiters.

5.22 Reconstruction for a Nonlinear Conservation Law

We consider now the case of the nonlinear conservation law

$$\frac{\partial u}{\partial t} + \frac{\partial}{\partial x} f(u) = 0,$$

approximated by the semi-discrete scheme

$$\Delta x \frac{dv_j}{dt} + h_{j+\frac{1}{2}} - h_{j-\frac{1}{2}} = 0. \tag{5.29}$$

The solution may now contain waves traveling in either direction. Using the Engquist–Osher splitting, we now construct the numerical flux as

$$h_{j+\frac{1}{2}} = f^+\left(v_{L,j+\frac{1}{2}}\right) + f^-\left(v_{R,j+\frac{1}{2}}\right), \tag{5.30}$$

where

$$f(u) = f^+(u) + f^-(u) \tag{5.31}$$

and

$$\frac{\partial f^+}{\partial u} = a^+(u) \geq 0, \quad \frac{\partial f^-}{\partial u} = a^-(u) \leq 0, \tag{5.32}$$

while

$$v_{L,j+\frac{1}{2}} = v_j + \frac{1}{2} L\left(\Delta v_{j+\frac{1}{2}}, \Delta v_{j-\frac{1}{2}}\right) \tag{5.33}$$

and

$$v_{R,j+\frac{1}{2}} = v_{j+1} - \frac{1}{2} L\left(\Delta v_{j+\frac{3}{2}}, \Delta v_{j+\frac{1}{2}}\right). \tag{5.34}$$

Thus,

$$\Delta x \frac{dv_j}{dt} = -\left(f^-\left(v_{R,j+\frac{1}{2}}\right) - f^-\left(v_{R,j-\frac{1}{2}}\right)\right) \\ -\left(f^+\left(v_{L,j+\frac{1}{2}}\right) - f^+\left(v_{L,j-\frac{1}{2}}\right)\right). \tag{5.35}$$

Now, by the mean value theorem,

$$\frac{f^-(v_{R,j+\frac{1}{2}}) - f^-(v_{R,j-\frac{1}{2}})}{v_{R,j+\frac{1}{2}} - v_{R,j-\frac{1}{2}}} = a^-(v_R^*) \leq 0$$

and

$$\frac{f^+(v_{L,j+\frac{1}{2}}) - f^+(v_{L,j-\frac{1}{2}})}{v_{L,j+\frac{1}{2}} - v_{L,j-\frac{1}{2}}} = a^+(v_L^*) \geq 0,$$

where v_R lies in the range between $v_{R,j+\frac{1}{2}}$ and $v_{R,j-\frac{1}{2}}$ and v_L in the range between $v_{L,j+\frac{1}{2}}$ and $v_{L,j-\frac{1}{2}}$. Denote the slope ratio as

$$r_j = \frac{\Delta v_{j+\frac{1}{2}}}{\Delta v_{j-\frac{1}{2}}}.$$

Then,

$$v_{R,j+\frac{1}{2}} - v_{R,j-\frac{1}{2}} = v_{j+1} - \frac{1}{2}L\left(\Delta v_{j+\frac{3}{2}}, \Delta v_{j+\frac{1}{2}}\right) - v_j + \frac{1}{2}L\left(\Delta v_{j+\frac{1}{2}}, \Delta v_{j-\frac{1}{2}}\right) \\ = \left(1 - \frac{1}{2}\phi(r_{j+1}) + \frac{1}{2}\phi\left(\frac{1}{r_j}\right)\right)\Delta v_{j+\frac{1}{2}}$$

and

$$v_{L,j+\frac{1}{2}} - v_{L,j-\frac{1}{2}} = v_j + \frac{1}{2}L\left(\Delta v_{j+\frac{1}{2}}, \Delta v_{j-\frac{1}{2}}\right) - v_{j-1} - \frac{1}{2}L\left(\Delta v_{j-\frac{1}{2}}, \Delta v_{j-\frac{3}{2}}\right) \\ = \left(1 + \frac{1}{2}\phi(r_j) - \frac{1}{2}\phi\left(\frac{1}{r_{j-1}}\right)\right)\Delta v_{j-\frac{1}{2}}.$$

Thus, (5.35) can be rearranged as

$$\Delta x \frac{dv_j}{dt} = -a^-(v_R^*)\left(1 - \frac{1}{2}\phi(r_{j+1}) + \frac{1}{2}\phi\left(\frac{1}{r_j}\right)\right)\Delta v_{j+\frac{1}{2}} \\ -a^+(v_L^*)\left(1 + \frac{1}{2}\phi(r_j) - \frac{1}{2}\phi\left(\frac{1}{r_{j-1}}\right)\right)\Delta v_{j-\frac{1}{2}}.$$

The coefficients of $(v_{j+1} - v_j)$ and $(v_{j-1} - v_j)$ are both non-negative, if for all r and s,

$$\phi(r) - \phi(s) \geq -2,$$

and accordingly the scheme is LED if the limiter satisfies

$$\phi(r) \geq -\alpha, \quad \phi(r) \leq 2 - \alpha,$$

where $0 \le \alpha \le 2$, as in the linear case. If we use a limiter satisfying

$$\phi(r) = 0, \quad r \le 0$$
$$0 \le \phi(r) \le 2, \quad r > 0,$$

then if v_j is a maximum or a minimum, $\Delta v_{j+\frac{1}{2}}$ and $\Delta v_{j-\frac{1}{2}}$ have opposite signs, so that

$$L\left(\Delta v_{j+\frac{1}{2}}, \Delta v_{j-\frac{1}{2}}\right) = 0$$

and

$$v_{L,j+\frac{1}{2}} = v_j.$$

If both $\Delta v_{j+\frac{1}{2}}$ and $\Delta v_{j-\frac{1}{2}}$ are positive,

$$0 \le L\left(\Delta v_{j+\frac{1}{2}}, \Delta v_{j-\frac{1}{2}}\right) \le 2\Delta v_{j+\frac{1}{2}}$$

and

$$v_j \le v_{L,j+\frac{1}{2}} \le v_{j+1},$$

while if both $\Delta v_{j-\frac{1}{2}}$ and $\Delta v_{j+\frac{1}{2}}$ are negative,

$$2\Delta v_{j+\frac{1}{2}} \le L\left(\Delta v_{j+\frac{1}{2}}, \Delta v_{j-\frac{1}{2}}\right) \le 0$$

and

$$v_{j+1} \le v_{L,j+\frac{1}{2}} \le v_j.$$

Thus, with this class of limiter, both $v_{L,j+\frac{1}{2}}$ and $v_{R,j+\frac{1}{2}}$ lie in the range between v_j and v_{j+1}:

$$\min(v_j, v_{j+1}) \le v_{L,j+\frac{1}{2}} \le \max(v_j, v_{j+1})$$

and

$$\min(v_j, v_{j+1}) \le v_{R,j-\frac{1}{2}} \le \max(v_j, v_{j+1}).$$

Consider also the fully discrete scheme

$$v_j^{n+1} = v_j^n - \frac{\Delta t}{\Delta x}\left(h_{j+\frac{1}{2}} - h_{j-\frac{1}{2}}\right), \tag{5.36}$$

where

$$h_{j+\frac{1}{2}} = f^-\left(v^n_{R,j+\frac{1}{2}}\right) + f^+\left(v^n_{L,j+\frac{1}{2}}\right).$$

Following the same argument,

$$v_j^{n+1} = A^-_{j+\frac{1}{2}}\frac{\Delta t}{\Delta x}v^n_{j+1} + \left(1 - \frac{\Delta t}{\Delta x}\left(A^-_{j+\frac{1}{2}} + A^+_{j-\frac{1}{2}}\right)\right)v_j^n + A^+_{j-\frac{1}{2}}\frac{\Delta t}{\Delta x}v^n_{j-1}, \tag{5.37}$$

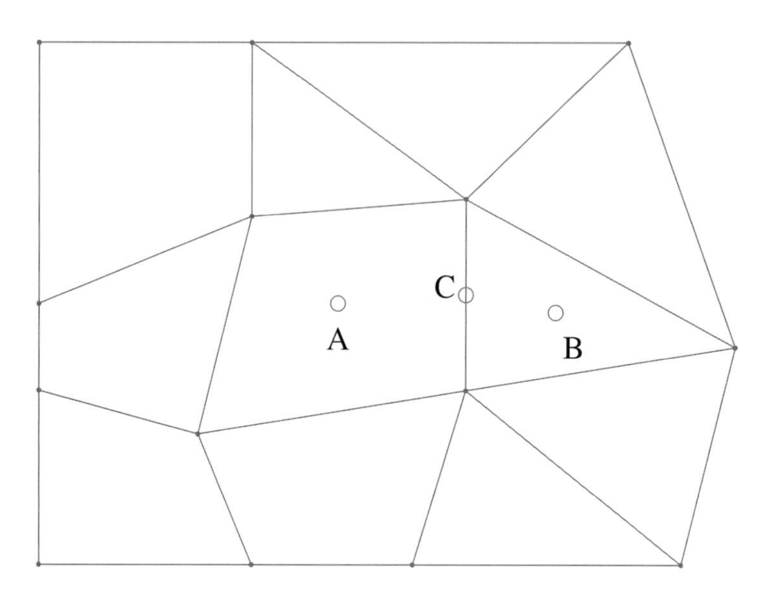

Figure 5.21 Reconstruction for an unstructured mesh.

where

$$A^{-}_{j+\frac{1}{2}} = -a^{-}(v^{*}_{R})\left(1 - \frac{1}{2}\phi(r_{j+1}) + \frac{1}{2}\phi\left(\frac{1}{r_j}\right)\right)$$

and

$$A^{+}_{j-\frac{1}{2}} = a^{+}(v^{*}_{L})\left(1 + \frac{1}{2}\phi(r_j) - \frac{1}{2}\phi\left(\frac{1}{r_{j-1}}\right)\right).$$

Under condition (5.36),

$$A^{-}_{j+\frac{1}{2}} \geq 0, \quad A^{+}_{j-\frac{1}{2}} \geq 0.$$

Also, they are both bounded, so we can choose Δt small enough that the coefficient of v^n_j is non-negative, and the scheme is LED.

The reconstruction approach proves to be particularly useful in the construction of high resolution schemes on unstructured meshes for multi-dimensional problems. Then, one can use an estimate of the gradient of the solution in each cell to recover left and right values at the center of the face separating any two cells, as illustrated in Figure 5.21. For example, at the edge center C,

$$v_L = v_A + \nabla v_A \cdot (\overrightarrow{x_C} - \overrightarrow{x_A})$$

$$v_R = v_B + \nabla v_B \cdot (\overrightarrow{x_C} - \overrightarrow{x_B}),$$

where ∇v_A and ∇v_B are appropriately limited estimates of the gradient in cells A and B.

5.23 SLIP Reconstruction

An alternative reconstruction formula can be derived from the SLIP scheme by setting

$$v_{L,j+\frac{1}{2}} = v_j + \frac{1}{2}L\left(\Delta v_{j+\frac{3}{2}}, \Delta v_{j-\frac{1}{2}}\right),$$
$$v_{R,j+\frac{1}{2}} = v_{j+1} - \frac{1}{2}L\left(\Delta v_{j+\frac{3}{2}}, \Delta v_{j-\frac{1}{2}}\right)$$

so that the difference

$$v_{R,j+\frac{1}{2}} - v_{L,j+\frac{1}{2}} = \Delta v_{j+\frac{1}{2}} - L\left(\Delta v_{j+\frac{3}{2}}, \Delta v_{j-\frac{1}{2}}\right)$$

is the same as that used in the construction of the artificial diffusion in the SLIP scheme, while the same slope estimate centered at $j + \frac{1}{2}$ is used to calculate both $v_{L,j+\frac{1}{2}}$ and $v_{R,j+\frac{1}{2}}$.

Defining the slope ratios

$$r^+ = \frac{\Delta v_{j+\frac{3}{2}}}{\Delta v_{j+\frac{1}{2}}}, \quad r^- = \frac{\Delta v_{j-\frac{3}{2}}}{\Delta v_{j-\frac{1}{2}}},$$

we now find that

$$\begin{aligned} v_{R,j+\frac{1}{2}} - v_{R,j-\frac{1}{2}} &= v_{j+1} - \frac{1}{2}L\left(\Delta v_{j+\frac{3}{2}}, \Delta v_{j-\frac{1}{2}}\right) - v_j + \frac{1}{2}L\left(\Delta v_{j+\frac{1}{2}}, \Delta v_{j-\frac{3}{2}}\right) \\ &= \Delta v_{j+\frac{1}{2}} - \frac{1}{2}\phi(r^+)\Delta v_{j-\frac{1}{2}} + \frac{1}{2}\phi(r^-)\Delta v_{j+\frac{1}{2}} \end{aligned}$$

and

$$\begin{aligned} v_{L,j+\frac{1}{2}} - v_{L,j-\frac{1}{2}} &= v_j + \frac{1}{2}L\left(\Delta v_{j+\frac{3}{2}}, \Delta v_{j-\frac{1}{2}}\right) - v_{j-1} - \frac{1}{2}L\left(\Delta v_{j+\frac{1}{2}}, \Delta v_{j-\frac{3}{2}}\right) \\ &= \Delta v_{j-\frac{1}{2}} + \frac{1}{2}\phi(r^+)\Delta v_{j-\frac{1}{2}} - \frac{1}{2}\phi(r^-)\Delta v_{j+\frac{1}{2}}. \end{aligned}$$

Thus, following the same argument that was used in the previous section, the semi-discrete scheme defined by (5.29)–(5.34) can be rearranged as

$$\begin{aligned} \Delta x \frac{dv_j}{dt} &= -a^-(v_R^*)\left(1 + \frac{1}{2}\phi(r^-) + \frac{1}{2}a^+(v_L^*)\phi(r^-)\right)\Delta v_{j+\frac{1}{2}} \\ &\quad - a^+(v_L^*)\left(1 + \frac{1}{2}\phi(r^+) - \frac{1}{2}a^-(v_L^*)\phi(r^+)\right)\Delta v_{j-\frac{1}{2}}. \end{aligned}$$

The coefficients of $v_{j+1} - v_j$ and $v_{j-1} - v_j$ are both nonnegative if for all r,

$$\phi(r) \geq 0,$$

and correspondingly the semi-discrete scheme is LED.

The fully discrete scheme described by (5.36) and (5.22) can be expressed in the form (5.37), where now

$$A^-_{j+\frac{1}{2}} = -a^-(v^*_R)\left(1 + \frac{1}{2}\phi(r^-)\right) + \frac{1}{2}a^+(v^*_L)\phi(r^-)$$

and

$$A^+_{j-\frac{1}{2}} = a^+(v^*_L)\left(1 + \frac{1}{2}\phi(r^+)\right) - \frac{1}{2}a^-(v^*_R)\phi(r^+).$$

Again, provided that $\phi(r)$ is nonnegative,

$$A^-_{j+\frac{1}{2}} \geq 0, \quad A^+_{j-\frac{1}{2}} \geq 0,$$

and if $\phi(r)$ is also bounded

$$0 \leq \phi(r) \leq M,$$

we can choose Δt small enough that the coefficient of v^n_j is positive, and the scheme is LED.

The SLIP reconstruction requires a less restrictive bound on the maximum value of $\phi(r)$ than the upwind reconstruction defined by (5.33) and (5.34). This has the advantage that it enables the use of a limiter that is closer to the arithmetic average. In particular one may use the switching function defined in Section 5.19 by (5.27).

6 Shock Capturing Schemes II: Systems of Equations and Gas Dynamics

6.1 Systems of Conservation Laws

In this chapter, we address the numerical solution of systems of conservation laws, considering first the case of a one-dimensional system

$$\frac{\partial \boldsymbol{w}}{\partial t} + \frac{\partial}{\partial x} \boldsymbol{f}(\boldsymbol{w}) = 0, \tag{6.1}$$

where $\boldsymbol{w}$ denotes the state vector and $\boldsymbol{f}(\boldsymbol{w})$ the flux vector, both of dimension n. In a smooth region, this may be written in quasilinear form as

$$\frac{\partial \boldsymbol{w}}{\partial t} + A(\boldsymbol{w}) \frac{\partial \boldsymbol{w}}{\partial x} = 0, \tag{6.2}$$

where

$$A(\boldsymbol{w}) = \frac{\partial \boldsymbol{f}}{\partial \boldsymbol{w}} \tag{6.3}$$

is the Jacobian matrix. The eigenvalues of A define the wave speeds of propagation, which may be in either direction.

We focus in particular on the equations of gas dynamics discussed in Chapter 2. Denoting the pressure, density, total energy, and velocity by p, ρ, E, and u, the state and flux vectors in (6.1) are now

$$\boldsymbol{w} = \begin{bmatrix} \rho \\ \rho u \\ \rho E \end{bmatrix}, \boldsymbol{f} = \begin{bmatrix} \rho u \\ \rho u^2 + p \\ \rho u H \end{bmatrix}, \tag{6.4}$$

where H is the total enthalpy

$$H = E + \frac{p}{\rho},$$

and the pressure p is determined from the internal energy as

$$p = (\gamma - 1)\rho \left(E - \frac{u^2}{2} \right). \tag{6.5}$$

Denoting the momentum and specific total energy by

$$m = \rho u, e = \rho E, \tag{6.6}$$

we can write the state and flux vectors as

$$w = \begin{bmatrix} \rho \\ m \\ e \end{bmatrix}, f = \begin{bmatrix} m \\ \frac{m^2}{\rho} + p \\ \frac{m}{\rho}(e + p) \end{bmatrix}, \tag{6.7}$$

and the Jacobian matrix (6.3) should then be evaluated from f and w in this form. The eigenvalues of A are

$$\lambda^{(1)} = u$$
$$\lambda^{(2)} = u + c$$
$$\lambda^{(3)} = u - c,$$

where c is the speed of sound.

This can be evaluated as

$$c = \sqrt{\gamma \frac{p}{\rho}},$$

where γ is the ratio of specific heats or equivalently from the relation

$$H = \frac{c^2}{\gamma - 1} + \frac{u^2}{2}.$$

In a subsonic flow, there will be acoustic waves traveling to the left and right with speeds $u - c$ and $u + c$ respectively, while there is a convective wave traveling with a speed u that may be of either sign.

6.2 Upwinding for a Linear System

Before considering the nonlinear case, it is instructive first to examine the case of a linear system

$$\frac{\partial w}{\partial t} + A \frac{\partial w}{\partial x} = 0$$

with a fixed matrix A. This corresponds to a conservation law with the flux

$$f(w) = Aw.$$

Suppose now that A can be decomposed as

$$A = V \Lambda V^{-1}, \tag{6.8}$$

where the columns of V are the right eigenvectors of A, and Λ is a diagonal matrix of the eigenvalues of A.

Now, multiplying by V^{-1} and setting

$$\tilde{w} = V^{-1} w,$$

(6.8) takes the form

$$\frac{\partial \tilde{\boldsymbol{w}}}{\partial t} + \Lambda \frac{\partial \tilde{\boldsymbol{w}}}{\partial x} = 0. \tag{6.9}$$

Thus, the equations are decoupled in terms of the characteristic variables $\tilde{w}^{(k)}, k = 1, n$, each satisfying a separate equation of the form

$$\frac{\partial \tilde{w}^{(k)}}{\partial t} + \lambda^{(k)} \frac{\partial \tilde{w}^{(k)}}{\partial x} = 0.$$

Now, suppose that this is discretized by a semi-discrete finite volume scheme of the form

$$\Delta x \frac{d\tilde{\boldsymbol{w}}_j}{dt} + \tilde{\boldsymbol{h}}_{j+\frac{1}{2}} - \tilde{\boldsymbol{h}}_{j-\frac{1}{2}} = 0, \tag{6.10}$$

where $\tilde{w}$ is the average value of the numerical solution in cell j, and $\tilde{h}_{j+\frac{1}{2}}$ is the numerical flux. With a pure upwind scheme,

$$\tilde{h}_{j+\frac{1}{2}} = \frac{1}{2}\lambda^{(k)}(\tilde{w}_{j+1} + \tilde{w}_j) - \frac{1}{2}|\lambda^{(k)}|(\tilde{w}_{j+1} - \tilde{w}_j).$$

Equations (6.10) can be reassembled as a vector equation for $\frac{d\tilde{\boldsymbol{w}}}{dt}$. Now, multiplying by V, they reduce to

$$\Delta x \frac{d\boldsymbol{w}_j}{dt} + \boldsymbol{h}_{j+\frac{1}{2}} - \boldsymbol{h}_{j-\frac{1}{2}} = 0, \tag{6.11}$$

where $\boldsymbol{w}_j$ now denotes the average value of the numerical solution for $\boldsymbol{w}$ in cell j, and $\boldsymbol{h}_{j+\frac{1}{2}}$ is the corresponding numerical flux, which can be expressed as

$$\boldsymbol{h}_{j+\frac{1}{2}} = \frac{1}{2}A(\boldsymbol{w}_{j+1} + \boldsymbol{w}_j) - \frac{1}{2}|A|(\boldsymbol{w}_{j+1} - \boldsymbol{w}_j), \tag{6.12}$$

where $|A|$ is the absolute Jacobian matrix

$$|A| = V|\Lambda|V^{-1},$$

obtained by replacing each eigenvalue $\lambda^{(k)}$ in Λ by its absolute value $|\lambda^{(k)}|$.

Note that the system matrix A could be split into parts corresponding to positive and negative eigenvalues as

$$A = A^+ + A^-,$$

where

$$A^+ = V\Lambda^+V^{-1}, \quad A^- = V\Lambda^-V^{-1},$$

and Λ^+ and Λ^- contain as their entries

$$\lambda^{(k)^+} = \max(\lambda^{(k)}, 0), \quad \lambda^{(k)^-} = \min(\lambda^{(k)}, 0).$$

Correspondingly, the flux vector can be split as

$$\boldsymbol{f} = \boldsymbol{f}^+ + \boldsymbol{f}^-,$$

where

$$f^+ = A^+ w, f^- = A^- w.$$

The positive and negative points f^+ and f^- contribute to solutions with right and left running waves, respectively.

Now, we can write the numerical flux as

$$\begin{aligned} h_{j+\frac{1}{2}} &= \frac{1}{2}A(w_{j+1} + w_j) - \frac{1}{2}|A|(w_{j+1} - w_j) \\ &= \frac{1}{2}(A^+ + A^-)(w_{j+1} + w_j) - \frac{1}{2}(A^+ - A^-)(w_{j+1} - w_j) \\ &= A^+ w_j + A^- w_{j+1} \\ &= f_j^+ + f_{j+1}^-, \end{aligned}$$

where $f_j^{\pm}$ denotes $f^{\pm}(w_j)$. Thus, the numerical flux is constructed by separately upwinding the parts corresponding to the left and right running waves.

6.3 Flux Vector Splitting for a Nonlinear System

The ideas of the previous section can be directly carried over to the case of a nonlinear system of conservation laws

$$\frac{\partial w}{\partial t} + \frac{\partial}{\partial x} f(w) = 0. \tag{6.13}$$

Again, denoting the average value of the numerical solution in cell j by w_j and the numerical interface flux by $h_{j+\frac{1}{2}}$, let (6.13) be approximated by the semi-discrete finite volume scheme

$$\Delta x \frac{\partial w_j}{\partial t} + h_{j+\frac{1}{2}} - h_{j-\frac{1}{2}} = 0. \tag{6.14}$$

Now, suppose that the flux vector is split as

$$f = f^+ + f^-$$

in any way such that $\frac{\partial f^+}{\partial w}$ and $\frac{\partial f^-}{\partial w}$ have nonnegative and nonpositive eigenvalues, respectively. Then, set

$$h_{j+\frac{1}{2}} = f_j^+ + f_{j+1}^-,$$

where $f_j^{\pm}$ denotes $f_j^{\pm}(w_j)$. The splitting of the flux into positive and negative parts is entirely nonunique. Many splittings are possible. The simplest is to set

$$f^{\pm}(w) = \frac{1}{2}(f(w) \pm \alpha w), \tag{6.15}$$

where if $\lambda^{(k)}(A)$ are the eigenvalues of $\frac{\partial f}{\partial w}$

$$\alpha \geq \max_k |\lambda^{(k)}(A(\boldsymbol{w}))|.$$

Then,

$$\frac{\partial \boldsymbol{f}^+}{\partial \boldsymbol{w}} = \frac{1}{2}(A(\boldsymbol{w}) + \alpha I)$$

with eigenvalues

$$\frac{1}{2}(\lambda^{(k)} + \alpha) \geq 0,$$

while the eigenvalues of $\frac{\partial f^-}{\partial w}$ are

$$\frac{1}{2}(\lambda^{(k)} - \alpha) \leq 0.$$

The corresponding numerical flux,

$$\begin{aligned} \boldsymbol{h}_{j+\frac{1}{2}} &= \frac{1}{2}(\boldsymbol{f}_j + \alpha_j \boldsymbol{w}_j) + \frac{1}{2}(\boldsymbol{f}_{j+1} - \alpha_{j+1}\boldsymbol{w}_{j+1}) \\ &= \frac{1}{2}(\boldsymbol{f}_{j+1} + \boldsymbol{f}_j) - \frac{1}{2}(\alpha_{j+1}\boldsymbol{w}_{j+1} - \alpha_j \boldsymbol{w}_j), \end{aligned}$$

represents a central difference scheme augmented by the artificial diffusive flux

$$\boldsymbol{d}_{j+\frac{1}{2}} = \frac{1}{2}(\alpha_{j+1}\boldsymbol{w}_{j+1} - \alpha_j \boldsymbol{w}_j).$$

In order to produce a higher order scheme, we can generate higher order estimate of the left and right states w_L and w_R at the interface $j + \frac{1}{2}$ in the same manner as was described for scalar conservation laws in Section 5.22. Then the interface flux is formed as

$$\boldsymbol{h}_{j+\frac{1}{2}} = \boldsymbol{f}^+(\boldsymbol{w}_L) + \boldsymbol{f}^-(\boldsymbol{w}_R).$$

As in the scalar case, limiters are used to prevent overshoots in the neighborhood of local extrema, such that the first order scheme is recovered with $\boldsymbol{w}_L = \boldsymbol{w}_j$ and $\boldsymbol{w}_R = \boldsymbol{w}_{j+1}$.

6.4 A Splitting with Full Upwinding in Supersonic Flow

Since all the waves travel in the same direction in a supersonic flow, it is natural then to use a fully upwind flux. The simple splitting (6.15) does not have this property, which can be recovered in the following manner. Define

$$\boldsymbol{f}^{\pm}(\boldsymbol{w}) = \epsilon^{\pm}\boldsymbol{f}(\boldsymbol{w}) \pm \frac{1}{2}\alpha^* c \boldsymbol{w}.$$

Here, the speed of sound is included in the coefficient of $\mathbf{w}$ so that α^* is dimensionless. For full upwinding when $|M| \geq 1$, we require

$$\alpha^* = 0\,,$$

$$\epsilon^+ = \begin{cases} 1, M \geq +1 \\ 0, M \leq -1 \end{cases}, \; \epsilon^- = \begin{cases} 0, M \geq +1 \\ 1, M \leq -1 \end{cases}.$$

In the range $-1 \leq M \leq 1$, a simple choice is

$$\epsilon^+ = \frac{1}{2}(1+M)\,,\; \epsilon^- = \frac{1}{2}(1-M).$$

Since the eigenvalues of the Jacobian matrix $A = \frac{\partial \mathbf{f}}{\partial \mathbf{w}}$ are $u, u+c$ and $u-c$, the smallest eigenvalue of f^+ is

$$\epsilon^+(u-c) + \frac{1}{2}\alpha^* c = \frac{1}{2}\left((1+M)(u-c) + \alpha^* c\right).$$

The requirement that this is nonnegative is satisfied if

$$\alpha^* \geq (1+M)(1-M) = 1 - M^2. \tag{6.16}$$

The largest eigenvalue of f^- is

$$\epsilon^-(u+c) - \frac{1}{2}\alpha^* c = \frac{1}{2}\left((1-M)(u+c) - \alpha^* c\right).$$

Similarly, this is nonpositive if α^* satisfies condition (6.16). Taking the minimum allowable value of α^* leads to the splitting

$$\mathbf{f}^+(\mathbf{w}) = \frac{1}{2}(1+M)\,(\mathbf{f}(\mathbf{w}) + (c-u)\mathbf{w})$$

$$\mathbf{f}^-(\mathbf{w}) = \frac{1}{2}(1-M)\,(\mathbf{f}(\mathbf{w}) - (c+u)\mathbf{w}), \tag{6.17}$$

when $|M| \leq 1$.

6.5 Characteristic Flux Splitting for Gas Dynamics

In the case of gas dynamics, a particularly natural and very widely used splitting is based on the characteristics. This was originally introduced by Steger and Warming (1981), and it takes advantage of the fact that the flux vector of gas dynamics given by (6.7) is homogeneous of degree one in the conservative variables as was discussed in Section 2.3, and accordingly, the flux vector is exactly equal to the state vector multiplied by the Jacobian matrix:

$$\mathbf{f}(\mathbf{w}) = A\mathbf{w}. \tag{6.18}$$

Now, as in the case of the linear system discussed in Section 6.2, we decompose A as

$$A = V \Lambda V^{-1},$$

where the columns of V are the right eigenvectors of A, and Λ is a diagonal matrix of the eigenvalues of A. Then, we define

$$A^+ = V\Lambda^+V^{-1}, A^- = V\Lambda^-V^{-1},$$

where the entries of Λ^+ and Λ^- are $\max(\lambda^{(k)},0)$ and $\min(\lambda^{(k)},0)$, respectively, and set

$$f^+ = A^+w, f^- = A^-w.$$

It may be verified that with this definition, the eigenvalues of $\frac{\partial f^+}{\partial w}$ and $\frac{\partial f^-}{\partial w}$ are in fact nonnegative and nonpositive, respectively, for standard gas properties (Witherden & Jameson 2018). The Jacobian matrices $A^\pm$ are most easily calculated by using the transformation to symmetric form given in Section 2.7. In the one-dimensional case,

$$A = \bar{M}\bar{A}\bar{M}^{-1},$$

where, setting

$$\bar{\gamma} = \frac{\gamma - 1}{c^2},$$

the transformation matrices are

$$\bar{M} = \begin{bmatrix} 1 & 0 & -1 \\ u & c & -u \\ H & uc & -\frac{u^2}{2} \end{bmatrix}, \quad \bar{M}^{-1} = \begin{bmatrix} \bar{\gamma}\frac{u^2}{2} & -\bar{\gamma}u & \bar{\gamma} \\ -\frac{u}{c} & \frac{1}{c} & 0 \\ \bar{\gamma}\frac{u^2}{2} - 1 & -\bar{\gamma}u & \bar{\gamma} \end{bmatrix},$$

and

$$\bar{A} = \begin{bmatrix} u & c & 0 \\ c & u & 0 \\ 0 & 0 & u \end{bmatrix}.$$

The eigenvector matrix of $\bar{A}$ is

$$\bar{V} = \begin{bmatrix} 0 & \frac{1}{\sqrt{2}} & \frac{1}{\sqrt{2}} \\ 0 & \frac{1}{\sqrt{2}} & -\frac{1}{\sqrt{2}} \\ 1 & 0 & 0 \end{bmatrix}.$$

and $\bar{V}^{-1} = \bar{V}^T$.

When the eigenvalues are replaced by q_1, q_2, and q_3, the modified Jacobian matrix takes the form

$$\bar{A}^* = \begin{bmatrix} r & s & 0 \\ s & r & 0 \\ 0 & 0 & q_1 \end{bmatrix},$$

where

$$r = \frac{q_2 + q_3}{2}, \quad s = \frac{q_2 - q_3}{2}.$$

In the case that $0 \leq u \leq c$, for example, we obtain $\bar{A}^+$ by setting

$$q_1 = u, \quad q_2 = u + c, \quad q_3 = 0,$$

$$q_1 + r = \frac{u + c}{2}, \quad q_2 = \frac{u + c}{2},$$

yielding

$$\bar{A}^+ = \begin{bmatrix} \frac{u+c}{2} & \frac{u+c}{2} & 0 \\ \frac{u+c}{2} & \frac{u+c}{2} & 0 \\ 0 & 0 & u \end{bmatrix}$$

with eigenvalues u, $u + c$, and zero. The flux vectors $f^\pm$ can actually be formed without directly calculating $A^\pm$ in the following manner. The eigenvalues of A are u, $u + c$, and $u - c$. The corresponding eigenvectors, scaled by the density, are

$$\boldsymbol{v}^{(1)} = \rho \begin{bmatrix} 1 \\ u \\ \frac{u^2}{2} \end{bmatrix}, \boldsymbol{v}^{(2)} = \rho \begin{bmatrix} 1 \\ u + c \\ H + uc \end{bmatrix}, \boldsymbol{v}^{(3)} = \rho \begin{bmatrix} 1 \\ u - c \\ H - uc \end{bmatrix}.$$

Then, the state vector can be expanded in terms of the eigenvectors as

$$\boldsymbol{w} = \frac{\gamma - 1}{\gamma} \boldsymbol{v}^{(1)} + \frac{1}{2\gamma} \boldsymbol{v}^{(2)} + \frac{1}{2\gamma} \boldsymbol{v}^{(3)}.$$

Since

$$A \boldsymbol{v}^{(k)} = \lambda^{(k)} \boldsymbol{v}^{(k)},$$

it follows from (6.18) that the flux vectors can also be expanded as

$$\boldsymbol{f}(\boldsymbol{w}) = \frac{u(\gamma - 1)}{\gamma} \boldsymbol{v}^{(1)} + \frac{u + c}{2\gamma} \boldsymbol{v}^{(2)} + \frac{u - c}{2\gamma} \boldsymbol{v}^{(3)}. \tag{6.19}$$

Now, if $|u| < c$ and $u > 0$,

$$\boldsymbol{f}^+(\boldsymbol{w}) = \frac{u(\gamma - 1)}{\gamma} \boldsymbol{v}^{(1)} + \frac{u + c}{2\gamma} \boldsymbol{v}^{(2)} \tag{6.20}$$

$$\boldsymbol{f}^-(\boldsymbol{w}) = \frac{u - c}{2\gamma} \boldsymbol{v}^{(3)}, \tag{6.21}$$

while if $v < 0$, the term containing the first eigenvector is shifted to $\boldsymbol{f}^-$.

Equations (6.19)–(6.21) enable characteristic splitting to be very efficiently implemented in computer programs. It is also evident from (6.19) that characteristic splitting results in perfect upwinding in locally supersonic flow.

The characteristic splitting scheme can be expressed in artificial viscosity form as follows. Denoting $A^{\pm}(w_j)$ by $A_j^{\pm}$ and $f^{\pm}(w_j)$ by $f_j^{\pm}$, the numerical flux can be expanded as

$$\begin{aligned}
\boldsymbol{h}_{j+\frac{1}{2}} &= \boldsymbol{f}_j^+ + \boldsymbol{f}_{j+1}^- \\
&= \frac{1}{2}(\boldsymbol{f}_{j+1}^+ + \boldsymbol{f}_j^+) - \frac{1}{2}(\boldsymbol{f}_{j+1}^+ - \boldsymbol{f}_j^+) + \frac{1}{2}(\boldsymbol{f}_{j+1}^- + \boldsymbol{f}_j^-) + \frac{1}{2}(\boldsymbol{f}_{j+1}^- - \boldsymbol{f}_j^-) \\
&= \frac{1}{2}(\boldsymbol{f}_{j+1} + \boldsymbol{f}_j) - \frac{1}{2}(\boldsymbol{f}_{j+1}^+ - \boldsymbol{f}_{j+1}^-) + \frac{1}{2}(\boldsymbol{f}_j^+ - \boldsymbol{f}_j^-) \\
&= \frac{1}{2}(\boldsymbol{f}_{j+1} + \boldsymbol{f}_j) - \frac{1}{2}(A_{j+1}^+ - A_{j+1}^-)\boldsymbol{w}_{j+1} + \frac{1}{2}(A_j^+ - A_j^-)\boldsymbol{w}_j \\
&= \frac{1}{2}(\boldsymbol{f}_{j+1} + \boldsymbol{f}_j) - \frac{1}{2}(|A|_{j+1}\boldsymbol{w}_{j+1} - |A|_j\boldsymbol{w}_j).
\end{aligned}$$

Moreover, the semi-discrete scheme can be expressed as

$$\begin{aligned}
\Delta x \frac{d\boldsymbol{v}_j}{dt} &= -\boldsymbol{h}_{j+\frac{1}{2}} + \boldsymbol{h}_{j-\frac{1}{2}} \\
&= -(\boldsymbol{f}_j^+ - \boldsymbol{f}_{j-1}^+) - (\boldsymbol{f}_{j+1}^- - \boldsymbol{f}_j^-).
\end{aligned} \tag{6.22}$$

6.6 Roe's Linearization and Flux Difference Splitting

An alternative way to extend the upwind scheme for a linear system defined by (6.11) and (6.12) to a nonlinear system was proposed by Roe (1981). Corresponding to the numerical definition of the wave speed for a scalar conservation law in Section 5.12 of the previous chapter, he introduced the idea of a mean value Jacobian matrix $A(\boldsymbol{w}_L, \boldsymbol{w}_R)$ depending on the states $\boldsymbol{w}_L$ and $\boldsymbol{w}_R$ such that

$$A(\boldsymbol{w}_L, \boldsymbol{w}_R)(\boldsymbol{w}_R - \boldsymbol{w}_L) = \boldsymbol{f}(\boldsymbol{w}_R) - \boldsymbol{f}(\boldsymbol{w}_L)$$

exactly. Define

$$\hat{\boldsymbol{w}}(\theta) = \boldsymbol{w}_L + \theta(\boldsymbol{w}_R - \boldsymbol{w}_L).$$

Then, we can evaluate $\boldsymbol{f}(\boldsymbol{w}_R) - \boldsymbol{f}(\boldsymbol{w}_L)$ as

$$\boldsymbol{f}(\boldsymbol{w}_R) - \boldsymbol{f}(\boldsymbol{w}_L) = \int_0^1 \frac{\partial \boldsymbol{f}}{\partial \boldsymbol{w}}(\hat{\boldsymbol{w}}(\theta))\frac{\partial \hat{\boldsymbol{w}}}{\partial \theta} d\theta = \int_0^1 \frac{\partial \boldsymbol{f}}{\partial \boldsymbol{w}}(\hat{\boldsymbol{w}}(\theta)) d\theta (\boldsymbol{w}_R - \boldsymbol{w}_L).$$

Thus, we can express the mean value Jacobian as

$$A(\boldsymbol{w}_L, \boldsymbol{w}_R) = \int_0^1 \frac{\partial \boldsymbol{f}}{\partial \boldsymbol{w}}(\hat{\boldsymbol{w}}(\theta)) d\theta.$$

In the case of gas dynamics, Roe found an elegant way to evaluate $A(w_L, w_R)$ in closed form, given in the next section. In order to produce an upwind scheme for the nonlinear system, we take $w_L = w_j$, $w_R = w_{j+1}$. Then, define

$$A_{j+\frac{1}{2}} = A(\boldsymbol{w}_L, \boldsymbol{w}_R)$$

and decompose $A_{j+\frac{1}{2}}$ as

$$A_{j+\frac{1}{2}} = V \Lambda V^{-1}$$

as in the case of the linear system, where the columns of V are the right eigenvectors of $A_{j+\frac{1}{2}}$, and Λ is a diagonal matrix containing its eigenvalues $\lambda^{(k)}$. Now, define the absolute Jacobian matrix as

$$|A_{j+\frac{1}{2}}| = V|\Lambda|V^{-1},$$

where the entries of $|\Lambda|$ are $|\lambda^{(k)}|$. Finally we approximate the conservation law by the semi-discrete scheme

$$\Delta x \frac{d\boldsymbol{w}_j}{dt} + \boldsymbol{h}_{j+\frac{1}{2}} - \boldsymbol{h}_{j-\frac{1}{2}} = 0, \tag{6.23}$$

where

$$\boldsymbol{h}_{j+\frac{1}{2}} = \frac{1}{2}(\boldsymbol{f}_{j+1} + \boldsymbol{f}_j) - \frac{1}{2}\left(|A|_{j+\frac{1}{2}}(\boldsymbol{w}_{j+1} - \boldsymbol{w}_j)\right). \tag{6.24}$$

The diffusive flux can be expressed as

$$\boldsymbol{d}_{j+\frac{1}{2}} = \frac{1}{2}\left(A^+_{j+\frac{1}{2}} - A^-_{j+\frac{1}{2}}\right)\Delta \boldsymbol{w}_{j+\frac{1}{2}},$$

where

$$A^+_{j+\frac{1}{2}} = V\Lambda^+ V^{-1}, A^-_{j+\frac{1}{2}} = V\Lambda^- V_{-1}$$

with $\Lambda^{\pm}$ containing the positive and negative eigenvalues of $A_{j+\frac{1}{2}}$, respectively. Now, defining

$$\Delta \boldsymbol{f}^+_{j+\frac{1}{2}} = A^+_{j+\frac{1}{2}} \Delta \boldsymbol{w}_{j+\frac{1}{2}}, \Delta \boldsymbol{f}^-_{j+\frac{1}{2}} = A^-_{j+\frac{1}{2}} \Delta \boldsymbol{w}_{j+\frac{1}{2}},$$

the flux difference $\Delta \boldsymbol{f}_{j+\frac{1}{2}}$ is split as

$$\Delta \boldsymbol{f}_{j+\frac{1}{2}} = \Delta \boldsymbol{f}^+_{j+\frac{1}{2}} + \Delta \boldsymbol{f}^-_{j+\frac{1}{2}}.$$

Then,

$$\boldsymbol{h}_{j+\frac{1}{2}} - \boldsymbol{h}_{j-\frac{1}{2}} = \frac{1}{2}\Delta \boldsymbol{f}_{j+\frac{1}{2}} + \frac{1}{2}\Delta \boldsymbol{f}_{j-\frac{1}{2}} - \frac{1}{2}\left(\Delta \boldsymbol{f}^+_{j+\frac{1}{2}} - \Delta \boldsymbol{f}^-_{j+\frac{1}{2}}\right) + \frac{1}{2}\left(\Delta \boldsymbol{f}^+_{j-\frac{1}{2}} - \Delta \boldsymbol{f}^-_{j-\frac{1}{2}}\right),$$

and accordingly the semi-discrete scheme (6.23) can be arranged as

$$\Delta x \frac{d\boldsymbol{w}_j}{dt} = -\Delta \boldsymbol{f}^+_{j-\frac{1}{2}} - \Delta \boldsymbol{f}^-_{j+\frac{1}{2}},$$

which may be compared with the characteristic flux split scheme 6.22. Note that one can also express $|A|$ as

$$|A| = V \text{ sgn } (\Lambda) \; \Lambda \; V^{-1} = SA,$$

where the sign matrix S is defined as

$$S = V \operatorname{sgn}(\Lambda) V^{-1}.$$

Now, the flux can be represented in the equivalent form

$$\boldsymbol{h}_{j+\frac{1}{2}} = \frac{1}{2}(\boldsymbol{f}_{j+1} + \boldsymbol{f}_j) - \frac{1}{2} S A(\boldsymbol{w}_{j+1} - \boldsymbol{w}_j) = \frac{1}{2}(\boldsymbol{f}_{j+1} + \boldsymbol{f}_j) - \frac{1}{2} S(\boldsymbol{f}_{j+1} - \boldsymbol{f}_j).$$

As in the case of flux vector splitting, a higher order scheme can be derived by replacing w_j and w_{j+1} by higher order estimates of the left and right states w_L and w_R at the interface $j + \frac{1}{2}$. Then, the numerical flux is defined as

$$\boldsymbol{h}_{j+\frac{1}{2}} = \frac{1}{2}\left(\boldsymbol{f}(\boldsymbol{w}_R) + \boldsymbol{f}(\boldsymbol{w}_L)\right) - \frac{1}{2}|A(\boldsymbol{w}_R, \boldsymbol{w}_L)|(\boldsymbol{w}_R - \boldsymbol{w}_L),$$

where

$$A(\boldsymbol{w}_R, \boldsymbol{w}_L)(\boldsymbol{w}_R - \boldsymbol{w}_L) = \boldsymbol{f}(\boldsymbol{w}_R) - \boldsymbol{f}(\boldsymbol{w}_L)$$

exactly.

6.7 The Roe Matrix for Gas Dynamics

Roe solved the problem of evaluating a mean value Jacobian $A(w_R, w_L)$ for gas dynamics that exactly satisfies the relation

$$A(\boldsymbol{w}_L, \boldsymbol{w}_R)(\boldsymbol{w}_R - \boldsymbol{w}_L) = \boldsymbol{f}(\boldsymbol{w}_R) - \boldsymbol{f}(\boldsymbol{w}_L)$$

by introducing the parameter vector

$$\boldsymbol{r} = \begin{bmatrix} \sqrt{\rho} \\ \sqrt{\rho} u \\ \sqrt{\rho} H \end{bmatrix}.$$

Then, the pressure can be expressed as

$$p = \frac{\gamma - 1}{\gamma} \rho \left(H - \frac{u^2}{2} \right) = \frac{\gamma - 1}{\gamma} \left(r_1 r_3 - \frac{r_2^2}{2} \right),$$

and the state and flux vectors take the form

$$\boldsymbol{w} = \begin{bmatrix} r_1^2 \\ r_1 r_2 \\ r_1 r_3 - \frac{\gamma-1}{\gamma}\left(r_1 r_3 - \frac{r_2^2}{2}\right) \end{bmatrix}, \boldsymbol{f} = \begin{bmatrix} r_1 r_2 \\ r_2^2 + \frac{\gamma-1}{\gamma}\left(r_1 r_3 - \frac{r_2^2}{2}\right) \\ r_2 r_3 \end{bmatrix}.$$

Thus, all the entries in both w and f are bilinear or quadratic in r_1, r_2, and r_3.

Now, consider the effect of changes Δa and Δb in a product

$$q = ab$$

of any two quantities a and b. Then

$$\Delta q = (a + \Delta a)(b + \Delta b) - ab$$
$$= \left(b + \frac{1}{2}\Delta b\right)\Delta a + \left(a + \frac{1}{2}\Delta a\right)\Delta b = \frac{\partial q}{\partial a}\Delta a + \frac{\partial q}{\partial b}\Delta b,$$

where $\frac{\partial q}{\partial a}$ and $\frac{\partial q}{\partial b}$ are evaluated using the mean values

$$\bar{a} = a + \frac{1}{2}\Delta a, \bar{b} = b + \frac{1}{2}\Delta b.$$

Applying this rule to the changes in $\boldsymbol{w}$ and $\boldsymbol{f}$ that result from a change $\Delta \boldsymbol{r}$ in the parameter vector, we find that

$$\Delta \boldsymbol{w} = B\Delta \boldsymbol{r}\,,\ \Delta \boldsymbol{f} = C\Delta \boldsymbol{r},$$

where

$$B = \frac{\partial \boldsymbol{w}}{\partial \boldsymbol{r}} = \begin{bmatrix} 2r_1 & 0 & 0 \\ r_2 & r_1 & 0 \\ \frac{1}{\gamma}r_3 & \frac{\gamma-1}{\gamma}r_2 & \frac{1}{\gamma}r_1 \end{bmatrix}$$

and

$$C = \frac{\partial \boldsymbol{f}}{\partial \boldsymbol{r}} = \begin{bmatrix} r_2 & r_1 & 0 \\ \frac{\gamma-1}{\gamma}r_3 & \frac{\gamma+1}{\gamma}r_2 & \frac{\gamma-1}{\gamma}r_1 \\ 0 & r_3 & r_2 \end{bmatrix},$$

and these are both evaluated with mean values $\bar{r}$.

Then, given $\Delta \boldsymbol{w}$ we can solve for $\Delta \boldsymbol{r}$ as

$$\Delta \boldsymbol{r} = B^{-1}\Delta \boldsymbol{w},$$

where

$$B^{-1} = \begin{bmatrix} \frac{1}{2r_1} & 0 & 0 \\ -\frac{1}{2}\frac{r_2}{r_1^2} & \frac{1}{r_1} & 0 \\ \frac{\gamma-1}{2}\frac{r_2^2}{r_1^3} - \frac{1}{2}\frac{r_3}{r_1^2} & -(\gamma-1)\frac{r_2}{r_1^2} & \frac{\gamma}{r_1} \end{bmatrix}.$$

Consequently,

$$\Delta \boldsymbol{f} = A\Delta \boldsymbol{w},$$

where

$$A = CB^{-1} = \frac{\partial \boldsymbol{f}}{\partial \boldsymbol{r}}\frac{\partial \boldsymbol{r}}{\partial \boldsymbol{w}} = \frac{\partial \boldsymbol{f}}{\partial \boldsymbol{w}}.$$

Multiplying out CB^{-1} using mean values of r_1, r_2, and r_3, we obtain the standard form for the Jacobian matrix

$$A = uI + \begin{bmatrix} 1 \\ u \\ H \end{bmatrix} [-u, 1, 0] + (\gamma - 1) \begin{bmatrix} 0 \\ 1 \\ u \end{bmatrix} \left[\frac{u^2}{2}, -u, 1 \right],$$

using the mean values

$$\bar{u} = \frac{\sqrt{\rho_L} u_L + \sqrt{\rho_R} u_R}{\sqrt{\rho_L} + \sqrt{\rho_R}}, \quad \bar{H} = \frac{\sqrt{\rho_L} H_L + \sqrt{\rho_R} H_R}{\sqrt{\rho_L} + \sqrt{\rho_R}}.$$

If the Jacobian is expressed in terms of u and c, then the mean value $\bar{c}$ should be evaluated from

$$\bar{c}^2 = (\gamma - 1)\left(\bar{H} - \frac{\bar{u}^2}{2}\right).$$

6.8 Godunov's Method

The semi-discete scheme (6.14) defines a procedure for updating the cell average values $\boldsymbol{w}_j$. In Godunov's approach, as was discussed in Section 5.16, the approximate solution is treated as piecewise constant with a value in each cell equal to the cell average. Then the interface flux $\boldsymbol{h}_{j+\frac{1}{2}}$ is calculated as the exact solution of the Riemann problem for the evaluation of the gas dynamics equation for an initial state consisting of constant left and right states $\boldsymbol{w}_L = \boldsymbol{w}_j$ and $\boldsymbol{w}_R = \boldsymbol{w}_{j+1}$ with a discontinuity at the interface. In general, this solution is a multiple wave pattern with a shock wave, a contact discontinuity and an expansion fan, as in the $x - t$ diagram of Figure 6.1. The numerical flux $\boldsymbol{h}_{j+\frac{1}{2}}$ has a constant value on the $x = 0$ axis, provided that the time step is restricted such that there is no interaction with the waves emanating from other interfaces. Actually the waves will travel into the cells on either side of the interface, but (6.14) still gives a correct update of the cell averages for a time step sufficiently small to prevent wave interactions. Referring to the control volume in Figure 6.8,

$$(\boldsymbol{w}_j^{n+1} - \boldsymbol{w}_j^n)\Delta x + \left(\boldsymbol{h}_{j+\frac{1}{2}} - \boldsymbol{h}_{j-\frac{1}{2}}\right)\Delta t = 0.$$

The cell averages at the end of the time step define a new piecewise constant state, and the process is repeated for the next time step.

The explicit assumption of a piecewise constant state enables the exact calculation of the numerical fluxes by an iterative procedure to solve the Riemann problem, such as that described by Toro (2009). In general, the true solution is, of course, not piecewise constant, and the assumption results in a first order accurate scheme. In this respect, Godunov's method is in the spirit of finite element methods, where the numerical solution is assumed to have an explicit spatial form, and it may be regarded as the lowest order finite element approximation.

The finite volume schemes described in the previous sections do not depend on any specific assumption about the form of the numerical solution. They simply use two

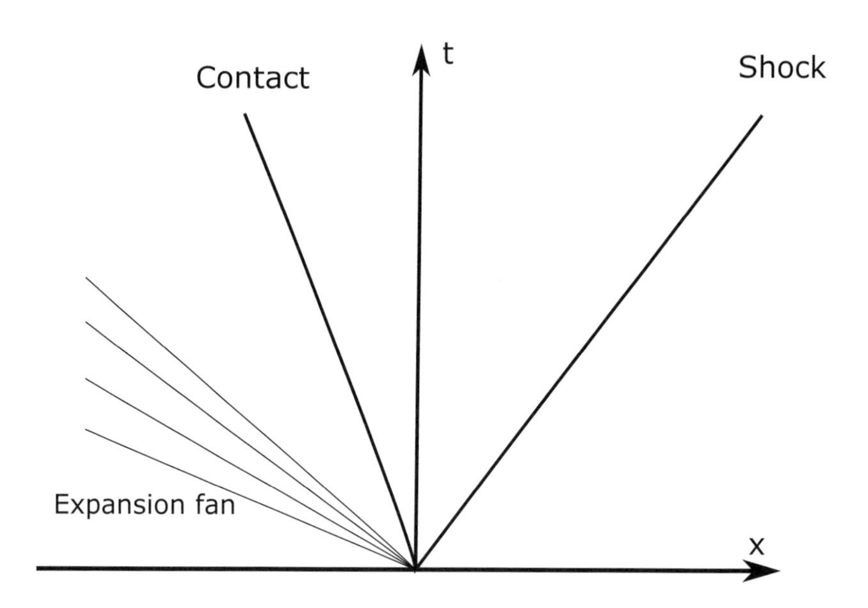

Figure 6.1 Solution of the Riemann problem for Godunov's method.

point flux formulas based on the average states to the left and right of each interface. If, however, one introduces the assumption of a piecewise constant state, then they may be, and often are, regarded as approximate solutions of the Riemann problem.

In the construction of higher order methods, this interpretation proves to be useful. Here, one uses reconstruction methods as described in Sections 5.21–5.23 to construct left and right states $\boldsymbol{w}_L$ and $\boldsymbol{w}_R$ at each interface, and then one may use whichever two point flux formula one prefers to evaluate the interface flux from $\boldsymbol{w}_L$ and $\boldsymbol{w}_R$. Reconstruction methods with an arbitrarily high order of accuracy can be formulated given a sufficiently large stencil. Alternatively, in the discontinuous Galerkin method, the numerical solution is represented by separate polynomials in each cell, which do not necessarily match at the cell interfaces. The values that the polynomials assume on either side of an interface then provide left and right states for calculation of the interface flux. These methods will be described in subsequent sections.

Godunov's work was popularized by Bram van Leer, who pioneered the use of reconstruction to devise higher order Godunov methods. In this case, the numerical flux for the higher order scheme is obtained by solving the Riemann problem exactly for the higher order estimates of the left and right states $\boldsymbol{w}_L$ and $\boldsymbol{w}_R$ at the interface $j + \frac{1}{2}$.

6.9 The Harten–Lax–Van Leer (HLL) Scheme

Another approach to calculating the interface flux was introduced by Harten, Lax, and van Leer in a celebrated paper (Harten, Lax, & van Leer 1983). They assumed an

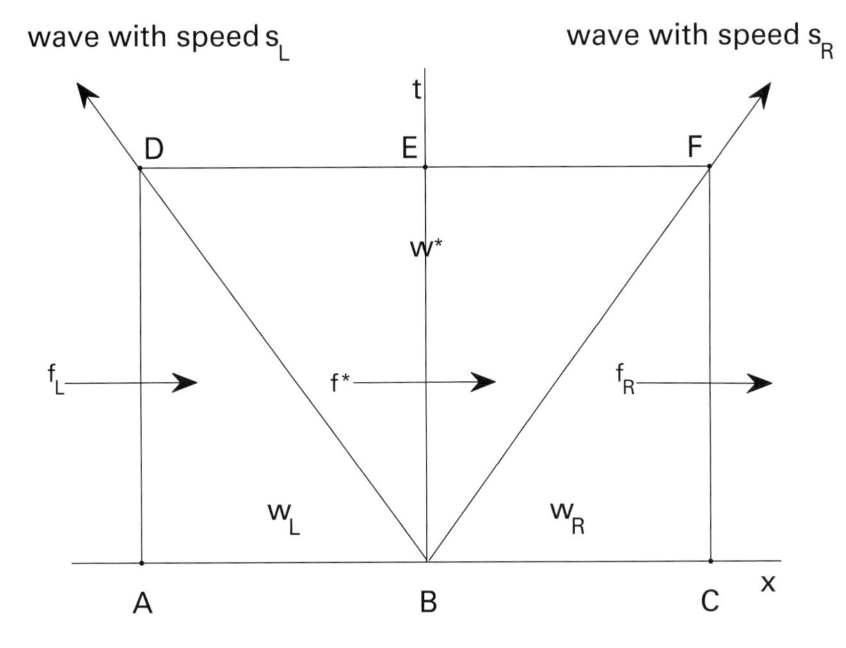

Figure 6.2 Flux balance for the HLL scheme.

intermediate state $\boldsymbol{w}^*$ separating the left and right states $\boldsymbol{w}_R$ and $\boldsymbol{w}_L$ by waves traveling to the left and right at speeds s_L and s_R, and a corresponding intermediate flux $\boldsymbol{f}^*$. Referring to the $x - t$ diagram in Figure 6.2 for the case of subsonic flow, the flux at the interface $x = 0$ is then $\boldsymbol{f}^*$ for all t, and this defines the numerical flux. The intermediate state $\boldsymbol{w}^*$ can be determined by considering the flux balance through the rectangle $ACFD$. According to the conservation law

$$\oint (\boldsymbol{w}dx - \boldsymbol{f}dt) = 0, \tag{6.25}$$

A and C are at $x = s_R\Delta t$ and $x = -s_L\Delta t$, respectively. Thus, (6.25) yields

$$(\boldsymbol{f}_R - \boldsymbol{f}_L)\Delta t + \boldsymbol{w}^*(s_R - s_L)\Delta t - (\boldsymbol{w}_R s_R - \boldsymbol{w}_L s_L)\Delta t = 0.$$

Accordingly,

$$\boldsymbol{w}^* = \frac{s_R\boldsymbol{w}_R - s_L\boldsymbol{w}_L - (\boldsymbol{f}_R - \boldsymbol{f}_L)}{s_R - s_L}. \tag{6.26}$$

Now, $\boldsymbol{f}^*$ may be determined by considering the flux balance through $BCFE$

$$(\boldsymbol{f}_R - \boldsymbol{f}^*)\Delta t + (\boldsymbol{w}^* - \boldsymbol{w}_R)s_R\Delta t = 0$$

or $ABED$

$$(\boldsymbol{f}^* - \boldsymbol{f}_L)\Delta t - (\boldsymbol{w}^* - \boldsymbol{w}_L)s_L\Delta t = 0.$$

Thus,

$$\boldsymbol{f}^* = \boldsymbol{f}_R + (\boldsymbol{w}^* - \boldsymbol{w}_R)s_R = \boldsymbol{f}_L + (\boldsymbol{w}^* - \boldsymbol{w}_L)s_L.$$

Substituting for $\boldsymbol{w}^*$, the numerical flux is

$$\begin{aligned}
\boldsymbol{f}^* &= \boldsymbol{f}_R + \frac{s_R\left(\boldsymbol{w}_R s_L - \boldsymbol{w}_L s_L - (\boldsymbol{f}_R - \boldsymbol{f}_L)\right)}{s_R - s_L} \\
&= \frac{s_R \boldsymbol{f}_L - s_L \boldsymbol{f}_R + s_R s_L(\boldsymbol{w}_R - \boldsymbol{w}_L)}{s_R - s_L} \\
&= \frac{1}{2}(\boldsymbol{f}_R + \boldsymbol{f}_L) - \frac{1}{2}\beta(\boldsymbol{f}_R - \boldsymbol{f}_L) - \frac{1}{2}\alpha^* c(\boldsymbol{w}_R - \boldsymbol{w}_L),
\end{aligned} \tag{6.27}$$

where

$$\alpha^* = \frac{2}{c}\frac{s_R s_L}{s_R - s_L}$$

and

$$\beta = \frac{s_R + s_L}{s_R - s_L}.$$

If s_R and s_L are taken as

$$s_R = u + c\,,\ s_L = u - c,$$

then

$$\alpha^* = \frac{u^2 - c^2}{c^2} = 1 - M^2, \tag{6.28}$$

and

$$\beta = \frac{u + c + u - c}{u + c - u + c} = M. \tag{6.29}$$

Note that in general, $\boldsymbol{f}^*$ is not equal to $\boldsymbol{f}(\boldsymbol{w}^*)$, since it is completely defined by the flux balance equations.

In the case of supersonic flow, both waves are on the same side of the interface, and

$$\boldsymbol{f}^* = \begin{cases} \boldsymbol{f}_L \text{ if } M \geq +1 \\ \boldsymbol{f}_R \text{ if } M \leq -1 \end{cases},$$

corresponding to

$$\alpha^* = 0$$

and

$$\beta = \operatorname{sgn}(M).$$

In the implementation of this scheme, it has to be decided what states should be used in evaluating the wave speeds. If one considers the left running wave to be emanating from the right state and the right running wave from the left state, one recovers the flux split scheme (6.17)

$$f = f_R^+ + f_L^-,$$

where

$$f^+ = \frac{1}{2}(1+M)\,(f + (c-u)w)$$
$$f^- = \frac{1}{2}(1-M)\,(f - (c+u)w)\,.$$

6.10 The HLLC Scheme

While Harten, Lax, and van Leer considered formulations with two intermediate states separated by three waves at each interface, the most widely accepted scheme of this type is the HLLC scheme proposed by Toro, Spruce, and Speares (1994), illustrated in Figure 6.3. They explicitly introduce a central wave with a speed s_M in the interface model to simulate a contact discontinuity.

Let w_L^* and w_R^* denote the intermediate states to the left and right of the middle wave. Now, conservation across the waves leads to the jump conditions

$$f_L^* = f_L + s_L(w_L^* - w_L) \tag{6.30}$$
$$f_R^* = f_L^* + s_M(w_R^* - w_L^*)$$
$$f_R^* = f_R + s_R(w_R^* - w_R). \tag{6.31}$$

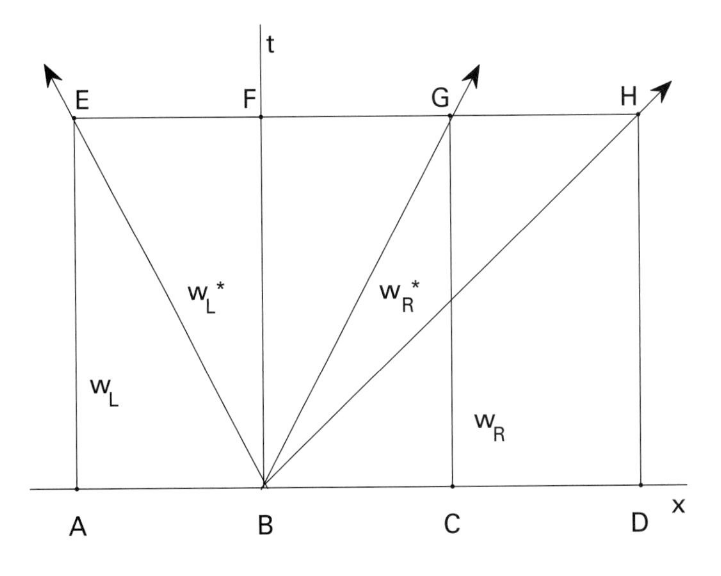

Figure 6.3 Flux balance for the HLLC scheme.

Because the middle wave represents a contact discontinuity, we impose the conditions

$$u_L^* = u_R^* = s_M$$
$$p_L^* = p_R^* = p^*.$$

Now, (6.30) and (6.31) can be rearranged as

$$s_L \boldsymbol{w}_L^* - \boldsymbol{f}_L^* = \boldsymbol{f}_L - s_L \boldsymbol{w}_L = \boldsymbol{q}_L$$
$$s_R \boldsymbol{w}_R^* - \boldsymbol{f}_R^* = \boldsymbol{f}_R - s_R \boldsymbol{w}_R = \boldsymbol{q}_R,$$

where $\boldsymbol{q}_R$ and $\boldsymbol{q}_L$ are known vectors depending on the left and right states. This enables the states $\boldsymbol{w}_L^*$ and $\boldsymbol{w}_R^*$ to be calculated as

$$\boldsymbol{w}_L^* = \frac{s_L - u_L}{s_L - s_M} = \begin{bmatrix} \rho_L \\ \rho_L\, s_M \\ \rho_L E_L + (s_M - u_L)\left(\rho_L\, s_M + \frac{p_L}{s_L - u_L}\right) \end{bmatrix}$$

$$\boldsymbol{w}_R^* = \frac{s_R - u_R}{s_R - s_M} = \begin{bmatrix} \rho_R \\ \rho_R\, s_M \\ \rho_R E_R + (s_M - u_R)\left(\rho_R\, s_M + \frac{p_R}{s_R - u_R}\right) \end{bmatrix},$$

which determine f_L^* and f_R^* via (6.30) and (6.31). Finally, the numerical flux is

$$\boldsymbol{f}_{HLLC} = \begin{cases} \boldsymbol{f}_L & \text{if} \quad 0 \le s_L \\ \boldsymbol{f}_L^* & \text{if} \quad s_L \le 0 \le s_M \\ \boldsymbol{f}_R^* & \text{if} \quad s_M \le 0 \le s_R \\ \boldsymbol{f}_R & \text{if} \quad s_R \le 0 \end{cases}.$$

These formulas require an estimate of the wave speed s_M. One way to do this is to consider the average state $\boldsymbol{w}^*$ between the left and right waves. According to (6.26) in the previous section,

$$\boldsymbol{w}^* = \frac{s_R \boldsymbol{w}_R - s_L \boldsymbol{w}_L - (\boldsymbol{f}_R - \boldsymbol{f}_L)}{s_R - s_L},$$

which gives

$$\rho^* = \frac{\rho_R(s_R - u_R) - \rho_L(s_L - u_L)}{s_R - s_L}$$

and

$$(\rho u)^* = \frac{\rho_R u_R(s_R - u_R) - \rho_L u_L(s_L - u_L) - p_R + p_L}{s_R - s_L}.$$

Then, we estimate

$$s_M = u^* = \frac{(\rho u)^*}{\rho^*} = \frac{\rho_R u_R(s_R - u_R) - \rho_L u_L(s_L - u_L) - p_R + p_L}{\rho_R(s_R - u_R) - \rho_L(s_L - u_L)}.$$

6.11 A Unified Model of Numerical Flux via Artificial Diffusion

Following the flux difference splitting approach, a variety of alternative numerical fluxes can be derived by introducing a generalized diffusive flux defined by a matrix coefficient. Suppose that the numerical flux is

$$\boldsymbol{h}_{j+\frac{1}{2}} = (\boldsymbol{f}_{j+1} + \boldsymbol{f}_j) - \boldsymbol{d}_{j+\frac{1}{2}},$$

where the diffusive flux is assumed to have the form

$$\boldsymbol{d}_{j+\frac{1}{2}} = \frac{1}{2}\alpha_{j+\frac{1}{2}} B_{j+\frac{1}{2}}(\boldsymbol{w}_{j+1} - \boldsymbol{w}_j).$$

The matrix $B_{j+\frac{1}{2}}$ determines the properties of the scheme, while the scaling factor $\alpha_{j+\frac{1}{2}}$ is included for convenience.

Introducing a Roe linearization, let $A_{j+\frac{1}{2}}(w_{j+1} - w_j)$ be an estimate of the Jacobian matrix with the property that

$$A_{j+\frac{1}{2}}(\boldsymbol{w}_{j+1} - \boldsymbol{w}_j) = \boldsymbol{f}_{j+1} - \boldsymbol{f}_j. \tag{6.32}$$

Scalar diffusion is produced by setting

$$B_{j+\frac{1}{2}} = I.$$

Then, in order to ensure that the upwind biasing terms dominate all the eigenvalues of $A_{j+\frac{1}{2}}$, it is natural to take

$$\alpha_{j+\frac{1}{2}} \geq \max_k \left|\lambda^{(k)}\left(A_{j+\frac{1}{2}}\right)\right|. \tag{6.33}$$

This scheme is variously referred to by different authors as the generalized Lax–Friedrichs flux or the Rusanov flux (Rusanov 1961). The choice

$$B_{j+\frac{1}{2}} = \left|A_{j+\frac{1}{2}}\right|$$

recovers the Roe scheme described in the last two sections.

A family of schemes is obtained by representing $B_{j+\frac{1}{2}}$ as a function of $A_{j+\frac{1}{2}}$, which could be expanded as a power series. But according to the Cayley–Hamilton theorem, a matrix satisfies its own characteristic equation. Therefore, the third and higher powers of $A_{j+\frac{1}{2}}$ can be eliminated, and there is no loss of generality in limiting $B_{j+\frac{1}{2}}$ to the quadratic form

$$B_{j+\frac{1}{2}} = \alpha_0 I + \alpha_1 A_{j+\frac{1}{2}} + \alpha_2 A^2_{j+\frac{1}{2}}.$$

This remains true even in the case of three-dimensional flow because the Jacobian matrix still has only three distinct eigenvalues. The characteristic upwind scheme (6.33) is obtained by substituting $A_{j+\frac{1}{2}} = V\Lambda V^{-1}$ and $A^2_{j+\frac{1}{2}} = V\Lambda^2 V^{-1}$. Then, α_0, α_1, and α_2 are determined from the three equations

$$\alpha_0 + \alpha_1\lambda^{(k)} + \alpha_2\lambda^{(k)^2} = \left|\lambda^{(k)}\right|, \; k = 1, 2, 3.$$

An intermediate scheme is obtaind by the two term expansion

$$B_{j+\frac{1}{2}} = \alpha_{j+\frac{1}{2}} I + \beta_{j+\frac{1}{2}} A_{j+\frac{1}{2}}. \tag{6.34}$$

Then, it follows from the definition (6.32) of the Roe matrix that the diffusive flux can be expressed as a combination of differences of the state and flux vectors

$$\boldsymbol{d}_{j+\frac{1}{2}} = \frac{1}{2}\alpha_{j+\frac{1}{2}}(\boldsymbol{w}_{j+1} - \boldsymbol{w}_j) + \frac{1}{2}\beta_{j+\frac{1}{2}}(\boldsymbol{f}_{j+1} - \boldsymbol{f}_j).$$

Schemes of this class are fully upwind in supersonic flow if one takes $\alpha_{j+\frac{1}{2}} = 0$ and $\beta_{j+\frac{1}{2}} = \operatorname{sgn}(M)$ when the absolute value of the Mach number M exceeds 1.

The next section examines the discrete shock structure associated with the choice of the diffusion matrix $B_{j+\frac{1}{2}}$. It turns out that stationary shocks with a single interior point are supported both by Roe's characteristic upwind scheme and also by the two term scheme, provided that the coefficients $\alpha_{j+\frac{1}{2}}$ and $\beta_{j+\frac{1}{2}}$ are chosen appropriately, leading to the convective upwind and split pressure (CUSP) scheme (Jameson 1995b). It is also shown in this paper that it is not possible to obtain a stationary shock with a single interior point using scalar diffusion.

6.12 Discrete Shock Structure

6.12.1 Conditions for a Stationary Shock

A discrete shock structure with a finite number of interior points has the advantage that it could be contained in an otherwise smooth solution without contaminating the accuracy of the smooth parts. In a one-dimensional flow, such a structure allows discrete solutions in exact agreement with the true solution outside the shock region. The model of a discrete shock, which will be examined, is illustrated in Figure 6.4. Suppose that $\boldsymbol{w}_L$ and $\boldsymbol{w}_R$ are left and right states that satisfy the jump conditions for a stationary shock, and that the corresponding fluxes are $\boldsymbol{f}_L = \boldsymbol{f}(\boldsymbol{w}_L)$ and $\boldsymbol{f}_R = \boldsymbol{f}(\boldsymbol{w}_R)$.

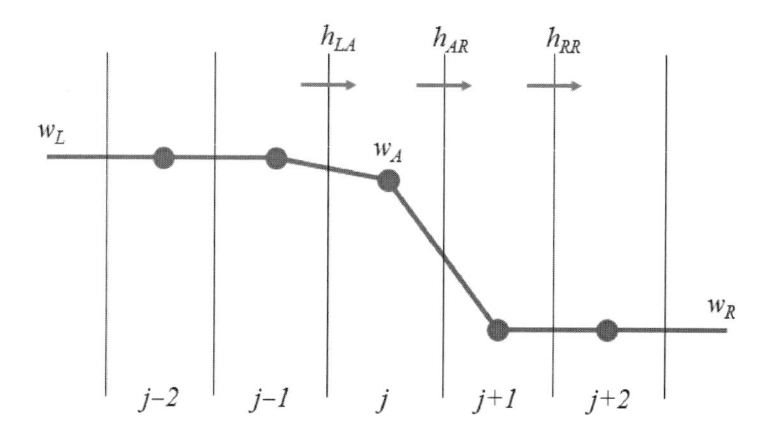

Figure 6.4 Discrete shock structure with a single interior point.

Since the shock is stationary, $\boldsymbol{f}_L = \boldsymbol{f}_R$. The ideal discrete shock has constant states $\boldsymbol{w}_L$ to the left and $\boldsymbol{w}_R$ to the right, and a single point with an intermediate value $\boldsymbol{w}_A$. The intermediate value is needed to allow the discrete solution to correspond to a true solution in which the shock wave does not coincide with an interface between two mesh cells. According to (6.1),

$$\int_0^L \boldsymbol{w}(T)dx = \int_0^L \boldsymbol{w}(0)dx - \int_0^T (\boldsymbol{f}_{RB} - \boldsymbol{f}_{LB})dt,$$

where $\boldsymbol{f}_{LB}$ and $\boldsymbol{f}_{RB}$ are the fluxes at the left and right boundaries. Assuming that the boundary conditions are compatible with a steady solution containing a stationary shock, the location x_s of the shock is fixed by this equation, since

$$\int_0^L \boldsymbol{w}(T)dx = x_s \boldsymbol{w}_L + (L - x_s)\boldsymbol{w}_R.$$

Similarly in the semi-discrete system,

$$\Delta x \sum_j \boldsymbol{w}_j(T) = \Delta x \sum_j \boldsymbol{w}_j(0) - \int_0^T (\boldsymbol{f}_{RB} - \boldsymbol{f}_{LB})dt.$$

Thus, $\sum_j \boldsymbol{w}_j(T)$ has a value that is determined by the initial and boundary conditions, and in general, it is not possible for this value to be attained by a discrete solution without an intermediate point, because then the sum would be quantized, increasing by $\Delta x(\boldsymbol{w}_R - \boldsymbol{w}_L)$ whenever the shock location is shifted one cell to the right.

6.12.2 Discrete Shock Structure of the Roe Scheme

It has been shown by Roe (1981) that the characteristic upwind scheme admits ideal stationary shocks with a single interior point. Assuming flow to the right with $u > 0$, the fluxes in the cell to the right of the shock are now

$$\boldsymbol{h}_{RR} = \boldsymbol{f}_R,$$

and

$$\boldsymbol{h}_{AR} = \frac{1}{2}(\boldsymbol{f}_R + \boldsymbol{f}_A) - \frac{1}{2}|A_{AR}|(\boldsymbol{w}_R - \boldsymbol{w}_A),$$

yielding equilibrium if

$$(A_{AR} + |A_{AR}|)(\boldsymbol{w}_R - \boldsymbol{w}_A) = V(\Lambda + |\Lambda|)V^{-1}(\boldsymbol{w}_R - \boldsymbol{w}_A) = 0.$$

With $u < c$, this is satisfied by the negative eigenvalue $u - c$, and since $w_R - w_A$ is the corresponding eigenvector, the Hugoniot equation

$$\boldsymbol{f}_R - \boldsymbol{f}_A = u_S(\boldsymbol{w}_R - \boldsymbol{w}_A)$$

is satisfied for the shock speed $u_S = u - c$. Thus, w_A lies on a Hugoniot curve. At the entrance to the shock, the transition from $\boldsymbol{w}_L$ to $\boldsymbol{w}_A$ is less than the full transition from $\boldsymbol{w}_L$ to $\boldsymbol{w}_R$ for which $u = c$. Thus, a structure is admitted in which $u > c$ in the transition from L to A, with the consequence that the flux is calculated from the upwind state

$$\boldsymbol{h}_{LA} = \frac{1}{2}(\boldsymbol{f}_L + \boldsymbol{f}_A) - \frac{1}{2}|A_{AR}|(\boldsymbol{w}_L - \boldsymbol{w}_A) = \boldsymbol{f}_L,$$

and equilibrium is maintained.

6.12.3 Discrete Shock Structure of the CUSP Scheme

In order to design a two term scheme of the type defined by (6.34) so that it supports an ideal stationary shock with a single interior point, it is convenient to redefine the coefficients so that

$$\boldsymbol{d}_{j+\frac{1}{2}} = \frac{1}{2}\alpha^* c(\boldsymbol{w}_{j+1} - \boldsymbol{w}_j) + \frac{1}{2}\beta(\boldsymbol{f}_{j+1} - \boldsymbol{f}_j),$$

where the factor c is included so that α^* is dimensionless. Let M be the Mach number $\frac{u}{c}$. If the flow is supersonic, an upwind scheme is obtained by setting

$$\alpha^* = 0\,,\ \beta = \text{sgn}(M).$$

Introducing the Roe linearization, the Mach number is calculated from u and c, and at the entrance to the shock, a transition to an intermediate value w_A is admitted with $u > c$ and

$$\boldsymbol{h}_{LA} = \frac{1}{2}(\boldsymbol{f}_L + \boldsymbol{f}_A) - \frac{1}{2}(\boldsymbol{f}_A - \boldsymbol{f}_L) = \boldsymbol{f}_L.$$

The fluxes leaving and entering the cell immediately to the right of the shock are now

$$\boldsymbol{h}_{RR} = \boldsymbol{f}_R$$

and

$$\boldsymbol{h}_{AR} = \frac{1}{2}(\boldsymbol{f}_R + \boldsymbol{f}_A) - \frac{1}{2}\alpha^* c(\boldsymbol{w}_R - \boldsymbol{w}_A) - \frac{1}{2}\beta(\boldsymbol{f}_R - \boldsymbol{f}_A).$$

These are in equilibrium if

$$\boldsymbol{f}_R - \boldsymbol{f}_A + \frac{\alpha^* c}{1+\beta}(\boldsymbol{w}_R - \boldsymbol{w}_A) = 0.$$

This is the Hugoniot equation for a shock moving to the left with a speed $u_S = \frac{\alpha^* c}{1+\beta}$. Also, introducing the Roe linearization,

$$\left(A_{AR} + \frac{\alpha^* c}{1+\beta} I\right)(\boldsymbol{w}_R - \boldsymbol{w}_A) = 0.$$

Thus, $\boldsymbol{w}_R - \boldsymbol{w}_A$ is an eigenvector of A_{AR}, and $-\frac{\alpha^* c}{1+\beta}$ is the corresponding eigenvalue. Since the eigenvalues are $u, u + c$, and $u - c$, the only choice that leads to positive diffusion when $u > 0$ is $u - c$, yielding the relationship

$$\alpha^* c = (1 + \beta)(c - u), \quad 0 < u < c.$$

Thus, β is uniquely determined once α^* is chosen, leading to a one-parameter family of schemes. The choice $\beta = M$ and $\alpha^* = 1 - M^2$ corresponds to the Harten–Lax–van Leer (HLL) scheme (6.27, 6.28, 6.29), which is extremely diffusive.

The term $\beta(f_R - f_A)$ contributes to the diffusion of the convective terms. Suppose that the convective terms are separated by splitting the flux as follows:

$$\boldsymbol{f} = u\boldsymbol{w} + \boldsymbol{f}_p,$$

where

$$\boldsymbol{f}_p = \begin{pmatrix} 0 \\ p \\ up \end{pmatrix}.$$

Then,

$$\boldsymbol{f}_{j+1} - \boldsymbol{f}_j = \bar{u}(\boldsymbol{w}_{j+1} - \boldsymbol{w}_j) + \bar{\boldsymbol{w}}(u_{j+1} - u_j) + \boldsymbol{f}_{p_{j+1}} - \boldsymbol{f}_{p_j},$$

where $\bar{u}$ and $\bar{\boldsymbol{w}}$ are the arithmetic averages

$$\bar{u} = \frac{1}{2}(u_{j+1} + u_j), \quad \bar{\boldsymbol{w}} = \frac{1}{2}(\boldsymbol{w}_{j+1} + \boldsymbol{w}_j).$$

Then, the total effective coefficient of convective diffusion is

$$\alpha c = \alpha^* c + \beta \bar{u}.$$

The choice $\alpha c = \bar{u}$ leads to low diffusion near a stagnation point and also leads to a smooth continuation of convective diffusion across the sonic line since $\alpha^* = 0$ and $\beta = 1$ when $|M| > 1$. The scheme must also be formulated so that the cases of $u > 0$ and $u < 0$ are treated symmetrically. Using the notation $M = \frac{u}{c}$, $\lambda^{\pm} = u \pm c$, this leads to the diffusion coefficients

$$\alpha = |M|, \tag{6.35}$$

$$\beta = \begin{cases} +\max(0, \frac{u+\lambda^-}{u-\lambda^-}) & \text{if} \quad 0 \le M \le 1 \\ -\max(0, \frac{u+\lambda^+}{u-\lambda^+}) & \text{if} \quad -1 \le M \le 0 \\ \operatorname{sgn}(M) & \text{if} \quad |M| \ge 1. \end{cases} \tag{6.36}$$

Near a stagnation point, α may be modified to $\alpha = \frac{1}{2}(\alpha_0 + \frac{|M|^2}{\alpha_0})$ if $|M|$ is smaller than a threshold α_0. The expression for β in subsonic flow can also be expressed as

$$\beta = \begin{cases} \max(0, 2M - 1) & \text{if} \quad 0 \le M \le 1 \\ \min(0, 2M + 1) & \text{if} \quad -1 \le M \le 0\,. \\ \operatorname{sgn}(M) & \text{if} \quad |M| \ge 1 \end{cases}$$

Equation (6.36) remains valid when the CUSP scheme is modified as described in Section 6.13 to allow solutions with constant stagnation enthalpy. The coefficients $\alpha(M)$ and $\beta(M)$ are displayed in Figure 6.5 for the case when $\alpha_0 = 0$. The cutoff of

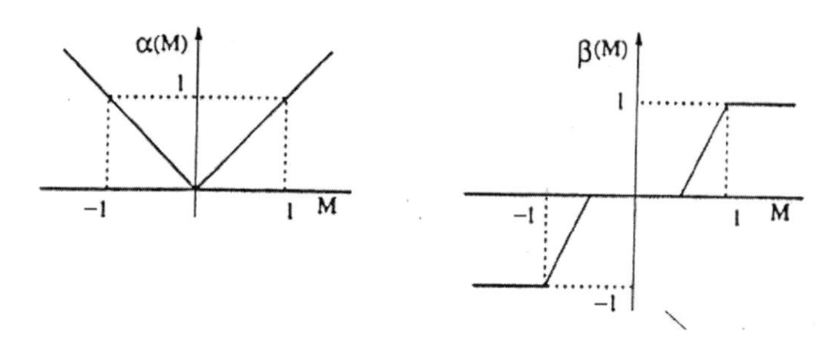

Figure 6.5 Diffusion Coefficients of the CUSP scheme.

β when $|M| < \frac{1}{2}$, together with α approaching zero as $|M|$ approaches zero, is also appropriate for the capture of contact discontinuities.

6.12.4 Criteria for a Single Point Stationary Shock

The analysis of these three cases shows that a discrete shock structure with a single interior point is supported by artificial diffusion that satisfies the two conditions that

(1) it produces an upwind flux if the flow is determined to be supersonic through the interface

(2) it satisfies a generalized eigenvalue problem for the exit from the shock of the form

$$(A_{AR} + B_{AR})(\boldsymbol{w}_R - \boldsymbol{w}_A) = 0,$$

where A_{AR} is the linearized Jacobian matrix, and B_{AR} is the matrix defining the diffusion for the interface AR. These two conditions are satisfied by both the charcteristic and CUSP schemes. Scalar diffusion does not satisfy the first condition.

6.13 Schemes Admitting Constant Total Enthalpy

In steady flow, the stagnation enthalpy H is constant, corresponding to the fact that the energy and mass equations are consistent when the constant factor H is removed from the energy equation. Discrete and semi-discete schemes do not necessarily satisfy this property. In the case of a semi-discrete scheme expressed in viscosity form,

$$\Delta x \frac{d\boldsymbol{w}_j}{dt} + \boldsymbol{h}_{j+\frac{1}{2}} - \boldsymbol{h}_{j-\frac{1}{2}} = 0,$$

where $\boldsymbol{h}_{j+\frac{1}{2}} = \frac{1}{2}(\boldsymbol{f}_{j+1} + \boldsymbol{f}_j) - \boldsymbol{d}_{j+\frac{1}{2}}$ and $\boldsymbol{d}_{j+\frac{1}{2}}$ is the added diffusive flux, a solution with constant H is admitted if the viscosity for the energy equation reduces to the viscosity for the continuity equation with ρ replaced by ρH. When the standard characteristic decomposition (6.24) is used, the viscous fluxes for ρ and ρH that result from composition of the fluxes for the characteristic variables do not have this property, and H is not constant in the discrete solution. In practice there is an excursion of H in the discrete shock structure that represents a local heat source. In very high speed flows, the corresponding error in the temperature may lead to a wrong prediction of associated effects such as chemical reactions.

The source of the error in the stagnation enthalpy is the discrepancy between the convective terms

$$u \begin{bmatrix} \rho \\ \rho u \\ \rho H \end{bmatrix}$$

in the flux vector, which contain ρH, and the state vector, which contains ρE. This may be remedied by introducing a modified state vector

$$\boldsymbol{w}_h = \begin{bmatrix} \rho \\ \rho u \\ \rho H \end{bmatrix}.$$

Isenthalpic formulations have been considered by Veuillot and Viviand (1979), and Lytton (1987).

One may now introduce the linearization

$$\boldsymbol{f}_R - \boldsymbol{f}_L = A_h(\boldsymbol{w}_{h_R} - \boldsymbol{w}_{h_L}).$$

Here, A_h may be calculated in the same way as the standard Roe linearization. On introducing the vector

$$\boldsymbol{v} = \begin{bmatrix} \sqrt{\rho} \\ \sqrt{\rho} u \\ \sqrt{\rho} H \end{bmatrix},$$

all quantities in both $\boldsymbol{f}$ and $\boldsymbol{w}_h$ are products of the form $v_j v_k$ that have the property that a finite difference $\Delta(v_j v_k)$ between left and right states can be expressed as

$$\Delta(v_j v_k) = \bar{v}_j \Delta v_k + \bar{v}_k \Delta v_j,$$

where $\bar{v}_j$ is the arithmetic mean $\frac{1}{2}(v_{j_R} + v_{j_L})$. Therefore,

$$\Delta \boldsymbol{w} = B \Delta \boldsymbol{v}, \quad \Delta \boldsymbol{f} = C \Delta \boldsymbol{v} = C B^{-1} \Delta \boldsymbol{w},$$

where B and C can be expressed in terms of appropriate mean values of the quantities v_j.

Define

$$u = \frac{\sqrt{\rho_R} u_R + \sqrt{\rho_L} u_L}{\sqrt{\rho_R} + \sqrt{\rho_L}}, \quad H = \frac{\sqrt{\rho_R} H_R + \sqrt{\rho_L} H_L}{\sqrt{\rho_R} + \sqrt{\rho_L}}$$

and

$$c = \sqrt{(\gamma - 1)\left(H - \frac{u^2}{2}\right)}.$$

Then,

$$A_h = \begin{bmatrix} 0 & 1 & 0 \\ -\frac{\gamma+1}{\gamma}\frac{u^2}{2} & \frac{\gamma+1}{\gamma}u & \frac{\gamma-1}{\gamma} \\ -uH & H & u \end{bmatrix}.$$

The eigenvalues of A_h are u, λ^+ and λ^-, where

$$\lambda^\pm = \frac{\gamma+1}{2\gamma}u \pm \sqrt{\left(\frac{\gamma+1}{2\gamma}u\right)^2 + \frac{c^2 - u^2}{\gamma}}. \tag{6.37}$$

Note that λ^+ and λ^- have the same sign as $u + c$ and $u - c$ and change sign at the sonic line $u = \pm c$. The corresponding eigenvectors of A_h are the columns of

$$V = \begin{bmatrix} 1 & 1 & 1 \\ u & \lambda^+ & \lambda^- \\ \frac{u^2}{2} & H & H \end{bmatrix}.$$

Also, the left eigenvectors of A_h are the rows of

$$V^{-1} = \frac{1}{D}\begin{bmatrix} (\lambda^+ - \lambda^-)H & 0 & (\lambda^+ - \lambda^-) \\ -\left(uH - \lambda^- \frac{u^2}{2}\right) & H - \frac{u^2}{2} & -(\lambda^- - u) \\ uH - \lambda^+ \frac{u^2}{2} & -\left(H - \frac{u^2}{2}\right) & \lambda^+ - u \end{bmatrix},$$

where

$$D = (\lambda^+ - \lambda^-)\left(H - \frac{u^2}{2}\right) = (\lambda^+ - \lambda^-)\frac{c^2}{\gamma - 1}.$$

Then

$$A_h = V \Lambda V^{-1},$$

where

$$\Lambda = \begin{bmatrix} u & & \\ & \lambda^+ & \\ & & \lambda^- \end{bmatrix}.$$

The same development can be carried out for multi-dimensional flows.

Using the modified linearization, both the characteristic upwind scheme and the CUSP scheme can be reformulated as follows to admit steady solutions with constant H.

6.13.1 H-Characteristic Upwind Scheme

The diffusion for the characteristic upwind scheme is now defined to be

$$\boldsymbol{d}_{j+\frac{1}{2}} = \frac{1}{2}\left|A_{h_{j+\frac{1}{2}}}\right|(\boldsymbol{w}_{j+1} - \boldsymbol{w}_j),$$

in which $|A_h|$ is defined to be

$$|A_h| = V \Lambda V^{-1},$$

where V is the eigenvector matrix of A_h, and

$$|\Lambda| = \begin{bmatrix} |u| & & \\ & |\lambda^+| & \\ & & |\lambda^-| \end{bmatrix}.$$

In order to show that the scheme admits a solution with constant H, split the diffusion into two parts:

$$\boldsymbol{d}_{j+\frac{1}{2}} = \boldsymbol{d}^{(1)}_{j+\frac{1}{2}} + \boldsymbol{d}^{(2)}_{j+\frac{1}{2}},$$

where $\boldsymbol{d}^{(1)}_{j+\frac{1}{2}}$ is the contribution from $|u|$ and where $\boldsymbol{d}^{(2)}_{j+\frac{1}{2}}$ is the contribution from $|\lambda^+|$ and $|\lambda^-|$. Then

$$\begin{aligned} \boldsymbol{d}^{(2)}_{j+\frac{1}{2}} &= \begin{bmatrix} 1 & 1 & 1 \\ u & \lambda^+ & \lambda^- \\ \frac{u^2}{2} & H & H \end{bmatrix} \begin{bmatrix} 0 & & \\ & |\lambda^+| & \\ & & |\lambda^-| \end{bmatrix} V^{-1} \Delta \boldsymbol{w}_h \\ &= \begin{bmatrix} 0 & |\lambda^+| & |\lambda^-| \\ 0 & |\lambda^+|\lambda^+ & |\lambda^-|\lambda^- \\ 0 & |\lambda^+|H & |\lambda^-|H \end{bmatrix} V^{-1} \Delta \boldsymbol{w}_h, \end{aligned}$$

and the third element of $d^{(1)}_{j+\frac{1}{2}}$ equals the first element multiplied by H. Also

$$\boldsymbol{d}^{(1)}_{j+\frac{1}{2}} = \frac{(\gamma-1)|u|}{c^2} \begin{bmatrix} 1 & 1 & 1 \\ u & \lambda^+ & \lambda^- \\ \frac{u^2}{2} & H & H \end{bmatrix} \begin{bmatrix} H\Delta\rho - \Delta(\rho H) \\ 0 \\ 0 \end{bmatrix} = \frac{(\gamma-1)\rho|u|\Delta H}{c^2} \begin{bmatrix} 1 \\ u \\ \frac{u^2}{2} \end{bmatrix},$$

and this is zero if H is constant. Thus, both contributions are consistent with a steady solution in which H is constant. The two variations of the characteristic splitting can conveniently be distinguished as the E- and H-characteristic schemes.

The property of admitting steady solutions with constant H is not automatically preserved by higher order constructions. If limiters are used to preserve monotonicity and these introduce comparisons of neighboring slopes of the characteristic variables, then the relationship between the diffusive fluxes for the mass and energy equations will no longer be consistent with constant stagnation enthalpy.

6.13.2 H-CUSP Scheme

The diffusive flux is now expressed as

$$\boldsymbol{d}_{j+\frac{1}{2}} = \frac{1}{2}\alpha^* c \Delta \boldsymbol{w}_h + \frac{1}{2}\beta \Delta \boldsymbol{f},$$

where Δ denotes the difference from $j+1$ to j. Again equilibrium at the entrance is established by upwinding, while equilibrium at the exit requires

$$\Delta \boldsymbol{f} + \frac{\alpha^* c}{1+\beta} \Delta \boldsymbol{w}_h = \left(A_h + \frac{\alpha^* c}{1+\beta} I \right) \Delta \boldsymbol{w}_h = 0.$$

Therefore, $-\frac{\alpha^* c}{1+\beta}$ must be an eigenvalue of A_h, and in the case $u > 0$, positive diffusion is obtained by taking

$$\alpha^* c = -(1+\beta)\lambda^-.$$

Now, the split is redefined as

$$\boldsymbol{f} = u\boldsymbol{w}_h + \boldsymbol{f}_p,$$

where

$$\boldsymbol{f}_p = \begin{bmatrix} 0 \\ p \\ 0 \end{bmatrix},$$

and the diffusive flux can be expressed as

$$\boldsymbol{d}_{j+\frac{1}{2}} = \frac{1}{2}\alpha c \Delta \boldsymbol{w}_h + \frac{1}{2}\beta \bar{\boldsymbol{w}}_h \Delta \boldsymbol{u} + \frac{1}{2}\beta \Delta \boldsymbol{f}_p.$$

Then, α and β are defined as before by (6.35) and (6.36), using the modified eigenvalues $\lambda^\pm$ defined in (6.37). This splitting corresponds to the Liou–Steffen splitting (Liou & Steffen 1993). The splitting in which the convective terms contain ρE corresponds to the wave particle splitting (Deshpande, Balakrishnan, & Raghurama Rao 1994). As in the case of characteristic splitting, the two variations can conveniently be distinguished as the E-CUSP and H-CUSP schemes.

6.14 The Advection Upstream Splitting Method (AUSM)

The advection upstream splitting method (AUSM) is an alternative splitting method designed to capture the best properties of both flux splitting and flux difference splitting. The original form of AUSM (Liou & Steffen 1993) has been further refined by Liou as AUSM$^+$ (Liou 1996) and also extended to low Mach flows as AUSM$^+$-up (Liou 2006). Here we shall describe AUSM$^+$. Both AUSM and AUSM$^+$ are based on the idea of splitting the flux into convective and pressure terms thus:

$$\boldsymbol{f} = u \begin{bmatrix} \rho \\ \rho u \\ \rho H \end{bmatrix} + \begin{bmatrix} 0 \\ p \\ 0 \end{bmatrix}.$$

These terms are then treated separately using blending formulas between the left and right states. First, upstream and downstream Mach numbers are defined as

$$M_j = \frac{u_j}{c_{j+\frac{1}{2}}}, \quad M_{j+1} = \frac{u_{j+1}}{c_{j+\frac{1}{2}}},$$

based on a common speed of sound $c_{j+\frac{1}{2}}$. Then, blending formulas

$$m_{j+\frac{1}{2}} = \mathcal{M}^+(M_j) + \mathcal{M}^-(M_{j+1})$$

and

$$p_{j+\frac{1}{2}} = \mathcal{P}^+(M_j) + \mathcal{P}^-(M_{j+1})$$

are introduced, and the interface flux is defined as

$$\boldsymbol{h}_{j+\frac{1}{2}} = c_{j+\frac{1}{2}} \left\{ \max\left(m_{j+\frac{1}{2}}, 0\right) \begin{bmatrix} \rho \\ \rho u \\ \rho H \end{bmatrix}_j + \min\left(m_{j+\frac{1}{2}}, 0\right) \begin{bmatrix} \rho \\ \rho u \\ \rho H \end{bmatrix}_{j+1} + \begin{bmatrix} 0 \\ p_{j+\frac{1}{2}} \\ 0 \end{bmatrix} \right\}.$$

The convective blending functions $\mathcal{M}^\pm$ are polynomials of the Mach number such that

$$\mathcal{M}^+ \geq 0, \quad \mathcal{M}^- \leq 0,$$

$$\mathcal{M}^+ = M, \quad \mathcal{M}^- = 0, \quad \text{when } M > 1,$$

$$\mathcal{M}^+ = 0, \quad \mathcal{M}^- = M, \quad \text{when } M < 1,$$

and

$$\mathcal{M}^\pm = \pm(M \pm 1)^2 \pm \beta(M^2 - 1)^2, \quad |M| < 1.$$

Liou recommends $\beta = \frac{1}{8}$, which yields a zero second derivative when $M = 0$. Correspondingly, the pressure blending functions $\mathcal{P}^\pm$ are polynomials such that

$$\mathcal{P}^\pm \geq 0,$$

$$\mathcal{P}^+ = 1, \quad \mathcal{P}^- = 0, \quad \text{when } M > 1,$$

$$\mathcal{P}^- = 1, \quad \mathcal{P}^+ = 0, \quad \text{when } M < 1,$$

and

$$\mathcal{P}^\pm = \frac{1}{4}(M \pm 1)^2(2 \mp M) \pm \alpha M(M^2 - 1)^2, \quad |M| < 1.$$

Liou recommends $\alpha = \frac{3}{16}$, which yields a zero second derivative at $M = \pm 1$.

Let c^* be the critical speed of sound at which $u = c^*$ in a steady flow for which

$$H = \frac{c^2}{\gamma - 1} + \frac{u^2}{2} = \text{constant}.$$

Then,

$$\frac{\gamma + 1}{\gamma - 1}\frac{c^{*2}}{2} = H.$$

Across a stationary shock, the left and right speeds u_L and u_R satisfy

$$u_L u_R = c^{*2}.$$

Liou shows that if $u_j > c_j$, the AUSM$^+$ scheme admits a solution with a stationary shock between cells j and $j+1$ if the interface speed of sound is evaluated as

$$c_{j+\frac{1}{2}} = \frac{c^{*2}_j}{u_j}.$$

To allow for other situations where u_j is subsonic or $u_j < 0$, he proposes

$$c_{j+\frac{1}{2}} = \min(\tilde{c}_j, \tilde{c}_{j+1}),$$

where

$$\tilde{c} = \begin{cases} \dfrac{c^{*2}}{|u|}, & \text{if } |u| > c^*, \\ c^*, & \text{if } |u| < c^*. \end{cases}$$

Liou also shows that the AUSM scheme satisfies H = constant in steady flow. Moreover, it can admit a stationary contact discontinuity between cells j and $j+1$, and it preserves positive density with a forward Euler time stepping scheme if Δt is restricted to be sufficiently small.

6.15 Positivity Preserving Schemes

The LED and TVD principles can be applied to a linear system when it is written in the characteristic form (6.9). The behavior of a nonlinear system is more complex: wave interactions could lead to new extrema in the conservative variables. In the case of gas dynamics, we know that the density and pressure must remain non-negative in a true solution. When the equations are solved in conservative form as defined by (6.1) to (6.7), the pressure is recovered from the internal energy via the relation

$$p = (\gamma - 1)\left(e - \frac{m^2}{2\rho}\right).$$

Here, e is the total energy, which is updated from the energy equation, while $\frac{m^2}{2\rho}$ is the kinetic energy, which is derived from the solution of the momentum equation.

Accordingly, over-prediction of the kinetic energy relative to the total energy can lead to a negative value of the pressure. This can be a real difficulty in simulations of hypersonic re-entry vehicles, such as the NASA space shuttle. These typically re-enter the atmosphere at a very high angle of attack, experiencing extremely high pressures and temperatures on their lower side but very low pressures on the upper lee side of their wings after the flow has expanded as it spills over the wing leading edge and tip.

In order to address these issues, we must examine precisely the conditions under which a numerical scheme can generate negative values of the pressure or density.

Conservative schemes that guarantee the non-negativity of these quantities are called "positivity preserving" or "positively conservative." Some preliminary results will be needed to enable the analysis of these schemes. We define a state

$$\mathbf{w} = \begin{bmatrix} \rho \\ m \\ e \end{bmatrix}$$

as physically realizable if $\rho > 0$ and the corresponding pressure $p \geq 0$.

First, it will be shown that a convex combination of physically realizable states is physically realizable. Suppose that

$$\mathbf{w} = \sum_{i=1}^{n} \alpha_i \mathbf{w}_i,$$

where

$$\mathbf{w}_i = \begin{bmatrix} \rho_i \\ m_i \\ e_i \end{bmatrix}$$

and

$$\alpha_i > 0, \quad p_i > 0, \quad \rho_i \geq 0.$$

Then,

$$\rho = \sum_{i=1}^{n} \alpha_i \rho_i > 0.$$

Also,

$$\begin{aligned} p &= (\gamma - 1)\left(\sum_{i=1}^{n} \alpha_i e_i - \frac{\left(\sum_{i=1}^{n} \alpha_i m_i\right)^2}{2\sum_{i=1}^{n} \alpha_i \rho_i}\right) \\ &= (\gamma - 1)\left(\sum_{i=1}^{n} \alpha_i \left(\frac{p_i}{\gamma - 1} + \frac{m_i^2}{2\rho_i}\right) - \frac{\left(\sum_{i=1}^{n} \alpha_i m_i\right)^2}{2\sum_{i=1}^{n} \alpha_i \rho_i}\right) \\ &= \sum_{i=1}^{n} \alpha_i p_i + \frac{\gamma - 1}{2} \frac{R}{\sum_{i=1}^{n} \alpha_i \rho_i}, \end{aligned}$$

where

$$R = \left(\sum_{i=1}^{n} \alpha_i \rho_i\right)\left(\sum_{i=1}^{n} \alpha_i \frac{m_i^2}{\rho_i}\right) - \left(\sum_{i=1}^{n} \alpha_i m_i\right)^2.$$

Also,

$$m_i = \rho_i u_i.$$

Set

$$r_i = \alpha_i \rho_i.$$

Then, $r_i > 0$, and

$$R = \left(\sum_{i=1}^{n} (\sqrt{r_i})^2\right) \sum \left(\sqrt{r_i} u_i\right)^2 - \left(\sum \sqrt{r_i}\sqrt{r_i} u_i\right)^2 \geq 0$$

by the Cauchy-Schwarz inequality. Hence,

$$p \geq \sum_{i=1}^{n} \alpha_i p_i \geq 0.$$

Next, observe that the eigenvectors of the Jacobian matrix A scaled by ρ,

$$\boldsymbol{v}^{(1)} = \rho \begin{bmatrix} 1 \\ v \\ \frac{u^2}{2} \end{bmatrix}, \quad \boldsymbol{v}^{(2)} = \rho \begin{bmatrix} 1 \\ u + c \\ H + uc \end{bmatrix}, \quad \boldsymbol{v}^{(3)} = \rho \begin{bmatrix} 1 \\ u - c \\ H - uc \end{bmatrix},$$

represent physically realizable states. In the case of $\boldsymbol{v}^{(1)}$, the total energy is

$$E = \frac{u^2}{2},$$

and

$$p = (\gamma - 1)\rho \left(E - \frac{u^2}{2}\right) = 0.$$

This represents a state at zero temperature. In the case of $\boldsymbol{v}^{(2)}$ or $\boldsymbol{v}^{(3)}$, the total energy is

$$E = H \pm uc = \frac{c^2}{\gamma - 1} + \frac{u^2}{2} \pm uc = \left(\frac{1}{\gamma - 1} - \frac{1}{2}\right) c^2 + \frac{1}{2}(u \pm c)^2,$$

while the velocity is $u \pm c$. Hence,

$$p = (\gamma - 1)\rho \left(E - \left(\frac{u \pm c}{2}\right)^2\right) = \rho \left(1 - \frac{\gamma - 1}{2}\right) c^2 > 0$$

if $\gamma < 3$, as is the case for a standard gas.

Now, using the fact that

$$\boldsymbol{f}(\boldsymbol{w}) = A\boldsymbol{w},$$

we can assess whether a discrete scheme is positively conservative by expressing it in terms of the eigenvectors corresponding to the state at each of the points in the stencil, and then determining whether the updated state $\boldsymbol{w}_j^{n+1}$ is a convex combination of the various eigenvectors.

6.16 Positivity for a Scheme with Scalar Diffusion

Consider first a scheme with scalar diffusion:

$$\boldsymbol{w}_j^{n+1} = \boldsymbol{w}_j^n - \frac{\Delta t}{\Delta x}\left(\boldsymbol{h}_{j+\frac{1}{2}}^n - \boldsymbol{h}_{j-\frac{1}{2}}^n\right),$$

where

$$\boldsymbol{h}_{j+\frac{1}{2}} = \frac{1}{2}(\boldsymbol{f}_{j+1} + \boldsymbol{f}_j) - \alpha_{j+\frac{1}{2}}(\boldsymbol{w}_{j+1} - \boldsymbol{w}_j).$$

Collecting the terms, we find that

$$\boldsymbol{w}_j^{n+1} = \frac{\Delta t}{\Delta x}\left(\alpha_{j+\frac{1}{2}} - \frac{1}{2}A_{j+1}\right)\boldsymbol{w}_{j+1}^n + \left(1 - \frac{\Delta t}{\Delta x}(\alpha_{j+\frac{1}{2}} + \alpha_{j-\frac{1}{2}})\right)\boldsymbol{w}_j^n$$

$$+ \frac{\Delta t}{\Delta x}(\alpha_{j-\frac{1}{2}} + \frac{1}{2}A_{j-1})\boldsymbol{w}_{j-1}^n.$$

Let $\boldsymbol{v}_j^{(k)}$ be the eigenvectors of A_j. Then,

$$\boldsymbol{w}_j^n = \sum_{k=1}^{3} \beta^{(k)}\boldsymbol{v}_j^{(k)},$$

where

$$\beta^{(1)} = \frac{\gamma - 1}{2\gamma}, \quad \beta^{(2)} = \frac{1}{2\gamma}, \quad \beta^{(3)} = \frac{1}{2\gamma}.$$

Now, the first term can be expressed as

$$\frac{\Delta t}{\Delta x}\sum_{k=1}^{3}\left(\alpha_{j+\frac{1}{2}} - \frac{1}{2}\lambda_{j+1}^{(k)}\right)\beta^{(k)}v_{j+1}^{(k)}$$

with non-negative coefficients if

$$\alpha_{j+\frac{1}{2}} \geq \frac{1}{2}\max_k |\lambda_{j+1}^{(k)}|.$$

The second term can be expressed as

$$\left(1 - \frac{\Delta t}{\Delta x}(\alpha_{j+\frac{1}{2}} + \alpha_{j-\frac{1}{2}})\right)\sum_{k=1}^{3}\beta^{(k)}\boldsymbol{v}_j^{(k)}.$$

The third term can be expressed as

$$\frac{\Delta t}{\Delta x}\sum_{k=1}^{3}\left(\alpha_{j-\frac{1}{2}} + \frac{1}{2}\lambda_{j-1}^{(k)}\right)\beta^{(k)}\boldsymbol{v}_{j-1}^{(k)}$$

with non-negative coefficients if

$$\alpha_{j-\frac{1}{2}} \geq \max_k |\lambda_{j-1}^{(k)}|.$$

Thus, the coefficients of every eigenvector are non-negative if for all j

$$\alpha_{j+\frac{1}{2}} \geq \max(\mu_{j+1}, \mu_j),$$

where

$$\mu_j = \frac{1}{2} \max_k(\lambda_j^{(k)})$$

and

$$\left(\alpha_{j+\frac{1}{2}} + \alpha_{j-\frac{1}{2}}\right) \frac{\Delta t}{\Delta x} \leq 1$$

or

$$\max(\mu_{j+1}, \mu_j, \mu_{j-1}) \frac{\Delta t}{\Delta x} \leq 1.$$

Accordingly, the scheme is positivity preserving under these conditions.

6.17 Positivity of the Flux Vector Splitting Scheme

Next, we consider the characteristic flux vector splitting scheme:

$$\boldsymbol{w}_j^{n+1} = w_j^n - \frac{\Delta t}{\Delta x}\left(\boldsymbol{h}_{j+\frac{1}{2}}^n - \boldsymbol{h}_{j-\frac{1}{2}}^n\right),$$

where

$$\boldsymbol{h}_{j+\frac{1}{2}} = \boldsymbol{f}_j^+ + \boldsymbol{f}_{j+1}^-.$$

Now,

$$\boldsymbol{w}_j^{n+1} = -\frac{\Delta t}{\Delta x} A_{j+1}^- \boldsymbol{w}_{j+1}^n + \left(I - \frac{\Delta t}{\Delta x}(A_j^+ - A_j^-)\right) \boldsymbol{w}_j^n + \frac{\Delta t}{\Delta x} A_{j-1}^+ \boldsymbol{w}_{j-1}^n.$$

Again, decomposing the state in cell j as

$$\boldsymbol{w}_j^n = \sum_{k=1}^{3} \beta^{(k)} \boldsymbol{v}_j^{(k)},$$

the first term is

$$-\frac{\Delta t}{\Delta x} \sum_{k=1}^{3} \left(\min(0, \lambda_{j+1}^{(k)}) \beta^{(k)} \boldsymbol{v}_{j+1}^{(k)}\right).$$

The second term is

$$1 - \frac{\Delta t}{\Delta x} \sum_{k=1}^{3} |\lambda^{(k)}| \beta^{(k)} \boldsymbol{v}_j^{(k)},$$

and the third term is

$$\frac{\Delta t}{\Delta x}\sum_{k=1}^{3}\left(\max(0,\lambda_{j-1}^{(k)})\beta^{(k)}\boldsymbol{v}_{j-1}^{(k)}\right).$$

The coefficients of every eigenvector are non-negative provided that

$$\max_k |\lambda^{(k)}|\frac{\Delta t}{\Delta x}\leq 1,$$

and correspondingly the scheme is positivity preserving under this restriction or the time step.

6.18 Second Order Accurate Schemes

The various upwind formulations just described lead to first order accuracy in their basic form. One might try to compensate for this by using very fine meshes, but this can lead to excessive computational costs. Assuming the use of affordable meshes, first order accuracy is generally not ever close to meeting the requirements of engineering analysis and design, as will be illustrated in some examples at the end of this chapter.

Second order accurate schemes using any of the fluxes discussed in the previous sections can be constructed by using reconstruction to obtain second order accurate estimates of the left and right values w_L and w_R of the state vector at each interface, as described in Sections 5.20 and 5.21, applied separately to each state variable. Then, one evaluates the interface flux as

$$\boldsymbol{h}_{j+\frac{1}{2}}=\frac{1}{2}(\boldsymbol{f}(\boldsymbol{w}_L)+\boldsymbol{f}(\boldsymbol{w}_R))-\boldsymbol{d}(\boldsymbol{w}_L,\boldsymbol{w}_R),$$

where the diffusive term $\boldsymbol{d}(\boldsymbol{w}_L,\boldsymbol{w}_R)$ is constructed according to whichever formula is preferred. If the mesh is smooth enough, one may alternatively restrict the use of reconstruction to the diffusive term and set

$$\boldsymbol{h}_{j+\frac{1}{2}}=\frac{1}{2}(\boldsymbol{f}_{j+1}+\boldsymbol{f}_j)-\boldsymbol{d}(\boldsymbol{w}_L,\boldsymbol{w}_R).$$

With this approach, the choice of a method of reconstruction is completely decoupled from the choice of a numerical flux.

6.19 The Jameson–Schmidt–Turkel Scheme for the Euler Equations

The Jameson–Schmidt–Turkel (JST) scheme is a particularly simple approach to the construction of a second order accurate scheme. Since this scheme has enjoyed considerable success, it seems useful to discuss its origin and some aspects of its implementation. The formulation of the scheme for a scalar conservation law has already been described in Section 5.17, where it was shown that the diffusive coefficients can be constructed in a manner that ensures that the scheme is LED. The scheme was actually first developed to solve the Euler equations, focusing particularly on the solution of steady state problems in transonic and supersonic flow. Its formulation was guided by a number of design principles as listed below:

1. The scheme should be in conservation form to ensure satisfaction of the shock jump conditions, according to the theorem of Lax and Wendroff (Section 4.10.3).
2. The scheme should be second order accurate in smooth regions of the flow.
3. Shock waves should be captured without overshoots or oscillations, at least in the steady state (but overshoots during the transient phase would be tolerated).
4. The steady state should be independent of the time evolution.
5. The scheme should be stable when using variable local time steps at a fixed CFL number, to accelerate convergence to a steady state.
6. The discrete steady state solution should have constant stagnation enthalpy, consistent with properties of the true steady state solutions.

These principles dictated several aspects of the scheme. In particular, in order to satisfy principle (4), the time integration scheme should be separated from the space discretization scheme by first constructing a semi-discrete approximation. This rules out the Lax–Wendroff and MacCormack schemes, for which the steady state solution depends on the time step Δt in the case of two- or three-dimensional problems. This issue is particularly important when variable local time steps are used in steady state calculations. When the mesh is stretched to a large distance from the body, the far field cells may become very large, with correspondingly large values of Δt.

The original JST scheme used scalar diffusion. Thus, the formula (5.16) for the diffusive flux was applied separately to each state variable, with the same coefficients $\epsilon^{(2)}_{j+\frac{1}{2}}$ and $\epsilon^{(4)}_{j+\frac{1}{2}}$. Thus, the diffusive flux for the density equation, for example, is

$$d_{j+\frac{1}{2}} = \epsilon^{(2)}_{j+\frac{1}{2}} \Delta\rho_{j+\frac{1}{2}} - \epsilon^{(4)}_{j+\frac{1}{2}} \left(\Delta\rho_{j+\frac{3}{2}} - 2\Delta\rho_{j+\frac{1}{2}} + \rho_{j-\frac{1}{2}} \right).$$

In order to ensure constant total enthalpy in the discrete steady state solution, however, the diffusive terms for the energy equation were constructed from ρH instead of ρE. The first term in the diffusive flux is needed to prevent oscillations near shock waves, but in order to maintain accuracy, it should be reduced in locations where the flow is smooth. The second term was found to be necessary to obtain consistent convergence to a steady state, but it must be turned off in the vicinity of shock waves, or else it will produce oscillations. This is consistent with the theory presented in Section 5.17. Both terms need to be conditioned to the local spectral radius

$$r_{j+\frac{1}{2}} = \max |\lambda(A)| = |u| + c.$$

This may be calculated directly at the cell interface or as the average between the cells j and $j+1$. It was found that the relative values of the coefficients $\epsilon^{(2)}_{j+\frac{1}{2}}$ and $\epsilon^{(4)}_{j+\frac{1}{2}}$ can be conveniently controlled by a pressure sensor of the form

$$s_j = \left| \frac{p_{j+1} - 2p_j + p_{j-1}}{p_{j+1} + 2p_j + p_{j-1}} \right|. \tag{6.38}$$

It is evident that $s_j \le 1$ and $s_j = \mathcal{O}(\Delta x^2)$ in a smooth region of the flow. The sensor at the interface is then calculated as

$$\bar{s}_{j+\frac{1}{2}} = \max(s_{j+1}, s_{j+1}, s_j, s_{j-1}).$$

Finally, we set

$$\epsilon^{(2)}_{j+\frac{1}{2}} = \min\left(\frac{1}{2}, k^{(2)}\bar{s}_{j+\frac{1}{2}}\right) r_{j+\frac{1}{2}} \tag{6.39}$$

and

$$\epsilon^{(4)}_{j+\frac{1}{2}} = \max\left(0, k^{(4)} - c^{(4)}\bar{s}_{j+\frac{1}{2}}\right) r_{j+\frac{1}{2}}, \tag{6.40}$$

where $k^{(2)}$, $k^{(4)}$, and $c^{(4)}$ are constants. Typically,

$$k^{(2)} = 1, \quad k^{(4)} = \frac{1}{32}, \quad c^{(4)} = 2.$$

With this construction, both diffusive terms have a magnitude proportional to Δx^3 in smooth regions of the flow because $\epsilon^{(2)}_{j+\frac{1}{2}} = \mathcal{O}(\Delta x^2)$, $\epsilon^{(4)}_{j+\frac{1}{2}} = \mathcal{O}(1)$, while the first difference of the flow variable is proportional to Δx and the third difference to Δx^3.

$\bar{r}_{j+\frac{1}{2}}$ generally reaches a maximum that is substantially less than unity, but with scalar diffusion, the magnitude of $\epsilon^{(2)}_{j+\frac{1}{2}}$ needed to prevent oscillations near shock waves is also substantially less than $\frac{1}{2}s_{j+\frac{1}{2}}$, so the choice $k^{(2)} = 1$ is a good compromise between accuracy and robustness. The coefficient $k^{(4)}$ can be optimized in conjunction with the time stepping scheme to maximize the rate of convergence to a steady state, as will be discussed in Chapters 9 and 10. The original JST scheme used a modified Runge–Kutta time stepping scheme in which the dissipative terms were frozen after the first stage.

In the case of multi-dimensional calculations with grids containing high aspect ratio cells, it has been found that the scheme is more robust if the coefficients $\epsilon^{(2)}_{j+\frac{1}{2}}$ and $\epsilon^{(4)}_{j+\frac{1}{2}}$ are conditioned on a blend of the spectral radii in the different coordinate directions (Martinelli 1987). Suppose that the spectral radii in the i, j, and k mesh directions are $s^{(i)}$, $s^{(j)}$, and $s^{(k)}$. Then, $s^{(i)}$ is replaced by

$$s^{(i)}\left(1 + \frac{s^{(j)}}{s^{(i)}} + \frac{s^{(k)}}{s^{(i)}}\right)^r,$$

where $0 \le r \le 1$, with corresponding modifications of $s^{(j)}$ and $s^{(k)}$. If $r = 1$, $s^{(i)}$, $s^{(j)}$, and $s^{(k)}$ are all replaced by their sum.

Alternative sensors may also be considered. A sensor closer to the formula proposed in Section 5.17 is

$$r_j = \left(\frac{\left|\Delta p_{j+\frac{1}{2}} - \Delta p_{j-\frac{1}{2}}\right|}{\max\left(\left(\left|\Delta p_{j+\frac{1}{2}}\right| + \left|\Delta p_{j-\frac{1}{2}}\right|\right), p_{\lim}\right)}\right)^q,$$

where $r_j = \mathcal{O}(\Delta x^2)$ if the power q is chosen equal to 2. This formula may be blended with the formula (6.38). Instead of constructing the sensor from the pressure, one may also construct the formulas (6.38) or (6.19) from the density or entropy (Martinelli 1987). In the true solution of the Euler equations, the entropy should only display discursions through shock waves, but in discrete solutions, entropy excursions generally occur along solid wall boundaries, particularly near stagnation point. The use of dissipative terms driven by excursions in the entropy has been analyzed by Guermond et al. (2011).

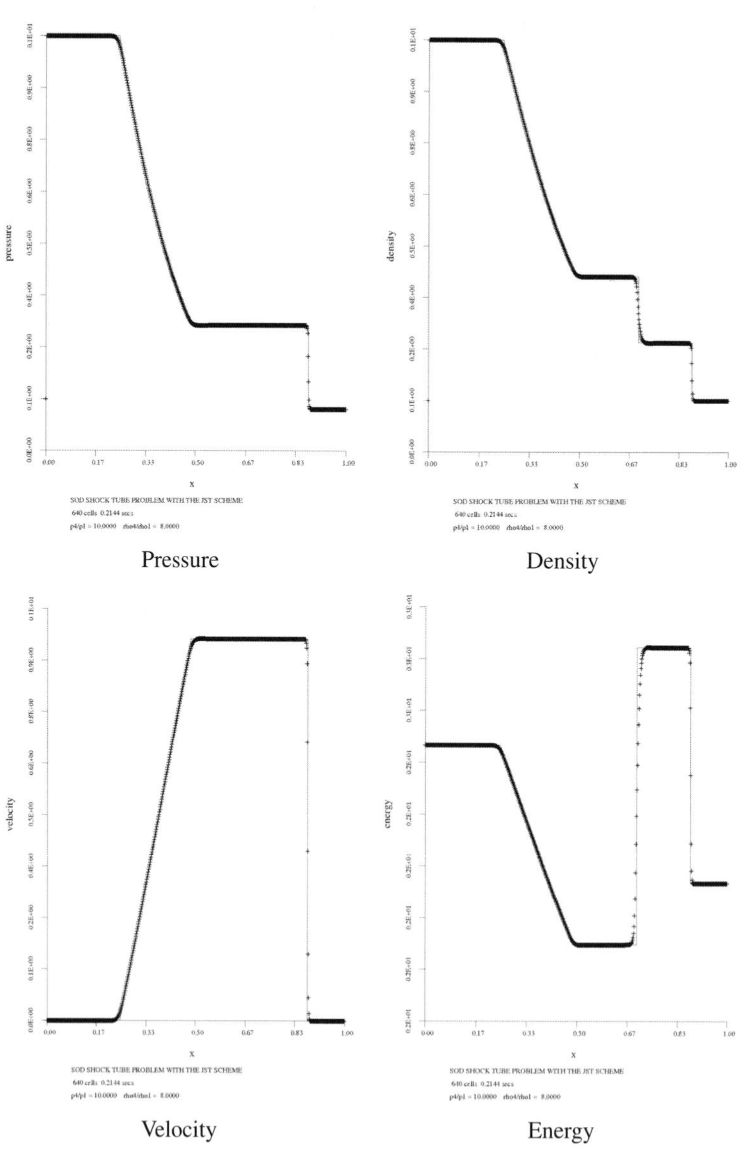

Figure 6.6 Sod shock tube problem with the JST scheme: 640 cells.

The use of a sensor based on the density enables the detection of contact discontinuities. While the JST scheme was originally targeted at steady flow calculations, it has also proved quite successful for unsteady flows. Figure 6.6 shows the result of a shock tube calculation for the Sod case (Sod 1978) on a mesh of 640 cells. In this case, the sensor (6.19) was calculated from the density with $q = 2$, while the JST constants were

$$k^{(2)} = 1, \quad k^{(4)} = \frac{1}{32}, \quad c^{(4)} = 4.$$

The reader is referred to Jameson (2017b) for a more detailed discussion of the origins and development of the JST scheme.

6.20 Some Representative Solutions

It can be seen from the previous section that there are a wide variety of flux formulations from which to choose. This section presents some representative solutions to illustrate the kinds of results which can be expected. Inviscid transonic flow over a NACA 0012 airfoil at Mach 0.8 and an angle of attack of 1.25 degrees has been chosen as a test case because it has been widely used as a benchmark and is representative of the kind of calculation that is needed to support aircraft design. While viscous simulations are ultimately needed, a high quality Euler solution is an essential ingredient of a good viscous calculation. Some hypersonic calculations over a blunt body are also presented to confirm the capability of the methods to calculate flows containing strong shock waves.

Beginning with the airfoil calculations, Figure 6.7 illustrates the mesh that was used, which has an "O-topology" containing $320 \times 64 = 20,480$ cells. The actual NACA 0012 airfoil has an open trailing edge, but for the purpose of these tests, the geometry was modified by extending it to a chord of 1.00893, at which it becomes closed. Figures 6.8–6.11 show solutions using the Roe scheme, the E- and H-CUSP schemes, and a characteristic flux split scheme, respectively. The figures show a plot of the pressure coefficients $C_p = \frac{p-p_\infty}{\frac{1}{2}\rho_\infty V_\infty^2}$, following the usual aeronautical convention with the negative direction upward, so that the upper pressure distribution appears above the lower pressure distribution in the case of a lifting solution. The entropy along the surface is also plotted in a direction normal to the profile, providing a measure of any numerically generated entropy ahead of the shock waves. The second panel on each figure shows Mach contours. The calculations were all converged to residual levels in the range of 10^{-9}, using acceleration techniques that will be described in a subsequent chapter. Plots of the convergence measured by the density residual are also included in the third panel of each figure. All the calculations are nominally second order accurate, using the SLIP reconstruction procedure described in Section 5.23. Table 6.1 shows the calculated values of the lift and drag coefficients.

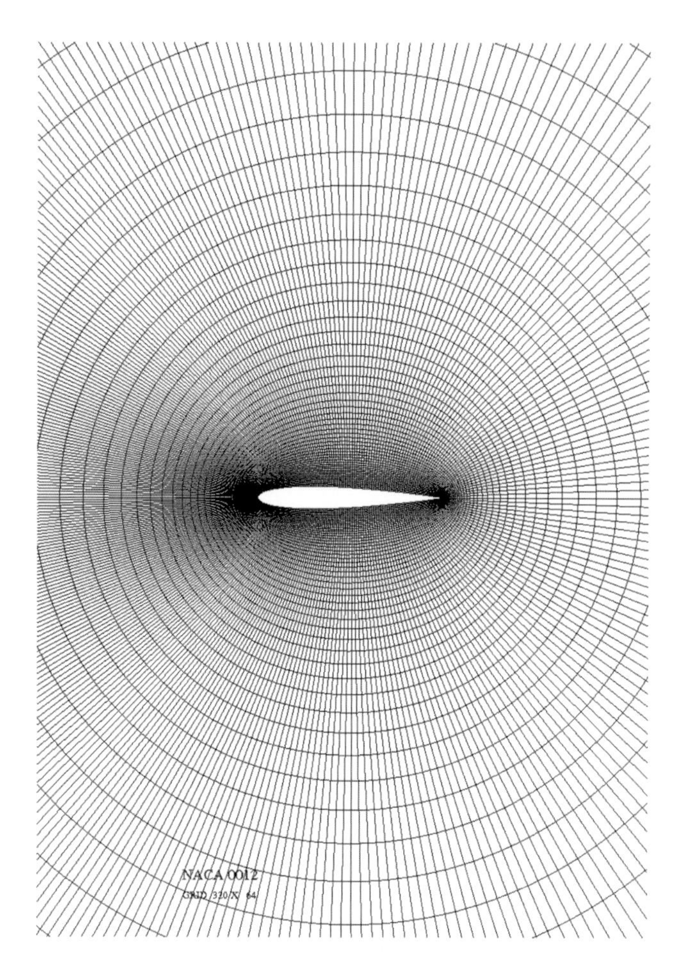

Figure 6.7 Mesh around NACA 0012 airfoil, 320 × 64 cells.

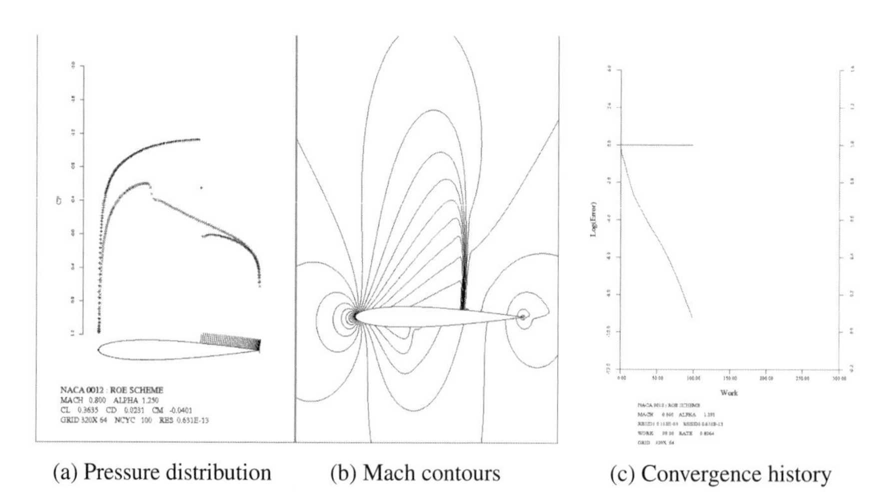

(a) Pressure distribution (b) Mach contours (c) Convergence history

Figure 6.8 NACA 0012: Roe scheme.

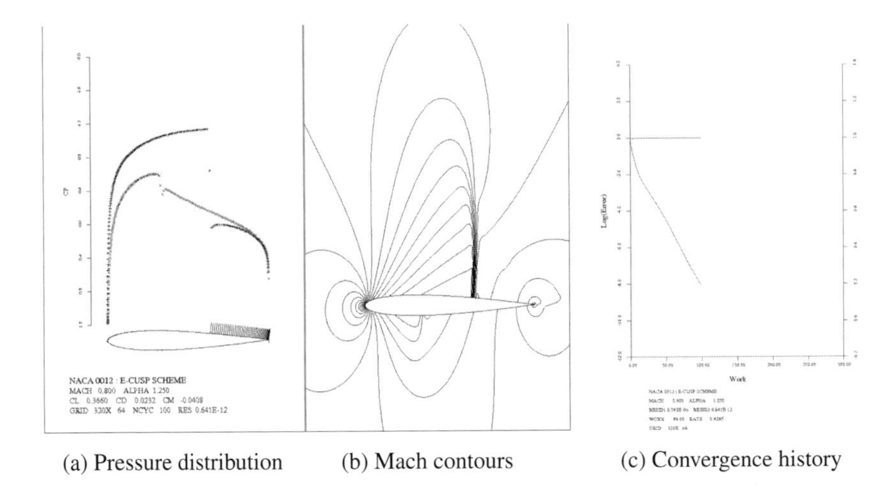

(a) Pressure distribution (b) Mach contours (c) Convergence history

Figure 6.9 NACA 0012: E-CUSP scheme.

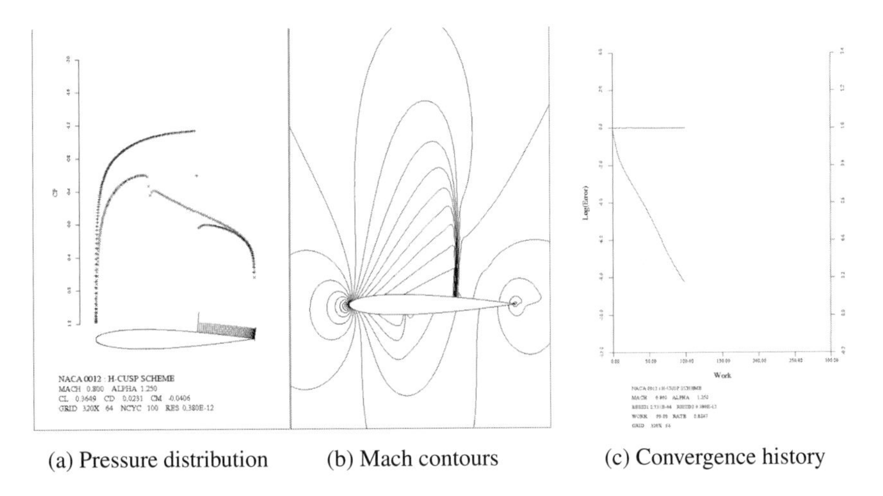

(a) Pressure distribution (b) Mach contours (c) Convergence history

Figure 6.10 NACA 0012: H-CUSP scheme.

The results of the first three schemes, which are variations of flux difference splitting, are extremely close. They all exhibit shocks with a single interior point on the upper surface and a minimal amount of numerical entropy in the vicinity of the leading edge. The flux split scheme has a less clean shock on the upper surface and significantly greater numerical entropy fluctuations at the leading edge. The H-CUSP scheme exhibits an excursion in entropy at the interior point of the shock but converges exactly to free stream total enthalpy everywhere, whereas the Roe and E-CUSP schemes have an excursion in enthalpy at the interior shock point, which is not shown in the figures.

Because the schemes revert to first order accuracy in the vicinity of shock waves, the asymptotic convergence with mesh refinement of the global solution is not necessarily second order. This issue was explored by Vassberg and Jameson in

Table 6.1. Table of calculated C_L and C_D from various schemes.

Scheme	C_L	C_D
Roe	0.3635	0.0231
E-CUSP	0.3660	0.0232
H-CUSP	0.3649	0.0231
Flux split	0.3716	0.0236

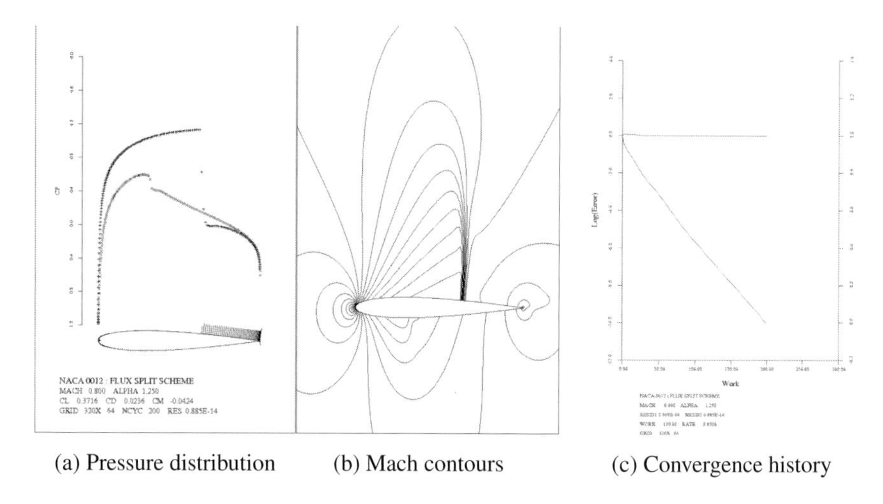

(a) Pressure distribution (b) Mach contours (c) Convergence history

Figure 6.11 NACA 0012: Flux splitting scheme.

mesh refinement studies all the way to a mesh with $4,096 \times 4,096 = 16,777,216$ cells (Vassberg & Jameson 2010). Second order asymptotic convergence rates were observed in subsonic solutions, but rates close to first order were observed in transonic solutions containing shock waves. The asymptotic values of the force coefficients were estimated to be $C_L = 0.3562$ and $C_D = 0.0227$, so the calculations in Figures 6.8–6.11 actually over-predict both C_L and C_D. This may be related to the numerical enforcement of the Kutta condition.

Figure 6.12 shows the result of a first order calculation without the reconstruction, in this case using the H-CUSP scheme. It can be seen that there is a huge amount of numerical entropy over the upper surface ahead of the shock, and also on the lower surface. Moreover, the calculated lift coefficient $C_L = 0.3105$ is in error by about 15 percent. This result confirms that in practice, it is essential to use a scheme that is nominally at least second order accurate.

Figure 6.13 shows the result for hypersonic flow over a blunt body at Mach 8 using the H-CUSP scheme. This provides evidence that although the E- and H-CUSP schemes actually have less numerical diffusion than the Roe scheme, they are very robust for simulations with strong shock waves.

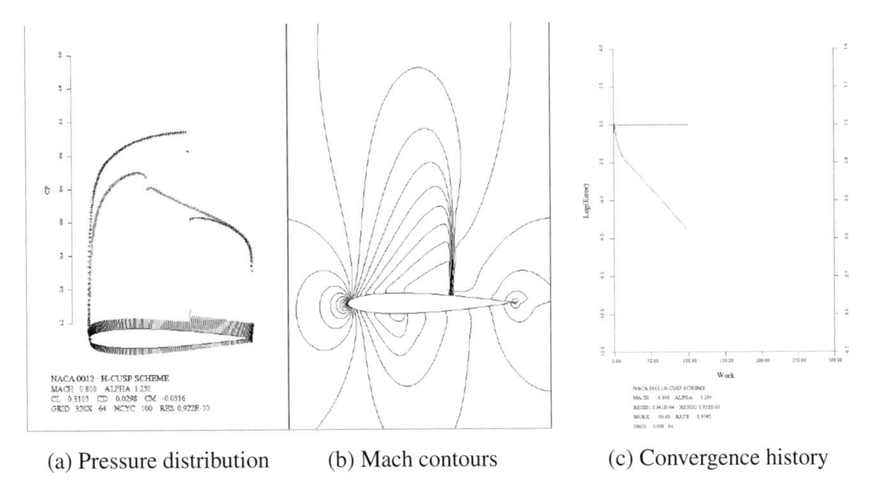

(a) Pressure distribution (b) Mach contours (c) Convergence history

Figure 6.12 NACA 0012: H-CUSP, first order scheme.

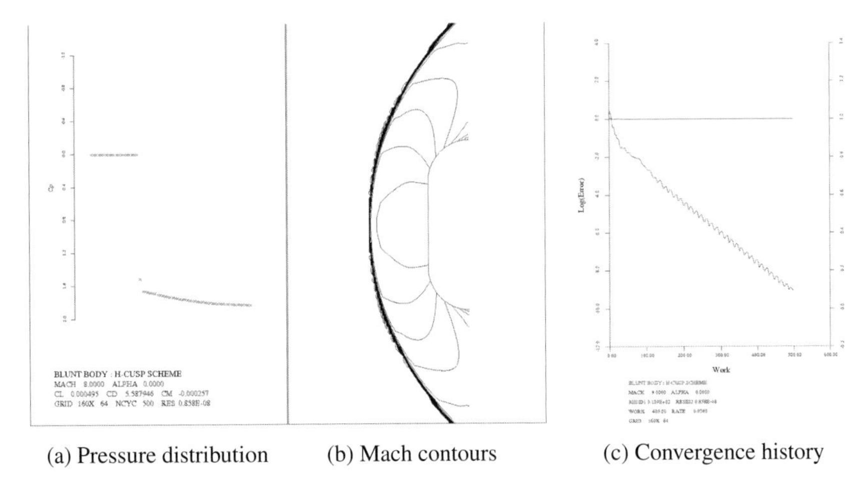

(a) Pressure distribution (b) Mach contours (c) Convergence history

Figure 6.13 Blunt body: H-CUSP scheme.

6.21 Anomalies

While the schemes described in this chapter have enjoyed tremendous success in a wide variety of applications, a number of anomalies have been discovered in their behavior for particular flows. These include the wall heating problem (Liou 2012), the problem of slow moving shock waves (Roberts 1990), and the carbuncle phenomenon, which is discussed in this section.

The carbuncle phenomenon appears to have been first encountered in calculations using the Roe scheme and was studied intensively by Quirk (1994). The bow shock wave in hypersonic flow over a blunt body becomes unstable and develops a protrusion or "carbuncle." Figure 6.14 shows result for flow over a semi-circular cylinder at Mach 6 using the Roe scheme, which develops a carbuncle. The solution is fully converged to

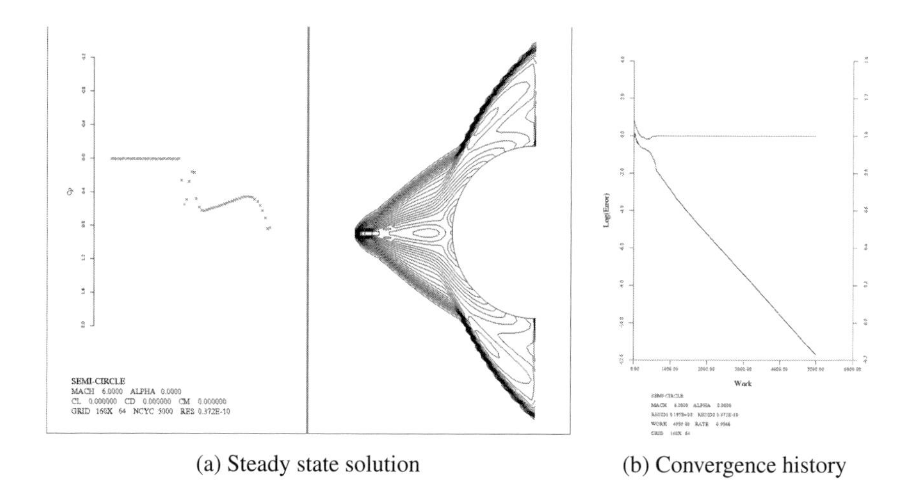

(a) Steady state solution (b) Convergence history

Figure 6.14 Semi-circle, carbuncle phenomenon: Roe scheme.

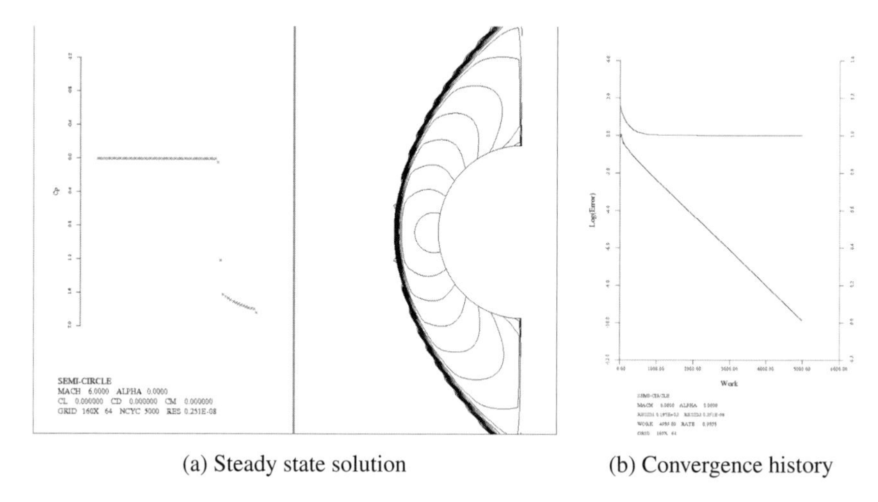

(a) Steady state solution (b) Convergence history

Figure 6.15 Semi-circle, solution with no carbuncle: Roe scheme using entropy fix.

a steady state and appears to be a legitimate solution of the discretized Euler equations. This phenomenon occurs when the entropy fix is reduced to a value less than 0.125 of the local speed of sound c. When it is increased to $0.5c$, it is suppressed, as can be seen in Figure 6.15. According to a recent study by Liou (2011), the carbuncle phenomenon is not limited to the Roe scheme but can occur with other schemes, including the Godunov scheme. To date, it has never been encountered with the E- and H-CUSP schemes, as evidenced by the solution shown in Figures 6.14 and 6.15. In numerical experiments by the author with the H-CUSP scheme, no carbuncle appeared even with the entropy fix reduced to zero.

7 Discretization Schemes for Flows in Complex Multi-dimensional Domains

7.1 Introduction

While the analysis of one-dimensional problems has laid the groundwork for a comprehensive theory of shock-capturing methods, practically all industrial applications require the simulation of flows in extremely complex, often multiply connected domains. Typical examples include simulations of complete aircraft with flow through engine nacelles, and Formula 1 racing cars. These applications require both the development of techniques to generate appropriate computational meshes and the development of discretization schemes compatible with whatever type of mesh is chosen. The principal alternatives are Cartesian meshes, body-fitted curvilinear meshes, and unstructured tetrahedral meshes. Each of these approaches has some advantages that have led to their use. The Cartesian mesh minimizes the complexity of the algorithm at interior points and facilitates the use of high order discretization procedures, at the expense of greater complexity, and possibly a loss of accuracy, in the treatment of boundary conditions at curved surfaces. This difficulty may be alleviated by using mesh refinement procedures near the surface. With their aid, schemes that use Cartesian meshes have recently been developed to treat very complex configurations (Samant, Bussoletti, Johnson et al. 1987, Berger & LeVeque 1989, Landsberg, Boris, et al. 1993, Melton, Pandya, & Steger 1993, Aftosmis, Melton, & Berger 1995).

Body-fitted meshes have been widely used and are particularly well suited to the treatment of viscous flow because they readily allow the mesh to be compressed near the body surface. With this approach, the problem of mesh generation itself has proved to be a major pacing item. The most commonly used procedures are algebraic transformations (Eiseman 1979, Eriksson 1982, Smith 1983, Baker 1986), methods based on the solution of elliptic equations, pioneered by Thompson (Thompson, Thames, & Mastin 1974, Thompson, Warsi, & Mastin 1982, Sorenson 1986, Sorenson 1988), and methods based on the solution of hyperbolic equations marching out from the body (Steger & Chaussee 1980). In order to treat very complex configurations, it generally proves expedient to use a multiblock procedure (Weatherill & Forsey 1985, Sawada & Takanashi 1987), with separately generated meshes in each block, which may then be patched at block faces, or allowed to overlap, as in the Chimera scheme (Benek, Buning, & Steger 1985, Benek, Donegan, & Suhs 1987). It remains both difficult and time consuming to generate a structured mesh for a complex configuration such as a complete aircraft, perhaps with its high-lift system deployed. The use of overset meshes facilitates the treatment of complex geometries, but automatic generation of structured meshes remains out of reach, and in current practice the definition of a sufficiently accurate geometry model and mesh generation pose a major bottleneck in the industrial application of CFD.

The alternative is to use unstructured tetrahedral or polyhedral meshes. This alleviates (but does not entirely eliminate) the difficulty of mesh generation and facilitates adaptive mesh refinement. This approach has been gaining acceptance as it is becoming apparent that it can lead to a speed-up and reduction in the cost of mesh generation that more than offsets the increased complexity and cost of the flow simulations. Two competing procedures for generating triangulations that have both proved successful are Delaunay triangulation (Delaunay 1934, Barth 1994), based on concepts introduced at the beginning of the century by Voronoi (Voronoi 1908), and the advancing front method (Löhner & Parikh 1988).

Mesh generation has evolved into a separate specialist discipline, which is outside the scope of this book. It is useful, however, to be aware of some of the principal characteristics of the meshes that might be used, in particular, the topological classification of meshes and the types of singularity they may contain. This is discussed in the following section.

7.2 Topological Classification of Structured Meshes

Structured meshes can be classified by their topological types. The three main types that have proved useful in two-dimensional calculations may be classified as H-, C-, and O-meshes. These are illustrated in Figures 7.1–7.3. In principle, they could all be generated by conformal mappings. An H-mesh can be mapped to a computational domain with flow past a horizontal slit. In the case of a conformal mapping, this leads to a singularity at the leading edge of a blunt profile where the smooth nose is mapped to a corner. At this singularity, the mapping modulus $h = \left|\frac{dz}{d\sigma}\right|$ becomes infinite. In the case of potential flow, defined by a potential ϕ in the computational domain, the velocity in the physical domain is

$$q = \frac{\nabla\phi}{h}.$$

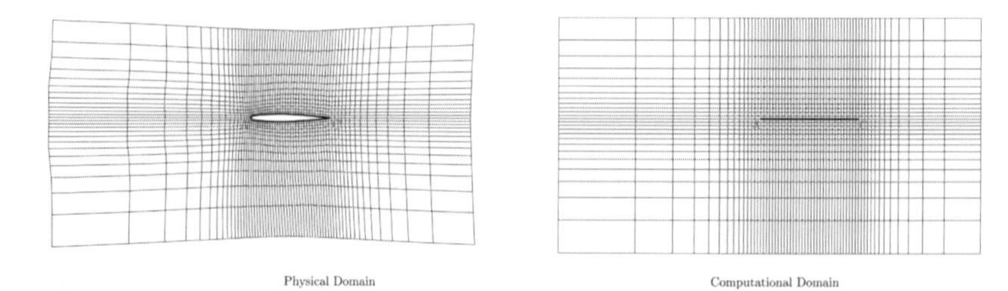

Figure 7.1 H-mesh.

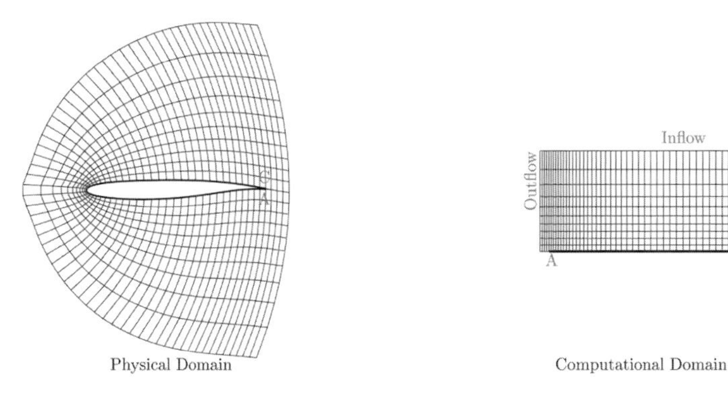

Figure 7.2 C-mesh.

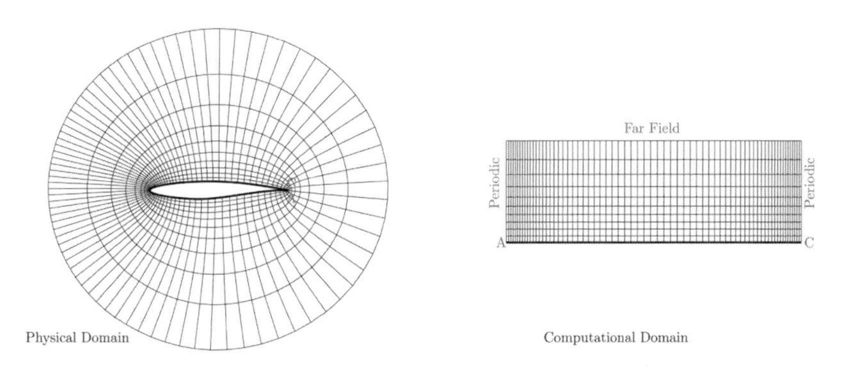

Figure 7.3 O-mesh.

Thus, the leading edge stagnation point is forced to coincide with the mesh point defined by the front end of the slit, which is generally not its true location. Consequently, H-meshes tend to produce inaccurate solutions in the neighborhood of the leading edge. They have the advantage, however, that the H-mesh concept is readily extended to three-dimensional geometries.

A C-mesh corresponds to a mapping of the physical domain to a half plane in the computational domain, where the profile covers a section of the lower boundary. The flow in the computational domain enters through the top boundary and exits through the two side boundaries. A mesh of this type can be generated by first mapping the profile to a shallow bump by the transformation

$$z = z_0 + \frac{1}{2}a\sigma^2, \quad \sigma = \sqrt{\frac{2(z - z_0)}{a}},$$

where the singular point z_0 lies inside the leading edge, as illustrated in Figure 7.4. Then, the bump is transformed to coincide with the boundary of the half plane by a shear

$$\zeta = \sigma - is,$$

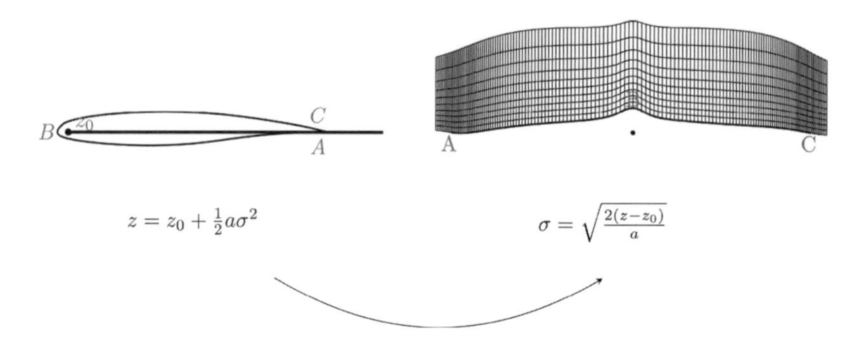

Figure 7.4 Generation of C-mesh by square root mapping and shear.

where s is the height of the bump. This enables the definition of a Cartesian mesh in the ζ plane, and finally the C-mesh is generated by the reverse mappings back to the physical plane.

In the case of a flat plate, this procedure is equivalent to the introduction of parabolic coordinates. In general, it leads to meshes with intervals that steadily increase away from the leading edge. Consequently additional stretches and shears are needed to improve the resolution in the neighborhood of the trailing edge. C-meshes are particularly suitable for viscous simulations because they can be adjusted to concentrate mesh lines in the vicinity of the wake.

An O-mesh can be generated by a conformal mapping of the profile to a circle. Then, polar coordinates in the computational domain are mapped back to generate the mesh in the physical plane. Exact conformal mappings can be calculated by the Theodorsen method (Abbott & Von Doenhoff 1959) and would produce a perfectly orthogonal mesh. A simpler procedure is to map the profile to a near circle by the Kármán–Trefftz transformation

$$\frac{z - z_1}{z - z_2} = \left(\frac{\sigma - \sigma_1}{\sigma - \sigma_2}\right)^{2-\dfrac{\epsilon}{\pi}},$$

where ϵ is the trailing edge corner angle, and σ_1 and σ_2 are singular points corresponding to the trailing edge point z_2 and a point z_1 located inside the leading edge. Then, the near circle in the σ plane is transformed to a perfect circle in the ζ plane by a shear, and finally polar coordinates in the ζ plane are mapped back to the physical plane by the reverse transformations.

O-meshes generated in this manner automatically concentrate the mesh near both the leading and trailing edges. Consequently, they are ideal for inviscid flow simulations. They are less suitable for viscous flow simulations because they do not allow the generation of mesh lines that smoothly follow the boundary layer into the wake.

Three-dimensional meshes can be generated by various combinations of the H-, C-, and O-mesh topologies. For example, a C–H-mesh around a wing would stack C-meshes around wing sections in vertical planes at successive span stations.

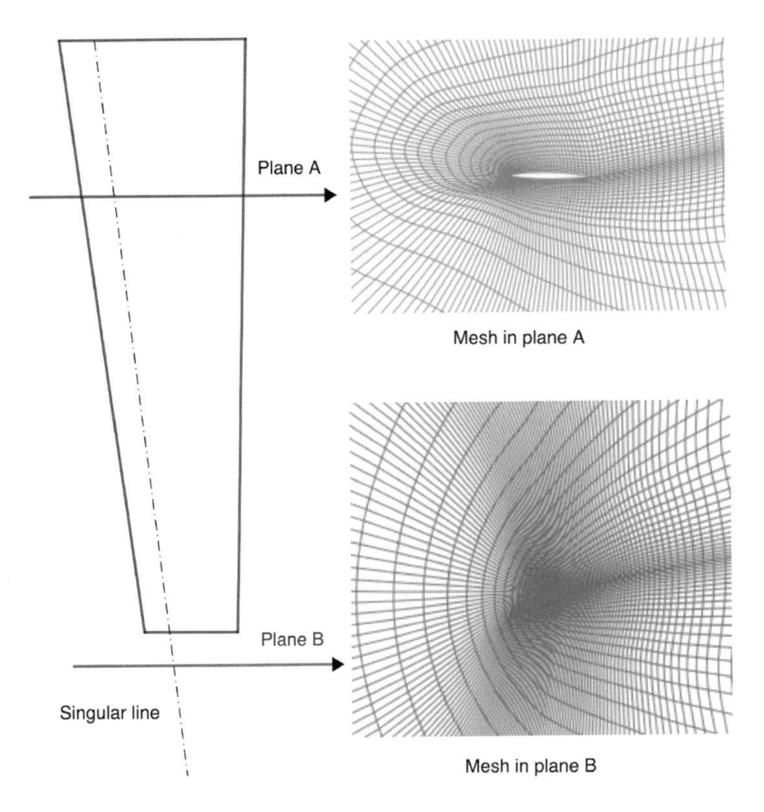

Figure 7.5 C-H mesh for a wing.

This would be continued beyond the wing tip with C-meshes containing a slit from a singular line emanating from the wing tip, as illustrated in Figure 7.5. H-meshes can be in principle be generated by compressing every component to vertical or horizontal slits aligned with the flow.

Three-dimensional structured meshes are bound to contain singularities of some kind, such as the slit beyond the wing tip in a C–H-mesh for a wing. This is a consequence of fundamental topological properties of three-dimensional space. For example, the hairy dog theorem states that if one brushes the hair of a dog, there will be at least two singular points at which the hair pattern is entirely inward or outward like a source or a sink, or entirely circumferential like a vortex. A dog could be continuously transformed to a sphere. Then, if one brushes the hair from the north to the south pole, the north pole will be a source and the south pole a sink, while if one brushes the hair circumferentially around the equator, each pole will be a vortex. If we wish to generate a mesh around a sphere, similar issues arise. For example, if we choose to generate a mesh with cylindrical coordinates aligned with the flow, there will be polar singularities along the axis of the coordinate system in front of and behind the sphere.

7.3 Cell-Centered Finite Volume Schemes on Structured Grids

The basic idea of the finite volume method has been discussed in Chapter 3. Here, we examine its application to gas dynamics in more detail for structured quadrilateral and hexahedral grids. The ideas of the last two chapters are easily adapted to these grids on a direction by direction basis.

We consider first the discretization of the two-dimensional conservation law

$$\frac{\partial \boldsymbol{w}}{\partial t} + \frac{\partial}{\partial x}\boldsymbol{f}(\boldsymbol{w}) - \frac{\partial}{\partial y}\boldsymbol{g}(\boldsymbol{w}) = 0, \tag{7.1}$$

which may represent the Euler equations for inviscid gas dynamics. For any subdomain $\mathcal{D}_h$ with boundary $\mathcal{B}_h$, we can write this in integral form as

$$\frac{d}{dt}\int_{\mathcal{D}_h} \boldsymbol{w}\, dS - \oint_{\mathcal{B}_h} (\boldsymbol{f}(\boldsymbol{w})\, dy - \boldsymbol{g}(\boldsymbol{w})\, dx) = 0.$$

The finite volume discretization follows from the evaluation of this equation for each grid cell, as was discussed in Section 3.4. Suppose that the cells are numbered by subscripts i and j. Referring to Figure 7.6, the discretization takes the form

$$S_{i,j}\frac{d}{dt}\boldsymbol{w}_{i,j} + \sum_{k=1}^{4}\left(\boldsymbol{f}_k \Delta y_k - \boldsymbol{g}_k \Delta x_k\right) = 0, \tag{7.2}$$

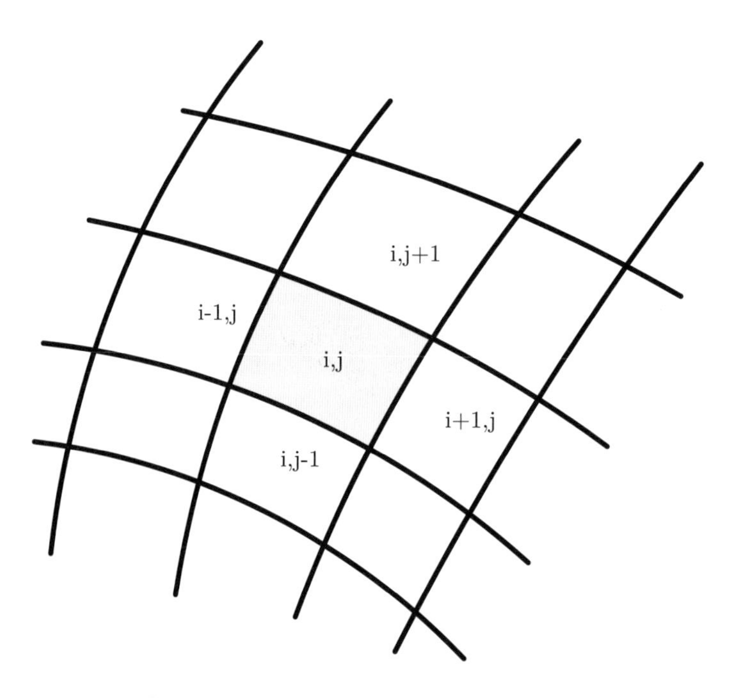

Figure 7.6 A finite volume cell in a smooth structured grid.

where $\boldsymbol{f}_k$ and g_k are estimates of the average value of the flux vectors $\boldsymbol{f}$ and $\boldsymbol{g}$ on side k, and Δx_k and Δy_k are the increments in x and y along side k. A simple central difference formula is obtained by setting

$$\boldsymbol{f}_1 = \frac{1}{2}\left(\boldsymbol{f}_{i+1,j} + \boldsymbol{f}_{i,j}\right), \quad \boldsymbol{g}_1 = \frac{1}{2}\left(\boldsymbol{g}_{i+1,j} + \boldsymbol{g}_{i,j}\right),$$

where

$$\boldsymbol{f}_{i,j} = \boldsymbol{f}(\boldsymbol{w}_{i,j}), \quad \boldsymbol{g}_{i,j} = \boldsymbol{g}(\boldsymbol{w}_{i,j}),$$

with corresponding formulas for the other three sides. We may also write (7.2) as

$$S_{i,j}\frac{d}{dt}\boldsymbol{w}_{i,j} + \boldsymbol{h}_{i+\frac{1}{2},j} - \boldsymbol{h}_{i-\frac{1}{2},j} + \boldsymbol{h}_{i,j+\frac{1}{2}} - \boldsymbol{h}_{i,j-\frac{1}{2}} = 0,$$

where $\boldsymbol{h}_{i+\frac{1}{2},j}$ is the numerical flux

$$\boldsymbol{h}_{i+\frac{1}{2},1} = \boldsymbol{f}_1\Delta y_1 - \boldsymbol{g}_1\Delta x_1,$$

across side 1, and so on.

According to the theory developed in Chapters 5 and 6, a central difference scheme of this kind cannot capture shock waves and contact discontinuities without oscillations. Assuming that the grid is smooth and regular, we may, however, apply any of the schemes that have been formulated in the last chapter separately in the i and j directions. For example, we can construct a first order accurate upwind scheme by introducing Roe averaged Jacobian matrices. On side 1, we construct a matrix $A_1(\boldsymbol{w}_{i+1,j}, \boldsymbol{w}_{i,j})$ such that

$$A_1(\boldsymbol{w}_{i+1,j} - \boldsymbol{w}_{i,j}) = (\boldsymbol{f}_{i+1,j} - \boldsymbol{f}_{i,j})\Delta y_1 - (\boldsymbol{g}_{i+1,j} - \boldsymbol{g}_{i,j})\Delta x_1.$$

Decomposing A_1 as

$$A_1 = V\Lambda V^{-1},$$

where the columns of V are the eigenvectors of A_1, and Λ is a diagonal matrix containing the eigenvalues of A_1, we define the absolute Jacobian matrix

$$|A_1| = V\,|\Lambda|\,V^{-1}.$$

Then, the upwind numerical flux across side 1 is constructed as

$$\boldsymbol{h}_{i+\frac{1}{2},j} = \frac{1}{2}(\boldsymbol{f}_{i+1,j} + \boldsymbol{f}_{i,j})\Delta y_1 - \frac{1}{2}(\boldsymbol{g}_{i+1,j} + \boldsymbol{g}_{i,j})\Delta x_1 - \frac{1}{2}|A_1|(\boldsymbol{w}_{i+1,j} - \boldsymbol{w}_{i,j}), \quad (7.3)$$

with corresponding formulas for the other three sides.

A scheme with scalar diffusion (the Rusanov or generalized Lax–Friedrichs flux) is obtained by setting

$$\boldsymbol{h}_{i+\frac{1}{2},j} = \frac{1}{2}(\boldsymbol{f}_{i+1,j} + \boldsymbol{f}_{i,j})\Delta y_1 - \frac{1}{2}(\boldsymbol{g}_{i+1,j} + \boldsymbol{g}_{i,j})\Delta x_1 - \epsilon_{i+\frac{1}{2},j}(\boldsymbol{w}_{i+1,j} - \boldsymbol{w}_{i,j}),$$

where

$$\epsilon_{i+\frac{1}{2},j} = \max|\lambda(A_1)|.$$

Other first order accurate schemes can be constructed by substituting different numerical fluxes. For example, the flux of the Godunov scheme may be calculated by solving Riemann problems between the states on either side of each of the four edges.

First order accurate schemes introduce too much numerical diffusion to be useful in practice. Any of the second order accurate formulations described in Chapter 4 can, however, be adapted to smooth and regular grids by applying them separately in the i and j directions. The simplest to implement are the Jameson–Schmidt–Turkel (JST) and SLIP schemes, but we may also use reconstruction with a variety of limiters.

7.4 Vertex-Centered Schemes on Quadrilateral Grids

Vertex-centered schemes for quadrilateral meshes can be defined in two ways. We can either take the control volume to be the union of the cells around a given node, as illustrated in Figure 7.7, or we can form a dual mesh by connecting the cell centers of the primary mesh and define the control volume to be the dual cell containing the node, as illustrated in Figure 7.8. Both approaches lead to central formulas that would be second order accurate on Cartesian or sufficiently smooth curvilinear meshes. While the time dependent updating scheme strictly describes the evolution of cell average values, these differ from the pointwise cell centered values by an amount proportional to the square of the mesh interval if the solution is smooth. Accordingly, we can use these values to estimate the interface fluxes without reducing the order of accuracy. The central formulas arising from either approach need to be stabilized by the addition of artificial diffusive terms, which directly emulate the upwind scheme in the case of the dual control volume.

In finite volume schemes that use the aggregated control volume consisting of the union of the neighboring cells around each node (Hall 1985, Jameson 1986b,

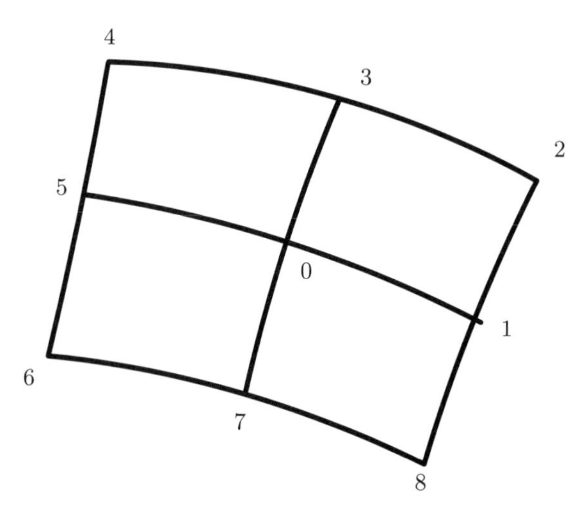

Figure 7.7 Aggregated control volume consisting of the union of cells containing node 0.

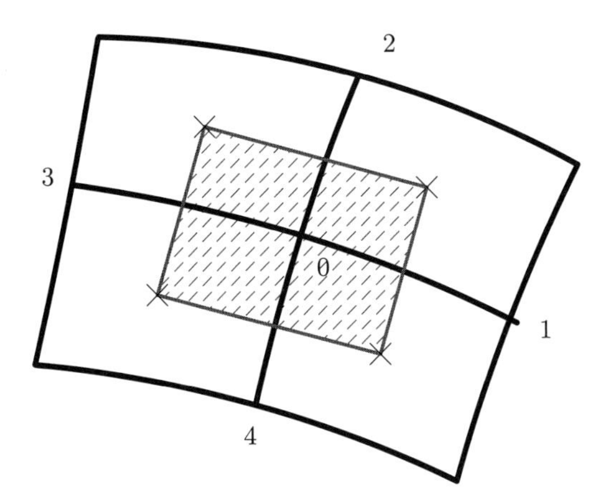

Figure 7.8 Dual control volume.

Radespiel, Rossow, & Swanson 1989), the flux balance is simply the sum of the flux balances of the neighboring cells. Referring to Figure 7.7, the flux balance for cell 0123 is evaluated by the trapezoidal integration as

$$\begin{aligned}\frac{1}{2}\{&(\boldsymbol{f}_1+\boldsymbol{f}_0)(y_1-y_0)-(\boldsymbol{g}_1+\boldsymbol{g}_0)(x_1-x_0)\\&+(\boldsymbol{f}_2+\boldsymbol{f}_1)(y_2-y_1)-(\boldsymbol{g}_2+\boldsymbol{g}_1)(x_2-x_1)\\&+(\boldsymbol{f}_3+\boldsymbol{f}_2)(y_3-y_2)-(\boldsymbol{g}_3+\boldsymbol{g}_2)(x_3-x_2)\\&+(\boldsymbol{f}_0+\boldsymbol{f}_3)(y_0-y_3)-(\boldsymbol{g}_0+\boldsymbol{g}_3)(x_0-x_3)\},\end{aligned}$$

where

$$\boldsymbol{f}_k=\boldsymbol{f}(\boldsymbol{w}_k),\quad \boldsymbol{g}_k=\boldsymbol{g}(\boldsymbol{w}_k),$$

with corresponding formulas for cells 0345, 0567, and 0781. On summing the contributions of the four cells, the fluxes across interior edges cancel. Thus, the state $\boldsymbol{w}_0$ is updated by the accumulated fluxes through the outer edges, corresponding to trapezoidal integration around the perimeter,

$$S_0\frac{d\boldsymbol{w}_0}{dt}+\frac{1}{2}\sum_{k=1}^{8}\{(\boldsymbol{f}_{k+1}+\boldsymbol{f}_k)(y_{k+1}-y_k)-(\boldsymbol{g}_{k+1}+\boldsymbol{g}_k)(x_{k+1}-x_k)\}=0, \tag{7.4}$$

where S_0 is the sum of the areas of the four cells, and node 9 is identified with node 1.

The flux balance for schemes using the dual control volume is calculated in the same way as the flux balance for the cell centered scheme – by averaging the flux vectors across each edge. Thus, referring to Figure 7.8, the central formula for updating the state w_0 is

$$S_0\frac{d\boldsymbol{w}_0}{dt}+\frac{1}{2}\sum_{k=1}^{4}(\boldsymbol{f}_{k0}\Delta y_{k0}-\boldsymbol{g}_{k0}\Delta x_{k0})=0,$$

where

$$f_{k0} = \frac{1}{2}(f_k + f_0), \quad g_{k0} = \frac{1}{2}(g_k + g_0),$$
$$f_k = f(w_k), \quad g_k = g(w_k).$$

Δx_{k0} and Δy_{k0} are the increments in x and y along the edge separating cell k and cell 0, and S_0 is the area of the dual cell.

Using the dual control volume, we can construct a first order upwind scheme in the usual manner by introducing Roe averaged Jacobian matrices A_{k0} that satisfy

$$(f_k - f_0)\Delta y_{k0} - (g_k - g_0)x_{k0} = A_{k0}(w_k - w_0). \tag{7.5}$$

Then, we set

$$S_0 \frac{dw0}{dt} + \sum_{k=1}^{4} h_{k0} = 0, \tag{7.6}$$

where the numerical flux across the edge $k0$ is

$$h_{k0} = f_{k0}\Delta y_{k0} - g_{k0}\Delta x_{k0} - \frac{1}{2}\left|A_{k0}\right|(w_k - w_0), \tag{7.7}$$

and the absolute Jacobian matrix is defined as

$$\left|A_{k0}\right| = V\,\left|\Lambda\right|\,V^{-1},$$

after decomposing the Jacobian matrix as

$$A_{k0} = V\Lambda V^{-1}.$$

In order to produce an upwind biased scheme using the aggregated control volume, we can rewrite the update (7.4) as

$$S_0 \frac{dw_0}{dt} + \frac{1}{2}\sum_{k=1}^{8}(f_k \Delta y_{k0} - g_k \Delta x_{k0}) = 0,$$

where

$$\Delta x_{k0} = x_{k+1} - x_{k-1}, \quad \Delta y_{k0} = y_{k+1} - y_{k-1},$$

and this is equivalent to

$$S_0 \frac{dw_0}{dt} + \frac{1}{2}\sum_{k=1}^{8}(f_k + f_0)\Delta y_{k0} - (g_k + g_0)\Delta x_{k0}) = 0,$$

since the sums of Δx_{k0} and Δy_{k0} are zero around a closed circuit. Now, we can introduce Roe matrices that satisfy (7.5) and define a scheme similar to (7.6) with fluxes defined by (7.7), where now the sum is from $k = 1$ to 8. A simpler alternative, which has proven to work well in practice (Jameson 1986b), is to add diffusive terms in the i and j coordinate directions similar to the JST scheme, where now the coefficients of the diffusive terms are scaled by the sum of the spectral radii of the two cells separated by the edges in each coordinate direction, such as cells 0123 and 0781 for edge 01 in Figure 7.7.

7.5 Wall Boundary Condition

At a solid boundary, inviscid flow solutions should satisfy the flow tangency condition

$$\boldsymbol{u} \cdot \boldsymbol{n} = 0, \tag{7.8}$$

where $\boldsymbol{n}$ is a vector normal to the wall.

Different treatments of the wall boundary condition are required for vertex and cell-centered schemes. In either case, assuming local body fitted coordinates, the tangency condition (7.8) for inviscid flow implies that there is no convective flux across the boundary, say $\xi_l = 0$. If the flow variables are stored at the vertices, their updated values at boundary vertices may no longer satisfy (7.8). A simple procedure is to correct the surface velocities by a projection in which the normal component of the velocity is set to zero.

In the case of a cell-centered scheme, the surface velocity does not explicitly appear in the discrete equations. While there is no convective flux across the boundary faces, sketched in Figure 7.9 for the two-dimensional case, it is necessary to determine the pressure on these faces. An accurate estimate can be obtained by using the normal pressure gradient to extrapolate the pressure at the wall from its value in the boundary cell. The tangency condition (7.8) implies that

$$\frac{\partial}{\partial t}(\rho \boldsymbol{u} \cdot \boldsymbol{n}) = 0.$$

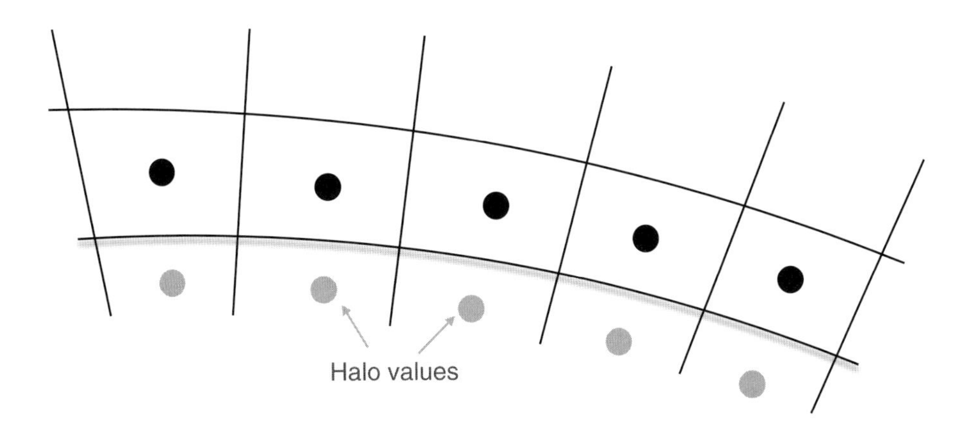

Figure 7.9 Boundary cells.

Combining the momentum equations, we find that

$$n_i \frac{\partial}{\partial x_j}(\rho u_i u_j) + n_i \frac{\partial p}{\partial x_i} = \frac{\partial}{\partial x_j}(\rho n_i u_i u_j) - \rho u_i u_j \frac{\partial n_i}{\partial x_j} + n_i \frac{\partial p}{\partial x_i} = 0,$$

where the first term vanishes by the tangency condition. Thus

$$\frac{\partial p}{\partial n} = \rho u_i u_j \frac{\partial n_i}{\partial x_j}.$$

Here, the normal pressure gradient balances the centrifugal terms induced by the curvature of the wall.

The computational grid is not necessarily orthogonal, so in the discrete equations one should estimate the pressure gradient in the direction of the radial mesh lines. Under a local coordinate transformation the equations become

$$\frac{\partial (J\rho u_i)}{\partial t} + \frac{\partial}{\partial \xi_j}\left(S_{jk}\rho u_i u_k + S_{ji}\, p\right) = 0,$$

where on a wall $\xi_l = 0$,

$$u_i S_{li} = 0. \tag{7.9}$$

Then,

$$\begin{aligned} 0 &= \frac{\partial \left(J\rho u_i S_{li}\right)}{\partial t} \\ &= S_{li}\frac{\partial}{\partial \xi_j}\left(S_{jk}\rho u_i u_k + S_{ji}\, p\right) \\ &= \frac{\partial}{\partial \xi_j}\left(S_{li} S_{jk}\rho u_i u_k\right) - S_{jk}\rho u_i u_k \frac{\partial S_{li}}{\partial \xi_j} + S_{li} S_{ji}\frac{\partial p}{\partial \xi_j} + S_{li}\, p\frac{\partial S_{ji}}{\partial \xi_j}, \end{aligned} \tag{7.10}$$

where the first term vanishes by (7.9), and the last term vanishes by the identity (2.2). Thus, the pressure gradients satisfy the relation

$$S_{li} S_{ji}\frac{\partial p}{\partial \xi_j} = \rho u_i u_k S_{jk}\frac{\partial S_{li}}{\partial \xi_j}, \tag{7.11}$$

where for $j = l, u_k S_{jk} = 0$. This allows the pressure gradient $\frac{\partial p}{\partial \xi_l}$ in the radial direction to be evaluated from the gradients in the tangential direction and the wall curvature. In order to allow the same equations to be used in calculating the flux across both interior and boundary faces, it is common practice to extrapolate the pressure to halo cells beyond the boundary.

7.6 Far Field Boundary Condition

The rate of convergence to a steady state will be impaired if outgoing waves are reflected back into the flow from the outer boundaries. The treatment of the far field boundary condition is based on the introduction of Riemann invariants for a one-dimensional flow normal to the boundary. Let subscripts ∞ and e denote free stream values and values extrapolated from the interior cells adjacent to the boundary, respectively, and let q_n and c be the velocity component normal to the boundary and the speed of sound. Assuming that the flow is subsonic at infinity, we introduce fixed and extrapolated Riemann invariants

$$R_\infty = q_{n_\infty} - \frac{2c_\infty}{\gamma - 1}$$

and

$$R_e = q_{n_e} + \frac{2c_e}{\gamma - 1},$$

corresponding to incoming and outgoing waves. These may be added and subtracted to give

$$q_n = \frac{1}{2}\left(R_e + R_\infty\right)$$

and

$$c = \frac{\gamma - 1}{4}\left(R_e - R_\infty\right),$$

where q_n and c are the actual normal velocity component and speed of sound to be specified in the far field. At an outflow boundary, the tangential velocity components and entropy are extrapolated from the interior, while at an inflow boundary they are specified as having free stream values. These quantities provide a complete definition of the flow in the far field. If the flow is supersonic in the far field, all the flow quantities are specified at an inflow boundary, and they are all extrapolated from the interior at an outflow boundary.

7.7 Discretization on Triangular Meshes

The successful use of unstructured meshes is contingent on the formulation of sufficiently accurate discretization procedures. It also poses a variety of problems related to data structures and indirect addressing. These issues are addressed for both triangular and tetrahedral meshes in the following sections. First, we discuss some general aspects of triangulated meshes in two and three dimensions. In both cases, there is no simple relationship between the number of vertices and the number of cells. Accordingly, data defining the connectivity has to be stored.

For a triangular mesh covering a planar domain,

$$V - E + C = 1, \tag{7.7}$$

where V is the number of vertices, E is the number of edges, and C is the number of cells. This can be proved by induction. The relation (7.7) holds for a single triangle. Then, the addition of each new triangle adds one vertex, one cell, and two edges, so (7.7) continues to hold. In a typical triangulation, there are roughly twice as many cells as vertices, corresponding to the fact that each rectangle of a Cartesian mesh could be

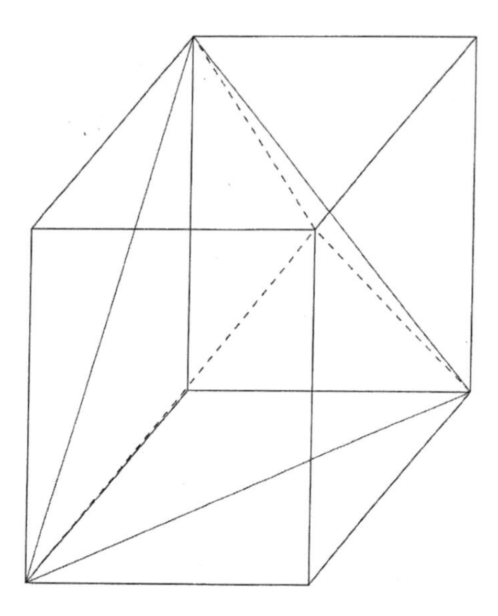

Figure 7.10 Decomposition of a cube into five tetrahedrons.

divided into two triangles. The plane can be perfectly tiled by equilateral triangles, and simple domains can, in general, be subdivided by very regular triangulations.

The situation is different for three-dimensional triangulations because equi-angular tetrahedrons cannot be packed to fill a solid domain. Thus, there will inevitably be some degree of irregularity in a three-dimensional triangulation. In general, a triangulation of a convex domain satisfies the Euler relation

$$V - E + F = 2,$$

where V, E, and F are the numbers of vertices, edges, and faces, respectively. A cube can be divided into six tetrahedra by first slicing it into two triangular prisms and then subdividing each prism into three tetrahedra. It can alternatively be divided into five tetrahedra by cutting off tetrahedra from opposing corners to leave one tetrahedron in the middle, as illustrated in Figure 7.10. In practice, the number of cells in a typical three-dimensional triangulation is around six times the number of vertices, while the number of edges is around seven times the number of vertices. Also, since each tetrahedron has four faces, and each interior face separates two tetrahedra, the number of faces is around twelve times the number of vertices.

7.7.1 Cell-Centered Finite Volume Schemes on Triangular Meshes

The finite volume method can be applied in a straightforward manner to triangulated meshes in two or three dimensions by taking the triangles or tetrahedrons themselves as the control volumes. In order to illustrate this, consider the discretization of the two-dimensional conservation law,

$$\frac{\partial \boldsymbol{w}}{\partial t} + \frac{\partial}{\partial x}\boldsymbol{f}(\boldsymbol{w}) + \frac{\partial}{\partial y}\boldsymbol{g}(\boldsymbol{w}) = 0, \tag{7.12}$$

in a triangulated domain $\mathcal{D}_h$ with boundary $\mathcal{B}_h$. The discretization of the integral form,

$$\frac{d}{dt}\int_{\mathcal{D}_h} \boldsymbol{w}dS + \oint_{\mathcal{B}_h}(\boldsymbol{f}dy - \boldsymbol{g}dx) = 0, \tag{7.13}$$

for a particular triangle is illustrated in Figure 7.11. For convenience, the triangle representing the central volume is labeled zero, with neighbors numbered $k = 1, 2, 3$. Labeling the edges $k0$, $k = 1, 2, 3$, with increments Δx_{k0} and Δy_{k0} in the coordinates along each edge, we discretize (7.13) as

$$S_0 \frac{d\boldsymbol{w}_0}{dt} + \sum_{k=1}^{3} \boldsymbol{h}_{k0} = 0, \tag{7.14}$$

where S_0 is the area of cell 0, $\boldsymbol{w}_0$ is cell average of the state vector, and $\boldsymbol{h}_{k0}$ is the numerical flux across the edge $k0$. We can evaluate $\boldsymbol{h}_{k0}$ as

$$\boldsymbol{h}_{k0} = \boldsymbol{f}_{k0}\Delta y_{k0} - \boldsymbol{g}_{k0}\Delta x_{k0}, \tag{7.15}$$

where $\boldsymbol{f}_{k0}$ and $\boldsymbol{g}_{k0}$ are estimates of the average values of the flux vectors $\boldsymbol{f}$ and $\boldsymbol{g}$ along the edge. A simple central difference scheme is obtained by taking $\boldsymbol{f}_{k0}$ and $\boldsymbol{g}_{k0}$ as arithmetic averages between cell k and cell 0,

$$\boldsymbol{f}_{k0} = \frac{1}{2}(\boldsymbol{f}_k + \boldsymbol{f}_0), \quad \boldsymbol{g}_{k0} = \frac{1}{2}(\boldsymbol{g}_k + \boldsymbol{g}_0), \tag{7.16}$$

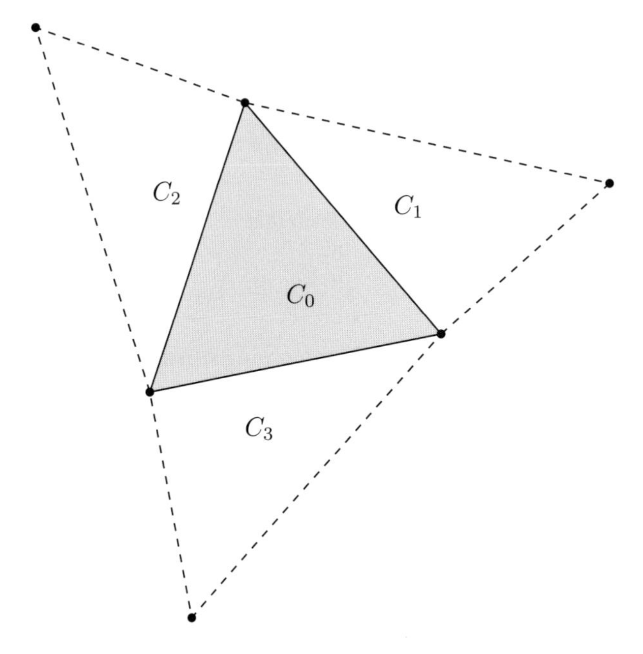

Figure 7.11 Cell-centered finite volume scheme on a triangular mesh.

where

$$\begin{aligned} \boldsymbol{f}_0 &= \boldsymbol{f}(\boldsymbol{w}_0), \quad \boldsymbol{f}_k = \boldsymbol{f}(\boldsymbol{w}_k), \\ \boldsymbol{g}_0 &= \boldsymbol{g}(\boldsymbol{w}_0), \quad \boldsymbol{g}_k = \boldsymbol{g}(\boldsymbol{w}_k). \end{aligned} \tag{7.17}$$

Alternative averaging procedures lead to alternative central difference schemes. In the case of the two-dimensional Euler equations, for example, the state and flux vectors are

$$\boldsymbol{w} = \begin{bmatrix} \rho \\ \rho u \\ \rho v \\ \rho E \end{bmatrix}, \quad \boldsymbol{f} = \begin{bmatrix} \rho u \\ \rho u^2 + p \\ \rho u v \\ \rho u H \end{bmatrix}, \quad \boldsymbol{g} = \begin{bmatrix} \rho v \\ \rho v u \\ \rho v^2 + p \\ \rho v H \end{bmatrix},$$

$$p = (\gamma - 1)\rho\left(E - \frac{1}{2}(u^2 + v^2)\right), \quad H = E + \frac{p}{\rho},$$

where ρ and p are the density and pressure, u and v are the velocity components, and E and H are the total energy and enthalpy, as described in Chapter 2. Then, we can define arithmetic averages ρ_{k0}, u_{k0}, v_{k0}, and E_{k0} as

$$\rho_{k0} = \frac{1}{2}(\rho_k + \rho_0),$$

and so on, and use these values to evaluate $\boldsymbol{f}_{k0}$ and $\boldsymbol{g}_{k0}$.

A first order accurate upwind scheme can be formulated as follows. Since the sums of Δy_{k0} and Δx_{k0} are zero around a closed loop, (7.14) is equivalent to

$$S_0 \frac{d\boldsymbol{w}_0}{dt} + \frac{1}{2}\sum_{k=1}^{3}\left\{(\boldsymbol{f}_k - \boldsymbol{f}_0)\Delta x_{k0} - (\boldsymbol{g}_k - \boldsymbol{g}_0)\Delta x_{k0}\right\} = 0. \tag{7.18}$$

Introduce Roe-averaged Jacobian matrices such that

$$(\boldsymbol{f}_k - \boldsymbol{f}_0)\Delta y_{k0} - (\boldsymbol{g}_k - \boldsymbol{g}_0)\Delta x_{k0} = A_{k0}(\boldsymbol{w}_k - \boldsymbol{w}_0), \tag{7.19}$$

following the definitions of Section 6.6. Decomposing A_{k0} as

$$A_{k0} = V \Lambda V^{-1}, \tag{7.20}$$

where the columns of $\mathbf{V}$ are the eigenvectors of A_{k0}, and Λ is a diagonal matrix containing the eigenvalues, we define the absolute Jacobian matrix as

$$|A_{k0}| = V|\Lambda|V^{-1}. \tag{7.21}$$

Now the upwind scheme is obtained by defining the numerical flux as

$$\boldsymbol{h}_{k0} = \boldsymbol{f}_{k0}\Delta y_{k0} - \boldsymbol{g}_{k0}\Delta x_{k0} - \frac{1}{2}\left|A_{k0}\right|(\boldsymbol{w}_k - \boldsymbol{w}_0). \tag{7.22}$$

Then, the scheme is equivalent to

$$S_0\frac{d\boldsymbol{w}}{dt} = \frac{1}{2}\left(\left|A_{k0}\right| - A_{k0}\right)(\boldsymbol{w}_k - \boldsymbol{w}_0).$$

This construction is essentially identical to the construction of the upwind scheme on a quadrilateral mesh defined by (7.3). As before, one may also use a scheme with scalar diffusion. Defining the coefficients

$$\epsilon_{k0} = \max\left|\lambda(A_{k0})\right|,$$

the numerical flux for the scheme with scalar diffusion is

$$\boldsymbol{h}_{k0} = \boldsymbol{f}_{k0}\Delta y_{k0} - \boldsymbol{g}_{k0}\Delta x_{k0} - \frac{1}{2}\epsilon_{k0}(\boldsymbol{w}_k - \boldsymbol{w}_0).$$

In the case of a scalar conservation law,

$$\frac{\partial u}{\partial t} + \frac{\partial}{\partial x}f(u) + \frac{\partial}{\partial y}g(u) = 0, \tag{7.23}$$

let v_k denote the solution in cell k. Define the coefficients

$$a_{k0} = \begin{cases} \dfrac{(f_k - f_0)\Delta y_{k0} - (g_k - g_0)\Delta x_{k0}}{v_k - v_0}, & v_k \neq v_0, \\ \left.\left(\dfrac{\partial f}{\partial u}\Delta y_{k0} - \dfrac{\partial g}{\partial u}\Delta x_{k0}\right)\right|_{u=v_0}, & v_k = v_0. \end{cases} \tag{7.24}$$

We can then formulate the semi-discrete scheme as

$$S_0\frac{dv_0}{dt} + \frac{1}{2}\sum_{k=1}^{3}(f_k + f_0) - |a_{k0}|(v_k - v_0) = 0,$$

which is equivalent to

$$S_0\frac{dv_0}{dt} + \frac{1}{2}\sum_{k=1}^{3}(|a_{k0}| - a_{k0})(v_k - v_0) = 0.$$

This has non-negative coefficients on the right hand side, thus satisfying the conditions given in Section 5.7 for a local extremum diminishing (LED) scheme.

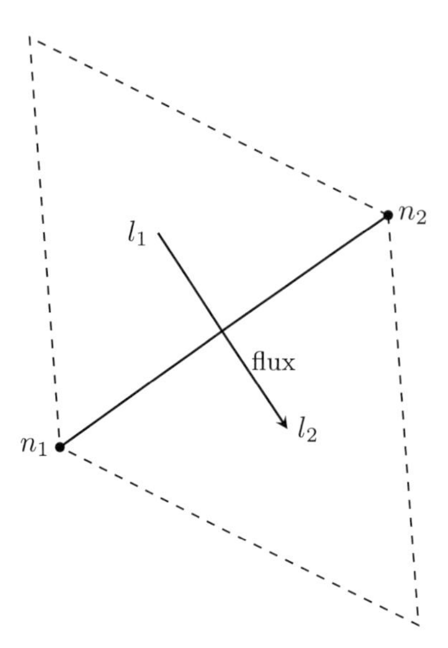

Figure 7.12 The vertices and cells associated with an edge.

7.7.2 Data Structure

If one were to evaluate the flux balance equation separately for each cell, each interface flux would have to be evaluated twice. A more efficient procedure, which avoids this duplication, is to preset the cell residuals to zero in a loop over the cells and then to calculate the interface fluxes in a loop over the faces and transfer the fluxes between the two cells on either side of each face. In order to implement this, we may introduce a data structure of the following kind. We associate with each edge the nodes $n1$ and $n2$, defining its vertices, and the cells $l1$ and $l2$ separated by the edge, as illustrated in Figure 7.12. Now the flux across the edge is evaluated as

$$\text{flux} = \frac{1}{2}\left\{(\boldsymbol{f}(l1) + \boldsymbol{f}(l2))\,(y(n2) - y(n1)) - (\boldsymbol{g}(l1) + \boldsymbol{g}(l2))\,(x(n2) - x(n1))\right\},$$

or the corresponding upwind formula. Then, the flux is subtracted from the residual at $l2$ and added to the residual at $l1$.

7.7.3 Second Order Upwind Schemes

Second order upwind schemes can be constructed following the procedures developed for one-dimensional conservation laws in Chapters 5 and 6. Instead of calculating the flux $\boldsymbol{h}_{k0}$ from the values $\boldsymbol{w}_k$ and $\boldsymbol{w}_0$, we estimate left and right states $\boldsymbol{w}_L$ and $\boldsymbol{w}_R$ at the center c of the edge $k0$, as illustrated in Figure 7.13, and evaluate the flux by substituting $\boldsymbol{w}_L$ and $\boldsymbol{w}_R$ for $\boldsymbol{w}_k$ and $\boldsymbol{w}_0$ in the formulas (7.16), (7.17), (7.19), and (7.22). Thus, we evaluate the flux as

$$\boldsymbol{h}_{k0} = \frac{1}{2}\{(\boldsymbol{f}_R + \boldsymbol{f}_L)\Delta y_{k0} - (\boldsymbol{g}_R + \boldsymbol{g}_L)\Delta x_{k0}\} - \frac{1}{2}|A_{RL}|(\boldsymbol{w}_k - \boldsymbol{w}_0),$$

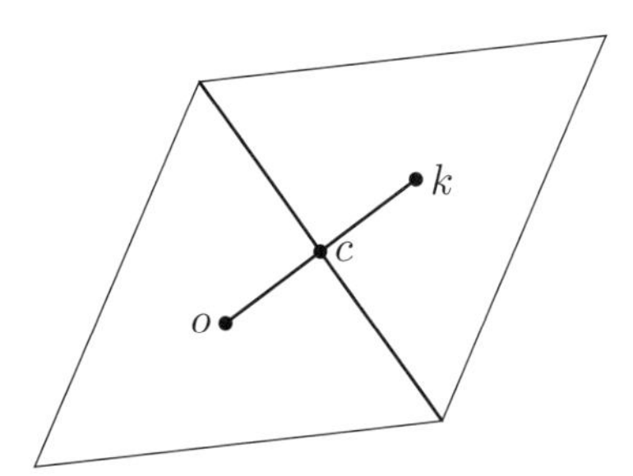

Figure 7.13 Left and right states at C are obtained by reconstruction from the solution values and gradients in the cells labeled 0 and k.

where

$$f_R = f(w_R), \quad f_L = f(w_L),$$
$$g_R = g(w_R), \quad g_L = g(w_L),$$

and

$$(f_R - f_L)\Delta y_{k0} - (g_R - g_L)\Delta x_{k0} = A_{RL}(w_k - w_0).$$

The states w_0 and w_k represent cell averages. In smooth regions of the flow, however, these differ from the centroidal values by $\mathcal{O}(h^2)$, where h is a distance representative of the cell width. Accordingly, we can estimate w_L and w_R with second order accuracy as

$$\begin{aligned} w_L &= w_0 + (x_c - x_0) \cdot \nabla w_0, \\ w_R &= w_0 + (x_c - x_k) \cdot \nabla w_k, \end{aligned} \tag{7.25}$$

where the vectors x_c, x_0, and x_k denote the center of the edge $k0$ and the centroids of cell k and cell 0, while ∇w_0 and ∇w_k are estimates of the gradients of the state vector at the centroids of cell k and cell 0. Moreover, we can use Gauss' theorem to estimate these gradients as in Chapter 3. According to Gauss' theorem, the average values of $\frac{\partial w}{\partial x}$ and $\frac{\partial w}{\partial y}$ in cell 0 are

$$\left.\frac{\partial w}{\partial x}\right|_0 = \frac{1}{S_0}\oint w dy,$$
$$\left.\frac{\partial w}{\partial y}\right|_0 = \frac{1}{S_0}\oint w dx,$$

where the integrals are around the boundary of the element. Now, using the arithmetic average of the states w_0 and w_k to estimate the state at the center of each edge, we can evaluate the average values of the gradients in cell 0 as

$$\left.\frac{\partial w}{\partial x}\right|_0 = \frac{1}{2S_0}\sum_{k=1}^{3}(w_k + w_0)\Delta y_{k0},$$
$$\left.\frac{\partial w}{\partial y}\right|_0 = -\frac{1}{2S_0}\sum_{k=1}^{3}(w_k + w_0)\Delta x_{k0}.$$

Finally, since the average and centroidal values of the gradients also differ by a quantity of second order, we can use these values of the gradients to evaluate $\boldsymbol{w}_L$ and $\boldsymbol{w}_R$ according to (7.25). This procedure for estimating the gradients corresponds closely to the formulas (7.15)–(7.17), which have been used to evaluate the flux balance.

7.7.4 Three-Dimensional Cell-Centered Finite Volume Scheme

The same procedure can be used to discretize the three-dimensional conservation law

$$\frac{\partial \boldsymbol{w}}{\partial t} + \frac{\partial}{\partial x_i} \boldsymbol{f}_i(\boldsymbol{w}) = 0$$

in a domain $\mathcal{D}_h$ with boundary $\mathcal{B}_h$, where the repeated index i denotes the sum over $i = 1, 2, 3$. Now, we discretize the integral form

$$\frac{d}{dt} \int_{\mathcal{D}} \boldsymbol{w} dV + \int_{\mathcal{B}} \boldsymbol{f}_i n_i dS = 0 \tag{7.26}$$

for a tetrahedron labeled zero, with neighbors labeled $k = 1, \ldots, 4$, as

$$V_0 \frac{d\boldsymbol{w}_0}{dt} + \sum_{k=1}^{4} \boldsymbol{h}_{k0} = 0,$$

where V_0 is the volume of cell 0, $\boldsymbol{w}_0$ is the cell average of the state vector $\boldsymbol{w}$ in cell 0, and $\boldsymbol{h}_{k0}$ is the flux through the face $k0$ separating cell k and cell 0, with area S_{k0}. Denoting the components of the unit normal to the face $k0$ by $n_{i_{k0}}$, we can evaluate the numerical flux as

$$\boldsymbol{h}_{k0} = n_i \boldsymbol{f}_{i_{k0}} S_{k0},$$

where $\boldsymbol{f}_{i_{k0}}$ is an estimate of the average value of the flux $\boldsymbol{f}_i$ over the face $k0$. As in the two-dimensional case, the flux of a central difference scheme can be evaluated by setting

$$\boldsymbol{f}_{i_{k0}} = \frac{1}{2}(\boldsymbol{f}_{i_k} + \boldsymbol{f}_{i_0}), \tag{7.27}$$

where

$$\boldsymbol{f}_{i_0} = \boldsymbol{f}_i(\boldsymbol{w}_0), \quad \boldsymbol{f}_{i_k} = \boldsymbol{f}_i(\boldsymbol{w}_k).$$

In order to define a first order accurate upwind scheme, we define Roe-averaged Jacobian matrices satisfying

$$n_i(\boldsymbol{f}_{i_k} - \boldsymbol{f}_{i_0}) S_{k0} = A_{k0}(\boldsymbol{w}_k - \boldsymbol{w}_0).$$

Then, defining the absolute Jacobian matrices $|A_{k0}|$ according to (7.20) and (7.21) as before, the upwind flux is constructed as

$$\boldsymbol{h}_{k0} = n_i \boldsymbol{f}_{i_{k0}} S_{k0} - \frac{1}{2}|A_{k0}|(\boldsymbol{w}_k - \boldsymbol{w}_0).$$

7.8 Vertex-Centered Schemes

And alternative approach is to define the solution by the values of the state vector at the vertices of the triangulation. Since the number of cells is typically around twice the number of vertices in a two-dimensional triangulation, and around six times the number of vertices in a three-dimensional triangulation, this leads to a substantial reduction in the number of degrees of freedom in the discretization on a given mesh, with a corresponding reduction in the memory needed to store the solution. It is no longer possible to satisfy the condition of a perfect flux balance for each cell, as there are not enough degrees of freedom. We can, however, define a control volume for each vertex as the union of the surrounding cells for which it is a common vertex.

We first illustrate this procedure for the case of the two-dimensional conservation law (7.12) with the integral form (7.13). Suppose that a particular interior node, labeled 0 for convenience, as illustrated in Figure 7.14, is surrounded by n nearest neighbors $k = 1, \ldots, n$, where node $n + 1$ is identified with node 1. These also form n surrounding cells. The area of the control volume is

$$S_0 = \sum_{k=1}^{n} S_k,$$

where S_k is the area of cell k. Then, using trapezoidal integration, we approximate the integral form (7.13) as

$$S_0 \frac{d\boldsymbol{f}_0}{dt} + \frac{1}{2} \sum_{k=1}^{n} \big((\boldsymbol{f}_{k+1} + \boldsymbol{f}_k)(y_{k+1} - y_k) - (\boldsymbol{g}_{k+1} + \boldsymbol{g}_k)(x_{k+1} - x_k) \big) = 0, \quad (7.28)$$

where

$$\boldsymbol{f}_k = \boldsymbol{f}(\boldsymbol{w}_k), \quad \boldsymbol{g}_k = \boldsymbol{g}(\boldsymbol{w}_k).$$

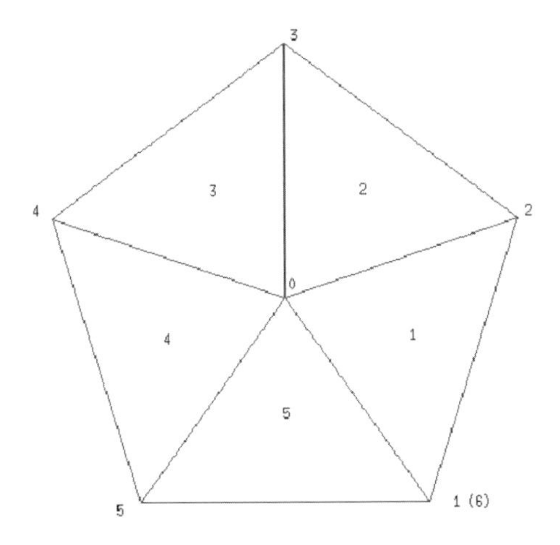

Figure 7.14 Cells and nodes surrounding the vertex 0.

Next, we show that this discretization closely corresponds to a discretization by the Galerkin method, using a piecewise linear trial solution. Moreover, we subsequently show that it is equivalent to a finite volume scheme using nonoverlapping median dual cells.

First, we show the correspondence to the Galerkin method. Following this method, as presented in Chapter 3, we approximate $\boldsymbol{w}$ by a piecewise linear trial solution

$$\boldsymbol{w}_h = \sum_{k=1}^{N} \boldsymbol{w}_k \phi_k, \tag{7.29}$$

where N is the total number of nodes, and each basis function has the value unity at a single node and zero at all other nodes, as was described in Section 3.3.5. In order to satisfy the conservation law (7.12), we require $\boldsymbol{w}_h$ to satisfy the orthogonality conditions

$$\iint_{\mathcal{D}_h} \phi_l \left(\frac{\partial \boldsymbol{w}_h}{\partial t} + \frac{\partial}{\partial x} \boldsymbol{f}(\boldsymbol{w}_h) + \frac{\partial}{\partial y} \boldsymbol{g}(\boldsymbol{w}_h) \right) dx\, dy = 0, \quad l = 1, \ldots, N. \tag{7.30}$$

Integrating by parts, we replace (7.30) by the weak form

$$\iint_{\mathcal{D}_h} \phi_l \frac{\partial \boldsymbol{w}_h}{\partial t}\, dx\, dy = R_l, \tag{7.31}$$

where

$$R_l = \iint_{\mathcal{D}_h} \left(\boldsymbol{f} \frac{\partial \phi_l}{\partial x} + \boldsymbol{g} \frac{\partial \phi_l}{\partial y} \right) dx\, dy - \int_{\mathcal{B}_h} \phi_l (n_x \boldsymbol{f} + n_y \boldsymbol{g}) dl.$$

Here s is the arc length along the boundary, while n_x and n_y are the components of the unit normal to the boundary. The boundary integral vanishes at every interior node because the corresponding test function is zero at every boundary node.

Consider now the particular interior node labeled zero, as illustrated in Figure 7.14. Since the basis function ϕ_0 is zero outside the immediately neighboring cells, only these cells contribute to the area integral in the residual R_0, which can thus be evaluated as

$$R_0 = \sum_{k=1}^{n} \left(\bar{\boldsymbol{f}}_k \frac{\partial \phi_0}{\partial x} + \bar{\boldsymbol{g}}_k \frac{\partial \phi_0}{\partial y} \right) S_k, \tag{7.32}$$

where $\bar{\boldsymbol{f}}_k$ and $\bar{\boldsymbol{g}}_k$ are the average values of $\boldsymbol{f}$ and $\boldsymbol{g}$ in cell k, S is its area, and $\frac{\partial \phi_0}{\partial x}$ and $\frac{\partial \phi_0}{\partial y}$ are constant in each cell because ϕ_0 is piecewise linear. As in Section 3.3.5, we can use Gauss' theorem to evaluate the derivatives of the test function. In cell k,

$$S_k \frac{\partial \phi_0}{\partial x} = \oint \phi_0\, dy,$$

$$S_k \frac{\partial \phi_0}{\partial y} = -\oint \phi_0\, dx,$$

where the integrals are taken in the counter-clockwise direction around the cell boundary. Using trapezoidal integration, which is exact for a linear function, we find that

$$S_k \frac{\partial \phi_0}{\partial x} = \frac{1}{2}(y_k - y_{k+1}),$$

$$S_k \frac{\partial \phi_0}{\partial y} = -\frac{1}{2}(x_k - x_{k+1}),$$

because ϕ_0 is zero at nodes k and $k+1$. With a piecewise linear trial solution v, $\boldsymbol{f}(v)$ and $\boldsymbol{g}(v)$ will, in general, be nonlinear. However, denoting $\boldsymbol{f}(\boldsymbol{w}_k)$ and $\boldsymbol{g}(\boldsymbol{w}_k)$ by $\boldsymbol{f}_k$ and $\boldsymbol{g}_k$, we can approximate the average values of $\boldsymbol{f}$ and $\boldsymbol{g}$ in cell k by the simple arithmetic averages

$$\bar{\boldsymbol{f}}_k = \frac{1}{3}(\boldsymbol{f}_0 + \boldsymbol{f}_k + \boldsymbol{f}_{k+1})$$

$$\bar{\boldsymbol{g}}_k = \frac{1}{3}(\boldsymbol{g}_0 + \boldsymbol{g}_k + \boldsymbol{g}_{k+1}).$$

Thus, the sum (7.32) reduces to

$$R_0 = -\frac{1}{6}\sum_{k=1}^{n}\left((\boldsymbol{f}_0 + \boldsymbol{f}_k + \boldsymbol{f}_{k+1})(y_{k+1} - y_k) - (\boldsymbol{g}_0 + \boldsymbol{g}_k + \boldsymbol{g}_{k+1})(x_{k+1} - x_k)\right).$$

However, $\boldsymbol{f}_0$ and $\boldsymbol{g}_0$ are multiplied by the sums of $y_{k+1} - y_k$ and $x_{k+1} - x_k$ over a closed loop, so they can be dropped. Thus, finally, we can write the Galerkin equation for node 0 as

$$\frac{d}{dt}\iint \phi_0 \boldsymbol{w}_k \, dx\, dy = -\frac{1}{6}\sum_{k=1}^{n}\left((\boldsymbol{f}_{k+1} + \boldsymbol{f}_k)(y_{k+1} - y_k) - (\boldsymbol{g}_{k+1} + \boldsymbol{g}_k)(x_{k+1} - x_k)\right). \tag{7.33}$$

The right hand side is just $\frac{1}{3}$ of the trapezoidal integration rule for $\oint(\boldsymbol{f}\, dy - \boldsymbol{g}\, dx)$. Also, carrying out the integrations on the left hand side,

$$\iint \phi_0 \frac{\partial \boldsymbol{w}_k}{\partial t}\, dx\, dy = M_0 \frac{d\boldsymbol{w}_0}{dt} + \sum_{k=1}^{n} M_k \frac{d\boldsymbol{w}_k}{dt},$$

where

$$M_0 = \frac{1}{6}\sum_{k=1}^{n} S_k$$

and

$$M_k = \frac{1}{12}(S_k + S_{k-1}).$$

Thus, the sum of the weights is

$$M_0 + \sum_{k=1}^{n} M_k = \frac{1}{3}\sum_{k=1}^{n} S_k,$$

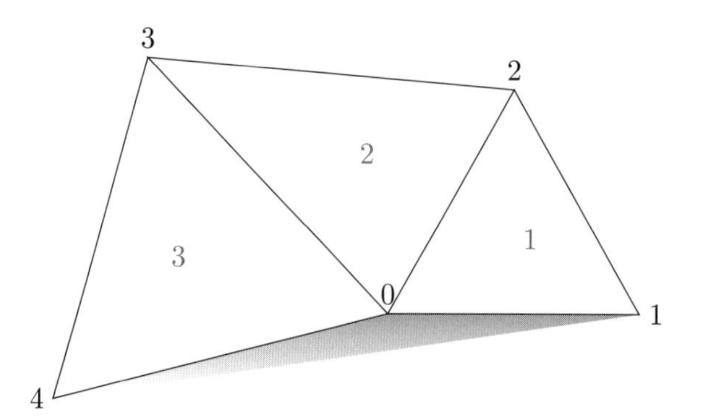

Figure 7.15 Control volume for a boundary node.

consistent with the fact that the right hand side is a discrete approximation to $\frac{1}{3}$ of the flux balance for the control volume comprised by the union of the neighboring triangles. If time accuracy is not needed, as in the case of a steady state calculation, the left hand side can be replaced by the lumped term $\left(\sum_{k=1}^{n} S_k\right)\frac{dw_0}{dt}$. Now, multiplying both sides of (7.33) by a factor of three, we recover the finite volume equation (7.28). At boundary nodes, a corresponding formula is obtained by trapezoidal integration over a one-sided control volume, as illustrated in Figure 7.15.

Summing the residual equation (7.28) over any subset of the nodes, possibly including some boundary nodes, the flux balance for each individual triangle appears three times in the sum of the flux balances, while the area of each triangle is multiplied by the rate of change of $\mathbf{w}_h$ at each of its three vertices, corresponding to three times the average rate of change of $\mathbf{w}_h$ in the triangle. Thus, the scheme is fully conservative. This is also apparent when the scheme is reformulated as an edge-based scheme, as is discussed in Section 7.8.2, where it is shown how the scheme may be modified to an upwind-biased scheme by the addition of artificial diffusive terms.

7.8.1 Data Structure and Evaluation of the Flux Balance

Referring to Figure 7.16, it can be seen that a given edge AB contributes only to the two control volumes P and Q centered at the vertices of the two triangles with the common base AB, since it is an interior edge of the control volumes centered at A and B. This suggests the use of a data structure in which each edge is associated with four nodes: the two vertices $n1$ and $n2$ of the edge and the outer vertices of the two of the two triangles based on that edge, as illustrated in Figure 7.17.

The flux balances for the entire mesh can then be calculated as follows. First, set the residuals at every node to zero in a loop over the nodes. Then, accumulate flux balances at the nodes in a loop over the edges. The flux across an edge is calculated using arithmetic averages between nodes $n1$ and $n2$:

$$\text{flux} = \frac{1}{2}\left((\boldsymbol{f}(n1)+\boldsymbol{f}(n2))(y(n2)-y(n1)) - (\mathbf{g}(n1)+\mathbf{g}(n2))(x(n2)-x(n1))\right).$$

Then, the flux is subtracted from the residual at $n3$ and added to the residual at $n4$. The residuals of the boundary nodes include contributions from the boundary edges,

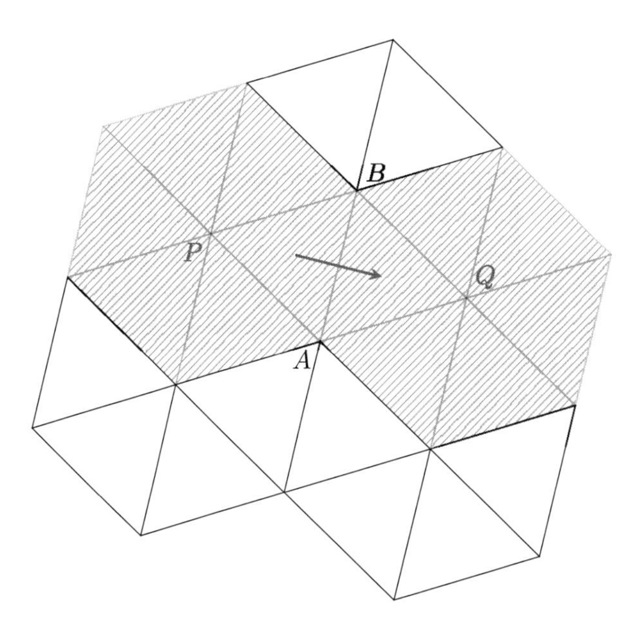

Figure 7.16 Control volumes affected by an edge.

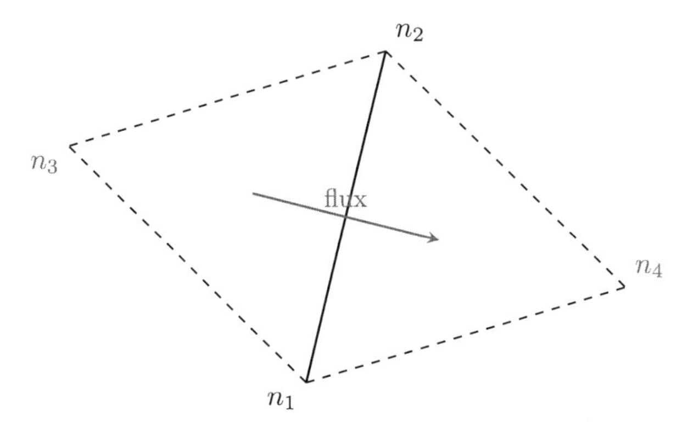

Figure 7.17 The four nodes associated with an edge.

as illustrated in Figure 7.15. However, we can introduce dummy nodes outside each boundary edge, as illustrated in Figure 7.18. Then, the flux across edges 01 and 40 can be transferred to the dummy nodes 5 and 6, and redistributed back to the nodes 0, 1, and 4 in an additional loop over the boundary edges.

7.8.2 Reformulation of the Finite Volume Scheme Based on Contributions along the Edges

The finite volume scheme (7.28) can be re-arranged in the following manner. Since $\boldsymbol{f}_k$ and $\boldsymbol{g}_k$ contribute to the flux across the edges from $k-1$ to k and k to $k+1$ in the trapezoidal integration rule, they actually make a contribution

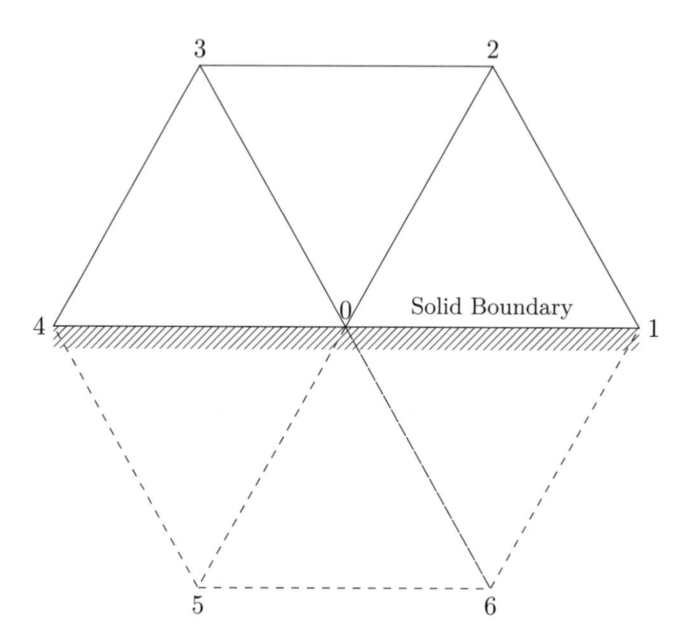

Figure 7.18 Accumulation of the flux balance at a boundary node.

$$\frac{1}{2}(\boldsymbol{f}_k(y_{k+1} - y_{k-1}) - \boldsymbol{g}_k(x_{k+1} - x_{k-1})).$$

Associating the intervals

$$\Delta y_{k0} = y_{k+1} - y_{k-1}, \quad \Delta x_{k0} = x_{k+1} - x_{k-1}$$

with the edge $k0$, (7.28) is thus equivalent to

$$S_0 \frac{d\boldsymbol{w}_0}{dt} + \frac{1}{2}\sum_{k=1}^{n}(\boldsymbol{f}_k \Delta y_{k0} - \boldsymbol{g}_k \Delta x_{k0}) = 0.$$

However, since the sums of Δy_{k0} and Δx_{k0} are zero around a closed loop, we can add or subtract $\boldsymbol{f}_0$ and $\boldsymbol{g}_0$ to obtain the equivalent radial edge based schemes

$$S_0 \frac{d\boldsymbol{w}_0}{dt} + \frac{1}{2}\sum_{k=1}^{n}((\boldsymbol{f}_k + \boldsymbol{f}_0)\Delta y_{k0} - (\boldsymbol{g}_k + \boldsymbol{g}_0)\Delta x_{k0}) = 0, \tag{7.34}$$

and

$$S_0 \frac{d\boldsymbol{w}_0}{dt} + \frac{1}{2}\sum_{k=1}^{n}((\boldsymbol{f}_k - \boldsymbol{f}_0)\Delta y_{k0} - (\boldsymbol{g}_k - \boldsymbol{g}_0)\Delta x_{k0}) = 0. \tag{7.35}$$

We can now construct a first order accurate upwind scheme in the same way that was used for the cell centered discretization in Section 7.7.1. Introducing

Roe-averaged Jacobian matrices that satisfy (7.19), we subtract an artificial diffusion term $\frac{1}{2}|A_{k0}|(\boldsymbol{w}_k - \boldsymbol{w}_0)$ along each edge to produce the upwind scheme

$$S_0 \frac{d\boldsymbol{w}_0}{dt} + \frac{1}{2} \sum_{k=1}^{n} ((\boldsymbol{f}_k + \boldsymbol{f}_0)\Delta y_{k0} - (\boldsymbol{g}_k + \boldsymbol{g}_0)\Delta x_{k0} - |A_{k0}|(\boldsymbol{w}_k - \boldsymbol{w}_0)) = 0, \quad (7.36)$$

which may equivalently be written as

$$S_0 \frac{d\boldsymbol{w}_0}{dt} = \frac{1}{2} \sum_{k=1}^{n} (|A_{k0}| - A_{k0})(\boldsymbol{w}_k - \boldsymbol{w}_0).$$

As in the case of the cell centered scheme, we can also formulate a scheme with scalar diffusion:

$$S_0 \frac{d\boldsymbol{w}_0}{dt} + \frac{1}{2} \sum_{k=1}^{n} ((\boldsymbol{f}_k + \boldsymbol{f}_0)\Delta y_{k0} + (\boldsymbol{g}_k - \boldsymbol{g}_0)\Delta x_{k0} - \epsilon_{k0}(\boldsymbol{w}_k - \boldsymbol{w}_0)) = 0, \quad (7.37)$$

where

$$\epsilon_{k0} = \max |\lambda(A_{k0})|.$$

7.8.3 Evaluation of the Flux Balance for Radial Edge-Based Scheme

As in the case of the first scheme defined by (3.4.1), the radial edge-based scheme can be implemented efficiently by accumulating the flux balances in a loop over the edges using the data structure illustrated in Figure 7.17. Now, however, the flux along the edge is evaluated as

$$\text{flux} = \frac{1}{2}((\boldsymbol{f}(n1) + \boldsymbol{f}(n2))(y(n4) - y(n3)) - (\boldsymbol{g}(n1) + \boldsymbol{g}(n2))(x(n4) - x(n3))) \quad (7.38)$$

and then exchanged between nodes $n1$ and $n2$. The artificial diffusive flux

$$\frac{1}{2}|A_{k0}|(\boldsymbol{w}_k - \boldsymbol{w}_0)$$

can be included in the same loop. The closure formulas needed to correct the flux balances at the boundary nodes are now, however, considerably more complicated.

In evaluating fluxes for the boundary edges according to formula (7.38), we may introduce dummy nodes ($n4$), defined as the midpoints of the edge

$$x(n4) = \frac{1}{2}(x(n1) + x(n2)),$$

$$y(n4) = \frac{1}{2}(y(n1) + y(n2)).$$

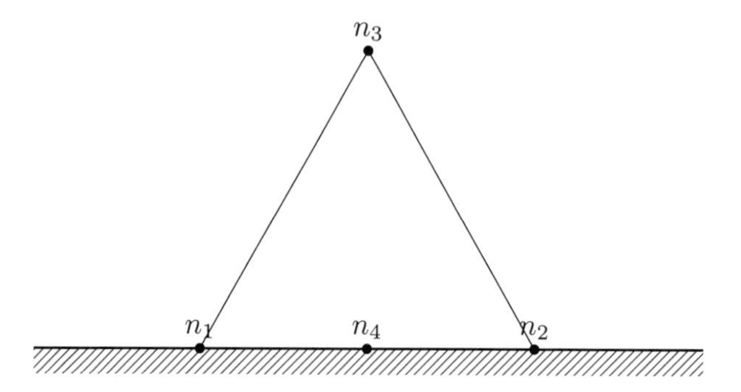

Figure 7.19 Nodes associated with a boundary edge.

Consider a boundary edge as illustrated in Figure 7.19. Let $\boldsymbol{R}_k,\ k=1,2$ be the residuals at each end of the edge that result from the edge loop. The equations for these nodes should be corrected as

$$\frac{d\boldsymbol{w}_k}{dt}+\boldsymbol{R}_k+\boldsymbol{C}_k=0,$$

where the correction $\boldsymbol{C}_k$ may be evaluated as follows. First, we calculate the flux across the edge from the solution values at each end as

$$\boldsymbol{F}_k=\boldsymbol{f}(\boldsymbol{w}_k)(y(n2)-y(n1))-\boldsymbol{g}(\boldsymbol{w}_k)(x(n2)-x(n1)).$$

Then, we set

$$\boldsymbol{Q}=\boldsymbol{F}_1+\boldsymbol{F}_2$$

and

$$\boldsymbol{C}_k=\frac{1}{4}\boldsymbol{Q}+\boldsymbol{F}_k.$$

Here, the flux through the edge is $\frac{1}{2}\boldsymbol{Q}$. At a solid wall, it follows from the flow tangency condition that there, the convective flux across the edge should be zero. Since the flux across the edge contributes not only to the residuals at each end of the edge but also to the residual at the interior vertex 3 of the triangle based on the edge, the convective part of the flux needs to be subtracted from the residuals at all three vertices.

7.8.4 Equivalence to a Finite Volume Scheme Using Median Dual Control Volumes

We now show that the finite volume scheme described by (7.27) or (7.31) is in fact exactly equivalent to a finite volume scheme using median dual control volumes. Referring to Figure 7.20, the median dual control volume is defined by connecting the centroids V_k of each neighboring cell to the midpoints of the edges $k0$. Since 1/3

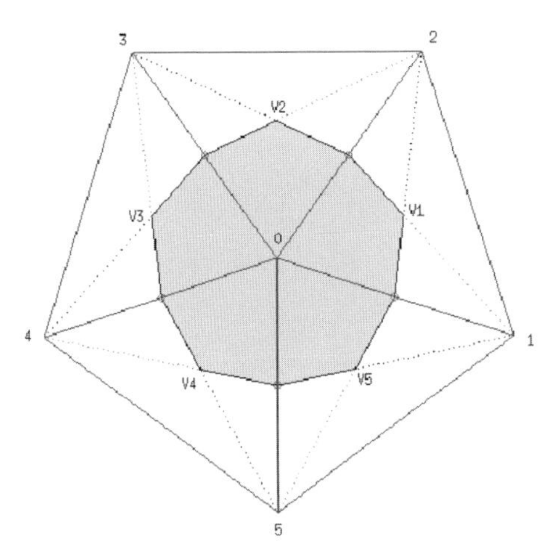

Figure 7.20 Median dual control volume.

of the area of each neighboring cell is assigned to the median dual control volume, its area is equal to $\frac{1}{3}\sum_{k=1}^{n} S_k$. Now, the flux along edge $k0$ is calculated as

$$\frac{1}{2}((\boldsymbol{f}_k + \boldsymbol{f}_0)(y(V_k) - y(V_{k-1})) - (\boldsymbol{g}_k + \boldsymbol{g}_0)(x(V_k) - x(V_{k-1}))).$$

Due to the construction from the medians, however,

$$y(V_k) - y(V_{k-1}) = \frac{1}{3}\Delta y_{k0},$$

$$x(V_k) - x(V_{k-1}) = \frac{1}{3}\Delta x_{k0}.$$

Thus, the formulas are exactly equivalent to (7.34) multiplied by 1/3. Since the median dual control volumes exactly cover the domain without overlapping, it is immediately evident that the scheme is conservative. Schemes of the type described in (7.28), (7.34), and (7.37) were first used by Jameson, Baker, and Weatherill (1986), who applied them directly to three-dimensional problems. A two-dimensional scheme directly based on the median dual control volume was apparently first proposed by Vijayasundaram (1986).

7.8.5 Second order Vertex-Centered Schemes

The schemes (7.36) and (7.37) are only first order accurate. In the case of the scheme (7.37) with scalar diffusion, a relatively simple way to obtain a more accurate scheme is to recycle the edge differencing procedure (Jameson et al. 1986). The accumulated dissipative term in (7.37) from all the edges may be written as

$$\boldsymbol{D}_0^{(1)} = \sum_{k=1}^{n} \epsilon_{k0}^{(1)}(\boldsymbol{w}_k - \boldsymbol{w}_0), \tag{7.39}$$

where now the superscript 1 is included to denote that this is the first order dissipative term. In order to define a higher order dissipative term, we set

$$\boldsymbol{E}_0 = \sum_{k=1}^{n} (\boldsymbol{w}_k - \boldsymbol{w}_0) \tag{7.40}$$

at every mesh point and then set

$$\boldsymbol{D}_0^{(2)} = \sum_{k=1}^{n} \epsilon_{k0}^{(2)} (\boldsymbol{E}_k - \boldsymbol{E}_0). \tag{7.41}$$

The Jameson–Schmidt–Turkel (JST) scheme ((5.17)–(5.26)) can now be emulated by blending $\boldsymbol{D}_0^{(1)}$ and $\boldsymbol{D}_0^{(2)}$ and adapting the coefficients $\epsilon_{k0}^{(1)}$ and $\epsilon_{k0}^{(2)}$ to the local pressure gradient. The resulting scheme has proven robust and has been found to have a good shock capturing capability.

7.8.6 SLIP Schemes on Multi-dimensional Unstructured Meshes

The SLIP construction may also be implemented along the edges of an unstructured mesh. For simplicity, we consider the case of the scalar conservation law (7.23). The first order vertex-centered semi-discrete scheme can be written as

$$S\frac{dv_o}{dt} + \sum_{k=1}^{n} h_{ko} = 0, \tag{7.42}$$

with the numerical flux

$$h_{k0} = \frac{1}{2}(f_k + f_0) - d_{k0} \tag{7.43}$$

along the edge $k0$, where d_{k0} is the diffusive flux. Defining coefficients a_{k0} according to (7.24), let

$$d_{k0} = \alpha_{k0}(v_k - v_0), \tag{7.44}$$

where

$$\alpha_{k0} \geq \frac{1}{2}|a_{k0}|. \tag{7.45}$$

Then, the scheme reduces to

$$S\frac{dv_o}{dt} = \sum_{k=1}^{n} (\alpha_{ko} - a_{k0})(v_k - v_0), \tag{7.46}$$

which has non-negative coefficients and is therefore LED.

Define

$$\Delta v_{k0} = v_k - v_0.$$

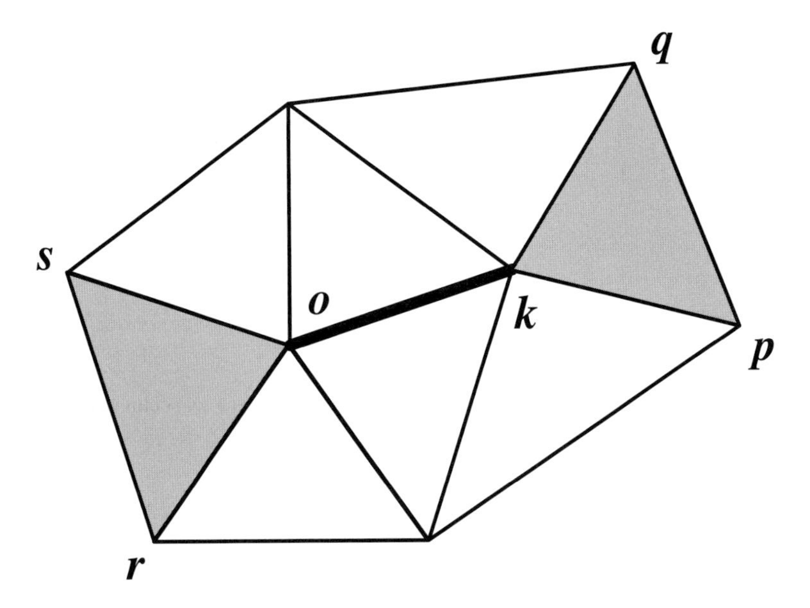

Figure 7.21 Edge $k0$ connecting adjacent control volumes.

Anti-diffusive terms may be added without violating the positivity condition (5.13) by the following generalization of the one-dimensional SLIP scheme. Let $\boldsymbol{l}_{ko}$ be the vector connecting the edge ko, and define the neighboring differences

$$\Delta^+ v_{ko} = \boldsymbol{l}_{ko} \cdot \nabla^+ v, \Delta^- v_{ko} = \boldsymbol{l}_{ko} \cdot \nabla^- v,$$

where $\nabla^{\pm} v$ are the gradients of v evaluated in the triangles out of which and into which $\boldsymbol{l}_{ko}$ points, as sketched in Figure 7.21. Arminjon and Dervieux have used a similar definition (Arminjon & Dervieux 1989). It may now be verified that

$$\Delta^+ v_{ko} = \epsilon_{pk}(v_p - v_k) + \epsilon_{qk}(v_q - v_k)$$

and

$$\Delta^- v_{ko} = \epsilon_{or}(v_o - v_r) + \epsilon_{os}(v_o - v_s),$$

where the coefficients $\epsilon_{pk}, \epsilon_{qk}, \epsilon_{ok}$, and ϵ_{os} are all non-negative. Now define the diffusive term for the edge ko as

$$d_{ko} = \alpha_{ko}\{\Delta v_{ko} - L(\Delta^+ v_{ko}, \Delta^- v_{ko})\}, \tag{7.47}$$

where $L(u, v)$ is a limited average with the properties (P1–P4) that were defined in Section 5.19. In considering the sum of the terms at the vertex o, write

$$L(\Delta^+ v_{ko}, \Delta^- v_{ko}) = \phi(r_{ko}^+)\Delta^- v_{ko},$$

where

$$r_{ko}^+ = \frac{\Delta^+ v_{ko}}{\Delta^- v_{ko}}.$$

Then, since the coefficients ϵ_{or} and ϵ_{os} are non-negative, and $\phi(r^+_{ko})$ is non-negative, the limited anti-diffusive term in (7.47) produces a contribution from every edge that reinforces the positivity condition (7.45). Similarly, in considering the sum of the terms at k, one writes

$$L(\Delta^+ v_{ko}, \Delta^- v_{ko}) = \phi(r^-_{ko})\Delta^+ v_{ko},$$

where

$$r^-_{ko} = \frac{\Delta^- v_{ko}}{\Delta^+ v_{ko}},$$

and again the discrete equation receives a contribution with the right sign. One may therefore deduce the following result:

THEOREM 7.8.1 (Positivity Theorem for Unstructured Meshes) *Suppose that the discrete conservation law (7.42) is augmented by flux limited dissipation following (7.46) and (7.47). Then, the positivity condition (7.45), together with the properties P1 to P4 in Section 5.19 for limited averages, are sufficient to ensure the LED property at every interior mesh point.*

Note also that if this construction is applied to any linear function v, then

$$\Delta v_{ko} = \Delta^+ v_{ko} = \Delta^- v_{ko},$$

with the consequence that the contribution of the diffusive terms is exactly zero. In the case of a smoothly varying function v, suppose that $\boldsymbol{l}_{ko} \cdot \nabla v \neq 0$ and the limiter is smooth in the neighborhood of $r^{\pm}_{ko} = 1$. Then, substitution of a Taylor series expansion indicates that the magnitude of the diffusive flux will be of second order. At an extremum, the anti-diffusive term is cut off, and the diffusive flux is of first order.

7.8.7 Three-Dimensional Vertex-Centered Schemes

The ideas of the preceding sections are easily extended to three-dimensional calculations. Let the tetrahedrons with a common vertex, labeled zero for convenience, be numbered $k = 1, \ldots, n$, with volume V_k. Figure 7.22 illustrates one of these tetrahedrons, with outer vertices $k1$, $k2$, and $k3$, and an outer face with area S_k and a unit normal with components n_{i_k}. We form a polyhedral control volume from a union of the surrounding cells, with a total volume

$$V_0 = \sum_{k=1}^{n} V_k.$$

Then, we discretize the integral form (7.26) of the conservation law as

$$V_0 \frac{d\boldsymbol{w}_0}{dt} + \frac{1}{3} \sum_{k=1}^{n} n_{i_k} S_k (\boldsymbol{f}_{i_{k1}} + \boldsymbol{f}_{i_{k2}} + \boldsymbol{f}_{i_{k3}}) = 0, \tag{7.48}$$

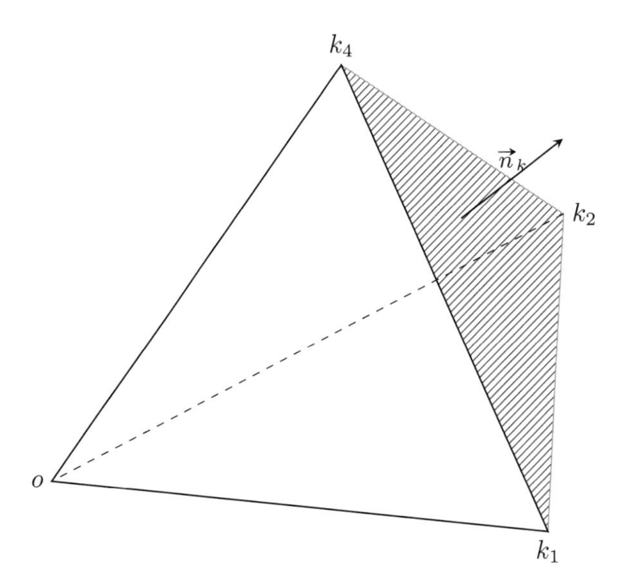

Figure 7.22 A tetrahedron with one vertex at node 0.

where

$$f_{i_{kl}} = f_i(w_{kl}), \quad l = 1, 2, 3.$$

Here, we approximate the average value of f_i over the face by the arithmetic average of its values at the three vertices.

As in the two-dimensional case, the finite volume equation (7.39) corresponds closely to a discretization by the Galerkin method with linear elements. Inserting a piecewise linear trial solution of the form (7.29), we now require it to satisfy the orthogonality conditions

$$\int_{\mathcal{D}_h} \phi_l \left(\frac{\partial \boldsymbol{w}_h}{\partial t} + \frac{\partial}{\partial x_i} \boldsymbol{f}_i(\boldsymbol{w}_h) \right) dV = 0, \quad l = 1, \dots, N.$$

Integrating by parts, we replace this by the weak form

$$\int_{\mathcal{D}_h} \phi_l \frac{\partial \boldsymbol{w}_h}{\partial t} dV = \boldsymbol{R}_l,$$

where at every interior node, the boundary integral vanishes, so that

$$\boldsymbol{R}_l = \int_{\mathcal{D}_h} \boldsymbol{f}_i \frac{\partial \phi_l}{\partial x_i} dV.$$

Figure 7.22 illustrates cell k. In this cell, the derivatives of the test function can be evaluated by Gauss' theorem as

$$\frac{\partial \phi_0}{\partial x_i} = \frac{n_i S_k}{3 V_k}.$$

Also, we may approximate the average value of f_i in cell k as

$$\frac{1}{4}\left(f_{i_0}+f_{i_{k1}}+f_{i_{k2}}+f_{i_{k3}}\right).$$

Now, multiplying the contribution of each cell by its volume, the residual is evaluated as

$$R_0=\frac{1}{12}\sum_{k=1}^{n} n_i S_k\left(f_{i_0}+f_{i_{k1}}+f_{i_{k2}}+f_{i_{k3}}\right).$$

Moreover,

$$\sum_{k=1}^{n} n_i S_k=0$$

because the faces cover a watertight volume. Thus, we can drop f_{i_0}, and finally,

$$R_0=\frac{1}{12}\sum_{k=1}^{n} n_i S_k\left(f_{i_{k1}}+f_{i_{k2}}+f_{i_{k3}}\right).$$

Carrying out the integrations on the left hand side,

$$\int \phi_0 \frac{\partial w_h}{\partial t} dV = M_0\frac{dw_0}{dt}+\sum_{k=1}^{n}\sum_{l=1}^{3} M_{kl}\frac{dw_{kl}}{dt},$$

where

$$M_0=\frac{1}{10}\sum_{k=1}^{n} V_k=\frac{1}{10}V_0$$

and

$$M_{kl}=\frac{1}{20}V_k,\quad l=1,2,3.$$

Thus, the sum of the weights is

$$\begin{aligned} M_0+\sum_{k=1}^{n}\sum_{l=1}^{3} M_{kl} &= \frac{1}{10}V_0+\frac{3}{20}\sum_{k=1}^{n} V_k \\ &= \frac{1}{4}V_0. \end{aligned}$$

After lumping the left hand side as $\frac{1}{4}V_0\frac{dw_0}{dt}$ and multiplying by four, we now recover the finite volume equation (7.48).

7.8.8 Face- and Edge-Based Decompositions of the Flux Balance

As in the two-dimensional case, the nodal flux balance in (7.48) can be broken down into contributions of fluxes through faces in an elegant manner. We associate five nodes with each face, its three vertices and the outer vertices of the two tetrahedrons

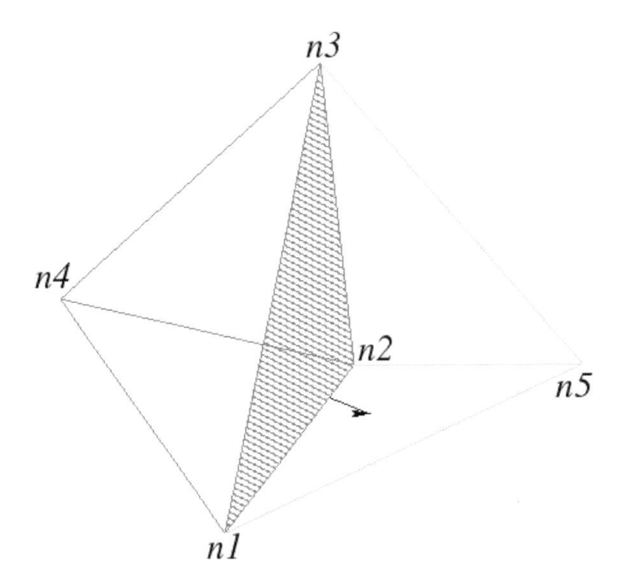

Figure 7.23 Flux through the face defined by nodes $n1$, $n2$, and $n3$ is out of the control volume centered at node $n4$ and into the control volume centered at node $n5$.

separated by the face, as illustrated in Figure 7.23. After first setting the nodal residuals to zero, we may then calculate the fluxes in a loop over the faces, using average values of the flux vectors $\boldsymbol{f}_i$ between the three vertices $n1$, $n2$, and $n3$ of each face and then transferring the fluxes between the other two nodes $n4$ and $n5$. To enable the same procedure to be used at boundary faces, we introduce dummy nodes ns outside each boundary face and then recover the contributions to the flux balances of its three vertices.

On a closer examination of this procedure, it may be observed that a vertex k of the polyhedral control volume surrounding an interior node 0 contributes to the average flux in each of the faces that meet at that vertex, as illustrated in Figure 7.24. Accordingly, we can associate the sum of the vector areas of the umbrella of faces around the vertex with the edge $k0$. Moreover, the contributions from node 0 to the flux balance at node k are similarly multiplied by the sum of the vector areas of the opposite umbrella of faces at node 0 around the edge $k0$, and this sum is the negative of the first sum, since the two umbrellas enclose a watertight volume consisting of all the tetrahedrons surrounding the edge $k0$. The finite volume equation (7.48) can now be reformulated as a sum of edge contributions for the m edges that meet at node 0,

$$V_0 \frac{d\boldsymbol{w}_0}{dt} + \frac{1}{3} \sum_{k=1}^{m} n_{i_{k0}} S_{k0} \boldsymbol{f}_k = 0,$$

where S_{k0} is the area associated with the edge $k0$, and $n_{i_{k0}}$ are the components of the corresponding unit normal. Moreover,

$$\sum_{k=1}^{n} n_{i_{k0}} S_{k0} = 0,$$

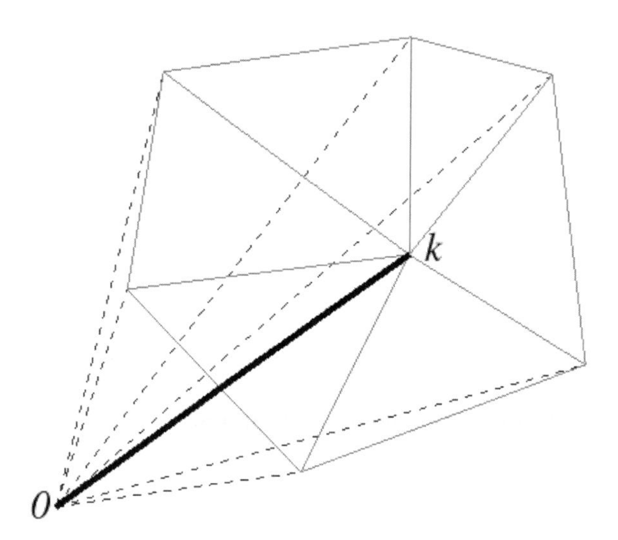

Figure 7.24 Umbrella of faces contributing fluxes along an edge.

since it is the sum of the vector face area of a watertight volume. Hence, we can write the finite volume equation in the symmetric form

$$V_0 \frac{d\boldsymbol{w}_0}{dt} + \frac{1}{3} \sum_{k=1}^{m} n_{i_{k0}} S_{k0} (\boldsymbol{f}_{i_k} + \boldsymbol{f}_{i_0}) = 0. \tag{7.49}$$

The edge-based decomposition enables the accumulation of the flux balances at every interior node via a loop over the edges, in which the flux along each edge is transferred between the two vertices of the edge. This is actually more efficient than the face-based procedure, because the number of edges in a typical triangulation is around 7 times the number of vertices – much fewer than the number of faces, which is around 12 times the number of vertices. It requires considerably more complicated closure formulas, however, to correct the flux balances of the boundary nodes.

The edge-based decomposition (7.49) also enables an upwind scheme to be formulated in the usual manner. Introduce Roe-averaged Jacobian matrices A_{k0} such that

$$A_{k0}(\boldsymbol{w}_k - \boldsymbol{w}_0) = n_{i_{k0}} S_{k0} (\boldsymbol{f}_{i_k} - \boldsymbol{f}_{i_0}).$$

Then, a first order upwind scheme can be written as

$$V_0 \frac{d\boldsymbol{w}_0}{dt} + \sum_{k=1}^{n} \boldsymbol{h}_{k0} = 0, \tag{7.50}$$

where the edge fluxes are constructed as

$$\boldsymbol{h}_{k0} = \frac{1}{3} (n_{i_{k0}} S_{k0} (\boldsymbol{f}_{i_k} + \boldsymbol{f}_{i_0}) - |A_{k0}| (\boldsymbol{w}_k - \boldsymbol{w}_0)), \tag{7.51}$$

and the absolute Jacobian matrix $|A_{k0}|$ is obtained by replacing the eigenvalues of A_{k0} by their absolute values in the diagonal decomposition as defined by (7.20) and (7.21). The edge flux for the corresponding scheme with scalar diffusion is

$$h_{k0} = \frac{1}{3}\left(n_{i_{k0}} S_{k0}(\boldsymbol{f}_{i_k} + \boldsymbol{f}_{i_0}) - \epsilon_{k0}(\boldsymbol{w}_k - \boldsymbol{w}_0)\right), \tag{7.52}$$

where

$$\epsilon_{k0} = \max |\lambda(A_{k0})| \,. \tag{7.53}$$

7.8.9 Equivalence of the Three-Dimensional Aggregated and Median Dual Finite Volume Schemes

It was shown in Section 7.8.4 that the two-dimensional aggregated and median dual finite volume schemes are equivalent at the interior nodes. In this section, we show that this is also true for three-dimensional calculations. Figure 7.25, supplied by Thomas Economon, shows the aggregated and median dual control volumes around a particular node of a representative mesh.

The semi-discrete equation for an interior node 0 using the dual control volume takes the form

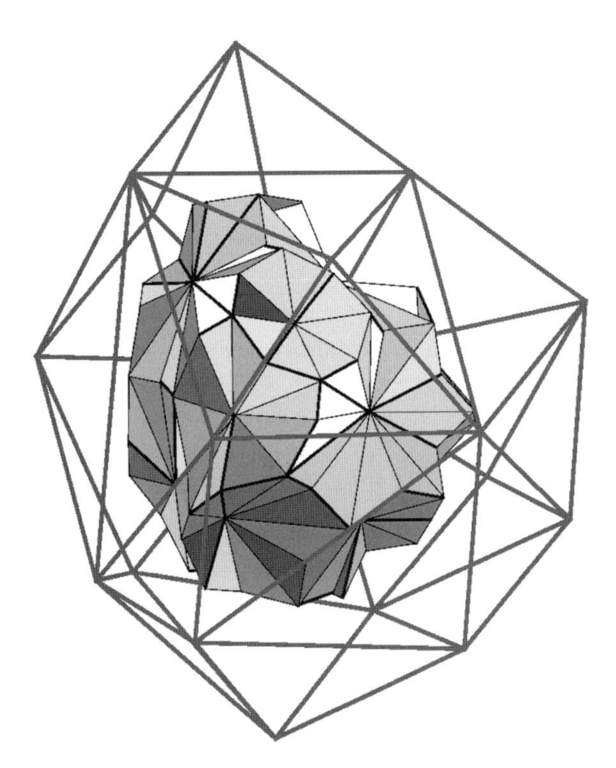

Figure 7.25 Aggregated and median dual control volumes.

$$V_0 \frac{d\boldsymbol{w}_0}{dt} + \frac{1}{2}\sum_{k=1}^{m} n_{i_{k0}} S_{k0}(\boldsymbol{f}_{i_k} + \boldsymbol{f}_{i_0}) = 0, \tag{7.54}$$

where the face areas S_{k0} and the cell volume V_0 are now the those of the dual control volume. This also differs from (7.49) by the factor 1/2 instead of 1/3 multiplying the sum.

In order to prove the equivalence of the schemes using the aggregated and dual control volumes, we first note that 1/4 the volume of each tetrahedron surrounding an interior node contributes to the median dual control volume, so that it has a total volume that is 1/4 that of the aggregated control volume. Next, we need to compare the face areas. For this purpose, it is convenient to use coordinates x, y, and z instead of $\boldsymbol{x}_i$, to avoid confusion with subscripts denoting the vertices. Consider an edge connecting a node k to the interior node 0, surrounded by n tetrahedrons with outer nodes $l = 1, \ldots, n$, as illustrated in Figure 7.26, which shows the tetrahedrons viewed along the edge. The edge umbrella area can be evaluated exactly by trapezoidal integration. Thus, the projected area in the z direction is

$$S_z^A = \frac{1}{2}\sum_{l=1}^{n}(x_{l+1} + x_l)(y_{l+1} - y_l), \tag{7.55}$$

where the node $n + 1$ is identified with node 1.

Next, we examine the contribution to the dual face area of a single tetrahedron, as illustrated in Figure 7.27, where the tetrahedron is viewed looking from k to 0, while the outer two nodes are numbered l and $l + 1$. The tetrahedron contains two facets

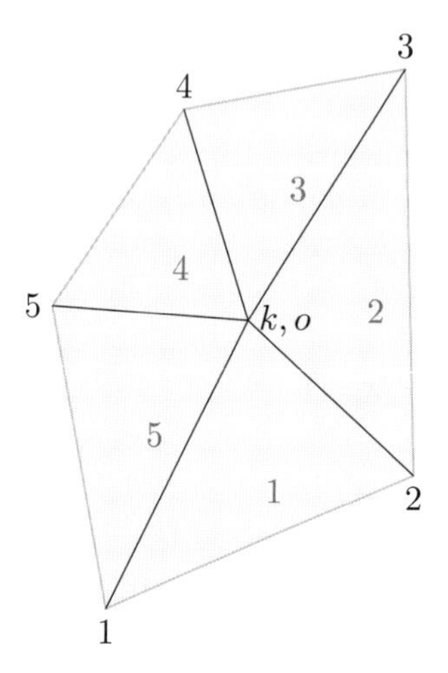

Figure 7.26 Tetrahedrons surrounding the edge $m0$ viewed along the edge looking from k to 0.

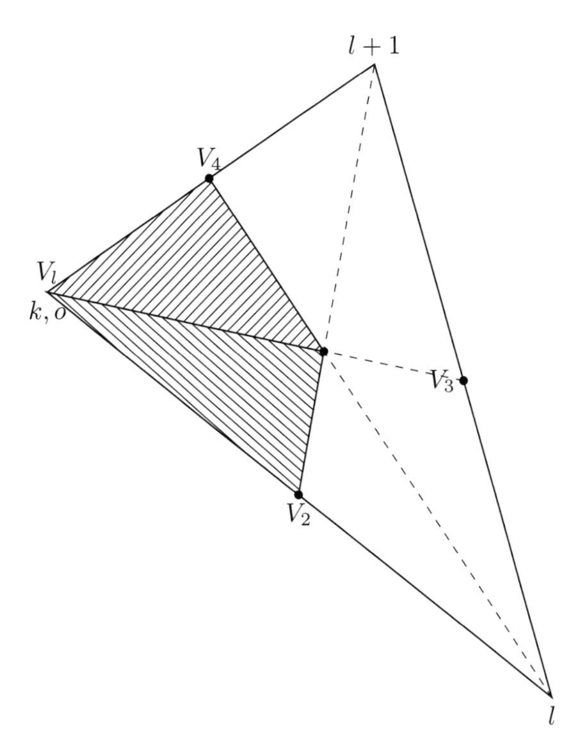

Figure 7.27 Contribution to the median dual face area of a single tetrahedron surrounding the edge $k, 0$, viewed along the edge, looking from k to 0.

of the dual control volume – with vertices labeled $V1$, $V2$, and $V3$ and $V1$, $V3$, and $V4$ – with the vector coordinates

$$\begin{aligned}
\boldsymbol{x}_{V1} &= \frac{1}{2}(\boldsymbol{x}_0 + \boldsymbol{x}_k) \\
\boldsymbol{x}_{V2} &= \frac{1}{3}(\boldsymbol{x}_0 + \boldsymbol{x}_k + \boldsymbol{x}_l) \\
\boldsymbol{x}_{V3} &= \frac{1}{4}(\boldsymbol{x}_0 + \boldsymbol{x}_k + \boldsymbol{x}_l + \boldsymbol{x}_{l+1}) \\
\boldsymbol{x}_{V4} &= \frac{1}{3}(\boldsymbol{x}_0 + \boldsymbol{x}_k + \boldsymbol{x}_{l+1}).
\end{aligned}$$

The combined area of the two facets projected in the z direction is

$$\begin{aligned}
S^D_{z_k} &= \frac{1}{2}\left[x_{V1}(y_{V2} - y_{V4}) + x_{V2}(y_{V3} - y_{V1}) + x_{V3}(y_{V4} - y_{V2}) + x_{V4}(y_{V3} - y_{V1})\right] \\
&= \frac{1}{2}\left[(x_{V3} - x_{V1})(y_{V4} - y_{V2}) + (x_{V4} - x_{V2})(y_{V3} - y_{V1})\right].
\end{aligned}$$

This is a well known formula for the projected area of a non-planar quadrilateral. Here,

$$x_{V3} - x_{V1} = \frac{1}{4}(x_{l+1} + x_l - x_0 - x_k)$$

$$x_{V4} - x_{V3} = \frac{1}{3}(x_{l+1} - x_l)$$

$$y_{V3} - y_{V1} = \frac{1}{4}(y_{l+1} + y_l - y_0 - y_k)$$

$$y_{V4} - y_{V3} = \frac{1}{3}(y_{l+1} - y_l).$$

Thus,

$$S^D_{z_k} = \frac{1}{24}\left[(x_{l+1} + x_l - x_0 - x_k)(y_{l+1} - y_l) - (x_{l+1} - x_l)(y_{l+1} + y_l - y_0 - y_k)\right].$$

Now, the total dual face area around the edge $k0$ is the sum of the contributions of the surrounding tetrahedrons

$$\begin{aligned} S^D_{z_k} &= \sum_{l=1}^{n} S^D_{z_l} \\ &= \frac{1}{24}\sum_{l=1}^{n}\left[(x_{l+1} + x_l - x_0 - x_k)(y_{l+1} - y_l) - (x_{l+1} - x_l)(y_{l+1} + y_l - y_0 - y_k)\right]. \end{aligned}$$

Here,

$$\sum_{l=1}^{n}(x_0 + x_k)(y_{l+1} - y_l) = 0,$$

$$\sum_{l=1}^{n}(x_{l+1} + x_l)(y_0 + y_k) = 0,$$

because the sums are around a closed loop. Also,

$$\sum_{l=1}^{n}(x_{l+1} - x_l)(y_{l+1} + y_l) = -\sum_{l=1}^{n}(x_{l+1} + x_l)(y_{l+1} - y_l).$$

Thus, finally, the edge dual face area projected in the z direction is

$$S^D_z = \frac{1}{12}\sum_{l=1}^{n}(x_{l+1} + x_l)(y_{l+1} - y_l). \tag{7.56}$$

This is exactly 1/6 of the edge umbrella area S^A_z of the aggregated control volume as given in (7.55). The same result holds for the areas projected in the y and z directions.

Now, comparing (7.49) and (7.54), the volume in (7.49) is four times the volume in (7.54), while the face areas are six times larger, but this is compensated by the factor 1/3 instead of 1/2, with the consequence that the equations are exactly equivalent.

7.8.10 Boundary Closure for the Three-Dimensional Vertex-Centered Scheme

An efficient procedure for evaluating the residuals in the flux balance equation (7.49) is first to set the residuals to zero in a loop over the nodes. Then, a loop is performed over the edges in which the fluxes along each edge are calculated according to the formulas in (7.49) and then exchanged between the nodes at each end of the edge. First or second order accurate artificial diffusive terms can be included in the same loop. The result after completion of the loop is a correct residual at every interior node. The residuals at the boundary nodes, however, remain incomplete. The control volume of a boundary node is one sided, containing faces in the boundary surface. The fluxes through these faces contribute to the residual, with the consequence that the accumulated residual from the edge loop needs to be corrected. In the case of a solid boundary, it follows from the flow tangency condition that there is no convective flux through the boundary, and one must allow for this, not only at the boundary nodes but also at interior nodes with control volumes containing faces in the boundary surfaces.

It is easiest to make these corrections in a loop over the boundary faces. Each face influences its three vertices and also the interior node that is the fourth vertex of the tetrahedron based on the face, as illustrated in Figure 7.28. The following is the correct closure formula. Let $\boldsymbol{R}_k$, $k = 1, 2, 3$, be the residuals calculated by the edge loop at the three vertices of a boundary face. Then, the equations are corrected as

$$\frac{d\boldsymbol{w}_k}{dt} + \boldsymbol{R}_k + \boldsymbol{C}_k = 0,$$

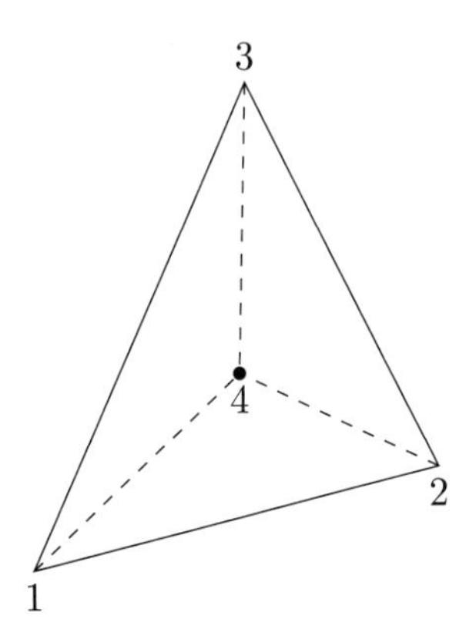

Figure 7.28 Boundary face influencing the boundary nodes 1, 2, and 3, and the interior node 4 that is the other vertex of the tetrahedron based on these nodes.

where in order to calculate the correction $\boldsymbol{C}_k$, we first calculate fluxes using the solution values at each vertex in turn:

$$\boldsymbol{F}_k = n_i S \boldsymbol{f}_i(\boldsymbol{w}_k),$$

where n_i are the components of the unit normal to the face and S is its area. Then, we set

$$\boldsymbol{Q} = \boldsymbol{F}_1 + \boldsymbol{F}_2 + \boldsymbol{F}_3$$

and

$$\boldsymbol{C}_k = \frac{1}{6}\boldsymbol{Q} + \frac{5}{6}\boldsymbol{F}_k.$$

Here $\frac{1}{3}\boldsymbol{Q}$ is the flux vector through the face. If the face is contained in a solid boundary, it is also necessary to subtract the convective part of the flux through the face from the residuals all four nodes of the tetrahedron. The convective part of the flux could be calculated by re-evaluating $\boldsymbol{F}_k$, $k = 1, 2, 3$, without the pressure terms, but it is less expensive to subtract the pressure contributions $\frac{1}{3} n_i S(p_1 + p_2 + p_3)$, $i = 1, 2, 3$ from the three momentum components in $\frac{1}{3}\boldsymbol{Q}$.

7.8.11 Code Fragment for a Three-Dimensional Euler Solution in a Tetrahedral Mesh

This section presents a code fragment, derived from the author's AIRPLANE code (see Section 7.9), which evaluates the convective flux balances for the three-dimensional Euler equations at the vertices of a tetrahedral mesh. It assumes the following data structure. The Cartesian coordinates of the vertices are stored as

$$\left.\begin{aligned} x &= \mathtt{x(1,i)} \\ y &= \mathtt{x(2,i)} \\ z &= \mathtt{x(3,i)} \end{aligned}\right\} \mathtt{i = 1,nnode.}$$

The components of the state vector are stored as

$$\left.\begin{aligned} \rho &= \mathtt{w(1,i)} \\ \rho u &= \mathtt{w(2,i)} \\ \rho v &= \mathtt{w(3,i)} \\ \rho w &= \mathtt{w(4,i)} \\ \rho E &= \mathtt{w(5,i)} \end{aligned}\right\} \mathtt{i = 1,nnode.}$$

The nodes connected by each edge are stored as

$$\left.\begin{aligned} \mathtt{n1} &= \mathtt{ndg(1,n)} \\ \mathtt{n2} &= \mathtt{ndg(2,n)} \end{aligned}\right\} \mathtt{n = 1,nedge.}$$

The components of the vector face area associated with each edge are pre-calculated and stored as

$$\left.\begin{aligned} \texttt{sx} &= \texttt{sg(1,n)} \\ \texttt{sy} &= \texttt{sg(2,n)} \\ \texttt{sz} &= \texttt{sg(3,n)} \end{aligned}\right\} \texttt{n} = \texttt{1,nedge}.$$

In order to calculate the boundary closure, a list of boundary faces is also required. The vertices of these faces are stored as

$$\left.\begin{aligned} \texttt{n1} &= \texttt{ndf(1,n)} \\ \texttt{n2} &= \texttt{ndf(2,n)} \\ \texttt{n3} &= \texttt{ndf(3,n)} \end{aligned}\right\} \texttt{n} = \texttt{1,nbnd},$$

with faces on the solid boundary listed first from $\texttt{n} = \texttt{1,nbod}$, and the outer boundary faces from $\texttt{n} = \texttt{nbod} + \texttt{1,nbnd}$.

The evaluation of the flux balances for the interior vertices requires only 32 lines of code. The boundary closure is more complicated and must be adjusted to allow for either a solid boundary or a far field boundary. This is accomplished by setting the flag s to zero or one.

Code fragment for the Euler flux balances at the nodes

```
c
c     *****************************************************************
c     *                                                               *
c     *   calculate the euler flux balances at the nodes              *
c     *                                                               *
c     *****************************************************************
c
c     broadmead,28 november 1987
c     (revised labor day weekend,3-5 september,1988)
c
c     *****************************************************************
c     *                                                               *
c     *   flow variables                                              *
c     *                                                               *
c     *****************************************************************
c
c     w(1,i)    = density
c     w(2,i)    = momentum in x direction
c     w(3,i)    = momentum in y direction
c     w(4,i)    = momentum in z direction
c     w(5,i)    = total energy
c     p(i)      = pressure
c
c     *****************************************************************
c
c     set the flux balances in each control volume to zero
c
      do n=1,5
      do i=1,nnode
```

```
            dw(n,i)     = 0.
         end do
         end do
c
c        calculate the convective fluxes across each edge
c        and accumulate the flux balances for each polyhedral subdomain
c
         do i=1,nedge

            n1          = ndg(1,i)
            n2          = ndg(2,i)

            qs1         = (w(2,n1)*sg(1,i)
     .                    +w(3,n1)*sg(2,i)
     .                    +w(4,n1)*sg(3,i))/w(1,n1)
            qs2         = (w(2,n2)*sg(1,i)
     .                    +w(3,n2)*sg(2,i)
     .                    +w(4,n2)*sg(3,i))/w(1,n2)
            pa          = p(n1)   +p(n2)

            fs1         = qs1*w(1,n1)   +qs2*w(1,n2)
            fs2         = qs1*w(2,n1)   +qs2*w(2,n2)   +pa*sg(1,i)
            fs3         = qs1*w(3,n1)   +qs2*w(3,n2)   +pa*sg(2,i)
            fs4         = qs1*w(4,n1)   +qs2*w(4,n2)   +pa*sg(3,i)
            fs5         = qs1*(w(5,n1)   +p(n1))   +qs2*(w(5,n2)   +p(n2))

            dw(1,n1)    = dw(1,n1)   +fs1
            dw(2,n1)    = dw(2,n1)   +fs2
            dw(3,n1)    = dw(3,n1)   +fs3
            dw(4,n1)    = dw(4,n1)   +fs4
            dw(5,n1)    = dw(5,n1)   +fs5

            dw(1,n2)    = dw(1,n2)   -fs1
            dw(2,n2)    = dw(2,n2)   -fs2
            dw(3,n2)    = dw(3,n2)   -fs3
            dw(4,n2)    = dw(4,n2)   -fs4
            dw(5,n2)    = dw(5,n2)   -fs5

         end do
         end do
c
c        correct the flux balances of boundary cells
c
         i1          = 1
         i2          = nbodf
         s           = 1.
         do i=i1,i2

            n1          = ndf(1,i)
            n2          = ndf(2,i)
            n3          = ndf(3,i)
            n4          = ndf(4,i)

            sx          = x(2,n1)*(x(3,n3)   -x(3,n2))
     .                   +x(2,n2)*(x(3,n1)   -x(3,n3))
     .                   +x(2,n3)*(x(3,n2)   -x(3,n1))
            sy          = x(3,n1)*(x(1,n3)   -x(1,n2))
```

```
.              +x(3,n2)*(x(1,n1)   -x(1,n3))
.              +x(3,n3)*(x(1,n2)   -x(1,n1))
      sz      = x(1,n1)*(x(2,n3)   -x(2,n2))
.              +x(1,n2)*(x(2,n1)   -x(2,n3))
.              +x(1,n3)*(x(2,n2)   -x(2,n1))

      pa      = p(n1)   +p(n2)   +p(n3)
      qs      = (w(2,n1)*sx   +w(3,n1)*sy   +w(4,n1)*sz)/w(1,n1)
      fs1     = qs*w(1,n1)
      fs2     = qs*w(2,n1)   +p(n1)*sx
      fs3     = qs*w(3,n1)   +p(n1)*sy
      fs4     = qs*w(4,n1)   +p(n1)*sz
      fs5     = qs*(w(5,n1)   +p(n1))

      qs      = (w(2,n2)*sx   +w(3,n2)*sy   +w(4,n2)*sz)/w(1,n2)
      gs1     = qs*w(1,n2)
      gs2     = qs*w(2,n2)   +p(n2)*sx
      gs3     = qs*w(3,n2)   +p(n2)*sy
      gs4     = qs*w(4,n2)   +p(n2)*sz
      gs5     = qs*(w(5,n2)   +p(n2))

      qs      = (w(2,n3)*sx   +w(3,n3)*sy   +w(4,n3)*sz)/w(1,n3)
      hs1     = qs*w(1,n3)
      hs2     = qs*w(2,n3)   +p(n3)*sx
      hs3     = qs*w(3,n3)   +p(n3)*sy
      hs4     = qs*w(4,n3)   +p(n3)*sz
      hs5     = qs*(w(5,n3)   +p(n3))

      q1      = fs1   +gs1   +hs1
      q2      = fs2   +gs2   +hs2
      q3      = fs3   +gs3   +hs3
      q4      = fs4   +gs4   +hs4
      q5      = fs5   +gs5   +hs5

      r1      = s*q1
      r2      = s*(q2   -pa*sx)
      r3      = s*(q3   -pa*sy)
      r4      = s*(q4   -pa*sz)
      r5      = s*q5

      dw(1,n1)  = dw(1,n1)   +.5*q1   +2.5*fs1   -r1
      dw(2,n1)  = dw(2,n1)   +.5*q2   +2.5*fs2   -r2
      dw(3,n1)  = dw(3,n1)   +.5*q3   +2.5*fs3   -r3
      dw(4,n1)  = dw(4,n1)   +.5*q4   +2.5*fs4   -r4
      dw(5,n1)  = dw(5,n1)   +.5*q5   +2.5*fs5   -r5

      dw(1,n2)  = dw(1,n2)   +.5*q1   +2.5*gs1   -r1
      dw(2,n2)  = dw(2,n2)   +.5*q2   +2.5*gs2   -r2
      dw(3,n2)  = dw(3,n2)   +.5*q3   +2.5*gs3   -r3
      dw(4,n2)  = dw(4,n2)   +.5*q4   +2.5*gs4   -r4
      dw(5,n2)  = dw(5,n2)   +.5*q5   +2.5*gs5   -r5

      dw(1,n3)  = dw(1,n3)   +.5*q1   +2.5*hs1   -r1
      dw(2,n3)  = dw(2,n3)   +.5*q2   +2.5*hs2   -r2
      dw(3,n3)  = dw(3,n3)   +.5*q3   +2.5*hs3   -r3
      dw(4,n3)  = dw(4,n3)   +.5*q4   +2.5*hs4   -r4
      dw(5,n3)  = dw(5,n3)   +.5*q5   +2.5*hs5   -r5
```

```
        dw(1,n4)   = dw(1,n4)   -r1
        dw(2,n4)   = dw(2,n4)   -r2
        dw(3,n4)   = dw(3,n4)   -r3
        dw(4,n4)   = dw(4,n4)   -r4
        dw(5,n4)   = dw(5,n4)   -r5

    end do
    end do

    if (i2.eq.nbndf2) return

    i1         = nbodf  +1
    i2         = nbndf
    s          = 0.
    go to 11
```

7.8.12 Finding the Edges

The edge-based algorithm (7.48) requires a list of the edges and their associated projected areas. Typically, the connectivity of a tetrahedral mesh is specified by a list of tetrahedrons with their nodes, such as

$$\left.\begin{aligned} \mathtt{n1} &= \mathtt{ndc(1,n)} \\ \mathtt{n2} &= \mathtt{ndc(2,n)} \\ \mathtt{n3} &= \mathtt{ndc(3,n)} \\ \mathtt{n4} &= \mathtt{ndc(4,n)} \end{aligned}\right\} \mathtt{n} = \mathtt{1,ncell}.$$

Then, the edges can be enumerated by looking at the six edges of each tetrahedron and adding any newly occurring edge to the existing list. If this is programmed naively, the $\mathtt{k}+\mathtt{1}^{\text{th}}$ edge can only be added after comparing it with $\mathtt{k}$ previous edges, leading to an operation count

$$1+2+3+\cdots+\mathtt{n} = \frac{1}{2}\mathtt{n}(\mathtt{n}+1)$$

to compile a list of $\mathtt{n}$ edges. This can be prohibitively expensive for very large meshes.

The following code fragment finds the edges and also calculates their associated areas with a linear operation count. The main loop ("do 20") over the cells contains a loop ("do 15") over the four vertices of each cell. This loop begins by calculating the area of the face with vertices n1, n2, and n3, projected in the coordinate directions. This face is opposite the vertex n4, and the projected areas contribute to the associated areas of the edges connecting n4 to n1, n2, and n3. Next, these edges are examined to find out whether they already appear in the existing edge list. Each edge that is added to the list is stored with its lower-numbered vertex first. Accordingly, it is only necessary to compare a candidate new edge with edges with the same lower-numbered vertex. For this purpose, two pointers are introduced. The first, ipoint(k), points to the first edge in the list with lower-numbered vertex `k`. The second, `npoint(i)`, points to the next edge after edge number `i` that has the same lower-numbered vertex. The three edges under consideration are examined in the loop over `l`, `l` $=$ `1,3`. The lower- and higher-numbered vertices are identified as `m1` and `m2`, and then `i` is set equal to `ipoint(m1)`. If `i` $=$ `0`, it must be a new edge and it is added to the list. Otherwise, it is necessary to check the second vertex of edge `i`. If it matches, the edge has been previously identified, and the areas are accumulated at statement 13. If it does not match, the search proceeds to the next edge `npoint(i)` and is continued until there is either a match or `npoint(i)` $=$ `0`, in which case a new edge is identified. At statement 14, the edges are permuted to allow reuse of the same code. Finally at the end of loop 15, the vertices are permuted to enable the entire procedure to be repeated for the other faces opposite the other three vertices.

Code fragment for finding the edges by searching the cells

```
c
c     *****************************************************************
c     *                                                               *
c     *   find the edges by searching the cells                       *
c     *                                                               *
c     *****************************************************************
c
c     broadmead,28 november 1987
c     revised labor day weekend,3-5 september,1988
c     rewritten to extract the edges from the cell list,28 april 1992
c
c     *****************************************************************
c
      nedge     = 0
c
c     set the initial edge pointers to zero
c
      do i=1,nnode
         ipoint(i)  = 0
      end do
c
c     loop over the cells
c
      do 20 n=1,ncell

      n1        = ndc(1,n)
      n2        = ndc(2,n)
      n3        = ndc(3,n)
      n4        = ndc(4,n)
c
c     loop over the vertices of the cell
c
      do 15 k=1,4
c
c     calculate the area of the face opposite n4
c
```

```
      sx          = x(2,n1)*(x(3,n3)  -x(3,n2))
     .             +x(2,n2)*(x(3,n1)  -x(3,n3))
     .             +x(2,n3)*(x(3,n2)  -x(3,n1))
      sy          = x(3,n1)*(x(1,n3)  -x(1,n2))
     .             +x(3,n2)*(x(1,n1)  -x(1,n3))
     .             +x(3,n3)*(x(1,n2)  -x(1,n1))
      sz          = x(1,n1)*(x(2,n3)  -x(2,n2))
     .             +x(1,n2)*(x(2,n1)  -x(2,n3))
     .             +x(1,n3)*(x(2,n2)  -x(2,n1))
      v           = (x(1,n4)   -x(1,n1))*sx
     .             +(x(2,n4)   -x(2,n1))*sy
     .             +(x(3,n4)   -x(3,n1))*sz
      t           = sign(1.,v)
      l2          = n4
c
c     loop over the vertices of the face
c
      do l=1,3

      l1          = n1
c
c     order the numbers of the vertices
c     of the edge joining l1 and l2
c
      m1          = min(l1,l2)
      m2          = max(l1,l2)
      s           = float(l2  -l1)
      s           = sign(.5,s)*t
c
c     find the first edge with low numbered vertex m1
c
      i           = ipoint(m1)
      if (i.eq.0) go to 12
c
c     check whether the other vertex matches m2
c
   11 if (m2.eq.ndg(2,i)) go to 13
c
c     find the next edge with low numbered vertex m1
c
      next        = npoint(i)
      if (next.eq.0) go to 12
      i           = next
      go to 11
c
c     new edge
c
   12 nedge       = nedge  +1
      ndg(1,nedge)    = m1
      ndg(2,nedge)    = m2
c
c     set the edge area equal to the face area
c
      sg(1,nedge)     = s*sx
      sg(2,nedge)     = s*sy
      sg(3,nedge)     = s*sz
      npoint(nedge)   = 0
```

```
      if (i.ne.0) npoint(i)  = nedge
      if (i.eq.0) ipoint(m1) = nedge
      go to 14
c
c     previously identified edge
c     add the face area to the accumulated edge area
c
   13 sg(1,i)    = sg(1,i)   +s*sx
      sg(2,i)    = sg(2,i)   +s*sy
      sg(3,i)    = sg(3,i)   +s*sz
c
c     permute the vertices of the face
c
   14 nn         = n1
      n1         = n2
      n2         = n3
      n3         = nn

      end do
c
c     permute the vertices of the cell
c
      nn         = n1
      n1         = n2
      n2         = n3
      n3         = n4
      n4         = nn

   15 continue

   20 continue
```

7.8.13 Edge Coloring for Implementation on Vector or Parallel Computers

When the nodal residuals are accumulated in a loop over the edges, the same node will appear repeatedly as a recipient of a contribution. This can lead to conflicts when the code is executed on vector or parallel machines. These conflicts can be avoided by first assigning a color to every edge such that no two edges of the same color meet at a common vertex. Figure 7.29 illustrates such a coloring for a two-dimensional mesh. The operations are then performed in a double loop, with the inner loop acting only on edges with the same color, while the outer loop is performed over the colors. Edge coloring algorithms are a topic in graph theory. The minimum number of colors needed for the coloring is known as the chromatic index of the graph. According to a theorem in graph theory, the chromatic index is at most $N+1$, where N is the largest number of edges meeting at a common vertex anywhere in the graph. For a dense graph in which any vertex might be connected to any other vertex, the graph coloring problem is known to be hard. The vertices of a triangulated mesh are only connected to their nearest neighbors, however, and in the author's experience, colorings can fairly easily be found using no more than N colors. Similar coloring procedures may also be used for loops over the faces.

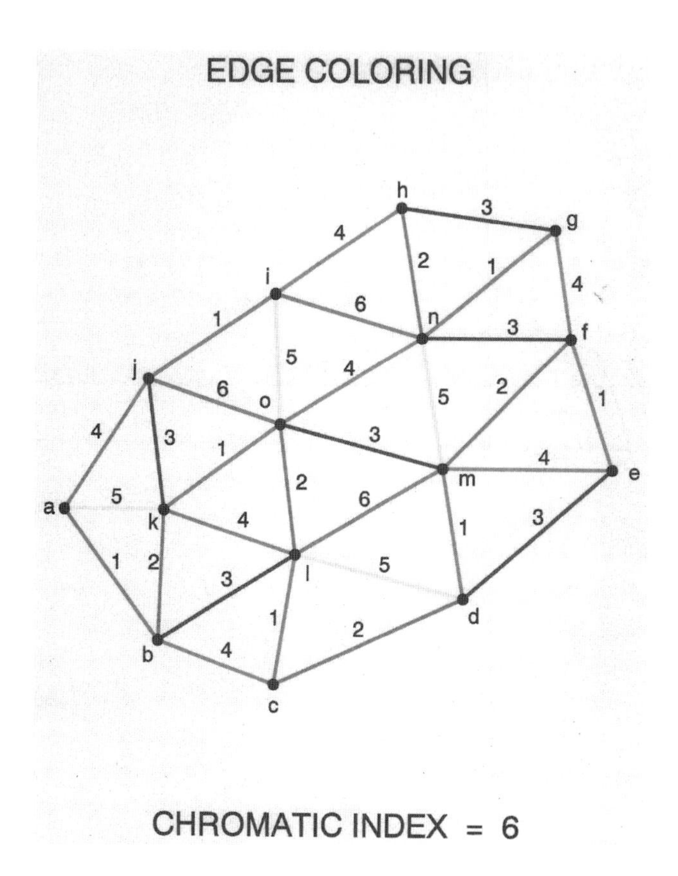

Figure 7.29 Edge coloring. A black and white version of this figure will appear in some formats. For the color version, refer to the plate section.

7.9 Some Representative Calculations on Unstructured Meshes

This section presents the results of some representative calculations on unstructured meshes. They were performed using the AIRPLANE code developed by Jameson, Baker, and Weatherill (1986) in the period 1987–1995.

Figure 7.30 shows the pressure distribution over a McDonnell Douglas MD11, flying at Mach 0.825, and the trailing vortex pattern in a vertical plane downstream. Figure 7.31 shows contours of the Mach number over the surface of an Airbus A320 in its cruising condition, and also in vertical cuts above the wing. Figure 7.32 shows the pressure distribution over the proposed NASA High Speed Civil Transport (HSCT) in supersonic cruise at Mach 2.4, and also pressure waves in a downstream vertical cut and on a horizontal cut below the aircraft. These would ultimately coalesce to create a sonic boom on the ground. Figure 7.33 shows the flow over the X33 lifting body design at Mach 2.0 and an angle of attack of 10 degrees.

Figure 7.30 Pressure distribution over MD11, $Ma = 0.825$. A black and white version of this figure will appear in some formats. For the color version, refer to the plate section.

Figure 7.31 Contours of Mach number over A320 wing surface. A black and white version of this figure will appear in some formats. For the color version, refer to the plate section.

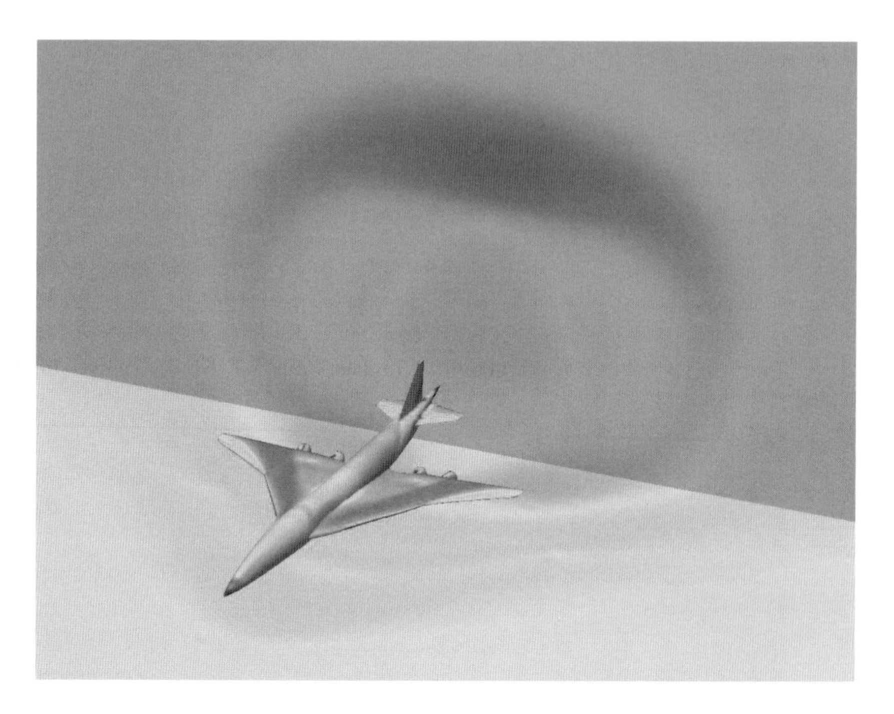

Figure 7.32 Pressure distribution over NASA High Speed Civil Transport, $Ma = 2.4$. A black and white version of this figure will appear in some formats. For the color version, refer to the plate section.

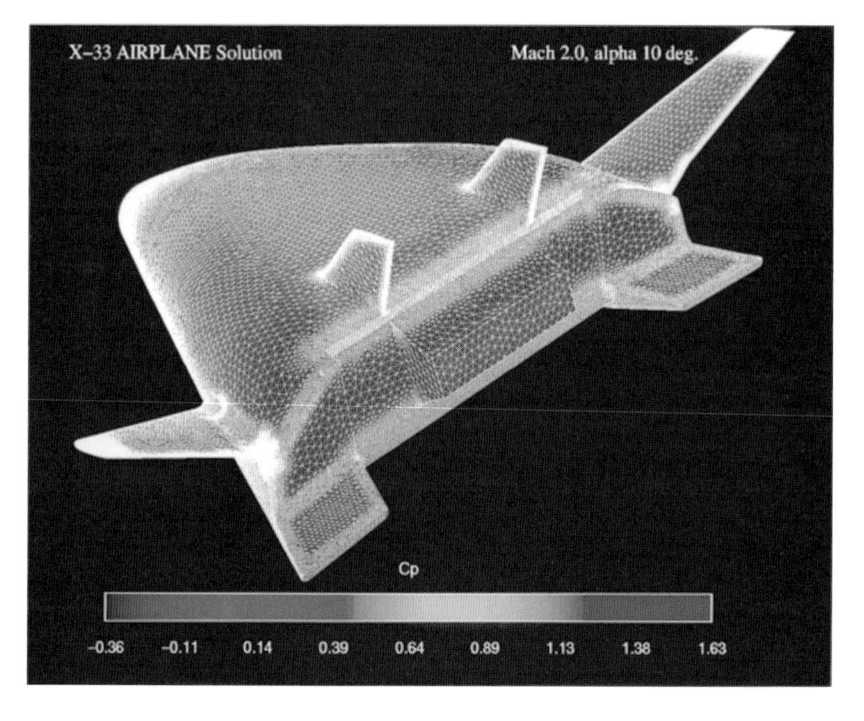

Figure 7.33 Pressure distribution over X33, $Ma = 2.0$. A black and white version of this figure will appear in some formats. For the color version, refer to the plate section.

7.10 Further Developments in Unstructured Mesh Simulations

Contemporaneously with the development of the AIRPLANE code, other groups were developing alternative finite element methods for incompressible and compressible flow simulations. Examples include the development of a finite element FCT scheme by Löhner, Morgan, et al. (1987) and an upwind characteristics based scheme (Peraire, Peiro, et al. 1988) by the Swansea group. Subsequently Hughes and his coworkers formulated and developed the streamline upwind Petrov Galerkin (SUPG) method in a series of papers and also advocated the use of entropy variables to symmetrize the Euler and Navier–Stokes equations (Hughes, Franca, & Mallet 1986a, Hughes, Mallet, & Akira 1986b). Barth and Jespersen (1989) presented a systematic approach to the formulation of upwind schemes on unstructured meshes. There has been ongoing research on residual distribution schemes that can provide truly multi-dimensional upwind biasing. Abgrall (2006) provides a comprehensive review of these developments.

Solvers for both structured and unstructured meshes have been incorporated in the German Megaflow software for aircraft design (Kroll & Fassbender 2006). Some of the most remarkable simulations to date have been performed by Rainald Löhner and his associates. Löhner's book (Löhner 2008) highlights the importance of data structures for an efficient implementation of computational algorithms for calculation on unstructured meshes. A particularly successful solver for unstructured grids is the NSU3D code primarily developed by Dimitri Mavriplis (Mavriplis & Long 2014), who has also given an excellent review of recent developments in the simulation of viscous flows (Mavriplis 2019). It may also be noted that solvers for unstructured meshes provide the basis for widely used commercial CFD software such as Fluent, Star-CCM, and CFD++.

8 The Calculation of Viscous Flow

8.1 Introduction

In this chapter, we address issues related to the calculation of viscous flows. The flows of interest in aeronautical science are generally characterized by high Reynolds numbers, of the order of 5–100 million in the case of flows over aircraft wings, and accordingly the emphasis here will be on this situation. Classical aerodynamic theory is based on Prandtl's insight that for flows at high Reynolds numbers, the viscous effects are confined to thin boundary layers over the surface in which there are large shearing velocity gradients to accommodate the no slip condition at the surface. Outside the boundary layers, the flow is well represented by the equations of inviscid flow. These may be either the Euler equations or, in the absence of strong shock waves, the potential flow equation. Many procedures have been devised for making boundary layer corrections to inviscid solutions, typically by modifying the geometry to allow for the displacement thickness of the boundary layer. However, these are hard to apply to complex shapes with intersections between solid surfaces, such as wing–fuselage or wing–pylon intersections.

The alternative is to solve the Navier–Stokes equations. Dating back to the pioneering investigations of pipe flows by Osborne Reynolds, it is well known that smooth laminar flows in shear layers become unstable at high Reynolds numbers and undergo transition to turbulent flow with eddies at multiple scales. Figures 8.1–8.6, which may be found in Milton Van Dyke's *Album of Fluid Motion*, illustrate the increasing complexity of viscous flows with the onset of separation and turbulence. According to the classical theory of Kolmogorov, based on dimensional analysis, the scale of the smallest eddies in a fully turbulent flow is proportional to $\frac{1}{Re^{\frac{3}{4}}}$, that Re is the Reynolds number based on a representative length scale of the global flow. This theory is reviewed in Section 8.2, while Section 8.3 presents a qualitative analysis of wall bounded flows. It follows from Kolmogorov's estimate that direct numerical simulations (DNS) of the full range of eddies in a three-dimensional turbulent flow would require a mesh with a number of cells proportional to $Re^{\frac{9}{4}}$. Moreover, the time scales that would need to be resolved have a similar order of magnitude, so the computational complexity of full DNS scales as Re^3, say 10^{21} for a flight Reynolds number of 10^7. This is well beyond the capability of any existing computer in the year 2020.

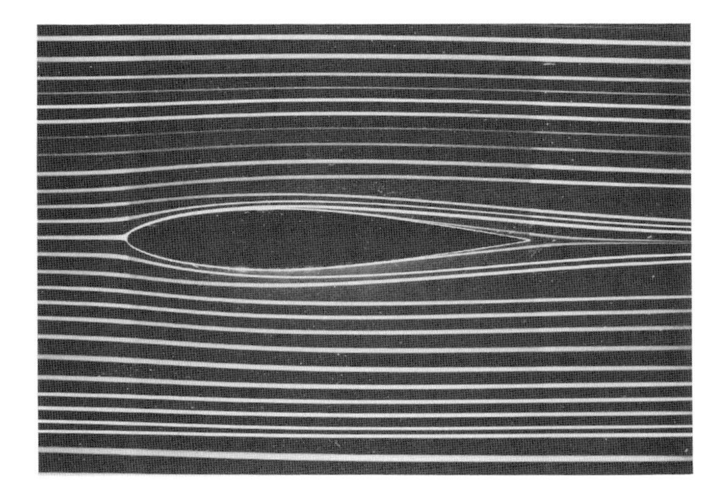

Figure 8.1 Symmetric plane flow past an airfoil. ONERA photograph, Werlé 1974, reproduced with permission

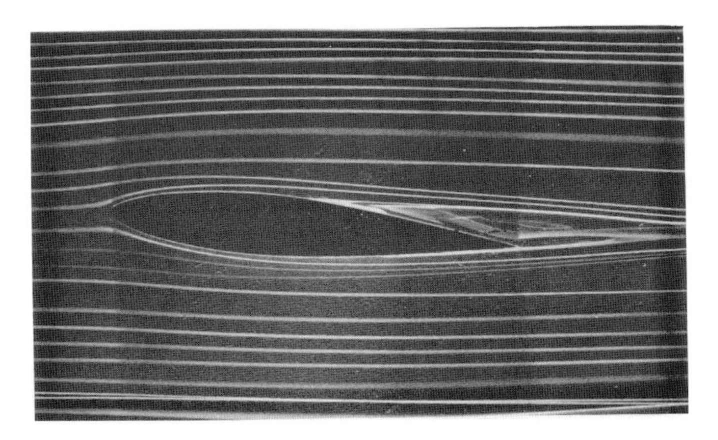

Figure 8.2 Boundary-layer separation on an inclined airfoil. ONERA photograph, Werlé 1974, reproduced with permission

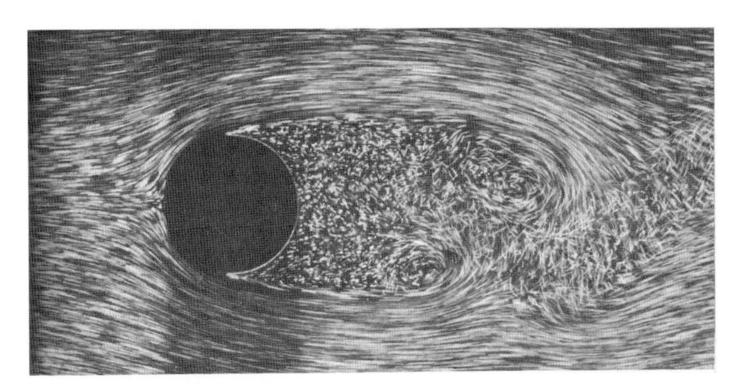

Figure 8.3 Circular cylinder at $Re = 2,000$. ONERA photograph, Werlé and Gallon 1972, reproduced with permission

Figure 8.4 Circular cylinder at $Re = 10,000$. Photograph by Thomas Corke and Hassan Nagib, reproduced with permission

Figure 8.5 Side view of a turbulent boundary layer. Photograph by Thomas Corke, Y. Guezennec and Hassan Nagib, reproduced with permission

Figure 8.6 Single lambda shock wave on an airfoil with a laminar boundary layer. Photograph by H.W. Liepmann, reproduced with permission

In the light of these considerations, the only feasible approach to the simulation of flows at flight Reynolds numbers is to introduce some kind of modeling of the small scales. The most widely used approach is Reynolds averaging, in which we compute time averaged quantities $\bar{u}_i$ of the instantaneous velocity components u_i. The Reynolds-averaged Navier–Stokes (RANS) equations are most easily derived for incompressible flow. The true Navier–Stokes equations contain derivatives of products such as $u_i u_j$.

The momentum equation for incompressible flow, for example is,

$$\rho\left(\frac{\partial u_i}{\partial t} + \frac{\partial u_i u_j}{\partial x_j}\right) + \frac{\partial p}{\partial x_i} = \mu \frac{\partial^2 u_i}{\partial x_i \partial x_j} = \rho\nu \frac{\partial^2 u_i}{\partial x_i \partial x_j},$$

where μ is the dynamic viscosity coefficient, and $\nu = \frac{\mu}{\rho}$ is the kinematic viscosity. Denoting the time average by an overbar, the time averaged equation is

$$\rho\left(\frac{\partial \bar{u}_i}{\partial t} + \frac{\partial \overline{u_i u_j}}{\partial x_j}\right) + \frac{\partial \bar{p}}{\partial x_i} = \mu \frac{\partial^2 \bar{u}_i}{\partial x_j \partial x_j},$$

where $\overline{u_i u_j}$ is not equal to $\bar{u}_i \bar{u}_j$. To recast the equation in its standard form, we introduce a residual stress tensor

$$\tau_{ij}^R = \rho\left(\bar{u}_i \bar{u}_j - \overline{u_i u_j}\right),$$

generally known as the Reynolds stress, to obtain

$$\rho\frac{\partial \bar{u}_i}{\partial t} + \frac{\partial \bar{u}_i \bar{u}_j}{\partial x_j} + \frac{\partial \bar{p}}{\partial x_i} = \mu \frac{\partial^2 \bar{u}_i}{\partial x_j \partial x_j} + \frac{\partial \tau_{ij}^R}{\partial x_j}.$$

In order to solve this equation, it is necessary to introduce a model for the Reynolds stress. The simplest models set

$$\tau_{ij}^R = \mu_t \left(\frac{\partial \bar{u}_i}{\partial x_j} + \frac{\partial \bar{u}_j}{\partial x_i}\right),$$

where $\mu_t = \rho\nu_t$ is the "eddy viscosity", which may be estimated using Prandtl's mixing length.

In principle, a higher level of fidelity can be attained at a correspondingly greater computational cost by large eddy simulation (LES), in which we attempt to directly compute the larger scale eddies, say those containing 90 percent of the turbulent kinetic energy, while modeling only the smaller scale eddies that cannot be resolved by the computational mesh. The modeled scales are then small enough that the assumption that the flow is locally isotropic becomes plausible.

The next two sections review the characteristics of fully developed turbulent and wall bounded flows. A discussion of the design of turbulence models for the RANS equations and subgrid models for LES is outside the scope of this book. The reader is referred to texts such as those of Wilcox (2006) and Sagaut (2002). In order to provide the basis for a discussion of the underlying numerical issues, however, Sections 8.3–8.5 present a summary of the RANS and Favre-averaged equations, while a brief description of several of the more widely used turbulence models is given in Appendix F. The elements of large eddy simulation are reviewed in Section 8.7.

Section 8.8 examines the advection-diffusion equation, which provides some useful insights into the discretization issues. Finally, Sections 8.9 and 8.10 discuss alternative discretization procedures for the viscous terms in the Navier–Stokes equations.

8.2 Fully Developed Turbulent Flows at Very High Reynolds Numbers

There is by now a fairly well established theory of the qualitative behavior of fully developed turbulent flows at very high Reynolds numbers, first proposed by the great Russian mathematician A. N. Kolmogorov (1941a, 1941b). This theory is based on the concept of the energy cascade suggested by Richardson (1922). According to Richardson, turbulent kinetic energy is created by the largest scales of the motion. These eddies are unstable and break up into smaller eddies, which in turn break up into yet smaller eddies. These eddies are still characterized by large Reynolds numbers, such that the local flow is dominated by inertial effects, and accordingly the energy is not dissipated. This process continues until finally eddies are generated at such small scales that they are characterized by Reynolds numbers ~ 1, and the turbulent kinetic energy is converted to heat by viscous dissipation. According to Richardson, cited in Pope (2000, p. 183),

> "Big whorls have little whorls,
> Which feed on their velocity,
> And little whorls have lesser whorls,
> and so on to viscosity."

Although the energy is finally dissipated by the smallest scale eddies, it is generated by the largest scale eddies, and a steady turbulent flow is only possible if it is sustained by a source of energy. Accordingly, the turbulent flow as a whole is characterized by the mean rate of dissipation of energy per unit mass, denoted by ϵ, and this must depend only on the largest eddies. These have a size of the order of magnitude of the length l of the region in which the turbulent flow is generated. Suppose that they have velocity fluctuations with a magnitude Δu. The dissipation rate ϵ has the dimensions $\frac{dE}{dt}$, where the energy E has the dimensions of velocity squared. Hence, the dimensions of ϵ are velocity cubed divided by length. Thus, for dimensional consistency,

$$\epsilon \approx \frac{\Delta u^3}{l}. \tag{8.1}$$

We now focus on the local properties of the small eddies with a size $\lambda \ll l$ that are generated in the cascade, assuming that they are separated from solid surfaces by distances $\gg \lambda$. We associate a local Reynolds number

$$Re_\lambda = \frac{v_\lambda \lambda}{\nu},$$

with the eddies at each scale where v_λ is the average magnitude of the velocity fluctuations. The cascade terminates with very small eddies for which $Re_\lambda \sim 1$, in which the energy is dissipated by viscosity. We denote this scale by λ_0.

Kolmogorov showed how this picture of an energy cascade can be systematically analyzed by dimensional arguments to produce a general description of fully developed turbulence, based on three main assumptions:

(1) As the eddies are repeatedly broken up, their motion should no longer depend on the directions of motion at the larger scales or the boundary conditions, and hence the small scale turbulent motions are statistically isotropic.
(2) The statistics of the smallest scale motions where the energy is dissipated have a universal form determined by the kinematic viscosity ν and the rate of turbulent energy dissipation ϵ.
(3) The statistics of the motions at intermediate scales for which $l \gg \lambda \gg \lambda_0$ have a universal form that depends only on the dissipation rate ϵ, independent of ν, because they are characterized by local Reynolds numbers $Re_\lambda \gg 1$.

We can use assumption (3) to determine the magnitude v_λ of the velocity fluctuations associated with eddies of a size λ. The only combination of ϵ and λ with the dimensions of velocity is $(\epsilon\lambda)^{\frac{1}{3}}$, and hence for dimensional consistency,

$$v_\lambda \sim (\epsilon\lambda)^{\frac{1}{3}}, \quad \epsilon \sim \frac{v_\lambda^3}{\lambda},$$

and substituting the estimate (8.1) for ϵ,

$$v_\lambda \sim \Delta u \left(\frac{\lambda}{l}\right)^{\frac{1}{3}}.$$

Also,

$$Re_\lambda = \frac{v_\lambda \lambda}{\nu} \sim \frac{\Delta u l}{\nu}\left(\frac{\lambda}{l}\right)^{\frac{4}{3}} = Re\left(\frac{\lambda}{l}\right)^{\frac{4}{3}}.$$

The scale λ_0 of the smallest eddies where energy is dissipated by viscous forces can now be estimated by setting $Re_{\lambda_0} = 1$. This yields

$$\frac{\lambda_0}{l} = \frac{1}{Re^{\frac{3}{4}}}. \tag{8.2}$$

An alternative derivation is to note that the only combinations of ϵ and ν with the dimensions of length, time, and velocity are

$$\lambda_0 = \left(\frac{\nu^3}{\epsilon}\right)^{\frac{1}{4}}, \quad \tau_0 = \left(\frac{\nu}{\epsilon}\right)^{\frac{1}{2}}, \quad v_0 = (\nu\epsilon)^{\frac{1}{4}}.$$

These are known as the Kolmogorov or inner scales for length, time, and velocity. Substituting the estimate (8.1) for ϵ, we obtain

$$\left(\frac{\lambda_0}{l}\right)^4 = \left(\frac{\nu}{\Delta u l}\right)^3,$$

thus recovering the estimate (8.2).

Another way to estimate the length scale at which dissipation becomes important is to note that the kinetic energy equation is obtained as the scalar product of the velocity with the momentum equation. Then, the dissipation due to the viscous terms contain terms such as $\nu u \frac{\partial^2 u}{\partial y^2}$. Accordingly, if velocity fluctuations of a magnitude Δu occur in a length scale of order λ_g,

$$\epsilon \sim \nu \left(\frac{\Delta u}{\lambda_g}\right)^2.$$

Now substituting the estimate (8.1) for ϵ leads to

$$\left(\frac{l}{\lambda_g}\right)^2 \sim \frac{\Delta u l}{\nu},$$

or

$$\frac{\lambda_g}{l} \sim \frac{1}{Re^{\frac{1}{2}}}.$$

This is known as the Taylor microscale and gives an indication of the length scale at the onset of the dissipation range.

Similar dimensional reasoning can be used to estimate the spectrum of turbulent kinetic energy $E(k)$ as a function of the wave number $k = \frac{1}{\lambda}$. The energy $E(k)$ per unit mass in the range dk has the dimensions of velocity squared times length. The only combination of ϵ and k with these dimensions is $\epsilon^{\frac{2}{3}} k^{-\frac{5}{3}}$, and hence for dimensional consistency,

$$E(k) \sim \epsilon^{\frac{2}{3}} k^{-\frac{5}{3}}.$$

Also the energy in scales $\leq \lambda$ is

$$\int_k^\infty E(k)dk \sim \left(\frac{\epsilon}{k}\right)^{\frac{2}{3}} \sim (\epsilon\lambda)^{\frac{2}{3}} \sim v_\lambda^2.$$

The overall description of the energy spectrum corresponding to this analysis is illustrated in Figure 8.7. Turbulent kinetic energy is produced by the very large eddies at small wave numbers. Then, the energy is transferred without being dissipated in a range in which $l \gg \lambda \gg \lambda_0$, known as the inertial range, with a slope of $-5/3$ on a log-log plot. Finally, the turbulent kinetic energy is converted to heat in the dissipation or viscous range characterized by the Kolmogorov scale.

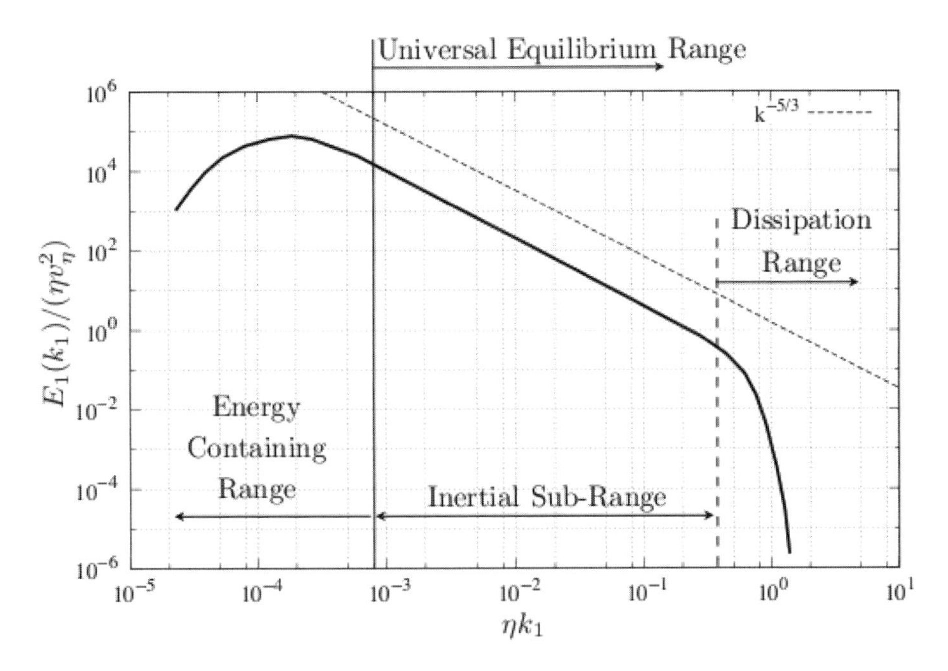

Figure 8.7 Description of the turbulent energy spectrum. Graph from G. Lodato

8.3 Wall Bounded Flow

Flows of interest in engineering science are generally wall bounded. The simplest examples are flows in pipes and channels, and the boundary layer over a flat plate, the latter being of particular relevance to aeronautical science as boundary layers will develop over the entire surface of an aircraft. In this section, we review the characteristics of wall bounded flows, focusing in particular on the nature of the flow very close to the solid boundary and how this impacts the computational requirements of numerical simulations.

Osborne Reynolds in his pioneering experiments showed that laminar flows in pipes are unstable and undergo transition to turbulence if the Reynolds number exceeds a critical value, which he measured as around 13,000. In general, laminar flow along any solid surface develops instabilities that result in transition to turbulence. The physics of transition are extremely complex. The initial instabilities can be treated by linearized theory, taking the form of Tollmien–Schlichting waves (Schlichting 1933). But as the disturbances grow, they are dominated by nonlinear processes such as vortex stretching, and no full mathematical description is as yet available. Transition is delayed when there is a favorable pressure gradient $\frac{\partial p}{\partial x} < 0$, and it is possible to design wing sections with suction peaks at around 40 or 50 percent of the chord, which achieve laminar flow in front of the suction peaks at moderate Reynolds numbers, representative of sailplanes and small aircraft, provided the surface is perfectly smooth and there are no excrescences such as rivets or leading edge devices. In general,

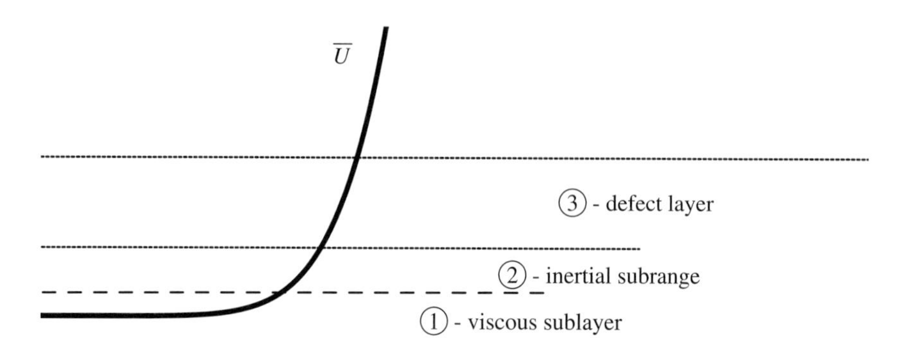

Figure 8.8 Velocity distribution in a turbulent boundary layer.

however, boundary layers undergo transition to turbulence very close to the leading edge on all larger aircraft.

The typical time-averaged velocity distribution in a turbulent boundary layer is illustrated in Figure 8.8. Momentum is transferred to the wall both directly by viscous shear stresses and by eddies. There can be no velocity fluctuations at the wall because of the no-slip boundary condition. Accordingly, there is a viscous sub-layer in which the viscous stresses are dominant. Above this, there is a layer where the transfer of momentum is dominated by the turbulent eddies, and the viscous effects are negligible, usually called the inertial sub-layer. Finally, there is an outer layer where the velocity in the boundary layer must finally conform to the velocity in the exterior flow, usually called the defect layer.

We consider first the viscous sublayer. Even in compressible flow, the velocities in this sublayer are so small that the effects of compressibility are negligible. The boundary layer equations for two-dimensional incompressible turbulent flow are

$$\frac{\partial u}{\partial x} + \frac{\partial v}{\partial y} = 0,$$

and

$$\rho \left(u \frac{\partial u}{\partial x} + v \frac{\partial v}{\partial y} \right) = -\frac{\partial p}{\partial x} + \frac{\partial}{\partial y} \left(\rho \nu \frac{\partial u}{\partial y} \right). \tag{8.3}$$

At the wall,

$$u = 0, \quad v = 0,$$

because of the no-slip condition. In the absence of a pressure gradient, the x-momentum equation (8.3) yields

$$\frac{\partial}{\partial y} \left(\rho \nu \frac{\partial u}{\partial y} \right) = 0 \text{ at } y = 0,$$

and hence, $\rho \nu \frac{\partial u}{\partial y}$ approaches a constant value equal to the wall shear stress τ_w. This can be conveniently stated in dimensionless form by introducing the friction velocity

$$u_\tau = \sqrt{\frac{\tau_w}{\rho}},$$

and dimensionless wall units of velocity and distance

$$u^+ = \frac{u}{u_\tau}, \quad y^+ = \frac{u_\tau y}{\nu}.$$

Then, at the wall,

$$\frac{\partial u}{\partial y} = \frac{u_\tau^2}{\nu},$$

which may be integrated to give

$$u = \frac{u_\tau^2 y}{\nu} + C,$$

in the vicinity of the wall, where the constant $C = 0$ since $u = 0$ at $y = 0$. Thus, we conclude that

$$u^+ = y^+,$$

in the viscous sublayer. Data from experiments and direct numerical simulations confirm that this result holds for $0 \leq y^+ \leq 5$.

Further away from the wall in the inertial sublayer, we expect eddies to form with a size proportional to the distance y to the walls and the inertial stresses to dominate the viscous stresses, so that $\frac{\partial \bar{u}}{\partial y}$ should be a function only of τ_w, ρ, and y. Then, for dimensional consistency,

$$\frac{\partial u}{\partial y} = \frac{u_\tau}{y} F\left(\frac{u_\tau y}{\nu}\right).$$

Experimental data indicates that a good approximation is

$$F\left(\frac{u_\tau y}{\nu}\right) = \frac{1}{\kappa},$$

for a wide range of boundary layers, where κ is the von Kármán constant (Von Kármán 1930). Then in wall units,

$$\frac{\partial u^+}{\partial y^+} = \frac{1}{\kappa y^+},$$

which may be integrated to give the log law of the wall,

$$u^+ = \frac{1}{\kappa} \log y^+ + B, \tag{8.4}$$

where B is a second constant. Typically, it has been found that good estimates of the constants are

$$\kappa = 0.41, \quad B = 5.2.$$

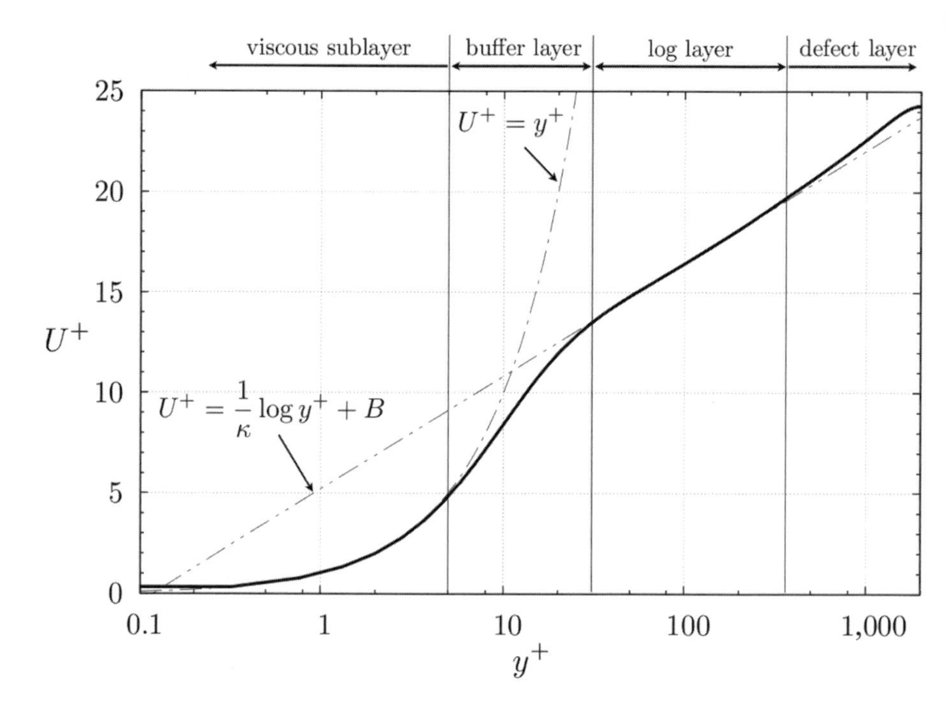

Figure 8.9 Log law. Graph from G. Lodato

The log law is generally a good approximation for $y^+ > 30$. It may also be derived by asymptotic matching of inner and outer solutions, as was first shown by Millikan (Millikan 1938, Pope 2000). Figure 8.9 displays the log law in a semi-log plot.

In the inertial sublayer, the Reynolds stresses can be modeled as

$$\tau_{xy} = \nu_t \frac{\partial \bar{u}}{\partial y},$$

where ν_t is the turbulent viscosity, which may be estimated according to Prandtl's mixing length hypothesis (Prandtl 1925) as

$$\nu_t = l_{\text{mix}}^2 \left| \frac{d\bar{u}}{dy} \right|$$

and

$$l_{\text{mix}} = \kappa y, \tag{8.5}$$

where κ is the von Kármán constant. This gives the wrong asymptotic behavior in the viscous sublayer. In fact, the linear law in the viscous sublayer $0 \leq y^+ \leq 5$ needs to be blended with the log law in the buffer region $5 \leq y^+ \leq 30$. Van Driest (2012) proposed that the estimate of the mixing length (8.5) should be modified by a damping function to take the form

$$l_{\text{mix}} = \kappa y \left(1 - e^{-y^+/A^+}\right),$$

where the constant A^+ has the value

$$A^+ = 26.$$

In order to obtain satisfactory estimates of the skin friction and surface heat transfer in numerical calculations, one should resolve the viscous sublayer. Accordingly, the generally accepted rule of good practice in RANS simulations is that the first mesh interval at the wall should have a size corresponding to a value of $y^+ = 1$ in wall units. Also, the total number of mesh intervals in the boundary layer should be of the order of 30 or more. Thus, RANS simulations require very fine meshes. These requirements can be alleviated by the introduction of wall models to represent the viscous sublayer.

8.4 The Reynolds-Averaged Navier–Stokes (RANS) Equations

We now examine Reynolds averaging for incompressible flow in more detail. In tensor notation, the equations for incompressible viscous flow are

$$\frac{\partial}{\partial t}(\rho u_i) + \frac{\partial}{\partial x_j}(\rho u_i u_j) + \frac{\partial p}{\partial x_i} = \mu \frac{\partial^2 u_i}{\partial x_j \partial x_j}, \tag{8.6}$$

where the density ρ is constant, and the continuity equation reduces to

$$\frac{\partial u_i}{\partial x_i} = 0.$$

Let the time average of a quantity z be defined over an interval T as

$$\bar{z}(t) = \frac{1}{T} \int_{t-\frac{T}{2}}^{t+\frac{T}{2}} z(s)\, ds.$$

Then, we can define the fluctuating part of z as

$$z' = z - \bar{z}.$$

Evidently,

$$\bar{z}' = 0.$$

Now, using this time averaging procedure for all the terms in the momentum equation (8.6),

$$\frac{\partial}{\partial t}(\rho \bar{u}_i) + \frac{\partial}{\partial x_j}(\rho \overline{u_i u_j}) + \frac{\partial \bar{p}}{\partial x_i} = \mu \frac{\partial^2 \bar{u}_i}{\partial x_j \partial x_j}, \tag{8.7}$$

where $\bar{u}_i$ satisfies the time averaged continuity equation

$$\frac{\partial \bar{u}_i}{\partial x_i} = 0.$$

Here, the average $\overline{u_i u_j}$ of the product $u_i u_j$ is not equal to the product of the averages $\bar{u}_i \bar{u}_j$. Substituting

$$u_i = \bar{u}_i + u_i',$$

we obtain

$$\overline{u_i u_j} = \bar{u}_i \bar{u}_j + \overline{u_i' u_j'},$$

and hence, (8.7) can be written as

$$\frac{\partial}{\partial t}(\rho \bar{u}_i) + \frac{\partial}{\partial x_j}(\rho \bar{u}_i \bar{u}_j) + \frac{\partial \bar{p}}{\partial x_i} = \mu \frac{\partial^2 \bar{u}_i}{\partial x_j \partial x_j} - \frac{\partial}{\partial x_j} \tau_{ij}^R, \tag{8.8}$$

where the term

$$\tau_{ij}^R = -\rho \overline{u_i' u_j'} = \rho \bar{u}_i \bar{u}_j - \rho \overline{u_i u_j},$$

is a residual stress tensor that cannot be determined from the mean velocities $\bar{u}_i$. Accordingly, the Reynolds-averaged Navier–Stokes (RANS) equation (8.8) is not closed, and the residual stress tensor, also known as the Reynolds stress tensor, has to be estimated from a turbulence model. To quote Wilcox (2006, p. 39), "In essence, Reynolds averaging is a brutal simplification that loses much of the information contained in the Navier–Stokes equation."

8.5 Remarks on Turbulence Models

The most widely used turbulence models are based on the Boussinesq hypothesis (Boussinesq 1877) that the Reynolds stress tensor is directly proportional to the mean rate of strain tensor with an "eddy" viscosity coefficient μ_t,

$$\tau_{ij}^R = \mu_t \left(\frac{\partial \bar{u}_i}{\partial x_j} + \frac{\partial \bar{u}_j}{\partial x_i} \right).$$

A large variety of models based on this assumption have been proposed. The simplest are "algebraic" models in which the eddy viscosity is directly calculated from mean flow quantities. In more complicated "one equation" and "two equation" models, the eddy viscosity is calculated from the solution of one or two advection-diffusion equations with source terms representing the production and destruction of turbulent energy.

The Boussinesq hypothesis is a gross simplification that is not true for complex flows. Nevertheless, models based on the hypothesis have proven to give quite reliable predictions of attached or weakly separated flows, such as those that are typically generated by properly designed wings.

More elaborate turbulence models that directly represent the full Reynolds stress tensor have also been intensively studied (Launder, Reece, & Rodi 1975). In the case of relatively simple flows, such as wings in the cruising condition, these have not so far been found to provide enough of an improvement in accuracy to justify their increased computational cost.

In order to provide a basis for the development of appropriate numerical solution methods, the detailed equations of some turbulence models that have been widely used in aerodynamic simulations are given in Appendix F. The main source of difficulty is the source terms, which are highly non linear and lead to stiff time dependent equations that may require implicit methods, as will be discussed later.

8.6 The Favre-Averaged Navier–Stokes (FANS) equations

Time averaging of the compressible Navier–Stokes equations is considerably more complicated because of the need to allow for fluctuations in the density and the appearance of triple products such as $\rho u_i u_j$ in both the momentum and energy equations. The time averaged equations can be expressed in a form similar to the incompressible equations through the use of Favre averaging (Favre 1965). Using the same definitions for the time average $\bar{z}$ of a quantity z as in the previous section, its Favre average is defined as

$$\tilde{z} = \frac{\overline{\rho z}}{\bar{\rho}}.$$

Then, the fluctuations z' and z'' are defined as

$$z' = z - \bar{z},$$
$$z'' = z - \tilde{z}.$$

It follows that

$$\bar{z}' = \overline{z - \bar{z}} = \bar{z} - \bar{z} = 0,$$
$$\tilde{z}'' = \widetilde{z - \tilde{z}} = \tilde{z} - \tilde{z} = 0,$$

and

$$\overline{\rho z''} = \bar{\rho}\tilde{z}'' = 0,$$

but

$$\bar{z}'' = \overline{z - \tilde{z}} = \bar{z} - \tilde{z} \neq 0.$$

Assuming that z is differentiable in both space and time, the averaging and differentiation operations commute

$$\widetilde{\frac{\partial z}{\partial x_i}} = \frac{\partial}{\partial x_i}\tilde{z}.$$

We now consider the Favre averages of the full Navier–Stokes equations, as given in Chapter 2:

$$\frac{\partial \rho}{\partial t} + \frac{\partial}{\partial x_i}(\rho u_i) = 0,$$

$$\frac{\partial}{\partial t}(\rho u_i) + \frac{\partial}{\partial x_j}(\rho u_i u_j) + \frac{\partial p}{\partial x_i} = -\frac{\partial}{\partial x_j}\tau_{ij},$$

$$\frac{\partial}{\partial t}(\rho E) + \frac{\partial}{\partial x_j}(\rho u_j H) = \frac{\partial}{\partial x_j}(u_i \tau_{ij}) - \frac{\partial q_j}{\partial x_j},$$

where ρ is the density, u_i are the velocity components, p is the pressure, E and H are the total energy and enthalpy, while τ_{ij} and q_j are the viscous stress tensor and heat flux. These are

$$\tau_{ij} = -\mu\left(\frac{\partial u_i}{\partial x_j} + \frac{\partial u_j}{\partial x_i}\right) - \lambda \delta_{ij}\frac{\partial u_k}{\partial x_k},$$

$$q_j = -\kappa \frac{\partial T}{\partial x_j},$$

where μ and λ are the viscosity coefficients, T is the temperature, and κ is the thermal conductivity. It is usually assumed that

$$\lambda = \frac{2}{3}\mu.$$

Also, in terms of the Prandtl number Pr and specific heat C_p,

$$\kappa = \mu \frac{C_p}{Pr}.$$

Time averaging the continuity equation,

$$\frac{\partial \bar{\rho}}{\partial t} + \frac{\partial}{\partial x_i}(\overline{\rho \tilde{u}_i + \rho u_i''}) = 0,$$

but

$$\overline{\rho \tilde{u}_i} = \bar{\rho}\tilde{u}_i,$$

$$\overline{\rho u_i''} = 0.$$

Hence,

$$\frac{\partial \bar{\rho}}{\partial t} + \frac{\partial}{\partial x_i}(\bar{\rho}\tilde{u}_i) = 0.$$

Time averaging the momentum equation,

$$\frac{\partial}{\partial t}\overline{\rho(\tilde{u}_i + u_i'')} + \frac{\partial}{\partial x_j}\overline{\rho(\tilde{u}_i + u_i'')(\tilde{u}_j + u_j'')} + \frac{\partial \bar{p}}{\partial x_i} = -\frac{\partial}{\partial x_j}\bar{\tau}_{ij},$$

where a term such as

$$\overline{\rho \tilde{u}_i u_j''} = \tilde{u}_i \overline{\rho u_j''} = 0.$$

Hence,

$$\frac{\partial}{\partial t}\left(\bar{\rho}\tilde{u}_i\right)+\frac{\partial}{\partial x_j}\left(\bar{\rho}\tilde{u}_i\tilde{u}_j\right)+\frac{\partial\bar{p}}{\partial x_i}=\frac{\partial}{\partial x_j}\left(\bar{\tau}_{ij}-\overline{\rho u_i''u_j''}\right).$$

In a similar manner, the time-averaged energy equation can be reduced to

$$\begin{aligned}\frac{\partial}{\partial t}(\bar{\rho}\tilde{E})+\frac{\partial}{\partial x_j}(\bar{\rho}\tilde{u}_j\tilde{H}_j)=&\frac{\partial}{\partial x_j}\left(\tilde{u}_i(\bar{\tau}_{ij}-\overline{\rho u_i''u_j''})\right)-\frac{\partial}{\partial x_j}\left(\bar{q}_j-\overline{\rho u_j''h''}\right)\\&+\frac{\partial}{\partial x_j}\left(\overline{u_i''\tau_{ij}}-\frac{1}{2}\overline{\rho u_i''u_i''u_j''}\right).\end{aligned}$$

Making the substitutions

$$\bar{\tau}_{ij}=\tilde{\tau}_{ij}+\overline{\tau_{ij}''},$$

$$\bar{q}_j=\kappa\left(\frac{\partial\tilde{T}}{\partial x_j}+\frac{\partial\overline{T''}}{\partial x_j}\right),$$

$$h=C_pT,$$

where h is the specific enthalpy, the momentum and energy equations reduce to

$$\frac{\partial}{\partial t}\left(\bar{\rho}\tilde{u}_i\right)+\frac{\partial}{\partial x_j}\left(\bar{\rho}\tilde{u}_i\tilde{u}_j\right)+\frac{\partial\bar{p}}{\partial x_i}=-\frac{\partial}{\partial x_j}\left(\tilde{\tau}_{ij}+\underbrace{\overline{\tau_{ij}''}}_{(1)}-\underbrace{\overline{\rho u_i''u_j''}}_{(2)}\right)$$

and

$$\begin{aligned}\frac{\partial}{\partial t}(\bar{\rho}\tilde{E})+\frac{\partial}{\partial x_j}(\bar{\rho}\tilde{u}_j\tilde{H}_j)=&\frac{\partial}{\partial x_j}\left(\tilde{u}_i\tilde{\tau}_{ij}+\underbrace{\tilde{u}_i\overline{\tau_{ij}''}}_{(3)}-\underbrace{\tilde{u}_i\overline{\rho u_i''u_j''}}_{(4)}\right)\\&-\frac{\partial}{\partial x_j}\left(\kappa\frac{\partial\tilde{T}}{\partial x_j}+\underbrace{\kappa\frac{\partial\overline{T''}}{\partial x_j}}_{(5)}+\underbrace{C_p\overline{\rho u_j''T''}}_{(6)}\right)\\&+\frac{\partial}{\partial x_j}\left(\underbrace{\overline{u_i''\tau_{ij}}}_{(7)}-\underbrace{\frac{1}{2}\overline{\rho u_i''u_i''u_j''}}_{(8)}\right).\end{aligned}$$

The terms labeled (1)–(8) have to be modeled. Here the stress tensor is augmented by the Reynolds stress tensor $\overline{\rho u_i''u_j''}$ in terms (2) and (4), while the heat flux is augmented by the Reynolds heat flux $C_p\overline{\rho u_j''T''}$ in term (6). It is usually assumed that

$$|\tilde{\tau}_{ij}|\gg\left|\overline{\tau_{ij}''}\right|,\qquad\left|\frac{\partial\tilde{\tau}}{\partial x_j}\right|\gg\left|\frac{\partial\overline{\tau''}}{\partial x_j}\right|,$$

so that terms (1), (3), and (5) can be neglected. Terms (7) and (8) represent the work done by the viscous stress due to turbulent velocity fluctuations and the transport of turbulent kinetic energy by turbulent velocity fluctuations. It is also usually assumed that these can be neglected, leaving the Reynolds stress tensor and Reynolds heat flux to be modeled.

Following the Boussinesq hypothesis, eddy viscosity models represent the Reynolds stress tensor as

$$\overline{\rho u_i'' u_j''} \approx -\mu_t \left(\frac{\partial \tilde{u}_i}{\partial x_j} + \frac{\partial \tilde{u}_j}{\partial x_j} \right) - \lambda_t \delta_{ij} \frac{\partial \tilde{u}_k}{\partial x_k},$$

where μ_t is the eddy viscosity and $\lambda_t = \frac{2}{3}\mu_t$. Also, the Reynolds heat flux is modeled as

$$C_p \overline{\rho u_j'' T''} \approx -C_p \frac{\mu_t}{Pr_t} \frac{\partial \tilde{T}}{\partial x_j},$$

where Pr_t is a turbulent Prandtl number, usually taken to be $Pr_t = 0.9$.

8.7 Large Eddy Simulations (LES)

8.7.1 Introduction

In large eddy simulation, we attempt to calculate directly the eddies larger than a specified cut off scale, while modeling eddies smaller than the cutoff scale. Ideally the cutoff scale should be in the inertial range where the small scale motions can be expected to be statistically isotropic, and their effects can be reasonably well represented by an isotropic model. In the usual practice, spatial filtering is used to separate the resolved scales above the cutoff from the modeled scales, although a combination of spatial and time filtering may be used. The ideal filter should be a high pass filter for the spatial scales, or equivalently a low pass filter in the frequency domain. As in the case of the RANS or FANS equations, the process of filtering the Navier–Stokes equations leads to unresolved terms that need to be modeled, but the aim is to achieve greater accuracy by restricting the modeling to very small scale motions. In the following sections, we present an overview of LES methods, focusing on the case of incompressible flow, which avoids the complexities of Favre averaging. In fact, LES methods have largely been derived for incompressible flow.

In incompressible flow, the continuity equation reduces to

$$\frac{\partial u_i}{\partial x_i} = 0.$$

Consequently, the viscous stress

$$\tau_{ij} = \mu \left(\frac{\partial u_i}{\partial x_j} + \frac{\partial u_j}{\partial x_i} \right),$$

contains no bulk viscosity term, and

$$\frac{\partial}{\partial x_j}\tau_{ij} = \mu\frac{\partial^2 u_i}{\partial x_j \partial x_j}.$$

Thus, the momentum equation may be written as

$$\rho\left(\frac{\partial u_i}{\partial t} + u_j\frac{\partial u_i}{\partial x_j}\right) + \frac{\partial p}{\partial x_i} = \mu\frac{\partial^2 u_i}{\partial x_j \partial x_j},$$

or equivalently as

$$\rho\left(\frac{\partial u_i}{\partial t} + \frac{\partial}{\partial x_j}(u_i u_j)\right) + \frac{\partial p}{\partial x_i} = \mu\frac{\partial^2 u_i}{\partial x_j \partial x_j}.$$

The pressure does not depend on an equation of state. Instead, it depends on the flow and may be recovered by taking the divergence of the momentum equation (8.7.1). This yields

$$\frac{\partial^2 p}{\partial x_i \partial x_i} = -\rho\frac{\partial^2}{\partial x_i \partial x_j}(u_i u_j). \tag{8.9}$$

These equations also have the property that they are Galilean invariant. Under a transformation to a coordinate frame moving at a constant speed with components V_i,

$$x_i^* = x_i - V_i t,$$

$$u_i^* = u_i - V_i,$$

it follows from the chain rule of differentiation that (8.7.1) and (8.7.1) retain the same form

$$\frac{\partial u_i^*}{\partial x_i^*} = 0,$$

$$\rho\left(\frac{\partial u_i^*}{\partial t} + u_j^*\frac{\partial u_i^*}{\partial x_j^*}\right) + \frac{\partial p}{\partial x_i^*} = \mu\frac{\partial^2 u_i^*}{\partial x_j^* \partial x_j^*}.$$

Accordingly, the filtering and modeling processes should be formulated to preserve Galilean invariance.

8.7.2 Filtering

In general, the filtered quantity $\bar{\phi}(\boldsymbol{x})$ derived from a space variable $\phi(\boldsymbol{x})$, symbolically

$$\bar{\phi} = G\phi,$$

may be represented by the integral

$$\bar{\phi}(\boldsymbol{x}) = \int_{-\infty}^{\infty} g(\boldsymbol{x},\boldsymbol{x}')\phi(\boldsymbol{x}')\, d\boldsymbol{x}',$$

where $g(\boldsymbol{x},\boldsymbol{x}')$ is the filter kernel associated with a cutoff scale Δ and is normalized such that

$$\int_{-\infty}^{\infty} g(\boldsymbol{x},\boldsymbol{x}')d\boldsymbol{x}' = 1.$$

We shall restrict the discussion to filters for which the kernel is a function of $\boldsymbol{x} - \boldsymbol{x}'$ only. In this case, the filtering operation commutes with differentiation:

$$\overline{\frac{\partial \phi}{\partial x_i}} = G\left(\frac{\partial \phi}{\partial x_i}\right) = \frac{\partial}{\partial x_i}(G\phi) = \frac{\partial \bar{\phi}}{\partial x_i}.$$

The residual field is defined as

$$\phi'(\boldsymbol{x}) = \phi(\boldsymbol{x}) - \bar{\phi}(\boldsymbol{x}).$$

In contrast to Reynolds averaging, in general, the filter is not a projector, $G^2 \neq G$, or

$$\bar{\bar{\phi}}(\boldsymbol{x}) \neq \bar{\phi}(\boldsymbol{x}),$$

and equivalently,

$$\bar{\phi}'(\boldsymbol{x}) \neq 0.$$

It is also useful to consider the properties of filters in the frequency domain. In the one-dimensional case, let $\hat{\phi}(k)$ be the Fourier transform of $\phi(\boldsymbol{x})$:

$$\hat{\phi}(k) = \int_{-\infty}^{\infty} \phi(\boldsymbol{x})e^{-ikx}d\boldsymbol{x},$$

$$\phi(\boldsymbol{x}) = \frac{1}{2\pi}\int_{-\infty}^{\infty} \hat{\phi}(k)e^{ikx}d\boldsymbol{x}.$$

According to the convolution theorem, the Fourier transform of

$$\bar{\phi} = \int_{-\infty}^{\infty} g(\boldsymbol{x} - \boldsymbol{x}')\phi(\boldsymbol{x}')d\boldsymbol{x}'$$

is

$$\hat{\bar{\phi}}(k) = \hat{g}(k)\hat{\phi}(k).$$

Thus, in order to cut off the high frequency modes, $\hat{g}(k)$ should be small beyond a cutoff frequency.

The three filters that have been most widely used are the top hat filter, the Gaussian filter, and the sharp spectral filter. The kernel of the top hat filter with cutoff length Δ is

$$g(\boldsymbol{x}) = \begin{cases} \dfrac{1}{\Delta} & \text{if } |\boldsymbol{x}| \leq \dfrac{\Delta}{2}, \\ 0 & \text{otherwise.} \end{cases}$$

Its transfer function is

$$\hat{g}(k) = \frac{\sin(k\Delta/2)}{k\Delta/2}.$$

The kernel and transfer function of the Gaussian filter are

$$g(\boldsymbol{x}) = \left(\frac{6}{\pi^2\Delta}\right)^{\frac{1}{2}} \exp\left(-\frac{6\boldsymbol{x}^2}{\Delta^2}\right)$$

and

$$\hat{g}(k) = \exp\left(-\frac{k^2\Delta^2}{24}\right).$$

The sharp spectral filter removes all modes above the cutoff frequency $k_c = \pi/\Delta$. Thus, its transfer function is

$$\hat{g}(k) = \begin{cases} 1 \text{ if } k \leq \pi/\Delta, \\ 0 \text{ if } k > \pi/\Delta. \end{cases}$$

Correspondingly, its kernel is

$$g(\boldsymbol{x}) = \frac{\sin(\pi\boldsymbol{x}/\Delta)}{\pi\boldsymbol{x}}.$$

Filters can also be defined by differential operators. For example, the inverse Helmholtz filter is defined as

$$\bar{\phi} - \Delta^2 \frac{\partial^2\bar{\phi}}{\partial x_k \partial x_k} = \phi.$$

This can be expressed as a convolution operator using the Green's function for the Helmholtz equation:

$$\bar{\phi} = \frac{1}{4\pi\Delta^2} \int \frac{\phi(\boldsymbol{x})}{|\boldsymbol{x} - \boldsymbol{x}'|} \exp\left(-\frac{\boldsymbol{x} - \boldsymbol{x}'}{\Delta}\right) d\boldsymbol{x}'.$$

In the one-dimensional case, it has the transfer function

$$\hat{G}(k) = \frac{1}{1 + 2\Delta^2(1 - \cos(k))}. \tag{8.10}$$

The inverse Helmholtz filter is an example of an invertible filter where we can write

$$\phi = Q\bar{\phi},$$

where $Q = G^{-1}$. An invertible filter cannot be a projector such that

$$G^2 = G.$$

If this were the case,

$$G^2 Q = GQ = I$$

and

$$G^2Q = G(GQ) = G,$$

so G reduces to the identity operator I and is not a filter. Invertible filters are of interest because in principle, the true solution can be recovered from the filtered solution by the inverse filter, a process that is often referred to as deconvolution.

8.7.3 The Filtered Navier–Stokes Equations

Filtering each term of the incompressible Navier–Stokes equations results in the filtered continuity equation

$$\frac{\partial \bar{u}_i}{\partial x_i} = 0, \tag{8.11}$$

and the filtered momentum equation

$$\rho\left(\frac{\partial \bar{u}_i}{\partial t} + \frac{\partial}{\partial x_j}\overline{u_i u_j}\right) + \frac{\partial \bar{p}}{\partial x_i} = \mu \frac{\partial^2}{\partial x_k \partial x_k}\bar{u}_i .$$

Introducing the subgrid scale (SGS) residual tensor

$$\tau_{ij}^R = \overline{u_i u_j} - \bar{u}_i \bar{u}_j,$$

we can rewrite the momentum equation in the standard form

$$\rho\left(\frac{\partial \bar{u}_i}{\partial t} + \frac{\partial}{\partial x_j}(\bar{u}_i \bar{u}_j)\right) + \frac{\partial \bar{p}}{\partial x_i} = \mu \frac{\partial^2 \bar{u}_i}{\partial x_j \partial x_j} - \frac{\partial}{\partial x_j}\tau_{ij}^R . \tag{8.12}$$

The SGS tensor τ_{ij}^R now needs to be modeled. Then, the filtered quantities $\bar{u}_i$ and $\bar{p}$ can be directly calculated by solving (8.11) and (8.12), where the filtered pressure $\bar{p}$ is determined from the Poisson equation

$$\frac{\partial^2 \bar{p}}{\partial x_i \partial x_i} = -\rho \frac{\partial}{\partial x_i}\frac{\partial}{\partial x_j}(\bar{u}_i \bar{u}_j) + \frac{\partial^2}{\partial x_i \partial x_j}\tau_{ij}^R,$$

which results from taking the divergence of (8.12).

8.7.4 The Leonard Decomposition

Following an idea first proposed by Leonard (1974), it is a common practice to decompose the SGS tensor into three parts as follows. Substituting

$$u_i = \bar{u}_i + u_i',$$

where u_i' is the SGS fluctuating velocity, we find that

$$\begin{aligned}\tau_{ij}^R &= \overline{(\bar{u}_i + u_i')(\bar{u}_j + u_j')} - \bar{u}_i \bar{u}_j \\ &= L_{ij} + C_{ij} + R_{ij},\end{aligned}$$

where the Leonard stress is

$$L_{ij} = \overline{\bar{u}_i \bar{u}_j} - \bar{u}_i \bar{u}_j,$$

the cross stress is

$$C_{ij} = \overline{\bar{u}_i u'_j} + \overline{u'_i \bar{u}_j},$$

and the SGS Reynolds stresses are

$$R_{ij} = \overline{u'_i u'_j}.$$

The Leonard stress represents the contribution from the resolved scales and can be recovered by explicitly filtering $\bar{u}_i \bar{u}_j$. The cross stress represents interactions between the resolved and the subgrid scale motions, while the SGS Reynolds stress represents interactions between the subgrid scale motions, and both these terms need to be modeled.

8.7.5 The Modified Leonard Decomposition

It can be verified by direct calculation that the terms L_{ij}, C_{ij}, and R_{ij} in the Leonard decomposition are not Galilean invariant. In order to remedy this, Germano (1986a, 1986b) proposed the modified decomposition

$$\tau_{ij}^R = L_{ij}^0 + C_{ij}^0 + R_{ij}^0,$$

where the modified Leonard stress is

$$L_{ij}^0 = \overline{\bar{u}_i \bar{u}_j} - \bar{\bar{u}}_i \bar{\bar{u}}_j, \tag{8.13}$$

the modified cross stress is

$$C_{ij}^0 = \overline{\bar{u}_i u'_j} + \overline{u'_i \bar{u}_j} - \bar{\bar{u}}_i \bar{u}'_j - \bar{u}'_i \bar{\bar{u}}_j, \tag{8.14}$$

and the modified SGS Reynolds stress is

$$R_{ij} = \overline{u'_i u'_j} - \bar{u}'_i \bar{u}'_j. \tag{8.15}$$

This decomposition preserves Galilean invariance in all three terms.

8.7.6 Exact SGS Stresses with Reverse Riltering

In the case of an invertible filter, we can apply reverse filtering

$$u_i = Q\bar{u}_i,$$

where $Q = G^{-1}$. In this case, we can derive an exact expression for the SGS tensor that depends only on the filtered quantities as follows, using the relation

$$Q\overline{u_i u_j} = u_i u_j = Q\bar{u}_i Q\bar{u}_j.$$

Then,

$$\tau_{ij}^R = \overline{u_i u_j} - \bar{u}_i \bar{u}_j$$
$$= Q^{-1}(Q\bar{u}_i Q\bar{u}_j - Q(\bar{u}_i \bar{u}_j)).$$

In the case of the inverse Helmholtz filter

$$Q = 1 - \Delta^2 \frac{\partial^2}{\partial x_k \partial x_k},$$

the residual stress tensor reduces to

$$\tau_{ij}^R = \Delta^2 Q^{-1} \left(2 \frac{\partial \bar{u}_i}{\partial x_k} \frac{\partial \bar{u}_j}{\partial x_k} + \Delta^2 \frac{\partial^2 \bar{u}_i}{\partial x_k \partial x_k} \frac{\partial^2 \bar{u}_j}{\partial x_l \partial x_l} \right)$$

and may be calculated by solving

$$\left(1 - \Delta^2 \frac{\partial^2}{\partial x_k \partial x_k}\right) \tau_{ij}^R = \Delta^2 \left(2 \frac{\partial \bar{u}_i}{\partial x_k} \frac{\partial \bar{u}_j}{\partial x_k} + \Delta^2 \frac{\partial^2 \bar{u}_i}{\partial x_k \partial x_k} \frac{\partial^2 \bar{u}_j}{\partial x_l \partial x_l} \right).$$

8.7.7 SGS Models

SGS models fall into two main classes – structural and functional. According to the Kolmogorov theory, the energy is dissipated at the small scales, beyond the inertial range. Functional models try to reproduce this by adding dissipative terms. Structural models try to reproduce the structure by assuming $\overline{u_i u_j} - \bar{u}_i \bar{u}_j$ is similar to the Leonard term $\overline{\bar{u}_i \bar{u}_j} - \bar{\bar{u}}_i \bar{\bar{u}}_j$. It has been found that these usually do not introduce enough dissipation, and accordingly mixed models that combine both approaches have been introduced.

8.7.8 Eddy Viscosity Models

The simplest models for the SGS stress tensor follow the Boussinesq hypothesis and set

$$\tau_{ij}^R = 2\nu_{\mathrm{SGS}} \bar{S}_{ij},$$

where $\bar{S}_{ij}$ is the filtered rate of strain

$$\bar{S}_{ij} = \frac{1}{2} \left(\frac{\partial \bar{u}_i}{\partial x_j} + \frac{\partial \bar{u}_j}{\partial x_i} \right).$$

With these models, there is no need to filter the equations explicitly because the filtered quantities are obtained as the solution of the filtered equations (8.12).

In incompressible flow, the trace

$$\bar{S}_{kk} = \frac{\partial \bar{u}_k}{\partial x_k} = 0,$$

while the trace τ_{kk}^{R} of the residual stress tensor is not necessarily zero. Accordingly, it is the usual practice to model the deviatoric residual stress as

$$\tau_{ij}^{d} = \tau_{ij}^{R} - \frac{1}{3}\delta_{ij}\tau_{kk}^{R},$$

which does have a zero trace. The isotropic residual stress is then included in the modified pressure

$$\bar{p}^{d} = \bar{p} + \frac{1}{3}\tau_{kk}^{R},$$

which can be determined by solving a Poisson equation in the usual manner.

The original SGS model introduced by Smagorinsky (1963) remains one of the most widely used eddy viscosity models. The eddy viscosity has the dimensions of length squared times the velocity gradient. In the Smagorinsky model, the length scale is taken as

$$l_s = c_s\Delta,$$

where c_s is the Smagorinsky constant, and Δ is the filter width, while the velocity gradient is taken as the magnitude

$$\bar{S} = \sqrt{2\bar{S}_{ij}\bar{S}_{ij}},$$

of the filtered rate of strain tensor, so the eddy viscosity is

$$\nu_{\mathrm{SGS}} = (c_s\Delta)^2\bar{S}. \qquad (8.16)$$

The Smagorinsky constant was estimated by Lilly (1968) to be

$$c_s \approx 0.17,$$

by comparing the resulting dissipation rate with the dissipation rate in the inertial range of the Kolmogorov spectrum. In practice, it has been found that in the presence of shear, smaller values in the range 0.065–0.1 should be used. Numerous variations of eddy viscosity models have been proposed. The most important are wall adapted large eddy (WALE) models, in which ν_{SGS} is reduced in the vicinity of the wall to produce the proper asymptotic behavior, and dynamic models, which estimate the Smagorinsky constant by comparing the subgrid scale tensor at two filter widths.

8.7.9 Scale Similarity and Mixed Models

Both experimental data and the results of direct numerical simulations confirm that the principal axes of the SGS residual stress τ_{ij}^{R} are not aligned with the principal axes of the filtered rate of strain tensor $\bar{S}_{ij}$. Thus, a serious deficiency of eddy viscosity models is that they force alignment of the modeled residual stress with the rate of strain. This motivates the introduction of more complex structural models.

Scale similarity models are based on the assumption that at scales far enough down in the inertial range, the turbulent motions are self similar, and accordingly, the motions of the largest SGS scales should be similar to the motions of the smallest resolved scales. Following this line of reasoning, the Bardina scale similarity model (Bardina et al., 1980) approximates τ_{ij}^R by the modified Leonard tensor L_{ij}^0 defined by (8.13). Since we can suppose that the filtered quantities are directly obtained as the computed solution of the filtered equations (8.12), we can obtain the required quantities by explicitly filtering the solution a single time, at a filter width somewhat larger than the grid interval, so the residual stress tensor is modeled as

$$L_{ij}^m = \widehat{\bar{u}_i \bar{u}_j} - \hat{\bar{u}}_i \hat{\bar{u}}_j,$$

where the hat indicates the explicit filtering operation. This model has been found to be insufficiently dissipative. In order to correct this, mixed models add an eddy viscosity term, which may be regarded as modeling the cross stresses and SGS residual stresses C_{ij}^0 and R_{ij}^0 defined by (8.14) and (8.15). As in the case of eddy viscosity models, it is the usual practice to model the deviatoric residual stress τ_{ij}^d. Thus, the mixed model becomes

$$\tau_{ij}^d = L_{ij}^m - \frac{1}{3}\delta_{ij} L_{kk}^m + 2\nu_{\mathrm{SGS}} \bar{S}_{ij},$$

where ν_{SGS} is defined by (8.16), with a suitable adjusted value of the constant c_s.

8.7.10 Deconvolution Models

Deconvolution models are based on the concept of recovering an estimate of the unfiltered quantities from the filtered quantities. In the one-dimensional case, we can write

$$\bar{\phi}(x) = \int G(\tilde{x}) \phi(x - \tilde{x})\, d\tilde{x},$$

and expand $\phi(x - \tilde{x})$ as a Taylor series,

$$\phi(x - \tilde{x}) = \phi(x) - \tilde{x}\frac{\partial \phi}{\partial x} + \frac{\tilde{x}^2}{2!}\frac{\partial^2 \phi}{\partial x^2} \cdots.$$

The filtered quantity can be expanded as

$$\bar{\phi}(x) = \left(1 + \sum^{\infty} \frac{(-1)^n}{n!} \Delta^n M_n \frac{\partial^n}{\partial x^n}\right) \phi(x),$$

where $M_n \Delta^n$ is the nth moment of the filter kernel, that is,

$$M_n = \frac{1}{\Delta^n} \int G(\tilde{x})\, \tilde{x}^n\, d\tilde{x}.$$

Hence, we can write

$$\phi = (I + F)^{-1} \bar{\phi},$$

where

$$F = \sum^{\infty} \frac{(-1)^n}{n!} \Delta^n M_n \frac{\partial^n}{\partial x^n},$$

and we can expand these formally as

$$\phi = (I - F + F^2 \ldots)\bar{\phi},$$

where the series is convergent if $\|F\| < 1$. For a symmetric filter, the odd moments are zero. Accordingly, a similar expansion in the three-dimensional case yields

$$u_i = \left(1 - \frac{1}{2} M_2 \Delta^2 \frac{\partial^2}{\partial x_k \partial x_k}\right) \bar{u}_i + \mathcal{O}(\Delta^4).$$

Thus, the inverse Helmholtz operator is an approximation to order Δ^4 for the inverse of filters satisfying the convergence condition $\|F\| < 1$. Substituting for u_i and u_j, we now find that

$$\begin{aligned} \tau_{ij}^R &= \overline{u_i u_j} - \bar{u}_i \bar{u}_j \\ &= M_2 \Delta^2 \frac{\partial \bar{u}_i}{\partial x_k} \frac{\partial \bar{u}_j}{\partial x_k} + \mathcal{O}(\Delta^4). \end{aligned}$$

The Clark deconvolution model takes only the first term of the expansion. Thus, the deviatoric stress is modeled as

$$\tau_{ij}^d = M_2 \Delta^2 \left(\frac{\partial \bar{u}_i}{\partial x_k} \frac{\partial \bar{u}_j}{\partial x_k} - \frac{1}{3} \frac{\partial \bar{u}_l}{\partial x_k} \frac{\partial \bar{u}_l}{\partial x_k} \delta_{ij} \right),$$

where in the case of a Gaussian or top-hat filter,

$$M_2 = \frac{1}{12}.$$

This may also be extended to a mixed model by adding an eddy viscosity term $2\nu_{\mathrm{SGS}} \bar{S}_{ij}$, as in the case of the mixed model.

8.7.11 Dynamic Models

Large eddy simulations of non-equilibrium flows have been significantly improved by the introduction of dynamic modeling techniques (Germano et al. 1991). In these, the model coefficients are computed dynamically during the course of the computation instead of being fixed a priori. The required information is obtained by filtering the computed solution with a test filter at a larger scale $\tilde{\Delta}$, typically twice the original filter width Δ. Then, we apply an identity due to Germano et al. (1991). Suppose that

$$\tau_{ij} = \overline{u_i u_j} - \bar{u}_i \bar{u}_j$$

and the corresponding residual stress after double filtering is

$$T_{ij} = \widetilde{\overline{u_i u_j}} - \tilde{\bar{u}}_i \tilde{\bar{u}}_j.$$

Then,

$$T_{ij} - \tau_{ij} = L_{ij}, \tag{8.17}$$

where

$$L_{ij} = \widetilde{\bar{u}_i \bar{u}_j} - \tilde{\bar{u}}_i \tilde{\bar{u}}_j,$$

and L_{ij} can be calculated from the computed solution $\bar{u}_i$. Suppose also that we model the deviatoric parts of τ_{ij} and T_{ij} by eddy viscosity models at the two different filter widths

$$\tau_{ij} = -2c_s \Delta^2 \bar{S}\bar{S}_{ij}, \quad T_{ij} = -2c_s \tilde{\Delta}^2 \tilde{\bar{S}}\tilde{\bar{S}}_{ij},$$

where the same constant c_s is used in each, and c_s replaces c_s^2 to allow for the possibility that c_s might be negative. Now we substitute for τ_{ij} and T_{ij} in the deviatoric part of (8.17) to obtain

$$L_{ij}^d = -2c_s \left(\tilde{\Delta}^2 \tilde{\bar{S}}\tilde{\bar{S}}_{ij} - \Delta^2 \widetilde{\bar{S}\bar{S}_{ij}} \right).$$

In general, this equation cannot be satisfied by the choice of the single coefficient c_s. But if we contract it with S_{ij}, it reduces to a scalar equation that can be solved for c_s to give

$$c_s = -\frac{1}{2} \frac{L_{ij}\bar{S}_{ij}}{\tilde{\Delta}^2 \tilde{\bar{S}}\tilde{\bar{S}}_{ij}\bar{S}_{ij} - \Delta^2 \widetilde{\bar{S}\bar{S}_{ij}}\bar{S}_{ij}}.$$

Then, we advance the solution using this value of c_s in the eddy viscosity model for τ_{ij}. In some circumstances, c_s proves to be negative, corresponding to the back scatter of energy from smaller to larger scales.

8.8 Analysis of the Advection-Diffusion Equation

In this section, we analyze discretization schemes for the advection-diffusion equation

$$\frac{\partial u}{\partial t} + a\frac{\partial u}{\partial x} = \nu \frac{\partial^2 u}{\partial x^2}, \tag{8.18}$$

where it is assumed that the diffusion coefficient $\nu \geq 0$. This serves to highlight some of the issues that arise in the discretization of the Navier–Stokes equations. The advection and diffusion equations have been separately analyzed in Chapters 3 and 4.

Let v_j^n denote the discrete solution of the advection-diffusion equation after n time steps at the mesh point j of a uniform mesh. Let

$$\lambda = a\frac{\Delta t}{\Delta x}, \quad \beta = \nu \frac{\Delta t}{\Delta x^2},$$

where Δt and Δx are the time step and mesh interval. In the absence of the diffusive term, the standard upwind scheme for the linear advection equation is recovered by adding artificial diffusion with a coefficient proportional to $\frac{1}{2}|a|\Delta x$, as has been discussed in Section 5.10. The scheme then takes the form

$$v_j^{n+1} = v_j^n - \frac{\lambda}{2}\left(v_{j+1}^n - v_{j-1}^n\right) + \frac{|\lambda|}{2}\left(v_{j+1}^n - 2v_j^n + v_{j-1}^n\right)$$

and is stable if Δt satisfies the CFL limit

$$|\lambda| \le 1.$$

In the absence of the convective terms, the use of a second order accurate central difference approximation to $\nu \frac{\partial^2 u}{\partial x^2}$ leads to the scheme

$$v_j^{n+1} = v_j^n + \beta\left(v_{j+1}^n - 2v_j^n + v_{j-1}^n\right).$$

In Sections 4.5 and 4.8, this scheme has been shown to be stable in both the L_∞ and L_2 norms if Δt satisfies the viscous limit

$$0 \le \beta \le \frac{1}{2}.$$

Following these guidelines, we consider a discretization of the advection-diffusion equation of the form

$$\begin{aligned} v_j^{n+1} &= v_j^n - \frac{\lambda}{2}(v_{j+1}^n - v_{j-1}^n) + (\beta + \alpha|\lambda|)(v_{j+1}^n - 2v_j^n + v_{j-1}^n) \\ &= \left(\beta + \alpha|\lambda| - \frac{\lambda}{2}\right) v_{j+1}^n + \left(1 - 2(\beta + \alpha|\lambda|)\right) v_j^n + \left(\beta + \alpha|\lambda| + \frac{\lambda}{2}\right) v_{j-1}^n, \end{aligned} \tag{8.19}$$

where the term $\beta(v_{j+1}^n - 2v_j^n + v_{j-1}^n)$ is a second order accurate discretization of the true diffusive term $\nu \frac{\partial^2 u}{\partial x^2}$, and artificial diffusion with a coefficient $\alpha|a|$ has been added to stabilize the convective part of the equation. According to the analysis of Sections 5.4 and 5.5, this scheme will be L_∞ stable, in fact local extremum diminishing (LED), if the coefficients of v_{j+1}^n, v_j^n, and v_{j-1}^n are all non-negative, and this will be the case if

$$0 \le \beta + \alpha|\lambda| \le \frac{1}{2}$$

and

$$\beta + \alpha|\lambda| \ge \frac{1}{2}|\lambda|.$$

Now, however, we also need to make sure that the artificial diffusion is small in comparison to the true diffusion, or

$$\alpha|\lambda| \ll \beta.$$

This suggests that Δt should be determined by the viscous limit $\beta \le \frac{1}{2}$ rather than the CFL limit $|\lambda| \le 1$. If we take $\alpha = 0$, the scheme is L_∞ stable and LED if

$$0 \le \beta \le \frac{1}{2}, \tag{8.20}$$

and

$$|\lambda| \leq 2\beta.$$

We can satisfy the second condition by taking Δx small enough. Actually,

$$\frac{|\lambda|}{\beta} = \frac{|a|\Delta x}{\nu},$$

which can be recognized as the cell Reynolds number Re_{cell}. Thus, the pure central difference scheme is stable if it satisfies the viscous limit (8.20) and

$$Re_{\text{cell}} \leq 2.$$

Moreover, if $Re_{\text{cell}} = 2$, the conditions $0 \leq \beta \leq \frac{1}{2}$ and $|\lambda| \leq 1$ are equivalent. If $Re_{\text{cell}} < 2$, the viscous limit $\beta \leq \frac{1}{2}$ dominates since then $|\lambda| = \beta Re_{\text{cell}} < 1$, while if $Re_{\text{cell}} > 2$, the CFL limit $|\lambda| \leq 1$ dominates since then $\beta = \frac{|\lambda|}{Re_{\text{cell}}} < \frac{1}{2}$. However, the scheme no longer has positive coefficients unless artificial diffusion is introduced. If we take $\alpha = \frac{1}{2}$, as is typical of upwind shock capturing schemes, we would need

$$\alpha \frac{|\lambda|}{\beta} = \alpha Re_{\text{cell}} \ll 1,$$

requiring a much finer mesh spacing. This indicates that first order artificial diffusion should be switched off in viscous dominated regions and be replaced by a higher order artificial diffusion such as that in the JST scheme described in Section 5.17.

We can gain some additional insights from a von Neumann test. Consider a mode of the form

$$v_j^n = g^n e^{i\omega x_j}.$$

Substituting this in the discrete equation (8.19), we obtain

$$g(\xi) = 1 - \lambda i \sin \xi - 2(\beta + \alpha|\lambda|)(1 - \cos \xi),$$

where ξ is the wave number $\omega \Delta x$. If we take $\alpha = 0$ and $\lambda = 2\beta$,

$$g(\xi) = 1 - 2\beta + 2\beta(\cos \xi - i \sin \xi),$$

which is a circle of radius of 2β centered at $1 - 2\beta$, as illustrated in Figure 8.10. Thus, the scheme is stable if $0 \leq 2\beta \leq 1$. Also, if this condition is satisfied and $|\lambda| < 2\beta$, $g(\xi)$ follows an ellipse inside the circle of radius 2β centered at $1 - 2\beta$, so the scheme remains stable.

The general conclusion of this analysis is that stable non-oscillatory discretizations can be assured by choosing a mesh interval small enough to ensure that $Re_{\text{cell}} \leq 2$. The implication for Navier–Stokes simulations is that the mesh interval normal to the wall in the boundary layer should satisfy a similar condition, and accordingly, a very fine mesh spacing will be needed for the simulation of flows at high Reynolds numbers.

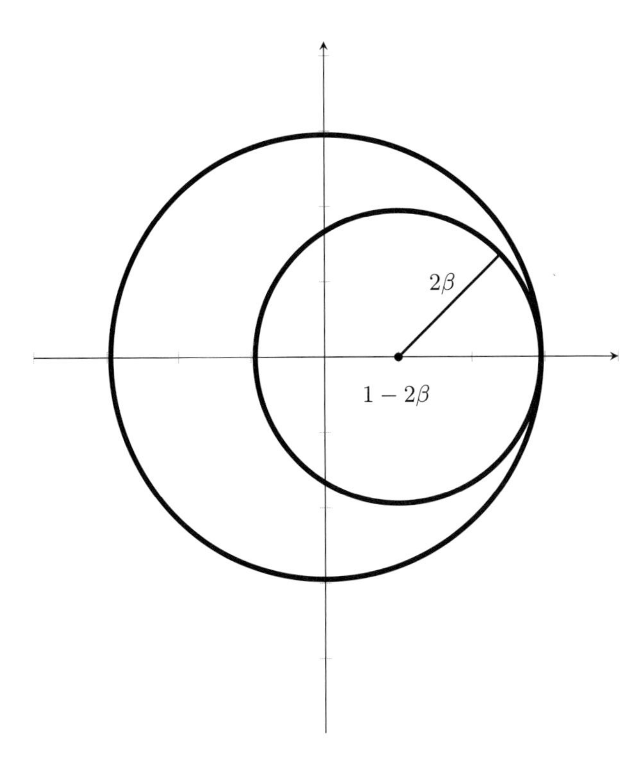

Figure 8.10 Locus of amplification factor $g(\xi)$ for the advection-diffusion equation.

8.9 Discretization of the Viscous Terms on Structured Meshes

The discretization of the viscous terms of the Navier–Stokes equations requires an approximation to the velocity derivatives $\frac{\partial u_i}{\partial x_j}$ in order to calculate the viscous stress tensor σ_{ij}, (2.4):

$$\frac{d}{dt}\boldsymbol{w}V + \sum_{\text{faces}} \boldsymbol{f}\cdot S = 0, \tag{8.21}$$

Then the viscous terms may be included in the flux balance (8.21). In order to evaluate the derivatives, one may apply the Gauss formula to a control volume V with the boundary S:

$$\int_V \frac{\partial u_i}{\partial x_j} dv = \int_S u_i n_j ds,$$

where n_j is the outward normal. For a polyhedral cell this gives

$$\overline{\frac{\partial u_i}{\partial x_j}} = \frac{1}{\text{vol}} \sum_{\text{faces}} \bar{u}_i \; n_j \; s, \tag{8.22}$$

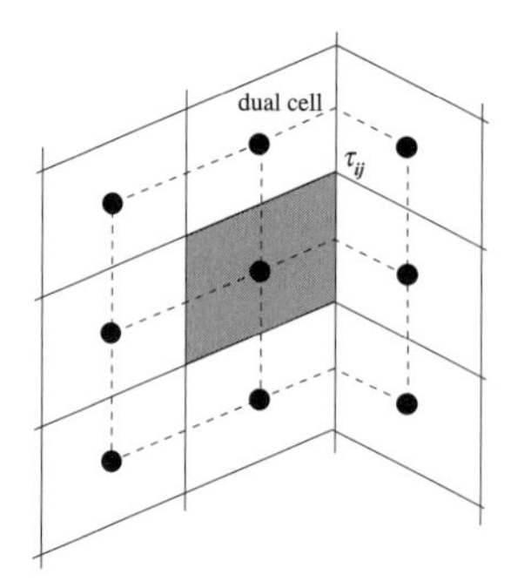

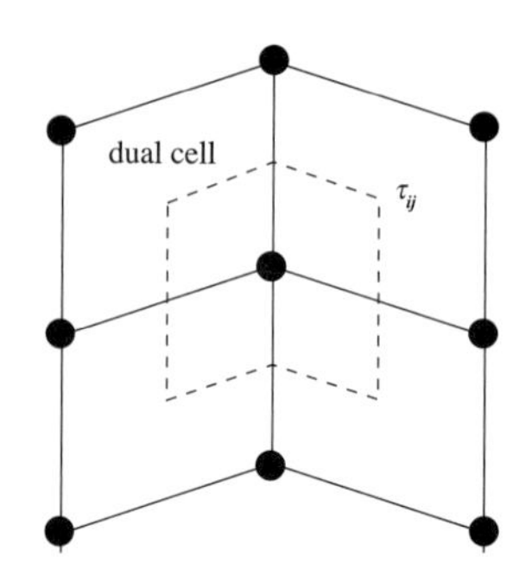

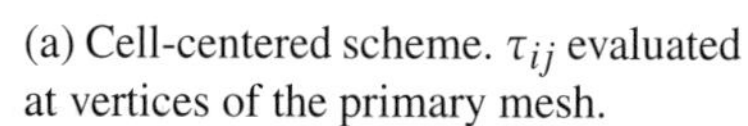

(a) Cell-centered scheme. τ_{ij} evaluated at vertices of the primary mesh.

(b) Cell-vertex scheme. τ_{ij} evaluated at cell centers of the primary mesh.

Figure 8.11 Viscous discretizations for cell-centered and cell-vertex algorithms.

where $\bar{u}_i$ is an estimate of the average of u_i over the face. If u varies linearly over a tetrahedral cell, this is exact. Alternatively, assuming a local transformation to computational coordinates ξ_j, one may apply the chain rule

$$\frac{\partial u}{\partial x} = \left[\frac{\partial u}{\partial \xi}\right]\left[\frac{\partial \xi}{\partial x}\right] = \frac{\partial u}{\partial \xi}\left[\frac{\partial x}{\partial \xi}\right]^{-1}. \tag{8.23}$$

Here the transformation derivatives $\frac{\partial x_i}{\partial \xi_j}$ can be evaluated with the same finite difference formulas as the velocity derivatives $\frac{\partial u_i}{\partial \xi_j}$. In this case, $\frac{\partial u}{\partial \xi}$ is exact if u is a linearly varying function.

For a cell-centered discretization (Figure 8.11a), $\frac{\partial u}{\partial \xi}$ is needed at each face. The simplest procedure is to evaluate $\frac{\partial u}{\partial \xi}$ in each cell and to average $\frac{\partial u}{\partial \xi}$ between the two cells on either side of a face (Jayaram & Jameson 1988). The resulting discretization does not have a compact stencil and supports undamped oscillatory modes. In a one-dimensional calculation, for example, $\frac{\partial^2 u}{\partial x^2}$ would be discretized as $\frac{u_{i+2}-2u_i+u_{i-2}}{4\Delta x^2}$. In order to produce a compact stencil, $\frac{\partial u}{\partial x}$ may be estimated from a control volume centered on each face, using formulas (8.22) or (8.23) (Rieger & Jameson 1988). This is computationally expensive because the number of faces is much larger than the number of cells. In a hexahedral mesh with a large number of vertices, the number of faces approaches three times the number of cells.

This motivates the introduction of dual meshes for the evaluation of the velocity derivatives and the flux balance, as sketched in Figure 8.11. The figure shows both cell-centered and cell-vertex schemes. The dual mesh connects cell centers of the primary mesh. If there is a kink in the primary mesh, the dual cells should be formed by assembling contiguous fractions of the neighboring primary cells. On smooth meshes, comparable results are obtained by either of these formulations (Martinelli & Jameson 1988, Martinelli, Jameson, & Malfa 1992, Liu & Jameson 1992). If the mesh has a kink, the cell-vertex scheme has the advantage that the derivatives $\frac{\partial u_i}{\partial x_j}$ are calculated in the interior of a regular cell, with no loss of accuracy.

A desirable property is that a linearly varying velocity distribution, as in a Couette flow, should produce a constant stress and hence an exact stress balance. This property is not necessarily satisfied in general by finite difference or finite volume schemes on curvilinear meshes. The characterization k-exact has been proposed for schemes that are exact for polynomials of degree k. The cell-vertex finite volume scheme is linearly exact if the derivatives are evaluated by (8.23), since then $\frac{\partial u_i}{\partial x_j}$ is exactly evaluated as a constant, leading to constant viscous stresses τ_{ij} and an exact viscous stress balance. This remains true when there is a kink in the mesh, because the summation of constant stresses over the faces of the kinked control volume sketched in Figure 8.11 still yields a perfect balance. The use of (8.23) to evaluate $\frac{\partial u_i}{\partial x_j}$, however, requires the additional calculation or storage of the nine metric quantities $\frac{\partial u_i}{\partial x_j}$ in each cell, whereas (8.22) can be evaluated from the same face areas that are used for the flux balance.

8.10 Discretization of the Viscous Terms on Unstructured Meshes

In the case of an unstructured mesh, the weak form

$$\frac{\partial}{\partial t}\iiint_{\Omega} \phi w d\Omega = \iiint_{\Omega} f \cdot \nabla\phi \, d\Omega - \iint_{\partial\Omega} \phi f \cdot dS$$

leads to a natural discretization with linear elements, in which the piecewise linear approximation yields a constant stress in each cell. This method yields a representation that is globally correct when averaged over the cells, as is proved by energy estimates for elliptic problems. It should be noted, however, that it yields formulas that are not necessarily locally consistent with the differential equations, as was observed in the discretization of the Laplacian $u_{xx} + u_{yy}$ in Section 3.3.6.

Anisotropic grids are needed in order to resolve the thin boundary layers that appear in viscous flows at high Reynolds numbers. Otherwise an excessively large number of grid cells may be required. The use of flat tetrahedra can have an adverse effect on both the accuracy of the solution and the rate of convergence to a steady state. This has motivated the use of hybrid prismatic-tetrahedral grids in which prismatic cells are used in the wall regions (Parthasarathy, Kallinderis, & Nakajima 1995). A review of many of the key issues in the design of flow solvers for unstructured meshes is given by Venkatakrishnan (1996).

9 Overview of Time Integration Methods

9.1 Introduction

If the space discretization procedure is implemented separately to produce a semi-discrete scheme, we have to solve a set of coupled ordinary differential equations (ODEs), which can be written in the form

$$\frac{d\boldsymbol{w}}{dt} + \boldsymbol{R}(\boldsymbol{w}) = 0, \tag{9.1}$$

where $\boldsymbol{w}$ now denotes the global vector of the flow variables at the mesh points, and $\boldsymbol{R}(\boldsymbol{w})$ is the vector of the residuals consisting of the flux balances defined by the space discretization scheme, together with the added dissipative terms. In the case of the finite volume schemes discussed in Chapter 7, this notation assumes that the local residuals have been divided by the cell area in the case of a two-dimensional calculation or the cell volume in a three-dimensional calculation. It now becomes necessary to choose a time stepping scheme to complete the discretization of the problem.

There is an extremely well developed theory of time integration methods for ordinary differential equations, which is thoroughly described in books by Lambert, Henrici, and Butcher, among others (Henrici 1962, Butcher 1987, Lambert 1991). The semi-discrete equations arising in fluid flow simulations are characterized, however, by a number of issues that distinguish them from ODEs in general. First, in many engineering applications, such as the analysis of flow past a wing, the primary objective is to calculate a steady flow, and details of the transient solution are immaterial. In fact, no one would wish to be a passenger in an aircraft that could not sustain essentially steady flow. In this situation, the time stepping scheme may be designed solely to maximize the rate of convergence to a steady state, without regard to the order of accuracy. If, on the other hand, the flow to be analyzed is inherently unsteady, the order of accuracy of the time stepping scheme may be critically important. Both situations will be considered in the following sections.

Second, the total number of equations to be integrated can be very large. In three-dimensional simulations of formula-one racing cars, for example, it is common practice to use meshes with 100 million cells. With a two equation turbulence model, there are 7 unknowns per cell, resulting in a total of 700 million nonlinear equations. This means that it is not realistic to assume that the equations of an implicit scheme can be exactly solved at each time step, as is commonly assumed in textbooks on ODEs.

Figure 9.1 Two dragons consuming each other: paradox of fast steady state solver.

Third, in time accurate simulations of flows with moving shock waves and contact discontinuities, one wants to prevent overshoots and oscillations triggered by the discretization scheme. The conditions for a local extremum diminishing (LED) scheme using a first order accurate forward Euler time stepping scheme were given in Section 5.5. Clearly, there is a need for higher order time stepping schemes that preserve this property. This has led to a whole new class of "strongly stability preserving" (SSP) schemes, which will be described later.

Fourth, there is an important class of periodic unsteady flows, such as the flow induced by a helicopter rotor in forward flight or a wind turbine in a nonuniform flow due to the boundary layer over the ground. In these cases, special techniques such as Fourier methods may be appropriate.

In the early days of CFD, it was commonly assumed that in order to obtain fast convergence to a steady state, it would be necessary to use an implicit scheme that allowed large time steps. Any implicit scheme, however, such as the backward Euler scheme

$$w^{n+1} = w^n - \Delta t \boldsymbol{R}(w^{n+1}),$$

with the superscript n denoting the time level, requires the solution of a large number of coupled nonlinear equations that have the same complexity as the steady state problem,

$$\boldsymbol{R}(w) = 0.$$

Accordingly, a fast steady state solver is an essential building block for an implicit scheme. This leads to a circular situation. We need an implicit scheme for fast convergence to a steady state, but we need a fast steady state solver to build an implicit scheme. This situation is reminiscent of the scene shown in Figure 9.1 of two dragons, each consuming the other's tail. In fact, one of the most successful approaches to building implicit schemes is the "dual time stepping" method (Jameson 1991), in which an explicit time stepping method marching in pseudo-time is used to solve the implicit equations at each time step. The efficiency of this method depends on the use of acceleration techniques such as multigrid to increase the rate of convergence of the inner time stepping scheme.

In the remainder of this chapter, we examine the properties of some general classes of time integration methods. The design of time-integration methods for steady state problems is discussed in the next chapter, including a variety of methods for accelerating convergence to a steady state. The following chapter addresses issues arising in the time-accurate simulation of unsteady flows.

9.2 Simple Time Stepping Schemes

Several simple time stepping schemes have already been discussed in Section 4.2. The prototype explicit scheme is the forward Euler scheme

$$\boldsymbol{w}^{n+1} = \boldsymbol{w}^n - \Delta t \boldsymbol{R}(\boldsymbol{w}^n).$$

When applied to hyperbolic equations, it was also shown in Section 4.3 that the allowable time step depends on the space discretization and must satisfy the CFL condition that the domain of dependence of the discrete scheme must contain the domain of dependence of the true solution. This restriction can be avoided by the use of an implicit time integration scheme.

9.3 Linear Multistep Schemes

In order to achieve higher order accuracy, one must either increase the number of time levels, as in linear multistep schemes, or increase the number of stages in a single time step, as in Runge–Kutta schemes. Linear multistep schemes are examined in this section, and Runge–Kutta schemes in the next section. Both classes of schemes can be either explicit, requiring only evaluations of the residual with known data values, or implicit, including evaluations of the residual with as yet unknown data values.

The general form of a linear multistep scheme with s steps covering $s+1$ time levels to solve (9.1) is

$$\sum_{j=0}^{s} \alpha_j \boldsymbol{w}^{n+j} = -\Delta t \sum_{j=0}^{s} \beta_j \boldsymbol{R}(\boldsymbol{w}^{n+j}).$$

It is conventional to normalize the scheme by setting $\alpha_s = 1$. If $\beta_s = 0$, the scheme is explicit, and if $\beta_s \neq 0$, it is implicit. The simplest examples are the forward Euler scheme and the implicit backward Euler scheme, which were discussed in Chapter 4. The more general implicit scheme

$$w^{n+1} = w^n - \Delta t\left((1-\epsilon)R(w^n) + \epsilon R(w^{n+1})\right),$$

will be examined in Section 9.8 to illustrate the issues in implementing implicit schemes in general. Higher order schemes require two or more steps. Consequently, they require initial data from more than one time level to start the time integration. One of the simplest is the leap frog scheme

$$w^{n+1} = w^{n-1} - 2\Delta t\, R(w^n).$$

This scheme is stable for the linear advection equation provided the time step satisfies the CFL condition, but not for the diffusion equation.

Linear multistep schemes can be constructed by starting from the integral

$$w^{n+s} - w^{n+s-1} = -\int_{(n+s-1)\Delta t}^{(n+s)\Delta t} R(w)\,\Delta t,$$

fitting a Lagrange interpolation polynomial through the data values of $R^{n+j} = R(w^{n+j})$ from $j = 0$ to $s-1$ for an explicit scheme or $j = 0$ to s for an implicit scheme, and integrating the polynomial. This leads to the explicit Adams–Bashforth and implicit Adams–Moulton methods. The fourth order Adams–Bashforth scheme is conveniently written as

$$w^{n+1} = w^n - \frac{\Delta t}{24}\left(55R^n - 59R^{n-1} + 37R^{n-2} - 9R^{n-3}\right).$$

The fourth order Adams–Moulton scheme is

$$w^{n+1} = w^n - \frac{\Delta t}{24}\left(9R^{n+1} + 19R^n - 5R^{n-1} + R^{n-2}\right).$$

Associated with any linear multistep scheme are the first and second characteristic polynomials

$$\rho(\zeta) = \sum_{j=0}^{s} \alpha_j \zeta^j,$$

and

$$\sigma(\zeta) = \sum_{j=0}^{s} \beta_j \zeta^j.$$

These arise naturally if we consider solutions of the linear problem

$$\frac{du}{dt} = \alpha u,$$

for which the scheme reduces to the recurrence equations

$$\sum_{j=0}^{s} \alpha_j u^{n+j} - \alpha\Delta t \sum_{j=0}^{s} \beta_j u^{n+j} = 0.$$

We seek solutions of the form

$$u(n\Delta t) = \zeta^n,$$

which lead to the polynomial equation

$$P(\zeta) = \rho(\zeta) - \alpha\Delta t\ \sigma(\zeta) = 0. \tag{9.2}$$

If $P(\zeta)$ has distinct roots ζ_k, a general solution is

$$u(n\Delta t) = \sum_{k=1}^{s} A_k \zeta_k^n.$$

If $\rho(\zeta)$ has a root of multiplicity q, then

$$P'(\zeta) = 0, \quad P''(\zeta) = 0, \ldots, P^{(q-1)}(\zeta) = 0,$$

or

$$s\alpha_s\zeta^{s-1} + \cdots + \alpha_1\zeta = 0,$$

$$s(s-1)\alpha_s\zeta^{s-2} + \cdots + \alpha_2\zeta = 0,$$

$$\ldots$$

leading to additional solutions of the form

$$n\zeta^n,$$

$$n(n-1)\zeta^n,$$

$$\ldots$$

Numerical solutions will not grow if all roots of $P(\zeta)$ satisfy the root condition

$$|\zeta_k| \le 1,$$

and if ζ_k is a multiple root,

$$|\zeta_k| < 1.$$

The stability region is the region in the complex plane $z = \alpha\Delta t$ for which this holds. The scheme is said to be A-stable if the stability region contains the entire left half-plane, as was discussed in Section 4.2. It is said to be L-stable if it is A-stable and also $|u^{n+1}|/|u^n| \to 0$ as $\alpha\Delta t \to \infty$, with the consequence that the discrete solution immediately decays to zero. As $\Delta t \to 0$, the roots depend only on the first characteristic polynomial $\rho(\zeta)$, and if the roots of $\rho(\zeta)$ satisfy the root condition, the scheme is said to be zero-stable.

For the Adams–Bashforth and Adams–Moulton schemes the first characteristic polynomial is

$$\rho(\xi) = \xi^s - \xi^{s-1} = \xi^{s-1}(\xi - 1),$$

with one root $\xi = 1$ and $s - 1$ roots $\xi = 0$. Thus, these schemes are zero–stable. It can also be seen from (9.2) that no explicit scheme can be A-stable because with $\alpha_s = 1$, the product of the roots given by $\alpha_0 - \alpha\Delta t\beta_0$ will become unbounded for large enough values of $\alpha\Delta t$ lying in the left half-plane if $\beta_0 \neq 0$. More generally, if some $\beta_k \neq 0$, a sum of products of roots will become unbounded.

The error of a linear multistep scheme is associated with the difference operator

$$L(y, \Delta t) = \sum_{j=0}^{\alpha} \alpha_j y(t + j\Delta t) - \Delta t \sum_{j=0}^{s} \beta_j y'(t + j\Delta t),$$

where $y(t)$ is an arbitrary differentiable function. If we assume y is sufficiently differentiable, we can expand $y(t + j\Delta t)$ and $y'(t + j\Delta t)$ in Taylor series about t and collect terms to obtain

$$L(y, \Delta t) = C_0 y(t) + C_1 y^{(1)}(t) + \cdots + C_q y^{(q)}(t) + \cdots,$$

where

$$C_0 = \sum_{j=0}^{s} \alpha_j = \rho(1),$$

$$C_1 = \sum_{j=0}^{s} (j\alpha_j - \beta_j) = \rho'(1) - \sigma(1),$$

$$C_q = \sum_{j=0}^{s} \left(\frac{j^q}{q!}\alpha_j - \frac{j^{q-1}}{(q-1)!}\beta_j \right), \quad q = 2, 3, \ldots.$$

The scheme is pth order accurate if

$$C_0 = C_1 = \cdots = C_p = 0, \quad C_{p+1} \neq 0$$

and consistent if it is at least first order accurate, or

$$\rho(1) = 0, \quad \rho'(1) = \sigma(1).$$

THEOREM 9.3.1 *A linear multistep scheme is actually* p*th order accurate if and only if the associated function*

$$\phi(\zeta) = \frac{\rho(\zeta)}{\log \zeta} - \sigma(\zeta) \tag{9.3}$$

has a p *fold zero at* $\zeta = 1$.

Proof

$$\begin{aligned} L(e^t, \Delta t) &= e^t \left(\rho\left(e^{\Delta t}\right) - \Delta t \sigma\left(e^{\Delta t}\right)\right) \\ &= e^t C_{p+1} \Delta t^{p+1} + \mathcal{O}\big(\Delta t^{p+2}\big); \end{aligned}$$

so

$$\frac{\rho\left(e^{\Delta t}\right)}{\Delta t} - \sigma\left(e^{\Delta t}\right) = C_{p+1}\Delta t^{p} + \cdots .$$

Setting $\zeta = e^{\Delta t}$ yields the condition. Conversely, if $\phi(\zeta)$ has a p fold zero at $\rho = 1$, then $\Delta t\phi\left(e^{\Delta t}\right)$ has a zero of order $p+1$ at $\Delta t = 0$, and hence $L(y, \Delta t)$ is of order Δt^{p+1} for $y = e^{x}$. But the order depends only on the coefficients, so the scheme is of order p. □

As in the case of partial differential equations discussed in Chapter 4, a time integration scheme is defined to be convergent if the numerical solution converges to the true solution in the limit $\Delta t \to 0$, $n \to \infty$, $n\Delta t \to t$. In 1956, Dahlquist proved the fundamental theorem that zero-stability and consistency are necessary and sufficient for convergence (Dahlquist 1956). In contrast to the Lax equivalence theorem discussed in Chapter 4, the proof holds for both linear and nonlinear ODEs. A detailed discussion of the proof is given in the text by Henrici (1962).

The linear multistep scheme has $2s+2$ coefficients, of which $\alpha_s = 1$, and if the scheme is explicit, $\beta_s = 0$. For a scheme of order p, the first $p+1$ terms in the Taylor expansion of $L(y, \Delta t)$ must vanish, placing $p+1$ conditions on the coefficients. Thus, the highest order that could be obtained by an s step method is $2s$ if it is implicit and $2s-1$ if it is explicit. In 1956, Dahlquist also proved, however, the following barrier theorem:

THEOREM 9.3.2 (Dahlquist) *The maximum order of a stable s step scheme is $s+1$ if s is odd, or $s+2$ if s is even.*

The proof depends on the analytic properties of the associated function 9.3 (Dahlquist 1956, Henrici 1962). In addition to these results Dahlquist subsequently proved the second barrier theorem (Dahlquist 1963):

THEOREM 9.3.3 (Dahlquist) *No A-stable linear multistep scheme can be better than second order accurate.*

9.4 Backward Difference Formulas

The backward difference formulas constitute a family of linear multistep schemes that have proved to be very useful for time-accurate CFD simulations. The first order scheme BDF1 is simply the backward Euler method

$$\frac{w^{n+1} - w^{n}}{\Delta t} + R(w^{n+1}) = 0.$$

The second order scheme BDF2 is

$$\frac{1}{2\Delta t}(3w^{n+1} - 4w^{n} + w^{n-1}) + R(w^{n+1}) = 0,$$

while the third order scheme BDF3 is

$$\frac{1}{6\Delta t}(11w^{n+1} - 18w^n + 9w^{n-1} - 2w^{n-2}) + \boldsymbol{R}(w^{n+1}) = 0.$$

Backward difference formulas with s^{th} order accuracy can be constructed as

$$D_t w^{(n+1)} + \boldsymbol{R}(w^{n+1}) = 0,$$

where

$$D_t = \frac{1}{\Delta t}\sum_{q=1}^{s}\frac{1}{q}(\Delta^-)^q$$

and

$$\Delta^- w^{n+1} = w^{n+1} - w^n.$$

Here, we have dropped the convention of setting $\alpha_s = 1$, because it is more natural to normalize the coefficients so that the left hand side represents $\frac{dw}{dt}$.

These schemes are complementary to the Adams schemes in the sense that they have the simplest possible second characteristic polynomial

$$\sigma(\zeta) = \zeta^s.$$

The first and second order schemes, BDF1 and BDF2, are both A- and L-stable. The first order scheme can be used as the basis of fast steady state solvers. The second order scheme realizes the Dahlquist limit of second order accuracy for an A-stable linear multistep scheme. The trapezoidal rule

$$w^{n+1} = w^n - \frac{\Delta t}{2}\left(\boldsymbol{R}(w^{n+1}) - \boldsymbol{R}(w^n)\right),$$

which is also the second order Adams–Moulton scheme, is A-stable but not L-stable, as was observed in Section 4.2. Thus, the second order backward difference formula may be preferred because it is L-stable. The third order backward difference formula is not A-stable, but its stability region covers almost all of the left half-plane, and in light of its higher order accuracy, it is also an attractive candidate for time-accurate simulations. Procedures for solving the nonlinear equations arising from implicit schemes are further discussed in Chapter 11.

9.5 Multistage Runge–Kutta Schemes

In contrast to linear multistep methods, Runge–Kutta methods require no special start up procedure. They can be tailored for higher order accuracy or, if this is not important, to give a large stability region, and they have proved extremely effective in practice as a method of solving the Euler or Navier–Stokes equations (Jameson et al. 1981, Jameson et al. 1986, Martinelli & Jameson 1988).

For a detailed analysis of Runge–Kutta (RK) schemes, the reader is referred to the books written by Butcher (1987, 2003). The classical fourth order RK scheme applied to (9.1) can be written as

$$\begin{aligned} w^{(1)} &= w^{(0)} - \frac{1}{2}\Delta t R(w^{(0)}), \\ w^{(2)} &= w^{(0)} - \frac{1}{2}\Delta t R(w^{(1)}), \\ w^{(3)} &= w^{(0)} - \Delta t R(w^{(2)}), \\ w^{(4)} &= w^{(0)} - \frac{1}{6}\Delta t\big(R(w^{(0)}) + 2R(w^{(1)}) + 2R(w^{(2)}) + R(w^{(3)})\big), \end{aligned} \tag{9.4}$$

where $w^{(0)}$ is the value w^n at the end of the previous time step, and the value w^{n+1} at the end of the time step is set equal to $w^{(4)}$. The scheme is fourth order accurate for both linear and nonlinear problems, and in general, it offers an excellent compromise between accuracy and computational costs. The stabilty region, illustrated later in Section 9.8, Figure 9.4, extends to $\pm i2\sqrt{2}$ along the imaginary axis and -2.8 along the real axis.

A general Runge–Kutta scheme with m stages can be written as

$$\begin{aligned} w^{(i)} &= w^{(0)} - \Delta t \sum_{j=0}^{m-1} a_{ij} R(w^{(j)}), \quad i = 1, \ldots, m-1, \\ w^{(m)} &= w^{(0)} - \Delta t \sum_{j=0}^{m-1} b_j R(w^{(i)}). \end{aligned} \tag{9.5}$$

This can be represented by the Butcher tableau where the vector $\mathbf{c}$ indicates the

$$\begin{array}{c|c} \mathbf{c} & \mathbf{A} \\ \hline & \mathbf{b}^T \end{array}$$

fraction of the time step corresponding to each stage, the elements of the matrix $\mathbf{A}$ show the dependence of each stage on the derivatives at other stages, and the vector $\mathbf{b}^T$ contains the quadrature weights for integrating $\frac{dw}{dt}$ over a complete time step. With this notation, the fourth order RK scheme given above has the tableau shown in Figure 9.2.

$$\begin{array}{c|cccc} 0 & & & & \\ \frac{1}{2} & \frac{1}{2} & & & \\ \frac{1}{2} & 0 & \frac{1}{2} & & \\ 1 & 0 & 0 & 1 & \\ \hline & \frac{1}{6} & \frac{1}{3} & \frac{1}{3} & \frac{1}{6} \end{array}$$

Figure 9.2 Butcher tableau for RK4.

This notation also allows for implicit RK schemes in which the derivative at any given stage may depend on its own value and possibly even the value of later stages. Schemes that include derivatives evaluated up to a given stage are called diagonal implicit RK (DIRK) schemes.

When an explicit Runge–Kutta scheme is applied to the linear model problem

$$\frac{du}{dt} = \alpha u,$$

it follows from repeated substitution that with $z = \alpha \Delta t$,

$$u^{(m)} = g(z) u^{(0)},$$

where $g(z)$ is a polynomial of degree m and accordingly has m zeros. The boundary of the stability region is the locus of $|g(z)| = 1$. Moreover, for k^{th} order accuracy, $u^{(m)}$ must match the first k terms of the Taylor series expansion

$$u(t + \Delta t) = u(t) + \Delta t \frac{du}{dt} + \cdots + \frac{\Delta t^k}{k!} \frac{d^k u}{dt^k} + \mathcal{O}(\Delta t^{k+1}).$$

Setting

$$\frac{du}{dt} = \alpha u, \quad \frac{d^2 u}{dt^2} = \alpha^2 u, \ldots,$$

this means that the first k terms of $g(z)$ match the first k terms of e^z,

$$1 + z + \cdots + \frac{z^k}{k!}.$$

RK schemes with k stages can deliver k^{th} order accuracy for both linear and nonlinear ODEs up to order four. For $k > 4$, it is known that more than k stages are needed for k^{th} order accuracy in the case of nonlinear ODEs, but k stages are sufficient in the case of linear ODEs.

9.6 Linear Analysis of Implicit Runge–Kutta Schemes

Analysis of the linear stability of implicit Runge–Kutta schemes is more complicated. Consider again the linear equation

$$\frac{\partial u}{\partial t} = \alpha u.$$

Setting $z = \alpha \Delta t$, the general s-stage Runge–Kutta scheme is

$$\xi_i = u^n + z \sum_{j=1}^{s} a_{ij} \xi_j,$$

$$u^{n+1} = u^n + z \sum_{i=1}^{s} b_i \xi_i.$$

Using vector notation, let

$$\xi = \begin{bmatrix} \xi_1 \\ \xi_2 \\ \vdots \\ \xi_s \end{bmatrix}, \quad \mathbf{e} = \begin{bmatrix} 1 \\ 1 \\ \vdots \\ 1 \end{bmatrix}.$$

Then,

$$\xi = u^n \mathbf{e} + zA\xi,$$

$$u^{n+1} = u^n \mathbf{e} + zb^T \xi,$$

where A is the Butcher matrix. The first equation may be solved to give

$$\xi = u^n (I - zA)^{-1} \mathbf{e}.$$

Hence,

$$u^{n+1} = r(z) u^n,$$

where $r(z)$ is the stability function defined as

$$r(z) = 1 + b^T (I - zA)^{-1} \mathbf{e}.$$

We can represent the inverse of $I - zA$ as

$$(I - zA)^{-1} = \frac{\operatorname{adj}(I - zA)}{\det(I - zA)},$$

where each entry of the adjunct matrix is a determinant of an $(s-1) \times (s-1)$ matrix and is hence a polynomial of degree $s-1$ in z. Thus, for the general s-stage implicit scheme,

$$r(z) = \frac{N(z)}{D(z)},$$

where $N(z)$ and $D(z)$ are both polynomials of degree s, and the stability region is the region of z in the complex plane for which

$$|r(z)| \leq 1.$$

In the case of an explicit scheme, A is strictly lower triangular, and $I - zA$ is lower triangular with diagonal elements $a_{jj} = 1$. Thus, $D(z) = 1$, and $r(z)$ reduces to a polynomial of degree s. Then $|r(z)|$ is unbounded as $z \to \infty$, and consequently no explicit Runge–Kutta scheme can be A-stable. On the other hand, A- and L-stable implicit Runge–Kutta schemes can be devised. One requirement is that $D(z)$ must have no zeros in the left half-plane. Moreover, to attain p^{th} order accuracy, $r(z)$ must be an approximation to $\mathbf{e}^z$ such that the first term in an expansion of $r(z) - \mathbf{e}^z$ must be a term in z^{p+1}. Implicit Runge–Kutta schemes will be further discussed in Chapter 11.

$$
\begin{array}{c|ccccccc}
0 & & & & & & & \\
\frac{1}{5} & \frac{1}{5} & & & & & & \\
\frac{3}{10} & \frac{3}{40} & \frac{9}{40} & & & & & \\
\frac{4}{5} & \frac{44}{45} & -\frac{56}{15} & \frac{32}{9} & & & & \\
\frac{8}{9} & \frac{19372}{6561} & -\frac{25360}{2187} & \frac{64448}{6561} & -\frac{212}{729} & & & \\
1 & \frac{9017}{3168} & -\frac{355}{33} & \frac{46732}{5247} & \frac{49}{176} & -\frac{5103}{18656} & & \\
1 & \frac{35}{384} & 0 & \frac{500}{1113} & \frac{125}{192} & -\frac{2187}{6784} & \frac{11}{84} & \\
\hline
 & \frac{35}{384} & 0 & \frac{500}{1113} & \frac{125}{192} & -\frac{2187}{6784} & \frac{11}{84} & 0 \\
 & \frac{5179}{57600} & 0 & \frac{7571}{16695} & \frac{393}{640} & -\frac{92097}{339200} & \frac{187}{2100} & \frac{1}{40}
\end{array}
$$

Figure 9.3 Butcher tableau for the Dormand–Price scheme. The first row of b coefficients gives the fifth order accurate solution and the second row gives the fourth order accurate solution.

9.7 Time Step Control with Embedded RK Pairs

In software packages for time integration, it is the usual practice to provide automatic control of the size of the time steps based on error estimates, which can be obtained by comparing the updates from two schemes with different orders of accuracy. This is particularly easy to implement with RK methods because each new step is completely independent of the previous steps. Suppose that the true solution when advancing Δt from w^n is w_{exact}, and we have two schemes of order p and $p+1$ such that

$$
\begin{aligned}
\boldsymbol{w}_{(1)}^{m+1} &= \boldsymbol{w}_{\text{exact}} + l_1 \Delta t^{p+1} + \mathcal{O}(\Delta t^{p+2}), \\
\boldsymbol{w}_{(2)}^{m+1} &= \boldsymbol{w}_{\text{exact}} + \mathcal{O}(\Delta t^{p+2}).
\end{aligned}
$$

Then we can estimate the error for one step of the lower order scheme as

$$
\boldsymbol{w}_{(1)}^{m+1} - \boldsymbol{w}_{(2)}^{n+1} + \mathcal{O}(\Delta t^{p+2}).
$$

If this error is larger than some specified threshold, Δt is reduced, and the step is repeated until the tolerance is satisfied. If the error is smaller than some fraction of the threshold, Δt may be increased. If the result of the lower order scheme is accepted for each time step, then the result is accompanied by the computed error estimate. It may be preferred, however, to accept the higher order estimate for continuation of the calculation because this should reduce the overall error, although we then no longer have a precise error estimate.

The error control procedure would be computationally very expensive if two completely independent schemes were used. It is possible, however, to devise embedded pairs of RK schemes that use the same fractional steps and recover schemes of different order by different combinations of the step values, so that the step control is achieved with a minimal computational cost. This was first shown by Fehlberg (1969). A widely used explicit embedded RK pair is due to Dormand and Prince (1980). This has seven stages but uses only six function evaluations because the last stage is evaluated at the same time instant as the first stage of the next step. The pair provides fourth and fifth order accurate solutions, and it is the usual practice to continue with the fifth order solution because Dormand and Prince optimized the coefficients to minimize the error of the higher order solution. The Butcher tableau for this scheme is displayed in Figure 9.3.

9.8 Model Problem for Stability Analysis

The following model problem provides a useful tool for the stability analysis of time integration methods for convection-dominated problems. Consider a semi-discretization of the linear advection equation

$$\frac{\partial u}{\partial t} + a\frac{\partial u}{\partial x} = 0, \tag{9.6}$$

on a uniform grid with mesh interval Δx. Using central differences, augmented by an artificial diffusive term $\sim \Delta x^3 \frac{\partial^3 u}{\partial x^3}$, this can be written as

$$\Delta t \frac{dv_j}{dt} = \frac{\lambda}{2}(v_{j+1} - v_{j-1}) - \lambda\mu(v_{j+2} - 4v_{j+1} + 6v_j - 4v_{j-1} + v_{j-2}),$$

where Δt is the assumed time step, and λ is the CFL number

$$\lambda = a\frac{\Delta t}{\Delta x}.$$

For a Fourier mode

$$v_j = \hat{v}(t)e^{i\omega x_j},$$

$$\Delta t \frac{d\hat{v}}{dt} = z\hat{v},$$

where z is the Fourier symbol of the space discretization. With $\xi = \omega\Delta x$,

$$z(\xi) = -\lambda\big(i \sin\xi + 4\mu(1 - \cos\xi)^2\big). \tag{9.7}$$

For stability of the fully discrete scheme, the locus of z, as ξ is varied, should be inside the stability region of the time integration method. The permissible CFL number thus depends on the stability interval along the imaginary axis as well as the negative real axis. Figure 9.4 shows the stability region of the standard fourth

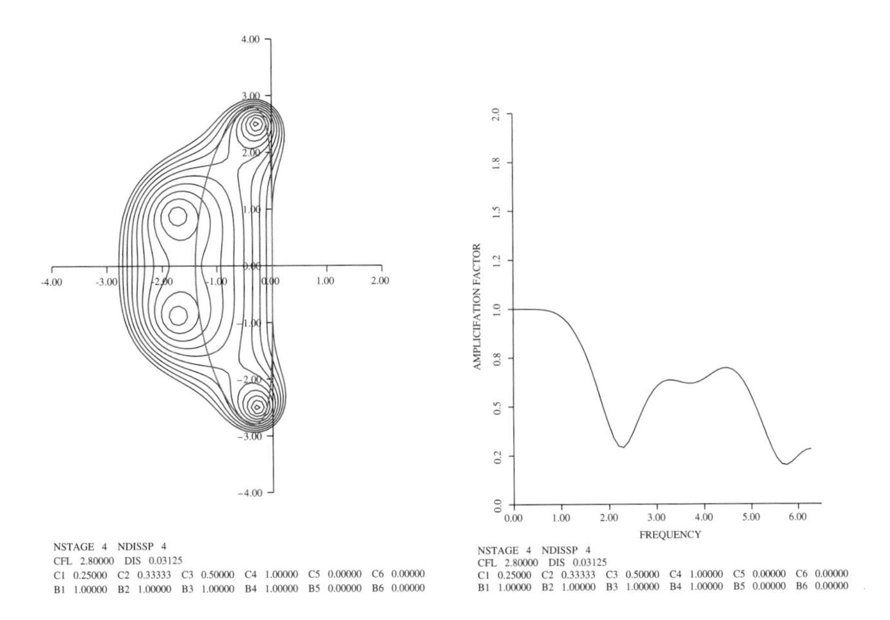

Figure 9.4 Stability region and amplification factor $|g|$ for the standard four-stage Runge–Kutta scheme (9.4). The oval curve is the locus of the Fourier symbol for linear advection with CFL number 2.8.

order RK4 scheme defined by (9.4) or the tableau in figure 9.2. It also shows the locus of the Fourier symbol for a CFL number of 2.8, $\mu = 1/32$ and the corresponding amplification factor.

9.9 Time Step Estimate for the Euler Equations

When an explicit scheme is used, it is necessary to estimate the maximum time step for which the scheme will be stable. In the case of the multi-dimensional gas dynamics equations, there is no simple formula that guarantees a bound on the time step. The analysis in this section provides an estimate that has proved useful in practice.

Consider central differencing for the two-dimensional Euler equations in Cartesian coordinates $x_j = j\Delta x, y_k = k\Delta y$, with velocity components u and v in the x and y directions.

$$\frac{d\boldsymbol{w}}{dt} + \frac{\partial \boldsymbol{f}}{\partial x} + \frac{\partial \boldsymbol{g}}{\partial y} = 0.$$

or

$$\frac{d\boldsymbol{w}}{dt} + A\frac{\partial \boldsymbol{w}}{\partial x} + B\frac{\partial \boldsymbol{w}}{\partial y} = 0.$$

where

$$A = \frac{\partial \boldsymbol{f}}{\partial \boldsymbol{w}}, \quad B = \frac{\partial \boldsymbol{g}}{\partial \boldsymbol{w}}.$$

Treating this as a linear equation, consider a Fourier mode of the form

$$\hat{\mathbf{w}} e^{i\omega_x x_j} e^{i\omega_y y_k}.$$

Then

$$\frac{d\hat{\mathbf{w}}}{dt} = Z\hat{\mathbf{w}},$$

where for wave numbers $\xi = \omega_x \Delta x$ and $\eta = \omega_x \Delta y$,

$$Z = i\left(\frac{A}{\Delta x}\sin\xi + \frac{B}{\Delta y}\sin\eta\right),$$

and for $\xi = \eta = \pi/2$,

$$Z = i\left(\frac{A}{\Delta x} + \frac{B}{\Delta y}\right).$$

With time integration by the standard RK4 defined in (9.4), we have

$$\hat{\mathbf{w}}^{n+1} = G\hat{\mathbf{w}}^n,$$

where

$$G = \left(I + \Delta t Z + \frac{(\Delta t Z)^2}{2} + \frac{(\Delta t Z)^3}{6} + \frac{(\Delta t Z)^4}{24}\right)$$

and

$$||G|| \leq 1 + \Delta t||Z|| + \frac{(\Delta t||Z||)^2}{2} + \frac{(\Delta t||Z||)^3}{6} + \frac{(\Delta t||Z||)^4}{24}.$$

Since the stability interval along the imaginary axis extends between $\pm i2\sqrt{2}$,

$$||G|| \leq 1 \quad \text{if} \quad \Delta t||Z|| \leq 2\sqrt{2}.$$

Using the symmetrizing variables defined in (2.19), A and B are both symmetric with real eigenvalues and a complete set of orthogonal eigenvectors. Thus, in the Euclidean norm,

$$||Z||_2 = \rho(Z),$$

where ρ is the spectral radius. Also

$$||Z||_2 \leq \left|\left|\frac{A}{\Delta x}\right|\right|_2 + \left|\left|\frac{B}{\Delta y}\right|\right|_2$$

by the triangle inequality, while

$$\left|\left|\frac{A}{\Delta x}\right|\right|_2 = \frac{\rho(A)}{\Delta x} = \frac{|u| + c}{\Delta x}$$

$$\left|\left|\frac{B}{\Delta y}\right|\right|_2 = \frac{\rho(B)}{\Delta y} = \frac{|v| + c}{\Delta y};$$

so the scheme will be stable if

$$\Delta t \leq \frac{2\sqrt{2}}{\frac{|u|+c}{\Delta x} + \frac{|v|+c}{\Delta y}}.$$

This estimate is quite conservative. It should also be remembered that for the nonlinear equation, A and B will change at each RK stage, but this change is generally small within a single time step. The analysis is easily extended to the three-dimensional case with mesh interval Δz and velocity component w by adding $\frac{|w|+c}{\Delta z}$ to the denominator.

The corresponding estimate for a finite volume scheme is

$$\Delta t \leq \frac{2\sqrt{2}V}{\sum_{k=1}^{3}(q_n + c)_k S_k},$$

where V is the cell volume, S_k is the averaged face area of the faces in the kth direction, and $(q_n + c)_k$ is the sum of the averaged normal velocity and speed of sound across the face cell in the kth direction.

Estimates can be derived for the Navier–Stokes equations by using the symmetrizing transformation for the inviscid and viscous Jacobian matrices developed by Gottlieb and Abarbanel (Abarbanel & Gottlieb 1981), as may be found in the thesis of L. Martinelli (1987).

9.10 Single Step Implicit Schemes

The prototype two-level implicit scheme can be formulated by estimating $\frac{\partial \boldsymbol{w}}{\partial t}$ at $t + \epsilon \Delta t$ as a linear combination of $\boldsymbol{R}(\boldsymbol{w}^n)$ and $\boldsymbol{R}(\boldsymbol{w}^{n+1})$. The resulting equation

$$\boldsymbol{w}^{n+1} = \boldsymbol{w}^n - \Delta t \left\{ (1-\epsilon)\, \boldsymbol{R}\left(\boldsymbol{w}^n\right) + \epsilon \boldsymbol{R}\left(\boldsymbol{w}^{n+1}\right) \right\}, \tag{9.8}$$

represents the first order accurate backward Euler scheme if $\epsilon = 1$ and the second order accurate trapezoidal rule if $\epsilon = \frac{1}{2}$. This equation calls for the solution of a large number of coupled nonlinear equations for the unknown state vector $\boldsymbol{w}^{n+1}$. For the purpose of analyzing these equations, it is convenient to consider the case of the two-dimensional conservation law

$$\frac{\partial \boldsymbol{w}}{\partial t} + \frac{\partial}{\partial x}\boldsymbol{f}(\boldsymbol{w}) + \frac{\partial}{\partial y}\boldsymbol{g}(\boldsymbol{w}) = 0,$$

discretized on a Cartesian grid. Then,

$$\boldsymbol{R}(\boldsymbol{w}) = D_x \boldsymbol{f}(\boldsymbol{w}) + D_y \boldsymbol{g}(\boldsymbol{w}),$$

where D_x and D_y denote the spatial difference operators approximating $\frac{\partial}{\partial x}$ and $\frac{\partial}{\partial y}$.

The implicit equation (9.8) can most readily be solved either via linearization or by resorting to an iterative method. Defining the correction vector

$$\delta \boldsymbol{w} = \boldsymbol{w}^{n+1} - \boldsymbol{w}^{n},$$

(9.8) can be linearized by approximating the local fluxes as

$$\boldsymbol{f}(\boldsymbol{w}^{n+1}) \sim \boldsymbol{f}(\boldsymbol{w}^{n}) + A\delta \boldsymbol{w}, \quad \boldsymbol{g}(\boldsymbol{w}^{n+1}) \sim \boldsymbol{g}(\boldsymbol{w}^{n}) + B\delta \boldsymbol{w},$$

where A and B are the Jacobian matrices,

$$A = \frac{\partial \boldsymbol{f}}{\partial \boldsymbol{w}}, \quad B = \frac{\partial \boldsymbol{g}}{\partial \boldsymbol{w}},$$

and the neglected terms are $\mathcal{O}(\|\delta w\|^2)$. This leads to a linearized implicit scheme that has the local form

$$\left\{I + \epsilon \Delta t (D_x A + D_y B)\right\} \delta \boldsymbol{w} + \Delta t \boldsymbol{R}(\boldsymbol{w}) = 0. \tag{9.9}$$

Here, we can recognize $D_x A + D_y B$ as $\frac{\partial \boldsymbol{R}}{\partial \boldsymbol{w}}$.

If one sets $\epsilon = 1$ and lets $\Delta t \to \infty$, this reduces to the Newton iteration for the steady state problem (9.1), which has been successfully used in two-dimensional calculations (Giles, Drela, & Thompkins 1985,Venkatakrishnan 1988). In the three-dimensional case with, say, an $N \times N \times N$ mesh, the bandwidth of the matrix that must be inverted is of order N^2. Direct inversion requires a number of operations proportional to the number of unknowns multiplied by the square of the bandwidth, resulting in a complexity of the order of N^7. This is prohibitive, and forces recourse to either an approximate factorization method or an iterative solution method.

The main possibilities for approximate factorization are the alternating direction and LU decomposition methods. The alternating direction method, which may be traced back to the work of Gourlay and Mitchell (1966), was given an elegant formulation for nonlinear problems by Beam and Warming (1976). In the two-dimensional case, (9.9) is replaced by

$$(I + \epsilon \Delta t D_x A)(I + \epsilon \Delta t D_y B)\delta \boldsymbol{w} + \Delta t \boldsymbol{R}(\boldsymbol{w}) = 0. \tag{9.10}$$

This may be solved in two steps:

$$(I + \epsilon \Delta t D_x A)\delta \boldsymbol{w}^* = -\Delta t \boldsymbol{R}(\boldsymbol{w})$$

$$(I + \epsilon \Delta t D_y B)\delta \boldsymbol{w} = \delta \boldsymbol{w}^*.$$

If we use a central difference approximation, each step requires block tridiagonal matrix inversions and may be performed in $\mathcal{O}(N^2)$ operations on an $N \times N$ mesh. The algorithm is amenable to vectorization by simultaneous solution of the tridiagonal system of equations along parallel coordinate lines. The alternating direction method has well balanced errors. If we take $\epsilon = \frac{1}{2}$, the time discretization error is $\mathcal{O}(\Delta t^2)$. Also assuming $||\delta \boldsymbol{w}|| = \mathcal{O}(\Delta t)$, the linearization error is $\mathcal{O}(\Delta t^2)$. Finally, the factorization error $\epsilon^2 \Delta t^2 D_x A \ D_y B$ is also $\mathcal{O}(\Delta t^2)$ in comparison with the linearized scheme. Thus, this method can be used as a relatively inexpensive implicit scheme for time-accurate calculations. It cannot be used with very large Δt, however, because then the factorization error becomes dominant and destroys the time accuracy. The method has been refined to a high level of efficiency by Pulliam and Steger (1985).

In the case of the scalar equation

$$\frac{\partial u}{\partial t} + A\frac{\partial u}{\partial x} + B\frac{\partial u}{\partial y} = 0,$$

a von Neumann analysis provides a useful insight of the properties of the linearized and alternating direction schemes using central differences. Substituting

$$u_{jk}^n = g^n e^{iw_x x_j} e^{iw_y y_k},$$

where g is the amplification factor, we find that for the linearized scheme,

$$(1 + i\epsilon(\lambda_x \sin\xi_x + \lambda_y \sin\xi_y))(g - 1) = -i(\lambda_x \sin\xi_x + \lambda_y \sin\xi_y),$$

where

$$\lambda_x = \frac{A\Delta t}{\Delta x},\ \lambda_y = \frac{B\Delta t}{\Delta y},\ \xi_x = w_x\Delta x,\ \xi_y = w_y\Delta y.$$

Hence,

$$g = \frac{1 - i(1-\epsilon)(\lambda_x \sin\xi_x + \lambda_y \sin\xi_y)}{1 + i\epsilon(\lambda_x \sin\xi_x + \lambda_y \sin\xi_y)}.$$

Evidently, $|g| = 1$ when $\epsilon = \frac{1}{2}$, while $|g| > 1$ if $\epsilon < \frac{1}{2}$, and $|g| < 1$ if $\epsilon > \frac{1}{2}$. For the alternating direction scheme,

$$\begin{aligned}(1 + i\epsilon(\lambda_x \sin\xi_x + \lambda_y \sin\xi_y) - \epsilon^2\lambda_x\lambda_y \sin\xi_x \sin\xi_y)(g - 1)\\ = -i(\lambda_x \sin\xi_x + \lambda_y \sin\xi_y),\end{aligned}$$

yielding

$$g = \frac{\alpha - i(1-\epsilon)(\lambda_x \sin\xi_x + \lambda_y \sin\xi_y)}{\alpha + i\epsilon(\lambda_x \sin\xi_x + \lambda_y \sin\xi_y)},$$

where

$$\alpha = 1 - \epsilon^2\lambda_x\lambda_y \sin\xi_x \sin\xi_y.$$

Again $|g| = 1$ when $\epsilon = \frac{1}{2}$, and the scheme is stable if $\epsilon > \frac{1}{2}$ and unstable if $\epsilon < \frac{1}{2}$. A similar von Neumann analysis indicates, however, that in the three-dimensional case, the alternating direction scheme with central differences is unstable. In practice, the scheme is quite successfully used when it is stabilized by the addition of artificial diffusive terms.

The idea of the LU decomposition method (Jameson & Turkel 1981) is to replace the operator in (9.9) by the product of lower and upper block triangular factors L and U,

$$LU\delta \boldsymbol{w} + \Delta t \boldsymbol{R}(\boldsymbol{w}) = 0.$$

Two factors are used independent of the number of dimensions, and the inversion of each can be accomplished by inversion of its diagonal blocks. The method can be conveniently illustrated by considering a one-dimensional example. Let the Jacobian matrix $A = \frac{\partial f}{\partial w}$ be split as

$$A = A^{+} + A^{-},$$

where the eigenvalues of A^{+} and A^{-} are positive and negative, respectively. Then, we can take

$$L = I + \epsilon \Delta t D_x^{-} A^{+}, \quad U = I + \epsilon \Delta t D_x^{+} A^{-},$$

where D_x^{+} and D_x^{-} denote forward and backward difference operators approximating $\frac{\partial}{\partial x}$. This leads to a factorization error $\epsilon^2 \Delta t^2 D_x^{-} A^{+} D_x^{+} A^{-}$. The reason for splitting A is to ensure the diagonal dominance of L and U, independent of Δt. Otherwise, stable inversion of both factors will only be possible for a limited range of Δt. A crude choice is

$$A^{\pm} = \frac{1}{2}(A \pm \alpha I), \tag{9.11}$$

where α is at least equal to the spectral radius of A. If flux splitting is used in the calculation of the residual, it is natural to use the corresponding splitting for L and U.

If one chooses to adopt an iterative technique to solve the implicit equations, the principal alternatives are variants of the Jacobi and Gauss–Seidel methods. These may be applied to either the nonlinear equation (9.8) or the linearized equation (9.9). A Jacobi method of solving (9.8) can be formulated by regarding it as an equation

$$\boldsymbol{w} - \boldsymbol{w}^{(0)} + \epsilon \Delta t \boldsymbol{R}(\boldsymbol{w}) + (1 - \epsilon) \Delta t \boldsymbol{R}(\boldsymbol{w}^{(0)}) = 0,$$

to be solved for $\boldsymbol{w}$. Here $\boldsymbol{w}^{(0)}$ is a fixed value obtained as the result of the previous time step. Now, using bracketed superscripts to denote the iterations, we have

$$\boldsymbol{w}^{(0)} = \boldsymbol{w}^{n}$$

$$\boldsymbol{w}^{(1)} = \boldsymbol{w}^{(0)} + \Delta t \boldsymbol{R}(\boldsymbol{w}^{(0)}),$$

and for $k > 1$,

$$\boldsymbol{w}^{(k+1)} = \boldsymbol{w}^{(k)} + \sigma_{k+1} \left\{ \left(\boldsymbol{w}^{(k)} - \boldsymbol{w}^{(0)} + \epsilon \Delta t \boldsymbol{R}(\boldsymbol{w}^{(k)}) + (1 - \epsilon) \Delta t \boldsymbol{R}(\boldsymbol{w}^{(0)}) \right) \right\},$$

where the parameters σ_k can be chosen to optimize convergence. Finally, if we stop after m iterations,

$$\boldsymbol{w}^{n+1} = \boldsymbol{w}^{(m)}.$$

We can express $\boldsymbol{w}^{(k+1)}$ as

$$\boldsymbol{w}^{(k+1)} = \boldsymbol{w}^{(0)} + (1 + \sigma_{k+1})(\boldsymbol{w}^{(k)} - \boldsymbol{w}^{(0)}) + \sigma_{k+1} \left\{ \left(\epsilon \Delta t \boldsymbol{R}(\boldsymbol{w}^{(k)}) + (1 - \epsilon) \Delta t \boldsymbol{R}(\boldsymbol{w}^{(0)}) \right) \right\}.$$

Since

$$\boldsymbol{w}^{(1)} - \boldsymbol{w}^{(0)} = \sigma_1 \Delta t \boldsymbol{R}(\boldsymbol{w}^{(0)}),$$

it follows that for all k, we can express $(\boldsymbol{w}^{(k)} - \boldsymbol{w}^{(0)})$ as a linear combination of $\boldsymbol{R}(\boldsymbol{w}^{(j)}), j < k$. Thus, we recover a scheme that belongs to the general class of explicit Runge–Kutta schemes described by (9.5). These have the advantages that they can be computed simultaneously and are readily amenable to vector and parallel processing.

Symmetric Gauss–Seidel schemes have proved to be particularly effective (Chakravarthy 1984, Hemker & Spekreijse 1984, MacCormack 1985, Yoon & Jameson 1987, Rieger & Jameson 1988). Following the analysis of Jameson (1986a), consider the case of a flux-split scheme in one dimension, for which

$$\boldsymbol{R}(\boldsymbol{w}) = D_x^+ \boldsymbol{f}^-(\boldsymbol{w}) + D_x^- \boldsymbol{f}^+(\boldsymbol{w}),$$

where the flux is split so that the Jacobian matrices

$$A^+ = \frac{\partial \boldsymbol{f}^+}{\partial \boldsymbol{w}} \quad \text{and} \quad A^- = \frac{\partial \boldsymbol{f}^-}{\partial \boldsymbol{w}}, \tag{9.12}$$

have positive and negative eigenvalues, respectively. Now, the linearized equation (9.9) becomes

$$\left\{I + \epsilon \Delta t \left(D_x^+ A^- + D_x^- A^+\right)\right\} \delta \boldsymbol{w} + \Delta t \boldsymbol{R}(\boldsymbol{w}) = 0.$$

At the jth mesh point, this is

$$\left\{I + \tau\left(A_j^+ - A_j^-\right)\right\}\delta \boldsymbol{w}_j + \tau A_{j+1}^- \delta \boldsymbol{w}_{j+1} - \tau A_{j-1}^+ \delta \boldsymbol{w}_{j-1} + \Delta t \boldsymbol{R}_j = 0,$$

where

$$\tau = \epsilon \frac{\Delta t}{\Delta x}.$$

Set $\delta \boldsymbol{w}_j^{(0)} = 0$. A two sweep symmetric Gauss–Seidel scheme is then

$$\left\{I + \tau\left(A_j^+ - A_j^-\right)\right\}\delta \boldsymbol{w}_j^{(1)} - \tau A_{j-1}^+ \delta \boldsymbol{w}_{j-1}^{(1)} + \Delta t \boldsymbol{R}_j = 0, \tag{9.13}$$

$$\left\{I + \tau\left(A_j^+ - A_j^-\right)\right\}\delta \boldsymbol{w}_j^{(2)} + \tau A_{j+1}^- \delta \boldsymbol{w}_{j+1}^{(2)} - \tau A_{j-1}^+ \delta \boldsymbol{w}_{j-1}^{(1)} + \Delta t \boldsymbol{R}_j = 0. \tag{9.14}$$

Subtracting (9.13) from (9.14), we find that

$$\left\{I + \tau\left(A_j^+ - A_j^-\right)\right\}\delta \boldsymbol{w}_j^{(2)} + \tau A_{j+1}^- \delta \boldsymbol{w}_{j+1}^{(2)} = \left\{I + \tau\left(A_j^+ - A_j^-\right)\right\}\delta \boldsymbol{w}_j^{(1)}.$$

Define the lower triangular, upper triangular, and diagonal operators L, U, and D as

$$\begin{aligned} L &= I - \tau A^- + \epsilon \Delta t D_x^- A^+, \\ U &= I + \tau A^+ + \epsilon \Delta t D_x^+ A^-, \\ D &= I + \tau(A^+ - A^-). \end{aligned}$$

It follows that the scheme can be written as

$$L D^{-1} U \, \delta \boldsymbol{w} = -\Delta t \, \boldsymbol{R}(\boldsymbol{w}).$$

Commonly the iteration is terminated after one double sweep. The scheme is then a variation of an LU implicit scheme, and hence, it has come to be known as the LUSGS method.

If we use the simple choice (9.11) for $A^{\pm}$, D reduces to a scalar factor, and the scheme requires no inversion. This is a significant advantage for the treatment of flows with chemical reactions including large numbers of species and a correspondingly large number of equations.

It is important to note, however, that the LUSGS scheme does not preserve time accuracy. To see this, it is convenient to express the scheme in terms of the shift operators

$$E^{+}w_j = \boldsymbol{w}_{j+1}, \quad E^{-}\boldsymbol{w}_j = \boldsymbol{w}_{j-1}.$$

Then, the factors take the form

$$L = D - \tau E^{-}A^{+},$$
$$U = D + \tau E^{+}A^{-}.$$

Multiplying out, we find that

$$\begin{aligned}&(D - \tau E^{-}A^{+})D^{-1}(D + \tau E^{+}A^{-})\\&\quad = D + \tau(E^{+}A^{-} - E^{-}A^{+}) - \tau^2 E^{-}A^{+}D^{-1}E^{+}A^{-},\end{aligned}$$

where the first two terms are an alternative representation of

$$I + \epsilon\frac{\Delta t}{\Delta x}(D_x^{+}A^{-} + D_x^{-}A^{+}),$$

and the third term is the factorization error. In contrast to the LU scheme, the magnitude of this is $\mathcal{O}(1)$ because $\alpha = \mathcal{O}\left(\frac{\Delta t}{\Delta x}\right)$, whereas in the LU scheme, the factor $\frac{1}{\Delta x^2}$ is absorbed in the difference operators D_x^{+} and D_x^{-}. Accordingly, the LUSGS scheme cannot be used directly for time-accurate calculations, but it yields comparatively fast convergence in steady state calculations, and it may be used for the inner iterations of a fully implicit time stepping scheme, as will be discussed in Chapter 11.

A similar modification may be made to the ADI scheme (9.10) to improve convergence to a steady state, yielding the diagonally dominant ADI (DDADI) scheme, first introduced by Bardina and Lombard (1987) and MacCormack (1997). Consider a two-dimensional flux split scheme with the residual

$$\boldsymbol{R}(\boldsymbol{w}) = D_x^{+}\boldsymbol{f}^{-} + D_x^{-}\boldsymbol{f}^{+} + D_y^{+}\boldsymbol{g}^{-} + D_y^{-}\boldsymbol{g}^{+},$$

where the superscripts denote forward and backward difference operators. The standard ADI scheme is

$$(I + \epsilon\Delta t(D_x^{+}A^{-} + D_x^{-}A^{+}))(I + \epsilon\Delta t(D_y^{+}B^{-} + D_y^{-}B^{+}))\delta\boldsymbol{w} = -\Delta t\boldsymbol{R}(\boldsymbol{w}),$$

where

$$A^{\pm} = \frac{\partial \boldsymbol{f}^{\pm}}{\partial \boldsymbol{w}}, \quad B^{\pm} = \frac{\partial \boldsymbol{g}^{\pm}}{\partial \boldsymbol{w}}.$$

In terms of the shift operators $E_x^{\pm}$ and $E_y^{\pm}$ in the x and y coordinate directions, the linearized implicit scheme may be written as

$$(D + \tau_x(E_x^+ A^- + E_x^- A^+) + \tau_y(E_y^+ B^- + E_y^- B^+))\delta \boldsymbol{w} = -\Delta t \boldsymbol{R}(\boldsymbol{w}),$$

where

$$\tau_x = \epsilon \frac{\Delta t}{\Delta x}, \quad \tau_y = \epsilon \frac{\Delta t}{\Delta y},$$

and

$$D = I + \tau_x(A^+ - A^-) + \tau_y(B^+ - B^-).$$

The DDADI scheme is obtained by removing a factor D before carrying out the factorization and then reintroducing it to obtain

$$(D + \tau_x(E_x^+ A^- + E_x^- A^+))D^{-1}(D + \tau_y(E_y^+ B^- + E_y^- B^+))\delta w = -\Delta t R(w).$$

Like the LUSGS scheme, this produces faster convergence to a steady state at the cost of losing time accuracy.

10 Steady State Problems

10.1 Introduction

It is a common practice to solve steady state problems by integrating the time dependent equations until they reach a steady state. The first question that must be addressed is whether the equations of gas dynamics necessarily reach a steady state. We know that small pressure disturbances are actually propagated as sound waves. These are well described by the acoustic equation, and in a confined domain, the waves might be reflected repeatedly. Solutions of the acoustic equation actually preserve the total energy in an infinite domain. Considering the two-dimensional case, the governing equation is

$$\phi_{tt} = c^2(\phi_{xx} + \phi_{yy}), \tag{10.1}$$

where c is the speed of sound and ϕ is the disturbance potential. This may be recovered from the unsteady potential flow equation

$$\phi_{tt} + 2u\phi_{xt} + 2v\phi_{yt} = (c^2 - u^2)\phi_{xx} - 2uv\phi_{xy} + (c^2 - v^2)\phi_{yy}, \tag{10.2}$$

by setting the velocity components u and v to zero. Also in this limit, we may take the speed of sound to be constant. To prove conservation of energy, we multiply (10.1) by ϕ_t and integrate by parts over the infinite domain to obtain

$$\int_{-\infty}^{\infty}\int_{-\infty}^{\infty} \phi_{tt}\phi_t \, dx \, dy = -c^2 \int_{-\infty}^{\infty}\int_{-\infty}^{\infty} (\phi_x\phi_{xt} + \phi_y\phi_{yt}) \, dx \, dy,$$

where ϕ is assumed to vanish in the far field so there are no boundary terms. It follows that

$$\frac{d}{dt}\int_{-\infty}^{\infty}\int_{-\infty}^{\infty} \frac{1}{2}(\phi_t^2 + c^2(\phi_x^2 + \phi_y^2)) \, dx \, dy = 0,$$

and the integral in this equation is a measure of energy.

For an external flow, the only mechanism for the near field to reach a steady state is by radiation of energy to the far field. Thus, in general, we can expect very slow convergence to a steady state, unless we modify the equations in a way that preserves the same steady state solution while increasing the rate of convergence.

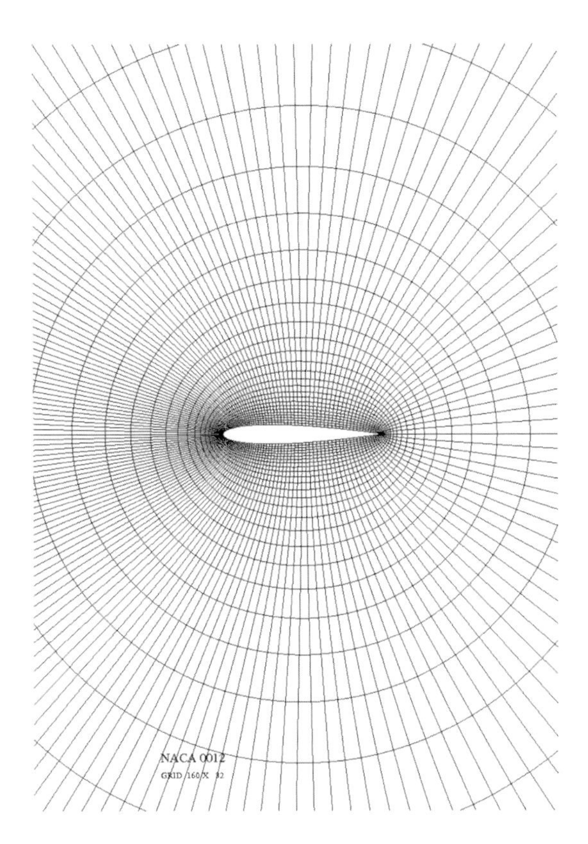

Figure 10.1 Mesh around NACA 0012 airfoil, 160×32 cells.

In order to illustrate what happens when the Euler equations are integrated to a steady state without any modification, Figures 10.1 and 10.2 show the results of a time-accurate calculation of the flow past an NACA 0012 airfoil at Mach 0.8 and an angle of attack 0f 1.25 degrees. This calculation was performed with the Jameson–Schmidt–Turkel (JST) scheme and the standard RK4 scheme defined by (9.4) or the tableau in Figure 10.2. The time step was defined by setting a maximum limit 2.0 for the CFL number anywhere in the computational domain. Figure 10.1 shows the grid with $160 \times 32 = 5,120$ cells. The O-topology leads to a concentration of very small cells at the trailing edge, which benefits the accuracy but restricts the allowable time step. Figures 10.2(a) and (b) shows the result after 200,000 time steps. Figure 10.2 (c) shows the convergence history measured by the density residual. It also shows the evolution of the lift coefficient as a function of its final value according to the scale on the right of the figure. It can be seen that more than 100,000 time steps are required for the lift coefficient to approach a steady state.

A variety of convergence acceleration techniques will be discussed in this chapter. First, however, we examine ways in which the time integration process itself may be optimized for steady state calculations. In this case, the accuracy of the time

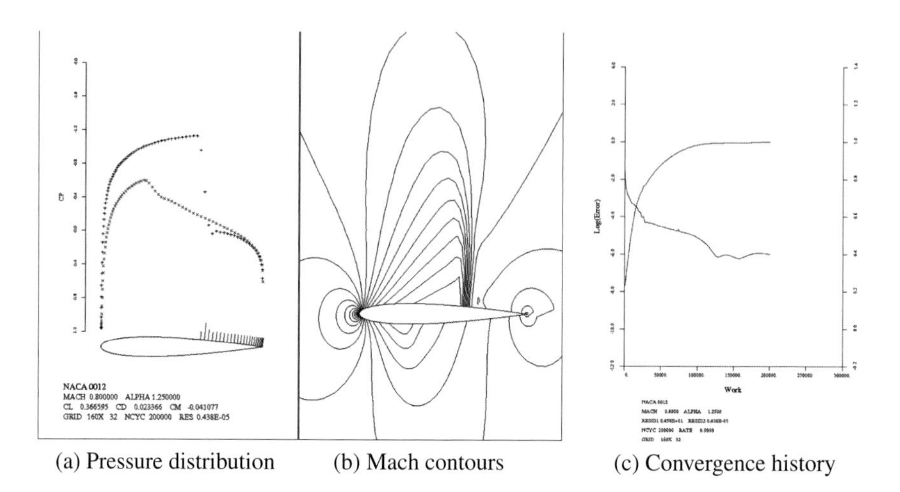

(a) Pressure distribution (b) Mach contours (c) Convergence history

Figure 10.2 Flow solution past an NACA 0012 airfoil using fixed Δt after 200,000 time steps.

integration scheme is no longer a consideration. This enables the use of modified RK schemes of reduced computational complexity. Moreover, the schemes may be tailored to increase the allowable time step, thereby promoting more rapid convergence to a steady state.

10.2 Linear Runge–Kutta Schemes with *m*th Order Accuracy

An immediate simplification of the time integration process is to introduce m-stage schemes of minimal complexity that are mth order accurate for the linear model problem

$$\frac{du}{dt} = \alpha u.$$

For a single time step, these have the form

$$\begin{aligned} u^{(1)} &= u^{(0)} + \frac{\Delta t}{m} \alpha u^{(0)}, \\ u^{(2)} &= u^{(0)} + \frac{\Delta t}{m-1} \alpha u^{(1)}, \\ \vdots &= \vdots \\ u^{(m)} &= u^{(0)} + \Delta t \, \alpha u^{(m-1)}. \end{aligned} \tag{10.3}$$

A Taylor series expansion gives

$$u(t + \Delta t) = u(t) + \Delta t \frac{du}{dt} + \cdots + \frac{\Delta t^m}{m!} \frac{d^m u}{dt^m} + \mathcal{O}(\Delta t)^{m+1}.$$

According to (10.3),

$$u^{(2)} = u^{(0)} + \frac{\Delta t}{m-1}\alpha\left(u^{(0)} + \frac{\Delta t}{m}\alpha u^{(0)}\right)$$
$$= u^{(0)} + \frac{\Delta t}{m-1}\alpha u^{(0)} + \frac{\Delta t^2}{(m-1)m}\alpha^2 u^{(0)},$$

and repeated substitution yields

$$u^{(m)} = u^{(0)} + \Delta t\ \alpha u^{(0)} + \frac{\Delta t}{2!}\alpha^2 u^{(0)} + \cdots + \frac{\Delta t^m}{m!}\alpha^m u^{(0)},$$

where

$$\alpha^k u^{(0)} = \frac{d^k u}{dt^k}.$$

Thus, $u^{(m)}$ matches the Taylor series expansion of $u(t+\Delta t)$ to m terms, and the scheme is m^{th} order accurate. Also, setting $z = \alpha\Delta t$,

$$u^{(m)} = g(z)u^{(0)},$$

where $g(z)$ is the amplification factor

$$g(z) = 1 + z + \frac{z^2}{2!} + \cdots + \frac{z^m}{m!},$$

which is a polynomial of degree m that matches the first m terms of e^z. Thus, $g(z)$ has m zeros.

For the nonlinear problem (9.1), the equations for a stage are

$$\begin{aligned}
\boldsymbol{w}^{(1)} &= \boldsymbol{w}^{(0)} - \frac{\Delta t}{m}\boldsymbol{R}\left(\boldsymbol{w}^{(0)}\right),\\
\boldsymbol{w}^{(2)} &= \boldsymbol{w}^{(0)} - \frac{\Delta t}{m-1}\boldsymbol{R}\left(\boldsymbol{w}^{(1)}\right),\\
\vdots &= \vdots\\
\boldsymbol{w}^{(m)} &= \boldsymbol{w}^{(0)} - \Delta t\ \boldsymbol{R}\left(\boldsymbol{w}^{(m-1)}\right).
\end{aligned} \tag{10.4}$$

The four stage scheme has the same stability region as the standard RK4 scheme shown in Figure 9.4.

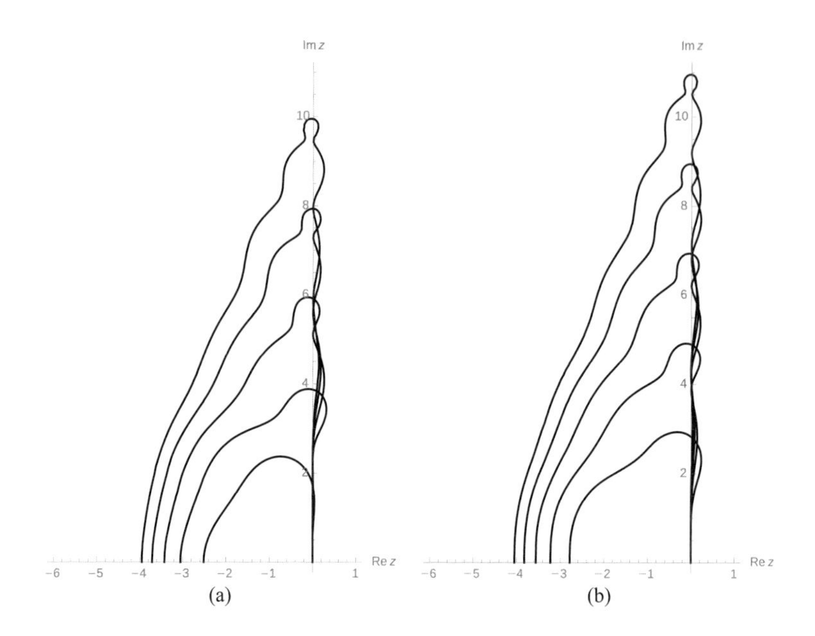

Figure 10.3 Stability regions of one step integration schemes with large stability limits for hyperbolic partial differential equations, I.P.E. Kinmark. Part (a) odd m = 3, 5, 7, 9, 11. Part (b) even m = 4, 6, 8, 10, 12.

10.3 Schemes with Maximal Stability Intervals Along the Imaginary Axis

If the order of accuracy is not a matter of concern, we may try to choose the coefficients of an m-stage RK scheme of the form

$$\begin{aligned} \boldsymbol{w}^{(1)} &= \boldsymbol{w}^{(0)} - \alpha_1 \Delta t\, \boldsymbol{R}\left(\boldsymbol{w}^{(0)}\right), \\ \boldsymbol{w}^{(2)} &= \boldsymbol{w}^{(0)} - \alpha_2 \Delta t\, \boldsymbol{R}\left(\boldsymbol{w}^{(1)}\right), \\ \vdots &= \vdots \\ \boldsymbol{w}^{(m)} &= \boldsymbol{w}^{(0)} - \Delta t\, \boldsymbol{R}\left(\boldsymbol{w}^{(m-1)}\right) \end{aligned}$$

to maximize the interval of stability along the imaginary axis, thereby in turn maximizing the permissible CFL number. This issue has been addressed by Van Leer (1974) among others. A complete solution was provided by Kinmark (1984). The maximum attainable stability interval with m stages is $m-1$, if the scheme is required to be no better than first order accurate. The stability interval is slightly reduced if one requires second order accuracy. Figure 10.3 shows the stability regions for these schemes. The coefficients of the four- and five-stage schemes with maximal stability intervals along the imaginary axis are

$$\alpha_1 = \frac{1}{3}$$

$$\alpha_2 = \frac{4}{15}$$

$$\alpha_3 = \frac{5}{9}$$
$$\alpha_4 = 1$$

and

$$\alpha_1 = \frac{1}{4}$$
$$\alpha_2 = \frac{1}{6}$$
$$\alpha_3 = \frac{3}{8}$$
$$\alpha_4 = \frac{1}{2}$$
$$\alpha_5 = 1$$

respectively.

10.4 Additive Runge–Kutta Schemes with Enlarged Stability Regions

In the case of steady state calculations, one may also use distinct time integration schemes to advance the convective and diffusive or dissipative parts of the equation. Then, the schemes for the two parts may be separately optimized to reduce the computational complexity or maximize the stability region. Such schemes belong to the class of additive RK schemes (Cooper & Sayfy 1980, 1983), which have been independently proposed in a different context.

Suppose that the residual is split as

$$\boldsymbol{R}(\boldsymbol{w}) = \boldsymbol{Q}(\boldsymbol{w}) + \boldsymbol{D}(\boldsymbol{w}), \tag{10.5}$$

where $\boldsymbol{Q}(\boldsymbol{w})$ is the convective part, and $\boldsymbol{D}(\boldsymbol{w})$ is the dissipative or diffusive part, which may constitute the artificial diffusion of the discretization scheme, together with the actual viscous terms in a Navier–Stokes simulation. The simplest scheme of this type is obtained by freezing the diffusive terms after the first stage. In the case of four stages, this yields the scheme

$$\boldsymbol{w}^{(1)} = \boldsymbol{w}^{(0)} - \frac{\Delta t}{4}\left(\boldsymbol{Q}(\boldsymbol{w}^{(0)}) + \boldsymbol{D}(\boldsymbol{w}^{(0)})\right),$$
$$\boldsymbol{w}^{(2)} = \boldsymbol{w}^{(0)} - \frac{\Delta t}{3}\left(\boldsymbol{Q}(\boldsymbol{w}^{(1)}) + \boldsymbol{D}(\boldsymbol{w}^{(0)})\right),$$
$$\boldsymbol{w}^{(3)} = \boldsymbol{w}^{(0)} - \frac{\Delta t}{2}\left(\boldsymbol{Q}(\boldsymbol{w}^{(2)}) + \boldsymbol{D}(\boldsymbol{w}^{(0)})\right),$$
$$\boldsymbol{w}^{(4)} = \boldsymbol{w}^{(0)} - \Delta t\left(\boldsymbol{Q}(\boldsymbol{w}^{(3)}) + \boldsymbol{D}(\boldsymbol{w}^{(0)})\right).$$

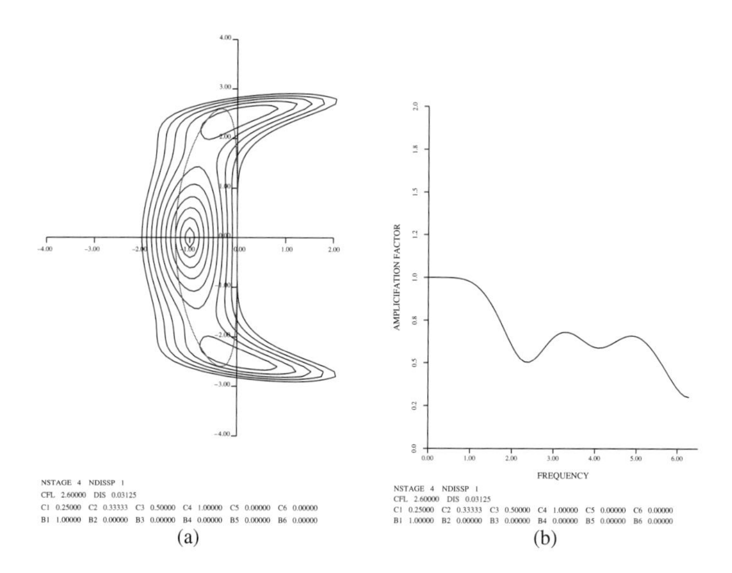

Figure 10.4 Stability region and amplification factor $|g|$ for the modified four-stage scheme. Contour lines $|g| = 1.0, 0.9, 0.8, \dots$ and locus of $z(\xi)$ for $\lambda = 2.6$, $\mu = \frac{1}{32}$, coefficients $\alpha_1 = \frac{1}{4}, \alpha_2 = \frac{1}{2}, \alpha_3 = \frac{1}{2}$. The oval curve is the locus of the Fourier symbol for linear advection with CFL number 2.4.

This is actually the scheme that was used in the original calculations using the Jameson–Schmidt–Turkel scheme (Jameson et al. 1981). The stability region is shown in the first plot of Figure 10.4. The figure also shows the locus of the Fourier symbol for the model problem described in Section 9.8 for a CFL number of 2.6 and $\mu = 1/32$. It can be seen that the stability region is comparable to that of the standard four-stage RK scheme, with a significant reduction in the computational cost. The second plot of Figure 10.4 shows the amplification factor g as a function of the wave number ξ of the mode.

One may also exploit the additional degrees of freedom with this larger class of schemes to increase the size of the stability region (Martinelli & Jameson 1988). With the split (10.5), these schemes have the form

$$\begin{aligned}
\boldsymbol{w}^{(1)} &= \boldsymbol{w}^{(0)} - \alpha_1 \Delta t \left(\boldsymbol{Q}^{(1)} + \boldsymbol{D}^{(1)} \right), \\
&\ \vdots \quad \vdots \\
\boldsymbol{w}^{(k)} &= \boldsymbol{w}^{(0)} - \alpha_k \Delta t \left(\boldsymbol{Q}^{(k-1)} + \boldsymbol{D}^{(k-1)} \right),
\end{aligned}$$

where $\alpha_m = 1$, and

$$\boldsymbol{Q}^{(0)} = \boldsymbol{Q}\big(\boldsymbol{w}^{(0)}\big), \quad \boldsymbol{D}^{(0)} = \boldsymbol{D}\big(\boldsymbol{w}^{(0)}\big)$$

and

$$\begin{aligned}
\boldsymbol{Q}^{(k)} &= \boldsymbol{Q}\left(\boldsymbol{w}^{(k)}\right), \\
\boldsymbol{D}^{(k)} &= \beta_k \boldsymbol{D}\left(\boldsymbol{w}^{(k)}\right) + (1 - \beta_k)\boldsymbol{D}^{(k-1)}.
\end{aligned}$$

The coefficients α_k are chosen to maximize the stability interval along the imaginary axis, while the coefficients β_k are chosen to increase the stability interval along the negative real axis.

Two schemes that have been found to be particularly effective are tabulated below. The first is a four-stage scheme with two evaluations of dissipation (4-2 scheme). Its coefficients are

$$\begin{aligned} \alpha_1 &= \tfrac{1}{3}, & \beta_1 &= 1, \\ \alpha_2 &= \tfrac{4}{15}, & \beta_2 &= \tfrac{3}{4}, \\ \alpha_3 &= \tfrac{5}{9}, & \beta_3 &= 0, \\ \alpha_4 &= 1, & \beta_4 &= 0. \end{aligned} \tag{10.6}$$

The second is a five-stage scheme with three evaluations of dissipation (5-3 scheme). Its coefficients are

$$\begin{aligned} \alpha_1 &= \tfrac{1}{4}, & \beta_1 &= 1, \\ \alpha_2 &= \tfrac{1}{6}, & \beta_2 &= 0, \\ \alpha_3 &= \tfrac{3}{8}, & \beta_3 &= 0.56, \\ \alpha_4 &= \tfrac{1}{2}, & \beta_4 &= 0, \\ \alpha_5 &= 1, & \beta_5 &= 0.44. \end{aligned} \tag{10.7}$$

Figure 10.5 shows the stability region of the 4-2 scheme, the locus of the Fourier symbol, and the corresponding amplication factor for the model problem defined in Section 9.8. In comparison with the standard RK4 scheme shown in Figure 9.4, the expansion of the stability region is apparent, while the high frequency damping is also significantly increased. The stability region can be expanded further to the left

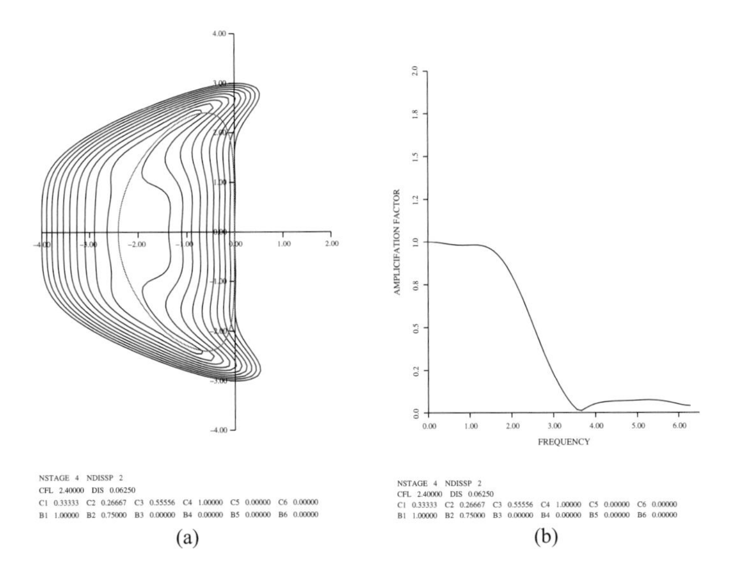

Figure 10.5 Stability region and amplification factor $|g|$ for the 4-2 scheme (10.6). The oval curve is the locus of the Fourier symbol for linear advection with CFL number 2.4.

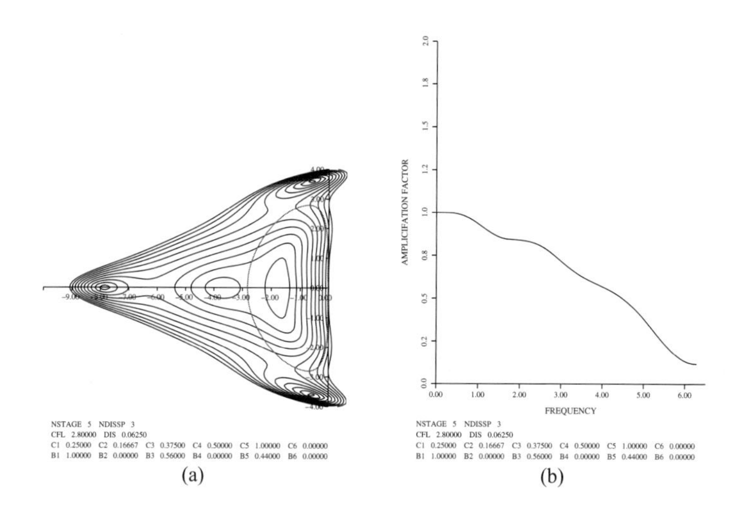

Figure 10.6 Stability region and amplification factor $|g|$ for the 5-3 scheme (10.7). The oval curve is the locus of the Fourier symbol for linear advection with CFL number 3.5.

by reducing β_2 to values less than 0.75, but this does not benefit the high frequency damping. Figure 10.6 shows the corresponding plots for the 5-3 scheme, with a representative choice of the coefficient of the third order artificial diffusion.

10.5 Newton–Krylov Methods

An alternative methodology for the treatment of both steady state problems and implicit schemes for time dependent simulations is to solve the resulting nonlinear equations by using Newton–Krylov methods. These are very briefly reviewed in this section, which summarizes the conclusions of Witherden, Jameson and Zingg in an article published in the Handbook of Numerical Analysis (Witherden, Jameson, & Zingg 2017).

The Newton method is a natural choice for the solution of nonlinear algebralc equations due to its quadratic convergence property if certain conditions are met. Application of the Newton method to large-scale problems in computational fluid dynamics has historically been limited by two issues. The first is that the method requires the solution of a large linear system at each iteration, and direct solution of these linear systems scales very badly with problem size and is generally infeasible for large-scale problems on current computing hardware. The second issue is that the Newton method will converge only if the initial guess is sufficiently close to the converged solution; in practical problems it is generally only possible to provide a suitable initial guess through a globalization procedure.

The first issue can be addressed through the use of iterative methods for linear algebraic systems, leading to the so-called inexact Newton method where the linear problem is solved iteratively to some tolerance at each Newton iteration. While this enables large-scale problems to be solved, it introduces the challenge of finding an iterative method that converges successfully for the linear systems encountered. Krylov methods for non symmetric linear systems, among which GMRES (Saad & Schultz 1986) is the most commonly used, have proven to be effective. The most common approach to globalization of the Newton method is pseudo-transient continuation, in which a non time-accurate time-dependent path is taken to partially converge the solution in order to provide a suitable initial condition for an inexact Newton method. This leads back to the time marching methods described in the previous sections. With this approach, the Newton–Krylov method is divided into two phases: a globalization phase followed by an inexact Newton phase.

Application of the implicit Euler method to (9.1) with a local time linearization gives

$$\left(\frac{I}{\Delta t} + A^{(n)}\right) \Delta \boldsymbol{w}^{(n)} = -\boldsymbol{R}(\boldsymbol{w}^{(n)}),$$

where

$$A = \frac{\partial \boldsymbol{R}}{\partial \boldsymbol{w}},$$

is the Jacobian of the discrete residual vector. It is easy to see that the Newton method is obtained in the limit as Δt goes to infinity. This linear system must be solved at each iteration. In the inexact Newton method, an iterative method is used for this purpose, and the system is solved to a specified finite relative tolerance, η_n, that is:

$$\left|\left|\left(T^{(n)} + A^{(n)}\right)\Delta \boldsymbol{w}^{(n)} + \boldsymbol{R}(\boldsymbol{w}^{(n)})\right|\right|_2 \leq \left|\left|\boldsymbol{R}(\boldsymbol{w}^{(n)})\right|\right|_2.$$

Here the term containing the time step has been modified to allow for a local time step as well as potentially a time step that varies between equations; for example, the turbulence model equation can have a different time step from the mean-flow equations. Strictly speaking, this term is zero in an inexact Newton method, but it is retained here as it is included in the pseudo-transient continuation approach to globalization. The sequence of relative tolerance values η_n affects both efficiency and robustness. If η_n is too small, the linear system will be over-solved, leading to an increase in computing time; if it is too large, the number of Newton iterations will increase, similarly leading to suboptimal performance.

One of the attributes of a Krylov method for solving linear systems of the form

$$A\mathbf{x} = \mathbf{b},$$

such as GMRES, is that the matrix A itself is never needed explicitly. What is needed is the product of A with an arbitrary vector $\boldsymbol{v}$. Therefore, the explicit formation and storage of the Jacobian matrix can be avoided by approximating these matrix–vector products by a finite difference approximation to a Fréchet derivative as follows:

$$A^{(n)}\boldsymbol{v} \approx \frac{\boldsymbol{R}(\boldsymbol{w}^{(n)} + \epsilon \boldsymbol{v}) - \boldsymbol{R}(\boldsymbol{w}^{(n)})}{\epsilon}.$$

The parameter ϵ must be chosen to balance truncation and round-off errors. For a detailed discussion of the Jacobian-free Newton–Krylov methods, the reader is referred to the review by Knoll and Keyes (2004).

The Jacobian matrices arising from the discretization of the Euler and RANS equations are typically poorly conditioned, such that GMRES will converge very slowly unless the system is first preconditioned. The choice of the preconditioner has a major impact on the overall speed and robustness of the algorithm. Pueyo and Zingg (1998) studied numerous preconditioners and showed that an incomplete lower-upper (ILU) factorization with some fill based on a level of fill approach, ILU(p), is an effective option, where p is the level of fill (Meijerink & van der Vorst 1977), which is an important parameter. Moreover, a block implementation of ILU(p) is preferred over a scalar implementation (Hicken & Zingg 2008). It is also possible to use multigrid solvers as preconditioners (Mavriplis & Brazell 2019).

Newton–Krylov methods provide an efficient option for problems with higher than usual stiffness. This can arise, for example, from the turbulence model, reacting flows, very high Reynolds numbers, or when high order spatial discretization is used.

10.6 Acceleration Methods

Acceleration methods fall into two main classes. The first class consists of modifications to the underlying differential equations that do not alter the steady state. Some ways of modifying the differential equations to accelerate the rate of convergence to a steady state are as follows:

1. to increase the speed at which disturbances are propagated through the domain;
2. to equalize the wave speeds of different types of disturbance, thereby enabling the use of larger time steps in the numerical scheme without violating the CFL condition;
3. to introduce terms which cause disturbances to be damped.

The second class of acceleration techniques consist of modifications to the numerical techniques. These range from modification to the time integration schemes to the adoption of general iterative methods for the solution of linear and nonlinear equations. The next sections explore both classes of acceleration methods.

10.7 Variable Local Time Stepping

An obvious way in which (9.1) can be modified without changing the steady state is to multiply the space derivatives by a preconditioning matrix P to produce

$$\frac{\partial \boldsymbol{w}}{\partial t} + P\left(\frac{\partial}{\partial x}\boldsymbol{f}(\boldsymbol{w}) + \frac{\partial}{\partial y}\boldsymbol{g}(\boldsymbol{w})\right) = 0.$$

The simplest choice of P is a diagonal matrix αI. Then, one can choose α locally so that the equations are advanced at the maximum CFL number permitted by the time integration scheme at every point in the domain. This is equivalent to using different local time steps at different points. Typically, one uses meshes with small cells adjacent to the body and increasingly large cells as the distance from the body increases, thus allowing increasingly large time steps in the far field.

As a model for this procedure, consider the wave equation in polar coordinates r and θ:

$$\phi_{tt} = c^2 \left\{ \frac{1}{r} \frac{\partial}{\partial r}(r\phi_r) + \frac{1}{r^2}\phi_{\theta\theta} \right\}.$$

Suppose that the wave speed c is proportional to the radius, say $c = \alpha r$. Then,

$$\phi_{tt} = \alpha^2 \left\{ r \frac{\partial}{\partial r}(r\phi_r) + \phi_{\theta\theta} \right\}.$$

This has solutions of the form

$$\phi = \frac{1}{r^n} e^{-\alpha n t},$$

indicating the possibility of exponential decay.

In practice, there are often very large variations in the size of the cells in body fitted meshes, such as the O-meshes typically used for airfoil calculations, as illustrated in Figure 6.7. Then, the use of variable local time steps with a fixed CFL number generally leads to an order of magnitude reduction in the number of time steps needed to reach a steady state. This is illustrated in Figure 10.7, which again shows a steady state calculation for an NACA 0012 airfoil at Mach 0.8 and an angle of attack $\alpha = 1.25°$, as was considered in Section 10.1, using the same grid as illustrated in

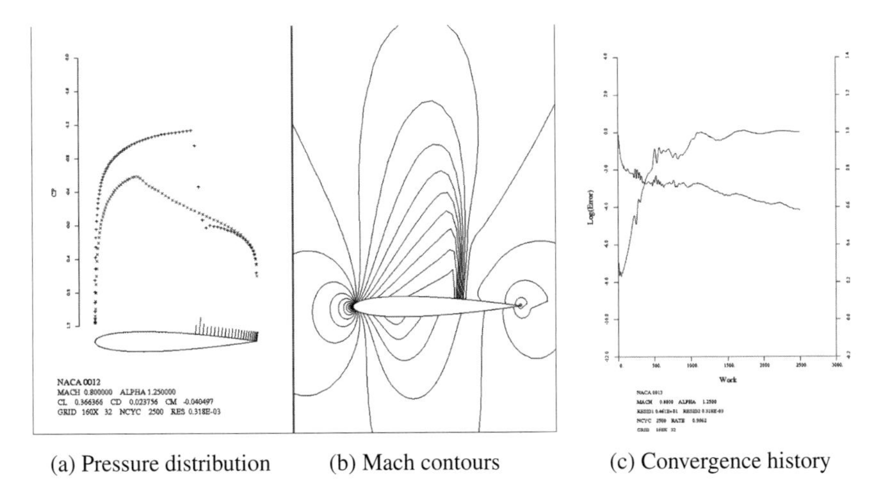

(a) Pressure distribution (b) Mach contours (c) Convergence history

Figure 10.7 NACA 0012. Steady state flow solution using a variable local time stepping method.

Figure 10.1. Now, a variable time step was used with a fixed nominal CFL number equal to 2.0 in every cell. Also, the simplified four-stage RK4-1 scheme defined by (10.4) was used since time accuracy was no longer required. Figurea 10.7 (a) and (b) shows the result after 2,500 steps, which is essentially the same as the result of the time-accurate calculation illustrated in Figures 10.2 (a) and (b) after 200,000 steps, while Figure 10.7 (c) shows the convergence history. It can be seen that this is not very smooth.

10.8 Enthalpy Damping

It was pointed out in Section 10.1 that solutions of the wave equation in an infinite domain have constant energy. If, on the other hand, the wave equation is modified by the addition of a term $\alpha\phi_t$, the energy decays. The modified equation takes the form

$$\phi_{tt} + \alpha\phi_t = c^2(\phi_{xx} + \phi_{yy}).$$

Now, multiplying by ϕ_t and integrating the right hand side by parts as before, we find that

$$\frac{d}{dt}\int_{-\infty}^{\infty}\int_{-\infty}^{\infty}\frac{1}{2}\left(\phi_t^2 + \phi_x^2 + \phi_y^2\right)\, dx\, dy = -\int_{-\infty}^{\infty}\int_{-\infty}^{\infty}\alpha\phi_t^2\, dx\, dy.$$

If $\alpha > 0$, the nonnegative integral on the left must continue to decrease as long as $\phi_t \neq 0$ anywhere in the domain. In fact, the over-relaxation method for solving Laplace's equation may be interpreted as a discretization of a damped wave equation, as was discussed in Section 3.2.5, and the rate of convergence may be optimized by a proper choice of α.

Provided that the flow is irrotational, as can be expected in subsonic flow, the Euler equations are satisfied by solutions of the unsteady potential flow equation (10.2). This does not contain a term in ϕ_t. However, in unsteady potential flow,

$$\frac{\partial\phi}{\partial t} + H - H_\infty = 0, \tag{10.8}$$

where H is the total enthalpy and H_∞ is the total enthalpy in the far field. This suggests that the difference $H - H_\infty$ can serve in the role of ϕ_t. Moreover, H is constant in steady solutions of the Euler equations, so such a term can be added without altering the steady state.

The following analysis applies to the three-dimensional case. The unsteady potential flow equation (10.1) and the enthalpy equation (10.8) may be derived as follows. Assuming there are no shock waves, consider the Euler equations in primitive form:

$$\frac{d\rho}{dt} + \rho\nabla\cdot\mathbf{u} = 0, \tag{10.9}$$

$$\frac{d\mathbf{u}}{dt} + \frac{1}{\rho}\nabla p = 0, \tag{10.10}$$

$$\frac{dH}{dt} - \frac{1}{\rho}\frac{\partial p}{\partial t} = 0, \tag{10.11}$$

where $\frac{d}{dt}$ is the substantial derivative

$$\frac{d}{dt} \equiv \frac{\partial}{\partial t} + \mathbf{u} \cdot \nabla,$$

and H is the total enthalpy

$$H = h + \frac{1}{2}u^2,$$

where h is the specific enthalpy and u is the magnitude of $\mathbf{u}$. The scalar product of (10.10) with $\mathbf{u}$ can be subtracted from (10.11) to yield

$$\frac{dh}{dt} - \frac{1}{\rho}\frac{dp}{dt} = 0,$$

which implies the isentropic relation

$$dh - \frac{dp}{\rho} = 0, \tag{10.12}$$

along particle paths. An initially homentropic flow therefore remains homentropic. In this case, we may write (10.9) as

$$\frac{d\mathbf{u}}{dt} + \nabla h = 0. \tag{10.13}$$

Let Γ be the circulation $\oint \mathbf{u} \cdot d\mathbf{l}$ around a closed material loop. Equation (10.13) yields Kelvin's theorem that

$$\frac{d\Gamma}{dt} = \oint \mathbf{u} \cdot d\mathbf{u} - \oint \nabla h \cdot d\mathbf{l} = 0.$$

Hence, a flow that is initially homentropic and irrotational remains so. In this case, we can set $\mathbf{u}$ equal to the gradient of a velocity potential

$$\mathbf{u} = \nabla\phi,$$

and (10.13) can be integrated to yield the enthalpy relation (10.8). Now, differentiating (10.8) with respect to time:

$$\phi_{tt} + u_i\frac{\partial u_i}{\partial t} + \frac{\partial h}{\partial t} = 0.$$

But it follows from the isentropic relation (10.12) that

$$\frac{\partial h}{\partial t} = \frac{dh}{d\rho}\frac{\partial \rho}{\partial t} = \frac{c^2}{\rho}\frac{\partial \rho}{\partial t},$$

and we can then recover the unsteady potential flow (10.1) by substituting for $\frac{\partial \rho}{\partial t}$. According to (10.9),

$$\frac{c^2}{\rho}\frac{\partial \rho}{\partial t} = -\frac{c^2}{\rho}u_i\frac{\partial \rho}{\partial x_i} - c^2\frac{\partial u_i}{\partial x_i}.$$

But since the flow is isentropic, we can replace $\frac{\partial p}{\partial x_i}$ in (10.10) by $c^2\frac{\partial \rho}{\partial x_i}$ to obtain

$$\frac{c^2}{\rho}\frac{\partial \rho}{\partial x_i} = -\frac{\partial u_i}{\partial t} - u_j\frac{\partial u_i}{\partial x_j}.$$

Thus,

$$\frac{c^2}{\rho}\frac{\partial \rho}{\partial t} = -c^2\frac{\partial u_i}{\partial x_i} + u_i\frac{\partial u_i}{\partial t} + u_iu_j\frac{\partial u_i}{\partial x_j},$$

and the unsteady potential flow equation follows on substituting $u_i = \frac{\partial \phi}{\partial x_i}$.

Suppose now that the continuity equation is modified by the addition of a term containing $H - H_\infty$ so that it takes the form

$$\frac{\partial \rho}{\partial t} + \frac{\partial}{\partial x_j}(\rho u_j) + \alpha\frac{\rho}{c^2}(H - H_\infty) = 0,$$

where α is a damping coefficient. Since the isentropic relation (10.12) is a consequence of the momentum and energy equations (10.10) and (10.11), it still holds. Accordingly, the entire argument can be repeated to give the unsteady potential flow equation with a damping term

$$\frac{\partial^2 \phi}{\partial t^2} + 2u_i\frac{\partial^2 \phi}{\partial x_i \partial t} + \alpha\frac{\partial \phi}{\partial t} = c^2\frac{\partial^2 \phi}{\partial x_i \partial x_i} - u_iu_j\frac{\partial^2 \phi}{\partial x_i \partial x_j}.$$

Thus, at least in subsonic flow, the added term can be expected to have the desired damping effect.

When the density equation is combined with the momentum equation to yield a system of equations in conservation form, the modified equations become

$$\frac{\partial \boldsymbol{w}}{\partial t} + \frac{\partial \boldsymbol{f}_i(\boldsymbol{w})}{\partial x_i} + \boldsymbol{s}(\boldsymbol{w}) = 0,$$

which differs from the standard equations (2.16) by the addition of the source term

$$\boldsymbol{s}(\boldsymbol{w}) = \frac{\rho}{c^2}(H - H_\infty)\begin{bmatrix} 1 \\ u_1 \\ u_2 \\ u_3 \\ H \end{bmatrix}.$$

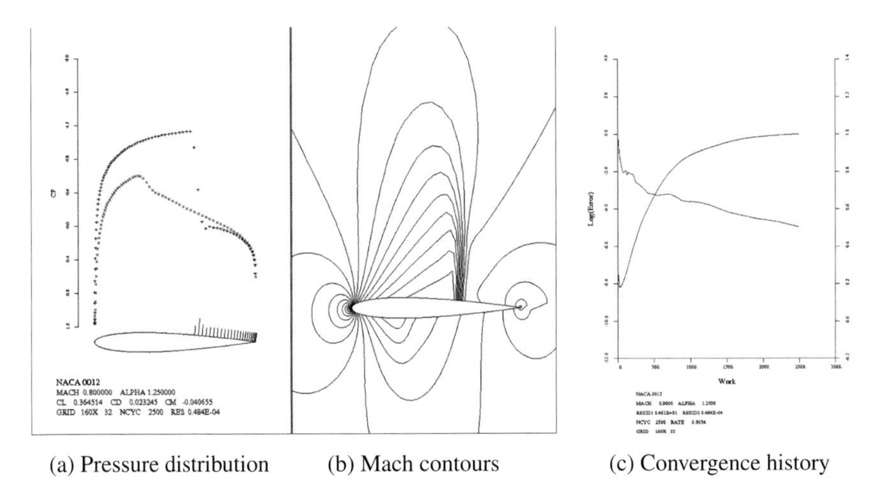

(a) Pressure distribution (b) Mach contours (c) Convergence history

Figure 10.8 NACA 0012. Steady state flow solution using a variable local time stepping method with enthalpy damping.

The energy equation has a quadratic term in H, like a Riccati equation, which could be destabilizing. An alternative is to modify the energy equation to the form

$$\frac{\partial}{\partial t}\rho E + \frac{\partial}{\partial x_i}(\rho u_i H) + \beta(H - H_\infty) = 0,$$

which tends to drive H towards H_∞ if the coefficient β is positive.

Figure 10.8 shows the effect of enthalpy damping on exactly the same steady state calculation for an NACA 0012 airfoil that was presented in the previous section, using the RK 4-1 time stepping scheme. It can be seen that the result after 2,500 time steps shown in Figures 10.8 (a) and (b) is essentially the same as before. It can also be seen from Figure 10.8 (c) that there is both a faster rate of convergence and a much smoother convergence history.

10.9 Preconditioning

In comparison with the use of variable local time steps discussed in Section 10.7, a more elaborate preconditioning technique is to multiply the spatial derivative by a matrix P designed to equalize the eigenvalues, so that all the waves can be advanced with optimal time steps. Van Leer, Lee, and Roe have proposed a symmetric preconditioner, which minimizes the ratio between the largest and smallest eigenvalues (Van Leer, Lee, & Roe 1991). When the equations are written in stream-aligned coordinates with the symmetrizing variables, which may be written in differential form as

$$d\tilde{w} = \left[\frac{dp}{\rho c}, du_1, du_2, du_3, dp - c^2 d\rho\right]^T,$$

it has the form

$$P = \begin{bmatrix} \frac{\tau}{\beta^2}M^2 & -\frac{\tau}{\beta}M & 0 & 0 & 0 \\ -\frac{\tau}{\beta}M & \frac{\tau}{\beta^2}+1 & 0 & 0 & 0 \\ 0 & 0 & \tau & 0 & 0 \\ 0 & 0 & 0 & \tau & 0 \\ 0 & 0 & 0 & 0 & 1 \end{bmatrix},$$

where

$$\beta = \tau = \sqrt{1-M^2}, \quad \text{if} \quad M < 1,$$

or

$$\beta = \sqrt{M^2-1}, \quad \tau = \sqrt{1-\frac{1}{M^2}}, \quad \text{if} \quad M \geq 1.$$

Turkel has proposed an asymmetric preconditioner designed to treat flow at low Mach numbers (Turkel 1987). A special case of the Turkel preconditioner advocated by Weiss and Smith (Smith & Weiss 1995) has the simple diagonal form

$$P = \begin{bmatrix} \epsilon^2 & 0 & 0 & 0 & 0 \\ 0 & 1 & 0 & 0 & 0 \\ 0 & 0 & 1 & 0 & 0 \\ 0 & 0 & 0 & 1 & 0 \\ 0 & 0 & 0 & 0 & 1 \end{bmatrix},$$

when written in terms of the symmetrizing variables. If ϵ^2 varies as M^2 as $M \to 0$, all the eigenvalues of PA depend only on the speed q. In order to improve the accuracy, the absolute Jacobian matrix $|A|$ apprearing in the artificial diffusion should be replaced by $P^{-1}|PA|$. In effect, this makes the diffusion depend on the modified wave speeds. The use of preconditioners of this type can lead to instability at stagnation points, where there is a zero eigenvalue that cannot be equalized with the eigenvalues $\pm c$. With a judiciously chosen limit on ϵ^2 as $M \to 0$, they have been proved effective in treating low speed flows.

10.10 Residual Averaging

Another approach to increasing the time step is to replace the residual at each point by a weighted average of residuals at neighboring points. Consider the multi-stage scheme described by (10.4). In the one-dimensional case, one might replace the residual R_j by the average

$$\bar{\boldsymbol{R}}_j = \epsilon \boldsymbol{R}_{j-1} + (1-2\epsilon)\boldsymbol{R}_j + \epsilon \boldsymbol{R}_{j+1},$$

at each stage of the scheme. This smooths the residuals and also increases the support of the scheme, thus relaxing the restriction on the time step imposed by the

Courant–Friedrichs–Lewy condition. In the case of the model problem (9.6), this would modify the Fourier symbol (9.7) by the factor

$$\beta = 1 - 2\epsilon(1 - \cos\xi).$$

As long as $c < \frac{1}{4}$, the absolute value $|\beta|$ decreases with increasing wave numbers ξ in the range $0 \le \xi \le \pi$. If $\epsilon = \frac{1}{4}$, however, $\bar{R}_j = 0$ for the odd-even mode $R_j = (-1)^j$.

In order to avoid a restriction on the smoothing coefficient, it is better to perform the averaging implicitly by setting

$$-\epsilon\bar{\boldsymbol{R}}_{j-1} + (1 - 2\epsilon)\bar{\boldsymbol{R}}_j - \epsilon\bar{\boldsymbol{R}}_{j+1} = \boldsymbol{R}_j. \tag{10.14}$$

This corresponds to a discretization of the inverse Helmholtz operator. For an infinite interval, this equation has the explicit solution

$$\bar{\boldsymbol{R}}_j = \frac{1-r}{1+r}\sum_{q=-\infty}^{\infty} r^{|q|}\boldsymbol{R}_{j+q},$$

where

$$\epsilon = \frac{r}{(1-r)^2}, \quad r < 1.$$

Thus, the effect of the implicit smoothing is to collect information from residuals at all points in the field, with an influence coefficient that decays by a factor r at each additional mesh interval from the point of interest.

Consider the model problem (9.6). According to (10.14), the Fourier symbol (9.7) will be replaced by

$$z = -\lambda\frac{i\sin\xi + 4\mu(1-\cos\xi)^2}{1 + 2\epsilon(1-\cos\xi)}.$$

In the absence of dissipation,

$$|z| = \lambda\left|\frac{\sin\xi}{1 + 2\epsilon(1-\cos\xi)}\right|.$$

Differentiating with respect to ξ, we find that $|z|$ reaches a maximum when

$$\cos\xi = \frac{2\epsilon}{1+2\epsilon}, \quad \sin\xi = \sqrt{1 - \left(\frac{2\epsilon}{1+2\epsilon}\right)^2}.$$

Then, $|z|$ attains the maximum value

$$|z|_{\max} = \frac{1}{\sqrt{1+4\epsilon}}.$$

Accordingly, if λ^* is the stability limit of the scheme, the CFL number λ should satisfy

$$\lambda \le \lambda^*\sqrt{1+4\epsilon}.$$

It follows that we can perform stable calculations at any desired CFL number λ by using a large enough smoothing coefficient such that

$$\epsilon \geq \frac{1}{4}\left(\left(\frac{\lambda}{\lambda*}\right)^2 - 1\right).$$

Implicit residual averaging was originally proposed by Jameson and Baker (1983). In practice, it has been found that the most rapid rate of convergence to a steady state is usually obtained with $\frac{\lambda}{\lambda*}$ in the range of 2 to 3, or $\lambda \approx 8$ for a four-stage Runge–Kutta scheme.

10.11 Multigrid Methods

Convergence acceleration by multigrid methods has been discussed in Chapter 3. These methods have proved very successful for elliptic problems. An early demonstration of the application of a multigrid method to a nonelliptic problem is due to South and Brandt, who obtained fast convergence for the transonic small disturbance equation (South & Brandt 1976). A successful multigrid method for the transonic potential equation was presented by Jameson in 1979 (1979). The analysis of multigrid methods for elliptic problems is based on the concept that the updating scheme acts as a smoothing operator at each grid level (Brandt 1977, Hackbusch 1978). This theory does not hold for hyperbolic systems. Nevertheless, it seems that it ought to be possible to accelerate the evolution of a hyperbolic system to a steady state by using large time steps on coarse grids so that disturbances will be more rapidly expelled through the outer boundary. Multigrid schemes to solve the Euler equations by taking advantage of this effect were independently developed by Ni (1982) and Jameson (1983). Subsequently, a variety of multigrid time stepping schemes have been developed (Hemker and Spekreijse 1984, Anderson, Thomas, & Whitfield 1986, Jameson 1986a, 1986b, Caughey 1987).

Ni's scheme was based on an adaptation of the Lax–Wendroff scheme to multiple grids. We focus here on the Runge–Kutta multigrid (RKMG) scheme proposed by Jameson, which can readily be adapted to alternative flux formulations of the type described in Chapter 6, and also to unstructured grids. The RKMG scheme is an adaptation of the full approximation scheme described in Chapter 3. It combines the full approximation scheme with a multi-stage time-stepping scheme of the type described in Sections 10.2–10.4, in conjunction with a finite volume space discretization scheme.

Consider a sequence of successively coarser quadrilateral grids generated by deleting alternate points in each coordinate direction as illustrated in Figure 10.9. The mesh cells at each coarser level roughly correspond to cells formed by agglomerating groups of four cells from the finer level, although they do not exactly coincide if the mesh is curved. In order to give a precise description of the multigrid scheme, we use subscripts to indicate the grid level, with the finest grid denoted as level 1.

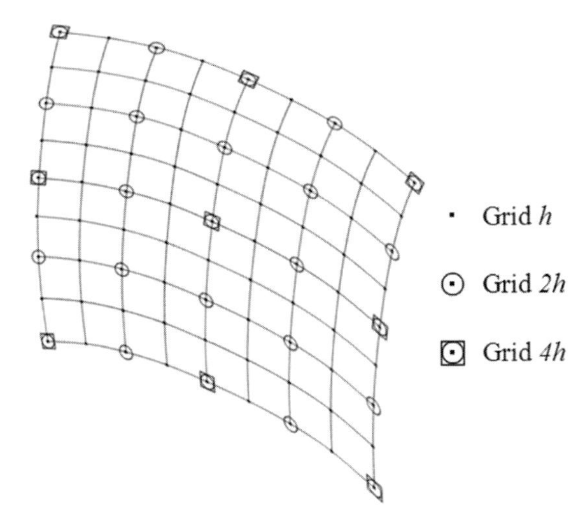

Figure 10.9 Nested grids with three levels for a multigrid scheme.

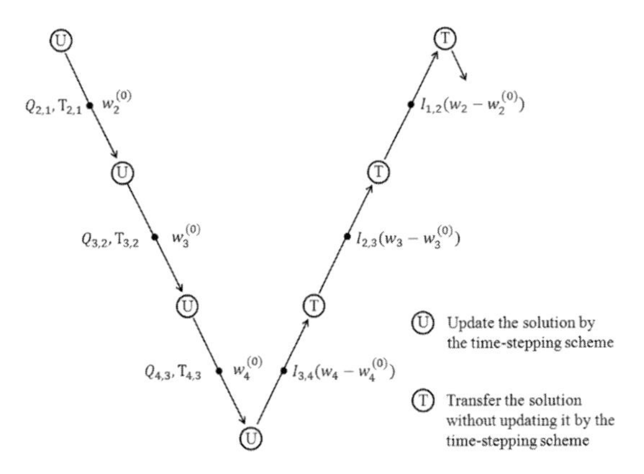

Figure 10.10 Sawtooth multigrid cycle.

The simplest multigrid cycle is a "sawtooth" cycle of the type illustrated in Figure 10.10, in which the solution is updated by the time stepping scheme at each grid level on the way down to the coarsest level, and then the corrections are successively interpolated to the next finer level on the way up to the finest grid, without any additional updates by the time-stepping scheme. The solution at level 1 is updated by the basic m-stage scheme

$$
\begin{aligned}
\boldsymbol{w}_1^{(1)} &= \boldsymbol{w}_1^{(0)} - \alpha_1 \Delta t \boldsymbol{R}_1^{(0)} \\
\vdots \quad & \quad \vdots \\
\boldsymbol{w}_1^{(q)} &= \boldsymbol{w}_1^{(0)} - \alpha_q \Delta t \boldsymbol{R}_1^{(q-1)} \\
\vdots \quad & \quad \vdots \\
\boldsymbol{w}_1^{(m)} &= \boldsymbol{w}_1^{(0)} - \Delta t \boldsymbol{R}_1^{(m-1)},
\end{aligned}
$$

where $w_1^{(0)}$ is the value at the beginning of the time step and $w_1^{(m)}$ is the final value at the end of the step, while the residual at any stage is calculated by the additive RK scheme defined by the equations in Section 10.4.

To complete the description of the multigrid scheme, several transfer operations need to be defined. First, the solution vector on grid k must be initialized as

$$\boldsymbol{w}_k^{(0)} = Q_{k,k-1}\boldsymbol{w}_{k-1},$$

where $\boldsymbol{w}_{k-1}$ is the current value on grid $k-1$, and $Q_{k,k-1}$ is a transfer operator. Next, it is necessary to calculate a residual forcing function such that the solution on grid k is driven by the residuals calculated on grid $k-1$. This can be accomplished by setting

$$\boldsymbol{P}_k = T_{k,k-1}\boldsymbol{R}_{k-1}(\boldsymbol{w}_{k-1}) - \boldsymbol{R}_k\left(\boldsymbol{w}_k^{(0)}\right),$$

where $T_{k,k-1}$ is a residual transfer operator, which aggregates the residuals from grid $k-1$. Note these residuals should be re-evaluated, using the final values $\boldsymbol{w}_{k-1}$, after the completion of the updating process on grid $k-1$. Now on grid k, we replace the residual $\boldsymbol{R}_k(w_k)$ by $\boldsymbol{R}_k(w_k) + \boldsymbol{P}_k$ at each stage of the time-stepping scheme. Thus, the multistage scheme is reformulated as

$$\begin{aligned}
\boldsymbol{w}_k^{(1)} &= \boldsymbol{w}_k^{(0)} - \alpha_1 \Delta t_k\left(\boldsymbol{R}_k^{(0)} + \boldsymbol{P}_k\right)\\
&\vdots\\
\boldsymbol{w}_k^{(q)} &= \boldsymbol{w}_k^{(0)} - \alpha_q \Delta t_k\left(\boldsymbol{R}_k^{(q-1)} + \boldsymbol{P}_k\right)\\
&\vdots\\
\boldsymbol{w}_k^{(m)} &= \boldsymbol{w}_k^{(0)} - \Delta t_k\left(\boldsymbol{R}_k^{(m-1)} + \boldsymbol{P}_k\right).
\end{aligned}$$

The result $\boldsymbol{w}_k^{(m)}$ provides the initial data for grid $k+1$. Finally, during the ascent back up the levels to the finest grid, the accumulated change $\boldsymbol{w}_k - \boldsymbol{w}_k^{(0)}$ is transferred to grid $k-1$ with the aid of an interpolation operator $I_{k-1,k}$.

It should be noted that at any grid level below the finest mesh, the residual calculated in the first stage is $\boldsymbol{R}_k^{(0)} = \boldsymbol{R}_k\left(\boldsymbol{w}_k^{(0)}\right)$, and this is cancelled by the second term in $\boldsymbol{P}_k$, with the result that the first stage update is determined purely by the aggregated residuals from grid $k-1$. If these are zero, $\boldsymbol{w}_k^{(1)} = \boldsymbol{w}_k^{(0)}$, and it follows that no change is made to the solution at any of the later stages. Thus, if the values of the solution on the finest grid satisfy the equations exactly, they will be left unchanged by the entire multigrid cycle. This is an essential requirement for a successful multigrid scheme.

For a sequence of nested grids of the type illustrated in Figure 10.9, the solution and residual transfer operators $Q_{k,k-1}$ and $T_{k,k-1}$ may be chosen as follows. The values of the flow variables are transferred to the next coarser grid by the rule

$$\boldsymbol{w}_{k,m}^{(0)} = \frac{\sum_n S_n \boldsymbol{w}_{k-1,n}}{\sum_n S_n},$$

where the sum is over the four cells on grid $k-1$, which are agglomerated to correspond to cell m on grid k, and S_n is the area of cell n on grid $k-1$. This rule conserves mass, momentum, and energy. The residual transfer operator is simply defined by summing the residuals of the cells on grid $k-1$ that correspond to cell m on grid k, so at cell m, on grid k:

$$\boldsymbol{P}_{k,m} = r \sum_{n} \boldsymbol{R}_{k-1,n} - \boldsymbol{R}_{k,m}\left(\boldsymbol{w}_k^{(0)}\right),$$

where r is a relaxation factor which is normally set equal to unity. In more difficult cases, however, such as hypersonic flow; or extremely stretched meshes, it is sometimes useful to reduce r to values in the range 0.5 to 0.75. The factor r is also useful for debugging, because we can set r to zero to simulate the case when the solution on grid k satisfies the equations exactly and produces zero residuals. In that case, none of the coarser grid levels should change the solution, and its evolution should be identical to the evolution of the solution with a single fine grid. This helps to identify possible conflicts between the coarse levels and the fine level, which may be introduced, for example, by boundary conditions on the coarse levels that are incompatible with the fine grid boundary conditions.

In order to transfer the corrections from grid k to grid $k-1$, we may use bilinear interpolation between the centers of the four coarse grid cells that contain the center of a particular fine grid cell. On a nonuniform curvilinear grid, this would require different interpolation formulas at every grid point. In practice, good results have been obtained by assuming that the mesh could be mapped to a Cartesian mesh in which each coarse mesh cell exactly corresponds to four fine mesh cells. The interpolation operator may then be based on bilinear interpolation on the corresponding Cartesian mesh, as illustrated in Figure 10.11. Accordingly, with the notation of the figure,

$$\begin{aligned}
\Delta \boldsymbol{w}_{f1} &= \frac{9}{16}\Delta \boldsymbol{w}_{c1} + \frac{3}{16}\Delta \boldsymbol{w}_{c2} + \frac{3}{16}\Delta \boldsymbol{w}_{c3} + \frac{1}{16}\Delta \boldsymbol{w}_{c4} \\
\Delta \boldsymbol{w}_{f2} &= \frac{9}{16}\Delta \boldsymbol{w}_{c2} + \frac{3}{16}\Delta \boldsymbol{w}_{c1} + \frac{3}{16}\Delta \boldsymbol{w}_{c4} + \frac{1}{16}\Delta \boldsymbol{w}_{c3} \\
\Delta \boldsymbol{w}_{f3} &= \frac{9}{16}\Delta \boldsymbol{w}_{c3} + \frac{3}{16}\Delta \boldsymbol{w}_{c1} + \frac{3}{16}\Delta \boldsymbol{w}_{c4} + \frac{1}{16}\Delta \boldsymbol{w}_{c2} \\
\Delta \boldsymbol{w}_{f4} &= \frac{9}{16}\Delta \boldsymbol{w}_{c4} + \frac{3}{16}\Delta \boldsymbol{w}_{c2} + \frac{3}{16}\Delta \boldsymbol{w}_{c3} + \frac{1}{16}\Delta \boldsymbol{w}_{c1},
\end{aligned}$$

where $\Delta \boldsymbol{w}_{f1}$, $\Delta \boldsymbol{w}_{f2}$, $\Delta \boldsymbol{w}_{f3}$ and $\Delta \boldsymbol{w}_{f4}$ denote the corrections in cells f_1, f_2, f_3, and f_4 on grid $k+1$; and $\Delta \boldsymbol{w}_{c1}$, $\Delta \boldsymbol{w}_{c2}$, $\Delta \boldsymbol{w}_{c3}$, and $\Delta \boldsymbol{w}_{c4}$ denote the accumulated change $\boldsymbol{w}_k - \boldsymbol{w}_k^{(0)}$ in the solution vector in cells c_1, c_2, c_3, and c_4 of grid k.

Care must be exercised in the application of boundary conditions at the coarse grid levels, because there is a possibility of introducing a conflict between the boundary conditions at different grid levels. This can occur when extrapolated values are set in halo cells outside a boundary, in order, for example, to satisfy an outflow boundary condition. In this situation, even if the solution satisfies the fine grid equations exactly, the extrapolated values at a coarser grid level may lead to corrections being interpolated back to the fine grid, as illustrated in Figure 10.12. A similar situation

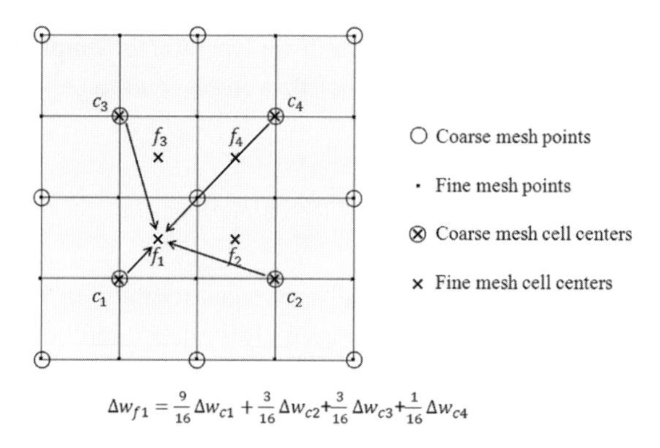

Figure 10.11 Interpolation scheme.

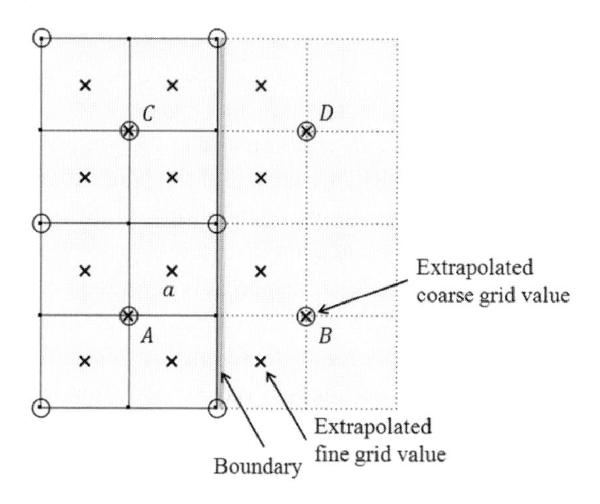

Figure 10.12 Influence of extrapolated boundary values: corrections to the interior fine grid point a depend on an extrapolated coarse grid values at B and D.

can occur when a no slip boundary condition is imposed at solid wall in a viscous flow simulation. One way to prevent this is to apply the boundary conditions on grid k immediately after the initial solution $\omega_k^{(0)}$ has been transferred from grid $k-1$, before updating the solution by the time-stepping scheme. Then, if the residuals transferred from grid $k-1$ are zero, the interior values of the solution on grid k will be left unchanged, and hence the extrapolated boundary values will also be left unchanged.

In nodal schemes, the solution values at a solid wall may be constrained to satisfy a flow tangency condition. In this situation, the wall normals calculated at coincident points on different grid levels may be different, leading to a conflict. This may be resolved by transferring the values of the normal calculated at the fine grid level to coincident points on the coarse grid levels instead of recalculating the normals at each level.

If the time step Δt_k is doubled at each coarser level, a 5-level sawtooth cycle could ideally advance the solution by an accumulated interval

$$\Delta t + 2\Delta t + 4\Delta t + 8\Delta t + 16\Delta t = 31\Delta t,$$

where Δt is the time step on the fine mesh. At the same time, the computational cost of the updates by the time-stepping scheme scales as

$$1 + \frac{1}{4} + \frac{1}{8} + \frac{1}{16} + \frac{1}{32} + \cdots < \frac{4}{3}$$

in a two-dimensional calculation, or

$$1 + \frac{1}{8} + \frac{1}{64} + \frac{1}{512} + \frac{1}{4096} + \cdots < \frac{8}{7}$$

in a three-dimensional calculation, to which there must be added a relatively small overhead due to the transfer operations.

In practice, the repeated interpolation operations generate high frequency errors on each higher level grid, and these need to be damped by the time-stepping scheme. Accordingly, the time-stepping scheme should be designed to behave like a low pass filter with the maximum possible attenuation of the high frequency modes with wave numbers in the band $\frac{\pi}{2} \leq \xi \leq \pi$. For this purpose, the additive Runge–Kutta schemes defined by (10.6) and (10.7) have proved particularly effective, since they both provide a large stability region, which allows the use of maximal time steps, and also attenuate the high frequency modes.

While the sawtooth cycle is quite effective, it has been found that W-cycles of the type illustrated in Figure 10.13 provide additional acceleration. Once the three-level W-cycle has been defined, W-cycles with more levels may be generated recursively as shown in the figure. In a W-cycle, the solution is updated twice on level 2, four times on level 3, and so on until the coarsest level is reached, where the number of updates is the same as the number on the second last level. Thus, if Δt_k is doubled at each coarser level, the effective time interval for a four-level W-cycle is

$$\Delta t + 4\Delta t + 16\Delta t + 64\Delta t = 85\Delta t,$$

while the computational cost of the updates by the time-stepping scheme scales as

$$1 + \frac{2}{8} + \frac{4}{64} + \frac{4}{512} < \frac{4}{3},$$

for a three-dimensional calculation.

The actual speed up realized in practice will be less than this because of the imperfections in the transfer operations, but it can still be very large. It should be noted, moreover, that multigrid methods simultaneously drive the solution toward equilibrium at all the grid levels. Consequently, global quantities such as the lift coefficient converge at the same rate as the local residuals, whereas on a single grid, the local residuals may be very small, while the global solution is still far from equilibrium.

Figure 10.14 shows the result of a multigrid calculation of the same benchmark test case that has been used in the previous sections: transonic flow past an NACA

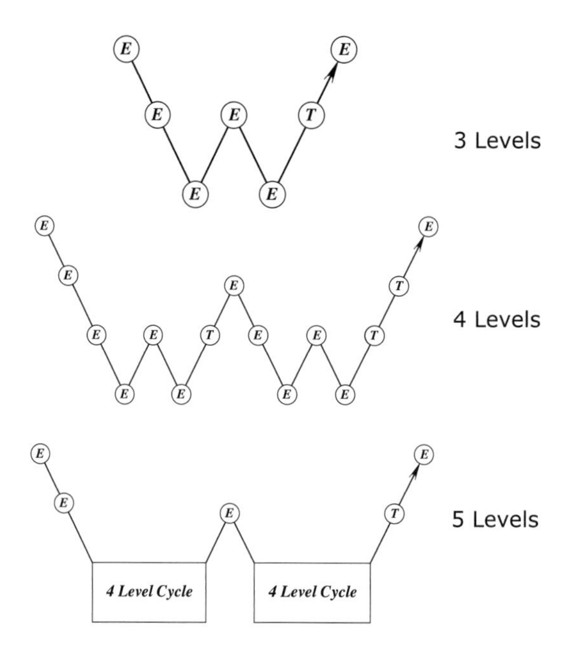

Figure 10.13 Multigrid W-cycle for managing the grid calculation. E: evaluate the change in the flow for one step; T: transfer the data without updating the solution.

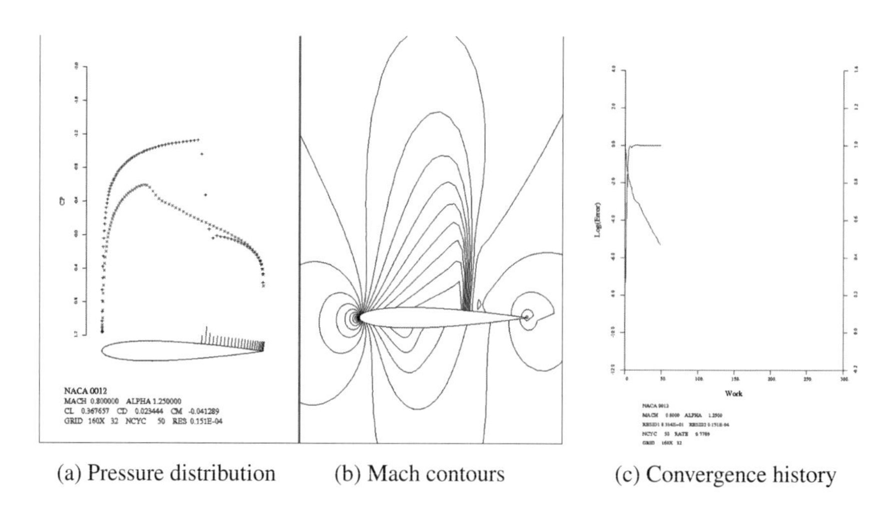

(a) Pressure distribution (b) Mach contours (c) Convergence history

Figure 10.14 NACA 0012. Steady state flow solution using multigrid acceleration method.

0012 airfoil at Mach 0.8 and an angle of attack of 1.25 degrees, discretized by the JST scheme. This calculation used a five-level W cycle with the 5-3 additive Runge–Kutta scheme defined by (10.6). The solution is well converged after 50 cycles. Moreover, it can be seen from the second curve in Figure 10.14 (c) that the lift coefficient reaches a steady state in about 25 cycles.

10.12 The Nonlinear SGS Multigrid Scheme

A further improvement in the convergence rate can be realized by substituting a symmetric Gauss–Seidel iterative scheme for the Runge–Kutta time stepping scheme as the driver of the multigrid cycle (Jameson & Caughey 2001). This scheme abandons any pretense of emulating a time dependent evolution toward a steady state and applies the symmetric Gauss–Seidel procedure directly to the nonlinear equations. It may be derived as follows.

The terms $\Delta t \boldsymbol{R}_j - \tau A^+_{j-1}\delta\tilde{\boldsymbol{w}}_{j-1}$ of (9.13) are a linearization of $\Delta t \boldsymbol{R}_j$ evaluated with $\tilde{\boldsymbol{w}}_{j-1} = \boldsymbol{w}_{j-1} + \delta\tilde{\boldsymbol{w}}_{j-1}$. Following this line of reasoning, the LU-SGS scheme can be recast (Jameson & Caughey 2001) as

$$\left\{I + \tau\left(A^+_j - A^-_j\right)\right\}\delta\tilde{\boldsymbol{w}}_j + \Delta t\tilde{\boldsymbol{R}}_j = 0, \tag{10.15}$$

$$\left\{I + \tau\left(A^+_j - A^-_j\right)\right\}\delta\tilde{\tilde{\boldsymbol{w}}}_j + \Delta t\tilde{\tilde{\boldsymbol{R}}}_j = 0, \tag{10.16}$$

where

$$\tilde{\boldsymbol{w}}_j = \boldsymbol{w}_j + \delta\tilde{\boldsymbol{w}}_j, \tilde{\boldsymbol{f}}^{\pm}_j = \boldsymbol{f}^{\pm}(\tilde{\boldsymbol{w}}_j), \tag{10.17}$$

$$\boldsymbol{w}^{n+1}_j = \tilde{\tilde{\boldsymbol{w}}}_j = \tilde{\boldsymbol{w}}_j + \delta\tilde{\tilde{\boldsymbol{w}}}_j, \tilde{\tilde{\boldsymbol{f}}}^{\pm}_j = \boldsymbol{f}^{\pm}(\tilde{\tilde{\boldsymbol{w}}}_j), \tag{10.18}$$

and

$$\tilde{\boldsymbol{R}}_j = \frac{1}{\Delta x}\left(\boldsymbol{f}^-_{j+1} - \boldsymbol{f}^-_j + \boldsymbol{f}^+_j - \tilde{\boldsymbol{f}}^+_{j-1}\right), \tag{10.19}$$

$$\tilde{\tilde{\boldsymbol{R}}}_j = \frac{1}{\Delta x}\left(\tilde{\tilde{\boldsymbol{f}}}^-_{j+1} - \tilde{\boldsymbol{f}}^-_j + \tilde{\boldsymbol{f}}^+_j - \tilde{\boldsymbol{f}}^+_{j-1}\right). \tag{10.20}$$

Using the simple splitting,

$$A^{\pm} = \frac{1}{2}A \pm \epsilon I,$$

where

$$\epsilon = \max|\lambda(A)|,$$

equations (10.15) and (10.16) can be written as

$$\delta\tilde{\boldsymbol{w}}_j = -\frac{\Delta t}{1+C}\tilde{\boldsymbol{R}}_j,$$

$$\delta\tilde{\tilde{\boldsymbol{w}}}_j = -\frac{\Delta t}{1+C}\tilde{\tilde{\boldsymbol{R}}}_j,$$

where $C = \frac{\epsilon\Delta t}{\Delta x}$ is the Courant number.

Alternatively, one may use the Jacobian splitting defined as

$$A^+ = \frac{1}{2}\left(A + |A|\right), \quad A^- = \frac{1}{2}\left(A - |A|\right),$$

where $|A| = V|\Lambda|V^{-1}$, and $|\Lambda|$ is the diagonal matrix whose entries are the absolute values of the eigenvalues of the Jacobian matrix A, while the columns of V are the right eigenvalues of A as defined in (6.6) and (6.6). Then (10.15) and (10.16) can be written

$$\{I + \alpha|A|\}\,\delta\tilde{\boldsymbol{w}}_j = -\Delta t\tilde{\boldsymbol{R}}_j,$$

$$\{I + \alpha|A|\}\,\delta\tilde{\tilde{\boldsymbol{w}}}_j = -\Delta t\tilde{\tilde{\boldsymbol{R}}}_j,$$

and, in the limit as the time step Δt goes to infinity, it follows from (9.10) that these equations represent the SGS Newton iteration

$$|A|\delta\tilde{\boldsymbol{w}}_j = -\Delta x\tilde{\boldsymbol{R}}_j, \tag{10.21}$$

$$|A|\delta\tilde{\tilde{\boldsymbol{w}}}_j = -\Delta x\tilde{\tilde{\boldsymbol{R}}}_j. \tag{10.22}$$

When the scheme corresponding to (10.21) and (10.22) is implemented for the finite-volume form of the equations of the equations, it can be represented (in two dimensions) as

$$\{|A| + |B|\}\,\delta\tilde{\boldsymbol{w}}_{i,j} = -\sigma\tilde{\boldsymbol{R}}_{i,j},$$

$$\{|A| + |B|\}\,\delta\tilde{\tilde{\boldsymbol{w}}}_{i,j} = -\sigma\tilde{\tilde{\boldsymbol{R}}}_{i,j},$$

where

$$\tilde{\boldsymbol{R}}_{i,j} = \boldsymbol{F}^{-}_{i+1,j} - \boldsymbol{F}^{-}_{i,j} + \boldsymbol{F}^{+}_{i,j} - \tilde{\boldsymbol{F}}^{+}_{i-1,j} + \boldsymbol{G}^{-}_{i,j+1} - \boldsymbol{G}^{-}_{i,j} + \boldsymbol{G}^{+}_{i,j} - \tilde{\boldsymbol{G}}^{+}_{i,j-1}, \tag{10.23}$$

$$\tilde{\tilde{\boldsymbol{R}}}_{i,j} = \tilde{\tilde{\boldsymbol{F}}}^{-}_{i+1,j} - \tilde{\boldsymbol{F}}^{-}_{i,j} + \tilde{\boldsymbol{F}}^{+}_{i,j} - \tilde{\boldsymbol{F}}^{+}_{i-1,j} + \tilde{\tilde{\boldsymbol{G}}}^{-}_{i,j+1} - \tilde{\boldsymbol{G}}^{-}_{i,j} + \tilde{\boldsymbol{G}}^{+}_{i,j} - \tilde{\boldsymbol{G}}^{+}_{i,j-1}, \tag{10.24}$$

and σ is a relaxation factor that can be used to optimize convergence rates. In these equations, $\boldsymbol{F}^{+}$, $\boldsymbol{F}^{-}$, $\boldsymbol{G}^{+}$, and $\boldsymbol{G}^{-}$ represent the split approximations to the cell area h times the contravariant components of the flux vectors in the corresponding mesh coordinate directions. The residual fluxes are approximated using either the scalar or Convective Upwind Split Pressure (CUSP) versions of the Symmetric LImited Positive (SLIP) approximations developed by (Jameson 1995a, 1995b).

The implementation of this procedure is made computationally very efficient by locally transforming the residuals to those corresponding to the equations written in symmetrizing variables, as described in Section 2.7, then transforming the corrections back to the conserved variables. Numerical experiments indicate that it can be beneficial to perform additional corrections in supersonic zones, when they are present in the solution. The CPU time required for these multiple sweeps is reduced by "freezing"

Table 10.1. Force coefficients for the fast, preconditioned multigrid solutions using CUSP spatial discretization.

Case	Fig.	MG Cycles	C_L	C_D
RAE 2822; $M_\infty = 0.75$; $\alpha = 3.00$	–	100	1.1417	0.04851
	15c	5	1.1429	0.04851
	15a	3	1.1451	0.04886
NACA 0012; $M_\infty = 0.80$; $\alpha = 1.25$	–	100	0.3725	0.02377
	15d	5	0.3746	0.02391
	15b	3	0.3770	0.02387

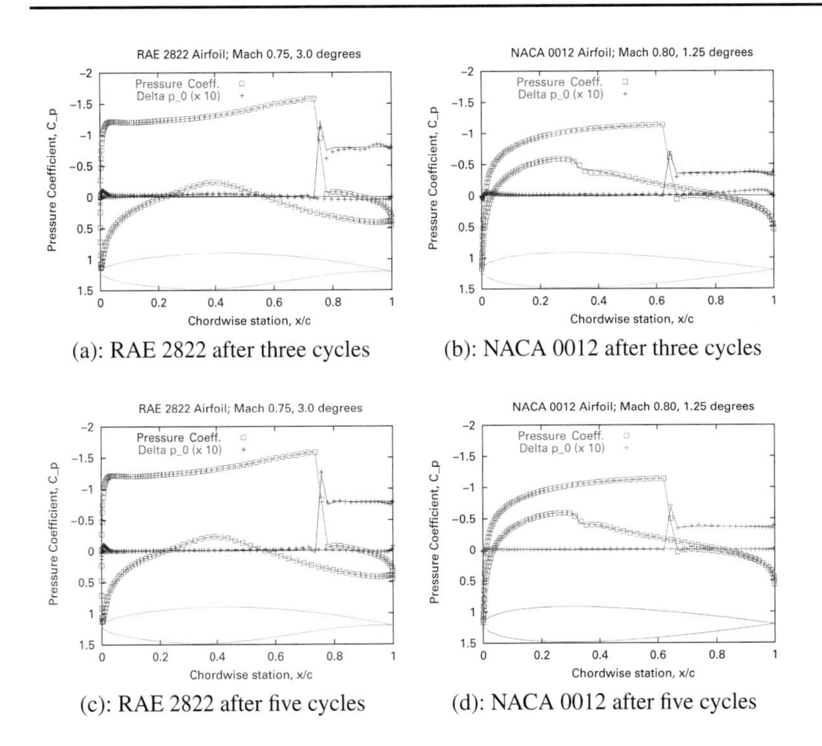

(a): RAE 2822 after three cycles (b): NACA 0012 after three cycles

(c): RAE 2822 after five cycles (d): NACA 0012 after five cycles

Figure 10.15 Pressure distribution for flow past the RAE 2822 and NACA 0012 airfoil after three cycles (a)-(b), and five cycles (c)-(d). Solid lines represent fully converged solution.

the matrix coefficients $|A|$ and $|B|$ that appear in (10.21) and (10.22). The additional memory required to store these coefficient matrices is minimized by storing only the symmetrized form of the Jacobians (which requires only seven additional quantities to be stored for each mesh cell). The SGS method has the disadvantage, however, that it cannot be implemented on massively parallel computers because it relies on successive updates.

The most dramatic results to date have been achieved by using the nonlinear SGS scheme to drive a multigrid procedure using W-cycles (Jameson & Caughey 2001). Figure 10.15 and Table 10.1 illustrate the results for two-dimensional transonic flow calculations. In Figure 10.15 the fully converged solution is shown by the solid lines, and it can be seen that the results are essentially fully converged after five cycles. This is an example of "text book" multigrid efficiency.

10.13 RANS Equations

Multigrid methods have so far proved less effective in calculations of turbulent viscous flows using the Reynolds, averaged Navier–Stokes equations. These require highly anisotropic grids with very fine mesh intervals normal to the wall to resolve the boundary layers. While simple multigrid methods still yield fast initial convergence, they tend to slow down as the calculation proceeds to a low asymptotic rate. This has motivated the introduction of semi-coarsening and directional coarsening methods (Mulder 1989, Mulder 1992, Allmaras 1993, Allmaras 1995, Allmaras 1997, Pierce & Giles 1997, Pierce, Giles, Jameson, & Martinelli 1997).

In 2007, Cord Rossow proposed using several iterations of an LUSGS scheme of the type described in Section 9.10 as a preconditioner at each stage of an explicit RK scheme (Rossow 2007). This hybrid RKSGS scheme was further developed by Swanson, Turkel, and Rossow (2007), and it has been proven to be an effective driver of a full approximation multigrid scheme for steady state RANS calculations, yielding rates of convergence comparable to those that have been achieved for Euler calculations. Schemes of this type have also been extensively studied by the present author, and the following paragraphs describe an RKSGS scheme that has proved effective. The scheme is generally similar to the Swanson–Turkel–Rossow implementation but differs in some significant details.

In order to solve the equation

$$\frac{d\boldsymbol{w}}{dt} + \boldsymbol{R}(\boldsymbol{w}) = 0,$$

where $\boldsymbol{R}(\boldsymbol{w})$ represents the residual of the space discretization, an n stage RKSGS scheme is formulated as

$$\begin{aligned}
\boldsymbol{w}^{(1)} &= \boldsymbol{w}^{(0)} - \alpha_1 \Delta t\, \boldsymbol{P}^{-1}\boldsymbol{R}^{(0)},\\
\boldsymbol{w}^{(2)} &= \boldsymbol{w}^{(0)} - \alpha_2 \Delta t\, \boldsymbol{P}^{-1}\boldsymbol{R}^{(1)},\\
\vdots &= \vdots\\
\boldsymbol{w}^{(m)} &= \boldsymbol{w}^{(0)} - \Delta t\, \boldsymbol{P}^{-1}\boldsymbol{R}^{(m-1)},
\end{aligned}$$

where $\mathbf{P}$ denotes the LUSGS preconditioner. As in the case of the basis additive RK scheme, the convective and dissipative parts of the residual are treated separately. Thus if $\boldsymbol{R}^{(k)}$ is split as

$$\boldsymbol{R}^{(k)} = \boldsymbol{Q}^{(k)} + \boldsymbol{D}^{(k)},$$

then

$$Q^{(0)} = Q\left(w^{(0)}\right), \quad D^{(0)} = D\left(w^{(0)}\right),$$

and for $k > 0$,

$$Q^{(k)} = Q\left(w^{(k)}\right),$$
$$D^{(k)} = \beta_k D\left(w^{(k)}\right) + (1 - \beta_k) D^{(k-1)}.$$

Two- and three-stage schemes that have proved effective are as follows. The coefficients of the two-stage scheme are

$$\alpha_1 = 0.24, \quad \beta_1 = 1,$$
$$\alpha_2 = 1, \quad \beta_2 = 2/3,$$

while those of the three-stage scheme are

$$\alpha_1 = 0.15, \quad \beta_1 = 1,$$
$$\alpha_2 = 0.40, \quad \beta_2 = 0.5,$$
$$\alpha_3 = 1, \quad \beta_3 = 0.5.$$

While the residual is evaluated with a second order accurate discretization, the preconditioner is based on a first order upwind discretization. Using a notation similar to Section 7.7.1 for a quadrilateral cell numbered zero, with neighbors $k = 1$ to 4, the residual of a central difference scheme is

$$R_0^c = \frac{1}{S_0} \sum_{k=1}^{4} h_{k0},$$

where S_0 is the cell area and

$$h_{k0} = \frac{1}{2}\left[(f_k + f_0)\Delta y_{k0} - (g_k + g_0)\Delta x_{k0}\right].$$

Introducing a Roe matrix A_{k0} satisfying

$$A_{k0}(w_k - w_0) = (f_k - f_0)\Delta y_{k0} - (g_k - g_0)\Delta x_{k0},$$

and noting that the sums

$$\sum_{k=1}^{4} \Delta y_{k0} = 0, \quad \sum_{k=1}^{4} \Delta x_{k0} = 0,$$

we can write

$$R_0^c = \frac{1}{S_0} \sum_{k=1}^{4} A_{k0}(w_k - w_0).$$

The residual of the first order upwind scheme is obtained by subtracting $|A_{k0}|(\boldsymbol{w}_k - \boldsymbol{w}_0)$ from $\boldsymbol{h}_{k0}$ to produce

$$\boldsymbol{R}_0 = \frac{1}{2S_0}\sum_{k=1}^{4}\left(A_{k0} - |A_{k0}|\right)(\boldsymbol{w}_k - \boldsymbol{w}_0).$$

Also noting that we can split A_{k0} as

$$A_{k0} = A_{k0}^{+} + A_{k0}^{-},$$

where

$$A_{k0}^{\pm} = \frac{1}{2}\left(A_{k0} \pm |A_{k0}|\right),$$

have positive and negative eigenvalues respectively, we can write

$$\boldsymbol{R}_0 = \frac{1}{S_0}\sum_{k=1}^{4} A_{k0}^{-}(\boldsymbol{w}_k - \boldsymbol{w}_0).$$

We now consider an implicit scheme of the form

$$\boldsymbol{w}^{n+1} = \boldsymbol{w}^{n} - \Delta t\Big(\epsilon \boldsymbol{R}(\boldsymbol{w}^{n+1}) + (1-\epsilon)\boldsymbol{R}(\boldsymbol{w}^{n})\Big),$$

and approximate $\mathbf{R}_0^{n+1}$ as

$$\boldsymbol{R}_0^{n+1} = \boldsymbol{R}_0^{n} + \frac{1}{S_0}\sum_{k=1}^{4} A_{k0}^{-}(\delta\boldsymbol{w}_k - \delta\boldsymbol{w}_0),$$

where $\delta\boldsymbol{w}$ denotes $\boldsymbol{w}^{n+1} - \boldsymbol{w}^{n}$. This yields the approximate implicit scheme

$$\left(I - \epsilon\frac{\Delta t}{S_0}\sum_{k=1}^{4} A_{k0}^{+}\right)\delta\boldsymbol{w}_0 + \epsilon\frac{\Delta t}{S_0}\sum_{k=1}^{4} A_{k0}^{-}\delta\boldsymbol{w}_k = -\Delta t\boldsymbol{R}_0^{n}. \tag{10.25}$$

The LUSGS preconditioner uses symmetric Gauss–Seidel forward and backward sweeps to approximately solve the equation using the latest available values for $\delta\mathbf{w}_k$, and starting from $\delta\mathbf{w} = 0$. In practice it has been found that a single forward and backward sweep is generally sufficient, and very rapid convergence of the overall multigrid scheme can be obtained with both the two- and the three-stage schemes. Moreover, a choice of the coefficient $\epsilon < 1$ is effectively a way to over-relax the iterations, and it turns out that the fastest rate of convergence is obtained with ϵ around 0.65 for the two-stage scheme and ϵ less than 0.5 for the three-stage scheme.

At each interface, A^{-} is modified by the introduction of an entropy fix to bound the absolute values of the eigenvalues away from zero. Denoting the components of the unit normal to the edge by n_x and n_y, the normal velocity is

$$q_n = n_x u + n_y v,$$

and the eigenvalues are

$$\lambda^{(1)} = q_n,$$
$$\lambda^{(2)} = q_n + c,$$
$$\lambda^{(3)} = q_n - c,$$

where c is the speed of sound. Velocity components, pressure, p and density ρ at the interface may be calculated by arithmetic averaging, and then $c = \sqrt{\frac{\gamma p}{\rho}}$.

The absolute eigenvalues $|\lambda^{(k)}|$ are then replaced by

$$e_k = \begin{cases} |\lambda^{(k)}|, & |\lambda^{(k)}| > a \\ \frac{1}{2}\left(a + \frac{\lambda^{(k)2}}{a}\right), & |\lambda^{(k)}| \le a, \end{cases}$$

where a is set as a fraction of the speed of sound:

$$a = d_1 c.$$

Without this modification, the scheme diverges. In numerical experiments it has been found that the overall scheme converges reliably for transonic flow with $d_1 \approx 0.10$. The modified eigenvalues of A^- are then

$$\lambda^{(1)} = q_n - e_1,$$
$$\lambda^{(2)} = q_n + c - e_2,$$
$$\lambda^{(3)} = q_n - c - e_3.$$

It also proves helpful to further augment the diagonal coefficients by a term proportional to a fraction d_2 of the normal velocity q_n. The interface matrix is also modified to provide for contributions from the viscous Jacobian. The final interface matrix $\mathbf{A}^{\pm}$ can be conveniently represented via a transformation to the symmetrizing variables,

$$d\hat{\mathbf{w}}^T = \left[\frac{dp}{c^2}, \quad \frac{\rho}{c}du, \quad \frac{\rho}{c}dv, \quad \frac{dp}{c^2} - d\rho\right],$$

as described in Section 2.7. Let s be the edge length. then the interface matrix can be expressed as

$$A^I = s\tilde{M}\left(\tilde{A}_c + \frac{s}{\rho S_0}\tilde{A}_v - d_2 q_n I\right)\tilde{M}^{-1},$$

where $\tilde{A}_c$ and $\tilde{A}_v$ are the convective and viscous contributions, $d_2 q_n I$ is the diagonal augmentation term, and $\tilde{M}$ is the transformation matrix from the conservative to the symmetrizing variables. Here

$$\tilde{M} = \begin{bmatrix} 1 & 0 & 0 & -1 \\ u & c & 0 & -u \\ v & 0 & c & -v \\ H & uc & vc & -\frac{q^2}{2} \end{bmatrix}$$

and

$$\tilde{M}^{-1} = \begin{bmatrix} \tilde{\gamma}\frac{q^2}{2} & -\tilde{\gamma}u & -\tilde{\gamma}v & \tilde{\gamma} \\ -\frac{u}{c} & \frac{1}{c} & 0 & 0 \\ -\frac{v}{c} & 0 & \frac{1}{c} & 0 \\ \tilde{\gamma}\frac{q^2}{2} - 1 & -\tilde{\gamma}u & -\tilde{\gamma}v & \tilde{\gamma} \end{bmatrix},$$

where

$$q^2 = u^2 + v^2, \quad \tilde{\gamma} = \frac{\gamma - 1}{c^2}.$$

Define

$$r_1 = \frac{1}{2}(q_2 + q_3 - 1),$$

$$r_2 = \frac{1}{2}(q_2 - q_3).$$

Then the modified convective Jacobian is

$$\tilde{A}_c = \begin{bmatrix} r_1 + q_1 & n_x r_2 & n_y r_2 & 0 \\ n_x r_2 & n_x^2 r_1 + q_1 & n_x n_y r_1 & 0 \\ n_y r_2 & n_x n_y r_1 & n_y^2 r_1 + q_1 & 0 \\ 0 & 0 & 0 & q_1 \end{bmatrix}.$$

Also let μ, λ, and κ be the viscosity, bulk viscosity, and conductivity coefficients. Then the viscous Jacobian is

$$\tilde{A} = \begin{bmatrix} -(\gamma - 1)\kappa & 0 & 0 & -\kappa \\ 0 & -(\mu + n_x^2 \mu^*) & -n_x n_y \mu^* & 0 \\ 0 & -n_x n_y \mu^* & -(\mu + n_y^2 \mu^*) & 0 \\ -(\gamma - 1)\kappa & 0 & 0 & -\kappa \end{bmatrix},$$

where

$$\mu^* = \mu + \lambda,$$

and typically $\lambda = -\frac{2}{3}\mu$. Here it is assumed that the Reynolds stress is modeled by an eddy viscosity. Then if μ_m and μ_t are the molecular and eddy viscosity,

$$\mu = \mu_m + \mu_t,$$

while it is the common practice to take

$$\kappa = \frac{\mu_m}{Pr} + \frac{\mu_t}{Pr_t},$$

where Pr and Pr_t are the molecular and turbulent Prandtl numbers.

An alternative implementation can be derived from a first order flux vector split scheme with the interface flux

$$\boldsymbol{h}_{k0} = (\boldsymbol{f}_0^+ + \boldsymbol{f}_k^-)\Delta y_{k0} - (\boldsymbol{g}_0^+ + \boldsymbol{g}_k^-)\Delta x_{k0},$$

where $\boldsymbol{f}$ and $\boldsymbol{g}$ are split as

$$\boldsymbol{f} = \boldsymbol{f}^+ + \boldsymbol{f}^-, \quad \boldsymbol{g} = \boldsymbol{g}^+ + \boldsymbol{g}^-,$$

and $\boldsymbol{f}^\pm$ and $\boldsymbol{g}^\pm$ have positive and negative eigenvalues respectively. In this case the implicit equation (10.25) should be replaced by

$$\left(I - \frac{\epsilon \Delta t}{S_0}\sum_{k=1}^{4} A_{k0}^+\right)\delta \boldsymbol{w}_0 + \frac{\epsilon \Delta t}{S_0}\sum_{k=1}^{4} A_{k0}^- \delta \boldsymbol{w}_k = -\Delta t \boldsymbol{R}_0^n,$$

where

$$A_{k0}^+ = \Delta y_{k0}\frac{\partial \boldsymbol{f}^+}{\partial \boldsymbol{w}} - \Delta x_{k0}\frac{\partial \boldsymbol{g}^+}{\partial \boldsymbol{w}},$$

and

$$A_{k0}^- = \Delta y_{k0}\frac{\partial \boldsymbol{f}^-}{\partial \boldsymbol{w}} - \Delta x_{k0}\frac{\partial \boldsymbol{g}^-}{\partial \boldsymbol{w}}.$$

This formula is similar to the scheme proposed by Swanson, Turkel, and Rossow (Swanson et al. 2007), but the consistent procedure is then to evaluate A_{k0}^+ at cell zero and A_{k0}^- at cell k. Numerical tests indicate that the alternative formulations work about equally well. In any case, these preconditioners may be applied with alternative second order schemes for the residual $\boldsymbol{R}_0$ on the right hand side. It turns out that the scheme works very well when $\boldsymbol{R}_0$ is evaluated by the Jameson–Schmidt–Turkel (JST) scheme, as presented in Section 6.19, provided that the artificial viscosity coefficients $k^{(2)}$ and $k^{(4)}$ in (6.39) and (6.40) and the blend factor r for the spectral radii in the different coordinate directions are all carefully tuned.

Figures 10.16–10.18 show the results of the RKSGS scheme for three standard benchmark cases: the RAE 2822 airfoil cases 1, 9, and 10. These calculations were performed with the Baldwin–Lomax turbulence model and the JST scheme.

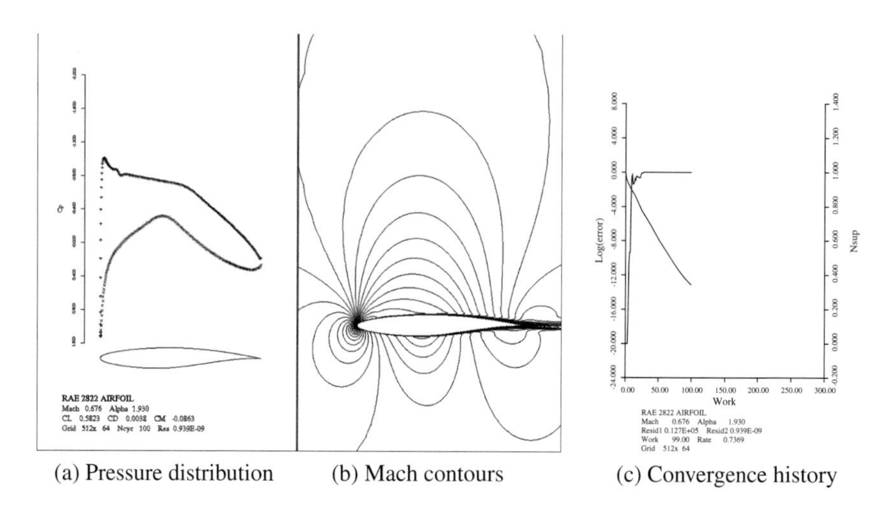

(a) Pressure distribution (b) Mach contours (c) Convergence history

Figure 10.16 RAE 2822 airfoil, Mach 0.676, α 1.930 degrees. RKSGS scheme using the Baldwin–Lomax turbulence model and the JST scheme.

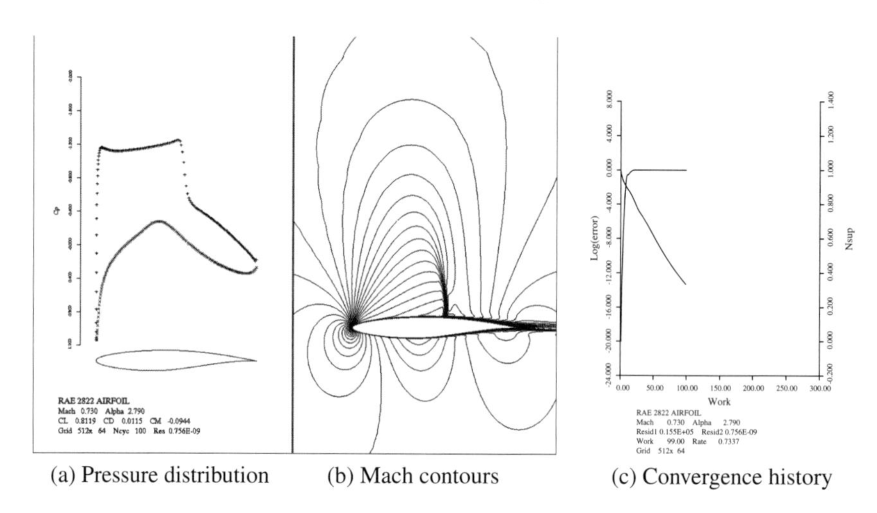

(a) Pressure distribution (b) Mach contours (c) Convergence history

Figure 10.17 RAE 2822 airfoil, Mach 0.730, α 2.790 degrees. RKSGS scheme using the Baldwin–Lomax turbulence model and the JST scheme.

10.14 Multigrid on Unstructured Meshes

The multigrid method can be applied on unstructured meshes by interpolating between a sequence of separately generated meshes with progressively increasing cell sizes (Jameson & Mavriplis 1987, Mavriplis & Jameson 1990, Mavriplis & Martinelli 1991, Peraire, Peirö, & Morgan 1992). In the case of a vertex centered scheme of the type described in Section 7.8, the grid transfer operators are defined by first locating the fine grid triangle or tetrahedron containing each coarse grid vertex and the coarse grid triangle or tetrahedron containing each fine grid vertex. Then flow variables at a coarse grid node P are taken as the linear interpolation of the corresponding values at nodes

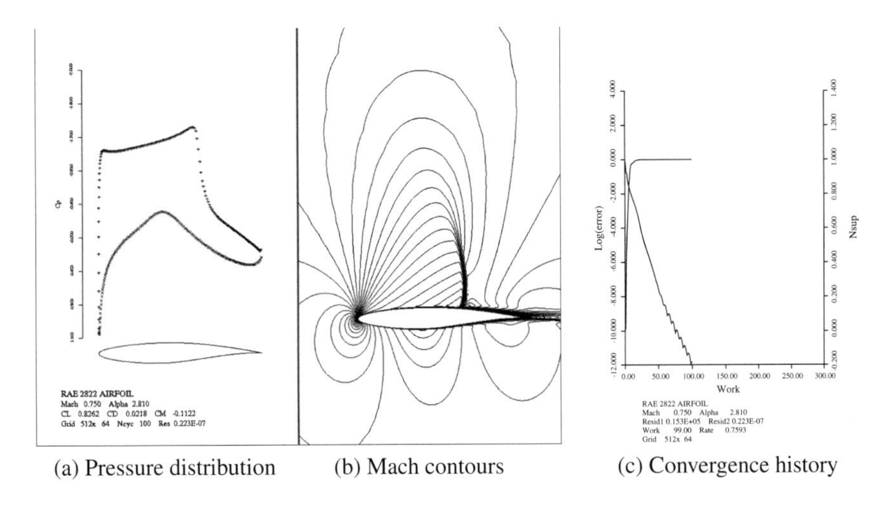

(a) Pressure distribution (b) Mach contours (c) Convergence history

Figure 10.18 RAE 2822 airfoil, Mach 0.750, α 2.810 degrees. RKSGS scheme using the Baldwin–Lomax turbulence model and the JST scheme.

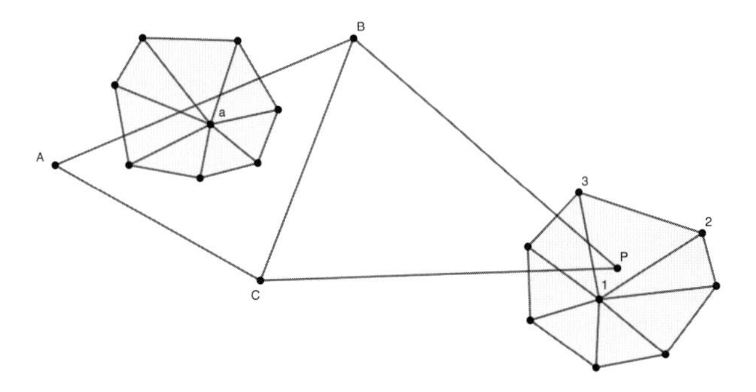

Figure 10.19 Grid transfers for the unstructured multigrid algorithm: residual at "a" is distributed to A, B, and C; flow variable at P is the linear interpolation of values at 1, 2, and 3.

1, 2, and 3, as shown in Figure 10.19, which are the vertices of the fine grid triangle enclosing P. These three nodes include the fine grid node that is closest to P, thus ensuring an accurate representation of the flow-field on the coarse grid. The fine grid residual $\boldsymbol{R}_a$ at "a" in Figure 10.19 is linearly distributed to the coarse grid nodes A, B, and C, which are the vertices of the coarse grid triangle enclosing "a." This linear distribution is accomplished by the use of shape functions that have the value 1 at one of the coarse grid triangle vertices and vanish at the other two vertices. This implies that the sum of the residual contribution to A, B, and C equals the residual at "a", and the weighting is such that, if "a" and A coincide, then the contribution at A is equal to R_a, and the contributions at B and C vanish. This type of transfer is conservative. When transferring the corrections from the coarse grid back to the fine grid, a simple linear interpolation formula is used. Thus, the correction at the fine grid node "a" is taken as the linear interpolation of the three corrections at nodes A, B, and C which enclose "a" on the coarse grid.

The remaining difficulty is the need to perform searches to locate the cells containing each vertex. In the case of the fine grid vertices, a naive search over all the coarse grid cells would require $O(N^2)$ operations, where N is the number of grid points, and thus would be prohibitively expensive, requiring more time than the flow solution itself. Hence. an efficient search algorithm is needed. It requires lists of the neighbors of each node and cell to be stored for both the coarse and fine grids. It is initiated by providing an initial guess IC_1 for the coarse grid cell and then testing IC_1 to see if it encloses the fine grid node NF. Since we are free to begin the search with any fine grid node and any coarse grid cell, we choose points whose locations are known (such as trailing edge values). If the test is negative, then the neighbors of IC_1 are tested. If these tests also fail, then the neighbors of these neighbors are tested. This process is continued until, after n tries, the address IC_n of the cell enclosing NF is located. This entire procedure is repeated for every node of the fine grid. The next fine grid node NF_2 is thus chosen as a neighbor of NF, and the initial guess for the enclosing cell is taken as IC_n, the coarse grid cell that is now known to enclose the previous NF. In this manner, we are assured of a good initial guess, since IC_n and NF_2 must be located in the same region of the computational domain. This type of search can be achieved in $O(NlogN)$ operations. In practice, of the order of 10 searches are required to locate an enclosing cell. An adjustment is required to treat boundary vertices in the case that they are not contained in any coarse grid cell.

It is not easy to generate very coarse meshes for complex configurations. Instead of generating the coarse meshes separately, one may generate successively coarser meshes by repeatedly collapsing the edges of a fine mesh to merge the vertices at each end of the edge (Crumpton & Giles 1995, May & Jameson 2005). This requires heuristic rules to determine the sequence of edges to be collapsed, and care should be taken to preserve the shape of the boundary.

An alternative approach that has proved popular is to agglomerate the control volumes of the fine mesh to form coarse grid control volumes (Lallemand & Dervieux 1987, Mavriplis & Venkatakrishnan 1996). This also requires heuristic rules to determine the sequence of control volumes to be agglomerated. It also requires a separate algorithm for the coarse grid solution to treat coarse grid control volumes of irregular shape.

Another way to simplify the definition of successively coarser discretizations is to resort to meshless discretizations at all coarser levels with a "multi-cloud" algorithm (Katz & Jameson 2009). Meshless discretizations have the disadvantage that they are generally not conservative, but this is not important when they are used solely on the coarser grids to accelerate convergence.

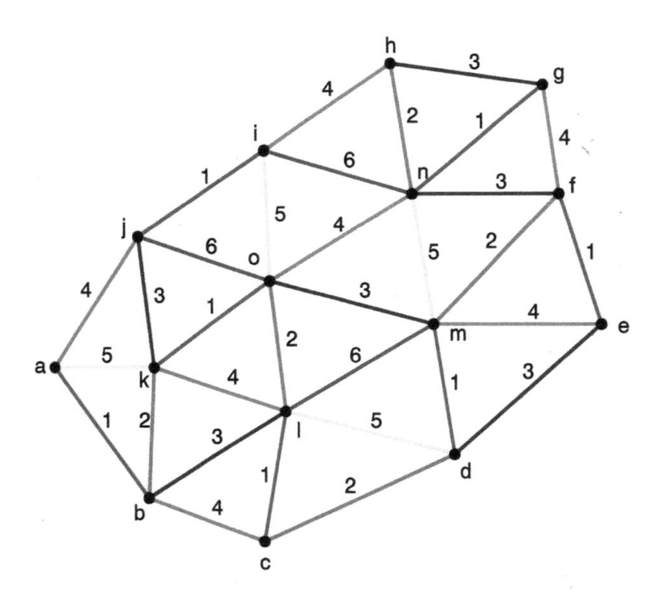

Figure 7.29 Edge coloring.

Figure 7.30 Pressure distribution over MD11, $Ma = 0.825$.

Figure 7.31 Contours of Mach number over A320 wing surface.

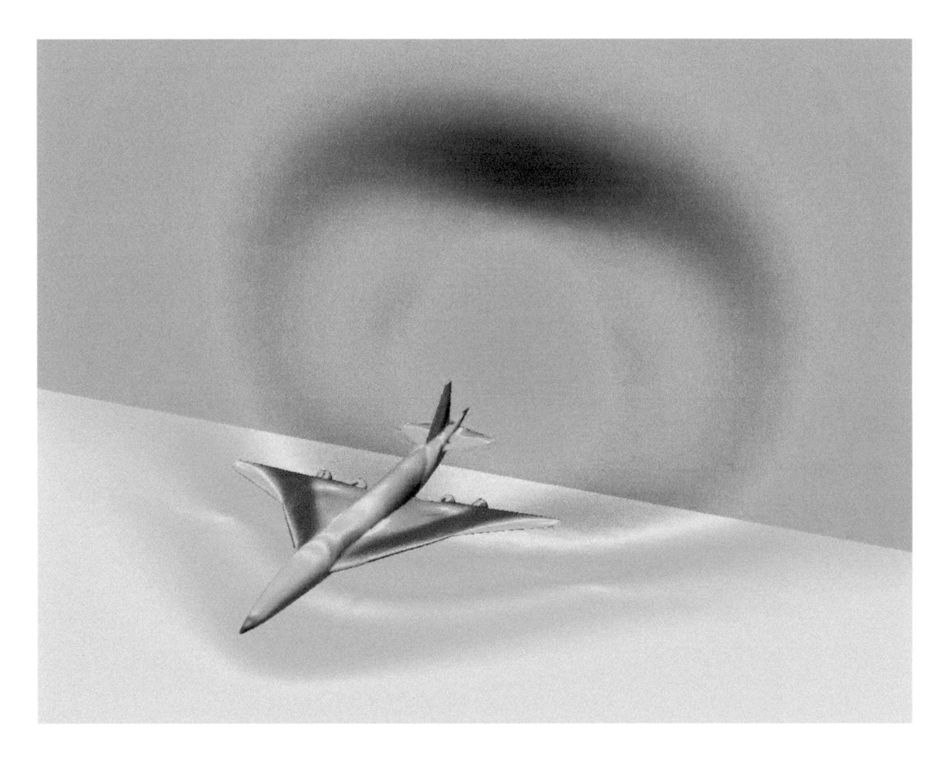

Figure 7.32 Pressure distribution over NASA High Speed Civil Transport, $Ma = 2.4$.

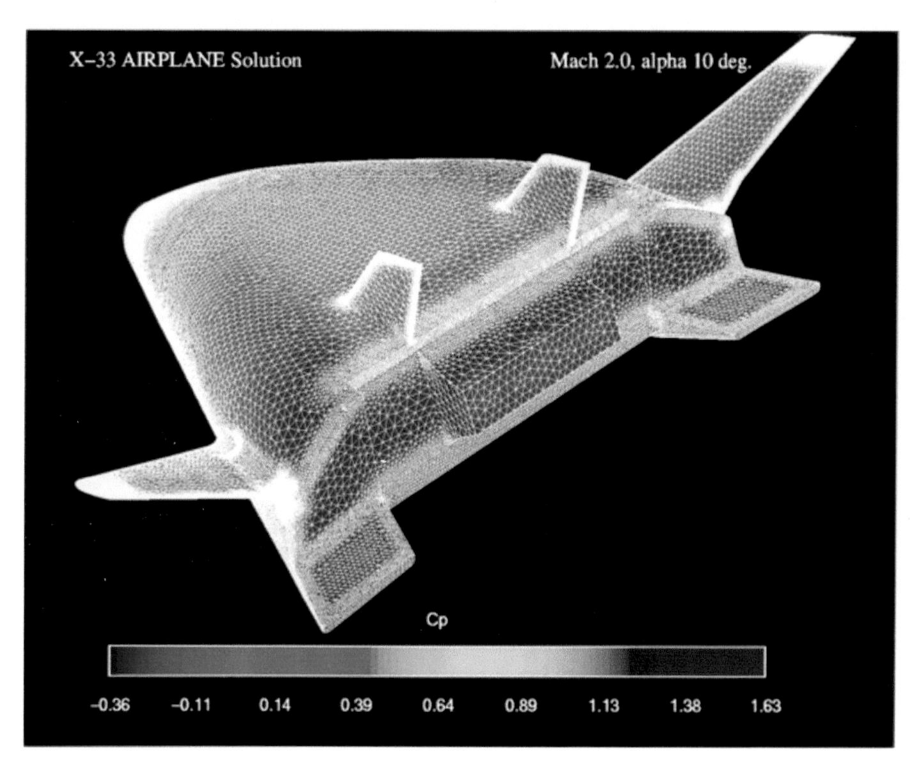

Figure 7.33 Pressure distribution over X33, $Ma = 2.0$.

11 Time-Accurate Methods for Unsteady Flow

11.1 Strong-Stability Preserving (SSP) Schemes

The general theory of shock capturing schemes for hyperbolic conservation laws has been presented in Chapters 5 and 6, where it is shown that schemes which satisfy the TVD or LED properties can capture discontinuities without any oscillations or overshoots. The analysis in these chapters is limited to semi-discrete schemes or schemes with a first order accurate forward Euler time discretization.

In order to enable accurate simulations of time dependent flows with moving shocks and contact discontinuities, there is a need for higher order accurate time discretization schemes that can preserve the TVD property. Higher order explicit time stepping schemes of this type have been developed during the last two decades. The first such schemes were introduced by Shu and Osher (1988) and were further developed by Gottlieb and Shu (1998). They were subsequently labeled strong-stability preserving (SSP) schemes in a review article by Gottlieb, Shu, and Tadmor (2001).

Suppose that the discrete solution of a nonlinear conservation law

$$\frac{d\boldsymbol{w}}{dt} + \boldsymbol{R}(\boldsymbol{w}) = 0$$

satisfies the strong-stability property

$$\|\boldsymbol{w}^{n+1}\| \leq \|\boldsymbol{w}^n\|$$

for some convex functional, such as the total variation, with a forward Euler time stepping scheme subject to the restriction

$$\Delta t \leq \Delta t_{\mathrm{FE}}.$$

A higher order time discretization scheme is strongly-stability preserving (SSP) if the strong stability property is satisfied whenever the time step satisfies the restriction

$$\Delta t \leq c\Delta t_{\mathrm{FE}},$$

where the ratio c of the allowable time step to the forward Euler time step is called the SSP coefficient.

The general form of an explicit SSP Runge–Kutta scheme is as follows. At time step n, a scheme with m stages is

$$
\begin{aligned}
\boldsymbol{w}^{(0)} &= \boldsymbol{w}^n, \\
\boldsymbol{w}^{(i)} &= \sum_{k=0}^{i-1} \left(\alpha_{ik}\boldsymbol{w}^{(k)} - \beta_{ik}\Delta t \boldsymbol{R}(\boldsymbol{w}^{(k)})\right), \quad i = 1,2,\ldots,m \\
&= \sum_{k=0}^{i-1} \alpha_{ik}\left(\boldsymbol{w}^{(k)} - \frac{\beta_{ik}}{\alpha_{ik}}\Delta t \boldsymbol{R}(\boldsymbol{w}^{(k)})\right), \quad i = 1,2,\ldots,m \\
\boldsymbol{w}^{n+1} &= \boldsymbol{w}^{(m)},
\end{aligned}
$$

where all $\alpha_{ik}, \beta_{ik} \geq 0$, and $\alpha_{ik} = 0$ only if $\beta_{ik} = 0$. Here for consistency

$$
\sum_{k=0}^{i-1} \alpha_{ik} = 1, \quad i = 1,2,\ldots,s,
$$

since if $\boldsymbol{R}(\boldsymbol{w}) = 0$, $\boldsymbol{w}$ should remain unchanged. Accordingly, each stage value $\boldsymbol{w}^{(i)}$ is a convex combination of forward Euler time steps with Δt replaced by $\frac{\beta_{ik}}{\alpha_{ik}}\Delta t$. It follows that the scheme is SSP if

$$
\max_{i,k} \frac{\beta_{ik}}{\alpha_{ik}} \Delta t \leq \Delta t_{\mathrm{FE}},
$$

and accordingly the SSP coefficient is

$$
c = \min \frac{\alpha_{ik}}{\beta_{ik}}
$$

over i and k for which $\beta_{ik} \neq 0$.

Shu and Osher (1988) identified the following two-stage second order accurate SSP scheme, SSPRK(2,2):

$$
\begin{aligned}
\boldsymbol{w}^{(1)} &= \boldsymbol{w}^{(0)} - \Delta t \boldsymbol{R}(\boldsymbol{w}^{(0)}), \\
\boldsymbol{w}^{(2)} &= \frac{1}{2}\boldsymbol{w}^{(0)} + \frac{1}{2}\left(\boldsymbol{w}^{(1)} - \Delta t \boldsymbol{R}(\boldsymbol{w}^{(1)})\right),
\end{aligned}
$$

for which the coefficients α_{ik} and β_{ik} are displayed in Figure 11.1. Shu and Osher also identified the following three stage third order accurate SSP scheme, SSPRK(3,3):

$$
\begin{aligned}
\boldsymbol{w}^{(1)} &= \boldsymbol{w}^{(0)} - \Delta t \boldsymbol{R}(\boldsymbol{w}^{(0)}), \\
\boldsymbol{w}^{(2)} &= \frac{3}{4}\boldsymbol{w}^{(0)} + \frac{1}{4}\left(\boldsymbol{w}^{(1)} - \Delta t \boldsymbol{R}(\boldsymbol{w}^{(1)})\right), \\
\boldsymbol{w}^{(3)} &= \frac{1}{3}\boldsymbol{w}^{(0)} + \frac{2}{3}\left(\boldsymbol{w}^{(2)} - \Delta t \boldsymbol{R}(\boldsymbol{w}^{(2)})\right),
\end{aligned}
$$

with the coefficients displayed in Figure 11.2. Subsequently, Gottlieb and Shu (1998) showed that these schemes are optimal in the sense that they maximize the SSP coefficients of two- and three-stage SSP schemes that are second and third order accurate, respectively.

α_{ik}	1	
	$\frac{1}{2}$	$\frac{1}{2}$
β_{ik}	1	
	0	$\frac{1}{2}$

Figure 11.1 SSPRK(2,2) coefficients.

α_{ik}	1		
	$\frac{3}{4}$	$\frac{1}{4}$	
	$\frac{1}{3}$	0	$\frac{2}{3}$
β_{ik}	1		
	0	$\frac{1}{4}$	
	0	0	$\frac{2}{3}$

Figure 11.2 SSPRK(3,3) coefficients.

Shu and Osher were not able to find a four-stage fourth order accurate SSP scheme with non-negative coefficients β_{ik}. However, they were able to construct four-stage fourth order accurate SSP schemes by allowing some negative coefficients β_{ik}, and whenever $\beta_{ik} < 0$, replacing $\boldsymbol{R}(\boldsymbol{w})$ by the corresponding operator $\tilde{\boldsymbol{R}}(\boldsymbol{w})$, which would be TVD or LED for a forward Euler step in reverse time. Such an operator can be constructed by substituting downwind differences for upwind differences. Shu and Osher identified schemes that required two evaluations of the reverse time operator, while Gottlieb and Shu were able to prove that a four-stage fourth order accurate SSP scheme must have at least one negative coefficient β_{ik}.

The introduction of the reverse time operator $\tilde{\boldsymbol{R}}(\boldsymbol{w})$ is an undesirable complication. Spiteri and Ruuth (2002) identified a five-stage fourth order accurate SSP scheme, SSPRK(5,4), with an SSP coefficient $c = 1.508$. This was maximized by numerical optimization. The results of recent research on SSP methods have been collected in a book by Gottlieb, Ketcheson, and Shu (2011). It is now known that explicit SSPRK methods are at most fourth order accurate (Ruuth & Spiteri 2002), and the scheme SSPRK(5,4) has been confirmed to be optimal. The efficiency of an SSP scheme can be measured by the effective SSP coefficient $\frac{c}{m}$, where m is the number of stages.

For the schemes SSPRK(2,2), SSPRK(3,3), and SSPRK(5,4), $\frac{c}{m} = \frac{1}{2}, \frac{1}{3}$, and 0.302, respectively.

If one accepts a lower order of accuracy, it is possible to attain larger values of the effective SSP coefficient. For example the three-stage second order accurate scheme, SSPRK(3,2),

$$\begin{aligned}
\boldsymbol{w}^{(0)} &= \boldsymbol{w}^n, \\
\boldsymbol{w}^{(1)} &= \boldsymbol{w}^{(0)} - \frac{\Delta t}{2}\boldsymbol{R}(\boldsymbol{w}^{(0)}), \\
\boldsymbol{w}^{(2)} &= \boldsymbol{w}^{(1)} - \frac{\Delta t}{2}\boldsymbol{R}(\boldsymbol{w}^{(1)}), \\
\boldsymbol{w}^{(3)} &= \frac{1}{3}\boldsymbol{w}^{(0)} + \frac{2}{3}\left(\boldsymbol{w}^{(2)} - \frac{\Delta t}{2}\boldsymbol{R}(\boldsymbol{w}^{(2)})\right),
\end{aligned}$$

has an effective coefficient $\frac{c}{3} = \frac{2}{3}$, while the four-stage third order accurate scheme, SSPRK(4,3),

$$\begin{aligned}
\boldsymbol{w}^{(0)} &= \boldsymbol{w}^n, \\
\boldsymbol{w}^{(1)} &= \boldsymbol{w}^{(0)} - \frac{\Delta t}{2}\boldsymbol{R}(\boldsymbol{w}^{(0)}), \\
\boldsymbol{w}^{(2)} &= \boldsymbol{w}^{(1)} - \frac{\Delta t}{2}\boldsymbol{R}(\boldsymbol{w}^{(1)}), \\
\boldsymbol{w}^{(3)} &= \frac{2}{3}\boldsymbol{w}^{(0)} + \frac{1}{3}\left(\boldsymbol{w}^{(2)} - \frac{\Delta t}{2}\boldsymbol{R}(\boldsymbol{w}^{(2)})\right), \\
\boldsymbol{w}^{(4)} &= \boldsymbol{w}^{(3)} - \frac{\Delta t}{2}\boldsymbol{R}(\boldsymbol{w}^{(3)}),
\end{aligned}$$

has an effective coefficient $\frac{c}{4} = \frac{1}{2}$. Moreover, these two schemes can be combined as an embedded pair for error estimation as

$$\begin{aligned}
\boldsymbol{w}^{(0)} &= \boldsymbol{w}^n, \\
\boldsymbol{w}^{(1)} &= \boldsymbol{w}^{(0)} - \frac{\Delta t}{2}\boldsymbol{R}(\boldsymbol{w}^{(0)}), \\
\boldsymbol{w}^{(2)} &= \boldsymbol{w}^{(1)} - \frac{\Delta t}{2}\boldsymbol{R}(\boldsymbol{w}^{(1)}), \\
\boldsymbol{w}_2^{n+1} &= \frac{1}{3}\boldsymbol{w}^{(0)} + \frac{2}{3}\left(\boldsymbol{w}^{(2)} - \frac{\Delta t}{2}\boldsymbol{R}(\boldsymbol{w}^{(2)})\right), \\
\boldsymbol{w}^{(3)} &= \frac{2}{3}\boldsymbol{w}^{(0)} + \frac{1}{3}\left(\boldsymbol{w}^{(2)} - \frac{\Delta t}{2}\boldsymbol{R}(\boldsymbol{w}^{(2)})\right), \\
\boldsymbol{w}_3^{n+1} &= \boldsymbol{w}^{(3)} - \frac{\Delta t}{2}\boldsymbol{R}(\boldsymbol{w}^{(3)}),
\end{aligned}$$

where $\boldsymbol{w}_2^{n+1}$ and $\boldsymbol{w}_3^{n+1}$ are second and third order accurate approximations.

By increasing the number of stages, larger effective SSP coefficients can be attained for a given order of accuracy. For example, the scheme SSPRK(m,2) due to Ketcheson achieves $\frac{c}{m} = 1 - \frac{1}{m}$, and it can be implemented in the low storage form

$$\boldsymbol{q}_1 = \boldsymbol{w}, \quad q_2 = 0,$$

for $i = 1, m-1$

$$\boldsymbol{q}_1 = \boldsymbol{q}_1 - \frac{\Delta t}{m-1}\boldsymbol{R}(\boldsymbol{q}_1)$$

end

$$\boldsymbol{q}_1 = (m-1)\boldsymbol{q}_1 + \boldsymbol{q}_2 - \frac{\Delta t}{m}\boldsymbol{R}(\boldsymbol{q}_1).$$

Gottlieb, Ketcheson, and Shu (2009) formulated a 10-stage fourth order accurate scheme with an effective SSP coefficient $\frac{c}{m} = 0.60$.

One might hope to obtain larger SSP coefficients by the introduction of implicit Runge–Kutta methods. It was already shown, however, by Gottlieb, Shu, and Tadmor (2001) that unfortunately any implicit Runge–Kutta scheme that is unconditionally SSP for arbitrarily large time steps is at most first order accurate. Recent results for implicit SSPRK schemes have been presented by Ketcheson, Macdonald, and Gottlieb (2009). It is now known that implicit SSPRK schemes are at most sixth order accurate. Implicit schemes can achieve higher values of the SSP coefficient c/m than explicit schemes, but the numerically optimized implicit schemes still have values of $c/m < 2$, and taking into account the computational cost of the implicit stages, they are not as efficient as explicit schemes. Accordingly, SSP schemes are primarily useful for simulations with fast time scales that can only be resolved by small time-steps corresponding to small CFL numbers.

11.2 Dual Time Stepping Schemes

Time dependent calculations are needed for a number of important applications, such as flutter analysis or the analysis of the flow past a helicopter rotor, in which the stability limit of an explicit scheme forces the use of much smaller time steps than would be needed for an accurate simulation. This motivates the "dual time stepping" scheme, in which a multigrid explicit scheme can be used in an inner iteration to solve the equations of a fully implicit time stepping scheme (Jameson 1991).

Suppose that (9.1) is approximated as

$$D_t \boldsymbol{w} + \boldsymbol{R}(\boldsymbol{w}) = 0.$$

Here D_t is the sth order accurate backward difference formula (BDF) given in Section 9.4, where it was observed that the first and second order forumlae are L-stable. Equation (9.1) is now treated as a modified steady state problem to be solved by a multigrid scheme using variable local time steps in a fictitious time τ. For example, in the case of the second order BDF, one solves

$$\frac{\partial \boldsymbol{w}}{\partial \tau} + \boldsymbol{R}^*(\boldsymbol{w}) = 0,$$

where

$$\boldsymbol{R}^*(w) = \frac{3}{2\Delta t}\boldsymbol{w} + \boldsymbol{R}(\boldsymbol{w}) - \frac{2}{\Delta t}\boldsymbol{w}^n + \frac{1}{2\Delta t}\boldsymbol{w}^{n-1},$$

and the last two terms are treated as fixed source terms. In the RK-BDF method, the modified Runge–Kutta scheme (described in Section 10.4) is used in the inner iterations with variable local time stepping and multigrid acceleration. The first term shifts the Fourier symbol of the equivalent model problem to the left in the complex plane. While this promotes stability, it may also require a limit to be imposed on the magnitude of the local time step $\Delta\tau$ relative to that of the implicit time step Δt. This may be relieved by a point-implicit modification of the multi-stage scheme (Melson, Sanetrik, & Atkins 1993). In the case of problems with moving boundaries, the equations must be modified to allow for movement and deformation of the mesh.

This method has proved effective for the calculation of unsteady flows that might be associated with wing flutter (Alonso & Jameson 1994, Alonso, Martinelli, & Jameson 1995) and also in the calculation of unsteady incompressible flows (Belov, Martinelli, & Jameson 1995). It has the advantage that it can be added as an option to a computer program that uses an explicit multigrid scheme, allowing it to be used for the efficient calculation of both steady and unsteady flows. A similar approach has been successfully adopted for unsteady flow simulations on unstructured grids by Venkatakrishnan and Mavriplis (1996).

Alternatively, in the SGS-BDF method, the nonlinear SGS multigrid scheme ((10.12)–(10.24)) is used in the inner iterations. The unsteady flow past a pitching airfoil provides a useful test case for time-integration methods. A case for which experimental data is also available is a NACA 64A010 at Mach 0.796, pitching about zero angle of attack with an amplitude ± 1.01 degrees, at a reduced frequency

$$\omega_r = \frac{\omega_{chord}}{2u_\infty} = 0.202.$$

This is test case CT6 in AGARD Report 702 (Davis, 1982). A snap shot of the resulting shock motion is shown in Figure 11.3, while Figure 11.4 shows the oval curve traced by the lift coefficient due to its phase lag as the angle of attack varies sinusoidally. These results were obtained using the nonlinear SGS scheme with the third order accurate BDF. It can be seen that the results over-plot when 24 or 36 time steps are used in each pitching cycle. In the case of 24 steps, the maximum CFL number is 4,153 in the vicinity of the trailing edge. It can also be seen that the results using three multigrid cycles in each time step are almost identical to those using nine multigrid cycles in each time step.

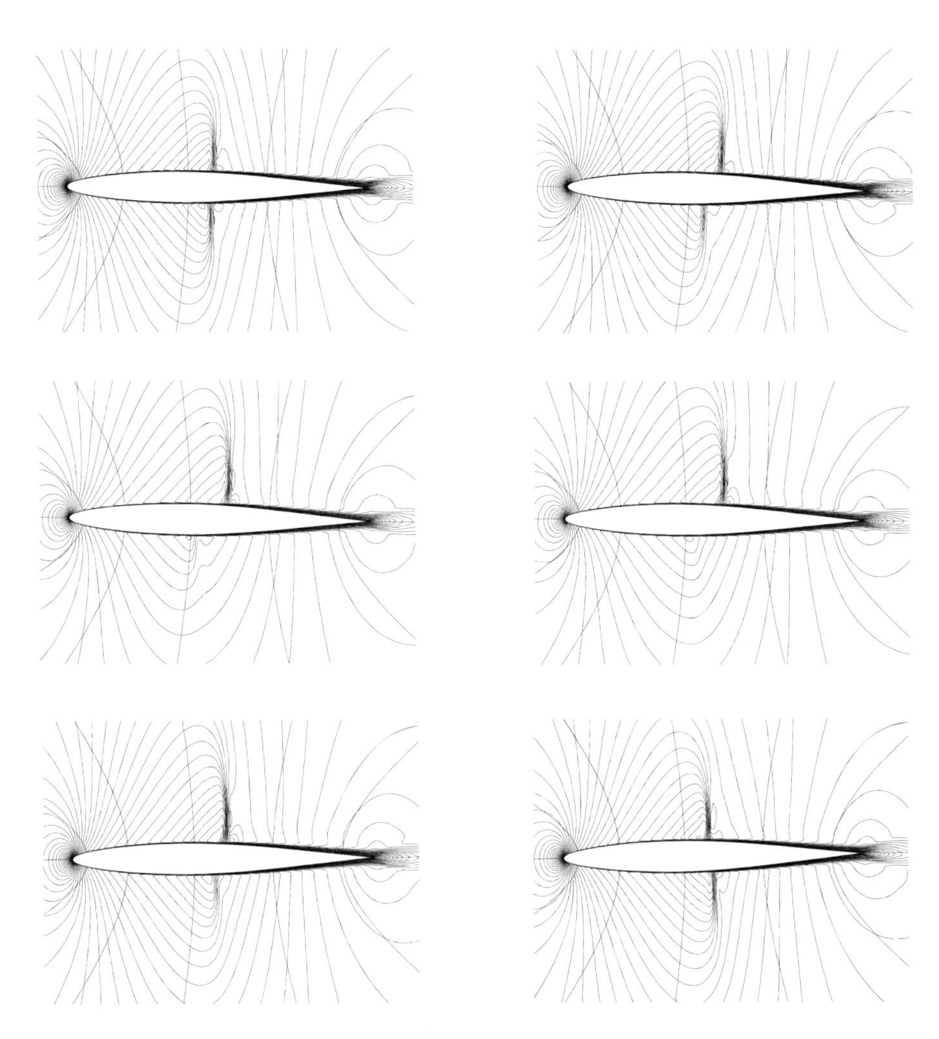

Figure 11.3 Pressure contours at various time instances: AGARD case CT6 showing the oscillating shock waves.

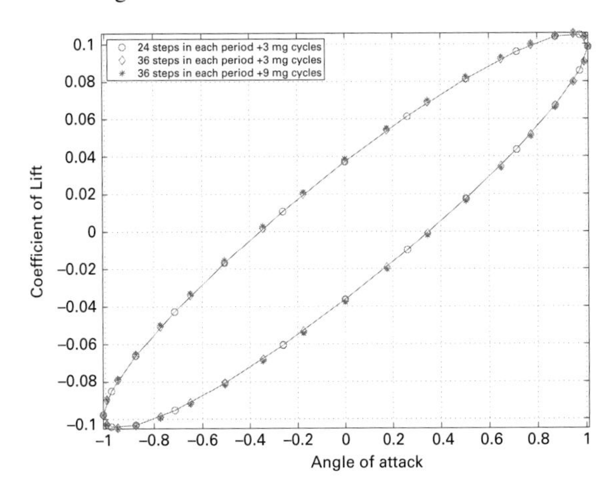

Figure 11.4 Results of the SGS-BDF Method for AGARD case CT6.

11.3 Implicit Runge–Kutta Schemes for Time-Accurate Simulations

In order to allow large time steps to be safely used, it is desirable that the time discretization scheme should be A-stable. Actually, they should preferably have the property of L-stability, meaning that solutions decay as $\alpha\Delta t \to \infty$ for $\alpha\Delta t$ in the left half-plane. The BDF2 scheme is L-stable, as has been discussed in Section 9.3, but in accordance with the Dahlquist barrier theorem (Dahlquist 1963), the higher order backward difference formulas are not A-stable

The limitation to second order accuracy can be overcome by resorting to implicit Runge–Kutta methods, pioneered by Butcher (1987). The cost of implicit Runge–Kutta methods is greatly reduced by limiting the implicit evaluations to $R(u^{(k)})$ in the kth stage – so-called diagonal implicit Runge–Kutta (DIRK) schemes. They can be further simplified by requiring the same diagonal coefficient at each stage to produce single diagonal implicit Runge–Kutta (SDIRK) schemes. The simplest SDIRK schemes are SDIRK1, which is the backward Euler scheme with the Butcher tableau

$$\begin{array}{c|c} 1 & 1 \\ & 1 \end{array},$$

and the second order accurate SDIRK2 scheme with the Butcher tableau

$$\begin{array}{c|cc} \alpha & \alpha & \\ 1 & 1-\alpha & \alpha \\ \hline & 1-\alpha & \alpha \end{array},$$

where $\alpha = 1 - \frac{1}{\sqrt{2}}$. These schemes are both L-stable.

The following third order accurate SDIRK scheme has been given by Butcher (2003). Its tableau is

$$\begin{array}{c|ccc} \alpha & \alpha & & \\ \frac{1}{2}(1-\alpha) & \frac{1}{2}(1-\alpha) & \alpha & \\ 1 & -\frac{1}{4}(6\alpha^2 - 16\alpha + 1) & \frac{1}{4}(6\alpha^2 - 20\alpha + 5) & \alpha \\ \hline & -\frac{1}{4}(6\alpha^2 - 16\alpha + 1) & \frac{1}{4}(6\alpha^2 - 20\alpha + 5) & \alpha \end{array},$$

with $\alpha = 0.4358665215$. This scheme is also L-stable, and it has been successfully used by Persson, in high order simulations of unsteady viscous flow (Persson, Willis, & Peraire 2010).

The search for even higher order A-stable and L-stable methods is a subject of ongoing research. It has been found that it is beneficial to include an initial explicit stage to produce a family of schemes known as ESDIRK schemes. A scheme that has been widely investigated is a six-stage fourth order accurate scheme due to Kennedy and Carpenter (Bijl, Carpenter, Vatsa, & Kennedy 2002). It was tested for unsteady Navier–Stokes simulations on unstructured meshes by Jothiprasad, Mavriplis, and Caughey (2003). They concluded that it could achieve a given level of accuracy at a lower computational cost than the BDF2 scheme because it allowed the use of larger time steps, which outweighed the additional cost of each time step. Recently, Boom and Zingg (2013) used numerical optimization to identify a six-stage fifth order accurate scheme, ESDIRK(6,5), which appears to offer some advantages over the ESDIRK(6,4) scheme of Kennedy and Carpenter.

Fully implicit Runge–Kutta schemes have not been widely used in CFD because of the complexity of solving the coupled equations. Recently the author showed that these schemes can be implemented using a dual time stepping method, provided that they are suitably preconditioned (Jameson 2017a). The following sections present the approach for two- and three-stage Gauss schemes (Butcher 1964b) and two- and three-stage Radau 2A schemes (Butcher 1964a).

There are typically very large variations of cell size in unsteady flow simulations, and this impacts the robustness of the Gauss schemes in particular. Accordingly the residual equation is written in the form

$$V\frac{d\boldsymbol{w}}{dt} + \boldsymbol{R}(\boldsymbol{w}) = 0, \tag{11.1}$$

where the cell area or volume V is explicitly displayed. For (11.1), the two-stage Gauss scheme takes the form

$$\begin{aligned}
\boldsymbol{\xi}_1 &= \boldsymbol{w}^n - \frac{\Delta t}{V}\left(a_{11}\boldsymbol{R}(\boldsymbol{\xi}_1) + a_{12}\boldsymbol{R}(\boldsymbol{\xi}_2)\right) \\
\boldsymbol{\xi}_2 &= \boldsymbol{w}^n - \frac{\Delta t}{V}t\left(a_{21}\boldsymbol{R}(\boldsymbol{\xi}_1) + a_{22}\boldsymbol{R}(\boldsymbol{\xi}_2)\right) \\
\boldsymbol{w}^{n+1} &= \boldsymbol{w}^n - \frac{\Delta t}{2V}\left(\boldsymbol{R}(\boldsymbol{\xi}_1) + \boldsymbol{R}(\boldsymbol{\xi}_2)\right),
\end{aligned} \tag{11.2}$$

where the matrix A of coefficients is

$$A = \begin{bmatrix} \frac{1}{4} & \frac{1}{4} - \frac{\sqrt{3}}{6} \\ \frac{1}{4} + \frac{\sqrt{3}}{6} & \frac{1}{4} \end{bmatrix},$$

and the stage values correspond to Gauss integration points inside the time step with the values $\left(\frac{1}{2} - \frac{\sqrt{3}}{6}\right)\Delta t$ and $\left(\frac{1}{2} + \frac{\sqrt{3}}{6}\right)\Delta t$.

The three-stage Gauss scheme takes the form

$$
\begin{aligned}
\boldsymbol{\xi}_1 &= \boldsymbol{w}^n - \frac{\Delta t}{V}\left(a_{11}\boldsymbol{R}(\boldsymbol{\xi}_1) + a_{12}R(\boldsymbol{\xi}_2) + a_{13}\boldsymbol{R}(\boldsymbol{\xi}_3)\right)\\
\boldsymbol{\xi}_2 &= \boldsymbol{w}^n - \frac{\Delta t}{V}\left(a_{21}\boldsymbol{R}(\boldsymbol{\xi}_1) + a_{22}\boldsymbol{R}(\boldsymbol{\xi}_2) + a_{23}\boldsymbol{R}(\boldsymbol{\xi}_3)\right)\\
\boldsymbol{\xi}_3 &= \boldsymbol{w}^n - \frac{\Delta t}{V}\left(a_{31}\boldsymbol{R}(\boldsymbol{\xi}_1) + a_{32}\boldsymbol{R}(\boldsymbol{\xi}_2) + a_{33}\boldsymbol{R}(\boldsymbol{\xi}_3)\right)\\
\boldsymbol{w}^{n+1} &= \boldsymbol{w}^n - \frac{\Delta t}{18V}\left(5\boldsymbol{R}(\boldsymbol{\xi}_1) + 8\boldsymbol{R}(\boldsymbol{\xi}_2) + 5\boldsymbol{R}(\boldsymbol{\xi}_3)\right),
\end{aligned}
$$

where the matrix A of coefficients is

$$
A = \begin{bmatrix}
\frac{5}{36} & \frac{2}{9} - \frac{\sqrt{15}}{15} & \frac{5}{36} - \frac{\sqrt{15}}{30}\\
\frac{5}{36} + \frac{\sqrt{15}}{30} & \frac{2}{9} & \frac{5}{36} - \frac{\sqrt{15}}{24}\\
\frac{5}{36} + \frac{\sqrt{15}}{30} & \frac{2}{9} + \frac{\sqrt{15}}{15} & \frac{5}{36}
\end{bmatrix},
$$

and the stage values correspond to the intermediate times $\left(\frac{1}{2} - \frac{\sqrt{15}}{10}\right)\Delta t, \frac{1}{2}\Delta t$, and $\left(\frac{1}{2} + \frac{\sqrt{15}}{10}\right)\Delta t$ within the time step.

The two-stage Gauss scheme is fourth order accurate, while the three-stage Gauss scheme is sixth order accurate. Both schemes are A-stable but not L-stable. The Radau 2A schemes include the end of the time interval as one of the integration points, corresponding to Radau integration. Consequently they have an order of accuracy $2s - 1$ for s stages. They have the advantages, on the other hand, that the last stage value is the final value, eliminating the need for an extra step to evaluate $\boldsymbol{w}^{n+1}$, and that they are L-stable.

For the solution of (11.1), the two-stage Radau 2A scheme takes the form

$$
\begin{aligned}
\boldsymbol{\xi}_1 &= \boldsymbol{w}^n - \frac{\Delta t}{V}\left(a_{11}\boldsymbol{R}(\boldsymbol{\xi}_1) + a_{12}\boldsymbol{R}(\boldsymbol{\xi}_2)\right)\\
\boldsymbol{\xi}_2 &= \boldsymbol{w}^n - \frac{\Delta t}{V}\left(a_{21}\boldsymbol{R}(\boldsymbol{\xi}_1) + a_{22}\boldsymbol{R}(\boldsymbol{\xi}_2)\right)\\
\boldsymbol{w}^{n+1} &= \boldsymbol{\xi}_2,
\end{aligned}
$$

where the matrix A of coefficients is

$$
A = \begin{bmatrix}
\frac{5}{12} & -\frac{1}{12}\\
\frac{3}{4} & \frac{1}{4}
\end{bmatrix},
$$

and the stage values correspond to Radau integration points at $\frac{1}{3}\Delta t$ and Δt.

The three-stage Radau 2A scheme takes the form

$$\xi_1 = w^n - \frac{\Delta t}{V}\left(a_{11}\boldsymbol{R}(\xi_1) + a_{12}\boldsymbol{R}(\xi_2) + a_{13}\boldsymbol{R}(\xi_3)\right)$$

$$\xi_2 = w^n - \frac{\Delta t}{V}\left(a_{21}\boldsymbol{R}(\xi_1) + a_{22}\boldsymbol{R}(\xi_2) + a_{23}\boldsymbol{R}(\xi_3)\right)$$

$$\xi_3 = w^n - \frac{\Delta t}{V}\left(a_{31}\boldsymbol{R}(\xi_1) + a_{32}\boldsymbol{R}(\xi_2) + a_{33}\boldsymbol{R}(\xi_3)\right)$$

$$w^{n+1} = \xi_3,$$

where the matrix A of coefficients is

$$A = \begin{bmatrix} \frac{88-7\sqrt{6}}{360} & \frac{296-169\sqrt{6}}{1800} & \frac{-2+3\sqrt{6}}{225} \\ \frac{296+169\sqrt{6}}{1800} & \frac{88+7\sqrt{6}}{360} & \frac{-2-3\sqrt{6}}{225} \\ \frac{16-\sqrt{6}}{36} & \frac{16+\sqrt{6}}{36} & \frac{1}{9} \end{bmatrix},$$

and the stage values correspond to the Radau integration points $\frac{4-\sqrt{6}}{10}\Delta t$, $\frac{4+\sqrt{6}}{10}\Delta t$, and Δt.

It is shown next that dual time stepping can be used to solve the equations of all these schemes with an inexpensive preconditioner. In order to clarify the issues, it is useful to consider first the application of the two-stage Gauss scheme to the scalar equation

$$\frac{du}{dt} = au,$$

where a is a complex coefficient lying in the left half-plane. A naive application of dual time stepping would simply add derivatives in pseudo time to produce the scheme

$$\begin{aligned} \frac{d\xi_1}{d\tau} &= a(a_{11}\xi_1 + a_{12}\xi_2) + \frac{u^n - \xi_1}{\Delta t} \\ \frac{d\xi_2}{d\tau} &= a(a_{21}\xi_1 + a_{22}\xi_2) + \frac{u^n - \xi_2}{\Delta t}, \end{aligned} \tag{11.3}$$

which may be written in vector form as

$$\frac{d\xi}{d\tau} = B\xi + c, \tag{11.4}$$

where

$$B = \begin{bmatrix} a_{11}a - \frac{1}{\Delta t} & a_{12}a \\ a_{21}a & a_{22}a - \frac{1}{\Delta t} \end{bmatrix}, \quad c = \frac{1}{\Delta t}\begin{bmatrix} u^n \\ u^n \end{bmatrix}.$$

For (11.4) to converge to a steady state, the eigenvalues of B should lie in the left half-plane. These are the roots of

$$\det(\lambda I - B) = 0$$

or

$$\lambda^2 - \lambda\left((a_{11} + a_{22})a - \frac{2}{\Delta t}\right) + a_{11}a_{22}a^2 - (a_{11} + a_{22})\frac{a}{\Delta t} + \frac{1}{\Delta t^2} - a_{12}a_{21}a^2 = 0.$$

Substituting the coefficient values for the Gauss scheme given in (11.3), we find that

$$\lambda = \frac{1}{4}a - \frac{1}{\Delta t} \pm ia\sqrt{\frac{1}{48}}.$$

Then if $a = p + iq$,

$$\lambda = \frac{1}{4}p \pm q\sqrt{\frac{1}{48}} - \frac{1}{\Delta t} + i\left(\frac{1}{4}q \pm p\sqrt{\frac{1}{48}}\right),$$

and for large Δt, one root could have a positive real part even when a lies in the left half-plane.

In order to prevent this, we can modify (11.3) by multiplying the right hand side by a preconditioning matrix. It is proposed here to take the inverse of the Runge–Kutta coefficient array A as the preconditioning matrix. Here

$$A^{-1} = \frac{1}{D}\begin{bmatrix} a_{22} & -a_{12} \\ a_{21} & a_{11} \end{bmatrix},$$

where the determinant of A is

$$D = a_{11}a_{22} - a_{12}a_{21}.$$

Setting

$$r_1 = a(a_{11}\xi_1 + a_{12}\xi_2) + \frac{u^n - \xi_1}{\Delta t}$$

$$r_2 = a(a_{21}\xi_1 + a_{22}\xi_2) + \frac{u^n - \xi_2}{\Delta t},$$

the preconditioned dual time stepping scheme now takes the form

$$\begin{aligned}\frac{d\xi_1}{d\tau} &= \frac{(a_{22}r_1 - a_{12}r_2)}{D} \\ &= a\xi_1 + \frac{a_{22}}{D\Delta t}\left(u^n - \xi_1\right) - \frac{a_{12}}{D\Delta t}\left(u^n - \xi_2\right) \\ \frac{d\xi_2}{d\tau} &= \frac{(a_{11}r_2 - a_{21}r_1)}{D} \\ &= a\xi_2 + \frac{a_{11}}{D\Delta t}\left(u^n - \xi_2\right) - \frac{a_{21}}{D\Delta t}\left(u^n - \xi_1\right),\end{aligned}$$

which may be written in the vector form (11.4) where now

$$B = \begin{bmatrix} a - \frac{a_{22}}{D\Delta t} & \frac{a_{12}}{D\Delta t} \\ \frac{a_{21}}{D\Delta t} & a - \frac{a_{11}}{D\Delta t} \end{bmatrix}, \quad c = \frac{1}{D\Delta t}\begin{bmatrix} (a_{22} - a_{12})u^n \\ (a_{11} - a - 21)u^n \end{bmatrix}.$$

Now the dual time stepping scheme will reach a steady state if the roots of

$$\det(\lambda I - B) = 0$$

lie in the left half-plane. Substituting the coefficients of B, the roots satisfy

$$\lambda^2 - \lambda\left(2a - \frac{a_{11} + a_{22}}{D\Delta t}\right) + a^2 - a\frac{a_{11} + a_{22}}{D\Delta t} + \frac{1}{D}\Delta t^2 = 0,$$

and using the coefficient values of the Gauss scheme, we now find that

$$\lambda^2 - \lambda\left(2a - \frac{6}{\Delta t}\right) + a^2 - \frac{6a}{\Delta t} + \frac{12}{\Delta t^2} = 0$$

yielding

$$\lambda = a - \frac{3}{\Delta t} \pm i\frac{\sqrt{3}}{\Delta t}.$$

Accordingly both roots lie in the left half-plane whenever a lies in the left half-plane, establishing the feasibility of the dual time stepping scheme.

In the case of the linear system

$$\frac{du}{dt} = Au,$$

a similar analysis may be carried out if A can be reduced to diagonal form by a similarity transformation

$$A = V\Lambda V^{-1}.$$

Then setting $v = V^{-1}u$,

$$\frac{dv}{dt} = \Lambda v,$$

and the same preconditioning scheme can be used separately for each component of v.

Following this approach, the proposed dual time stepping scheme for the nonlinear equations (11.2) is

$$\boldsymbol{r}_1 = \frac{V}{\Delta t}(\boldsymbol{w}^n - \boldsymbol{\xi}_1) - a_{11}\boldsymbol{R}(\boldsymbol{\xi}_1) - a_{12}\boldsymbol{R}(\boldsymbol{\xi}_2)$$

$$\boldsymbol{r}_2 = \frac{V}{\Delta t}(\boldsymbol{w}^n - \boldsymbol{\xi}_2) - a_{21}\boldsymbol{R}(\boldsymbol{\xi}_1) - a_{22}\boldsymbol{R}(\boldsymbol{\xi}_2)$$

and

$$\frac{d\boldsymbol{\xi}_1}{d\tau} = (a_{22}\boldsymbol{r}_1 - a_{12}\boldsymbol{r}_2)/D$$

$$\frac{d\boldsymbol{\xi}_2}{d\tau} = (a_{11}\boldsymbol{r}_2 - a_{21}\boldsymbol{r}_1)/D.$$

A more general approach, which facilitates the analysis of dual time stepping for implicit Runge–Kutta schemes, is as follows. Using vector notation, a naive application of dual time stepping yields the equations

$$\frac{d\xi}{d\tau} = aA\xi + \frac{1}{\Delta t}\left(w^n - \xi\right),$$

and the eigenvalues of the matrix

$$B = aA - \frac{1}{\Delta t}I$$

do not necessarily lie in the left half-plane. Introducing A^{-1} as a preconditioning matrix, the dual time stepping equations become

$$\frac{d\xi}{d\tau} = a\xi + \frac{1}{\Delta t}A^{-1}\left(w^n - \xi\right),$$

so we need the eigenvalues of

$$B = aI - \frac{1}{\Delta t}A^{-1}$$

to lie in the left half-plane for all values of a in the left half-plane.

The eigenvalues of B are

$$a - \frac{1}{\Delta t}\frac{1}{\lambda_k}, \quad k = 1, 2, 3,$$

where λ_k are the eigenvalues of A. Thus they will lie in the left half-plane for all values of a in the left half-plane if the eigenvalues of A lie in the right half-plane. The characteristic polynomials of A for the two-stage Gauss and Radau 2A schemes are

$$\lambda^2 - \frac{1}{2}\lambda + \frac{1}{12} = 0$$

and

$$\lambda^2 - \frac{2}{3}\lambda + \frac{1}{6} = 0,$$

with roots

$$\lambda = \frac{1}{4} \pm i\sqrt{\frac{1}{48}}$$

and

$$\lambda = \frac{1}{3} \pm i\sqrt{\frac{1}{18}}$$

respectively, which in both cases lie in the right half-plane. It may be determined by a rather lengthy calculation that the characteristic polynomials for the three-stage Gauss and Radau 2A schemes are

$$\lambda^3 - \frac{1}{2}\lambda^2 + \frac{1}{10}\lambda - \frac{1}{120} = 0$$

and

$$\lambda^3 - \frac{6}{10}\lambda^2 + \frac{3}{20}\lambda - \frac{1}{60} = 0.$$

Rather than calculating the roots directly, it is simpler to use the Routh–Hurwitz criterion, which states that the roots of

$$a_3\lambda^3 + a_2\lambda^2 + a_1\lambda + a_0$$

lie in the left half-plane if all the coefficients are positive and

$$a_2a_1 > a_3a_0.$$

The roots of A will lie in the right half-plane if the roots of $-A$ are in the left half-plane. Here, the characteristic polynomials of $-A$ for the two three-stage schemes are

$$\lambda^3 + \frac{1}{2}\lambda^2 + \frac{1}{10}\lambda + \frac{1}{120}$$

and

$$\lambda^3 + \frac{6}{10}\lambda^2 + \frac{3}{20}\lambda + \frac{1}{60},$$

and it is easily verified that the Routh–Hurwitz condition is satisfied in both cases. Thus, it may be concluded that the preconditioned dual time stepping scheme will work for all four of the implicit Runge–Kutta schemes under considerations.

In the present analysis it has been directly proved that for each of the four schemes, the eigenvalues of A lie in the right half-plane. It may be noted, however, that for an implicit Runge–Kutta scheme with coefficient matrix A and row vector $\boldsymbol{b}^T$ of the final stage coefficients, the stability function is

$$r(z) = 1 + \boldsymbol{b}^T(I - zA)^{-1}\boldsymbol{e},$$

where

$$\boldsymbol{e}^T = [1, 1, \ldots].$$

When this is expanded as

$$r(z) = \frac{N|z|}{D|z|},$$

$\det(I - zA)$ appears in the denominator. For an A-stable scheme, $D(z)$ cannot have any zeros in the left half-plane. Accordingly it may be concluded the eigenvalues of A must lie in the right half-plane for any A-stable implicit Runge–Kutta schemes.

The proposed dual time stepping scheme for the three-stage schemes is finally

$$\boldsymbol{r}_1 = \frac{V}{\Delta t}(\boldsymbol{w}^n - \boldsymbol{\xi}_1) - a_{11}\boldsymbol{R}(\boldsymbol{\xi}_1) - a_{12}\boldsymbol{R}(\boldsymbol{\xi}_2) - a_{13}\boldsymbol{R}(\boldsymbol{\xi}_3)$$

$$\boldsymbol{r}_2 = \frac{V}{\Delta t}(\boldsymbol{w}^n - \boldsymbol{\xi}_2) - a_{21}\boldsymbol{R}(\boldsymbol{\xi}_1) - a_{22}\boldsymbol{R}(\boldsymbol{\xi}_2) - a_{23}\boldsymbol{R}(\boldsymbol{\xi}_3)$$

$$\boldsymbol{r}_3 = \frac{V}{\Delta t}(\boldsymbol{w}^n - \boldsymbol{\xi}_3) - a_{31}\boldsymbol{R}(\boldsymbol{\xi}_1) - a_{32}\boldsymbol{R}(\boldsymbol{\xi}_2) - a_{33}\boldsymbol{R}(\boldsymbol{\xi}_3)$$

and

$$\frac{d\xi_1}{d\tau} = d_{11}\boldsymbol{r}_1 + d_{12}\boldsymbol{r}_2 + d_{13}\boldsymbol{r}_3$$

$$\frac{d\xi_2}{d\tau} = d_{21}\boldsymbol{r}_1 + d_{22}\boldsymbol{r}_2 + d_{23}\boldsymbol{r}_3$$

$$\frac{d\xi_3}{d\tau} = d_{31}\boldsymbol{r}_1 + d_{32}\boldsymbol{r}_2 + d_{33}\boldsymbol{r}_3,$$

where the coefficients d_{jk} are the entries of A^{-1}.

These schemes have been tested for the AGARD case CT6 of a pitching airfoil at Mach 0.796 discussed in Section 11.2 (Jameson 2017a). The equations were solved using the RKSGS scheme described in Section 13.3). It was found that the Gauss schemes require the equations to be solved to a very small tolerance in each time step, and were not very robust. Insufficient convergence could result in a failure to satisfy the Kutta condition, leading to a low pressure spike and possibly a vacuum state at the trailing edge. It appears that the most dangerous operation in the Gauss scheme is the calculation of the final update values from the stage values, which involves multiplication by $\Delta t/V$. Here Δt may be quite large, while the cell volume V becomes very small in the vicinity of the trailing edge.

The Radau schemes eliminate this operation because the last stage value is the final updated value, and have the advantage that they are L-stable. In practice they have proved extremely robust, and preserve high accuracy without requiring deep convergence of the inner iterations. Figures 11.5 and 11.6 show the results of a calculation with the 3-stage Radau scheme using only 3 time steps per pitching cycle in comparison with a reference solution calculated using the same scheme with 360 time steps per pitching cycle. It can be seen that the results for both lift and drag are very close to the reference solution.

11.4 Time-Spectral Methods

There are many unsteady flows in engineering devices such as turbomachinery or helicopter rotors, in which the flow is periodic. In this situation there is the opportunity to gain spectral accuracy by using a Fourier representation in time (Hall 1985, McMullen, Jameson, & Alonso 2002). Suppose the period T is divided into N time steps $\Delta t = T/N$. Let $\hat{w}_k$ be the discrete Fourier transform of w^n,

$$\hat{\boldsymbol{w}}_k = -\sum_{n=0}^{N-1} \boldsymbol{w}^n e^{-ikn\Delta t}.$$

Then the semi-discretization (9.1) is discretized as the pseudo-spectral scheme

$$D_t\boldsymbol{w}^n + \boldsymbol{R}(\boldsymbol{w}^n) = 0, \tag{11.5}$$

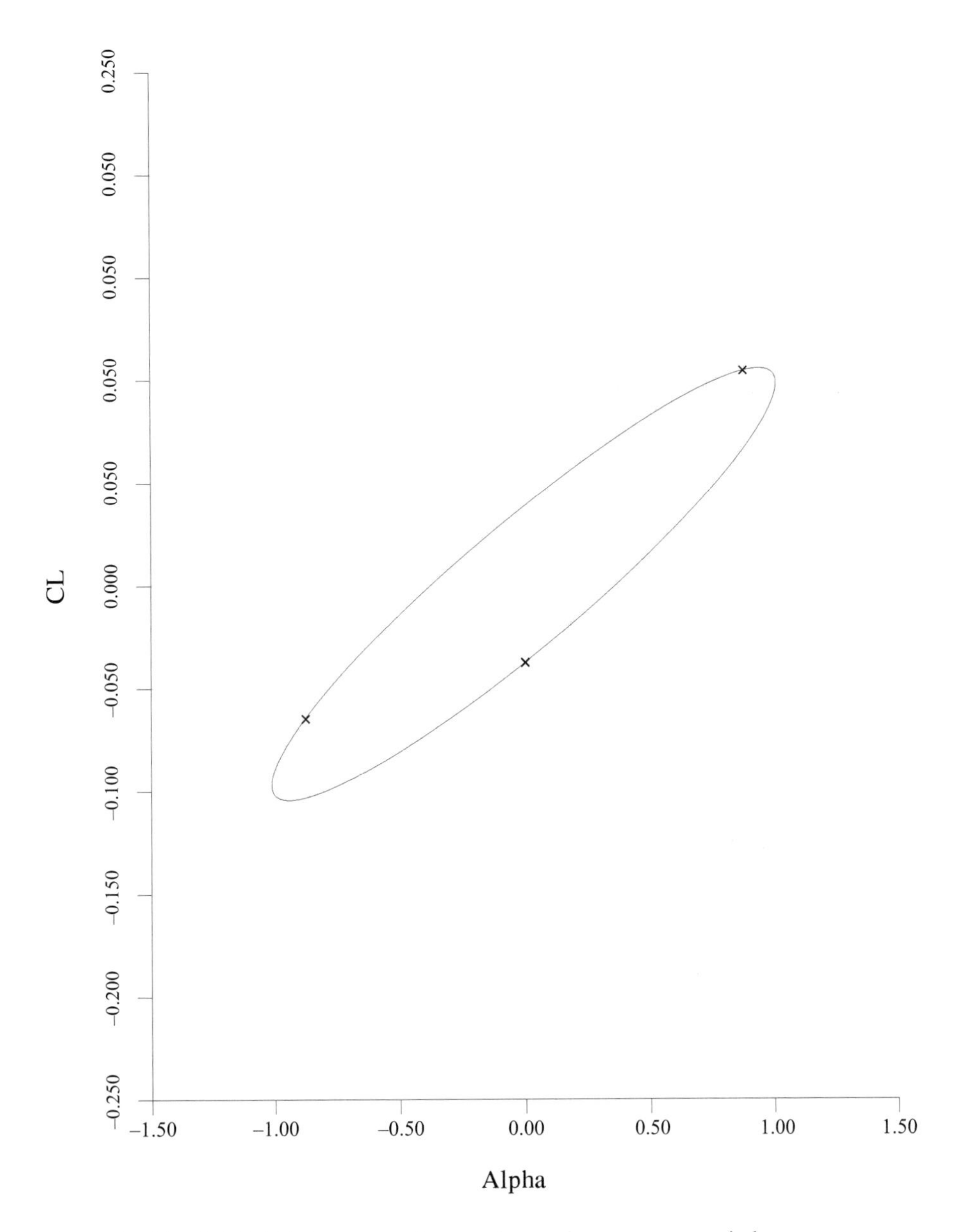

Figure 11.5 NACA 64A010, three-stage Radau scheme, three steps per period.

where

$$D_t w^n = \sum_{k=-\frac{N}{2}}^{\frac{N}{2}-1} ik\hat{w}_k e^{ikn\Delta t}.$$

Here D_t is a central difference formula connecting all the time levels, so equation (11.5) is an integrated space-time formulation that requires the simultaneous solution

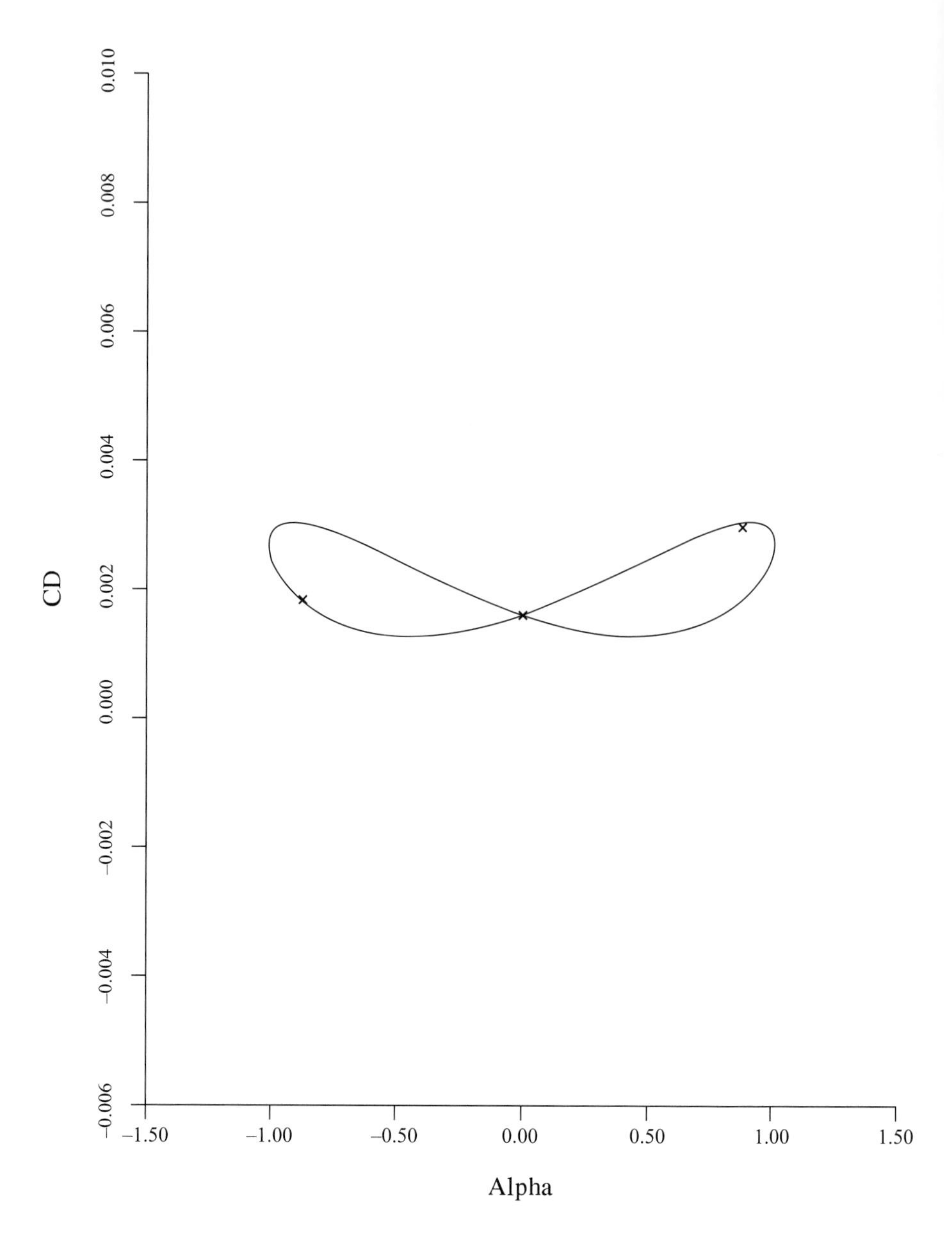

Figure 11.6 NACA 64A010, three-stage Radau scheme, three steps per period.

of the equations for all the time levels. Provided, however, that the solution is sufficiently smooth, (11.5) should yield spectral accuracy (exponential convergence with increasing N).

The time-spectral equation (11.5) may be solved by dual time stepping as

$$\frac{d\boldsymbol{w}^n}{d\tau} + D_t \boldsymbol{w}^n + \boldsymbol{R}(\boldsymbol{w}^n) = 0,$$

in pseudo-time τ, as in the case of the BDF. Alternatively it may be solved in the frequency domain. In this case we represent (11.5) as

$$\hat{\boldsymbol{R}}_k^* = ik\hat{\boldsymbol{w}}_k + \hat{\boldsymbol{R}}_k = 0, \tag{11.6}$$

where $\hat{\boldsymbol{R}}_k$ is the Fourier transform of $\boldsymbol{R}(\boldsymbol{w}(t))$. Because $\boldsymbol{R}(\boldsymbol{w})$ is nonlinear, $\hat{\boldsymbol{R}}_k$ depends on all the modes $\hat{\boldsymbol{w}}_k$. We now solve (11.6) by time evolution in pseudo-time:

$$\frac{d\boldsymbol{w}_k}{d\tau} + \hat{\boldsymbol{R}}_k^* = 0.$$

At each iteration in pseudotime, $\hat{\boldsymbol{R}}_k$ is evaluated indirectly. First $\boldsymbol{w}(t)$ is obtained as the reverse transform of $\hat{\boldsymbol{w}}_k$. Then we calculate the corresponding time history of the residual,

$$\boldsymbol{R}(t) = \boldsymbol{R}(\boldsymbol{w}(t)),$$

and obtain $\hat{\boldsymbol{R}}_k$ as the Fourier transform of $\boldsymbol{R}(t)$, as shown in Figure 11.7. McMullen (2003) has shown that the modified RK method with multigrid acceleration achieves

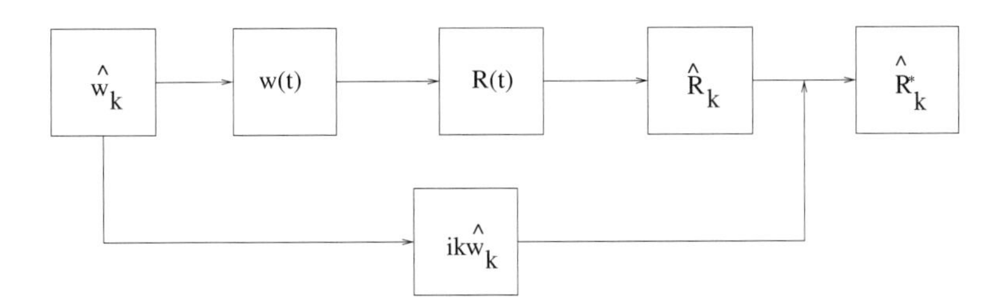

Figure 11.7 Solution of the time-spectral method in the frequency domain.

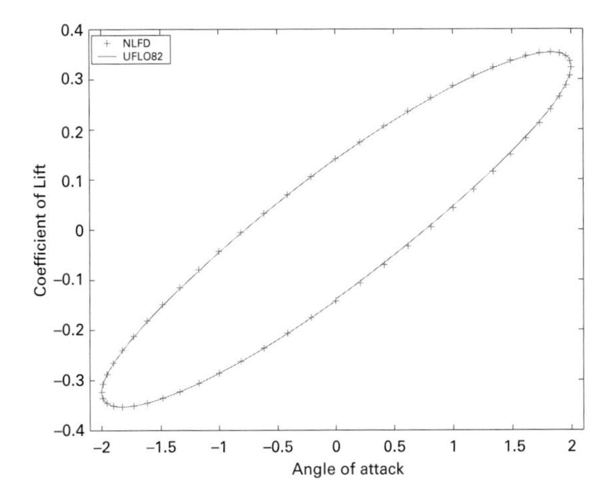

Figure 11.8 Comparison of the time-spectral and dual time stepping schemes.

essentially the same rate of convergence for the solution of the frequency-domain equation (11.6) as it does for the BDF.

While the time-spectral method should make it possible to achieve spectral accuracy, numerical tests have shown that it can give the accuracy required for practical applications ("engineering accuracy") with very small numbers of modes. Figure 11.8 shows perfect agreement between the dual time stepping and time-spectral method for the pitching airfoil considered in Section 11.2.

12 Energy Stability for Nonlinear Problems

12.1 Introduction

The use of energy estimates to establish the stability of discrete approximations to initial value problems has a long history. The energy method is discussed in the classical book by Richtmyer & Morton (1994), and it has been emphasized by the Uppsala school under the leadership of Kreiss and Gustafsson. Consider a well posed initial value problem of the form

$$\frac{du}{dt} = Lu, \tag{12.1}$$

where u is a state vector, and L is a linear differential operator in space with appropriate boundary conditions. Then, forming the inner product with u,

$$\left(u, \frac{du}{dt}\right) = \frac{1}{2}\frac{d}{dt}(u,u) = (u, Lu).$$

If L is skew self-adjoint, $L^* = L^T = -L$, and the right hand side is

$$\frac{1}{2}(u, Lu) + \frac{1}{2}(u, L^* u) = 0.$$

Then, the energy $\frac{1}{2}(u,u)$ cannot increase.

If (12.1) is approximated in semi-discrete form on a mesh as

$$\frac{dv}{dt} = Av,$$

where v is the vector of the solution values of the mesh points, the corresponding energy balance is

$$\frac{1}{2}\frac{d}{dt}(v^T v) = v^T Av,$$

and stability is established if

$$v^T Av \leq 0. \tag{12.2}$$

A powerful approach to the formulation of discretizations with this property is to construct A in a manner that allows summation by parts (SBP) of $v^T Av$, annihilating all interior contributions and leaving only boundary terms. Then, one seeks boundary operators such that (12.2) holds. In particular, suppose that A is split as

$$A = D + B,$$

where D is an interior operator and B is a boundary operator. Then, if D is skew-symmetric, $D^T = -D$, the contribution $\boldsymbol{v}^T D\boldsymbol{v}$ vanishes leaving only the boundary terms.

12.2 Burgers' Equation

Burgers' equation, which has been discussed in Section 4.10, is the simplest example of a nonlinear equation that supports wave motion in opposite directions and the formation of shock waves, and consequently it provides a very useful example for the analysis of the energy method. Expressed in conservation form, the inviscid Burgers' equation is

$$\frac{\partial u}{\partial t} + \frac{\partial}{\partial x} f(u) = 0, \qquad a \leq x \leq b, \tag{12.3}$$

where

$$f(u) = \frac{u^2}{2}$$

and the wave speed is

$$a(u) = \frac{\partial f}{\partial u} = u.$$

Boundary conditions specifying the value of u at the left or right boundaries should be imposed if the direction of u is towards the interior at the boundary.
Smooth solutions of (12.3) remain constant along characteristics

$$x - ut = \xi.$$

If a faster moving wave over-runs a slower moving wave, this would indicate a multi-valued solution. Instead, following the exposition of Section 4.10, a proper weak solution is obtained by assuming the formation of a shock wave across which there is a discontinuous transition between left and right states u_L and u_R. In order to satisfy the conservation law (12.3) in integral form, u_L and u_R must satisfy the jump condition

$$f(u_R) - f(u_L) = s(u_R - u_L), \tag{12.4}$$

where s is the shock speed. For Burgers' equation, this gives a shock speed

$$s = \frac{1}{2}(u_R + u_L). \tag{12.5}$$

Provided that the solution remains smooth, (12.3) can be multiplied by u^{k-1} and rearranged to give an infinite set of invariants of the form

$$\frac{\partial}{\partial t}\left(\frac{u^k}{k}\right)+\frac{\partial}{\partial x}\left(\frac{u^{k+1}}{k+1}\right)=0.$$

Here, we focus on the first of these,

$$\frac{\partial}{\partial t}\left(\frac{u^2}{2}\right)+\frac{\partial}{\partial x}\left(\frac{u^3}{3}\right)=0. \tag{12.6}$$

This may be integrated over x from a to b to determine the rate of change of the energy

$$E=\int_a^b \frac{u^2}{2}dx$$

in terms of the boundary fluxes as

$$\frac{dE}{dt}=\frac{u_a^3}{3}-\frac{u_b^3}{3}. \tag{12.7}$$

This equation fails in the presence of shock waves, as can easily be seen by considering the initial data $u=-x$ in the interval $[-1,1]$. Then a wave moves inwards from each boundary at unit speed toward the center until a stationary shock wave is formed at $t=1$, after which the energy remains constant. Thus,

$$E(t)=\begin{cases}\frac{1}{3}+\frac{2t}{3}, & 0\le t\le 1\\ 1, & t>1.\end{cases}$$

In order to correct (12.7) in the presence of a shock wave with left and right states u_L and u_R, (12.6) should be integrated separately on each side of the shock. If the shock is moving at a speed S, there is an additional contribution to $\frac{dE}{dt}$ of

$$S\left(\frac{u_L^2}{2}-\frac{u_R^2}{2}\right).$$

Substituting (12.5) for the shock speed,

$$\frac{dE}{dt}=\frac{u_a^3}{3}-\frac{u_L^3}{3}+\frac{u_R^3}{3}-\frac{u_b^3}{3}+\frac{1}{2}(u_L+u_R)\left(\frac{u_L^2}{2}-\frac{u_R^2}{2}\right),$$

which can be simplified to

$$\frac{dE}{dt}=\frac{u_a^3}{3}-\frac{u_b^3}{3}-\frac{1}{12}(u_L-u_R)^3.$$

In the presence of multiple shocks, each will remove energy at the rate $\frac{1}{12}(u_L-u_R)^3$.

As was already observed by Morton and Richtmyer (Richtmyer & Morton, 1994, page 142), a skew-symmetric difference operator consistent with (12.3) for smooth data can be constructed by splitting it between conservation and quasilinear form as

$$\frac{\partial u}{\partial t}+\frac{2}{3}\frac{\partial}{\partial x}\left(\frac{u^2}{2}\right)+\frac{1}{3}u\frac{\partial u}{\partial x}=0.$$

Suppose this is discretized on a uniform mesh $x_j = j\Delta x$, $j = 0, 1, \ldots n$. Denoting the discrete solution by v_j at the location j, central differencing of both spatial derivatives at interior points yields the semi-discrete scheme

$$\frac{dv_j}{dt} = \frac{1}{6\Delta x}\left(v_{j+1}^2 - v_{j-1}^2\right) + \frac{1}{6\Delta x} v_j \left(v_{j+1} - v_{j-1}\right) = 0, \qquad j = 1, n-1. \quad (12.8)$$

The skew symmetric operator is completed by the use of one sided schemes at each boundary:

$$\begin{aligned} \frac{dv_0}{dt} &= \frac{1}{3\Delta x}\left(v_1^2 - v_0^2\right) + \frac{1}{3\Delta x} v_0(v_1 - v_0) \\ \frac{dv_n}{dt} &= \frac{1}{3\Delta x}\left(v_n^2 - v_{n-1}^2\right) + \frac{1}{3\Delta x} v_n(v_n - v_{n-1}). \end{aligned} \quad (12.9)$$

Rewriting the quasilinear term as $\frac{1}{6\Delta x}(v_{j+1}v_j - v_j v_{j-1})$, (12.8) and (12.9) can be expressed in the conservation form

$$\frac{dv_j}{dt} + \frac{1}{\Delta x}\left(f_{j+\frac{1}{2}} - f_{j-\frac{1}{2}}\right) = 0, \qquad j = 1, n-1, \quad (12.10)$$

where

$$f_{j+\frac{1}{2}} = \frac{1}{6}\left(v_{j+1}^2 + v_{j+1}v_j + v_j^2\right) \quad (12.11)$$

and

$$\begin{aligned} \frac{dv_0}{dt} + \frac{2}{\Delta x}\left(f_{\frac{1}{2}} - f_0\right) &= 0 \\ \frac{dv_n}{dt} + \frac{2}{\Delta x}\left(f_n - f_{n-\frac{1}{2}}\right) &= 0, \end{aligned} \quad (12.12)$$

where

$$f_0 = \frac{v_0^2}{2}, \qquad f_n = \frac{v_n^2}{2}.$$

Now, let the discrete energy be represented by trapezoidal integration as

$$E = \frac{\Delta x}{2}\left(\frac{v_0^2}{2} + \frac{v_n^2}{2}\right) + \Delta x \sum_{j=1}^{n-1} \frac{v_j^2}{2}.$$

Multiplying (12.10) by v_j and summing by parts,

$$\Delta x \sum_{j=1}^{n-1} v_j \frac{dv_j}{dt} = -\sum_{j=1}^{n-1} v_j \left(f_{j+\frac{1}{2}} - f_{j-\frac{1}{2}}\right) = f_{\frac{1}{2}} v_0 - f_{n+\frac{1}{2}} v_n.$$

Hence, including the boundary points, we find that

$$\frac{dE}{dt} = \frac{v_0^3}{3} - \frac{v_n^3}{3}, \quad (12.13)$$

which is the exact discrete analog of the continuous energy evolution equation (12.7).

The formulation so far does not include the boundary conditions. Suppose that boundary data $u = g_0$ should be imposed at x_0 if the left boundary x_0 is an inflow boundary, and correspondingly, $u = g_n$ should be imposed at x_n if the right boundary x_n is an inflow boundary. It is convenient to introduce the positive and negative wave speeds

$$a^+(u) = \max(a(u), 0), \qquad a^-(u) = \min(a(u), 0).$$

Then, we modify the equations at the boundary points by adding simultaneous approximation terms (SAT), so that instead of (12.12) we solve

$$\frac{dv_0}{dt} + \frac{2}{\Delta x}\left(f_{\frac{1}{2}} - f_0\right) + \frac{\tau}{\Delta x}a_0^+(v_0 - g_0) = 0$$

$$\frac{dv_n}{dt} + \frac{2}{\Delta x}\left(f_n - f_{n-\frac{1}{2}}\right) - \frac{\tau}{\Delta x}a_0^-(v_n - g_n) = 0, \tag{12.14}$$

where the parameter τ determines the amount of the penalty if the boundary condition is not satisfied exactly. The linear case has been analyzed by Mattsson (2003). Here, we wish to ensure stability in the nonlinear case. It is evident from (12.13) that outflow boundaries ($u_0 < 0$ or $u_n > 0$) promote energy decay. Thus, we need only consider the effect of inflow boundary conditions.

For this purpose suppose that $\frac{d}{dt}E_{\text{true}}$ is the rate of change of energy that would result if the boundary conditions were exactly satisfied. Then we wish to choose τ so that $\frac{dE}{dt}$ is bounded from above by $\frac{d}{dt}E_{\text{true}}$:

$$\frac{dE}{dt} \le \frac{d}{dt}E_{\text{true}}. \tag{12.15}$$

Consider the construction at the left boundary, assuming it is an inflow boundary. Suppose that a_0^+ is evaluated as $\frac{1}{2}(v_0 + g_0)$. Omitting for the moment the contribution of the right boundary, we find that

$$\frac{d}{dt}(E - E_{\text{true}}) = \frac{u_0^3}{3} - \frac{g_0^3}{3} - \frac{\tau}{4}v_0(v_0^2 - g_0^2).$$

Suppose now that v_0 has the value αg_0. Then, the rate of change of the energy is modified by the cubic expression $F(\alpha)g_0^3$, where

$$F(\alpha) = \frac{\alpha^3}{3} - \frac{1}{3} - \frac{\tau}{4}\alpha(\alpha^2 - 1).$$

Here, g_0 should be positive if it is truly an inflow boundary condition, so we require $F(\alpha)$ to be nonpositive in the range of α corresponding to inflow, $\alpha > -1$. However, $F(\alpha) = 0$ when $\alpha = 1$ and $v_0 = g_0$, so its sign will change at $\alpha = 1$ unless this is a double root. Here,

$$F'(\alpha) = \alpha^2 - \frac{\tau}{4}(3\alpha^2 - 1),$$

and the condition $F'(1) = 0$ yields

$$\tau = 2. \tag{12.16}$$

Then,

$$F(\alpha) = -\frac{1}{6}(\alpha - 1)^2(\alpha + 2)$$

and is non positive whenever $\alpha > -2$. A similar analysis at the right boundary confirms that condition (12.16) is sufficient to assure the favorable energy comparison (12.15) whenever either boundary is an inflow boundary.

Numerical experiments have been conducted to verify the stability of the semi-discrete scheme (12.10–12.11) with the boundary conditions (12.14–12.16). The SSPRK(3,3) scheme of Shu and Osher described in Section 11.1 was used for time integration.

Figure 12.1 displays snapshots of the solution with initial data $u = -x$ in $[-1, 1]$ at times $t = 0, 0.5, 1, 1.5$ using a grid with 256 intervals. The true solution is a straight line connecting wave fronts moving inwards from both boundaries at unit speed until a shock is formed at $t = 1$. It can be seen that the discrete solution closely tracks the true solution prior to the formation of the shock. After the shock is formed, the discrete

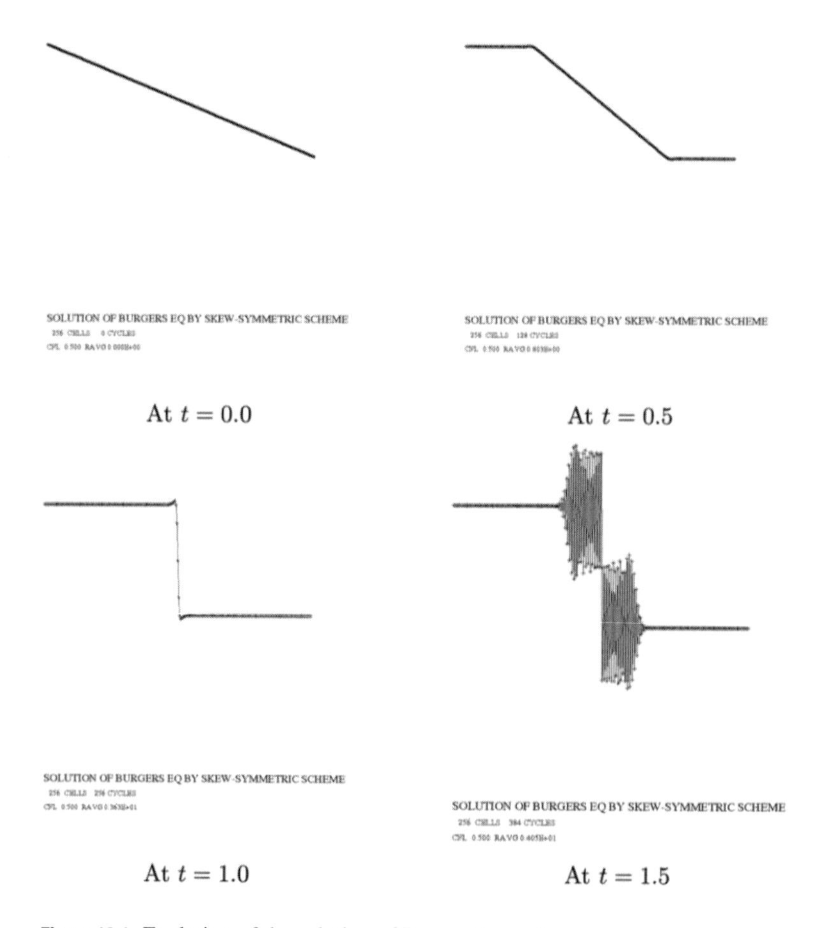

Figure 12.1 Evolution of the solution of Burgers' equation.

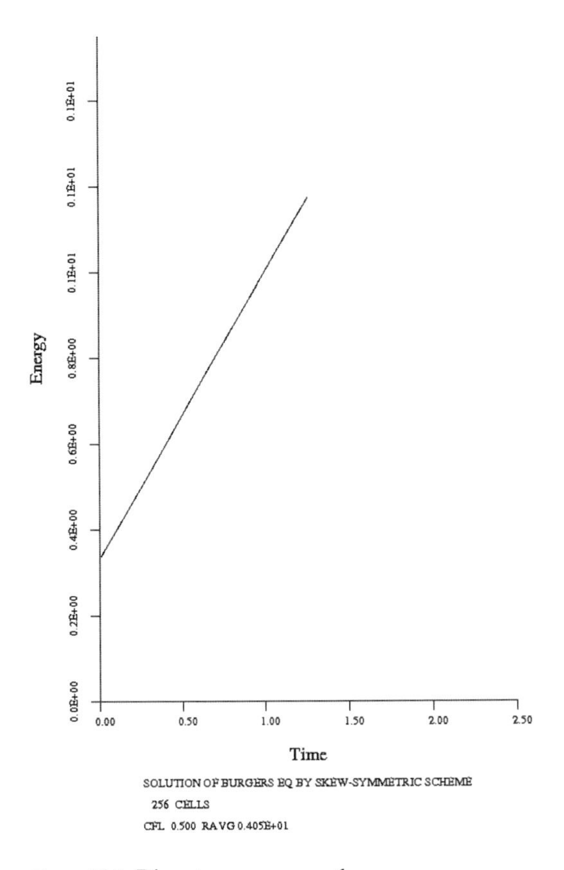

Figure 12.2 Discrete energy growth.

solution develops strong oscillations in a zone expanding outward from the shock toward both boundaries. This is consistent with the fact that according to (12.13), the discrete energy continues to grow at the rate $\frac{2}{3}t$ when $t > 1$, as illustrated in Figure 12.2, and the energy must be absorbed in the solution. It is evident that the scheme must be modified to preserve stability in the presence of shock waves.

The conditions for a semi-discrete equation to be local extremum diminishing (LED) or total variation diminishing (TVD) have been discussed in Sections 5.5 and 5.13. In the case of the inviscid Burgers' equation, these conditions are satisfied by the upwind schemes in which the numerical flux (12.11) is replaced by

$$f_{j+\frac{1}{2}} = \begin{cases} v_j^2 & \text{if} \quad a_{j+\frac{1}{2}} > 0 \\ v_{j+1}^2 & \text{if} \quad a_{j+\frac{1}{2}} < 0 \\ \frac{1}{2}(v_{j+1}^2 + v_j^2) & \text{if} \quad a_{j+\frac{1}{2}} = 0, \end{cases} \tag{12.17}$$

where the numerical wave speed is evaluated as

$$a_{j+\frac{1}{2}} = \frac{1}{2}(v_{j+1} + v_j). \tag{12.18}$$

Moreover, the upwind scheme (12.17) admits a stationary numerical shock structure with a single interior point, as was shown in Section 5.14.

The LED condition only needs to be satisfied in the neighborhoods of local extrema, which may be detected by a change of sign in the first differences $\Delta v_{j+\frac{1}{2}} = v_{j+1} - v_j$. A shock operator that meets these requirements can be constructed as follows. The numerical flux (12.11) can be converted to the upwind flux (12.17) by subtracting a diffusive term of the form

$$d_{j+\frac{1}{2}} = \alpha_{j+\frac{1}{2}} \Delta v_{j+\frac{1}{2}}.$$

The required coefficient is

$$\alpha_{j+\frac{1}{2}} = \frac{1}{4}\left|v_{j+1} + v_j\right| - \frac{1}{12}\left(v_{j+1} - v_j\right). \tag{12.19}$$

In order to detect an extremum, we can use the same strategy as the JST scheme described in Section 5.17. We introduce the function

$$R(u,v) = \left|\frac{u-v}{|u|+|v|}\right|^q,$$

where q is an integer power. $R(u,v) = 1$ whenever u and v have opposite signs. When $u = v = 0$, $R(u,v)$ should be assigned the value zero. Now, set

$$s_{j+\frac{1}{2}} = R\left(\Delta v_{j+\frac{3}{2}}, \Delta v_{j-\frac{1}{2}}\right) \tag{12.20}$$

so that $s_{j+\frac{1}{2}} = 1$ when $\Delta v_{j+\frac{3}{2}}$ and $\Delta v_{j-\frac{1}{2}}$ have opposite signs, which will generally be the case if either v_{j+1} or v_j is an extremum. In a smooth region where $\Delta v_{j+\frac{3}{2}}$ and $\Delta v_{j-\frac{1}{2}}$ are not both zero, $s_{j+\frac{1}{2}}$ is of the order Δx^q, since $\Delta v_{j+\frac{3}{2}} - \Delta v_{j-\frac{1}{2}}$ is an undivided difference. In order to avoid activating the switch at smooth extrema, and also to protect against division by zero, $R(u,v)$ may be redefined as

$$R(u,v) = \left|\frac{u-v}{\max\{(|u|+|v|), \epsilon\}}\right|, \tag{12.21}$$

where ϵ is a tolerance (Jameson 1995a).

Finally, the diffusive term is modified to

$$d_{j+\frac{1}{2}} = \max\left(s_{j+\frac{3}{2}}, s_{j+\frac{1}{2}}, s_{j-\frac{1}{2}}\right) \alpha_{j+\frac{1}{2}} \Delta u_{j+\frac{1}{2}}.$$

Thus, the coeffcient is reduced to a magnitude of order Δx^q in smooth regions, while it has the value $\alpha_{j+\frac{1}{2}}$ in the neighborhood of a shock. The value $q = 8$ has proved satisfactory in numerical experiments.

Figure 12.3 shows the evolution of the discrete solution for the same case as Figure 12.1, with initial data $u = -x$ in $[-1, 1]$, using the shock operator defined by (12.19–12.21). A stationary shock with a single interior point is formed when $t = 1$, as expected. Figure 12.4 confirms that the discrete energy grows at the rate $\frac{2t}{3}$ until the shock forms and then remains constant. The difference between the discrete energy

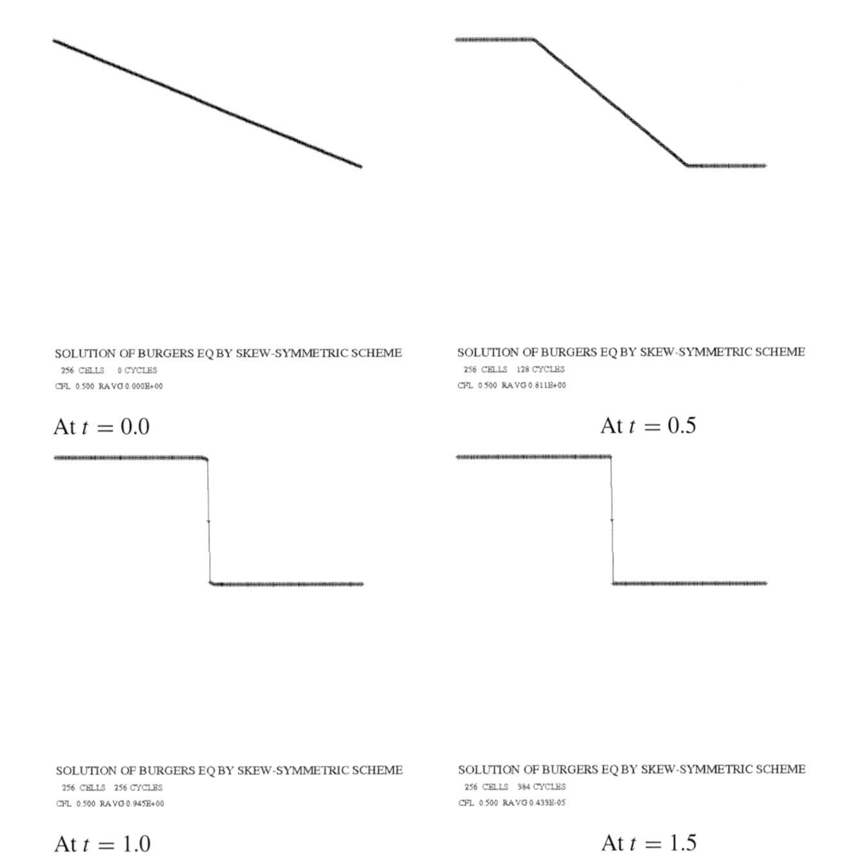

Figure 12.3 Evolution of the solution of Burgers' equation with a switch.

and the true energy of the stationary solution is $-\frac{1}{2}\Delta x$ because of the zero value in the middle of the discrete shock. Once the numerical shock structure is established, the additional diffusive terms only contribute to $\frac{dv_j}{dt}$ at the three points $s-1$, s, and $s+1$ comprising the shock, for which

$$v_{s-1} = 1, \qquad u_s = 0, \qquad v_{s+1} = -1.$$

The only non-zero values of $\Delta v_{j+\frac{1}{2}}$ are

$$\Delta v_{s-\frac{1}{2}} = -1, \qquad \Delta v_{s+\frac{1}{2}} = -1.$$

Also,

$$a_{s-\frac{1}{2}} = -\frac{1}{2}, \qquad \alpha_{s-\frac{1}{2}} = \frac{1}{3}$$

$$a_{s+\frac{1}{2}} = -\frac{1}{2}, \qquad \alpha_{s+\frac{1}{2}} = \frac{1}{3}.$$

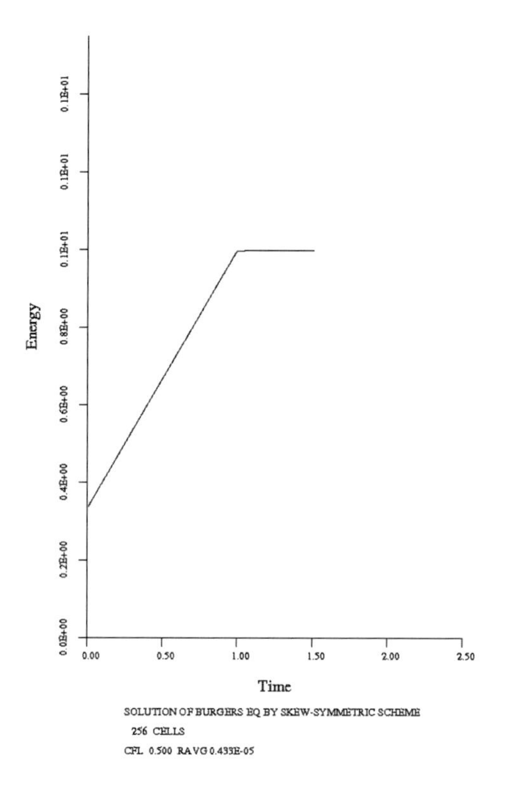

Figure 12.4 Discrete energy growth with a limiter.

Thus, the additional contribution to $\frac{dE}{dt}$ due to the shock is

$$\sum_{s=-1}^{1} v_s \left(\alpha_{s+\frac{1}{2}} \Delta v_{s+\frac{1}{2}} - \alpha_{s-\frac{1}{2}} \Delta v_{s-\frac{1}{2}}\right) = \alpha_{s-\frac{1}{2}} v_{s-1} \Delta v_{s-\frac{1}{2}} - \alpha_{s+\frac{1}{2}} v_{s+1} \Delta v_{s+\frac{1}{2}}$$
$$= -\frac{2}{3}.$$

This exactly cancels the contribution of $\frac{2}{3}$ from the boundaries, so that the total rate of change of the discrete energy is zero.

In the case of the viscous Burgers' equation with the viscosity coefficient ν,

$$\frac{\partial u}{\partial t} + \frac{\partial}{\partial x}\left(\frac{u^2}{2}\right) = \nu \frac{\partial^2 u}{\partial x}, \tag{12.22}$$

the energy balance is modified by the viscous dissipation. Multiplying by u and integrating the right hand side by parts with $\frac{\partial u}{\partial x} = 0$ at each boundary, the energy balance equation assumes the form

$$\frac{dE}{dt} = \frac{u_a^3}{3} - \frac{u_b^3}{3} - \nu \int_a^b \left(\frac{\partial u}{\partial x}\right)^2 dx \tag{12.23}$$

instead of (12.7). Suppose that $\frac{\partial^2 u}{\partial x^2}$ is discretized by a central difference operator at interior points with one sided formulas at the boundaries corresponding to $\frac{\partial u}{\partial x} = 0$,

$$\begin{aligned} &\frac{1}{\Delta x^2}(v_{j+1} - 2v_j + v_{j-1}), && j = 2, n-1 \\ &\frac{1}{\Delta x^2}(v_1 - v_0) && \text{at the left boundary,} \\ &\frac{1}{\Delta x^2}(v_n - v_{n-1}) && \text{at the right boundary,} \end{aligned} \tag{12.24}$$

as proposed by Mattsson (2003). Then, summing by parts with the convective flux evaluated by (12.11) as before, the discrete energy balance is found to be

$$\frac{dE}{dt} = \frac{v_0^3}{3} - \frac{v_n^3}{3} - \nu \sum_{j=0}^{n-1} (v_{j+1} - v_j)^2. \tag{12.25}$$

This enables the possibility of fully resolving shock waves without the need to add any additional numerical diffusion via shock operators. The convective flux difference $f_{j+\frac{1}{2}} - f_{j-\frac{1}{2}}$ can be factored as

$$\frac{1}{3\Delta x}(v_{j+1} + v_j + v_{j-1})(v_{j+1} - v_{j-1}).$$

Accordingly, the semi-discrete approximation to (12.22) can written as

$$\frac{du_j}{dt} = a_{j+\frac{1}{2}}(v_{j+1} - v_j) + a_{j-\frac{1}{2}}(v_{j-1} - v_j),$$

where

$$a_{j+\frac{1}{2}} = \frac{\nu}{\Delta x^2} - \frac{v_{j+1} + v_j + v_{j-1}}{3\Delta x}$$

and

$$a_{j-\frac{1}{2}} = \frac{\nu}{\Delta x^2} + \frac{v_{j+1} + v_j + v_{j-1}}{3\Delta x}.$$

According to the analysis in Section 5.7, the semi-discrete approximation satisfies the conditions for a local extremum diminishing scheme if $a_{j+\frac{1}{2}} \geq 0$ and $a_{j-\frac{1}{2}} \geq 0$. This establishes Theorem 12.2.1:

THEOREM 12.2.1 *The semi-discrete approximation (12.10) using the numerical flux (12.11) and the central difference operator (12.24) for $\frac{\partial^2 u}{\partial x^2}$ is local extremum diminishing if the cell Reynolds number satisfies the condition*

$$\frac{\bar{u}\Delta x}{\nu} \leq 2, \tag{12.26}$$

where the local speed is evaluated as

$$\bar{u} = \frac{1}{3}\left|v_{j+1} + v_j + v_{j-1}\right|.$$

It has been confirmed by numerical experiments that shock waves are indeed fully resolved with no oscillation if the cell Reynolds number satisfies condition (12.26). Figure 12.5 shows the evolution of the discrete viscous Burgers' equation using this scheme for the same initial data as before, $u = -x$ in $[-1, 1]$. The Reynolds number $\frac{uL}{\nu}$ was 2,048, based on the boundary velocity $u = \pm 1$ and the length of the interval, and the solution was calculated on a uniform mesh with 1,024 intervals, so that the cell Reynolds number condition (12.26) was satisfied in the entire domain. It can be seen that a stationary shock wave is formed at the time $t = 1$, and it is finally resolved with three interior points. Correspondingly, the energy becomes constant after the shock wave is formed, as can be seen in Figure 12.6.

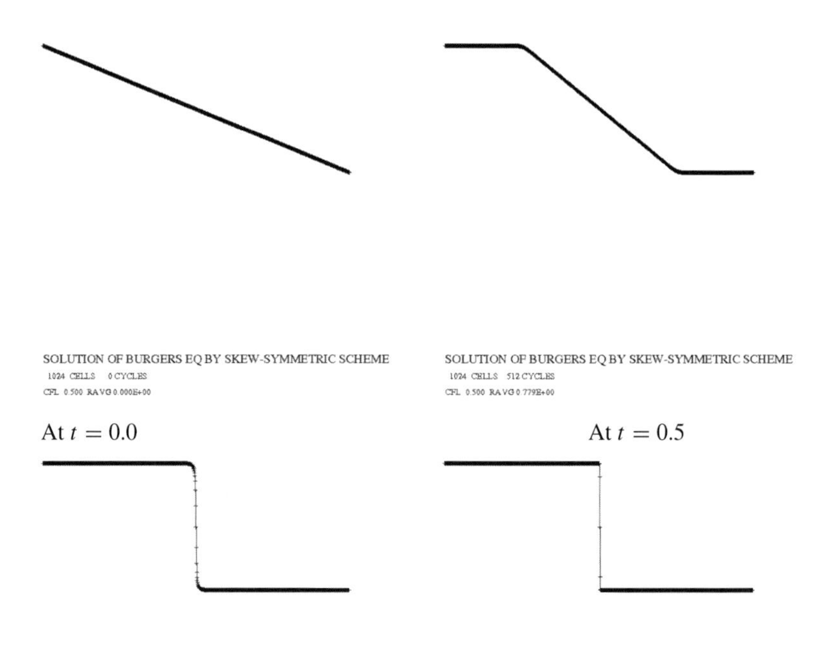

SOLUTION OF BURGERS EQ BY SKEW-SYMMETRIC SCHEME
1024 CELLS 1024 CYCLES
CFL 0.500 RAVG 0.231E+01

At $t = 1.0$

SOLUTION OF BURGERS EQ BY SKEW-SYMMETRIC SCHEME
1024 CELLS 1536 CYCLES
CFL 0.500 RAVG 0.923E-14

At $t = 1.5$

Figure 12.5 Evolution of the solution of the viscous Burgers' equation.

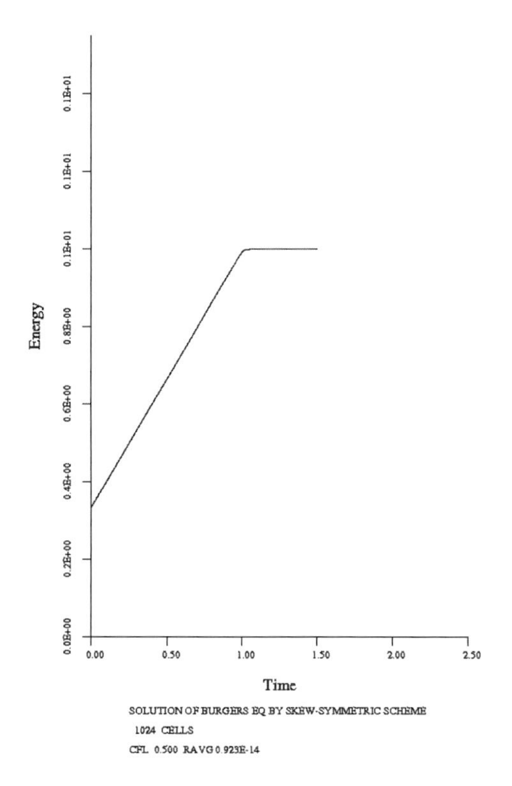

Figure 12.6 Discrete energy growth for the viscous Burgers' equation.

12.3 The General One-Dimensional Scalar Conservation Law

Consider the scalar conservation law

$$\frac{\partial u}{\partial t} + \frac{\partial}{\partial x} f(u) = 0 \tag{12.27}$$

$$u(x,0) = u_0(x),$$

$$u \qquad \text{specified at inflow boundaries.}$$

Correspondingly, smooth solutions of (12.27) also satisfsy

$$\frac{\partial}{\partial t}\left(\frac{u^2}{2}\right) + \frac{\partial}{\partial x} F(u) = 0, \tag{12.28}$$

where

$$F_u = u f_u,$$

since multiplying (12.27) by u yields

$$u\frac{\partial u}{\partial t} + u f_u \frac{\partial u}{\partial x} = 0.$$

Defining the energy as

$$E = \int_a^b \frac{u^2}{2} dx,$$

it follows from (12.28) that smooth solutions of (12.27) satisfy the energy equation

$$\frac{dE}{dt} = F(u_a) - F(u_b). \tag{12.29}$$

Introducing the function $G(u)$ such that

$$G_u = f$$

and multiplying (12.27) by u, we obtain

$$\begin{aligned} u\frac{\partial u}{\partial t} + u\frac{\partial f}{\partial x} &= \frac{\partial}{\partial t}\left(\frac{u^2}{2}\right) + \frac{\partial}{\partial x}(uf) - f\frac{\partial u}{\partial x} \\ &= \frac{\partial}{\partial t}\left(\frac{u^2}{2}\right) + \frac{\partial}{\partial x}(uf) - G_u\frac{\partial u}{\partial x} \\ &= \frac{\partial}{\partial t}\left(\frac{u^2}{2}\right) + \frac{\partial}{\partial x}(uf - G) \\ &= 0. \end{aligned}$$

Thus, F and G can be identified as

$$F = uf - G, \qquad G = uf - F.$$

For the inviscid Burgers' equation,

$$F = \frac{u^3}{3}, \qquad G = \frac{u^3}{6}.$$

Suppose now that (12.27) is discretized on a uniform grid over the range $j = 0, n$. Consider a semi-discrete conservative scheme of the form

$$\frac{dv_j}{dt} + \frac{1}{\Delta x}\left(f_{j+\frac{1}{2}} - f_{j-\frac{1}{2}}\right) = 0 \tag{12.30}$$

at every interior point, where the numerical flux $f_{j+\frac{1}{2}}$ is a function of v_i over a range bracketing v_j such that $f_{j+\frac{1}{2}} = f(u)$ whenever u is substituted for the v_i, thus satisfying Lax's consistency condition. Multiplying (12.27) by v_j and summing by parts over the interior points, we obtain

$$\begin{aligned}
\Delta x \sum_{j=1}^{n-1} v_j \frac{dv_j}{dt} &= -\sum_{j=1}^{n-1} v_j \Big(f_{j+\frac{1}{2}} - f_{j-\frac{1}{2}}\Big) \\
&= -v_1 f_{\frac{3}{2}} - v_2 f_{\frac{5}{2}} \ldots - v_{n-2} f_{n-\frac{3}{2}} - v_{n-1} f_{n-\frac{1}{2}} \\
&\quad + v_1 f_{\frac{1}{2}} + v_2 f_{\frac{3}{2}} + v_3 f_{\frac{5}{2}} \ldots + v_{n-1} f_{n-\frac{3}{2}} \\
&= v_1 f_{\frac{1}{2}} - v_{n-1} f_{n-\frac{1}{2}} + \sum_{j=1}^{n-2} f_{j+\frac{1}{2}} (v_{j+1} - v_j).
\end{aligned}$$

Suppose now that

$$f_{j+\frac{1}{2}} = G_{u_{j+\frac{1}{2}}}, \tag{12.31}$$

where $G_{v_{j+\frac{1}{2}}}$ is the mean value of G_u in the range from v_j to v_{j+1} such that

$$G_{v_{j+\frac{1}{2}}} (v_{j+1} - u_j) = G(v_{j+1}) - G(v_j). \tag{12.32}$$

Then, denoting $G(v_j)$ by G_j,

$$\begin{aligned}
\Delta x \sum_{j=1}^{n-1} v_j \frac{du_j}{dt} &= v_1 f_{\frac{1}{2}} - v_{n-1} f_{n-\frac{1}{2}} + \sum_{j=1}^{n-2} (G_{j+1} - G_j) \\
&= v_1 f_{\frac{1}{2}} - v_{n-1} f_{n-\frac{1}{2}} - G_1 + G_{n-1}.
\end{aligned}$$

Now, let (12.27) be discretized at the end points as

$$\frac{dv_0}{dt} + \frac{2}{\Delta x}\left(f_{\frac{1}{2}} - f_0\right), \qquad \frac{dv_n}{dt} + \frac{2}{\Delta x}\left(f_n - f_{n-\frac{1}{2}}\right), \tag{12.33}$$

where

$$f_0 = f(v_0), \qquad f_n = f(v_n),$$

and define the discrete approximation to the energy as

$$E = \frac{\Delta x}{2}\left(\frac{v_0^2}{2} + \frac{v_n^2}{2}\right) + \Delta x \sum_{j=1}^{n-1} \frac{v_j^2}{2}.$$

Then,

$$\begin{aligned}
\frac{dE}{dt} &= v_0 f_0 - v_n f_n - G_0 + G_n \\
&= F_0 - F_n.
\end{aligned}$$

Thus, the energy balance (12.29) is exactly recovered by the discrete scheme. Equations (12.31) and (12.32) are satisfied by evaluating the numerical flux as

$$f_{j+\frac{1}{2}} = \int_0^1 f(\hat{v}(\theta)) d\theta, \tag{12.34}$$

where

$$\hat{v}(\theta) = v_j + \theta(v_{j+1} - v_j), \tag{12.35}$$

since then

$$\begin{aligned} G_{j+1} - G_j &= \int_0^1 G_u(\hat{v}(\theta))v_\theta d\theta \\ &= \int_0^1 G_u(\hat{v}(\theta)d\theta(v_{j+1} - v_j). \end{aligned}$$

Thus, we have established Theorem 12.3.1:

THEOREM 12.3.1 *If the scalar conservation law (12.27) is approximated by the semi-discrete conservative scheme (12.30), it also satisfies the semi-discrete energy conservation law (12.33) if the numerical flux $f_{j+\frac{1}{2}}$ is evaluated by (12.34) and (12.35).*

In the case of Burgers' equation, formulas (12.34) and (12.35) yield the same numerical flux that was defined in Section 12.2:

$$f_{j+\frac{1}{2}} = \frac{v_{j+1}^2 + v_{j+1}v_j + v_j^2}{6}.$$

If (12.34) cannot be evaluated in closed form, one may approximate it by numerical integration. The Lobatto quadrature rule, which uses the end points and $n-2$ interior points, is suitable for this purpose, giving an exact result for polynomials of degree up to $2n-3$. Taking $n=3$ yields Simpson's rule:

$$f_{j+\frac{1}{2}} = \frac{1}{6}(f(v_{j+1}) + 4f\left(\frac{1}{2}(v_{j+1} + v_j)\right) + f(v_j)),$$

which is exact for Burgers' equation.

In order to enforce appropriate inflow and outflow boundary conditions, we introduce simultaneous approximation terms (SAT). Denote the wave speed by

$$a(u) = \frac{\partial f}{\partial u},$$

and let

$$a^+ = \frac{1}{2}(a + |a|), \qquad a^- = \frac{1}{2}(a - |a|).$$

Then, we replace (12.33) by

$$\left.\begin{aligned} \frac{dv_0}{dt} + \frac{2}{\Delta x}\left(f_{\frac{1}{2}} - f_0\right) + \frac{\tau}{\Delta x}a^+(v_0 - g_0) = 0 \\ \frac{dv_n}{dt} + \frac{2}{\Delta x}\left(f_n - f_{\frac{1}{2}}\right) - \frac{\tau}{\Delta x}a^-(v_n - g_n) = 0 \end{aligned}\right\}, \tag{12.36}$$

where g_0 and g_n denote the boundary values that should be imposed at inflow boundaries, and the magnitude of the penalty for not exactly satisfying the boundary conditions is determined by the parameter τ.

12.4 Discrete Conservation of Energy or entropy for a System

In the treatment of a scalar conservation law, we expressed $f(u)$ as a derivative $\frac{\partial G}{\partial u}$ of a scalar function $G(u)$ in order to construct a numerical flux that ensures the satisfaction of a discrete energy conservation law. In the case of a system

$$\frac{\partial \boldsymbol{w}}{\partial t} + \frac{\partial}{\partial x} \boldsymbol{f}(\boldsymbol{w}) = 0, \tag{12.37}$$

if we could express $\boldsymbol{f}(\boldsymbol{w})$ in the form $\frac{\partial R}{\partial \boldsymbol{w}}$ for some scalar function $R(\boldsymbol{w})$, it would imply that the Jacobian

$$A_{ij} = \frac{\partial f_i}{\partial w_j} = \frac{\partial^2 R}{\partial w_i \partial w_j}$$

is symmetric. In order to generalize the procedure described in Section 12.3 to discretely satisfy an additional invariant, we must therefore treat a system in symmetric form.

It is shown in Chapter 2, Section 2.10, that if we can find a convex scalar function $U(\boldsymbol{w})$ such that

$$\frac{\partial U}{\partial \boldsymbol{w}} \frac{\partial f}{\partial \boldsymbol{w}} = \frac{\partial G}{\partial \boldsymbol{w}}$$

for some scalar function $G(\boldsymbol{w})$, then multiplying (12.37) by $\frac{\partial U}{\partial \boldsymbol{w}}$, we obtain

$$\frac{\partial U}{\partial \boldsymbol{w}} \frac{\partial \boldsymbol{w}}{\partial t} = \frac{\partial U}{\partial t} = -\frac{\partial U}{\partial \boldsymbol{w}} \frac{\partial \boldsymbol{f}}{\partial x} = -\frac{\partial U}{\partial \boldsymbol{w}} \frac{\partial f}{\partial \boldsymbol{w}} \frac{\partial \boldsymbol{w}}{\partial x} = -\frac{\partial G}{\partial \boldsymbol{w}} \frac{\partial \boldsymbol{w}}{\partial x} = -\frac{\partial G}{\partial x},$$

so that U satisfies the additional conservation law

$$\frac{\partial}{\partial t} U(\boldsymbol{w}) + \frac{\partial}{\partial x} G(\boldsymbol{w}) = 0. \tag{12.38}$$

Here, $U(\boldsymbol{w})$ is a generalized entropy function, and $G(\boldsymbol{w})$ is the corresponding entropy flux. Moreover, if we introduce dependent variables

$$v^T = \frac{\partial U}{\partial \boldsymbol{w}},$$

the equations are symmetrized as

$$\frac{\partial \boldsymbol{w}}{\partial \boldsymbol{v}} \frac{\partial \boldsymbol{v}}{\partial t} + \frac{\partial \boldsymbol{f}}{\partial \boldsymbol{v}} \frac{\partial \boldsymbol{v}}{\partial x} = 0,$$

where $\frac{\partial \boldsymbol{w}}{\partial \boldsymbol{v}}$ and $\frac{\partial \boldsymbol{f}}{\partial \boldsymbol{v}}$ are both symmetric. Also, multiplying (12.37) by $\boldsymbol{v}^T$, we recover the entropy equation (12.38). We can also now construct a discrete scheme that satisfies a discrete form of the entropy equation (12.38).

Suppose that (12.37) is approximated in semi-discrete form on a uniform grid over the range $j = 0, n$ as

$$\frac{d\boldsymbol{w}_j}{dt} + \frac{1}{\Delta x}\left(\boldsymbol{f}_{j+\frac{1}{2}} - \boldsymbol{f}_{j-\frac{1}{2}}\right) = 0, \tag{12.39}$$

where the numerical flux $\boldsymbol{f}_{j+\frac{1}{2}}$ is a function of $\boldsymbol{w}_i$ over a range bracketing $\boldsymbol{w}_j$. Then, we can construct a scheme that discretely satisfies the energy or entropy conservation law in the same manner as for the scalar case. Multiplying (12.39) by $\boldsymbol{v}_j^T$ and summing by parts over the interior points:

$$\Delta x \sum_{j=1}^{n-1} \boldsymbol{v}_j^T \frac{d\boldsymbol{w}_j}{dt} = \sum_{j=1}^{n-1} U_{\boldsymbol{w}_j} \frac{\partial \boldsymbol{w}_j}{\partial t} = \sum_{j=1}^{n-1} \frac{dU_j}{dt}$$

$$= \boldsymbol{v}_1^T \boldsymbol{f}_{\frac{1}{2}} - \boldsymbol{v}_{n-1}^T \boldsymbol{f}_{n-\frac{1}{2}} + \sum_{j=1}^{n-2} \boldsymbol{f}_{j+\frac{1}{2}}^T (\boldsymbol{v}_{j+1} - \boldsymbol{v}_j).$$

Now, suppose that

$$\boldsymbol{f}_{j+\frac{1}{2}}^T = R_{\boldsymbol{v}_{j+\frac{1}{2}}}, \tag{12.40}$$

where $R_{\boldsymbol{v}_{j+\frac{1}{2}}}$ is a mean value of $R_{\boldsymbol{v}}$ between $\boldsymbol{v}_j$ and $\boldsymbol{v}_{j+1}$ in the sense of Roe, such that

$$R_{\boldsymbol{v}_{j+\frac{1}{2}}} (\boldsymbol{v}_{j+1} - \boldsymbol{v}_j) = R_{j+1} - R_j, \tag{12.41}$$

where R_j denotes $R(\boldsymbol{v}_j)$. Then,

$$\Delta x \sum_{j=1}^{n-1} \frac{dU_j}{dt} = v_1^T f_{\frac{1}{2}} - v_{n-1}^T f_{n-\frac{1}{2}} - R_1 + R_{n-1}.$$

Now, let (12.37) be discretized at the end points as

$$\frac{d\boldsymbol{w}_0}{dt} + \frac{2}{\Delta x}(\boldsymbol{f}_{\frac{1}{2}} - \boldsymbol{f}_0) = 0, \qquad \frac{d\boldsymbol{w}_n}{dt} + \frac{2}{\Delta x}(\boldsymbol{f}_n - \boldsymbol{f}_{n-\frac{1}{2}}) = 0,$$

where

$$\boldsymbol{f}_0 = \boldsymbol{f}(\boldsymbol{w}_0), \qquad \boldsymbol{f}_n = \boldsymbol{f}(\boldsymbol{w}_n).$$

Then, we obtain the discrete conservation law

$$\frac{\Delta x}{2}\left(\frac{dh_0}{dt} + \frac{dU_n}{dt}\right) + \Delta x \sum_{j=1}^{n-1} \frac{dU_j}{dt} = \boldsymbol{v}_0^T \boldsymbol{f}_n - \boldsymbol{v}_n^T \boldsymbol{f}_n - R_0 + R_n$$

$$= G_0 - G_n, \tag{12.42}$$

where G is the entropy flux

$$G = \boldsymbol{v}^T \boldsymbol{f} - R.$$

$R_{v_{j+\frac{1}{2}}}$ can be constructed to satisfy (12.41) exactly in the form

$$R_{\boldsymbol{v}_{j+\frac{1}{2}}} = \int_0^1 R_v\left(\hat{\boldsymbol{v}}(\theta)\right) d\theta,$$

where

$$\hat{\boldsymbol{v}}(\theta) = \boldsymbol{v}_j + \theta(\boldsymbol{v}_{j+1} - \boldsymbol{v}_j), \tag{12.43}$$

since then

$$R_{j+1} - R_j = \int_0^1 R_v\left(\hat{v}(\theta)\right) v_\theta \, d\theta$$
$$= \int_0^1 R_v\left(\hat{v}(\theta)\right) d\theta (v_{j+1} - v_j).$$

Thus we can state Theorem 12.4.1:

THEOREM 12.4.1 *The semi-discrete conservation law (12.39) satisfies the semi-discrete entropy conservation law (12.42) if the numerical flux $f_{j+\frac{1}{2}}$ is constructed as*

$$f_{j+\frac{1}{2}} = \int_0^1 f\left(\hat{v}(\theta)\right) d\theta, \tag{12.44}$$

where $\hat{v}(\theta)$ is defined by (12.43).

For some systems, it may not be possible to express the integral (12.44) in closed form. Then, one may rescale the interval of integration for θ to $(-1,1)$ so that

$$f_{j+\frac{1}{2}} = \frac{1}{2}\int_{-1}^{1} f\left(\tilde{v}(\theta)\right) d\theta,$$

where

$$\tilde{v}(\theta) = \frac{1}{2}(v_{j+1} + v_j) + \frac{1}{2}\theta(v_{j+1} - v_j)$$

and apply the n point Lobatto rule.

In general, neither boundary is necessarily purely inflow or outflow. Consequently, in order to impose proper boundary conditions, it is essential to distinguish ingoing and outgoing waves at the boundaries. For this purpose, we can split the Jacobian matrix $A = f_u$ into positive and negative parts $A^{\pm}$. Suppose that A is decomposed as

$$A = V\Lambda V^{-1},$$

where the columns of V are the right eigenvectors of A, and Λ is a diagonal matrix comprising the eigenvalues. Then,

$$A^{\pm} = V\Lambda^{\pm}V^{-1},$$

where Λ^{+} and Λ^{-} contain the positive and negative eigenvalues respectively. Now, the boundary conditions maybe imposed by adding simultaneous approximation terms (SAT) at the boundaries. Accordingly, we set

$$\frac{dw_0}{dt} + \frac{1}{\Delta x}\left(f_{\frac{1}{2}} - f_0\right) + \frac{\tau}{\Delta x}A^{+}(w_0 - g_0) = 0$$
$$\frac{dw_n}{dt} + \frac{1}{\Delta x}\left(f_n - f_{n-\frac{1}{2}}\right) - \frac{\tau}{\Delta x}A^{-}(w_n - g_n) = 0,$$

where g_0 and g_n define the exterior data, and the parameter τ determines the magnitude of the penalty when the solution is not consistent with the incoming waves. Appropriate values of A_0 and A_n may be obtained by taking them to be the mean valued Jacobian matrices in the sense of Roe (1981), such that

$$A_0(\boldsymbol{w}_0 - \boldsymbol{g}_0) = \boldsymbol{f}(\boldsymbol{w}_0) - \boldsymbol{f}(\boldsymbol{g}_0),$$

$$A_n(\boldsymbol{w}_n - \boldsymbol{g}_n) = \boldsymbol{f}(\boldsymbol{w}_n) - \boldsymbol{f}(\boldsymbol{g}_n).$$

If shock waves appear in the solution, the scheme needs to be modified by shock operators, since by construction the basic scheme prevents entropy generation. For this purpose, we can use any of the upwind biased numerical fluxes discussed in Chapter 7. Let $\boldsymbol{f}_{C_{j+\frac{1}{2}}}$ be the central flux defined by (12.44), and let $\boldsymbol{f}_{U_{j+\frac{1}{2}}}$ be the upwind biased flux. Then, we construct the flux throughout the domain as

$$\boldsymbol{f}_{j+\frac{1}{2}} = \left(1 - S_{j+\frac{1}{2}}\right) \boldsymbol{f}_{C_{j+\frac{1}{2}}} + S_{j+\frac{1}{2}} \boldsymbol{f}_{U_{j+\frac{1}{2}}},$$

where $S_{j+\frac{1}{2}}$ is a switching function with values in the range $0 \leq S_{j+\frac{1}{2}} \leq 1$, of the order of a high power of Δx except in the neighborhood of a shock wave, where it should have a value of unity. The switching function can be constructed in a manner similar to the switch used for the Burgers' equation, (12.20) and (12.21). The same formulas may be used to detect extrema in either the pressure or the entropy. Alternatively, we can use these formulas to identify extrema in the the characteristic variables by applying them to

$$\Delta \boldsymbol{v}_{j+\frac{1}{2}} = V^{-1}_{j+\frac{1}{2}} \Delta \boldsymbol{u}_{j+\frac{1}{2}},$$

where the Jacobian matrix is decomposed in terms of its eigenvectors and eigenvalues as

$$A_{j+\frac{1}{2}} = V \Lambda V^{-1}.$$

Note that in order to implement this scheme, we can still solve the equations in the original conservation form (12.37). The entropy variables only need to be introduced in the evaluation of the numerical flux $\boldsymbol{f}_{j+\frac{1}{2}}$. In the case of the gas dynamics equations, we can take

$$U(\boldsymbol{w}) = -\frac{\rho S}{\gamma - 1}, \quad G(\boldsymbol{w}) = -\frac{\rho U S}{\gamma - 1},$$

as given by (12.10) and (12.10) in Section 2.10.

If the solution is discontinuous, (12.38) can be integrated over an arbitrary domain to give a valid integral form that is nonpositive for physically relevant solutions, corresponding to a decrease in the generalized entropy, and in the case of gas dynamics to an increase in the physical entropy. The reader is referred to Fisher and Carpenter (2013) for a comprehensive discussion of entropy consistent and entropy stable schemes.

12.5 Kinetic Energy Preserving Schemes

While entropy conserving schemes are theoretically interesting, they still require the use of special operators to capture shock waves and contact discontinuities, and it is not clear that they lead to any substantial improvement. An interesting alternative is to construct kinetic energy preserving schemes, which exactly satisfy a discrete form of the kinetic energy conservation law, thus ensuring that the discrete kinetic energy evolves in a manner that exactly corresponds to the true equation for kinetic energy. Since the total energy is directly conserved by discretizing the energy conservation law, a bound on kinetic energy should limit the possible excursions of the internal energy. Correct simulation of the evolution of kinetic energy is also a crucial requirement for accurate simulations of turbulence, where there is an energy cascade between the different eddy scales.

It is shown in this section that kinetic energy preserving schemes can be obtained by an appropriate construction of the numerical flux. It can be seen in general that the latitude in the formulation of the numerical flux enables the satisfaction of an additional conservation law for free. The construction of a kinetic energy preserving (KEP) scheme requires an approach in which the fluxes of the continuity and momentum equations are separately constructed in a compatible manner. Denoting the specific kinetic energy by k,

$$k = \rho \frac{u^2}{2}, \quad \frac{\partial k}{\partial u} = \left[-\frac{u^2}{2}, u, 0 \right].$$

Thus,

$$\begin{aligned} \frac{\partial k}{\partial t} &= u \frac{\partial}{\partial t}(\rho u) - \frac{u^2}{2} \frac{\partial \rho}{\partial t} \\ &= -\frac{\partial}{\partial x} \left\{ u \left(p + \rho \frac{u^2}{2} \right) \right\} + p \frac{\partial u}{\partial x}. \end{aligned}$$

Suppose that the semi-discrete conservation scheme (12.39) is written separately for the continuity and momentum equations as

$$\Delta x_j \frac{d\rho_j}{dt} + (\rho u)_{j+\frac{1}{2}} - (\rho u)_{j-\frac{1}{2}} = 0 \tag{12.45}$$

$$\Delta x_j \frac{d}{dt}(\rho u)_j + (\rho u^2)_{j+\frac{1}{2}} - (\rho u^2)_{j-\frac{1}{2}} + p_{j+\frac{1}{2}} - p_{j-\frac{1}{2}} = 0. \tag{12.46}$$

Now, multiplying (12.45) by $\frac{u_j^2}{2}$ and (12.46) by u_j, adding them, and summing by parts,

$$\begin{aligned} \sum_{j=1}^{n} \Delta x_j \left(u_j \frac{d}{dt}(\rho u)_j - \frac{u_j^2}{2} \frac{d\rho_j}{dt} \right) &= \sum_{j=1}^{n} \Delta x_j \frac{d}{dt} \left(\rho_j \frac{u_j^2}{2} \right) \\ = \sum_{j=1}^{n} \frac{u_j^2}{2} \left((\rho u_j)_{j+\frac{1}{2}} - (\rho u_j)_{j-\frac{1}{2}} \right) &- \sum_{j=1}^{n} u_j \left((\rho u^2)_{j+\frac{1}{2}} - (\rho u^2)_{j-\frac{1}{2}} \right) \end{aligned}$$

$$
\begin{aligned}
&- \sum_{j=1}^{n} u_j \left(p_{j+\frac{1}{2}} - p_{j-\frac{1}{2}} \right) \\
&= -\frac{u_1^2}{2}(\rho u)_{\frac{1}{2}} + u_1(\rho u^2)_{\frac{1}{2}} + u_1 p_{\frac{1}{2}} + \frac{u_n^2}{2}(\rho u)_{n+\frac{1}{2}} - u_n(\rho v^2)_{n+\frac{1}{2}} - u_n p_{n+\frac{1}{2}} \\
&+ \sum_{j=1}^{n-1} \left\{ \frac{1}{2}(\rho u)_{j+\frac{1}{2}} (u_{j+1}^2 - u_j^2) - \frac{1}{2}(\rho u^2)_{j+\frac{1}{2}} (u_{j+1} - u_j) \right\} \\
&+ \sum_{j=1}^{n-1} p_{j+\frac{1}{2}} \left(u_{j+1} - u_j \right). \qquad (12.47)
\end{aligned}
$$

Each term in the first sum containing the convective terms can be expanded as

$$
\left\{ (\rho u)_{j+\frac{1}{2}} \frac{u_{j+1} + u_j}{2} - (\rho u^2)_{j+\frac{1}{2}} \right\} \left(u_{j+1} - u_j \right)
$$

and will vanish if

$$
(\rho u^2)_{j+\frac{1}{2}} = (\rho u)_{j+\frac{1}{2}} \frac{u_{j+1} + u_j}{2}. \qquad (12.48)
$$

Now, evaluating the boundary fluxes as

$$
\begin{aligned}
(\rho u)_{\frac{1}{2}} &= \rho_1 u_1, & (\rho u^2)_{\frac{1}{2}} &= \rho_1 u_1^2, & p_{\frac{1}{2}} &= p_1, \\
(\rho u)_{n+\frac{1}{2}} &= \rho_n u_n, & (\rho u^2)_{n+\frac{1}{2}} &= \rho_n u_n^2, & p_{n+\frac{1}{2}} &= p_n,
\end{aligned} \qquad (12.49)
$$

(12.47) reduces to the semi-discrete kinetic energy conservation law

$$
\begin{aligned}
\sum_{j=1}^{n} \Delta x_j \left(\rho_j \frac{u_j^2}{2} \right) &= u_1 \left(p_1 + \rho_1 \frac{u_1^2}{2} \right) - u_n \left(p_n + \rho_n \frac{u_n^2}{2} \right) \\
&+ \sum_{j=1}^{n} p_{j+\frac{1}{2}} \left(u_{j+1} + u_j \right). \qquad (12.50)
\end{aligned}
$$

Denoting the arithmetic average of any quantity q between $j+1$ and j as

$$
\bar{q} = \frac{1}{2} \left(q_{j+1} + q_j \right),
$$

the interface pressure may be evaluated as

$$
p_{j+\frac{1}{2}} = \bar{p}. \qquad (12.51)
$$

Also, if one sets

$$
(\rho u)_{j+\frac{1}{2}} = \bar{\rho}\bar{u}, \qquad (12.52)
$$

$$
(\rho u^2)_{j+\frac{1}{2}} = \bar{\rho}\bar{u}^2, \qquad (12.53)
$$

condition (12.48) is satisfied. Consistently, one may set

$$\left(\rho u H\right)_{j+\frac{1}{2}} = \bar{\rho}\bar{u}\bar{H}. \tag{12.54}$$

The foregoing argument establishes

THEOREM 12.5.1 *The semi-discrete conservation law (12.39) satisfies the semi-discrete kinetic energy global conservation law (12.50) if the fluxes for the continuity and momentum equations satisfy condition (12.48) and the boundary fluxes are calculated by (12.49).*

A fully discrete kinetic energy conserving scheme has been developed by Subbareddy and Candler 2009. Numerical experiments with both entropy preserving (EP) and kinetic energy preserving (KEP) schemes confirm that they can simulate viscous compressible flows with shock waves provided that the mesh is fine enough to resolve the viscous shock structure (Jameson 2008). Thus, the use of upwind schemes or schemes with added numerical diffusion is no longer needed if sufficient computational resources are available to enable simulations on very fine meshes.

Condition (12.48) allows some latitude in the construction of the fluxes. For example, it is also satisfied if one sets

$$\begin{aligned}
\left(\rho u\right)_{j+\frac{1}{2}} &= \overline{\rho u}, \\
\left(\rho u^2\right)_{j+\frac{1}{2}} &= \overline{\rho u}\bar{u}, \\
\left(\rho u H\right)_{j+\frac{1}{2}} &= \overline{\rho u}\bar{H},
\end{aligned}$$

instead of (12.53–12.54). It has been realized that this latitude enables the construction of conservative schemes that are both kinetic energy and entropy conserving (Chandrashekar 2013; Kuya, Totani, & Kawai 2018). In order to treat flows with shocks waves with schemes of this type, it is necessary to introduce a shock operator to enable entropy production (Ray & Chandrashekar 2013).

13 High order Methods for Structured Meshes

13.1 High Order Finite Difference Methods

The finite difference method enables the formulation of comparatively simple high order schemes for Cartesian meshes. For example, we can approximate $\frac{\partial f}{\partial x}$ to fourth order accuracy on a uniform mesh with interval $\Delta x = h$ as

$$\frac{1}{12h}(-f_{j+2} + 8f_{j+1} - 8f_{j-1} + f_{j+2}) = \frac{\partial f}{\partial x} + \mathcal{O}(h^4) \tag{13.1}$$

and to sixth order accuracy as

$$\frac{1}{60h}(f_{j+3} - 9f_{j+2} + 45f_{j+1} - 45f_{j-1} + 9f_{j+2} - f_{j-3}) = \frac{\partial f}{\partial x} + \mathcal{O}(h^6). \tag{13.2}$$

In general, a stencil of $n+1$ points is needed to derive a formula with nth order accuracy. In fact, both centered and one-sided or offset formulas can be derived by differentiating the Lagrange interpolation polynomial of degree n over a stencil of $n+1$ points. Offset formulas are needed in the vicinity of boundaries.

In the case of a one-dimensional system of conservation laws,

$$\frac{\partial \boldsymbol{w}}{\partial t} + \frac{\partial}{\partial x}\boldsymbol{f}(\boldsymbol{w}) = 0,$$

the fourth and sixth order accurate formulas (13.1) and (13.2) can be recast to produce the conservation form

$$\frac{d\boldsymbol{w}_j}{dt} + \frac{1}{\Delta x}\left(\boldsymbol{h}_{j+\frac{1}{2}} - \boldsymbol{h}_{j-\frac{1}{2}}\right) = 0,$$

where

$$\boldsymbol{h}_{j+\frac{1}{2}} = \frac{1}{12}(-\boldsymbol{f}_{j+2} + 7\boldsymbol{f}_{j+1} + 7\boldsymbol{f}_j - \boldsymbol{f}_{j-1})$$

and

$$\boldsymbol{h}_{j+\frac{1}{2}} = \frac{1}{60}(\boldsymbol{f}_{j+3} - 8\boldsymbol{f}_{j+2} + 37\boldsymbol{f}_{j+1} + 37\boldsymbol{f}_{j-1} - 8\boldsymbol{f}_{j-2} + \boldsymbol{f}_{j-3})$$

respectively. In order to treat flows with shock waves, a stable non-oscillatory scheme may be obtained by an extension of the Jameson–Schmidt–Turkel scheme in which the numerical flux is augmented by a diffusive flux of the form

$$\boldsymbol{d}_{j+\frac{1}{2}} = (\epsilon^{(2)} - \epsilon^{(4)}\Delta^+\Delta^- + \epsilon^{(6)}(\Delta^+\Delta^-)^2 \ldots)\Delta^+ w_j,$$

where $\Delta+$ and $\Delta-$ are the forward and backward difference operators

$$\Delta^+\boldsymbol{w}_j = \boldsymbol{w}_{j+1} - \boldsymbol{w}_j, \quad \Delta^-\boldsymbol{w}_j = \boldsymbol{w}_j - \boldsymbol{w}_{j-1}.$$

These coefficients should be scaled to the spectral radius of the Jacobian matrix $A = \frac{\partial f}{\partial w}$,

$$\rho = \max_k \left|\lambda^{(k)}(A)\right|,$$

as has been described in the discussion of shock capturing schemes in Chapter 7. In a smooth region of the flow, only the highest term $\epsilon^{(k)}$ is needed to provide background dissipation of order Δx^{k-1}. In the neighborhood of an extremum, the highest terms should be switched off, while $\epsilon^{(2)}$ should have a value $\geq \frac{1}{2}\rho$. Schemes of this type have been developed and successfully used by Pulliam (2011).

For the multi-dimensional conservation law,

$$\frac{\partial \boldsymbol{w}}{\partial t} + \frac{\partial}{\partial x_i}\boldsymbol{f}_i(w) = 0,$$

these formulas can be applied separately in each coordinate direction. In the case of a sufficiently smooth curvilinear mesh with coordinates ξ_i derived by a global transformation, one may discretize the strong conservation form derived in Section 2.2,

$$\frac{\partial}{\partial t}(J\boldsymbol{w}) + \frac{\partial}{\partial \xi_i}\boldsymbol{F}_i(\boldsymbol{w}) = 0,$$

where

$$K_y = \frac{\partial \xi_i}{\partial x_j}, \quad J = \det K, \quad S = JK^{-1}$$

and

$$\boldsymbol{F}_i = S_{ij}\boldsymbol{f}_j(w).$$

Now, the metric derivatives S_{ij} may be calculated by difference formulas of the same order of accuracy as those used to approximate the flux derivatives.

It should be noted that this procedure does not yield zero residuals when it is applied to a uniform flow, because the terms arising from the metric derivatives do not necessarily cancel. In order to correct this defect, one may subtract the residual for a uniform flow from the calculated residuals of the nonuniform flow.

13.2 High Order Compact Difference Schemes

13.2.1 Overview

In order to reduce the size of the stencil for a high order formula, one may solve for a linear combination of the derivatives at neighboring mesh points. Schemes of this type are generally known as compact difference schemes, because they enable a given order of accuracy to be obtained with a narrower stencil. For example, a fourth order accurate formula for the first derivative is provided by the scheme

$$\frac{1}{6}v_{j-1} + \frac{2}{3}v_j + \frac{1}{6}v_{j+1} = \frac{f_{j+1} - f_{j-1}}{2h}, \tag{13.3}$$

where v_j is a numerical approximation to $\frac{df}{dx}$. Expanding both sides as truncated Taylor series, the two sides differ by terms of order h^4. This formula is actually identical to the formula for the first derivative of a cubic spline on a uniform mesh.

The Pade difference schemes are a well known example of compact difference formulas (Kreiss & Oliger 1973, Orszag & Israeli 1974). Here, we shall follow the approach of Lele (1992) to derive a general family of compact difference schemes for the first and second derivatives f' and f'' with orders of accuracy up to 10.

13.2.2 Derivation of Compact Schemes for the First Derivative

In order to calculate a numerical approximation v_j to the first derivative f'_j at the mesh point j on a uniform mesh with interval h, we shall solve an equation of the form

$$\begin{aligned} v_j + \alpha_1(v_{j+1} + v_{j-1}) + \alpha_2(v_{j+2} + v_{j-2}) \\ = a_1\frac{f_{j+1} - f_{j-1}}{2h} + a_2\frac{f_{j+2} - f_{j-2}}{4h} + a_3\frac{f_{j+3} - f_{j-3}}{6h}, \end{aligned} \tag{13.4}$$

where α_1, α_2, and a_1, a_2, and a_3 are coefficients to be determined, depending on the desired order of accuracy. If $\alpha_2 \neq 0$, the left side has a five-point stencil, and if $a_3 \neq 0$, the right side has a seven-point stencil. Suppose that

$$v_j = f'_j + e_j,$$

where e_j is the error between v_j and the true derivative f'_j. Now, we expand the various terms as truncated Taylor series of the appropriate order, using the remainder theorem that the error for a series truncated at the nth term is $\frac{f^{(n+1)}}{(n+1)!}$, where $f^{(n+1)}$ is evaluated somewhere in the interval of the expansion. Then,

$$\begin{aligned} v_{j+1} + v_{j-1} &= v_j + hv'_j + \frac{h^2}{2!}v''_j + \frac{h^3}{3!}v'''_j + \cdots + v_j - hv'_j + \frac{h^2}{2!}v''_j - \frac{h^3}{3!}v'''_j + \cdots \\ &= 2\left(v_j + \frac{h^2}{2!}v''_j + \frac{h^4}{4!}v^{(4)}_j + \frac{h^6}{6!}v^{(6)}_j + \frac{h^8}{8!}v^{(8)}_j\right) + \mathcal{O}(h^{10}), \end{aligned}$$

while

$$v_{j+2} + v_{j-2} = 2\left(v_j + \frac{(2h)^2}{2!}v''_j + \frac{(2h)^4}{4!}v^{(4)}_j + \frac{(2h)^6}{6!}v^{(6)}_j + \frac{(2h)^8}{8!}v^{(8)}_j\right) + \mathcal{O}(h^{10}).$$

Hence, the left hand side can be expanded as

$$
\begin{aligned}
e_j + \alpha_1(e_{j+1} + e_{j-1}) + \alpha_2(e_{j+2} + e_{j-2}) &+ (1 + 2\alpha_1 + 2\alpha_2)f_j' \\
&+ 2(\alpha_1 + 2^2\alpha_2)\frac{h^2}{2!}f_j''' \\
&+ 2(\alpha_1 + 2^4\alpha_2)\frac{h^4}{4!}f_j^5 \\
&+ 2(\alpha_1 + 2^6\alpha_2)\frac{h^6}{6!}f_j^7 \\
&+ 2(\alpha_1 + 2^8\alpha_2)\frac{h^8}{8!}f_j^9 \\
&+ \mathcal{O}(h^{10}).
\end{aligned}
$$

Also,

$$
\frac{f_{j+1} - f_{j-1}}{2h} = f_j' + \frac{h^2}{3!}f_j''' + \frac{h^4}{5!}f_j^{(5)} + \frac{h^6}{7!}f_j^{(7)} + \frac{h^8}{9!}f_j^{(9)} + \mathcal{O}(h^{10}),
$$

with corresponding expansions for $\frac{f_{j+2}-f_{j-2}}{4h}$ and $\frac{f_{j+3}-f_{j-3}}{6h}$. Hence, the right hand side can be expanded as

$$
\begin{aligned}
&(a_1 + a_2 + a_3)f_j' \\
&+ (a_1 + 2^2a_2 + 3^2a_3)\frac{h^2}{3!}f_j''' \\
&+ (a_1 + 2^4a_2 + 3^4a_3)\frac{h^4}{5!}f_j^{(5)} \\
&+ (a_1 + 2^6a_2 + 3^6a_3)\frac{h^6}{7!}f_j^{(7)} \\
&+ (a_1 + 2^8a_2 + 3^8a_3)\frac{h^8}{9!}f_j^{(9)} \\
&+ \mathcal{O}(h^{10}).
\end{aligned}
$$

Now, schemes of various orders of accuracy can be obtained by choosing α_1, α_2, a_1, a_2, and a_3 to satisfy constraints such that the terms of successively higher powers of h are matched in the left and right sides of the equation. The constraints are as follows:

$$
a_1 + a_2 + a_3 = 1 + 2\alpha_1 + 4\alpha_2 \quad \text{(second order)}, \tag{13.5}
$$

$$
a_1 + 2^2a_2 + 3^2a_3 = 6(\alpha_1 + 2^2\alpha_2) \quad \text{(fourth order)},
$$

$$
a_1 + 2^4a_2 + 3^4a_3 = 10(\alpha_1 + 2^4\alpha_2) \quad \text{(sixth order)},
$$

$$
a_1 + 2^6a_2 + 3^6a_3 = 14(\alpha_1 + 2^6\alpha_2) \quad \text{(eighth order)},
$$

$$
a_1 + 2^8a_2 + 3^8a_3 = 18(\alpha_1 + 2^8\alpha_2) \quad \text{(tenth order)}. \tag{13.6}
$$

In any case, the error equation at points sufficiently far from the boundary reduces to

$$e_j + \alpha_1(e_{j+1} + e_{j-1}) + \alpha_2(e_{j+2} + e_{j-2}) = r_j, \tag{13.7}$$

where

$$|r_j| \le Mh^k,$$

and k is the power of the first unmatched term and M is a bound on $\left|f^{(k+1)}\right|$. Depending on the treatment of points in the vicinity of the boundaries, it may be possible to obtain a corresponding global bound on the error.

13.2.3 Summary of Formulas for the First Derivative

The constraints (13.5) to (13.6) lead to progressively more accurate and complex schemes. Taking $\alpha_2 = 0$ leads to a variety of tridiagonal systems. Taking also $a_3 = 0$ leads to a one parameter family of fourth order accurate schemes,

$$\alpha_2 = 0, \quad a_1 = \frac{2}{3}(\alpha_1 + 2), \quad a_2 = \frac{1}{3}(4\alpha_1 - 1), \quad a_3 = 0.$$

The standard fourth order central difference scheme is obtained by setting $\alpha_1 = 0$. The Pade scheme (13.3), which has a three-point stencil on both sides, is obtained by setting $\alpha_1 = \frac{1}{4}$. However, the extra degree of freedom in α_1 allows a sixth order accurate scheme with $\alpha_1 = \frac{1}{3}$. For this scheme, which has a three-point stencil on the left and a five-point stencil on the right,

$$\alpha_1 = \frac{1}{3}, \quad \alpha_2 = 0, \quad a_1 = \frac{14}{9}, \quad a_2 = \frac{1}{9}, \quad a_3 = 0. \tag{13.8}$$

Keeping a three-point stencil on the left while allowing a seven-point stencil on the right, we can obtain a one parameter family of sixth order accurate schemes with

$$\alpha_2 = 0, \quad a_1 = \frac{1}{6}(\alpha_1 + 9), \quad a_2 = \frac{1}{15}(32\alpha_1 - 9), \quad a_3 = \frac{1}{10}(1 - 3\alpha_1).$$

The choice $\alpha_1 = \frac{1}{3}$, $a_3 = 0$ recovers the scheme (13.8). The choice $\alpha_1 = \frac{3}{8}$ leads to an eighth order accurate scheme for which

$$\alpha_1 = \frac{3}{8}, \quad \alpha_2 = 0, \quad a_1 = \frac{25}{12}, \quad a_2 = \frac{1}{5}, \quad a_3 = -\frac{1}{80}.$$

Allowing a five-point stencil on the left, we can obtain a one parameter family of eighth order accurate schemes with

$$\alpha_2 = \frac{1}{20}(8\alpha_1 - 3), \quad a_1 = \frac{1}{6}(12 - 7\alpha_1),$$

$$a_2 = \frac{1}{150}(568\alpha_1 - 183), \quad a_3 = \frac{1}{50}(9\alpha_1 - 4),$$

Table 13.1. Highest order compact schemes for the first derivative.

Order of accuracy	Left stencil	Right stencil	α_1	α_2	a_1	a_2	a_3
4	3	3	$\frac{1}{4}$	0	$\frac{3}{2}$	0	0
6	3	5	$\frac{1}{3}$	0	$\frac{14}{9}$	$\frac{1}{9}$	0
8	5	5	$\frac{4}{9}$	$\frac{1}{36}$	$\frac{40}{27}$	$\frac{25}{54}$	0
10	5	7	$\frac{1}{2}$	$\frac{1}{20}$	$\frac{17}{12}$	$\frac{101}{150}$	$\frac{1}{100}$

which is a scheme given by Collatz (1966) and which has five-point stencils on both sides with

$$\alpha_1 = \frac{4}{9}, \quad \alpha_2 = \frac{1}{36}, \quad a_1 = \frac{40}{27}, \quad a_2 = \frac{25}{54}, \quad a_3 = 0.$$

Finally, there is a unique tenth order accurate scheme with

$$\alpha_1 = \frac{1}{2}, \quad \alpha_2 = \frac{1}{20}, \quad a_1 = \frac{17}{12}, \quad a_2 = \frac{101}{150}, \quad a_3 = \frac{1}{100}.$$

The most accurate schemes for each combination of stencils are summarized in Table 13.1.

13.2.4 Global Error Estimates for the First Derivative

In the case of a finite domain with non-periodic boundary conditions, the centered formula (13.4) has to be supplemented with offset and one-sided formulas in the vicinity of the boundaries. Introducing boundary operators, the error equation can be expressed in matrix vector form as

$$A\mathbf{e} = \mathbf{r},$$

where $\mathbf{r}$ and $\mathbf{e}$ are the residual and error vectors, and for interior nodes, the matrix A is tridiagonal or pentadiagonal. Accordingly, we obtain the global error bound

$$\|\mathbf{e}\| \leq \|A^{-1}\| \|\mathbf{r}\|.$$

Assuming that $\|A^{-1}\|$ is bounded, $\|\mathbf{e}\| = \mathcal{O}(h^k)$ if $\|\mathbf{r}\| = \mathcal{O}(h^k)$. However, it is evident that the scheme may not attain the formal order of accuracy of the interior points if the points in the vicinity of the boundaries have a lower order of accuracy.

If the boundary conditions are periodic, both the defining equation (13.4) and the error equation (13.7) can be solved by a discrete Fourier transform. Suppose that the domain ranges from 0 to 2π with $2n$ equally spaced intervals of width $h = \frac{2\pi}{n}$. Let $\hat{\mathbf{e}}$ be the transform of $\mathbf{e}$ defined as

$$\hat{e}_k = \frac{1}{2n} \sum_{j=0}^{2n-1} e_j e^{-ikx_j},$$

where

$$e_j = \sum_{k=-n}^{n-1} \hat{e}_k e^{ikx_j}, \tag{13.9}$$

and let $\hat{\mathbf{r}}$ be the transform of $\mathbf{r}$. Then

$$\hat{v}_k + \alpha_1(e^{ikh} + e^{-ikh})\hat{v}_k + \alpha_2(e^{2ikh} + e^{-2ikh})\hat{v}_k = \hat{r}_k.$$

Hence,

$$\hat{e}_k = \frac{\hat{r}_k}{1 + 2\alpha_1 \cos \xi_k + 2\alpha_2 \cos 2\xi_k},$$

where

$$\xi = kh = \frac{\pi k}{n},$$

and e_j is given by the reverse transform (13.9). Also, in the Euclidean norm,

$$\|\mathbf{e}\| = \sqrt{2}n\|\hat{\mathbf{e}}\|, \quad \|\mathbf{r}\| = \sqrt{2}n\|\hat{\mathbf{r}}\|,$$

where

$$\|\hat{\mathbf{e}}\| = \left(\sum_{k=-n}^{n-1} \left| \frac{\hat{r}_k}{1 + 2\alpha_1 \cos \xi + 2\alpha_2 \cos 2\xi} \right|^2 \right)^{\frac{1}{2}}.$$

Hence,

$$\|\hat{\mathbf{e}}\| \leq C\|\hat{\mathbf{r}}\|, \quad \|\mathbf{e}\| < C\|\mathbf{r}\|,$$

where

$$C = \frac{1}{\min\limits_k (1 + 2\alpha_1 \cos \xi + 2\alpha_2 \cos 2\xi)}.$$

Here, ξ lies in the range from 0 to π. Consider the function

$$S(\xi) = 1 + 2\alpha_1 \cos \xi + 2\alpha_2 \cos 2\xi.$$

Then,

$$\begin{aligned} S'(\xi) &= -2(\alpha_1 \sin \xi + 2\alpha_2 \sin 2\xi) \\ &= -2 \sin \xi(\alpha_1 + 4\alpha_2 \cos \xi). \end{aligned}$$

If $\alpha_1 > 4\alpha_2$, $S'(\xi) = 0$ only at $0,\pi$, and $S(\xi)$ reaches a minimum at π. The results for the schemes listed in Table 13.1 are shown in Table 13.2.

In all cases, $S_{\min} > 0$, so C is bounded, confirming a global error estimate of the same order as the formal order of accuracy at interior points.

Table 13.2. Error indicators for compact schemes.

Order of accuracy	Stencil	α_1	α_2	$S_{\min}$
4	3 - 3	$\frac{1}{4}$	0	$\frac{1}{2}$
6	3 - 5	$\frac{1}{3}$	0	$\frac{1}{3}$
8	5 - 5	$\frac{4}{9}$	$\frac{1}{36}$	$\frac{1}{6}$
10	5 - 7	$\frac{1}{2}$	$\frac{1}{20}$	$\frac{1}{10}$

In the non-periodic case, a bound can be determined on $\|A^{-1}\|$ if it is strictly diagonally dominant

$$|a_{ii}| > \sum_{j \neq 1} |a_{ij}|$$

for all rows. This is the case for the 4th, 6th, and 8th order accurate schemes in Table 13.1, provided diagonal dominance is maintained in the boundary operators. Then, using the maximum norm,

$$\|\mathbf{r}\|_\infty = \|A\mathbf{e}\|_\infty = \max_i \left| \sum_{j=1}^{n} a_{ij} e_j \right| \geq \sum_{j=1}^{n} \left| a_{kj} e_j \right|,$$

where $\|\mathbf{e}\|_\infty = |e_k|$. Thus,

$$\begin{aligned}
\|\mathbf{r}\|_\infty &\geq |a_{kk}|\ |e_k| + \sum_{j \neq k} |a_{kj}|\ |e_j| \\
&\geq \left\{ |a_{kk}| - \sum_{j \neq k} |a_{kj}| \right\} \|\mathbf{e}\|_\infty \\
&\geq \min_i \left\{ |a_{ii}| - \sum_{j \neq i} |a_{ij}| \right\} \|\mathbf{e}\|_\infty,
\end{aligned}$$

and hence,

$$\|\mathbf{e}\|_\infty \leq \left[\min_i \left\{ |a_{ii}| - \sum_{j \neq i} |a_{ij}| \right\} \right]^{-1} \|\mathbf{r}\|_\infty.$$

Equivalently,

$$\|A^{-1}\|_\infty \leq \frac{1}{\min_i C_i},$$

Table 13.3. Diagonal dominance of the schemes listed in Table 13.1.

Order of accuracy	C_i
4	$\frac{1}{2}$
6	$\frac{1}{3}$
8	$\frac{1}{18}$
10	(Not diagonally dominant)

Table 13.4. One sided schemes at first boundary point.

Order of accuracy	b_1	b_2	b_3	b_4	b_5	b_6	b_7
5	$-\frac{137}{60}$	5	-5	$\frac{10}{3}$	$-\frac{5}{4}$	$\frac{1}{5}$	0
6	$-\frac{49}{20}$	6	$-\frac{15}{2}$	$\frac{20}{3}$	$-\frac{15}{4}$	$\frac{6}{5}$	$-\frac{1}{6}$

where

$$C_i = |a_{ii}| - \sum_{j \neq i} |a_{ij}|.$$

At the interior point, C_i has the values displayed in Table 13.3.

The boundary formulas, proposed by Lele (1992), do not by and large preserve diagonal dominance. Gaitonde and Visbal (1998) propose formulas of up to 6th order accuracy at the first and second boundary points, some of which do preserve diagonal dominance.

At the first boundary point, we can use explicit schemes of the form

$$f_1' = \sum_{r=1}^{q} b_r f_r + \mathcal{O}(h^{q-1}),$$

which can be derived by differentiating one-sided Lagrange interpolation polynomials. Schemes of 5th and 6th order accuracy are displayed in Table 13.4.

At the second boundary point, we can use formulas that are symmetric and tridiagonal on the left of the form

$$v_2 + \alpha_1(v_1 + v_3) = \sum_{r=1}^{q} c_r f_r$$

Table 13.5. Asymmetric schemes at second boundary point.

Order of accuracy	α_1	c_1	c_2	c_3	c_4	c_5	c_6
5	$\frac{3}{14}$	$-\frac{19}{28}$	$-\frac{5}{42}$	$\frac{6}{7}$	$-\frac{1}{14}$	$\frac{1}{84}$	0
6	$\frac{2}{11}$	$-\frac{20}{33}$	$-\frac{35}{132}$	$\frac{34}{33}$	$-\frac{7}{33}$	$\frac{2}{33}$	$-\frac{1}{132}$

where, as before, v_i is the discrete estimate of f_i'. Schemes of 5th and 6th order accuracy are displayed in Table 13.5.

According to the analysis of the preceding paragraphs, a combination of the 6th order scheme of Table 13.1 with the sixth order boundary schemes of Tables 13.4 and 13.5 would preserve global accuracy of 6th order in the maximum norm. Also, the matrix A is symmetric with any of these boundary closures. Then, A^{-1} is also symmetric, and hence its 1 and ∞ norms are the same.

$$\|A^{-1}\|_1 = \|A^{-1}\|_\infty.$$

Now, the average error $\frac{\|\mathbf{e}\|_1}{N}$ on a mesh with N points using the 6th order scheme of Table 13.1 with 5th order boundary closures of Tables 13.4 and 13.5 will be of order h^6 as $h \to 0$, because $N = \frac{L}{h}$ for a domain of length L, and accordingly the contributions from the boundary points will be weighted by $\frac{h}{L}$, while the average error of the interior points remains of order h^6.

13.2.5 Phase Error Analysis of Compact Schemes

Additional insights into the properties of the high order schemes can be obtained by phase error analysis. Consider the linear advection equation

$$\frac{\partial u}{\partial t} + a\frac{\partial u}{\partial x} = 0$$

in the interval $0 \le x \le 2\pi$, with periodic boundary conditions, and initial data

$$u(x,0) = e^{ikx}.$$

The true solution is the traveling wave

$$u(x,t) = e^{ik(x-at)}$$

with phase speed a. We now consider semi-discrete schemes on a uniform grid $x = jh$, $j = 0, \dots, 2n$ with $h = \frac{2\pi}{n}$, where the spatial derivatives are determined by formulas of the form (13.4). The simplest is the second order central difference scheme

$$\frac{dv_j}{dt} + \frac{a}{2h}(v_{j+1} - v_{j-1}) = 0.$$

Introducing the discrete Fourier transform as in the previous section,

$$v_j(t) = \sum_{k=-n}^{n-1} \hat{v}_k(t) e^{ikx_j},$$

where

$$\hat{v}_k(t) = \frac{1}{2n} \sum_{j=0}^{2n-1} v_j(t) e^{-ikx_j}.$$

Then,

$$\begin{aligned} \frac{d\hat{v}_k}{dt} &= -\frac{a}{2h}\left(e^{ikh} - e^{-ikh}\right)\hat{v}_k \\ &= -ia\frac{\sin kh}{h}\hat{v}_k, \end{aligned}$$

and

$$\hat{v}_k(t) = \hat{v}(0) e^{-ia_h kt},$$

where, setting $\xi = kh = \frac{k\pi}{n}$,

$$a_h(\xi) = a\frac{\sin \xi}{\xi}. \tag{13.10}$$

Now,

$$v_j(t) = \sum_{k=-n}^{n} \hat{v}_k e^{ik(x_j - a_h t)},$$

where each mode represents a wave traveling with a numerical wave speed a_h defined by the dispersion relation (13.10) for ξ in the range 0 to π. With central difference formulas, the dispersion relation is real, with the consequence that there is an error in the phase speed of each mode but not in the amplitude. Offset or one-sided schemes result in a dispersion relation with both real and imaginary parts. If the imaginary part is positive, the amplitude decays, and the scheme is dissipative. In the case of the general compact scheme (13.4), a similar analysis shows that the numerical wave speed satisfies the dispersion relation

$$a_h(\xi) = a\frac{D(\xi)}{\xi},$$

where

$$D(\xi) = \frac{a_1 \sin \xi + \frac{1}{2}a_2 \sin 2\xi + \frac{1}{3}a_3 \sin 3\xi}{1 + 2\alpha_1 \cos \xi + 2\alpha_2 \cos 2\xi},$$

and there would be no phase error if $D(\xi) = \xi$. Figure 13.1 shows plots of $D(\xi)$ versus ξ for the 4th, 6th, 8th, and 10th order schemes in Table 13.1.

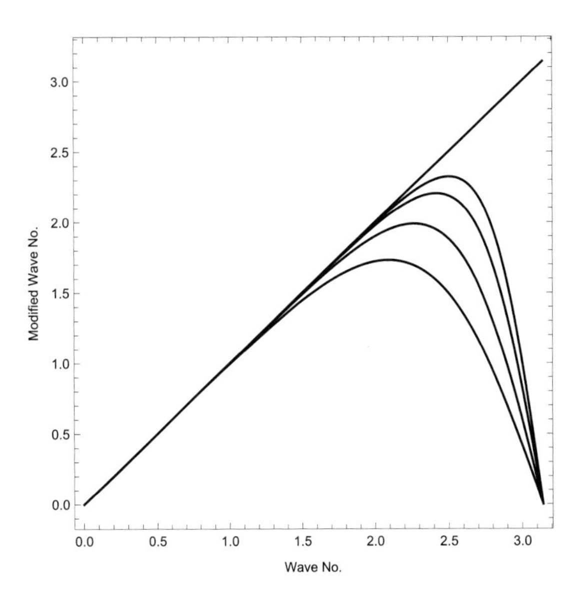

Figure 13.1 Plot of modified wave number vs. wave number for first derivative approximations: derivative approximations for the schemes of Table 13.1. Curves for 4th to 10th order schemes progressively to the right.

13.2.6 Derivation of Compact Schemes for the Second Derivative

Compact formulas for the second derivative can be obtained in a similar manner. Now, we solve

$$\begin{aligned} v_j + \alpha_1(v_{j+1} + v_{j-1}) + \alpha_2(v_{j+2} + v_{j-2}) = \frac{a_1}{h^2}(f_{j+1} - 2f_j - f_{j-1}) \\ + \frac{a_2}{4h^2}(f_{j+2} - 2f_j - f_{j-2}) \\ + \frac{c}{9h^2}(f_{j+3} - 2f_j - f_{j-3}), \end{aligned}$$

where v_j is the numerical approximation to f_j''. Now, setting

$$v_j = f_j'' + e_j,$$

where e_j is the error between v_j and the true second derivative f_j'', we obtain a penta-diagonal error equation of the form (13.7), where the residual r_j is of order h^k, and the constraints for successively higher orders of accuracy are

$$\begin{aligned} a_1 + a_2 + a_3 &= 1 + 2\alpha_1 + 2\alpha_2 \quad \text{(second order)}, \\ a_1 + 2^2 a_2 + 3^2 a_3 &= 3.4(\alpha_1 + 2^2\alpha_2) \quad \text{(fourth order)}, \\ a_1 + 2^4 a_2 + 3^4 a_3 &= 5.6(\alpha_1 + 2^4\alpha_2) \quad \text{(sixth order)}, \end{aligned}$$

$$a_1 + 2^6 a_2 + 3^6 a_3 = 7.8(\alpha_1 + 2^6 \alpha_2) \quad \text{(eighth order)},$$

$$a_1 + 2^8 a_2 + 3^8 a_3 = 9.10(\alpha_1 + 2^8 \alpha_2) \quad \text{(tenth order)}.$$

13.2.7 Filtering

The centered compact difference formulas described in the previous sections are non-dissipative. In practical problems involving mappings to nonuniform curvilinear coordinates and nonlinear equations, instabilities may occur due to the growth of high frequency modes. It has been shown by Gaitonde and Visbal that these can be controlled by the application of high order filters after each time step. They propose a family of schemes with a tridiagonal stencil on the left of the form

$$\bar{v}_j + \alpha_f(\bar{v}_{j+1} - \bar{v}_{j-1}) = \frac{1}{2}\sum_{r=0}^{q} a_r(v_{j+r} + v_{j-r}),$$

where $\bar{v}_j$ is the filtered quantity derived from v_j, and for diagonal dominance, $-\frac{1}{2} \leq \alpha_f \leq \frac{1}{2}$. Applied to a Fourier mode

$$v_j = \hat{v}e^{i\omega x_j},$$

this leads to

$$\hat{\bar{v}} = g\bar{v}$$

with the amplification factor

$$g(\xi) = \frac{\sum_{r=0}^{q} a_r \cos(r\xi)}{1 + 2\alpha_f \cos(\xi)},$$

where $\xi = \omega h$. The constraint $g(\pi) = 0$ is applied to achieve complete damping of the highest frequency mode. Schemes of 6th, 8th, and 10th order accuracy are displayed in Table 13.6, in which α_f is treated as a free parameter.

Table 13.6. Filters of 6th, 8th, and 10th order accuracy.

Order	a_0	a_1	a_2	a_3	a_4	a_5
6	$\frac{11+10\alpha_f}{16}$	$\frac{15+34\alpha_f}{32}$	$\frac{-3+6\alpha_f}{16}$	$\frac{1-2\alpha_f}{32}$	0	0
8	$\frac{93+70\alpha_f}{128}$	$\frac{7+18\alpha_f}{16}$	$\frac{-7(1-2\alpha_f)}{32}$	$\frac{1-2\alpha_f}{16}$	$\frac{-(1-2\alpha_f)}{128}$	0
10	$\frac{193+126\alpha_f}{256}$	$\frac{105+302\alpha_f}{256}$	$\frac{-15(1-2\alpha_f)}{64}$	$\frac{45(1-2\alpha_f)}{512}$	$\frac{-5(1-2\alpha_f)}{256}$	$\frac{1-2\alpha_f}{512}$

13.3 ENO and WENO Schemes

Schemes using reconstruction still require the use of limiters near extrema to maintain the LED or TVD property. In order to alleviate this difficulty, Harten, Osher, Engquist, and Chakaravarthy proposed the idea of "essentially non-oscillatory" (ENO) schemes (Harten et al. 1987). These use an adaptive stencil for the reconstruction of v_L and v_R designed to find the least oscillatory interpolant of a given order of accuracy. To construct $v_{L_{j+\frac{1}{2}}}$, for example, with rth order accuracy, one would examine the interpolants of r stencils, each spanning r cells, symmetrically arranged around cell j, as illustrated in Figure 13.2. In order to measure which interpolant is least oscillatory, we can introduce the divided differences for each stencil. For a function $V(x)$ defined at the points $x_0, x_1, \ldots, x_r$, the divided differences are defined recursively as

$$V[x_0] = V(x_0)$$

$$V[x_0, x_1] = \frac{V[x_1] - V[x_0]}{x_1 - x_0}$$

$$V[x_0, x_1, x_2] = \frac{V[x_1, x_2] - V[x_0, x_1]}{x_2 - x_0}.$$

Then, if $V(x)$ is smooth in the stencil $[x_0, x_2]$, the rth divided difference is proportional to the rth derivative at a point ξ within the stencil

$$V[x_0, \ldots, x_r] = \frac{1}{r!} V^{(r)}(\xi),$$

while if $V(x)$ is discontinuous,

$$V[x_0, \ldots, x_r] = \mathcal{O}\frac{1}{(\Delta x^r)}.$$

Accordingly, the divided differences provide a measure of the smoothness of $V(x)$ in the stencil.

The ENO scheme builds up successively higher order interpolants to the primitive function $V(x)$ defined as

$$V_{x+\frac{1}{2}} = \int_{\infty}^{x_{j+\frac{1}{2}}} v(\xi) d\xi = \sum_{i=-\infty}^{j} \bar{v}_i \Delta x_i,$$

where $\bar{v}_i$ is the average value of $v(x)$ in cell i. Then the derivative of the Lagrange interpolation polynomial $P(x)$ over an r-cell stencil satisfies

$$P'(x) = V'(x) + \mathcal{O}(\Delta x^r),$$

with the result that $P'(x_{j+\frac{1}{2}})$ provides the desired rth order reconstruction of $v_{j+\frac{1}{2}}$.

The reconstruction procedure starts with a stencil consisting of the single cell j. At each step, a higher order interpolant is formed by adding a cell to the left or right of the previous stencil. We choose whichever of the two new candidate stencils has the smaller absolute value of the divided difference, thus arriving at the least

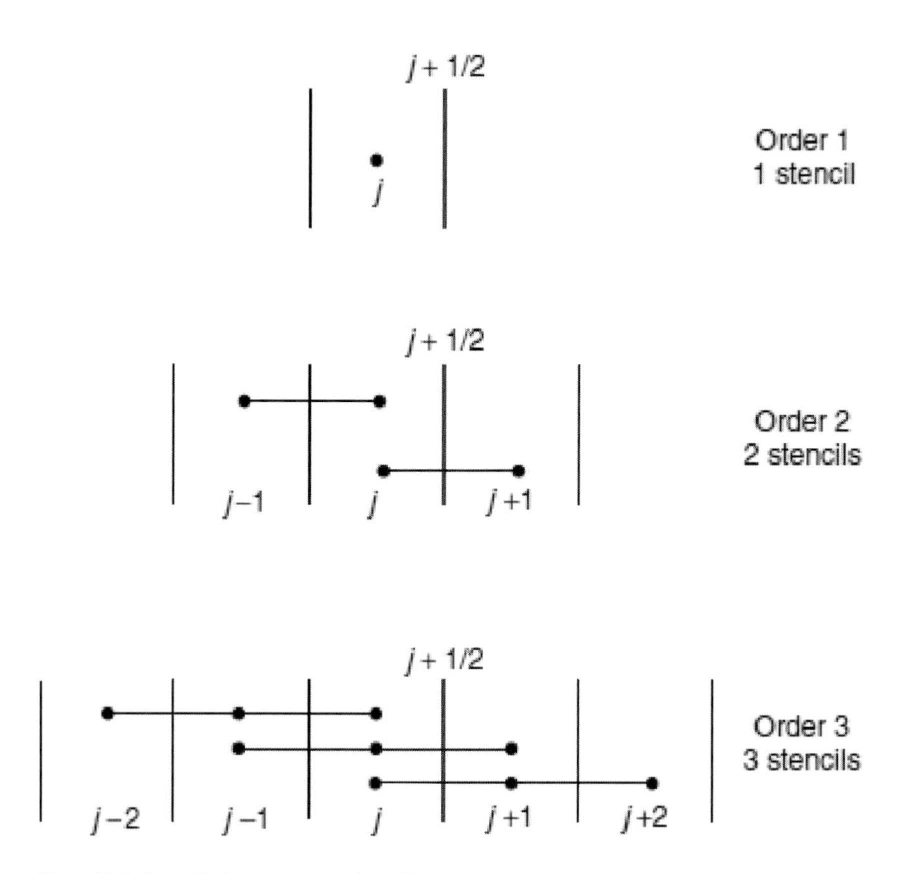

Figure 13.2 Stencils for reconstruction of $v_{L_{j+\frac{1}{2}}}$.

oscillatory interpolants of any given order. Alternative reconstruction procedures to generate interpolants from point values have also been worked out.

The original ENO schemes have the drawback that very small changes in the solution could cause sudden changes in the adaptive stencils. In steady flow calculations, this could lead to limit cycles that prevent complete convergence. If the initial data was an odd-even mode $v_j = (-1)^j$, for example, all divided differences would be equal in magnitude, and the choice of adaptive stencils would be random.

These deficiencies are remedied by the weighted essentially non-oscillatory (WENO) schemes originally developed by Liu, Osher, and Chan (1994), and Jiang and Shu (1996). Instead of picking one stencil at each stage, the WENO schemes use a weighted average of the r possible r-cell stencils symmetrically arranged about cell j for the reconstruction of v_L at the interface $j + \frac{1}{2}$ and about cell $j + 1$ for the reconstruction of v_R. Referring to Figure 13.2, it can be seen that these span $2r - 1$ cells, with the consequence that the weights can be chosen to recover $(2r - 1)$th order accurate estimates of v_L and v_R.

Suppose that the r cell interpolants for the stencils $(j + k - r + 1, \ldots, j + k)$, $k = 0, \ldots, r - 1$ are q_k^r, and the optimal $(2r - 1)$th order accurate combination is

$$\hat{v}_L = \sum_{k=0}^{r-1} C_k^r q_k^r$$

with weights C_k^r such that

$$\sum_{k=0}^{r-1} C_k^r = 1.$$

In order to find a less oscillatory interpolant when the solution is not smooth, we replace these weights with weights ω_k, which are deflated if the interpolant q_k^r is oscillatory. For this purpose, we introduce a measure IS_k of the smoothness of q_k^r. Then, the weights ω_k may be defined as

$$\omega_k = \frac{\alpha_k}{\alpha_0 + \cdots + \alpha_{r-1}},$$

where

$$\alpha_k = \frac{C_k^r}{(\epsilon + IS_k)^p}, \quad k = 0, 1, \ldots, r-1. \tag{13.11}$$

These definitions ensure that

$$\sum_{k=0}^{r-1} \omega_k = 1.$$

The quantity ϵ is a small tolerance (>0) to ensure that α_k is defined as $IS_k = 0$. Jiang and Shu recommend a formula based on the divided differences for the smoothness indicator IS_k and the value $p = 2$ for the power in (13.11).

WENO schemes have been widely adopted and shown to give excellent results for unsteady flows with shock waves and contact discontinuities, such as flows in shock tubes (Jiang & Shu 1996). Their extension to multi-dimensional flows on unstructured grids has been worked out, but is very complex.

13.4 Spectral Methods

Like finite element methods, spectral methods represent the solution as an expansion in basis functions

$$u_h = \sum u_j \phi_j,$$

but now the basis functions have global support, with the consequence that the solution at any location depends on data throughout the domains. The classical example is the use of Fourier expansions for problems with periodic boundary conditions. Consider the linear advection equation

$$\frac{\partial u}{\partial t} + a \frac{\partial u}{\partial x} = 0$$

in the interval $[0, 2\pi]$ with periodic boundary conditions. Assuming that N is even, we substitute the truncated Fourier expansion

$$u_h = \sum_{k=-N/2}^{N/2} \hat{u}_k(t) e^{ikx},$$

where

$$\hat{u}_k = \frac{1}{2\pi} \int_0^{2\pi} u(x) e^{-ikx} dx.$$

In a Fourier–Galerkin method, we then require the residual equation to be orthogonal to the expansion functions,

$$\int_0^{2\pi} \left(\frac{\partial u_h}{\partial t} + a \frac{\partial u_h}{\partial x} \right) e^{-imx} dx = 0,$$

for $|m| \leq N/2$. Since

$$\int_0^{2\pi} e^{ikx} e^{-imx} = 0$$

when $m \neq k$, we obtain a separate equation

$$\frac{d\hat{u}_k}{dt} + iak\hat{u}_k = 0 \quad \text{for } |k| \leq N/2$$

for each mode $\hat{u}_k$. These can be integrated to give

$$\hat{u}_k(t) = \hat{u}_k(0) e^{-iakt},$$

and hence

$$\begin{aligned} u_h(x,t) &= \sum_{k=-N/2}^{N/2} u_k(0) e^{ik(x-at)} \\ &= g_h(x - at), \end{aligned} \tag{13.12}$$

where

$$g_h(x) = u_h(x, 0).$$

Thus the initial projection is propagated exactly.

The Fourier–Galerkin method proves, however, to be cumbersome for nonlinear problems. In the case of the nonlinear conservation law

$$\frac{\partial u}{\partial t} + \frac{\partial}{\partial x} f(u) = 0,$$

the same substitution leads to

$$\int_0^{2\pi} \left(\frac{\partial u_h}{\partial t} + \frac{\partial}{\partial x} f(u_h) \right) e^{-imx} dx = 0 \quad \text{for } |m| \leq \frac{N}{2},$$

which yields the frequency domain equation

$$\frac{d\hat{u}_k}{dt} + \int_0^{2\pi} \frac{\partial}{\partial x} f(u_h) e^{-ikx} dx = 0.$$

This can be simplified by replacing $f(u_h)$ by

$$f_h = \sum_{k=-N/2}^{N/2} \hat{f}_k e^{ikx},$$

where now in general, the coefficients

$$\hat{f}_k = \frac{1}{\pi} \int_0^{2\pi} f(u_h) e^{-ikx} dx$$

must be evaluated by numerical quadrature. Then we obtain the frequency domain equation

$$\frac{d\hat{u}_k}{dt} + ik\hat{f} = 0,$$

which may be integrated numerically, where now the modes are all coupled since $\hat{f}_k$ depends on all the modes defining $u_h(x)$.

The complexity of spectral methods for nonlinear problems is significantly reduced by simply requiring the residual equations to be satisfied at a set of collocation points defined by a grid

$$x_j = \frac{2\pi j}{N}, \quad j = 0, 1, \ldots, N-1.$$

This method, alternatively known as the Fourier collocation or pseudo-spectral method, can be viewed as a Petrov–Galerkin method with shifted Dirac delta functions as the test functions. One now writes the residual equations in the physical domain. Assuming N to be even, the linear advection equation is now approximated as

$$\frac{du_j}{dt} + \sum_{k=-N/2}^{N/2} iak\hat{u}_k e^{ix_j} = 0$$

and the nonlinear conservation law as

$$\frac{du_j}{dt} + \sum_{k=-N/2}^{N/2} ik\hat{f}_k e^{ikx_j} = 0.$$

In general the expansion coefficients $\hat{u}_k$ and $\hat{f}_k$ need to be evaluated by numerical quadrature. In the case of Fourier series, numerical quadrature by the trapezoidal rule is equivalent to trigonometric interpolation, as discussed in Appendix B, and accordingly it is natural to reformulate the spectral method in terms of the trigonometric interpolants. Let u_j and f_j be the values of $u(x)$ and $f(u(x))$ at the grid points. Then in the case that N is even, one sets

$$u_h = \sum_{k=-N/2}^{N/2-1} \tilde{u}_k e^{ikx} \tag{13.13}$$

and

$$f_h = \sum_{k=-N/2}^{N/2-1} \tilde{f}_k e^{ikx}, \tag{13.14}$$

such that

$$u_h(x_j) = u_j$$

and

$$f_h(x_j) = f_j.$$

These are the inverse discrete Fourier transforms of $\tilde{u}_k$ and $\tilde{f}_k$, which are now evaluated by the forward discrete transforms

$$\tilde{u}_k = \frac{1}{N} \sum_{j=0}^{N-1} u_j e^{-ikx_j} \tag{13.15}$$

and

$$\tilde{f}_k = \frac{1}{N} \sum_{j=0}^{N-1} f_j e^{-ikx_j}. \tag{13.16}$$

When N is even, $\frac{N}{2} x_j = j\pi$ and hence $\tilde{f}_{N/2} = \tilde{f}_{-N/2}$, so some writers prefer to replace (13.13) and (13.15) by

$$u_h = \sum_{k=-N/2}^{N/2} \tilde{u}_k e^{ikx}$$

and

$$\tilde{u}_k = \frac{1}{c_k N} \sum_{j=0}^{N-1} u_j e^{-ikx_j},$$

where

$$c_k = \begin{cases} 1 & \text{if } |k| < \frac{N}{2} \\ 2 & \text{if } k = \frac{N}{2} \end{cases}$$

with corresponding modifications of equations (13.14) and (13.16).

In the linear case, the Fourier–Galerkin method is now implemented by requiring discrete orthogonality of the residuals and the expansion functions

$$\sum_{j=0}^{N-1}\left(\frac{\partial u_h}{\partial t}+a\frac{\partial u_h}{\partial x}\right)e^{-imx_j}=0$$

for $m=-N/2$ to $N/2-1$. Then, since

$$\sum_{j=0}^{N-1}e^{ikx_j}e^{-imx_j}=\begin{cases}0 & \text{if } k\neq m\\ N & \text{if } k=m\end{cases},$$

the model equation reduces to

$$\frac{d\tilde{u}_k}{dt}+iak\tilde{u}_k=0,$$

yielding

$$\begin{aligned}\tilde{u}_k(t)&=\tilde{u}_k(0)e^{ik(x-at)}\\&=g_h(x-at),\end{aligned}$$

where

$$g_k(x)=u_h(x,0)$$

and the initial interpolant is propagated exactly.

Using trigonometric interpolants, the Fourier collocation method for the nonlinear conservation law reduces to the grid point equation

$$\frac{du_j}{dt}+\sum_{k=-N/2}^{N/2-1}ik\tilde{f}_k e^{ikx_j}=0,$$

where

$$\left.\frac{\partial}{\partial x}f_h\right|_{x_j}=\sum_{k=-N/2}^{N/2-1}ik\tilde{f}_k e^{ikx_j},$$

and the interpolation coefficients $\tilde{f}_k$ are calculated by the discrete Fourier transform equations 13.16. This amounts to evaluating $\frac{\partial}{\partial x}f_h\big|_{x_j}$ by a forward discrete transform and multiplication by ik, followed by the inverse discrete transform. We can use fast Fourier transform (FFT) methods to perform these calculations with $\mathcal{O}(N\log N)$ operations. When $k=-N/2$, $x_j=j\pi$, and $\tilde{f}_{-N/2}$ is real for real data f_j. Consequently the contribution of $\tilde{f}_{-N/2}$ to $\frac{\partial}{\partial x}f_h\big|_{x_j}$ is purely imaginary and can be neglected. We can then represent the numerical differentiation process as a matrix vector multiplication

$$\left.\frac{\partial}{\partial x}f_h\right|_{x_j}=\sum_{l=0}^{N-1}D_{jl}f_l, \tag{13.17}$$

where

$$D_{jl} = \frac{1}{N} \sum_{k=-N/2+1}^{N/2-1} ike^{ik(x_j - x_l)}$$

$$= \frac{1}{N} \sum_{k=-N/2+1}^{N/2-1} ike^{2\pi \frac{ik}{N}(j-l)}. \tag{13.18}$$

When $j \neq l$, this sum can be expressed in closed form as $\frac{1}{N}\frac{dS}{dx}$, where the geometric series

$$S = \sum_{k=-N/2+1}^{N/2-1} e^{ikx}$$

$$= e^{-i(N/2-1)x}\left(1 + e^{ix} \cdots + e^{i(N-2)x}\right)$$

$$= e^{-i(N/2-1)x}\frac{1 - e^{i(N-1)x}}{1 - e^{ix}}$$

$$= \frac{e^{i(N-1)x/2} - e^{-i(N-1)x/2}}{e^{ix/2} - e^{-ix/2}}$$

$$= \frac{\sin\left(\frac{N-1}{2}x\right)}{\sin\left(\frac{x}{2}\right)}.$$

Then

$$\frac{dS}{dx} = \frac{\frac{N-1}{2}\cos\left(\frac{N-1}{2}x\right)\sin\left(\frac{x}{2}\right) - \frac{1}{2}\cos\left(\frac{x}{2}\right)\sin\left(\frac{N-1}{2}x\right)}{\sin^2\left(\frac{x}{2}\right)}.$$

Now substituting $x = \frac{2\pi}{N}(j-l)$ into $\frac{dS}{dx}$, expanding the terms, and noting that

$$\cos\left(\frac{N-1}{2}x\right) = \cos\left(\frac{Nx}{2}\right)\cos\left(\frac{x}{2}\right) + \sin\left(\frac{Nx}{2}\right)\sin\left(\frac{x}{2}\right) = (-1)^{j-l}\cos\left(\frac{x}{2}\right),$$

and

$$\sin\left(\frac{N-1}{2}x\right) = \sin\left(\frac{Nx}{2}\right)\cos\left(\frac{x}{2}\right) - \cos\left(\frac{Nx}{2}\right)\sin\left(\frac{x}{2}\right) = -(-1)^{j-l}\sin\left(\frac{x}{2}\right),$$

the differentiation matrix can be expressed as

$$D_{jl} = \begin{cases} \frac{1}{2}(-1)^{j-l}\cot(j-l)\frac{\pi}{N} & \text{if } j \neq l, \\ 0 \quad \text{if } j = l. \end{cases} \tag{13.19}$$

This is an anti-symmetric circulant matrix.

Any Fourier mode e^{ikx} for $|k| < N/2$ is represented exactly by its trigonometric interpolant since the interpolant is unique and e^{ikx} interpolates itself. Correspondingly the differentiation formulas (13.17) and (13.18) or (13.19) exactly differentiate e^{ikx} for modes $k < |N/2|$, and since

$$\frac{d}{dx}e^{ikx} = ike^{ikx},$$

the discrete modes $v_j^{(k)} = e^{ikx_j}$, $|k| < N/2$, are eigenvectors of the differentiation matrix (13.19) with eigenvalues $\lambda^{(k)} = ik$. By reasons of symmetry, it can be seen that the odd-even mode $v^{(N/2)} = e^{i\frac{N}{2}x_j} = e^{ij\pi}$ satisfies

$$\sum_{l=0}^{N-1} D_{kl}\, v_l^{(N/2)} = 0.$$

It may now be recognized that the formula (13.18) for the differentiation matrix is simply its eigenvector decomposition

$$D = V\Lambda V^{-1},$$

where the eigenvectors of D are the columns of V, Λ is a diagonal matrix containing the eigenvalues, and $V^{-1} = \frac{1}{N}V^T$.

It is shown in Appendix B that the interpolation coefficients $\tilde{u_k}$ are related to the continuous Fourier coefficients $\hat{u_k}$ by the formula

$$\tilde{u_k} = \hat{u_k} + \sum_{\substack{m=-\infty \\ m\neq 0}}^{\infty} \hat{u}_{k+Nm}, \quad k = -N/2, \ldots, N/2-1.$$

Moreover if u has q continuous derivatives, there is a constant κ such that

$$|u_h - u| \leq \frac{\kappa}{N^q},$$

and by a similar argument if $p < q$, there is a constant κ_p such that

$$\left|\frac{d^p}{dx^p}u_h - \frac{d^p u}{dx^p}\right| \leq \frac{\kappa_p}{N^{q-p}},$$

with the consequence that the error decays faster than any power of N (exponentially) for infinitely differentiable data.

In the case that the number of grid points N is odd, the forward and reverse transforms can be expressed as

$$\tilde{f_k} = \frac{1}{N}\sum_{j=0}^{N-1} f_j e^{-ikx_j}$$

and

$$f_j = \sum_{k=-\frac{N-1}{2}}^{\frac{N-1}{2}} \tilde{f_k} e^{ikx_j},$$

and the differentiation formula is

$$\left.\frac{\partial f}{\partial x}\right|_{x_j} = \sum_{j=0}^{N-1} D_{jl} f_l,$$

where

$$D_{jl} = \frac{1}{N} \sum_{k=-\frac{N-1}{2}}^{\frac{N-1}{2}} ike^{\frac{2\pi ik}{N}(j-l)}.$$

Following a procedure similar to that for N even, this sum can be expressed as $\frac{1}{N}\frac{dS}{dx}$, where

$$\begin{aligned} S &= \sum_{k=-\frac{N-1}{2}}^{\frac{N-1}{2}} e^{ikx} \\ &= e^{-i\frac{N-1}{2}x}\left(1 + e^{ix} \cdots + e^{i(N-1)x}\right) \\ &= e^{-i\frac{N-1}{2}x}\frac{1-e^{iNx}}{1-e^{ix}} \\ &= \frac{\sin\left(\frac{N}{2}x\right)}{\sin\left(\frac{x}{2}\right)}. \end{aligned}$$

and

$$\frac{dS}{dx} = \frac{\frac{N}{2}\cos\left(\frac{N}{2}x\right)\sin\left(\frac{x}{2}\right) - \frac{1}{2}\cos\left(\frac{x}{2}\right)\sin\left(\frac{N}{2}x\right)}{\sin^2\left(\frac{N}{2}x\right)}.$$

Now substituting $x = \frac{2\pi}{N}(j-l)$ and noting that $\sin\left(\frac{N}{2}\pi\right) = 0$, $\cos\left(\frac{N}{2}\pi\right) = (-1)^{j-l}$, we find that

$$D_{jl} = \begin{cases} \frac{1}{2}(-1)^{j-l}\csc\left(\frac{\pi(j-l)}{N}\right) & \text{if } j \neq l, \\ 0 \quad \text{if } j = l. \end{cases}$$

It can be shown that this is the limit of a finite difference scheme with nth order accuracy as $n \to \infty$ (Hesthaven, Gottlieb, & Gottlieb 2007, pp.15–16).

A historically important exposition of spectral methods was given by Gottlieb and Orszag (1977). For a comprehensive study of spectral methods, the reader is referred to the books by Canuto, Hussaini, Quarteroni, and Zang (1988, 2007a, 2007b).

14 High Order Methods for Unstructured Meshes

14.1 Introduction

The foundation of much modern research on high order methods stems from a series of papers by Cockburn and Shu in which they reformulated the discontinuous Galerkin (DG) method to treat hyperbolic conservation laws. Their formulation provides a theoretical basis for rigorous stability proofs and error estimates, while accommodating the techniques of upwind and Godunov type schemes in a very natural manner through the use of Riemann or approximate Riemann solutions to resolve the discontinuities in the numerical solution at the cell interfaces. The computational complexity of this approach grows quite rapidly with the order of the scheme, and this has motivated the development of a variety of approaches that aim to eliminate the cost of repeated quadratures or otherwise reduce the computational complexity. A particularly simple and efficient family of DG schemes utilize high order Lagrange polynomial basis functions inside each element, with the solution defined by values at the corresponding distinct nodal points. With each element mapped to a universal reference element, the required quadratures can be pre-calculated and stored, reducing the computational operations other than the Riemann solutions to a sequence of matrix vector multiplications within each element. An exposition of nodal DG (NDG) schemes of this type can be found in the recent textbook by Hesthaven and Warburton (2008), as well as various articles by the same authors. Wang's spectral volume scheme (Wang 2002) is another approach that has proved successful but still suffers a rapid growth of complexity with increasing order.

In the last few years spectral difference (SD) methods, which directly approximate the differential form of the equations, have emerged as a promising alternative. In SD methods the solution u is represented in each element by a polynomial of degree p defined by values at $p + 1$ interior nodes, while the flux $f(u)$ is represented by a polynomial of degree $p + 1$ defined by values of $f(u)$ at p interior nodes interspersed with the solution nodes and values at the left and right boundaries of the element defined by an exact or approximate Riemann solution for the discontinuity between the element and its left or right neighbor.

In 2007 Huynh first presented his flux reconstruction (FR) method (Huynh 2007, 2009), which further simplifies the treatment of the equations in differential form. Instead of calculating the flux at a separate set of flux collocation points, he proposed simply to modify the flux $f(u)$ calculated from the solution at the interior nodal points by corrections from the left and right boundaries based on the difference between the Riemann flux f^* at the interface and the value $f(u)$ calculated from the internal solution polynomial in the element. These corrections are propagated from each boundary by polynomials of degree $p+1$ that vanish at the opposite boundary. Thus the corrected flux is represented by a polynomial of degree $p+1$, so that its derivative $\frac{\partial f}{\partial x}$ is a polynomial of degree p, consistent with the polynomial representing the solution. For the linear case, Huynh was able to show that by appropriate choices of the correction polynomials, he could recover both the standard NDG scheme and the SD scheme as well as a variety of hitherto unexplored variations that might have some potential advantages. He also used Fourier analysis to verify the stability of some of these schemes for third order accuracy. In the nonlinear case, the FR schemes with appropriate correction polynomials are no longer exactly equivalent to the corresponding NDG and SD schemes. However, the FR methodology provides a rich framework for the design of high order schemes of minimal complexity.

Sections 14.2–14.4 review recent developments in the formulation and analysis of high order schemes for unstructured meshes, with an emphasis on the flux reconstruction method and energy stability. The exposition focuses on the treatment of scalar conservation laws to illustrate the basic principles.

For a more complete survey of high order methods for unstructured meshes, the reader is referred to the recent articles by Wang (2007), and Vincent and Jameson (2011). The extension of discontinuous finite element methods to treat advection-diffusion problems and the Navier–Stokes equations is subject to ongoing research. Section 14.5 provides citations to some relevant recent works.

14.2 The Discontinuous Galerkin (DG) Method

14.2.1 Overview

The discontinuous Galerkin (DG) method appears to have been originally proposed in 1961 by Reed and Hill (1973) as a method to solve the Neutron transport equation. Subsequently it was widely studied for elliptic problems. The focus here will be on the DG method for hyperbolic conservation laws, for which the theoretical basis has been provided in a series of papers by Cockburn and Shu (Cockburn & Shu 1989, Cockburn, Lin, & Shu 1989, Cockburn, Hou, & Shu 1990, Cockburn & Shu 1998b, Cockburn & Shu 2001), and its extension to the Navier–Stokes equations.

Consider the one-dimensional scalar conservation law

$$\frac{\partial u}{\partial t} + \frac{\partial}{\partial x} f(u) = 0 \tag{14.1}$$

in a domain from A to B. Suppose that the domain is divided into K nonoverlapping elements $D^k = [x_L^k, x_R^k]$ with $x_R^{k-1} = x_L^k$ as illustrated in Figure 14.1.

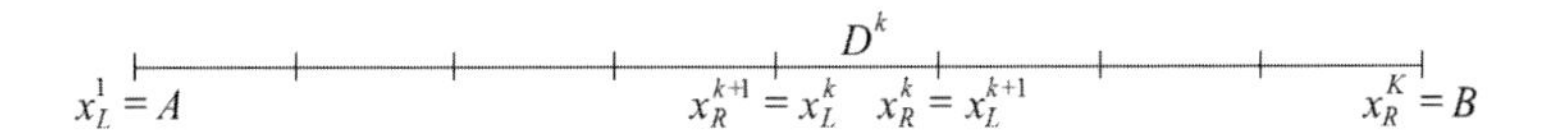

Figure 14.1 One-dimensional domain illustration.

Suppose also that the discrete solution u_h is a represented in D^k as an expansion in a set of basis functions $\Phi_j^k(x)$ defined within the element

$$u_h^k = \sum_{j=0}^{p} u_j^k \phi_j^k(x) \tag{14.2}$$

with $n = p + 1$ coefficients. Then we require the residual

$$R_h^k = \frac{\partial}{\partial t} u_h^k + \frac{\partial}{\partial x} f(u_h^k)$$

to be orthogonal to a set of test functions, which are taken to be the basis functions. Accordingly the n coefficients u_j^k are required to satisfy the n conditions

$$\int \left(\frac{\partial u_h^k}{\partial t} + \frac{\partial}{\partial x} f(u_h^k) \right) \phi_j^k(x) dx = 0, \quad j = 0, \dots, p. \tag{14.3}$$

At this point the elemental solutions $u_h^k(x)$ are independent of each other. In order to couple them to obtain a global solution, we integrate (14.3) by parts. Thus in each element we require

$$\int_{x_L}^{x_R} \frac{\partial u_h^k}{\partial t} \phi_j^k(x) dx - \int_{x_R}^{x_L} f(u_h^k) \frac{\partial \phi_j^k}{\partial x} dx + f^* \phi_j^k \Big|_{x_L}^{x_R} = 0 \tag{14.4}$$

to hold for $j = 0, \dots, p$, where f^* is a single valued numerical flux that is uniquely defined at the left and right interfaces and is also used in the corresponding equations for the neighboring elements. At each interface f^* is calculated from the left and right states u_L and u_R, defined by the solutions in the elements on either side of the interface between elements k and $k + 1$:

$$x_R^k = x_L^{k+1}, \quad u_L = u_h^k(x_R^k), \quad u_R = u_h^{k+1}(x_L^{k+1}).$$

Then the interface flux

$$f^* = f^*(u_L, u_R)$$

is defined as a function of the left and right states u_L and u_R in exactly the same way as it is in a finite volume scheme. Accordingly the elemental solutions are not only coupled to comprise a global solution, but we are also able to introduce an upwind bias in the interface flux. For this purpose we may use any of the standard upwind flux formulas.

Equation 14.4 defines the DG scheme in weak form. A reverse integration by parts recovers the strong form of the DG scheme:

$$\int_{x_L}^{x_R} \frac{\partial u_h^k}{\partial t} \phi_j^k(x) dx + \int_{x_L}^{x_R} \frac{\partial}{\partial x} f(u_h^k) \phi_j^k dx + \left(f^* - f(u_h^k)\right) \phi_j^k \Big|_{x_L}^{x_R} = 0. \qquad (14.5)$$

Both the weak and the strong forms (14.4) and (14.5) are completed by defining appropriate inflow or outflow boundary values of the numerical flux f^* at the outer boundaries of the domain $x = A$ and $x = B$. The semi-discrete equations can then be integrated forward in time to obtain the numerical solution. Assuming exact quadratures, it is possible to prove stability for both linear and nonlinear problems.

14.2.2 Quadrature Free DG Schemes

Without further simplification, both the weak and strong forms of the DG scheme are computationally expensive because of the need for repeated quadratures. In order to reduce the computational cost for nonlinear problems, one may replace $f(u_h^k)$ in (14.4) and (14.5) by an expansion similar to that of the solution

$$f_h^k = \sum_{j=0}^{p} f_j^k \phi_j^k(x).$$

In the linear case for which $f = au$, this representation is exact. Now on inserting the expansions for u_h^k and f_h^k in (14.4) and (14.5), we obtain the equations

$$M^k \frac{d\mathbf{u}^k}{dt} - {S^k}^T \mathbf{f}^k + f^* \boldsymbol{\Phi} \Big|_{x_R}^{x_L} = 0$$

and

$$M^k \frac{d\mathbf{u}^k}{dt} + S^k \mathbf{f}^k + \left(f^* - f_h^k\right) \boldsymbol{\Phi} \Big|_{x_R}^{x_L} = 0 \qquad (14.6)$$

in weak and strong forms for the local solution vectors

$$\begin{aligned} \mathbf{u}^T &= [u_1 \ \cdots \ u_n] \\ \mathbf{f}^T &= [f_1 \ \cdots \ f_n] \\ \Phi^T &= [\phi_1 \ \cdots \ \phi_n], \end{aligned} \qquad (14.7)$$

where M^k and S^k are the local mass and stiffness matrices

$$M_{ij}^k = \int_{x_L}^{x_R} \phi_i(x) \phi_j(x) dx$$

$$S_{ij}^k = \int_{x_L}^{x_R} \phi_i(x) {\phi'}_j(x) dx.$$

14.2.3 Nodal and Modal Representations

DG schemes can be defined using any set of basis functions, which might appear to offer some advantages. The following discussion will be restricted to the use of piecewise continuous polynomial expansions defined by polynomial bases in each element. One could simply use the monomial basis $1, x, x^2, \ldots$, but this leads to poorly conditioned mass and stiffness matrices, as discussed by Hesthaven and Warburton (2008, pp. 45–51).

Two representations are particularly attractive, modal and nodal. In the modal representation the solution in each element is represented by an expansion in orthogonal polynomials, with the consequence that the mass matrix is diagonal. However, the coefficients do not correspond to the value of the solution at any particular location, and in order to calculate a nonlinear flux $f(u_h^k)$, the expansion must be evaluated explicitly. In the nodal representation we introduce n collocation points x_j^k in each element and define the local solution by the Lagrange polynomial of degree $p = n-1$:

$$u_h^k = \sum_{j=0}^{p} u_j l_j(x),$$

where

$$l_j(x) = \frac{\prod_{i \neq j} (x - x_i)}{\prod_{i \neq j} (x_j - x_i)}$$

satisfying

$$\begin{aligned} l_j(x_i) &= 0, \quad i \neq j \\ l_j(x_j) &= 1, \end{aligned}$$

and u_j is now the solution value at the point x_j^k.

With either representation it is convenient to make a local transformation of each element to a reference element covering the interval $[-1, 1]$ by the mapping

$$x = x_L^k + \frac{1}{2}(1 + \xi)(x_R^k - x_L^k). \tag{14.8}$$

Now

$$\begin{aligned} M_{ij}^k &= \frac{x_R - x_L}{2} M_{ij} \\ S_{ij}^k &= S_{ij}, \end{aligned}$$

where **M** and **S** are the reference mass and stiffness matrices with the universal form

$$\begin{aligned} M_{ij} &= \int_{-1}^{1} \phi_i \phi_j d\xi \\ S_{ij} &= \int_{-1}^{1} \phi_i {\phi'}_j d\xi. \end{aligned}$$

An orthogonal basis for the reference element is now provided by the Legendre polynomials $L_p(\xi)$, which may be generated by the recurrence formula

$$L_{p+1}(\xi) = \frac{2p+1}{p+1}\xi L_p(\xi) + \frac{p}{p+1} L_{p-1}(\xi)$$

with

$$L_0(\xi) = 1$$

$$L_1(\xi) = \xi.$$

They are alternately odd and even with

$$L_p(1) = 1, \quad L_p(-1) = (-1)^p.$$

Also

$$\int_{-1}^{1} L_p(\xi)L_q(\xi)dt = 0, \quad p \neq q,$$

$$\int_{-1}^{1} L_p^2(\xi)d\xi = \frac{2}{2p+1},$$

which defines a mass matrix $\hat{M}$. Correspondingly the scaled Legendre polynomials

$$\tilde{L}_p(\xi) = \sqrt{\frac{2p+1}{2}} L_p(\xi)$$

provide an orthonormal basis.

The transformation between the modal and nodal representations can be expressed as follows. Writing the nodal representation as

$$u_k(\xi) = \sum_{j=0}^{p} \hat{u}_j L_j(\xi),$$

we consider a Lagrange representation with collocation points ξ_i and collocation values $u_i = u(\xi_i)$,

$$u_i = \sum_{j=0}^{p} \hat{u}_j L_j(\xi_i)$$

$$= \sum_{j=0}^{p} V_{ij}\hat{u}_j,$$

where $\mathbf{V}$ is the Vandermonde matrix with entries

$$V_{ij} = L_j(\xi_i).$$

In matrix vector notation:

$$\mathbf{u} = V\hat{\mathbf{u}}, \quad \hat{\mathbf{u}} = V^{-1}\mathbf{u},$$

where

$$\mathbf{u}^T = [u_0, u_1, \ldots]$$
$$\hat{\mathbf{u}}^T = [\hat{u}_0, \hat{u}_1, \ldots].$$

Moreover, since $L_i(\xi)$ is a polynomial of degree $i \leq p$, it can be represented exactly as

$$L_i(\xi) = \sum_{j=0}^{p} L_i(\xi_j) l_j(\xi) = \sum_{j=0}^{p} V_{ji} l_j(\xi)$$

or

$$\mathbf{L} = V^T \mathbf{l}, \quad \mathbf{l} = V^{T^{-1}} \mathbf{L},$$

where

$$\mathbf{L}^T = [L_0, L_1, \ldots, L_p],$$

and

$$\mathbf{l} = [l_0, l_1, \ldots, l_p]^T.$$

Now the mass matrix for the Lagrange representation has entries

$$\begin{aligned}
M_{ij} &= \int_{-1}^{1} l_i(\xi) l_j(\xi) d\xi \\
&= \int_{-1}^{1} \left(\sum_{k=0}^{p} V_{ik}^{T^{-1}} L_k \right) \left(\sum_{l=0}^{p} V_{jl}^{T^{-1}} L_l \right) d\xi \\
&= \sum_{k=0}^{p} \sum_{l=0}^{p} V_{ik}^{T^{-1}} V_{jl}^{T^{-1}} \hat{M}_{kl} \\
&= \sum_{k=0}^{p} \sum_{l=0}^{p} V_{ik}^{T^{-1}} \hat{M}_{kl} V_{lj}^{-1}
\end{aligned}$$

or

$$M = V^{T^{-1}} \hat{M} V^{-1}, \quad \hat{M} = V^T M V.$$

If f_h^k is also represented as a Lagrange polynomial

$$f_h^k = \sum_{j=0}^{p} f_j l_j(x), \tag{14.9}$$

we can evaluate its derivative as

$$\frac{d}{d\xi}f(\xi) = \frac{d}{d\xi}\sum_{j=0}^{p} f_j l_j(\xi)$$
$$= \sum_{j=0}^{p} f_j l_j'(\xi).$$

Thus

$$f'(\xi_i) = \sum_{j=0}^{p} D_{ij} f_j,$$

where

$$D_{ij} = l_j'(\xi_i),$$

or in vector notation

$$\mathbf{f}' = D\mathbf{f},$$

where D is the differentiation matrix with respect to ξ in the reference element. Correspondingly the differentiation matrix with respect to x in the kth element is

$$D^k = \frac{2}{x_R^k - x_L^k} D.$$

Also, multiplying by the mass matrix,

$$(MD)_{ij} = \sum_{k=0}^{p} \left(\int_{-1}^{1} l_i(\xi) l_j(\xi) d\xi \right) l_k'(\xi_j)$$
$$= \int_{-1}^{1} l_i(\xi) l_j'(\xi) d\xi.$$

These are the entries of the stiffness matrix. Thus for the reference element

$$MD = S, \quad D = M^{-1}S, \tag{14.10}$$

and correspondingly for the kth element

$$M^k D^k = S^k, \quad D^k = M^{k^{-1}} S^k.$$

Finally, multiplying the strong form (14.6) by $M^{k^{-1}}$, we obtain the differential form

$$\frac{d\mathbf{u}^k}{dt} + D^k \mathbf{f} + (f^* - f_h) M^{k^{-1}} \mathbf{l} \Big|_{x_L}^{x_R} = 0,$$

which maps to the reference element as

$$\frac{x_R - x_L}{2}\frac{d\mathbf{u}}{dt} + D\mathbf{f} + (f^* - f_h)M^{-1}\mathbf{l}\Big|_{-1}^{1} = 0. \qquad (14.11)$$

This equation can be expressed in modal form by substituting

$$\mathbf{u} = V\hat{\mathbf{u}}, \quad \mathbf{f} = V\hat{\mathbf{f}}, \quad \mathbf{l} = V^{T^{-1}}L, \quad M^{-1} = V\hat{M}^{-1}V^T.$$

Then on multiplying 14.11 by V^{-1}, we obtain

$$\frac{x_R - x_L}{2}\frac{d\hat{\mathbf{u}}}{dt} + \hat{D}\hat{\mathbf{f}} + (f^* - f_h)\hat{M}^{-1}\mathbf{L}\Big|_{-1}^{1} = 0,$$

where

$$\hat{D} = V^{-1}DV, \quad D = V\hat{D}V^{-1}.$$

In the modal basis the stiffness matrix element

$$\hat{S}_{jk} = \int_{-1}^{1} L_j L_k' d\xi = 0$$

if $j \geq k$ because L_j is orthogonal to polynomials of lower degree. Also if $j < k$, integrating by parts:

$$\int_{-1}^{1} L_j L_k d\xi = L_j L_k\Big|_{-1}^{1} - \int_{-1}^{1} L_k L_j' d\xi,$$

where the last term is zero by orthogonality, and

$$L_j(1) = 1, \quad L_j(-1) = (-1)^j.$$

Hence

$$\hat{S}_{jk} = \begin{cases} 0 & \text{if } j+k \text{ is even,} \\ 2 & \text{if } j+k \text{ is odd.} \end{cases}$$

Thus $\hat{S}$ is upper triangular with alternate bands 2 and 0. When $n = 5$,

$$\hat{S} = \begin{bmatrix} 0 & 2 & 0 & 2 & 0 \\ 0 & 0 & 2 & 0 & 2 \\ 0 & 0 & 0 & 2 & 0 \\ 0 & 0 & 0 & 0 & 2 \\ 0 & 0 & 0 & 0 & 0 \end{bmatrix}.$$

Also in this case,

$$\hat{M} = \begin{bmatrix} 2 & 0 & 0 & 0 & 0 \\ 0 & \frac{2}{3} & 0 & 0 & 0 \\ 0 & 0 & \frac{2}{5} & 0 & 0 \\ 0 & 0 & 0 & \frac{2}{7} & 0 \\ 0 & 0 & 0 & 0 & \frac{2}{9} \end{bmatrix},$$

and the differentiation matrix is

$$\hat{D} = \hat{M}^{-1}\hat{S} = \begin{bmatrix} 1 & 0 & 0 & 0 & 0 \\ 0 & 3 & 0 & 0 & 0 \\ 0 & 0 & 5 & 0 & 0 \\ 0 & 0 & 0 & 7 & 0 \\ 0 & 0 & 0 & 0 & 9 \end{bmatrix} \begin{bmatrix} 0 & 1 & 0 & 1 & 0 \\ 0 & 0 & 1 & 0 & 1 \\ 0 & 0 & 0 & 1 & 0 \\ 0 & 0 & 0 & 0 & 1 \\ 0 & 0 & 0 & 0 & 0 \end{bmatrix}.$$

14.2.4 Choice of Collocation Points

While the same polynomial expansion of degree $n-1$ can be represented by Lagrange polynomials with any set of n collocation points, the choice of these points impacts the computational cost. Moreover, a poor choice can lead to a badly conditioned mass matrix. For example, it is well known that with equally spaced points, the Lagrange polynomials became highly oscillatory as n is increased, leading to the Runge phenomenon (Davis 1975).

Hesthaven and Warburton (2008, pp. 47–50) favor the Legendre–Gauss–Lobato (LGL) quadrature points, which are the zeros of

$$(1-\xi^2)L'_{n-1}(\xi),$$

on the grounds that they minimize the Lebesque constant

$$A = \max_{\xi} \sum_{i=0}^{p} |l_i(\xi)|\,.$$

This choice also eliminates the need to extrapolate the expansion to the element boundaries $r = \pm 1$ to provide the states $u_h^k(x_R)$ and $u_h^k(x_L)$ needed to calculate the common flux f^*, but it leads to a mass matrix that is not diagonal.

An alternative choice that avoids a badly conditioned mass matrix is simply to use the Legendre–Gauss quadrature points, which are the zeros of the Legendre polynomial $L_n(\xi)$. This results in a diagonal mass matrix because if $i \neq j$, the product $l_i(\xi)l_j(\xi)$ of degree $2n-2$ contains all the factors $\xi - \xi_i$, $i = 0, \ldots, n-1$ of $L_n(\xi)$. Hence it can be written as the product of $L_n(\xi)$ and a polynomial $p_{n-2}(\xi)$ of degree $n-2$. Then

$$\int l_i(\xi)l_j(\xi)d\xi = \int L_n(\xi)p_{n-2}(\xi)d\xi = 0$$

because $L_n(\xi)$ is orthogonal to all polynomials of lower degree.

The use of the Legendre–Gauss quadrature points is equivalent to calculating the coefficients $\hat{u}_j$ of the expansion in Legendre polynomials

$$\hat{u}_j = \frac{2j+1}{2}\int_{-1}^{1} u_h(\xi)L_j(\xi)d\xi$$

using Gauss integration. In the nonlinear case, if the interior flux $f(u_h)$ is approximated as

$$f_h(\xi) = \sum_{j=0}^{p} \hat{f}_j L_j(\xi)$$

with

$$\hat{f}_j = \frac{2j+1}{2} \int_{-1}^{1} f(u_h(\xi)) L_j(\xi) d\xi,$$

this maximizes the degree of the polynomial that could be integrated exactly, and this may be beneficial in minimizing aliasing errors, as will be discussed in a later section.

Tables 14.1 and 14.2 show the Vandermonde, mass, stiffness, and differentiation matrices for the quadratic case, $p = 2$ and $n = 3$. It can be seen that the entries for the differentiation matrix correspond exactly to the second order forward, central, and backward difference formulas that approximate $\frac{du}{d\xi}$.

Table 14.1. Matrices for Legendre Gauss Lobatto points with $n = 3$ (quadratic case).

$\xi_0 = -1$	$\xi_1 = 0$	$\xi_2 = 1$
$\tilde{L}_0 = \frac{1}{\sqrt{2}}$	$\tilde{L}_1 = \sqrt{\frac{3}{2}}\xi$	$\tilde{L}_2 = \sqrt{\frac{5}{8}}(3\xi^2 - 1)$

$$V^T = \begin{bmatrix} \frac{1}{\sqrt{2}} & \frac{1}{\sqrt{2}} & \frac{1}{\sqrt{2}} \\ -\sqrt{\frac{3}{2}} & 0 & \sqrt{\frac{3}{2}} \\ \sqrt{\frac{5}{2}} & -\sqrt{\frac{5}{8}} & \sqrt{\frac{5}{2}} \end{bmatrix}$$

$$VV^T = M^{-1} = \begin{bmatrix} \frac{9}{2} & -\frac{3}{4} & \frac{3}{2} \\ -\frac{3}{4} & \frac{9}{8} & -\frac{3}{4} \\ \frac{3}{2} & -\frac{3}{4} & \frac{9}{2} \end{bmatrix}$$

$$M = \frac{1}{15} \begin{bmatrix} 4 & 2 & -1 \\ 2 & 16 & 2 \\ -1 & 2 & 4 \end{bmatrix}$$

$$S = \begin{bmatrix} -\frac{1}{2} & \frac{2}{3} & -\frac{1}{6} \\ -\frac{2}{3} & 0 & \frac{2}{3} \\ \frac{1}{6} & -\frac{2}{3} & \frac{1}{2} \end{bmatrix}$$

$$D = M^{-1}S = \frac{1}{2} \begin{bmatrix} -3 & 4 & -1 \\ -1 & 0 & 1 \\ 1 & -4 & 3 \end{bmatrix}$$

Table 14.2. Matrices for Legendre Gauss points with $n = 3$ (quadratic case).

$\xi_0 = -\sqrt{\frac{3}{5}}$ $\quad \xi_1 = 0$ $\quad \xi_2 = \sqrt{\frac{3}{5}}$

$\tilde{L}_0 = \frac{1}{\sqrt{2}}$ $\quad \tilde{L}_1 = \sqrt{\frac{3}{2}}\xi$ $\quad \tilde{L}_2 = \sqrt{\frac{5}{8}}(3\xi^2 - 1)$

$$V^T = \begin{bmatrix} \frac{1}{\sqrt{2}} & \frac{1}{\sqrt{2}} & \frac{1}{\sqrt{2}} \\ -\sqrt{\frac{9}{10}} & 0 & \sqrt{\frac{9}{10}} \\ \sqrt{\frac{2}{5}} & -\sqrt{\frac{5}{8}} & \sqrt{\frac{2}{5}} \end{bmatrix}$$

$$VV^T = M^{-1} = 9\begin{bmatrix} \frac{1}{5} & 0 & 0 \\ 0 & \frac{1}{8} & 0 \\ 0 & 0 & \frac{1}{5} \end{bmatrix}$$

$$M = \frac{1}{9}\begin{bmatrix} 5 & 0 & 0 \\ 0 & 8 & 0 \\ 0 & 0 & 5 \end{bmatrix}$$

$$S = \frac{5}{18}\sqrt{\frac{5}{3}}\begin{bmatrix} -3 & 4 & -1 \\ -\frac{8}{5} & 0 & \frac{8}{5} \\ 1 & -4 & 3 \end{bmatrix}$$

$$D = M^{-1}S = \frac{1}{2}\sqrt{\frac{5}{3}}\begin{bmatrix} -3 & 4 & -1 \\ -1 & 0 & 1 \\ 1 & -4 & 3 \end{bmatrix}$$

14.2.5 Accuracy of DG Schemes

The accuracy of DG schemes has been extensively investigated in theoretical studies. A summary of the conclusions is given by Hesthaven and Warburton (2008, pp. 85–88). For polynomial expansions of degree $p = n - 1$, it has been proved that for linear problems

$$\|u - u_h\| \leq Ch^{p+1},$$

where h is a bound on the element size, and C grows linearly with time. For nonlinear problems, a similar error bound has generally been found to be satisfied in numerical experiments.

14.2.6 Energy Stability Proof for the Nodal DG Scheme

For linear problems, the DG scheme can be proved to be stable with an appropriate choice of the numerical flux f^* by the following argument. With the flux $f = au$, the conservation law (14.1) reduces to the scalar advection equation

$$\frac{\partial u}{\partial t} + a\frac{\partial u}{\partial x} = 0$$

in the domain from A to B. For convenience assume that $a > 0$, corresponding to a right running wave. Multiplying by u and integrating over the domain,

$$\int_A^B u \frac{\partial u}{\partial t} dx = -a \int_A^B u \frac{\partial u}{\partial x} = -a \int_A^B \frac{\partial}{\partial x} \left(\frac{u^2}{2} \right) dx.$$

Thus the solution satisfies the energy estimate

$$\frac{d}{dt} \int_A^B \frac{u^2}{2} dx = \frac{1}{2} a \left(u_A^2 - u_B^2 \right).$$

In order to prove the stability of the DG scheme, we wish to prove that the discrete solution satisfies a similar estimate. For this purpose we can take the local solution, which is a linear combination of the basis polynomials, as the test function. Using the strong form this yields

$$\mathbf{u}^T M^k \frac{d\mathbf{u}}{dt} + \mathbf{u}^T S^k \mathbf{f} + \mathbf{u}^T \Phi(f^* - f) \Big|_{x_L}^{x_R} = 0$$

for each element. Using the fact that M^k and S^k have been pre-integrated exactly, this is equivalent to

$$\frac{d}{dt} \int_{x_L}^{x_R} \frac{u_h^2}{2} dx + a \int_{x_L}^{x_R} u_h \frac{\partial u_h}{\partial x} + u_h(f^* - a u_h) \Big|_{x_L}^{x_R} = 0,$$

where the middle term can be integrated and combined with the last term to give

$$\frac{d}{dt} \int_{x_L}^{x_R} \frac{u_h^2}{2} dx = - \left(u_h f^* - a \frac{u_h^2}{2} \right) \Big|_{x_L}^{x_R}. \tag{14.12}$$

Let u_L and u_R be the values of u_h on the left and right sides of a cell interface. For the numerical flux we now take

$$f^* = \frac{1}{2} a (u_R + u_L) - \frac{1}{2} \alpha |a| (u_R - u_L), \quad 0 \le \alpha \le 1,$$

where if $\alpha = 0$ we have a central flux, and if $\alpha = 1$ we have the upwind flux. Now on summing (14.12) over the elements, the left side yields

$$\frac{d}{dt} \int_A^B \frac{u_h^2}{2} dx,$$

while at each interior interface, collecting the contributions from the elements on the left and right sides, there is a total contribution

$$u_R f^* - a \frac{u_R^2}{2} - \left(u_L f^* - a \frac{u_L^2}{2} \right) = \frac{1}{2} a (u_R^2 - u_L^2) - \frac{1}{2} \alpha |a| (u_R - u_L)^2 - \frac{1}{2} a (u_R^2 - u_L^2)$$
$$= -\frac{1}{2} \alpha |a| (u_R - u_L)^2.$$

If we set the numerical flux to the true value $a u_A$ at the inflow boundary and to the extrapolated upwind value $a u_h$ at the outflow boundary, it now follows that, if $\alpha > 0$, there is a negative contribution at every element boundary except the inflow boundary, where the contribution is

$$au_A u_h - \frac{1}{2}au_h^2 = \frac{1}{2}au_A^2 - \frac{1}{2}a(u_A - u_h)^2,$$

which is strictly less than the boundary contribution $a\frac{u_a h^2}{2}$ in the true solution. This completes the proof that the DG scheme is energy stable for the linear advection equation.

14.2.7 Conditions for Nonlinear Stability

In the case of the scalar nonlinear conservation law

$$\frac{\partial u}{\partial t} + \frac{\partial}{\partial x} f(u) = 0,$$

the energy $h(u) = \frac{u^2}{2}$ can be regarded as an entropy function satisfying the auxiliary entropy conservation law

$$\frac{\partial h}{\partial t} + \frac{\partial F}{\partial x} = 0, \tag{14.13}$$

where $F(u)$ is the entropy flux. Multiplying the conservation law by u,

$$u\frac{\partial u}{\partial t} = \frac{\partial h}{\partial t} = -u\frac{\partial f}{\partial u}\frac{\partial u}{\partial x}.$$

Also, in the absence of a shock wave, the entropy conservation law can be expanded as

$$\frac{\partial h}{\partial t} = -\frac{\partial F}{\partial u}\frac{\partial u}{\partial x}.$$

Accordingly it will be recovered if

$$\frac{\partial F}{\partial u} = u\frac{\partial f}{\partial u} = \frac{\partial}{\partial u}(uf) - f$$

or

$$f = \frac{\partial G}{\partial u}, \tag{14.14}$$

where

$$G = uf - F, \quad F = uf - G.$$

Thus in general, F can be determined by integrating $f(u)$ with respect to u to obtain $G(u)$. Now the energy statement becomes

$$\frac{d}{dt}\int_A^B \frac{u^2}{2}dx = -\int_A^B \frac{\partial F}{\partial x}dx = F(u_A) - F(u_B).$$

In the case of the inviscid Burgers' equation, for example,

$$f(u) = \frac{u^2}{2}, \quad G(u) = \frac{u^3}{6}, \quad F(u) = \frac{u^3}{3}.$$

In the presence of shock waves, the entropy equation is modified to

$$\frac{\partial h}{\partial t} + \frac{\partial}{\partial x} F \leq 0. \tag{14.15}$$

Suppose that the interior flux $f(u_h)$ in the DG equation is approximated by a polynomial of degree $p = n - 1$ in each element of the form

$$f_h = \sum_{j=0}^{p} f(u_j) l_j(\xi) = \sum_{j=0}^{p} \hat{f}_j L_j(\xi).$$

Then it will be shown that the common flux f^* can be constructed to yield energy stability provided that f_h^k satisfies the condition

$$\int_{x_L}^{x_R} (f_h - f(u_h)) p_k(x) dx = 0 \tag{14.16}$$

for all polynomials of degree $k < p$.

Consider the weak form (14.4) of the DG equation. Since u_h^k is a polynomial of degree $p = n - 1$, we can take it as the test function. Substituting the approximate flux f_h^k for $f(u_h)$, we obtain

$$\int_{x_L}^{x_R} u_h^k \frac{\partial u_h^k}{\partial t} dx - \int_{x_L}^{x_R} f_h^k \frac{\partial u_h^k}{\partial x} dx + f^* u_h^k \Big|_{x_L}^{x_R} = 0.$$

Since $\frac{\partial u_h^k}{\partial x}$ is a polynomial of degree $p - 1$, condition (14.16) allows this to be rewritten as

$$\frac{d}{dt} \int_{x_L}^{x_R} \frac{u_h^{k\,2}}{2} dx - \int_{x_L}^{x_R} f(u_h^k) \frac{\partial u_h^k}{\partial x} dx + f^* u_h^k \Big|_{x_L}^{x_R} = 0. \tag{14.17}$$

Replacing $f(u)$ by $\frac{\partial G}{\partial u}$, the middle term becomes

$$\int_{x_L}^{x_R} \frac{\partial G}{\partial u} \frac{\partial u_h^k}{\partial x} dx = \int_{x_L}^{x_R} \frac{\partial G}{\partial x} dx,$$

so (14.17) can be integrated to obtain

$$\frac{d}{dt} \int_{x_L}^{x_R} \frac{u_h^{k\,2}}{2} dx - \left(f^* u_h^k - G \right) \Big|_{x_L}^{x_R} = 0.$$

Now summing over the elements, the contribution from each interior interface reduces to

$$f^*(u_R - u_L) - G(u_R) + G(u_L) = f^*(u_R - u_L) - G_u(u)(u_R - u_L),$$

where u lies between u_R and u_L. Hence the interface contribution is nonpositive if

$$(f(u) - f^*(u_R, u_L))(u_R - u_L) \geq 0.$$

This is the condition for a monotone or E flux (Osher 1984), which is satisfied, for example, by

$$f^*(u_R, u_L) = \frac{1}{2}(f(u_R) + f(u_L)) - \frac{1}{2}|a|(u_R - u_L),$$

where

$$a(u_R, u_L) = \max |f'(u)|,$$

for u in the interval between u_R and u_L. Accordingly with appropriate inflow and outflow boundary conditions, the DG scheme satisfies the energy equation

$$\frac{d}{dt}\int_A^B u_h^2 dx \leq F(u_A) - F(u_B).$$

If the flux approximation f_h^k is obtained as the least squares projection of $f(u_h)$

$$f_h(\xi) = \sum_{j=0}^{p} \hat{f}_j L_j(\xi)$$

with coefficients

$$\hat{f}_j = \frac{2j+1}{2}\int_{-1}^{1} f(u_h(\xi))L_j(\xi)d\xi, \tag{14.18}$$

the condition (14.16) will be satisfied for $k \leq p$. However, this would require the integrals (14.18) to be evaluated exactly, which in practice is not always possible. By using the Legendre–Gauss points as the collocation points, the coefficients can be evaluated using Gauss quadrature. This will maximize the degree of the polynomial for which the integration is exact.

14.3 The Spectral Difference (SD) and Flux Reconstruction (FR) Schemes

14.3.1 Introduction

The rapid growth of the computational complexity of DG methods with increasing order has spurred the search for simpler or more efficient variants. It was shown in the previous section that after pre-integration of the mass and stiffness matrices, the DG scheme can be reduced to the differential form (DF). Various schemes that directly treat the differential form of the equations have emerged as competitive alternatives. These include the spectral difference (SD) and flux reconstruction (FR) methods, which will now be discussed.

14.3.2 The Spectral Difference Scheme

The basic idea of the SD method was first proposed in 1996 by Kopriva and Kolias (1996) under the name "staggered grid Chebyshev multi-domain" method. In 2006,

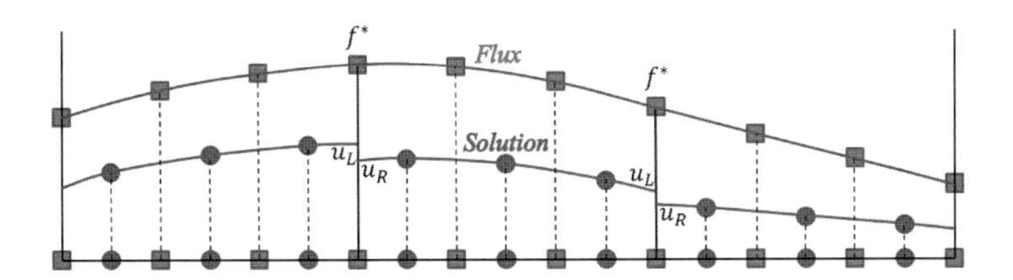

Figure 14.2 The spectral difference scheme in one dimension.

Liu, Vinokur, and Wang (2006) presented a more general formulation for both triangular and quadrilateral elements, and this name has generally been accepted.

Figure 14.2 illustrates the SD method for a one-dimensional problem. As in the DG method, the computational domain is divided into elements in each of which the discrete solution u_h is represented by a local polynomial of degree p. These polynomials do not necessarily match at the element boundaries, so the global representation is piecewise continuous. As before it is convenient to map each element to a reference element by the mapping (14.8), for which the transformed conservation laws

$$J\frac{\partial u}{\partial t} + \frac{\partial}{\partial \xi} f(u) = 0 \tag{14.19}$$

hold where

$$J = \frac{dx}{d\xi} = \frac{x_R - x_L}{2}.$$

In the reference element, the discrete solution u_h is expanded as a Lagrange polynomial of degree p defined by the values of u_h at $n = p + 1$ collocation points ξ_j,

$$u_h(\xi) = \sum_{j=0}^{p} u_j l_j(\xi),$$

where $u_j = u_h(\xi_j)$ and

$$l_j(\xi) = \frac{\prod_{i \neq j}(\xi - \xi_i)}{\prod_{i \neq j}(\xi_j - \xi_i)}.$$

The discrete flux is then represented by another Lagrange polynomial of degree $p + 1$ at $n + 1 = p + 2$ flux collocation points $\hat{\xi}_j$,

$$f_h(\xi) = \sum_{k=0}^{n} f_j \hat{l}_k(\xi),$$

where $f_k = f_h(\xi_k)$ and

$$\hat{l}_k(\xi) = \frac{\prod_{i \neq k}(\xi - \hat{\xi}_i)}{\prod_{i \neq k}(\hat{\xi}_k - \hat{\xi}_i)}.$$

The flux points consist of the element boundaries $\hat{\xi}_0 = -1$, $\hat{\xi}_n = 1$, and p interior points interspersed with the solution points. At each boundary, f_h is set to a common interface value f^* shared between the element and its neighbor. This is determined from the left and right solution values u_L and u_R at the interface in exactly the same way as in the DG scheme.

At interior flux points f_h is determined from the interpolated solutions

$$f_k = f(u_h(x_k)).$$

Then to advance the solution in time, we simply differentiate the flux polynomial to obtain a polynomial of degree p, consistent with the degree of the solution polynomial, and evaluate it at the solution points

$$\frac{x_R - x_L}{2}\frac{du_j}{dt} + \frac{d}{d\xi} f_h(\xi) = 0,$$

allowing for the mapping from the reference element.

The SD method is natural and intuitive and easy to apply to a system of conservation laws. It can easily be extended to quadrilateral and hexahedral elements by using tensor products of one-dimensional Lagrange polynomials in each coordinate direction. The local elements may be mapped to square or cubic reference elements by bilinear or trilinear mappings.

14.3.3 The Flux Reconstruction (FR) Method

The flux reconstruction method further simplifies the treatment of the equations in differential form. It was first presented in 2007 by Huynh (2007), who later extended it to the diffusion equation in 2009 (Huynh 2009). The solution is represented as a Lagrange polynomial $f_h^k(x)$ of degree p defined by the values $f_j = f(u_j)$ at the solution points as described by (14.9). It is then modified by corrections from the left and right boundaries of the element based on the difference between the common interface flux f^* and the boundary value of f_h^k. The common flux f^* is evaluated from the left and right values u_L and u_R of the solution at the interface as in the DG and SD schemes. The corrections are propagated from each boundary by polynomials of degree $p + 1$ that vanish at the opposite boundary.

Figure 14.3 illustrates the flux reconstruction method for a one-dimensional problem. Assuming that each element is mapped to the reference element by the mapping (14.8), so that the solution is governed locally by the transformed conservation law (14.19), the discrete solution is represented by a Lagrange polynomial of degree p as in the nodal DG method:

$$u_h(\xi) = \sum_{j=0}^{p} u_j l_j(\xi),$$

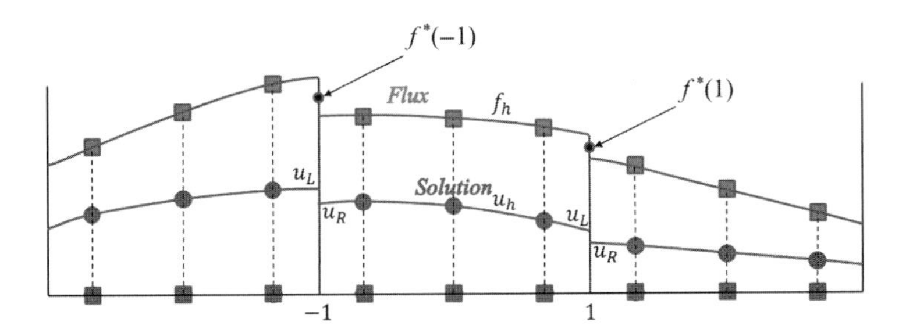

Figure 14.3 The flux reconstruction scheme in one dimension.

where $u_j = u_h(\xi_j)$ and

$$l_j(\xi) = \frac{\prod_{i \neq j}(\xi - \xi_i)}{\prod_{i \neq j}(\xi_j - \xi_i)}$$

as before. Correspondingly, the uncorrected flux is represented as

$$f_h(\xi) = \sum_{j=0}^{p} f_j l_j(\xi),$$

where $f_j = f(u_h(\xi_j))$. The common fluxes f^* at the boundaries are evaluated as

$$f^* = f^*(u_L, u_R),$$

where u_L and u_R are the left and right states evaluated from the solution polynomials on either side of the interface.

Next we introduce left and right correction polynomials $g_L(\xi)$ and $g_R(\xi)$ of degree $p + 1$ satisfying

$$\begin{aligned} g_L(-1) = 1, \quad g_L(1) = 0 \\ g_R(1) = 1, \quad g_R(-1) = 0. \end{aligned} \tag{14.20}$$

A correction polynomial of degree 3 is illustrated in Figure 14.4. Now we augment the discrete flux $f_h(\xi)$ by corrections $f_{CL} g_L(\xi)$ and $f_{CR} g_R(\xi)$, where

$$f_{CL} = f^*(-1) - f_h(-1), \quad f_{CR} = f^*(1) - f_h(1),$$

to produce the corrected flux of degree $p + 1$:

$$\tilde{f}_h(\xi) = f_h(\xi) + f_{CL}\, g_L(\xi) + f_{CR}\, g_R(\xi).$$

This is continuous across the interface between any two elements since for each element,

$$\tilde{f}(-1) = f^*(-1), \quad \tilde{f}(1) = f^*(1).$$

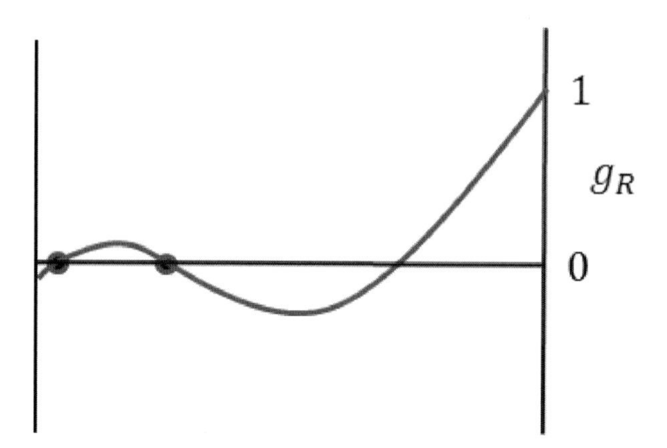

Figure 14.4 Right correction polynomial.

Now we advance the solution in time by setting

$$\frac{x_R - x_L}{2}\frac{du_j}{dt} = -\frac{\partial}{\partial \xi}\tilde{f}_h(\xi),$$

as in the SD scheme. This is equivalent to

$$\frac{x_R - x_L}{2}\frac{\partial u_h}{\partial t} + \frac{\partial}{\partial \xi}(f_h + f_{CL}\, g_L + f_{CR}\, g_R) = 0. \tag{14.21}$$

A variety of schemes can be obtained by different choices of the correction polynomials $g_L(\xi)$ and $g_R(\xi)$. Since they have one zero at $\xi = \pm 1$, they will be fully defined by the location of their p remaining zeros. In the linear case, the SD scheme is recovered by choosing these zeros to coincide with the interior flux collocation points. Then the FR flux is identical to the SD flux, since they are both polynomials of degree $p + 1$ that have the same values at $p + 2$ collocation points, and accordingly they must both be identical to the unique Lagrange polynomials through those points. In the nonlinear case, this choice of correction functions does not recover the SD scheme exactly because the interior fluxes are calculated from the interpolated values of u_h.

A natural approach to the choice of the correction functions is to require them to be as small as possible, subject to the two constraints imposed by the boundary conditions (14.20). The best that can be achieved is to satisfy $p - 1$ conditions that they are orthogonal to all polynomials of degree $\leq p - 1$. These can be realized by choosing g_L and g_R to be linear combinations of the Legendre polynomials L_p and L_{p+1}. Then the boundary conditions are also satisfied by choosing

$$g_R = \tfrac{1}{2}(L_p + L_{p+1})$$
$$g_L = \tfrac{(-1)^p}{2}(L_p - L_{p+1}).$$

These are the left and right Radau polynomials.

This choice recovers the nodal DG scheme. In order to demonstrate this, we multiply the FR equation (14.21) by $l_i(\xi)$ and integrate over the element to obtain

$$\frac{x_R - x_L}{2} \int_{-1}^{1} l_i \frac{\partial u_h}{\partial t} \, d\xi + \int_{-1}^{1} l_i \, f_h' \, d\xi + f_{CL} \int_{-1}^{1} l_i \, g_L' \, d\xi + f_{CR} \int_{-1}^{1} l_i \, g_R' \, d\xi = 0. \tag{14.22}$$

Integrating by parts,

$$f_{CR} \int_{-1}^{1} l_i \, g_R' d\xi = g_R \, l_i \Big|_{-1}^{1} - \int_{-1}^{1} g_R \, l_i' \, d\xi.$$

Here, $g_R(1) = 1$ and $g_R(-1) = 0$, while l_i' is a polynomial of degree $p - 1$. Hence, the last term vanishes because g_R is orthogonal to all polynomials of degree $\leq p - 1$, and

$$f_{CR} \int_{-1}^{1} l_i \, g_R' d\xi = f_{CR} \, l_i(1).$$

Similarly,

$$f_{CL} \int_{-1}^{1} l_i \, g_L' d\xi = -f_{CL} \, l_i(-1).$$

Now, using the definition of f_{CL} and f_{CR}, (14.22) reduces to

$$\frac{x_R - x_L}{2} \int_{-1}^{1} l_i \frac{\partial u_h}{\partial t} \, d\xi + \int_{-1}^{1} l_i \, f_h' \, d\xi + l_i \, (f^* - f_h)\Big|_{-1}^{1} = 0.$$

This is precisely the strong form of the nodal DG scheme.

Using Fourier analysis, Huynh found that schemes using correction polynomials orthogonal to polynomials of degree $\leq m$ lead to super accuracy of order $p + m + 2$ for linear problems on a uniform mesh. Accordingly, the DG scheme is super accurate to the order $2p + 1$. He also found that the CFL limit increases as the zeros of the correction function are moved away from the boundary, and he identified a particularly interesting scheme in which the correction functions have a double zero at the far boundary, which he labeled g_2. This is stable for a larger CFL limit than the DG and SD schemes. The correction functions for the g_2 scheme can be represented as

$$g_R = \frac{1}{2}\left(L_p + \frac{(p+1)L_{p-1} + pL_{p+1}}{2p+1}\right)$$

$$g_L = \frac{(-1)p}{2}\left(L_p - \frac{(p+1)L_{p-1} + pL_{p+1}}{2p+1}\right),$$

which are orthogonal to polynomials of degree $< p - 1$.

The next section presents proofs that both the SD scheme and a class of FR schemes are stable in an energy norm for linear problems. In comparison with the Fourier method of stability analysis, the energy method is general and rigorous, enabling proofs of stability for all orders of accuracy for entire classes of schemes. Moreover, these proofs are valid for nonuniform meshes. While the Fourier method requires a separate analysis for each case and normally assumes a uniform mesh, it remains useful because it provides more detailed information about the distribution of dispersive and diffusive errors, and it also enables the identification of the property of super accuracy for linear problems.

14.3.4 Stability of the Spectral Difference Method

The stability of the SD scheme for linear problems was proved by Jameson (2010) by regarding it as a special case of the FR scheme. Restricting our attention to the case of linear advection, $f = au$, the first step is to rewrite the flux at each boundary, following the FR procedure, as

$$f^*(-1) = au_h(-1) + f_{CL}, \quad f^*(1) = au_h(1) + f_{CR},$$

where f_{CL} and f_{CR} are boundary corrections

$$f_{CL} = f^*(-1) - au_h(-1), \quad f_{CR} = f^*(1) - au_h(1).$$

Now

$$f_h(\xi) = f_{CL}\hat{l}_1(\xi) + f_{CR}\hat{l}_{n+1}(\xi) + a\sum_{j=0}^{n} u_h(\hat{\xi}_j)\hat{l}_j(\xi).$$

But since $u_h(\xi)$ is a polynomial of degree p, it is exactly represented by the sum. Hence

$$f_h(\xi) = f_{CL}\hat{l}_1(\xi) + f_{CR}\hat{l}_{n+1}(\xi) + au_h(\xi).$$

We can now rewrite the SD scheme as

$$\frac{\partial u_h}{\partial t} = -a\frac{\partial u_h}{\partial \xi} - f_{CL}\hat{l}'_1 - f_{CR}\hat{l}'_{n+1}.$$

Evaluating this at the solution points:

$$\frac{du_i}{dt} = -a\sum_{j=0}^{p} D_{ij}u_j - f_{CL}\hat{l}'_1(\xi_i) - f_{CR}\hat{l}'_{n+1}(\xi_i), \tag{14.23}$$

where D is the differentiation matrix associated with the solution collocation points and is uniquely determined by the location of these points and the polynomial degree p. Thus D is represented by (14.10). In the case of an upwind numerical flux, there will only be a correction from the left boundary, and in order to simplify the analysis this will now be assumed.

Equation (14.23) can be converted to a form that resembles the nodal DG method by multiplying it by the mass matrix to produce

$$\sum_j M_{ij}\frac{du_j}{dt} + a\sum_j S_{ij}u_j = -f_{CL}\sum_j M_{ij}\hat{l}_1'(\xi_j).$$

Now since $\hat{l}_0(-1) = 1$ and $\hat{l}_0(1) = 0$,

$$\begin{aligned}\sum_{j=0}^{p} M_{ij}\hat{l}_0'(\xi_j) &= \int_{-1}^{1} l_i(\xi)\sum_{j=1}^{n}\hat{l}_0'(\xi_j)l_j(\xi)d\xi \\ &= \int_{-1}^{1} l_i(\xi)\hat{l}_0'(\xi)d\xi \\ &= \hat{l}_0 l_i\Big|_{-1}^{1} - \int_{-1}^{1} l_i'(\xi)\hat{l}_0(\xi)d\xi \\ &= -l_i(-1) - \int_{-1}^{1} l_i'(\xi)\hat{l}_0(\xi)d\xi.\end{aligned}$$

Thus

$$\sum_j M_{ij}\frac{du_j}{dt} + a\sum_j S_{ij}u_j = f_{CL}\left(l_i(-1) + \int_{-1}^{1} l_i'(\xi)\hat{l}_0(\xi)d\xi\right). \tag{14.24}$$

This differs from the corresponding nodal DG equation only in the last term. In order to compensate for this, we can replace the mass matrix M by a matrix $Q > 0$ such that

$$QD = S. \tag{14.25}$$

This will be the case if Q has the form $M + C$, where

$$CD = 0.$$

Thus each row of C must be orthogonal to every column of D. Because $DR_p = R_p'$ for any polynomial $R_p(\xi)$ of degree p, the coefficients of each row of D must sum to zero, so the rank of D is no greater than $n - 1$. In order to find a row vector that is orthogonal to every column of D, consider the pth difference operator d^T, which gives

$$\sum_{j=0}^{p} d_j R_p(\xi_j) = R_p^{(p)}$$

for any polynomial of degree p. Then

$$\sum_i d_i \sum_j D_{ij}R_p(\xi_j) = R_p^{(p+1)} = 0.$$

Thus the matrix

$$Q = M + cdd^T,$$

where c is an arbitrary parameter, satisfies (14.25). Also since any polynomial $R_p(\xi)$ of degree p can be represented exactly as

$$R_p = \sum_i R_p(\xi_i) l_i(\xi),$$

it follows that if $l_i(\xi)$ is expanded as

$$l_i(\xi) = a_i \xi^p + \cdots,$$

where a_i is the leading coefficient, then

$$d_i = l_i^{(p)} = p!\, a_i.$$

Now on multiplying (14.23) by Q instead of M, we obtain the extra term

$$-c\ f_{CL}\ d_i \sum_j d_j \hat{l}_1'(\xi_j) = -c\ f_{CL}\ \hat{l}_1^{(p+1)} l_i^{(p)}$$

on the right, so that (14.24) is replaced by

$$\sum_j Q_{ij} \frac{du_j}{dt} + a \sum_j S_{ij} u_j = f_{CL}\left(l_i(-1) + \int_{-1}^{1} l_i'(\xi) \hat{l}_0(\xi) d\xi - c\, \hat{l}_0^{(p+1)} l_i^{(p)} \right).$$

Now if we can choose c so that the last two terms on the right cancel, we can attain an energy estimate with the norm $\mathbf{u}^T Q \mathbf{u}$ replacing $\mathbf{u}^T M \mathbf{u}$ in each element. For this purpose we can choose the interior flux collocation points as the zeros of the Legendre polynomial $L_p(\xi)$ of degree p. Then

$$\hat{l}_0(\xi) = (-1)^p \frac{1}{2} (1 - \xi) L_p(\xi)$$

and

$$\int_{-1}^{1} \hat{l}_0(\xi) l_i'(\xi) d\xi = (-1)^{p+1} \frac{1}{2} \int_{-1}^{1} \xi L_p(\xi) l_i'(\xi) d\xi,$$

since $l_i'(\xi)$ is a polynomial of degree $p-1$ and $L_p(\xi)$ is orthogonal to all polynomials of degree $< p$. Moreover only the leading term in $\xi l_i'(\xi)$ contributes to the integral for the same reason. Let

$$L_p(\xi) = c_p \xi^p + \cdots$$

where the leading coefficient

$$c_p = \frac{1 \cdot 3 \cdot 5 \ldots \cdot (2p-1)}{p!} \xi^p + \cdots.$$

Also

$$\xi l_i'(\xi) = p a_i x^p + \cdots,$$

where a_i is the leading coefficient in $l_i(\xi)$. Noting that

$$\int_{-1}^{1} L_p^2 d\xi = \frac{2}{2p+1},$$

we obtain

$$\int_{-1}^{1} \xi L_p(\xi) l_i'(\xi) d\xi = \frac{2p}{2p+1} \frac{a_i}{c_p}.$$

Also

$$\hat{l}^{(p+1)} = (-1)^{p+1} \frac{1}{2} (p+1)! \, c_p$$

and

$$l_i^{(p)} = p! \; a_i.$$

Thus the desired cancellation is obtained by setting

$$c = \frac{2p}{2p+1} \frac{1}{c_p^2} \frac{1}{p! \; (p+1)!} > 0.$$

In the case that the interface flux is not fully upwind, a similar calculation shows that the convection from the right boundary is correspondingly reduced, so that finally

$$\sum_j Q_{ij} \frac{du_j}{dt} + a \sum_j S_{ij} u_j = f_{CL} l_i(-1) - f_{CR} l_i(1).$$

Since u_h is a polynomial of degree p,

$$\sum_i d_i u_i = u_h^{(p)},$$

and in each element, allowing for the scaling factor $J = \frac{\xi_R - \xi_L}{2}$ in the transformation from the reference element,

$$\sum_i \sum_j u_i Q_{ij} u_j = \frac{1}{J} \int_{x_L}^{x_R} (u_h^2 + J^{2p} \, c \, {u_h^{(p)}}^2) dx.$$

Now the same argument that was used to prove the energy stability of the nodal DG scheme establishes the energy stability of the SD scheme with the norm

$$\int_A^B (u_h^2 + J^{2p} \, c \, {u_h^{(p)}}^2) dx$$

with the piecewise constant scaling factor J, for the case of solution polynomials of degree p, provided that the interior flux collocation points are the zeros of $L_p(\xi)$. It can be seen that c decreases very rapidly with increasing p, as is illustrated by the following table of values of c.

p	c
1	$\frac{1}{3}$
2	$\frac{4}{135}$
3	$\frac{1}{1050}$

14.3.5 Energy Stable Flux Reconstruction Schemes

The analysis of the previous section can be generalized to prove the linear stability of a broad class of FR schemes, usually referred to as energy stable FR (ESFR) schemes, as was shown by Vincent, Castonguay, and Jameson (2011). Since it has been observed that in the linear case the SD scheme can be recovered as an FR scheme, it is natural to consider whether other FR schemes may be linearly stable for a similar norm of the form

$$\int_{-1}^{1}\left(u_h^2+\frac{c}{2}u_h^{(p)^2}\right)d\xi$$

in each element after it has been mapped to the reference element. For the linear flux $f=au$, the FR scheme can be expressed in the reference element as

$$\frac{\partial u_h}{\partial t}=-a\frac{\partial u_h}{\partial \xi}-(f_L^*-au_L)g_L'-(f_R^*-au_R)g_R', \tag{14.26}$$

where f_L^* and f_R^* are the common interface fluxes at the left and right boundaries, and the prime denotes differentiation with respect to r. Noting that the $(p+1)$th derivative of u_h is zero, this may be differentiated p times to yield

$$\frac{\partial u_h^{(p)}}{\partial t}=-(f_L^*-au_L)g_L^{(p+1)}-(f_R^*-au_R)g_R^{(p+1)}, \tag{14.27}$$

where the superscripts (p) and $(p+1)$ denote the pth and $(p+1)$th derivatives. Multiplying (14.26) by u_h and integrating, we obtain

$$\begin{aligned}\frac{d}{dt}\int_{-1}^{1}\frac{u_h^2}{2}d\xi&=-a\int u_h\frac{\partial u_h}{\partial r}d\xi\\&\quad-(f_L^*-au_L)\int_{-1}^{1}u_hg_L'd\xi\\&\quad-(f_R^*-au_R)\int_{-1}^{1}u_hg_R'd\xi.\end{aligned}$$

Integrating by parts and noting that $g_L(-1) = 1$, $g_L(1) = 0$, $g_R(-1) = 0$, and $g_R(1) = 1$, this reduces to

$$\begin{aligned}\frac{d}{dt}\int_{-1}^{1} u_h^2 d\xi = &-a(u_R^2 - u_L^2)\\ &- 2(f_L^* - au_L)\left(-u_L - \int_{-1}^{1} g_L \frac{\partial u_h}{\partial \xi} d\xi\right)\\ &- 2(f_R^* - au_R)\left(-u_R - \int_{-1}^{1} g_R \frac{\partial u_h}{\partial \xi} d\xi\right).\end{aligned}$$

Also, multiplying (14.27) by $u_h^{(p)}$ and integrating, using the fact that $u_h^{(p)}$, $g_L^{(p+1)}$ and $g_R^{(p+1)}$ are constant, we obtain

$$\begin{aligned}\frac{1}{2}\frac{d}{dt}\int_{-1}^{1} u_h^{(p)^2} d\xi = &- 2(f_L^* - au_L)u_h^{(p)} g_L^{(p+1)}\\ &- 2(f_R^* - au_R)u_h^{(p)} g_R^{(p+1)}.\end{aligned}$$

Hence,

$$\begin{aligned}\frac{d}{dt}\int_{-1}^{1}\left(u_h^2 + \frac{c}{2}u_h(p)^2\right) d\xi = &- a(u_R^2 - u_L^2)\\ &- 2(f_L^* - au_L)\left[-u_L - \int_{-1}^{1} g_L \frac{\partial u_h}{\partial \xi} d\xi + cu_h^{(p)} g_L^{(p+1)}\right]\\ &- 2(f_R^* - au_R)\left[-u_R - \int_{-1}^{1} g_R \frac{\partial u_h}{\partial \xi} d\xi + cu_h^{(p)} g_R^{(p+1)}\right].\end{aligned}$$

The right hand side reduces to the same form as the DG scheme if

$$\int_{-1}^{1} g_L \frac{\partial u_h}{\partial \xi} d\xi = cu_h^{(p)} g_L^{(p+1)}$$

and

$$\int_{-1}^{1} g_R \frac{\partial u_h}{\partial \xi} d\xi = cu_h^{(p)} g_R^{(p+1)},$$

and hence energy stability follows by the same argument when the interface fluxes are evaluated with the same upwind bias.

Substituting

$$u_h = \sum u_i l_i(\xi),$$

these conditions will be satisfied if

$$\int_{-1}^{1} g_L l_i' d\xi = c\, l_i^{(p)} g_L^{(p+1)} \tag{14.28}$$

and

$$\int_{-1}^{1} g_R l_i' d\xi = c\, l_i^{(p)} g_R^{(p+1)}.$$

Since $g_L(r) = g_R(r)$, and assuming the collocation points are symmetric, these conclusions are equivalent, and we need only consider the first. Expressing

$$l_i = a_i \xi^p + \text{lower powers of } \xi,$$

$$l_i' = p a_i \xi^{p-1} + \text{lower powers of } \xi,$$

and

$$l_i^{(p)} = p!\ a_i,$$

we choose g_L to be orthogonal to all polynomials of degree $\leq p - 2$, with the consequence that (14.28) reduces to

$$p \int g_L \xi^{p-1} d\xi = c\ p!\ g_L^{(p+1)}. \tag{14.29}$$

The orthogonality requirement can be satisfied by choosing g_L to be a linear combination of Legendre polynomials

$$g_L + r_{p-1} L_{p-1} + r_p L_p + r_{p+1} L_{p+1}, \tag{14.30}$$

where the coefficients must be chosen such that

$$g_L(-1) = 1, \quad g_L(1) = 0.$$

Since

$$L_p(-1) = (-1)^p, \quad L_p(1) = 1,$$

this leads to the conditions

$$r_{p-1} - r_p + r_{p+1} = (-1)^{p+1},$$

$$r_{p-1} + r_p + r_{p+1} = 0,$$

whence

$$r_p = \frac{1}{2}(-1)^p$$

and

$$r_{p-1} + r_{p+1} = \frac{1}{2}(-1)^{p+1}.$$

Also,

$$L_p = c_p \xi^p + \text{lower powers of } \xi,$$

where the leading coefficient

$$c_p = \frac{1 \cdot 3 \cdot 5 \ldots \cdot (2p-1)}{p!} = \frac{2p-1}{p} c_{p-1}.$$

Thus substituting (14.30) in (14.29), we find that

$$\begin{aligned}
\int_{-1}^{1} g_L \xi^{p-1} d\xi &= r_{p-1} \int_{-1}^{1} L_{p-1} \xi^{p-1} d\xi \\
&= \frac{r_{p-1}}{a_{p-1}} \int_{-1}^{1} L_{p-1}^2 d\xi \\
&= \frac{r_{p-1}}{a_{p-1}} \frac{2}{2p-1} \\
&= \frac{2}{p a_p} r_{p-1}
\end{aligned}$$

yielding

$$r_{p-1} = \eta_p r_{p+1},$$

where

$$\begin{aligned}
\eta_p &= \frac{c}{2}\, p!\ (p+1)!\ c_p\, c_{p+1} \\
&= \frac{c}{2}\, (2p+1)\, (c_p\, p!)^2.
\end{aligned}$$

Thus finally,

$$r_{p-1} = \frac{1}{2} \frac{\eta_p}{1+\eta_p} (-1)^{p+1}, \quad r_{p+1} = \frac{1}{2} \frac{(-1)^{p+1}}{1+\eta_p},$$

and

$$g_L = \frac{1}{2} (-1)^p \left(L_p - \frac{\eta_p L_{p-1} + L_{p+1}}{1+\eta_p} \right),$$

while by symmetry

$$g_R = \frac{1}{2} \left(L_p + \frac{\eta_p L_{p-1} + L_{p+1}}{1+\eta_p} \right).$$

The best known schemes are recovered as follows:

Nodal DG Scheme

By the choice of c, if the flux correction functions are reduced to the right and left Radau polynomials, then a nodal scheme is recovered. The corresponding flux correction functions are

$$g_L = \frac{(-1)^p}{2}(L_p - L_{p+1})$$
$$g_R = \frac{(+1)^p}{2}(L_p - L_{p+1}),$$

which is a result of picking $c = 0$, $\eta_p = 0$.

SD Scheme

The SD scheme has a set of flux collocation points within the element. At those points, the flux correction functions should assume zero values. By choosing $c = \frac{2p}{(2p+1)(p+1)(c_p p!)^2}$ and $\eta_p = \frac{p}{p+1}$, the resulting flux correction functions are

$$g_L = \frac{(-1)^p}{2}(1 - \xi)L_p$$
$$g_R = \frac{(+1)^p}{2}(1 + \xi)L_p.$$

Huynh g_2 Scheme

If the value of c is set equal to $\frac{2(p+1)}{(2p+1)p(c_p p!)^2}$ corresponding to $\eta_p = \frac{p+1}{p}$, the resulting flux correction functions are

$$g_L = \frac{(-1)^p}{2}\left[L_p - \left(\frac{(p+1)L_{p-1} + pL_{p+1}}{2p+1}\right)\right]$$
$$g_R = \frac{(+1)^p}{2}\left[L_p + \left(\frac{(p+1)L_{p-1} + pL_{p+1}}{2p+1}\right)\right].$$

This recovers a particular scheme that was proposed by Huynh, which he found to be very stable.

14.4 DG, SD, and FR Schemes for Multi-dimensional Problems

14.4.1 Background

The flux reconstruction family of schemes (including NDG and SD) can readily be extended to quadrilateral and hexahedral elements using tensor product formulations as suggested by Huynh (2007). These had already been used by Kopriva and Kolias (1996), and it has proved successful in practice (Ou, Liang, & Jameson 2010, Ou, Liang, Premasuthan, & Jameson 2009, Premasuthan, Liang, Jameson, & Wang 2009).

Figure 14.5 illustrates the SD scheme in the two-dimensional case with $p = 2$. There are 9 solution points and 24 flux points. In general, there are n^2 solution points and $2n(n-1)$ flux points, where $n = p + 1$. Let ξ and η be the coordinates in the reference element, $-1 \leq \xi \leq 1$, $-1 \leq \eta \leq 1$. The conservation law

$$\frac{\partial \boldsymbol{w}}{\partial t} + \frac{\partial}{\partial x}\boldsymbol{f}(\boldsymbol{w}) + \frac{\partial}{\partial y}\boldsymbol{g}(\boldsymbol{w}) = 0$$

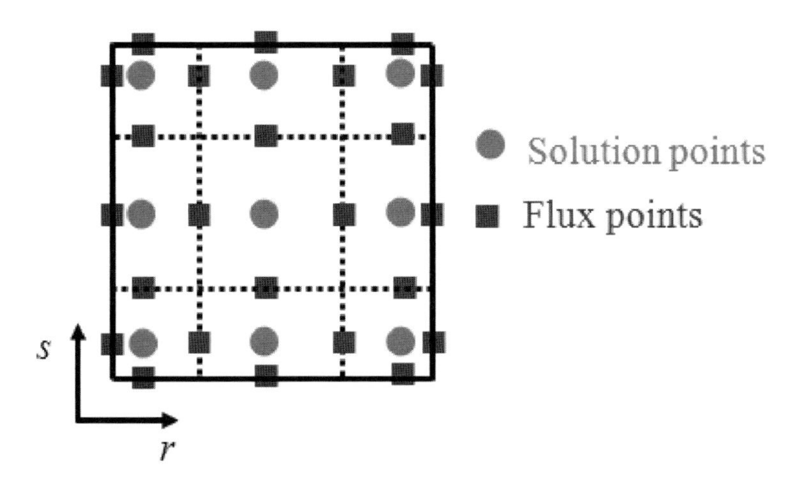

Figure 14.5 Distribution of flux and solution points for a two-dimensional case with $p = 2$.

is transformed as in Section 2.2 to

$$\frac{\partial \boldsymbol{w}}{\partial t} + \frac{\partial}{\partial \xi}\boldsymbol{F}(\boldsymbol{w}) + \frac{\partial}{\partial \eta}\boldsymbol{G}(\boldsymbol{w}) = 0,$$

where

$$K = \begin{bmatrix} x_\xi & x_\eta \\ y_\xi & y_\eta \end{bmatrix}, \quad J = \det(K)$$

$$\begin{bmatrix} \xi_x & \xi_y \\ \eta_x & \eta_y \end{bmatrix} = K^{-1} = \frac{1}{J}\begin{bmatrix} y_\eta & -x_\eta \\ -y_\xi & x_\xi \end{bmatrix}$$

and

$$\boldsymbol{W} = J\boldsymbol{w}, \quad \begin{bmatrix} \boldsymbol{F} \\ \boldsymbol{G} \end{bmatrix} = JK^{-1}\begin{bmatrix} \boldsymbol{f} \\ \boldsymbol{g} \end{bmatrix}.$$

Then we represent the solution as

$$\boldsymbol{w}_h(\xi,\eta) = \sum_{i=0}^{p}\sum_{j=0}^{p}\frac{\boldsymbol{W}_{ij}}{J_{ij}}l_i(\xi)l_j(\eta)$$

and the fluxes as

$$\boldsymbol{F}_h(\xi,\eta) = \sum_{i=0}^{n}\sum_{j=0}^{p}\boldsymbol{F}_{ij}\hat{l}_i(\xi)l_j(\eta)$$

$$\boldsymbol{G}_h(\xi,\eta) = \sum_{i=0}^{p}\sum_{j=0}^{n}\boldsymbol{G}_{ij}l_i(\xi)\hat{l}_j(\eta).$$

The stability of tensor product FR schemes for linear problems has recently been established for an extended energy norm by Sheshadri and Jameson (2016).

14.4.2 DG, SD, and FR Schemes for Simplex Elements

While the formulation of the nodal DG method for simplex elements is well established (Hesthaven & Warburton 2008), the original SD scheme for simplex elements (Liu et al. 2006) was found to be unstable when the order was increased. This has motivated the search for stable variants of the FR formulation for simplex elements, and three approaches are currently being pursued. Wang and Gao have proposed the LCP method (Wang & Gao 2009), while Balan, May, and Schöberl (2012) have reformulated the SD scheme using Raviart Thomas basis polynomials and have used Fourier analysis to verify stability for the case of third order accuracy. Recently Castonguay, Vincent and Jameson have extended the formulation of energy stable flux reconstruction schemes to multi-dimensional problems, including the nodal DG scheme as a special case (Castonguay, Vincent, & Jameson 2012).

14.5 Advection-Diffusion Problems and the Navier–Stokes Equations

The application of FR schemes to advection-diffusion problems and the Navier–Stokes equations has been described by Castonguay et al. (2013) and by Williams et al. (Williams, Castonguay, Vincent, & Jameson 2013, Williams and Jameson 2014). The equations are first expressed as a first order system. For example, the one-dimensional conservation law

$$\frac{\partial u}{\partial t} + \frac{\partial f}{\partial x} = 0,$$

where $f = f\left(u, \frac{\partial u}{\partial x}\right)$, is rewritten as

$$\frac{\partial u}{\partial t} + \frac{\partial f(u,q)}{\partial x} = 0$$

$$q - \frac{\partial u}{\partial x} = 0,$$

and q is an auxiliary variable. To evaluate the second equation, the discontinuous discrete approximation u_h is replaced by a continuous value u^* at each cell interface, and the difference $u_h - u^*$ is propagated to the interior of each cell by correction functions of degree $p+1$, in a manner similar to the construction of a continuous flux in Section 14.3.3. Then the numerical flux $f^*(u,q)$ is formed from left and right values of u and q at each interface. There are a variety of approaches for the construction of common interface values of the diffusive part of the flux, including the central flux (CF), the local discontinuous Galerkin (LDG), the compact discontinuous Galerkin (CDG), interior penalty (IP), Bassi Rebay 1 (BR1), and Bassi Rebay 2 (BR2) approaches. The CF approach uses central averaging for both u and q, while the LDG approach uses one sided values in the left and right direction respectively for u and q. The reader is referred to Bassi and Rebay (1997), Cockburn and Shu (1998a), Peraire and Persson (2008), Arnold (1982), and Bassi and Rebay (2000) for more details.

15 Aerodynamic Shape Optimization

15.1 Introduction

This chapter reviews the formulation and application of optimization techniques based on control theory for aerodynamic shape design in both inviscid and viscous compressible flow. The theory is applied to a system defined by the partial differential equations of the flow, with the boundary shape acting as the control. The Frechet derivative of the cost function is determined via the solution of an adjoint partial differential equation, and the boundary shape is then modified in a direction of descent. This process is repeated until an optimum solution is approached. Each design cycle requires the numerical solution of both the flow and the adjoint equations, leading to a computational cost roughly equal to the cost of two flow solutions. Representative results are presented for viscous optimization of transonic wing–body combinations.

15.2 Aerodynamic Design

The definition of the aerodynamic shapes of modern aircraft relies heavily on computational simulation to enable the rapid evaluation of many alternative designs. Wind tunnel testing is then used to confirm the performance of designs that have been identified by simulation as promising to meet the performance goals. In the case of wing design and propulsion system integration, several complete cycles of computational analysis followed by testing of a preferred design may be used in the evolution of the final configuration. Wind tunnel testing also plays a crucial role in the development of the detailed loads needed to complete the structural design and in gathering data throughout the flight envelope for the design and verification of the stability and control system. The use of computational simulation to scan many alternative designs has proved extremely valuable in practice, but it still suffers from the limitation that it does not guarantee the identification of the best possible design. Generally one has to accept the best so far by a given cutoff date in the program schedule. To ensure the realization of the true best design, the ultimate goal of computational simulation methods should not just be the analysis of prescribed shapes but the automatic determination of the true optimum shape for the intended application.

This is the underlying motivation for the combination of computational fluid dynamics with numerical optimization methods. Some of the earliest studies of such an approach were made by Hicks and Henne (Hicks, Murman, & Vanderplaats 1974, Hicks & Henne 1979). The principal obstacle was the large computational cost of determining the sensitivity of the cost function to variations of the design parameters by repeated calculation of the flow. Another way to approach the problem is to formulate aerodynamic shape design within the framework of the mathematical theory for the control of systems governed by partial differential equations (Lions 1971). In this view the wing is regarded as a device to produce lift by controlling the flow, and its design is regarded as a problem in the optimal control of the flow equations by changing the shape of the boundary. If the boundary shape is regarded as arbitrary within some requirements of smoothness, then the full generality of shapes cannot be defined with a finite number of parameters, and one must use the concept of the Frechet derivative of the cost with respect to a function. Clearly such a derivative cannot be determined directly by separate variation of each design parameter, because there are now an infinite number of these.

Using techniques of control theory, however, the gradient can be determined indirectly by solving an adjoint equation that has coefficients determined by the solution of the flow equations. This directly corresponds to the gradient technique for trajectory optimization pioneered by Bryson and Ho (1975). The cost of solving the adjoint equation is comparable to the cost of solving the flow equations, with the consequence that the gradient with respect to an arbitrarily large number of parameters can be calculated with roughly the same computational cost as two flow solutions. Once the gradient has been calculated, a descent method can be used to determine a shape change that will make an improvement in the design. The gradient can then be recalculated, and the whole process can be repeated until the design converges to an optimum solution, usually within 10–50 cycles. The fast calculation of the gradients makes optimization computationally feasible even for designs in three-dimensional viscous flow. There is a possibility that the descent method could converge to a local minimum rather than the global optimum solution. In practice this has not proved a difficulty, provided care is taken in the choice of a cost function that properly reflects the design requirements. Conceptually, with this approach the problem is viewed as infinite-dimensional, with the control being the shape of the bounding surface. Eventually the equations must be discretized for a numerical implementation of the method. For this purpose the flow and adjoint equations may either be separately discretized from their representations as differential equations, or, alternatively, the flow equations may be discretized first and the discrete adjoint equations then derived directly from the discrete flow equations.

The effectiveness of optimization as a tool for aerodynamic design also depends crucially on the proper choice of cost functions and constraints. One popular approach is to define a target pressure distribution and then solve the inverse problem of finding the shape that will produce that pressure distribution. Since such a shape does not necessarily exist, direct inverse methods may be ill-posed. The problem of designing a two-dimensional profile to attain a desired pressure distribution was studied by Lighthill, who solved it for the case of incompressible flow with a conformal mapping of the profile to a unit circle (Lighthill 1945). The speed over the profile is

$$q = \frac{1}{h} |\nabla\phi|,$$

where ϕ is the potential, which is known for incompressible flow, and h is the modulus of the mapping function. The surface value of h can be obtained by setting $q = q_d$, where q_d is the desired speed, and since the mapping function is analytic, it is uniquely determined by the value of h on the boundary. A solution exists for a given speed q_∞ at infinity only if

$$\frac{1}{2\pi} \oint q \, d\theta = q_\infty,$$

and there are additional constraints on q if the profile is required to be closed.

The difficulty that the target pressure may be unattainable may be circumvented by treating the inverse problem as a special case of the optimization problem, with a cost function that measures the error in the solution of the inverse problem. For example, if p_d is the desired surface pressure, one may take the cost function to be an integral over the body surface of the square of the pressure error,

$$I = \frac{1}{2} \int_{\mathcal{B}} (p - p_d)^2 d\mathcal{B},$$

or possibly a more general Sobolev norm of the pressure error. This has the advantage of converting a possibly ill-posed problem into a well-posed one. It has the disadvantage that it incurs the computational costs associated with optimization procedures.

The inverse problem still leaves the definition of an appropriate pressure architecture to the designer. One may prefer to directly improve suitable performance parameters, for example, to minimize the drag at a given lift and Mach number. In this case it is important to introduce appropriate constraints. For example, if the span is not fixed, the vortex drag can be made arbitrarily small by sufficiently increasing the span. In practice, a useful approach is to fix the planform and optimize the wing sections subject to constraints on minimum thickness.

Studies of the use of control theory for optimum shape design of systems governed by elliptic equations were initiated by Pironneau (1984). The control theory approach to optimal aerodynamic design was first applied to transonic flow by Jameson, who formulated the method for inviscid compressible flows with shock waves governed by both the potential flow and the Euler equations (Jameson 1988). Numerical results showing the method to be extremely effective for the design of airfoils in transonic potential flow were presented in Jameson (1989, 1990), and for three-dimensional wing design using the Euler equations in Jameson (1994). More recently the method has been employed for the shape design of complex aircraft configurations (Reuther, Jameson, Alonso, Rimlinger, & Saunders 1999, Reuther, Alonso, Vassberg, Jameson, & Martinelli 1997), using a grid perturbation approach to accommodate the geometry modifications. The method has been used to support the aerodynamic design studies of several industrial projects, including the Beech Premier, the McDonnell Douglas MDXX, blended wing–body projects, and the Gulfstream G650. The application to the MDXX is described in Jameson (1997). The experience gained in these industrial applications made it clear that the viscous effects cannot be ignored in transonic wing design, and the method has therefore been extended to treat the Reynolds Averaged Navier–Stokes equations (Jameson, Martinelli, & Pierce 1998). Adjoint methods have also been the subject of studies by a number of other authors, including Baysal and Eleshaky (1992), Huan and Modi (1994), Desai and Ito (1994), Anderson and Venkatakrishnan (1997), and Peraire and Elliot (1996). Ta'asan, Kuruvila, and Salas (1992) have implemented a one shot approach in which the constraint represented by the flow equations is only required to be satisfied by the final converged solution. In their work, computational costs are also reduced by applying multigrid techniques to the geometry modifications as well as the solution of the flow and adjoint equations.

15.3 Formulation of the Design Problem as a Control Problem

The simplest approach to optimization is to define the geometry through a set of design parameters, which may, for example, be the weights α_i applied to a set of shape functions $b_i(x)$ so that the shape is represented as

$$f(x) = \sum \alpha_i b_i(x).$$

Then a cost function I is selected, which might, for example, be the drag coefficient or the lift to drag ratio, and I is regarded as a function of the parameters α_i. The sensitivities $\frac{\partial I}{\partial \alpha_i}$ may now be estimated by making a small variation $\delta\alpha_i$ in each design parameter in turn and recalculating the flow to obtain the change in I. Then

$$\frac{\partial I}{\partial \alpha_i} \approx \frac{I(\alpha_i + \delta\alpha_i) - I(\alpha_i)}{\delta\alpha_i}.$$

The gradient vector $\frac{\partial I}{\partial \alpha}$ may now be used to determine a direction of improvement. The simplest procedure is to make a step in the negative gradient direction by setting

$$\alpha^{n+1} = \alpha^n - \lambda\delta\alpha,$$

so that to first order,

$$I + \delta I = I - \frac{\partial I^T}{\partial \alpha}\delta\alpha = I - \lambda\frac{\partial I^T}{\partial \alpha}\frac{\partial I}{\partial \alpha}.$$

More sophisticated search procedures may be used, such as quasi-Newton methods, which attempt to estimate the second derivative $\frac{\partial^2 I}{\partial \alpha_i \partial \alpha_j}$ of the cost function from changes in the gradient $\frac{\partial I}{\partial \alpha}$ in successive optimization steps. These methods also generally introduce line searches to find the minimum in the search direction, which is defined at each step. The main disadvantage of this approach is the need for a number of flow calculations proportional to the number of design variables to estimate the gradient. The computational costs can thus become prohibitive as the number of design variables is increased.

Using techniques of control theory, however, the gradient can be determined indirectly by solving an adjoint equation that has coefficients defined by the solution of the flow equations. The cost of solving the adjoint equation is comparable to that of solving the flow equations. Thus the gradient can be determined with roughly the computational costs of two flow solutions, independently of the number of design variables, which may be infinite if the boundary is regarded as a free surface. The underlying concepts are clarified by the following abstract description of the adjoint method.

For flow about an airfoil or wing, the aerodynamic properties that define the cost function are functions of the flow-field variables (w) and the physical location of the boundary, which may be represented by the function $\mathcal{F}$, say. Then

$$I = I\left(\boldsymbol{w}, \mathcal{F}\right),$$

and a change in $\mathcal{F}$ results in a change

$$\delta I = \left[\frac{\partial I^T}{\partial \boldsymbol{w}}\right]_I \delta \boldsymbol{w} + \left[\frac{\partial I^T}{\partial \mathcal{F}}\right]_{II} \delta \mathcal{F} \tag{15.1}$$

in the cost function. Here, the subscripts I and II are used to distinguish the contributions due to the variation $\delta \boldsymbol{w}$ in the flow solution from the change associated directly with the modification $\delta \mathcal{F}$ in the shape. This notation assists in grouping the numerous terms that arise during the derivation of the full Navier–Stokes adjoint operator, outlined later, so that the basic structure of the approach as it is sketched in the present section can easily be recognized.

Suppose that the governing equation $\boldsymbol{R}$, which expresses the dependence of $\boldsymbol{w}$ and $\mathcal{F}$ within the flowfield domain D, can be written as

$$R\left(\boldsymbol{w}, \mathcal{F}\right) = 0. \tag{15.2}$$

Then δw is determined from the equation

$$\delta \boldsymbol{R} = \left[\frac{\partial \boldsymbol{R}}{\partial \boldsymbol{w}}\right]_I \delta \boldsymbol{w} + \left[\frac{\partial \boldsymbol{R}}{\partial \mathcal{F}}\right]_{II} \delta \mathcal{F} = 0.$$

Since the variation δR is zero, it can be multiplied by a Lagrange multiplier ψ and subtracted from the variation δI without changing the result. Thus equation (15.1) can be replaced by

$$\begin{aligned} \delta I &= \frac{\partial I^T}{\partial \boldsymbol{w}} \delta \boldsymbol{w} + \frac{\partial I^T}{\partial \mathcal{F}} \delta \mathcal{F} - \psi^T \left(\left[\frac{\partial \boldsymbol{R}}{\partial \boldsymbol{w}}\right] \delta \boldsymbol{w} + \left[\frac{\partial \boldsymbol{R}}{\partial \mathcal{F}}\right] \delta \mathcal{F} \right) \\ &= \left\{ \frac{\partial I^T}{\partial \boldsymbol{w}} - \psi^T \left[\frac{\partial \boldsymbol{R}}{\partial \boldsymbol{w}}\right] \right\}_I \delta \boldsymbol{w} + \left\{ \frac{\partial I^T}{\partial \mathcal{F}} - \psi^T \left[\frac{\partial \boldsymbol{R}}{\partial \mathcal{F}}\right] \right\}_{II} \delta \mathcal{F}. \end{aligned}$$

Choosing $\boldsymbol{\psi}$ to satisfy the adjoint equation

$$\left[\frac{\partial \boldsymbol{R}}{\partial \boldsymbol{w}}\right]^T \psi = \frac{\partial I}{\partial \boldsymbol{w}}, \tag{15.3}$$

the first term is eliminated, and we find that

$$\delta I = \mathcal{G} \delta \mathcal{F}, \tag{15.4}$$

where

$$\mathcal{G} = \frac{\partial I^T}{\partial \mathcal{F}} - \psi^T \left[\frac{\partial \boldsymbol{R}}{\partial \mathcal{F}} \right].$$

The advantage is that (15.4) is independent of δw, with the result that the gradient of I with respect to an arbitrary number of design variables can be determined without the need for additional flow-field evaluations. In the case that (15.2) is a partial differential equation, the adjoint equation (15.3) is also a partial differential equation, and determination of the appropriate boundary conditions requires careful mathematical treatment.

Jameson (1988) derived the adjoint equations for transonic flows modeled by both the potential flow equation and the Euler equations. The theory was developed in terms of partial differential equations, leading to an adjoint partial differential equation. In order to obtain numerical solutions, both the flow and the adjoint equations must be discretized. Control theory might be applied directly to the discrete flow equations that result from the numerical approximation of the flow equations by finite element, finite volume or finite difference procedures. This leads directly to a set of discrete adjoint equations with a matrix which is the transpose of the Jacobian matrix of the full set of discrete nonlinear flow equations. On a three-dimensional mesh with indices i,j,k, the individual adjoint equations may be derived by collecting together all the terms multiplied by the variation $\delta \boldsymbol{w}_{i,j,k}$ of the discrete flow variable $\boldsymbol{w}_{i,j,k}$. The resulting discrete adjoint equations represent a possible discretization of the adjoint partial differential equation. If these equations are solved exactly, they can provide an exact gradient of the inexact cost function that results from the discretization of the flow equations. The discrete adjoint equations derived directly from the discrete flow equations become very complicated when the flow equations are discretized with higher order upwind biased schemes using flux limiters. On the other hand, any consistent discretization of the adjoint partial differential equation will yield the exact gradient in the limit as the mesh is refined.

The true optimum shape belongs to an infinite-dimensional space of design parameters. One motivation for developing the theory for the partial differential equations of the flow is to provide an indication in principle of how such a solution could be approached if sufficient computational resources were available. It displays the character of the adjoint equation as a hyperbolic system with waves travelling in the reverse direction to those of the flow equations, and the need for correct wall and far-field boundary conditions. It also highlights the possibility of generating ill-posed formulations of the problem. For example, if one attempts to calculate the sensitivity of the pressure at a particular location to changes in the boundary shape, there is the possibility that a shape modification could cause a shock wave to pass over that location. Then the sensitivity could become unbounded. The movement of the shock, however, is continuous as the shape changes. Therefore a quantity such as the drag coefficient, which is determined by integrating the pressure over the surface, also depends continuously on the shape. The adjoint equation allows the sensitivity of the drag coefficient to be determined without the explicit evaluation of pressure sensitivities, which would be ill-posed. Another benefit of the continuous adjoint formulation is that it allows grid sensitivity terms to be eliminated from the gradient, which can finally be expressed purely in terms of the boundary displacement, as will be shown in Section 15.5. This greatly simplifies the implementation of the method for overset or unstructured grids.

The discrete adjoint equations, whether they are derived directly or by discretization of the adjoint partial differential equation, are linear. Therefore they could be solved by direct numerical inversion. In three-dimensional problems on a mesh with, say, n intervals in each coordinate direction, the number of unknowns is proportional to n^3 and the bandwidth to n^2. The complexity of direct inversion is proportional to the number of unknowns multiplied by the square of the bandwidth, resulting in a complexity proportional to n^7. The cost of direct inversion can thus become prohibitive as the mesh is refined, and it becomes more efficient to use iterative solution methods. Moreover, because of the similarity of the adjoint equations to the flow equations, the same iterative methods that have been proved to be efficient for the solution of the flow equations are efficient for the solution of the adjoint equations.

15.4 Design Using the Euler Equations

The application of control theory to aerodynamic design problems is illustrated in this section for the case of three-dimensional wing design using the compressible Euler equations as the mathematical model. In order to simplify the derivation of the adjoint equations and boundary conditions, we first express the flow equations in a fixed computational domain by a transformation of coordinates from (x_1, x_2, x_3) to (ξ_1, ξ_2, ξ_3). Following the analysis of Section 2.2, the Euler equations can be written in the combined vector form

$$J\frac{\partial \boldsymbol{w}}{\partial t} + \boldsymbol{R}(\boldsymbol{w}) = 0, \tag{15.5}$$

where

$$\boldsymbol{R}(\boldsymbol{w}) = S_{ij}\frac{\partial \boldsymbol{f}_j}{\partial \xi_i} = \frac{\partial}{\partial \xi_i}\left(S_{ij}\boldsymbol{f}_j\right). \tag{15.6}$$

We can write the transformed fluxes in terms of the scaled contravariant velocity components

$$U_i = S_{ij}u_j$$

as

$$\boldsymbol{F}_i = S_{ij}\boldsymbol{f}_j = \begin{bmatrix} \rho U_i \\ \rho U_i u_1 + S_{i1}p \\ \rho U_i u_2 + S_{i2}p \\ \rho U_i u_3 + S_{i3}p \\ \rho U_i H \end{bmatrix}.$$

For convenience, the coordinates ξ_i describing the fixed computational domain are chosen so that each boundary conforms to a constant value of one of these coordinates. Variations in the shape then result in corresponding variations in the mapping derivatives defined by K_{ij}. Suppose that the performance is measured by a cost function

$$I = \int_{\mathcal{B}} \mathcal{M}(\boldsymbol{w}, S)\, dB_\xi + \int_{\mathcal{D}} \mathcal{P}(\boldsymbol{w}, S)\, dD_\xi,$$

containing both boundary and field contributions, where dB_ξ and dD_ξ are the surface and volume elements in the computational domain. In general, $\mathcal{M}$ and $\mathcal{P}$ will depend on both the flow variables w and the metrics S defining the computational space. The design problem is now treated as a control problem where the boundary shape represents the control function, which is chosen to minimize I subject to the constraints defined by the flow equations (15.5). A shape change produces a variation in the flow solution $\delta \boldsymbol{w}$ and the metrics δS, which in turn produce a variation in the cost function

$$\delta I = \int_{\mathcal{B}} \delta\mathcal{M}(\boldsymbol{w}, S)\, dB_\xi + \int_{\mathcal{D}} \delta\mathcal{P}(\boldsymbol{w}, S)\, dD_\xi. \tag{15.7}$$

This can be split as

$$\delta I = \delta I_I + \delta I_{II}, \tag{15.8}$$

with

$$\begin{aligned} \delta\mathcal{M} &= [\mathcal{M}_w]_I\, \delta\boldsymbol{w} + \delta\mathcal{M}_{II}, \\ \delta\mathcal{P} &= [\mathcal{P}_w]_I\, \delta\boldsymbol{w} + \delta\mathcal{P}_{II}, \end{aligned} \tag{15.9}$$

where we continue to use the subscripts I and II to distinguish between the contributions associated with the variation of the flow solution $\delta\boldsymbol{w}$ and those associated with the metric variations δS. Thus $[\mathcal{M}_w]_I$ and $[\mathcal{P}_w]_I$ represent $\frac{\partial \mathcal{M}}{\partial \boldsymbol{w}}$ and $\frac{\partial \mathcal{P}}{\partial \boldsymbol{w}}$ with the metrics fixed, while $\delta\mathcal{M}_{II}$ and $\delta\mathcal{P}_{II}$ represent the contribution of the metric variations δS to $\delta\mathcal{M}$ and $\delta\mathcal{P}$.

In the steady state, the constraint equation (15.5) specifies the variation of the state vector δw by

$$\delta\boldsymbol{R} = \frac{\partial}{\partial \xi_i}\delta\boldsymbol{F}_i = 0. \tag{15.10}$$

Here also, $\delta\boldsymbol{R}$ and $\delta\boldsymbol{F}_i$ can be split into contributions associated with $\delta\boldsymbol{w}$ and δS using the notation

$$\begin{aligned} \delta\boldsymbol{R} &= \delta\boldsymbol{R}_I + \delta\boldsymbol{R}_{II} \\ \delta\boldsymbol{F}_i &= \left[\boldsymbol{F}_{iw}\right]_I \delta\boldsymbol{w} + \delta\boldsymbol{F}_{i\,II}, \end{aligned} \tag{15.11}$$

where

$$\left[\boldsymbol{F}_{iw}\right]_I = S_{ij}\frac{\partial f_i}{\partial w}.$$

Multiplying by a co-state vector $\boldsymbol{\psi}$, which will play an analogous role to the Lagrange multiplier introduced in (15.3), and integrating over the domain produces

$$\int_{\mathcal{D}} \boldsymbol{\psi}^T \frac{\partial}{\partial \xi_i} \delta \boldsymbol{F}_i d\mathcal{D}_\xi = 0.$$

Assuming that $\boldsymbol{\psi}$ is differentiable, the terms with subscript I may be integrated by parts to give

$$\int_{\mathcal{B}} n_i \boldsymbol{\psi}^T \delta \boldsymbol{F}_{i_I} d\mathcal{B}_\xi - \int_{\mathcal{D}} \frac{\partial \boldsymbol{\psi}^T}{\partial \xi_i} \delta \boldsymbol{F}_{i_I} d\mathcal{D}_\xi + \int_{\mathcal{D}} \boldsymbol{\psi}^T \delta \boldsymbol{R}_{II} d\mathcal{D}_\xi = 0. \tag{15.12}$$

This equation results directly from taking the variation of the weak form of the flow equations, where ψ is taken to be an arbitrary differentiable test function. Since the left hand expression equals zero, it may be subtracted from the variation in the cost function (15.7) to give

$$\delta I = \delta I_{II} - \int_{\mathcal{D}} \boldsymbol{\psi}^T \delta \boldsymbol{R}_{II} d\mathcal{D}_\xi - \int_{\mathcal{B}} \left[\delta \mathcal{M}_I - n_i \boldsymbol{\psi}^T \delta \boldsymbol{F}_{i_I} \right] d\mathcal{B}_\xi$$
$$+ \int_{\mathcal{D}} \left[\delta \mathcal{P}_I + \frac{\partial \boldsymbol{\psi}^T}{\partial \xi_i} \delta \boldsymbol{F}_{i_I} \right] d\mathcal{D}_\xi. \tag{15.13}$$

Now, since $\boldsymbol{\psi}$ is an arbitrary differentiable function, it may be chosen in such a way that δI no longer depends explicitly on the variation of the state vector $\delta \boldsymbol{w}$. The gradient of the cost function can then be evaluated directly from the metric variations without having to recompute the variation $\delta \boldsymbol{w}$ resulting from the perturbation of each design variable.

Comparing (15.9) and (15.11), the variation $\delta \boldsymbol{w}$ may be eliminated from (15.13) by equating all field terms with subscript "I" to produce a differential adjoint system governing $\boldsymbol{\psi}$:

$$\frac{\partial \boldsymbol{\psi}^T}{\partial \xi_i} \left[\boldsymbol{F}_{i w} \right]_I + [\mathcal{P}_w]_I = 0 \quad \text{in } \mathcal{D}. \tag{15.14}$$

Taking the transpose of (15.14), in the case that there is no field integral in the cost function, the inviscid adjoint equation may be written as

$$C_i^T \frac{\partial \boldsymbol{\psi}}{\partial \xi_i} = 0 \quad \text{in } \mathcal{D},$$

where the inviscid Jacobian matrices in the transformed space are given by

$$C_i = S_{ij} \frac{\partial \boldsymbol{f}_j}{\partial \boldsymbol{w}}.$$

The corresponding adjoint boundary condition is produced by equating the subscript "I" boundary terms in (15.13) to produce

$$n_i \boldsymbol{\psi}^T \left[\boldsymbol{F}_{i w} \right]_I = [\mathcal{M}_w]_I \quad \text{on } \mathcal{B}. \tag{15.15}$$

The remaining terms from (15.13) then yield a simplified expression for the variation of the cost function that defines the gradient,

$$\delta I = \delta I_{II} + \int_{\mathcal{D}} \psi^T \delta \boldsymbol{R}_{II} d\mathcal{D}_\xi,$$

which consists purely of the terms containing variations in the metrics, with the flow solution fixed. Hence an explicit formula for the gradient can be derived once the relationship between mesh perturbations and shape variations is defined.

The details of the formula for the gradient depend on the way in which the boundary shape is parameterized as a function of the design variables and the way in which the mesh is deformed as the boundary is modified. Using the relationship between the mesh deformation and the surface modification, the field integral is reduced to a surface integral by integrating along the coordinate lines emanating from the surface. Thus the expression for δI is finally reduced to the form of (15.4):

$$\delta I = \int_{\mathcal{B}} \mathcal{G} \delta \mathcal{F} \, d\mathcal{B}_\xi,$$

where $\mathcal{F}$ represents the design variables, and $\mathcal{G}$ is the gradient, which is a function defined over the boundary surface.

The boundary conditions satisfied by the flow equations restrict the form of the left hand side of the adjoint boundary condition (15.15). Consequently, the boundary contribution to the cost function $\mathcal{M}$ cannot be specified arbitrarily. Instead, it must be chosen from the class of functions that allow cancellation of all terms containing δw in the boundary integral of (15.13). On the other hand, there is no such restriction on the specification of the field contribution to the cost function $\mathcal{P}$, since these terms may always be absorbed into the adjoint field equation (15.14) as source terms.

For simplicity, it will be assumed that the portion of the boundary that undergoes shape modifications is restricted to the coordinate surface $\xi_2 = 0$. Then (15.13) and (15.15) may be simplified by incorporating the conditions

$$n_1 = n_3 = 0, \quad n_2 = 1, \quad d\mathcal{B}_\xi = d\xi_1 d\xi_3,$$

so that only the variation δF_2 needs to be considered at the wall boundary. The condition that there is no flow through the wall boundary at $\xi_2 = 0$ is equivalent to

$$U_2 = 0,$$

so that

$$\delta U_2 = 0$$

when the boundary shape is modified. Consequently the variation of the inviscid flux at the boundary reduces to

$$\delta \boldsymbol{F}_2 = \delta p \begin{bmatrix} 0 \\ S_{21} \\ S_{22} \\ S_{23} \\ 0 \end{bmatrix} + p \begin{bmatrix} 0 \\ \delta S_{21} \\ \delta S_{22} \\ \delta S_{23} \\ 0 \end{bmatrix}.$$

Since $\delta \boldsymbol{F}_2$ depends only on the pressure, it is now clear that the performance measure on the boundary $\mathcal{M}(\boldsymbol{w}, S)$ may only be a function of the pressure and metric terms. Otherwise, complete cancellation of the terms containing $\delta \boldsymbol{w}$ in the boundary integral would be impossible. One may, for example, include arbitrary measures of the forces and moments in the cost function, since these are functions of the surface pressure.

In order to design a shape that will lead to a desired pressure distribution, a natural choice is to set

$$I = \frac{1}{2}\int_{\mathcal{B}} (p - p_d)^2 \, dS,$$

where p_d is the desired surface pressure, and the integral is evaluated over the actual surface area. In the computational domain this is transformed as in Section 2.2 to

$$I = \frac{1}{2}\iint_{\mathcal{B}_w} (p - p_d)^2 \, |S_2| \, d\xi_1 d\xi_3,$$

where the quantity

$$|S_2| = \sqrt{S_{2j} S_{2j}}$$

denotes the face area corresponding to a unit element of face area in the computational domain. Now, to cancel the dependence of the boundary integral on δp, the adjoint boundary condition reduces to

$$\psi_j n_j = p - p_d, \tag{15.16}$$

where n_j are the components of the surface normal

$$n_j = \frac{S_{2j}}{|S_2|}.$$

This amounts to a transpiration boundary condition on the co-state variables corresponding to the momentum components. Note that it imposes no restriction on the tangential component of ψ at the boundary.

We find finally that

$$\begin{aligned} \delta I = &- \int_{\mathcal{D}} \frac{\partial \boldsymbol{\psi}^T}{\partial \xi_i} \delta S_{ij} \boldsymbol{f}_j d\mathcal{D} \\ &- \iint_{B_W} \left(\delta S_{21}\psi_2 + \delta S_{22}\psi_3 + \delta S_{23}\psi_4\right) p \, d\xi_1 d\xi_3. \end{aligned}$$

Here the expression for the cost variation depends on the mesh variations throughout the domain, which appear in the field integral. However, the true gradient for a shape variation should not depend on the way in which the mesh is deformed but only on the true flow solution. In the next section we show how the field integral can be eliminated to produce a reduced gradient formula that depends only on the boundary movement (Jameson and Kim 2003).

15.5 The Reduced Gradient Formulation

Consider the case of a mesh variation with a fixed boundary. Then,

$$\delta I = 0,$$

but there is a variation in the transformed flux,

$$\delta \boldsymbol{F}_i = C_i \delta \boldsymbol{w} + \delta S_{ij} \boldsymbol{f}_j.$$

Here the true solution is unchanged. Thus, the variation δw is due to the mesh movement δx at each mesh point. Therefore

$$\delta \boldsymbol{w} = \nabla \boldsymbol{w} \cdot \delta x = \frac{\partial \boldsymbol{w}}{\partial x_j} \delta x_j \left(= \delta \boldsymbol{w}^*\right),$$

and since

$$\frac{\partial}{\partial \xi_i} \delta \boldsymbol{F}_i = 0,$$

it follows that

$$\frac{\partial}{\partial \xi_i} \left(\delta S_{ij} \boldsymbol{f}_j\right) = -\frac{\partial}{\partial \xi_i} \left(C_i \delta \boldsymbol{w}^*\right). \tag{15.17}$$

It is verified below that this relation holds in the general case with boundary movement. Now

$$\begin{aligned}
\int_{\mathcal{D}} \boldsymbol{\psi}^T \delta \boldsymbol{R} d\mathcal{D} &= \int_{\mathcal{D}} \boldsymbol{\psi}^T \frac{\partial}{\partial \xi_i} C_i \left(\delta \boldsymbol{w} - \delta \boldsymbol{w}^*\right) d\mathcal{D} \\
&= \int_{\mathcal{B}} \boldsymbol{\psi}^T C_i \left(\delta \boldsymbol{w} - \delta \boldsymbol{w}^*\right) d\mathcal{B} \\
&\quad - \int_{\mathcal{D}} \frac{\partial \boldsymbol{\psi}^T}{\partial \xi_i} C_i \left(\delta \boldsymbol{w} - \delta \boldsymbol{w}^*\right) d\mathcal{D}.
\end{aligned}$$

Here on the wall boundary

$$C_2 \delta \boldsymbol{w} = \delta \boldsymbol{F}_2 - \delta S_{2j} \boldsymbol{f}_j.$$

Thus, by choosing ϕ to satisfy the adjoint equation (15.4) and the adjoint boundary condition (15.15), we reduce the cost variation to a boundary integral that depends only on the surface displacement:

$$\begin{aligned}
\delta I = &\int_{\mathcal{B}_{\mathcal{W}}} \boldsymbol{\psi}^T \left(\delta S_{2j} \boldsymbol{f}_j + C_2 \delta \boldsymbol{w}^*\right) d\xi_1 d\xi_3 \\
&- \iint_{B_W} \left(\delta S_{21} \boldsymbol{\psi}_2 + \delta S_{22} \psi_3 + \delta S_{23} \psi_4\right) p \, d\xi_1 d\xi_3.
\end{aligned}$$

For completeness the general derivation of (15.17) is presented here. Using the formula (2.12), and the property (2.2):

$$\begin{aligned}\frac{\partial}{\partial \xi_i}\left(\delta S_{ij}\boldsymbol{f}_j\right) &= \frac{1}{2}\frac{\partial}{\partial \xi_i}\left\{\epsilon_{jpq}\epsilon_{irs}\left(\frac{\partial \delta x_p}{\partial \xi_r}\frac{\partial x_q}{\partial \xi_s}+\frac{\partial x_p}{\partial \xi_r}\frac{\partial \delta x_q}{\partial \xi_s}\right)\boldsymbol{f}_j\right\}\\
&= \frac{1}{2}\epsilon_{jpq}\epsilon_{irs}\left(\frac{\partial \delta x_p}{\partial \xi_r}\frac{\partial x_q}{\partial \xi_s}+\frac{\partial x_p}{\partial \xi_r}\frac{\partial \delta x_q}{\partial \xi_s}\right)\frac{\partial \boldsymbol{f}_j}{\partial \xi_i}\\
&= \frac{1}{2}\epsilon_{jpq}\epsilon_{irs}\left\{\frac{\partial}{\partial \xi_r}\left(\delta x_p\frac{\partial x_q}{\partial \xi_s}\frac{\partial \boldsymbol{f}_j}{\partial \xi_i}\right)\right\}\\
&\quad + \frac{1}{2}\epsilon_{jpq}\epsilon_{irs}\left\{\frac{\partial}{\partial \xi_s}\left(\delta x_q\frac{\partial x_p}{\partial \xi_r}\frac{\partial \boldsymbol{f}_j}{\partial \xi_i}\right)\right\}\\
&= \frac{\partial}{\partial \xi_r}\left(\delta x_p\epsilon_{pqj}\epsilon_{rsi}\frac{\partial x_q}{\partial \xi_s}\frac{\partial \boldsymbol{f}_j}{\partial \xi_i}\right).\end{aligned}$$

Now express δx_p in terms of a shift in the original computational coordinates

$$\delta x_p = \frac{\partial x_p}{\partial \xi_k}\delta \xi_k.$$

Then we obtain

$$\frac{\partial}{\partial \xi_i}\left(\delta S_{ij}\boldsymbol{f}_j\right) = \frac{\partial}{\partial \xi_r}\left(\epsilon_{pqj}\epsilon_{rsi}\frac{\partial x_p}{\partial \xi_k}\frac{\partial x_q}{\partial \xi_s}\frac{\partial \boldsymbol{f}_j}{\partial \xi_i}\delta \xi_k\right).$$

The term in $\frac{\partial}{\partial \xi_1}$ is

$$\epsilon_{123}\epsilon_{pqj}\frac{\partial x_p}{\partial \xi_k}\left(\frac{\partial x_q}{\partial \xi_2}\frac{\partial \boldsymbol{f}_j}{\partial \xi_3}-\frac{\partial x_q}{\partial \xi_3}\frac{\partial \boldsymbol{f}_j}{\partial \xi_2}\right)\delta \xi_k.$$

Here the term multiplying $\delta\xi_1$ is

$$\epsilon_{jpq}\left(\frac{\partial x_p}{\partial \xi_1}\frac{\partial x_q}{\partial \xi_2}\frac{\partial f_j}{\partial \xi_3}-\frac{\partial x_p}{\partial \xi_1}\frac{\partial x_q}{\partial \xi_3}\frac{\partial f_j}{\partial \xi_2}\right).$$

According to the formula (2.12) this may be recognized as

$$S_{2j}\frac{\partial \boldsymbol{f}_1}{\partial \xi_2}+S_{3j}\frac{\partial \boldsymbol{f}_1}{\partial \xi_3},$$

or, using the quasi-linear form(15.6) of the equation for steady flow, as

$$-S_{1j}\frac{\partial \boldsymbol{f}_1}{\partial \xi_1}.$$

The terms multiplying $\delta\xi_2$ and $\delta\xi_3$ are

$$\epsilon_{jpq}\left(\frac{\partial x_p}{\partial \xi_2}\frac{\partial x_q}{\partial \xi_2}\frac{\partial \boldsymbol{f}_j}{\partial \xi_3}-\frac{\partial x_p}{\partial \xi_2}\frac{\partial x_q}{\partial \xi_3}\frac{\partial \boldsymbol{f}_j}{\partial \xi_2}\right) = -S_{1j}\frac{\partial \boldsymbol{f}_1}{\partial \xi_2}$$

and

$$\epsilon_{jpq}\left(\frac{\partial x_p}{\partial \xi_3}\frac{\partial x_q}{\partial \xi_2}\frac{\partial \boldsymbol{f}_j}{\partial \xi_3} - \frac{\partial x_p}{\partial \xi_3}\frac{\partial x_q}{\partial \xi_3}\frac{\partial \boldsymbol{f}_j}{\partial \xi_2}\right) = -S_{1j}\frac{\partial \boldsymbol{f}_1}{\partial \xi_3}.$$

Thus the term in $\frac{\partial}{\partial \xi_1}$ is reduced to

$$-\frac{\partial}{\partial \xi_1}\left(S_{1j}\frac{\partial \boldsymbol{f}_1}{\partial \xi_k}\delta \xi_k\right).$$

Finally, with similar reductions of the terms in $\frac{\partial}{\partial \xi_2}$ and $\frac{\partial}{\partial \xi_3}$, we obtain

$$\frac{\partial}{\partial \xi_i}\left(\delta S_{ij}\boldsymbol{f}_j\right) = -\frac{\partial}{\partial \xi_i}\left(S_{ij}\frac{\partial \boldsymbol{f}_j}{\partial \xi_k}\delta \xi_k\right) = -\frac{\partial}{\partial \xi_i}\left(C_i \delta \boldsymbol{w}^*\right)$$

as was to be proved.

15.6 Optimization Procedure

15.6.1 The Need for a Sobolev Inner Product in the Definition of the Gradient

Another key issue for successful implementation of the continuous adjoint method is the choice of an appropriate inner product for the definition of the gradient. It turns out that there is an enormous benefit from the use of a modified Sobolev gradient, which enables the generation of a sequence of smooth shapes. This can be illustrated by considering the simplest case of a problem in the calculus of variations.

Suppose that we wish to find the path $y(x)$ that minimizes

$$I = \int_a^b F(y, y')dx$$

with fixed end points $y(a)$ and $y(b)$. Under a variation $\delta y(x)$,

$$\begin{aligned}\delta I &= \int_a^b \left(\frac{\partial F}{\partial y}\delta y + \frac{\partial F}{\partial y'}\delta y'\right)dx \\ &= \int_a^b \left(\frac{\partial F}{\partial y} - \frac{d}{dx}\frac{\partial F}{\partial y'}\right)\delta y dx.\end{aligned}$$

Thus defining the gradient as

$$g = \frac{\partial F}{\partial y} - \frac{d}{dx}\frac{\partial F}{\partial y'}$$

and the inner product as

$$(u, v) = \int_a^b uv dx,$$

we find that

$$\delta I = (g, \delta y).$$

If we now set

$$\delta y = -\lambda g, \qquad \lambda > 0,$$

we obtain a improvement

$$\delta I = -\lambda(g, g) \leq 0$$

unless $g = 0$, the necessary condition for a minimum.

Note that g is a function of y, y', y'',

$$g = g(y, y', y'').$$

In the well known case of the Brachistrone problem, for example, which calls for the determination of the path of quickest descent between two laterally separated points when a particle falls under gravity,

$$F(y, y') = \sqrt{\frac{1 + y'^2}{y}}$$

and

$$g = -\frac{1 + y'^2 + 2yy''}{2\left(y(1 + y'^2)\right)^{3/2}}.$$

It can be seen that each step

$$y^{n+1} = y^n - \lambda^n g^n$$

reduces the smoothness of y by two classes. Thus the computed trajectory becomes less and less smooth, leading to instability.

In order to prevent this we can introduce a weighted Sobolev inner product (Jameson, Martinelli, & Vassberg 2003):

$$\langle u, v \rangle = \int (uv + \epsilon u'v')dx,$$

where ϵ is a parameter that controls the weight of the derivatives. We now define a gradient $\overline{g}$ such that

$$\delta I = \langle \overline{g}, \delta y \rangle.$$

Then we have

$$\begin{aligned}
\delta I &= \int (\overline{g}\delta y + \epsilon \overline{g}'\delta y')dx \\
&= \int \left(\overline{g} - \frac{\partial}{\partial x}\epsilon\frac{\partial \overline{g}}{\partial x}\right)\delta y dx \\
&= (g, \delta y),
\end{aligned}$$

where

$$\overline{g} - \frac{\partial}{\partial x}\epsilon\frac{\partial \overline{g}}{\partial x} = g,$$

and $\overline{g} = 0$ at the end points. Thus $\overline{g}$ can be obtained from g by a smoothing equation. Now the step

$$y^{n+1} = y^n - \lambda^n \overline{g}^n$$

gives an improvement

$$\delta I = -\lambda^n \langle \overline{g}^n, \overline{g}^n \rangle,$$

but y^{n+1} has the same smoothness as y^n, resulting in a stable process.

15.6.2 Sobolev Gradient for Shape Optimization

In applying control theory to aerodynamic shape optimization, the use of a Sobolev gradient is equally important for the preservation of the smoothness class of the redesigned surface. Accordingly, using the weighted Sobolev inner product defined above, we define a modified gradient $\bar{\mathcal{G}}$ such that

$$\delta I = < \bar{\mathcal{G}}, \delta \mathcal{F} > .$$

In the one-dimensional case, $\bar{\mathcal{G}}$ is obtained by solving the smoothing equation

$$\bar{\mathcal{G}} - \frac{\partial}{\partial \xi_1}\epsilon\frac{\partial}{\partial \xi_1}\bar{\mathcal{G}} = \mathcal{G}. \tag{15.18}$$

In the multi-dimensional case, the smoothing is applied in product form. Finally we set

$$\delta \mathcal{F} = -\lambda \bar{\mathcal{G}} \tag{15.19}$$

with the result that

$$\delta I = -\lambda < \bar{\mathcal{G}}, \bar{\mathcal{G}} > \quad < 0,$$

unless $\bar{\mathcal{G}} = 0$, and correspondingly $\mathcal{G} = 0$.

When second order central differencing is applied to (15.18), the equation at a given node, i, can be expressed as

$$\bar{\mathcal{G}}_i - \epsilon\left(\bar{\mathcal{G}}_{i+1} - 2\bar{\mathcal{G}}_i + \bar{\mathcal{G}}_{i-1}\right) = \mathcal{G}_i, \qquad 1 \leq i \leq n,$$

where $\mathcal{G}_i$ and $\bar{\mathcal{G}}_i$ are the point gradients at node i before and after the smoothing respectively, and n is the number of design variables equal to the number of mesh points in this case. Then,

$$\bar{\mathcal{G}} = A\mathcal{G},$$

where A is the $n \times n$ tri-diagonal matrix such that

$$A^{-1} = \begin{bmatrix} 1+2\epsilon & -\epsilon & 0 & . & 0 \\ \epsilon & . & . & & \\ 0 & . & . & . & \\ . & & . & . & -\epsilon \\ 0 & & & \epsilon & 1+2\epsilon \end{bmatrix}.$$

Using the steepest descent method in each design iteration, a step, $\delta\mathcal{F}$, is taken such that

$$\delta\mathcal{F} = -\lambda A \mathcal{G}.$$

As can be seen from the form of this expression, implicit smoothing may be regarded as a preconditioner that allows the use of much larger steps for the search procedure and leads to a large reduction in the number of design iterations needed for convergence.

15.6.3 Outline of the Design Procedure

The design procedure can finally be summarized as follows (Figure 15.1):

1. Solve the flow equations for ρ, u_1, u_2, u_3, p.
2. Solve the adjoint equations for ψ subject to appropriate boundary conditions.
3. Evaluate $\mathcal{G}$, and calculate the corresponding Sobolev gradient $\bar{\mathcal{G}}$.

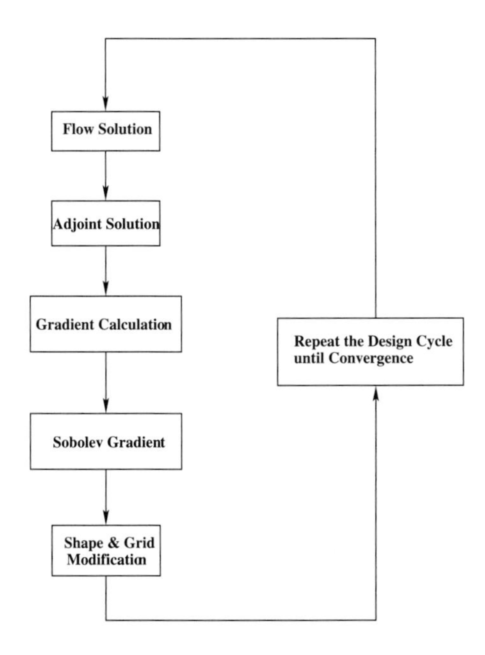

Figure 15.1 Design cycle.

4. Project $\bar{\mathcal{G}}$ into an allowable subspace that satisfies any geometric constraints.
5. Update the shape based on the direction of steepest descent.
6. Return to step 1 until convergence is reached.

Practical implementation of the design method relies heavily upon fast and accurate solvers for both the state (w) and co-state (ψ) systems. The results obtained in Section 15.9 have been obtained using well-validated software for the solution of the Euler and Navier–Stokes equations developed over the course of many years (Jameson et al. 1981, Martinelli and Jameson 1988, Tatsumi, Martinelli, & Jameson 1995). For inverse design, the lift is fixed by the target pressure. In drag minimization it is also appropriate to fix the lift coefficient, because the induced drag is a major fraction of the total drag, and this could be reduced simply by reducing the lift. Therefore the angle of attack is adjusted during each flow solution to force a specified lift coefficient to be attained, and the influence of variations of the angle of attack is included in the calculation of the gradient. The vortex drag also depends on the span loading, which may be constrained by other considerations such as structural loading or buffet onset. Consequently, the option is provided to force the span loading by adjusting the twist distribution as well as the angle of attack during the flow solution.

15.7 Design using the Navier-Stokes Equations

15.7.1 The Navier–Stokes Equations in the Computational Domain

The next sections present the extension of the adjoint method to the Navier–Stokes equations (Jameson, Martinelli, & Pierce 1998). It proves convenient to express the viscous fluxes separately, denote the viscous stress tensor by sigma, and write the equations in the form

$$\frac{\partial \boldsymbol{w}}{\partial t} + \frac{\partial \boldsymbol{f}_i}{\partial x_i} = \frac{\partial \boldsymbol{f}_{vi}}{\partial x_i} \quad \text{in } \mathcal{D},$$

where the state vector $\boldsymbol{w}$, inviscid flux vector $\boldsymbol{f}$, and viscous flux vector $\boldsymbol{f}_v$ are described respectively by

$$\boldsymbol{w} = \begin{bmatrix} \rho \\ \rho u_1 \\ \rho u_2 \\ \rho u_3 \\ \rho E \end{bmatrix},$$

$$\boldsymbol{f}_i = \begin{bmatrix} \rho u_i \\ \rho u_i u_1 + p\delta_{i1} \\ \rho u_i u_2 + p\delta_{i2} \\ \rho u_i u_3 + p\delta_{i3} \\ \rho u_i H \end{bmatrix}, \qquad \boldsymbol{f}_{vi} = \begin{bmatrix} 0 \\ \sigma_{ij}\delta_{j1} \\ \sigma_{ij}\delta_{j2} \\ \sigma_{ij}\delta_{j3} \\ u_j\sigma_{ij} + k\frac{\partial T}{\partial x_i} \end{bmatrix}.$$

The viscous stresses may be written as

$$\sigma_{ij} = \mu \left(\frac{\partial u_i}{\partial x_j} + \frac{\partial u_j}{\partial x_i} \right) + \lambda \delta_{ij} \frac{\partial u_k}{\partial x_k},$$

where μ and λ are the first and second coefficients of viscosity. The coefficient of thermal conductivity and the temperature are computed as

$$k = \frac{c_p \mu}{Pr}, \qquad T = \frac{p}{R\rho}, \tag{15.20}$$

where Pr is the Prandtl number, c_p is the specific heat at constant pressure, and R is the gas constant.

Using a transformation to a fixed computational domain as before, the Navier–Stokes equations can be written in the transformed coordinates as

$$\frac{\partial (J\boldsymbol{w})}{\partial t} + \frac{\partial (\boldsymbol{F}_i - \boldsymbol{F}_{vi})}{\partial \xi_i} = 0 \quad \text{in } \mathcal{D},$$

where the viscous terms have the form

$$\frac{\partial \boldsymbol{F}_v}{\partial \xi_i}{}_i = \frac{\partial}{\partial \xi_i}(S_{ij} \boldsymbol{f}_{vj}).$$

Computing the variation $\delta \boldsymbol{w}$ resulting from a shape modification of the boundary, introducing a co-state vector $\boldsymbol{\psi}$, and integrating by parts, following the steps outlined by (15.10) to (15.12), we obtain

$$\int_{\mathcal{B}} \boldsymbol{\psi}^T (\delta S_{2j} \boldsymbol{f}_{vj} + S_{2j} \delta \boldsymbol{f}_{vj}) d\mathcal{B}_\xi - \int_{\mathcal{D}} \frac{\partial \boldsymbol{\psi}^T}{\partial \xi_i} (\delta S_{ij} \boldsymbol{f}_{vj} + S_{ij} \delta \boldsymbol{f}_{vj}) d\mathcal{D}_\xi,$$

where the shape modification is restricted to the coordinate surface $\xi_2 = 0$ so that $n_1 = n_3 = 0$, and $n_2 = 1$. Furthermore, it is assumed that the boundary contributions at the far field may either be neglected or else eliminated by a proper choice of boundary conditions as previously shown for the inviscid case (Jameson 1990, 1995c).

The viscous terms will be derived under the assumption that the viscosity and heat conduction coefficients μ and k are essentially independent of the flow and that their variations may be neglected. This simplification has been successfully used for may aerodynamic problems of interest. However, if the flow variations could result in significant changes in the turbulent viscosity, it may be necessary to account for its variation in the calculation.

15.7.2 Transformation to Primitive Variables

The derivation of the viscous adjoint terms can be simplified by transforming to primitive variables

$$\tilde{\boldsymbol{w}}^T = \left[\rho, u_1, u_2, u_3, p\right]^T,$$

because the viscous stresses depend on the velocity derivatives $\frac{\partial u_i}{\partial x_j}$, while the heat flux can be expressed as

$$\kappa \frac{\partial}{\partial x_i}\left(\frac{p}{\rho}\right),$$

where $\kappa = \frac{k}{R} = \frac{\gamma\mu}{Pr(\gamma-1)}$. The relationship between the conservative and primitive variations is defined by the expressions

$$\delta w = M\delta\tilde{w}, \quad \delta\tilde{w} = M^{-1}\delta w,$$

which make use of the transformation matrices $M = \frac{\partial w}{\partial \tilde{w}}$ and $M^{-1} = \frac{\partial \tilde{w}}{\partial w}$. These matrices are provided in transposed form for future convenience

$$M^T = \begin{bmatrix} 1 & u_1 & u_2 & u_3 & \frac{u_i u_i}{2} \\ 0 & \rho & 0 & 0 & \rho u_1 \\ 0 & 0 & \rho & 0 & \rho u_2 \\ 0 & 0 & 0 & \rho & \rho u_3 \\ 0 & 0 & 0 & 0 & \frac{1}{\gamma-1} \end{bmatrix}$$

$$M^{-1^T} = \begin{bmatrix} 1 & -\frac{u_1}{\rho} & -\frac{u_2}{\rho} & -\frac{u_3}{\rho} & \frac{(\gamma-1)u_i u_i}{2} \\ 0 & \frac{1}{\rho} & 0 & 0 & -(\gamma-1)u_1 \\ 0 & 0 & \frac{1}{\rho} & 0 & -(\gamma-1)u_2 \\ 0 & 0 & 0 & \frac{1}{\rho} & -(\gamma-1)u_3 \\ 0 & 0 & 0 & 0 & \gamma-1 \end{bmatrix}.$$

The conservative and primitive adjoint operators L and $\tilde{L}$ corresponding to the variations δw and $\delta\tilde{w}$ are then related by

$$\int_{\mathcal{D}} \delta w^T L\psi \, d\mathcal{D}_\xi = \int_{\mathcal{D}} \delta\tilde{w}^T \tilde{L}\psi \, d\mathcal{D}_\xi,$$

with

$$\tilde{L} = M^T L,$$

so that after determining the primitive adjoint operator by direct evaluation of the viscous portion of (15.14), the conservative operator may be obtained by the transformation $L = M^{-1^T}\tilde{L}$. Since the continuity equation contains no viscous terms, it makes no contribution to the viscous adjoint system. Therefore, the derivation proceeds by first examining the adjoint operators arising from the momentum equations.

15.7.3 Contributions from the Momentum Equations

In order to make use of the summation convention, it is convenient to set $\psi_{j+1} = \phi_j$ for $j = 1,2,3$ and $\psi_5 = \theta$. Then the contribution from the momentum equations is

$$\int_{\mathcal{B}} \phi_k \left(\delta S_{2j}\sigma_{kj} + S_{2j}\delta\sigma_{kj}\right) d\mathcal{B}_\xi - \int_{\mathcal{D}} \frac{\partial \phi_k}{\partial \xi_i} \left(\delta S_{ij}\sigma_{kj} + S_{ij}\delta\sigma_{kj}\right) d\mathcal{D}_\xi. \tag{15.21}$$

The velocity derivatives can be expressed as

$$\frac{\partial u_i}{\partial x_j} = \frac{\partial u_i}{\partial \xi_l}\frac{\partial \xi_l}{\partial x_j} = \frac{S_{lj}}{J}\frac{\partial u_i}{\partial \xi_l}$$

with corresponding variations

$$\delta\frac{\partial u_i}{\partial x_j} = \left[\frac{S_{lj}}{J}\right]_I \frac{\partial}{\partial \xi_l}\delta u_i + \left[\frac{\partial u_i}{\partial \xi_l}\right]_{II} \delta\left(\frac{S_{lj}}{J}\right).$$

The variations in the stresses are then

$$\delta\sigma_{kj} = \left\{\mu\left[\tfrac{S_{lj}}{J}\tfrac{\partial}{\partial \xi_l}\delta u_k + \tfrac{S_{lk}}{J}\tfrac{\partial}{\partial \xi_l}\delta u_j\right] + \lambda\left[\delta_{jk}\tfrac{S_{lm}}{J}\tfrac{\partial}{\partial \xi_l}\delta u_m\right]\right\}_I$$
$$+\left\{\mu\left[\delta\left(\tfrac{S_{lj}}{J}\right)\tfrac{\partial u_k}{\partial \xi_l} + \delta\left(\tfrac{S_{lk}}{J}\right)\tfrac{\partial u_j}{\partial \xi_l}\right] + \lambda\left[\delta_{jk}\delta\left(\tfrac{S_{lm}}{J}\right)\tfrac{\partial u_m}{\partial \xi_l}\right]\right\}_{II}.$$

As before, only those terms with subscript I, which contain variations of the flow variables, need be considered further in deriving the adjoint operator. The field contributions that contain δu_i in (15.21) appear as

$$-\int_{\mathcal{D}} \frac{\partial \phi_k}{\partial \xi_i} S_{ij}\left\{\mu\left(\frac{S_{lj}}{J}\frac{\partial}{\partial \xi_l}\delta u_k + \frac{S_{lk}}{J}\frac{\partial}{\partial \xi_l}\delta u_j\right) + \lambda\delta_{jk}\frac{S_{lm}}{J}\frac{\partial}{\partial \xi_l}\delta u_m\right\} d\mathcal{D}_\xi.$$

This may be integrated by parts to yield

$$\int_{\mathcal{D}} \delta u_k \frac{\partial}{\partial \xi_l}\left(S_{lj}S_{ij}\frac{\mu}{J}\frac{\partial \phi_k}{\partial \xi_i}\right) d\mathcal{D}_\xi$$
$$+\int_{\mathcal{D}} \delta u_j \frac{\partial}{\partial \xi_l}\left(S_{lk}S_{ij}\frac{\mu}{J}\frac{\partial \phi_k}{\partial \xi_i}\right) d\mathcal{D}_\xi$$
$$+\int_{\mathcal{D}} \delta u_m \frac{\partial}{\partial \xi_l}\left(S_{lm}S_{ij}\frac{\lambda\delta_{jk}}{J}\frac{\partial \phi_k}{\partial \xi_i}\right) d\mathcal{D}_\xi,$$

where the boundary integral has been eliminated by noting that $\delta u_i = 0$ on the solid boundary. By exchanging indices, the field integrals may be combined to produce

$$\int_{\mathcal{D}} \delta u_k \frac{\partial}{\partial \xi_l} S_{lj}\left\{\mu\left(\frac{S_{ij}}{J}\frac{\partial \phi_k}{\partial \xi_i} + \frac{S_{ik}}{J}\frac{\partial \phi_j}{\partial \xi_i}\right) + \lambda\delta_{jk}\frac{S_{im}}{J}\frac{\partial \phi_m}{\partial \xi_i}\right\} d\mathcal{D}_\xi,$$

which is further simplified by transforming the inner derivatives back to Cartesian coordinates:

$$\int_{\mathcal{D}} \delta u_k \frac{\partial}{\partial \xi_l} S_{lj}\left\{\mu\left(\frac{\partial \phi_k}{\partial x_j} + \frac{\partial \phi_j}{\partial x_k}\right) + \lambda\delta_{jk}\frac{\partial \phi_m}{\partial x_m}\right\} d\mathcal{D}_\xi. \tag{15.22}$$

The boundary contributions that contain δu_i in (15.21) may be simplified using the fact that

$$\frac{\partial}{\partial \xi_l}\delta u_i = 0 \quad \text{if} \quad l = 1,3$$

on the boundary $\mathcal{B}$ so that they become

$$\int_{\mathcal{B}} \phi_k S_{2j} \left\{ \mu \left(\frac{S_{2j}}{J} \frac{\partial}{\partial \xi_2} \delta u_k + \frac{S_{2k}}{J} \frac{\partial}{\partial \xi_2} \delta u_j \right) + \lambda \delta_{jk} \frac{S_{2m}}{J} \frac{\partial}{\partial \xi_2} \delta u_m \right\} d\mathcal{B}_\xi. \quad (15.23)$$

Together, (15.22) and (15.23) comprise the field and boundary contributions of the momentum equations to the viscous adjoint operator in primitive variables.

15.7.4 Contributions from the Energy Equation

In order to derive the contribution of the energy equation to the viscous adjoint terms, it is convenient to set

$$\psi_5 = \theta, \quad Q_j = u_i \sigma_{ij} + \kappa \frac{\partial}{\partial x_j} \left(\frac{p}{\rho} \right),$$

where the temperature has been written in terms of pressure and density using (15.20). The contribution from the energy equation can then be written as

$$\int_{\mathcal{B}} \theta \left(\delta S_{2j} Q_j + S_{2j} \delta Q_j \right) d\mathcal{B}_\xi - \int_{\mathcal{D}} \frac{\partial \theta}{\partial \xi_i} \left(\delta S_{ij} Q_j + S_{ij} \delta Q_j \right) d\mathcal{D}_\xi. \quad (15.24)$$

The field contributions that contain δu_i,δp, and $\delta \rho$ in (15.24) appear as

$$-\int_{\mathcal{D}} \frac{\partial \theta}{\partial \xi_i} S_{ij} \delta Q_j d\mathcal{D}_\xi$$

$$= -\int_{\mathcal{D}} \frac{\partial \theta}{\partial \xi_i} S_{ij} \left\{ \delta u_k \sigma_{kj} + u_k \delta \sigma_{kj} + \kappa \frac{S_{lj}}{J} \frac{\partial}{\partial \xi_l} \left(\frac{\delta p}{\rho} - \frac{p}{\rho} \frac{\delta \rho}{\rho} \right) \right\} d\mathcal{D}_\xi. \quad (15.25)$$

The term involving $\delta \sigma_{kj}$ may be integrated by parts to produce

$$\int_{\mathcal{D}} \delta u_k \frac{\partial}{\partial \xi_l} S_{lj} \left\{ \mu \left(u_k \frac{\partial \theta}{\partial x_j} + u_j \frac{\partial \theta}{\partial x_k} \right) + \lambda \delta_{jk} u_m \frac{\partial \theta}{\partial x_m} \right\} d\mathcal{D}_\xi,$$

where the conditions $u_i = \delta u_i = 0$ are used to eliminate the boundary integral on $\mathcal{B}$. Notice that the other term in (15.25) that involves δu_k need not be integrated by parts and is merely carried on as

$$-\int_{\mathcal{D}} \delta u_k \sigma_{kj} S_{ij} \frac{\partial \theta}{\partial \xi_i} d\mathcal{D}_\xi.$$

The terms in expression (15.25) that involve δp and $\delta \rho$ may also be integrated by parts to produce both a field and a boundary integral. The field integral becomes

$$\int_{\mathcal{D}} \left(\frac{\delta p}{\rho} - \frac{p}{\rho} \frac{\delta \rho}{\rho} \right) \frac{\partial}{\partial \xi_l} \left(S_{lj} S_{ij} \frac{\kappa}{J} \frac{\partial \theta}{\partial \xi_i} \right) d\mathcal{D}_\xi,$$

which may be simplified by transforming the inner derivative to Cartesian coordinates

$$\int_{\mathcal{D}} \left(\frac{\delta p}{\rho} - \frac{p}{\rho} \frac{\delta \rho}{\rho} \right) \frac{\partial}{\partial \xi_l} \left(S_{lj} \kappa \frac{\partial \theta}{\partial x_j} \right) d\mathcal{D}_\xi. \quad (15.26)$$

The boundary integral becomes

$$\int_{\mathcal{B}} \kappa \left(\frac{\delta p}{\rho} - \frac{p}{\rho}\frac{\delta \rho}{\rho} \right) \frac{S_{2j} S_{ij}}{J} \frac{\partial \theta}{\partial \xi_i} d\mathcal{B}_\xi .$$

This can be simplified by transforming the inner derivative to Cartesian coordinates

$$\int_{\mathcal{B}} \kappa \left(\frac{\delta p}{\rho} - \frac{p}{\rho}\frac{\delta \rho}{\rho} \right) \frac{S_{2j}}{J} \frac{\partial \theta}{\partial x_j} d\mathcal{B}_\xi$$

and identifying the normal derivative at the wall

$$\frac{\partial}{\partial n} = S_{2j} \frac{\partial}{\partial x_j}, \tag{15.27}$$

and the variation in temperature

$$\delta T = \frac{1}{R} \left(\frac{\delta p}{\rho} - \frac{p}{\rho}\frac{\delta \rho}{\rho} \right),$$

to produce the boundary contribution

$$\int_{\mathcal{B}} k \delta T \frac{\partial \theta}{\partial n} d\mathcal{B}_\xi . \tag{15.28}$$

This term vanishes if T is constant on the wall but persists if the wall is adiabatic.

There is also a boundary contribution left over from the first integration by parts (15.24), which has the form

$$\int_{\mathcal{B}} \theta \delta \left(S_{2j} Q_j \right) d\mathcal{B}_\xi, \tag{15.29}$$

where

$$Q_j = k \frac{\partial T}{\partial x_j},$$

since $u_i = 0$. If the wall is adiabatic,

$$\frac{\partial T}{\partial n} = 0,$$

so that using (15.27),

$$\delta \left(S_{2j} Q_j \right) = 0,$$

and both the δw and δS boundary contributions vanish.

On the other hand, if T is constant, $\frac{\partial T}{\partial \xi_l} = 0$ for $l = 1, 3$, so that

$$Q_j = k \frac{\partial T}{\partial x_j} = k \left(\frac{S_{lj}}{J} \frac{\partial T}{\partial \xi_l} \right) = k \left(\frac{S_{2j}}{J} \frac{\partial T}{\partial \xi_2} \right).$$

Thus, the boundary integral (15.29) becomes

$$\int_{\mathcal{B}} k \theta \left\{ \frac{{S_{2j}}^2}{J} \frac{\partial}{\partial \xi_2} \delta T + \delta \left(\frac{{S_{2j}}^2}{J} \right) \frac{\partial T}{\partial \xi_2} \right\} d\mathcal{B}_\xi . \tag{15.30}$$

Therefore, for constant T, the first term corresponding to variations in the flow field contributes to the adjoint boundary operator, and the second set of terms corresponding to metric variations contribute to the cost function gradient.

Finally the contributions from the energy equation to the viscous adjoint operator are the three field terms (15.7.4), (15.7.4), and (15.26), and either of two boundary contributions (15.28) or (15.30), depending on whether the wall is adiabatic or has constant temperature.

15.7.5 The Viscous Adjoint Field Operator

Collecting together the contributions from the momentum and energy equations, the viscous adjoint operator in primitive variables can be expressed as

$$
\begin{aligned}
(\tilde{L}\psi)_1 &= -\frac{p}{\rho^2}\frac{\partial}{\partial \xi_l}\left(S_{lj}\kappa\frac{\partial \theta}{\partial x_j}\right)\\
(\tilde{L}\psi)_{i+1} &= \frac{\partial}{\partial \xi_l}\left\{S_{lj}\left[\mu\left(\frac{\partial \phi_i}{\partial x_j}+\frac{\partial \phi_j}{\partial x_i}\right)+\lambda\delta_{ij}\frac{\partial \phi_k}{\partial x_k}\right]\right\}\\
&\quad+\frac{\partial}{\partial \xi_l}\left\{S_{lj}\left[\mu\left(u_i\frac{\partial \theta}{\partial x_j}+u_j\frac{\partial \theta}{\partial x_i}\right)+\lambda\delta_{ij}u_k\frac{\partial \theta}{\partial x_k}\right]\right\} \qquad \text{for} \quad i=1,2,3\\
&\quad-\sigma_{ij}S_{lj}\frac{\partial \theta}{\partial \xi_l}\\
(\tilde{L}\psi)_5 &= \frac{1}{\rho}\frac{\partial}{\partial \xi_l}\left(S_{lj}\kappa\frac{\partial \theta}{\partial x_j}\right).
\end{aligned}
$$

The conservative viscous adjoint operator may now be obtained by the transformation

$$
L = M^{-1^T}\tilde{L}.
$$

15.8 Viscous Adjoint Boundary Conditions

It was recognized in Section 15.4 that the boundary conditions satisfied by the flow equations restrict the form of the performance measure that may be chosen for the cost function. There must be a direct correspondence between the flow variables for which variations appear in the variation of the cost function and those variables for which variations appear in the boundary terms arising during the derivation of the adjoint field equations. Otherwise it would be impossible to eliminate the dependence of δI on δw through proper specification of the adjoint boundary conditions. Consequently the contributions of the pressure and viscous stresses need to be merged. As in the derivation of the field equations, it proves convenient to consider the contributions from the momentum equations and the energy equation separately.

15.8.1 Boundary Conditions Arising from the Momentum Equations

The boundary term that arises from the momentum equations including both the δw and δS components (15.21) takes the form

$$\int_{\mathcal{B}} \phi_k \delta \left(S_{2j} \left(\delta_{kj} p + \sigma_{kj} \right) \right) d\mathcal{B}_\xi .$$

Replacing the metric term with the corresponding local face area S_2 and unit normal n_j defined by

$$|S_2| = \sqrt{S_{2j} S_{2j}}, \quad n_j = \frac{S_{2j}}{|S_2|}$$

then leads to

$$\int_{\mathcal{B}} \phi_k \delta \left(|S_2| \, n_j \left(\delta_{kj} p + \sigma_{kj} \right) \right) d\mathcal{B}_\xi .$$

Defining the components of the total surface stress as

$$\tau_k = n_j \left(\delta_{kj} p + \sigma_{kj} \right)$$

and the physical surface element

$$dS = |S_2| \, d\mathcal{B}_\xi ,$$

the integral may then be split into two components,

$$\int_{\mathcal{B}} \phi_k \tau_k \, |\delta S_2| \, d\mathcal{B}_\xi + \int_{\mathcal{B}} \phi_k \delta \tau_k dS, \tag{15.31}$$

where only the second term contains variations in the flow variables and must consequently cancel the δw terms arising in the cost function. The first term will appear in the expression for the gradient.

A general expression for the cost function that allows cancellation with terms containing $\delta \tau_k$ has the form

$$I = \int_{\mathcal{B}} \mathcal{N}(\tau) dS, \tag{15.32}$$

corresponding to a variation

$$\delta I = \int_{\mathcal{B}} \frac{\partial \mathcal{N}}{\partial \tau_k} \delta \tau_k dS,$$

for which cancellation is achieved by the adjoint boundary condition

$$\phi_k = \frac{\partial \mathcal{N}}{\partial \tau_k}.$$

Natural choices for $\mathcal{N}$ arise from force optimization and as measures of the deviation of the surface stresses from desired target values.

The force in a direction with cosines q_i has the form

$$C_q = \int_{\mathcal{B}} q_i \tau_i dS.$$

If we take this as the cost function (15.32), this quantity gives

$$\mathcal{N} = q_i \tau_i .$$

Cancellation with the flow variation terms in (15.31) therefore mandates the adjoint boundary condition

$$\phi_k = q_k.$$

Note that this choice of boundary condition also eliminates the first term in (15.31) so that it need not be included in the gradient calculation.

In the inverse design case, where the cost function is intended to measure the deviation of the surface stresses from some desired target values, a suitable definition is

$$\mathcal{N}(\tau) = \frac{1}{2} a_{lk} \left(\tau_l - \tau_{dl}\right) \left(\tau_k - \tau_{dk}\right),$$

where τ_d is the desired surface stress, including the contribution of the pressure, and the coefficients a_{lk} define a weighting matrix. For cancellation,

$$\phi_k \delta \tau_k = a_{lk} \left(\tau_l - \tau_{dl}\right) \delta \tau_k.$$

This is satisfied by the boundary condition

$$\phi_k = a_{lk} \left(\tau_l - \tau_{dl}\right). \tag{15.33}$$

Assuming arbitrary variations in $\delta \tau_k$, this condition is also necessary.

In order to control the surface pressure and normal stress, one can measure the difference

$$n_j \left\{\sigma_{kj} + \delta_{kj} \left(p - p_d\right)\right\},$$

where p_d is the desired pressure. The normal component is then

$$\tau_n = n_k n_j \sigma_{kj} + p - p_d,$$

so that the measure becomes

$$\begin{aligned} \mathcal{N}(\tau) &= \frac{1}{2} \tau_n^2 \\ &= \frac{1}{2} n_l n_m n_k n_j \left\{\sigma_{lm} + \delta_{lm} \left(p - p_d\right)\right\} \left\{\sigma_{kj} + \delta_{kj} \left(p - p_d\right)\right\}. \end{aligned}$$

This corresponds to setting

$$a_{lk} = n_l n_k$$

in (15.33). Defining the viscous normal stress as

$$\tau_{vn} = n_k n_j \sigma_{kj},$$

the measure can be expanded as

$$\begin{aligned} \mathcal{N}(\tau) &= \frac{1}{2} n_l n_m n_k n_j \sigma_{lm} \sigma_{kj} + \frac{1}{2} \left(n_k n_j \sigma_{kj} + n_l n_m \sigma_{lm}\right) \left(p - p_d\right) + \frac{1}{2} \left(p - p_d\right)^2 \\ &= \frac{1}{2} \tau_{vn}^2 + \tau_{vn} \left(p - p_d\right) + \frac{1}{2} \left(p - p_d\right)^2. \end{aligned}$$

For cancellation of the boundary terms,

$$\phi_k \left(n_j \delta \sigma_{kj} + n_k \delta p\right) = \left\{n_l n_m \sigma_{lm} + n_l^2 \left(p - p_d\right)\right\} n_k \left(n_j \delta \sigma_{kj} + n_k \delta p\right),$$

leading to the boundary condition

$$\phi_k = n_k \left(\tau_{vn} + p - p_d\right).$$

In the case of a high Reynolds number, this is well approximated by the equations

$$\phi_k = n_k \left(p - p_d\right), \tag{15.34}$$

which should be compared with the single scalar equation derived for the inviscid boundary condition (15.16). In the case of an inviscid flow, choosing

$$\mathcal{N}(\tau) = \frac{1}{2} \left(p - p_d\right)^2$$

requires

$$\phi_k n_k \delta p = \left(p - p_d\right) n_k^2 \delta p = \left(p - p_d\right) \delta p,$$

which is satisfied by (15.34) but which represents an overspecification of the boundary condition since only the single condition (15.16) needs be specified to ensure cancellation.

15.8.2 Boundary Conditions Arising from the Energy Equation

The form of the boundary terms arising from the energy equation depends on the choice of temperature boundary condition at the wall. For the adiabatic case, the boundary contribution is

$$\int_{\mathcal{B}} k \delta T \frac{\partial \theta}{\partial n} d\mathcal{B}_\xi,$$

while for the constant temperature case, the boundary term is (15.30). One possibility is to introduce a contribution into the cost function that depends on T or $\frac{\partial T}{\partial n}$ so that the appropriate cancellation would occur. Since there is little physical intuition to guide the choice of such a cost function for aerodynamic design, a more natural solution is to set

$$\theta = 0$$

in the constant temperature case or

$$\frac{\partial \theta}{\partial n} = 0$$

in the adiabatic case. Note that in the constant temperature case, this choice of θ on the boundary would also eliminate the boundary metric variation terms in (15.29).

15.9 Some Representative Results

15.9.1 Redesign of the Boeing 747 Wing

Here the optimization of the wing of the Boeing 747-200 is presented to illustrate the kinds of benefits that can be obtained. In these calculations the flow was modeled by the Reynolds Averaged Navier–Stokes equations. A Baldwin–Lomax turbulence model was considered sufficient, since the optimization is for the cruise condition with attached flow. The calculations were performed to minimize the drag coefficient at a fixed lift coefficient, subject to the additional constraints that the span loading should not be altered, and the thickness should not be reduced. It might be possible to reduce the induced drag by modifying the span loading to an elliptic distribution, but this would increase the root bending moment and consequently require an increase in the skin thickness and structure weight. A reduction in wing thickness would not only reduce the fuel volume, but it would also require an increase in skin thickness to support the bending moment. Thus these constraints assure that there will be no penalty in either structure weight or fuel volume.

Figure 15.2 displays the result of an optimization at a Mach number of 0.86, which is roughly the maximum cruising Mach number attainable by the existing design before the onset of significant drag rise. The lift coefficient of 0.42 is the contribution of the exposed wing. Allowing for the contribution of the fuselage, the total lift coefficient is about 0.47. It can be seen that the redesigned wing is essentially shock-free, and the drag coefficient is reduced from 0.01269 (127 counts) to 0.01136 (114 counts). The total drag coefficient of the aircraft at this lift coefficient is around 270 counts, so this would represent a drag reduction of the order of five percent.

Figure 15.3 displays the result of an optimization at Mach 0.90. In this case the shock waves are not eliminated, but their strength is significantly weakened, while the drag coefficient is reduced from 0.01819 (182 counts) to 0.01293 (129 counts). Thus the redesigned wing has essentially the same drag at Mach 0.9 as the original wing at Mach 0.86. The Boeing 747 wing could apparently be modified to allow such an increase in the cruising Mach number because it has a higher sweep-back than later designs, and a rather thin wing section with a thickness to chord ratio of eight percent.

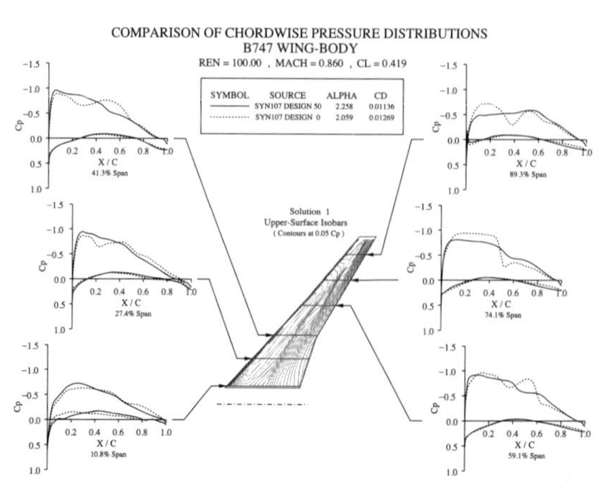

Figure 15.2 Redesigned Boeing 747 wing at Mach 0.86, Cp distributions.

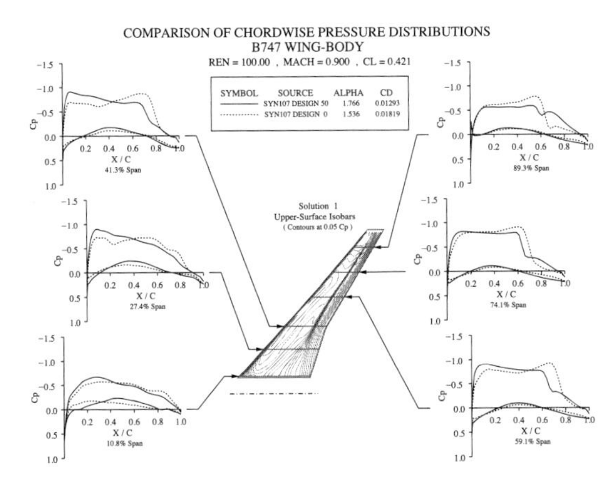

Figure 15.3 Redesigned Boeing 747 wing at Mach 0.90, Cp distributions.

15.9.2 Inverse Design

Figures 15.4–15.7 show the result of an inverse design calculation using the Euler equations, where the initial geometry was a wing made up of NACA 0012 sections, and the target pressure distribution was the pressure distribution over the Onera M6 wing. It can be seen from these plots that the target pressure distribution is well recovered in 50 design cycles, verifying that the design process is capable of recovering pressure distributions that are significantly different from the initial distribution. This is a particularly challenging test because it calls for the recovery of a smooth symmetric profile from an asymmetric pressure distribution containing a triangular pattern of shock waves.

An example of inverse design with the RANS equations uses the wing from an airplane (code named SHARK) (Ahlstrom, Gregg, Vassberg, & Jameson 2000), which was designed for the Reno Air Races. The initial and final pressure distributions are shown in Figure 15.8. As can be seen from these plots, the initial pressure distribution has a weak shock in the outboard sections of the wing, while the final pressure distribution is shock-free. The final pressure distributions are compared with the target distributions along three sections of the wing in Figures 15.9, 15.10, and 15.11. Again the design process captures the target pressure with good accuracy in about 50 design cycles.

15.10 Further Developments

The further development of aerodynamic shape optimization techniques is a topic of widespread ongoing research. Theoretical studies of continuous adjoint methods have been presented by Castro et al. (Castro, Lozano, Palacios, & Zuazua 2007), and Lozano (2012, 2019). Representative examples of state of the art multidisciplinary optimization software include SU2 (Economon, Palacios, Copeland, Lukaczyk, & Alonso 2016) and ADflow (Mader, Kenway, Yildirim, & Martins 2020, Bons & Martins 2020).

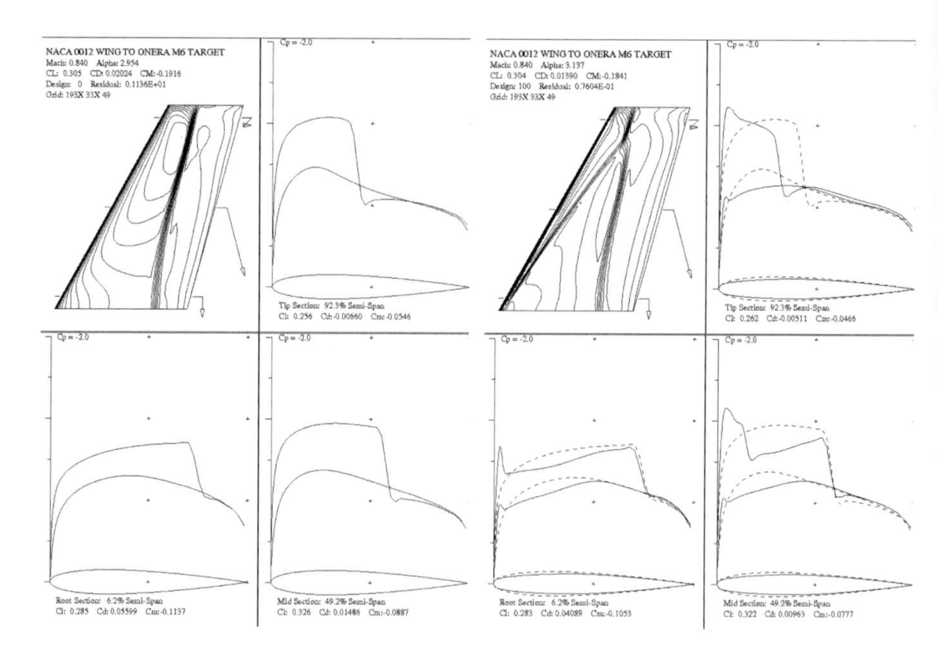

Figure 15.4 Initial geometry with NACA 0012 wing sections and C_p distribution over the planform and at three span stations.

Figure 15.5 Final geometry and C_p distribution over the planform and at three span stations recovering the target C_p distribution.

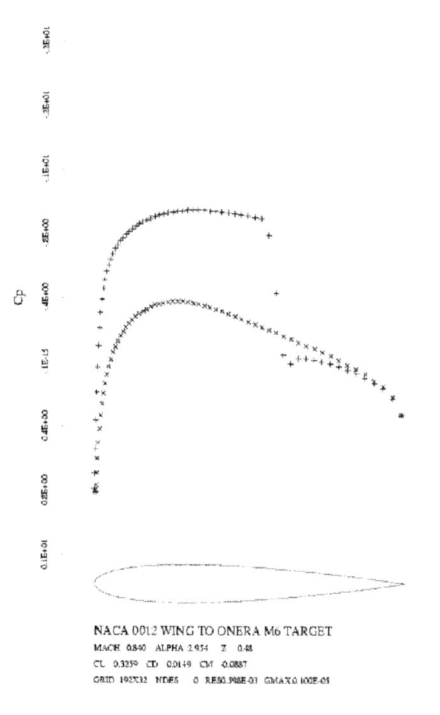

Figure 15.6 NACA 0012 section and C_p distribution of the initial wing at midspan.

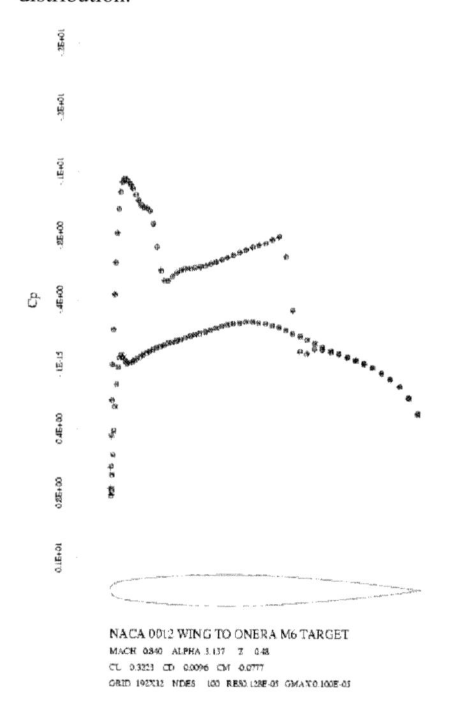

Figure 15.7 Section and C_p distribution of the final wing at midspan with the C_p distribution of the ONERA M6 wing displayed by open circles.

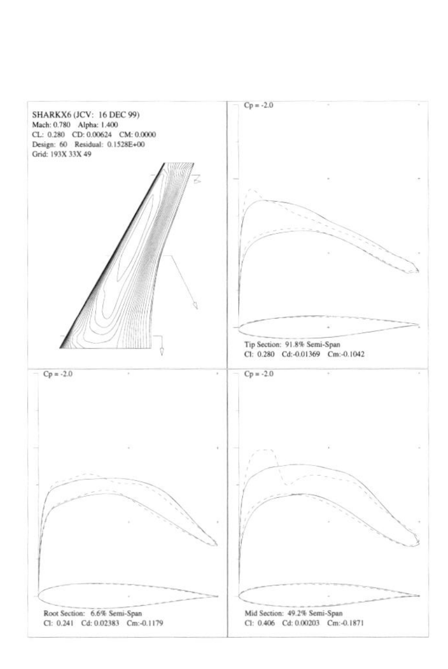

Figure 15.8 Initial and final pressure and section geometries.

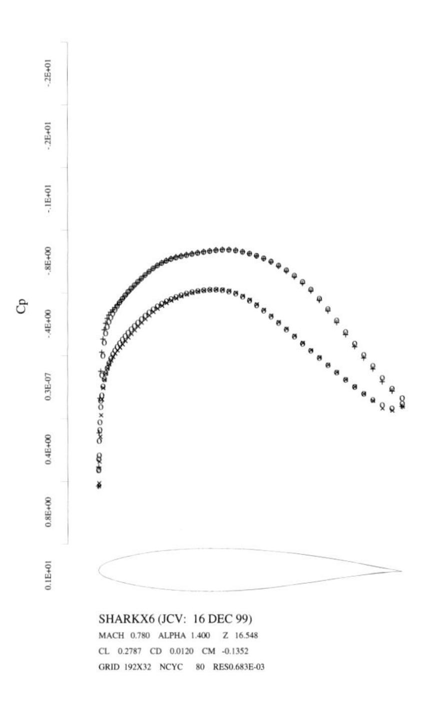

Figure 15.9 Initial and final pressure distributions at 5% of the wing span.

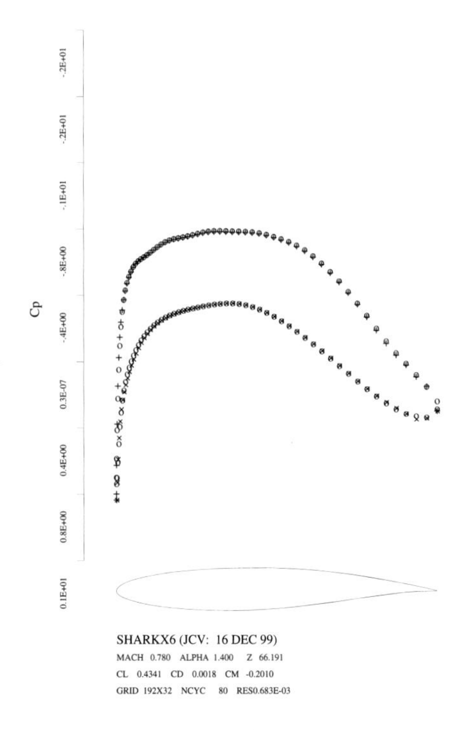

Figure 15.10 Initial and final pressure distributions at 50% of the wing span.

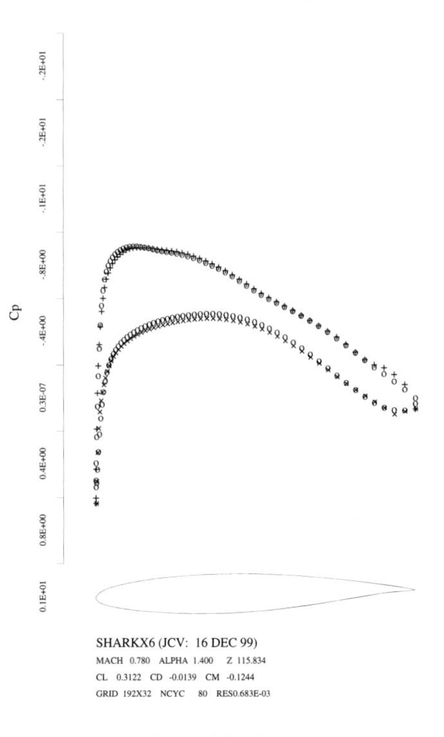

Figure 15.11 Initial and final pressure distributions at 95% of the wing span.

Appendix A Vector and Function Spaces

A.1 Basic Concepts

The well known concepts of Euclidean space can be generalized to n-dimensional vectors and also to functions. These concepts are extremely useful in the analysis of numerical methods and are briefly reviewed in this appendix. The idea of function space was one of the great achievements of nineteenth century mathematics, particularly due to Banach and Hilbert. Hilbert was the first to introduce spaces with an abstract inner product. As a preliminary, it is useful to derive some basic inequalities due to Jensen, Hölder, Cauchy, and Schwarz.

A.2 Convex and Concave Functions: Jensen's Theorem

A function of $f(x)$ of a real variable x is convex if

$$f(tx_1 + (1-t)x_2) \leq tf(x_1) + (1-t)f(x_2)$$

and concave if

$$f(tx_1 + (1-t)x_2) \geq tf(x_1) + (1-t)f(x_2),$$

whenever $0 < t < 1$. It is strictly convex or concave if equality in these expressions implies $x_1 = x_2$.

If f is twice differentiable, it is convex if $f'' \geq 0$ and concave if $f'' \leq 0$, and strictly convex or concave if $f'' > 0$ or $f'' < 0$.

JENSEN'STHEOREM *If $f(x)$ is a concave function,*

$$\sum_{i=1}^{n} t_i f(x_i) \leq f\left(\sum_{i=1}^{n} t_i x_i\right),$$

whenever $t_1, \ldots, t_n \in (0,1)$ and $\sum_{i=1}^{n} t_i = 1$.

Proof This may be proved by induction. It is true for $n = 2$ by the definition of a concave function. Suppose $n \geq 3$ and the assertion holds for smaller values of n. Then, for $i = 2, \ldots, n$, set $t_i' = t_i/(1-t_1)$, so that $\sum_{i=2}^{n} t_i' = 1$. Then,

$$\begin{aligned}
\sum_{i=1}^{n} t_i f(x_1) &= t_1 f(x_1) + (1-t_1)\sum_{i=2}^{n} t_i' f(x_i) \\
&\le t_1 f(x_1) + (1-t_1) f\left(\sum_{i=2}^{n} t_i' x_i\right) \\
&\le f\left(t_1 x_1 + (1-t_1)\sum_{i=2}^{n} t_i' x_i\right) \\
&= f\left(\sum_{i=1}^{n} t_i x_i\right). \qquad \square
\end{aligned}$$

A.3 The Generalized AM-GM Inequality

The geometric mean (GM) of n real numbers does not exceed the arithmetic mean (AM). This result follows from Jensen's theorem applied to the function $\log x$, which is strictly concave. Thus if $p_1, \ldots, p_n > 0$ and $\sum_{i=1}^{n} p_i = 1$,

$$\sum_{i=1}^{n} p_i \log a_i \le \log\left(\sum_{i=1}^{n} p_i a_i\right)$$

and hence

$$\prod_{i=1}^{n} a_i^{p_i} \le \sum_{i=1}^{n} p_i a_i.$$

This is known as the generalized AM-GM inequality. Setting $p_i = \frac{1}{n}$ recovers the basic AM-GM inequality

$$\left(\prod_{i=1}^{n} a_i\right)^{\frac{1}{n}} \le \frac{1}{n}\sum_{i=1}^{n} a_i.$$

A.4 Hölder's Inequality and the Cauchy–Schwarz Inequality

Suppose $p, q > 1$ and $\frac{1}{p} + \frac{1}{q} = 1$. Then

$$\left|\sum_{k=1}^{n} a_k b_k\right| \le \left(\sum_{k=1}^{n} |a_k|^p\right)^{\frac{1}{p}} \left(\sum_{k=1}^{n} |b_k|^q\right)^{\frac{1}{q}}.$$

Set $x_1 = a^p$, $x_2 = b^q$, $p_1 = \frac{1}{p}$, $p_2 = \frac{1}{q}$. By the generalized GM-AM theorem,

$$ab = x_1^{p_1} x_2^{p_2} \le p_1 x_1 + p_2 x_2 = \frac{a^p}{p} + \frac{b^q}{q}.$$

Take vectors such that

$$\sum_{k=1}^{n} |a_k|^p = \sum_{k=1}^{n} |b_k|^q = 1.$$

Then

$$\left|\sum_{k=1}^{n} a_k b_k\right| \leq \sum_{k=1}^{n} |a_k b_k| \leq \sum_{k=1}^{n} \left(\frac{|a_k|^p}{p} + \frac{|b_k|^q}{q}\right) = \frac{1}{p} + \frac{1}{q} = 1.$$

When $p = 2$ this reduces to the Cauchy–Schwartz inequality:

$$\left|\sum_{k=1}^{n} a_k b_k\right| \leq \left(\sum_{k=1}^{n} a_k^2\right)^{\frac{1}{2}} \left(\sum_{k=1}^{n} b_k^2\right)^{\frac{1}{2}}.$$

This may be proved directly by noting that for any value of ρ

$$\sum_{k=1}^{n} (a_k + \rho b_k)(a_k + \rho b_k) \geq 0$$

and consequently

$$\sum_{k=1}^{n} a_k^2 + 2\rho \left|\sum_{k=1}^{n} a_k b_k\right| + \rho^2 \sum_{k=1}^{n} b_k^2 \geq 0.$$

Then setting

$$\rho = -\frac{\left|\sum_{k=1}^{n} a_k b_k\right|}{\sum_{k=1}^{n} b_k^2}$$

and multiplying by $\sum_{k=1}^{n} b_k^2$:

$$\left(\sum_{k=1}^{n} a_k^2\right) \left(\sum_{k=1}^{n} b_k^2\right) \geq \left|\sum_{k=1}^{n} a_k b_k\right|^2.$$

A.5 Vector Norms

The size of an n-dimensional vector can be conveniently represented by a variety of measures. Such measures are called norms, which are required to satisfy the following axioms:

To each vector $\mathbf{x}$ assign a number $||\mathbf{x}||$ where

1. $||\mathbf{x}|| \geq 0$
2. $||\alpha \mathbf{x}|| = |\alpha| ||\mathbf{x}||$ for scalar α
3. $||\mathbf{x} + \mathbf{y}|| \leq ||\mathbf{x}|| + ||\mathbf{y}||$ (triangle inequality)
4. $||\mathbf{x}|| = 0$ if and only if $\mathbf{x} = \mathbf{0}$

Some widely used norms are

- $||\mathbf{x}||_p = \left(\sum |x_i|^p\right)^{\frac{1}{p}}$ for $p = 1, 2, 3, \ldots \infty$
- $||\mathbf{x}||_1 = \sum |x_i|$
- $||\mathbf{x}||_2 = \left(\sum |x_i|^2\right)^{\frac{1}{2}}$
- $||\mathbf{x}||_\infty = \max_i |x_i|$

Axioms (1) and (2) are evident. For (3):

- $||\mathbf{x}+\mathbf{y}||_1 = \sum |x_i + y_i| \leq \sum (|x_i| + |y_i|) \leq ||\mathbf{x}||_1 + ||\mathbf{y}||_1$
- $||\mathbf{x}+\mathbf{y}||_\infty = \max_i |x_i + y_i| \leq \max_i |x_i| + \max_i |y_i| \leq ||\mathbf{x}||_\infty + ||\mathbf{y}||_\infty$

To verify (3) for $||\mathbf{x}||_2$, use the Cauchy–Schwartz inequality

$$|\mathbf{x}^T\mathbf{y}| \leq ||\mathbf{x}||\ ||\mathbf{y}||.$$

Then

$$\begin{aligned}||\mathbf{x}+\mathbf{y}||_2 &= \left((\mathbf{x}+\mathbf{y})^T(\mathbf{x}+\mathbf{y})\right)^{\frac{1}{2}} \\ &\leq \left(\mathbf{x}^T\mathbf{x} + 2|\mathbf{y}^T\mathbf{x}| + \mathbf{y}^T\mathbf{y}\right)^{\frac{1}{2}} \leq \left(\mathbf{x}^T\mathbf{x} + 2(\mathbf{x}^T\mathbf{x})^{\frac{1}{2}}(\mathbf{y}^T\mathbf{y})^{\frac{1}{2}} + \mathbf{y}^T\mathbf{y}\right)^{\frac{1}{2}} \\ &= ||\mathbf{x}||_2 + ||\mathbf{y}||_2.\end{aligned}$$

A.5.1 Minkowski's Inequality – Triangle Inequality for the *p* Norm

$$||\mathbf{x}+\mathbf{y}||_p \leq ||\mathbf{x}||_p + ||\mathbf{y}||_p.$$

Proof

$$\sum |x_k + y_k|^p \leq \sum |x_k + y_k|^{p-1}|x_k| + \sum |x_k + y_k|^{p-1}|y_k|.$$

Then by Hölder's inequality:

$$\begin{aligned}\sum |x_k + y_k|^p &\leq \left(\sum |x_k + y_k|^{(p-1)q}\right)^{\frac{1}{q}} \left(\left(\sum |x_k|^p\right)^{\frac{1}{p}} + \left(\sum |y_k|^p\right)^{\frac{1}{p}}\right) \\ &= \left(\sum |x_k + y_k|^p\right)^{\frac{1}{q}} \left(\left(\sum |x_k|^p\right)^{\frac{1}{p}} + \left(\sum |y_k|^p\right)^{\frac{1}{p}}\right)\end{aligned}$$

since $p - 1 = \frac{p}{q}$. Divide both sides by

$$\left(\sum |x_k + y_k|^p\right)^{\frac{1}{q}}$$

to obtain the triangle inequality.

A.5.2 Equivalence of Norms for Finite-Dimensional Vectors

For finite-dimensional vectors, all norms are equivalent in the sense that for any two norms $M(\mathbf{x})$ and $N(\mathbf{x})$, there exist constants c_1, c_2 such that for all $\mathbf{x}$,

$$c_1 M(\mathbf{x}) \leq N(\mathbf{x}) \leq c_2 M(\mathbf{x})$$

$$\frac{1}{c_2} N(\mathbf{x}) \leq M(\mathbf{x}) \leq \frac{1}{c_1} N(\mathbf{x}).$$

This need only be proved for $M(\mathbf{x}) = N_\infty(\mathbf{x})$ since if M and N are both equivalent to N_∞ then they are also equivalent to each other. This may be verified as follows. Suppose that

$$c_1 N_\infty \leq M \leq c_2 N_\infty$$

and

$$d_1 N_\infty \leq N \leq d_2 N_\infty.$$

Then

$$c_1 N \leq d_2 M$$

and

$$d_1 M \leq c_2 N,$$

whence

$$\frac{c_1}{d_2} N \leq M \leq \frac{c_2}{d_1} N.$$

Consider the unit ball S of vectors for which $N_s(\mathbf{x}) = 1$. Let $\mathbf{x}_0$ and $\mathbf{x}_1$ be elements such that

$$N(\mathbf{x}_0) = \min_{\mathbf{x} \in S} N(\mathbf{x}), \; N(\mathbf{x}_1) = \max_{\mathbf{x} \in S} N(\mathbf{x}).$$

(These exist since $N(\mathbf{x})$ is a continuous function of the elements of $\mathbf{x}$.)

Now for any $\mathbf{y}$, $\dfrac{\mathbf{y}}{N_\infty(\mathbf{y})}$ is in S, so

$$N(\mathbf{x}_0) \leq N\left(\frac{\mathbf{y}}{N_\infty(\mathbf{y})}\right) \leq N(\mathbf{x}_1)$$

or

$$N(\mathbf{x}_0) N_\infty(\mathbf{y}) \leq N(\mathbf{y}) \leq N(\mathbf{x}_1) N_\infty(\mathbf{y}).$$

This is the desired result with

$$c_1 = N(\mathbf{x}_0),\ c_2 = N(\mathbf{x}_1).$$

A.6 Matrix Norms

The induced norm of a matrix is defined as $||A|| = \sup\limits_{\mathbf{x}\neq 0} \frac{||A\mathbf{x}||}{||\mathbf{x}||}$. $||A\mathbf{x}||$ is a continuous function of $||\mathbf{x}||$, so with $||\mathbf{x}|| = 1$,

$$||A|| = \sup ||A\mathbf{x}||,$$

where the maximum is attained.

Let $\mathbf{x}$ be such that with $||\mathbf{x}|| = 1$,

$$||A + B|| = ||(A + B)\mathbf{x}|| \leq ||A\mathbf{x}|| + ||B\mathbf{x}|| \leq ||A|| + ||B||.$$

Thus the triangle inequality is satisfied. Also note that

$$||AB\mathbf{x}|| \leq ||A||\ ||B\mathbf{x}|| \leq ||A||\ ||B||\ ||\mathbf{x}||$$

so that

$$||AB|| \leq ||A||\ ||B||.$$

A.6.1 Infinity Norm

Corresponding to $||\mathbf{x}||_\infty$ we have

$$||A||_\infty = \max_i \sum_j |a_{ij}| \text{ (max absolute row sum)}$$

since

$$\begin{aligned} ||A\mathbf{x}||_\infty = \max_i \left| \sum_j a_{ij}x_j \right| &\leq \max_i \left| \sum_j a_{ij} \right| \max_j |x_j| \\ &\leq \max_i \left| \sum_j a_{ij} \right| ||\mathbf{x}||_\infty \end{aligned}$$

where the max is attained with $x_j = \text{sgn}(a_{ij})$.

A.6.2 1 Norm

Corresponding to $||\mathbf{x}||_1$ we have

$$||A||_1 = \max_j \sum_i |a_{ij}| \text{ (max absolute column sum)}$$

since

$$
\begin{aligned}
||A\mathbf{x}||_1 &= \sum_i \left| \sum_j a_{ij} x_j \right| \leq \sum_i \sum_j |a_{ij}||x_j| \\
&= \sum_j \left(\sum_i |a_{ij}| \right) |x_j| \leq \max_j \left(\sum_i |a_{ij}| \right) \sum_j |x_j| \\
&= \max_j \left(\sum_i |a_{ij}| \right) ||\mathbf{x}||_1,
\end{aligned}
$$

with the maximum attained when $\mathbf{x} = e_j$, where j is the index of the max sum.

A.6.3 2 Norm

Corresponding to $||\mathbf{x}||_2$ we have

$$
\frac{||A\mathbf{x}||_2^2}{||\mathbf{x}||_2^2} = \frac{\mathbf{x}^H A^H A\mathbf{x}}{\mathbf{x}^H \mathbf{x}},
$$

where A^H is the Hermitian transpose of A.

Let u_i be an eigenvector of $A^H A$ with real eigenvalue σ_i^2. Set $\mathbf{x} = \sum \alpha_i u_i$, then

$$
\frac{\mathbf{x}^H A^H A\mathbf{x}}{\mathbf{x}^H \mathbf{x}} = \frac{\sum \alpha_i^2 \sigma_i^2}{\sum \alpha_i^2} \leq \sigma_1^2,
$$

where σ_1^2 is the largest eigenvalue of $A^H A$. The decomposition is possible because the u_i are independent.

A.7 Norms of Functions

A.7.1 Function Space

In order to consider errors, we need to measure differences between functions. It is convenient to regard functions as vectors in an infinite-dimensional space. It turns out that many of the familiar concepts of three-dimensional Euclidean geometry carry over to function spaces. In particular we can introduce norms to measure the size of a function and an abstract definition of a generalized inner product of two functions, corresponding to the scalar product of two vectors. This leads to the concept of orthogonal functions. The introduction of axiomatic definitions leads to proofs that hold for a variety of different norms and inner products. In order to enable standard arguments of real analysis to be carried over, function spaces are required to satisfy an axiom of completeness: that the limit of a sequence of elements in a given space is contained in the space. Spaces with an inner product are known as Hilbert spaces in honor of their inventor.

A.7.2 Norms

Norms will be a measure of the size of a function, regarded as a vector. For example

$$||f|| = \left(\int_0^1 f^2 dx\right)^{\frac{1}{2}}$$

is a generalization of

$$||f|| = \left(\sum_1^3 f_i^2 dx\right)^{\frac{1}{2}}$$

for Euclidean space.

To be more precise, we require the norm of a function to satisfy the following axioms:

1. $||f|| \geq 0$
2. $||\alpha f|| = |\alpha|||f||$ for scalar α
3. $||f+g|| \leq ||f|| + ||g||$
4. $||f|| = 0$ if and only if $f = 0$

If a function is given by a table

$$f_i = f(x_i) \text{ for } i = 0, 1, 2, \ldots, n,$$

it may be convenient to use as a measure

$$||f|| = \left(\sum_0^n f_i^2 dx\right)^{\frac{1}{2}}.$$

Then we can have

$$||f|| = 0 \text{ for } f \neq 0.$$

This measure satisfies the first three axioms only. Such a measure is called a **semi-norm**.

Examples of norms are:

$$||f||_\infty = \max_{x\in[a,b]} |f(x)| \quad \text{(Maximum norm)}$$

$$||f||_2 = \left(\int_a^b f^2(x)dx\right)^{\frac{1}{2}} \quad \text{(Euclidean norm)}$$

$$||f||_1 = \int_a^b |f(x)|dx.$$

These are special cases of

$$||f||_p = \left(\int_a^b |f(x)|^p dx\right)^{\frac{1}{p}} \text{ for } p \geq 1.$$

Note that, for example, we can regard $||f||_2$ as the limit of

$$||f|| = \left(\frac{1}{n}\sum_{0}^{n} f_i^2 dx\right)^{\frac{1}{2}},$$

where f_i is the table of values representing $f(x)$.

We can also introduce weighted norms such as the weighted Euclidean norm

$$||f|| = \left(\int_a^b f^2(x)w(x)dx\right)^{\frac{1}{2}},$$

where $w(x)$ is a non-negative weight function.

All these norms can be shown to satisfy the axioms.

A.7.3 Non-equivalence of Norms for Functions

For functions, norms are not equivalent. If $p > q$ then

$$||f||_p \to 0 \text{ implies } ||f||_q \to 0$$

but not the other way round.

For example, consider

$$f_n = \begin{cases} 1 & \text{if } -\frac{1}{2n} \le x \le \frac{1}{2n} \\ 0 & \text{otherwise.} \end{cases}$$

Then

$$\int_{-1}^{1} f_n^2 dx = \frac{1}{n}$$

and

$$||f_n||_2 = \frac{1}{\sqrt{n}} \to 0 \text{ as } n \to \infty$$

while

$$||f_n||_\infty = 1.$$

On the other hand if

$$||f||_\infty = \epsilon$$

then

$$\int_a^b f^2 dx \le \epsilon^2(b-a)$$

so

$$||f||_2 \le \epsilon\sqrt{b-a} \to 0 \text{ as } \epsilon \to 0.$$

A.8 Inner Products of Functions: Orthogonality

A.8.1 Orthogonality

It is also convenient to introduce the idea of orthogonality in function space. Given a nonnegative weight function $w(x)$ define the inner product as

$$\begin{aligned}(f,g) &= \int_a^b f(x)\, g(x)\, w(x)\, dx \quad \text{in the continuous case} \\ &= \sum_{i=0}^{n} f_i\ g_i\ w_i \quad \text{in the discrete case.}\end{aligned}$$

Then if (f,g) vanishes, the functions are said to be orthogonal.

The inner product satisfies the following conditions:

$$\begin{aligned}&(f,g) = (g,f) && \text{(commutativity)} \\ &(c_1 f + c_2 g, \phi) = c_1(f,\phi) + c_2(g,\phi) && \text{(linearity)} \\ &(f,f) \geq 0 && \text{(positivity).}\end{aligned}$$

These can be taken as axioms for an abstract inner product.

A.8.2 Triangle Inequality for the ∞ and 1 Norms

We have

$$\max|f+g| \leq \max(|f|+|g|) \leq \max|f| + \max|g|$$

and

$$\int |f+g|dx \leq \int (|f|+|g|)dx.$$

Thus the ∞ and 1 norms satisfy the triangle inequality. They also already satisfy axioms (1), (2), and (4).

A.8.3 Cauchy–Schwarz Inequality

$$(f,g)^2 \leq (f,f)(g,g).$$

Proof

$$0 \leq (f+\rho g, f+\rho g) \leq (f,f) + 2\rho|(f,g)| + \rho^2(g,g).$$

But since this is a quadratic function of ρ, it has imaginary roots or coincident real roots. The discriminant yields the desired inequality.

Or set

$$\rho = -\frac{|(f,g)|}{(g,g)}$$

and multiply by (g,g) to get

$$0 \leq (f,f)\,(g,g) - 2|(f,g)|^2 + |(f,g)|^2. \qquad \square$$

A.8.4 Triangle Inequality for the Euclidean Norm

$$
\begin{aligned}
||f+g||_2 &= (f+g, f+g)^{\frac{1}{2}} \\
&\le ((f,f) + 2|(f,g)| + (g,g))^{\frac{1}{2}} \\
&\le \left((f,f) + 2(f,f)^{\frac{1}{2}}(g,g)^{\frac{1}{2}} + (g,g)\right)^{\frac{1}{2}} \\
&= ||f|| + ||g||
\end{aligned}
$$

using the Schwartz inequality.

A.8.5 Pythagorean Theorem

In the Euclidean norm, if $(f,g) = 0$ then

$$||f+g||^2 = ||f||^2 + ||g||^2.$$

Proof

$$||f+g||^2 = (f+g, f+g) = (f,f) + (f,g) + (g,f) + (g,g). \qquad \square$$

A.8.6 Linear Independence

A set of functions $\phi_0, \phi_1, \ldots, \phi_n$ is said to be linearly independent if

$$\left|\left|\sum c_i\, \phi_i\right|\right| = 0$$

implies

$$c_i = 0 \text{ for } i = 0, 1, \ldots, n.$$

Orthogonal functions are independent since then from the Pythagorean theorem

$$\left|\left|\sum_{i=0}^{n} c_i\, \phi_i\right|\right|^2 = \sum_{i=0}^{n} c_i^2 ||\phi_i||^2.$$

A.8.7 Weierstrass Theorem

Let $f(x)$ be given in interval $[a,b]$. Let the lower bound of the error in the maximum norm for all polynomials of order n be

$$E_n(f) = \min_{p_n(x)} ||f - p_n(x)||_\infty.$$

Then if f is continuous,

$$\lim_{n\to\infty} E_n(f) = 0.$$

That is, a continuous function can be arbitrarily well approximated in a closed interval by a polynomial of sufficiently high order.

The proof is by construction of the required $p_n(x)$ (Isaacson and Keller, p183).

A.9 Portraits of Courant and Hilbert

Figure A.1 Richard Courant (1888–1972).

Figure A.2 David Hilbert (1862–1943).

Appendix B Approximation Theory

B.1 Least Squares Approximation and Projection

Suppose that we wish to approximate a given function f by a linear combination of independent basis functions $\phi_j, j = 0, .., n$. In general we wish to choose coefficients c_j to minimize the error

$$\left\| f - \sum_{j=0}^{n} c_j \phi_j \right\|$$

in some norm. Ideally we might use the infinity norm, but it turns out that it is computationally very expensive to do this, requiring an iterative process. The best known method for finding the best approximation in the infinity norm is the exchange algorithm (Powell 1981). On the other hand, it is very easy to calculate the best approximation in the Euclidean norm as follows. Choose c_j to minimize the least squares integral

$$J(c_0, c_1, c_2, \ldots c_n) = \int \left(\sum_{j=0}^{n} c_j \phi_j - f \right)^2 dx,$$

equivalent to minimizing the Euclidean norm. Then we require

$$0 = \frac{\partial J}{\partial c_i} = \int 2 \left(\sum_{j=0}^{n} c_j \phi_j - f \right) \phi_i dx.$$

Thus the best approximation is

$$f^* = \sum_{j=0}^{n} c_j^* \phi_j,$$

where

$$\left(\sum_{j=0}^{n} c_j^* \phi_j - f, \phi_i \right) = (f^* - f, \phi_i) = 0, \text{ for } i = 0, 1, 2, \ldots, n$$

or

$$\sum_{j=0}^{n} a_{ij}c_j^* = b_i,$$

where

$$a_{ij} = (\phi_i, \phi_j),\ b_i = (f, \phi_i).$$

Notice that since f^* is a linear combination of the ϕ_j, these equations state that

$$(f - f^*, f^*) = 0.$$

Thus the least squares approximation f^* is orthogonal to the error $f - f^*$. This means that f^* is the projection of f onto the space spanned by the basis functions ϕ_j, in the same way that a three-dimensional vector might be projected onto a plane.

If the ϕ_j form an orthonormal set,

$$a_{ij} = \delta_{ij},$$

where $\delta_{ij} = 1$ if $i = j$ and 0 if $i \neq j$, and we have directly

$$c_i^* = (f, \phi_i).$$

B.2 Alternative Proof of Least Squares Solution

Let

$$f^* = \sum c_j^* \phi_j$$

be the best approximation, and consider another approximation

$$\sum c_j \phi_j.$$

Then if $f - f^*$ is orthogonal to all ϕ_j,

$$\begin{aligned}\left|\left|\sum c_j\phi_j - f\right|\right|^2 &= \left|\left|\sum (c_j - c_j^*)\phi_j - (f - f^*)\right|\right|^2 \\ &= \left|\left|\sum (c_j - c_j^*)\phi_j\right|\right|^2 + \left|\left|(f - f^*)\right|\right|^2 \\ &\geq \left|\left|(f - f^*)\right|\right|^2.\end{aligned}$$

B.3 Bessel's Inequality

Since f^* is orthogonal to $f - f^*$, the Pythagorean theorem gives

$$||f||^2 = ||f^* + f - f^*||^2 = ||f^*||^2 + ||f - f^*||^2.$$

Also if the ϕ_j are orthonormal,

$$||f^*||^2 = \sum c_j^{*2}$$

so we have Bessel's inequality

$$\sum c_j^{*2} \leq ||f||^2.$$

Thus the series $\sum c^{*2}_j$ is convergent, and if $||f^* - f|| \to 0$ as $n \to \infty$ we have Parseval's formula

$$\sum c_j^{*2} = ||f||^2.$$

This is the case for orthogonal polynomials because of the Weiesstrass theorem.

B.4 Orthonormal Systems

The functions ϕ_i form an orthonormal set if they are orthogonal and $||\phi_i|| = 1$. We can write

$$(\phi_i, \phi_j) = \delta_{ij},$$

where $\delta_{ij} = 1$ if $i = j$ and 0 if $i \neq j$.

An example of an orthogonal set is

$$\phi_j = \cos(jx), \text{ for } j = 0, 1, \ldots, n$$

on the ineral $[0, \pi]$. Then if $j \neq k$,

$$(\phi_j, \phi_k) = \int_0^{\pi} \cos(jx)\cos(kx)\, dx = \int_0^{\pi} \frac{1}{2}\left(\cos(j-k)x + \cos(j+k)x\right)\, dx = 0.$$

Also

$$(\phi_j, \phi_j) = \int_0^{\pi} \cos^2(jx) dx = \int_0^{\pi} \frac{1}{2}\left(1 + \cos(2jx)\right)\, dx = \frac{\pi}{2}.$$

B.5 Construction of Orthogonal Functions

This can be done by Gram Schmidt orthogonalization. Define the inner product

$$(f, g) = \int_a^b f(x)g(x)dx.$$

Functions g_i are called independent if

$$\sum_{i=0}^{n} \alpha_i g_i = 0 \text{ implies } \alpha_i = 0 \text{ for } i = 0, \ldots, n.$$

Let $g_i(x), i = 0, 1, \ldots, n$, be any set of independent functions. Then set

$$
\begin{aligned}
f_0(x) &= d_0 g_0(x) \\
f_1(x) &= d_1 \left(g_1(x) - c_{01} f_0(x)\right) \\
&\vdots \\
f_n(x) &= d_n \left(g_n(x) - c_{0n} f_0(x) \ldots - c_{n-1,n} f_{n-1}(x)\right).
\end{aligned}
$$

We require

$$(f_i, f_j) = \delta_{ij}.$$

This gives

$$d_0 = \frac{1}{\sqrt{(g_0, g_0)}}.$$

Then $0 = (f_0, f_1) = d_1 \left((f_0, g_1) - c_{01}\right)$, and in general

$$c_{jk} = (f_j, g_k),$$

while the d_k are easily obtained from

$$(f_k, f_k) = 1.$$

B.6 Orthogonal Polynomials

A similar approach can be used to construct orthogonal polynomials. Given orthogonal polynomials, $\phi_j(x)$, $j = 0, 1, \ldots, n$, we can generate a polynomial of degree $n + 1$ by multiplying $\phi_n(x)$ by x, and we can make it orthogonal to the previous $\phi_j(x)$ by subtracting a linear combination of these polynomials. Thus, we set

$$\phi_{n+1}(x) = \alpha_n x \phi_n(x) - \sum_{i=0}^{n} c_{ni} \phi_i(x).$$

Now,

$$\alpha_n (x\phi_n, \phi_j) - \sum_{i=0}^{n} c_{ni} (\phi_i, \phi_j) = 0.$$

But $(\phi_i, \phi_j) = 0$ for $i \neq j$, so

$$c_{nj} ||\phi_j||^2 = \alpha_n (x\phi_n, \phi_j) = \alpha_n (\phi_n, x\phi_j).$$

Since $x\phi_j$ is a polynomial of order $j + 1$, this vanishes except for $j = n - 1, n$. Thus

$$\phi_{n+1}(x) = \alpha_n x \phi_n(x) - c_{nn} \phi_n(x) - c_{n,n-1} \phi_{n-1}(x),$$

where

$$c_{nn} = \frac{\alpha_n(\phi_n, x\phi_n)}{||\phi_n||^2}, \quad c_{n,n-1} = \frac{\alpha_n(\phi_n, x\phi_{n-1})}{||\phi_{n-1}||^2}$$

and α_n is arbitrary.

B.7 Zeros of Orthogonal Polynomials

An orthogonal polynomial $P_n(x)$ for the interval $[a,b]$ has n distinct zeros in $[a,b]$. Suppose it had only $k < n$ zeros $x_1, x_2, \ldots, x_k$ at which $P_n(x)$ changes sign. Then consider

$$Q_k(x) = (x - x_1) \ldots (x - x_k),$$

where for $k = 0$ we define $Q_0(x) = 1$. Then $Q_k(x)$ changes sign at the same points as $P_n(x)$, and

$$\int_a^b P_n(x)Q_k(x)w(x)dx \neq 0.$$

But this is impossible since $P_n(x)$ is orthogonal to all polynomials of degree $< n$.

B.8 Legendre Polynomials

These are orthogonal in the interval $[-1,1]$ with a constant weight function $w(x) = 1$. They can be generated from Rodrigues' formula:

$$L_0(x) = 1$$

$$L_n(x) = \frac{1}{2^n n!}\frac{d^n}{dx^n}\left[(x^2-1)^n\right], \text{ for } n = 1, 2, \ldots.$$

The first few are

$$L_0 = 1$$

$$L_1(x) = x$$

$$L_2(x) = \frac{1}{2}(3x^2 - 1)$$

$$L_3(x) = \frac{1}{2}(5x^3 - 3x)$$

$$L_4(x) = \frac{1}{8}(35x^4 - 30x^2 + 3)$$

$$L_5(x) = \frac{1}{8}(63x^5 - 70x^3 + 15x),$$

where they can be successively generated for $n \geq 2$ by the recurrence formula

$$L_{n+1} = \frac{2n+1}{n+1} x L_n(x) - \frac{n}{n+1} L_{n-1}(x).$$

They are alternately odd and even and are normalized so that

$$L_n(1) = 1\ ,\ L_n(-1) = (-1)^n,$$

while for $|x| < 1$,

$$|L_n(x)| < 1.$$

Also

$$(L_n, L_j) = \int_{-1}^{1} L_n(x) L_j(x) dx = \begin{cases} 0 & \text{if } n \neq j \\ \frac{2}{2n+1} & \text{if } n = j \end{cases}$$

Writing the recurrence relation as

$$L_n(x) = \frac{2n-1}{n} x L_{n-1}(x) - \frac{n-1}{n} L_{n-2}(x),$$

it can be seen that the leading coefficient of $L_n(x)$ is multiplied by $(2n-1)/n$ when $L_n(x)$ is generated. Hence the leading coefficient is

$$a_n = \frac{1 \cdot 3 \cdot 5 \cdot \ldots \cdot (2n-1)}{1 \cdot 2 \cdot 3 \cdot \ldots \cdot n} = \frac{1}{2^n} \frac{2n!}{(n!)^2}.$$

$L_n(x)$ also satisfies the differential equation

$$\frac{d}{dx}\left[\left(1-x^2\right)\frac{d}{dx} L_n(x)\right] + n(n+1) L_n(x) = 0.$$

B.9 Chebyshev Polynomials

Let

$$x = \cos\ \phi,\ 0 \leq \phi \leq \pi$$

and define

$$T_n(x) = \cos(n\phi)$$

in the interval $[-1,1]$. Since

$$\cos(n+1)\phi + \cos(n-1)\phi = 2\cos\phi \cos n\phi,\ n \geq 1,$$

we have

$$\begin{aligned}T_0(x) &= 1\\ T_1(x) &= x\\ &\vdots\\ T_{n+1}(x) &= 2xT_n(x) - T_{n-1}(x).\end{aligned}$$

These are the Chebyshev polynomials.
Property 1:
Leading coefficient is 2^{n-1} for $n \geq 1$.

Property 2:

$$T_n(-x) = (-1)^n T_n(x).$$

Property 3:
$T_n(x)$ has n zeros at

$$x_k = \cos\left(\frac{2k-1}{n}\frac{\pi}{2}\right) \text{ for } k = 1, \ldots, n$$

and $n+1$ extrema in $[-1,1]$ with the values $(-1)^k$, at

$$x'_k = \cos\frac{k\pi}{n} \text{ for } k = 0, \ldots, n.$$

Property 4 (Continuous orthogonality):
They are orthogonal in the inner product

$$(f,g) = \int_{-1}^{1} f(x)g(x)\frac{dx}{\sqrt{1-x^2}}$$

since

$$(T_r, T_s) = \int_0^{\pi} \cos r\theta \cos s\theta \, d\theta = \begin{cases} 0 & \text{if } r \neq s \\ \frac{\pi}{2} & \text{if } r = s \neq 0 \\ \pi & \text{if } r = s = 0 \end{cases}$$

Property 5 (Discrete orthogonality):
They are orthogonal in the discrete inner product

$$(f,g) = \sum_{j=0}^{n} f(x_j)g(x_j),$$

where x_j are the zeros of $T_{n+1}(x)$ and

$$\arccos x_j = \theta_j = \frac{2j+1}{2n+1}\frac{\pi}{2}, \; j = 0, 1, \ldots, n.$$

Then

$$(T_r, T_s) = \sum_{j=0}^{n} \cos r\theta_j \cos s\theta_j = \begin{cases} 0 & \text{if } r \neq s \\ \frac{n+1}{2} & \text{if } r = s \neq 0 \\ n+1 & \text{if } r = s = 0 \end{cases}$$

Property 6 (Minimax):
$\left(1/2^{n-1}\right) T_n(x)$ has smallest maximum norm in $[-1, 1]$ of all polynomials with leading coefficient unity.

Proof Suppose $||p_n(x)||_\infty$ were smaller. Now at the $n+1$ extrema $x_0, \ldots, x_n$ of $T_n(x)$, which all have the same magnitude of unity,

$$p_n(x_n) < \frac{1}{2^{n-1}} T_n(x_n)$$
$$p_n(x_{n-1}) > \frac{1}{2^{n-1}} T_n(x_{n-1})$$
$$\vdots$$

Thus $p_n(x) - \left(1/2^{n-1}\right) T_n(x)$ changes sign n times in $[-1, 1]$, but this is impossible since it is of degree $n-1$ and has $n-1$ roots. □

B.10 Best Approximation Polynomial

Let $f(x) - P_n(x)$ have maximum deviations, $+e$, $-e$, alternately at $n+2$ points $x_0, \ldots, x_{n+1}$ in $[a, b]$. Then $P_n(x)$ minimizes $||f(x) - P_n(x)||_\infty$.

Proof Suppose $||f(x) - Q_n(x)||_\infty < |e|$; then we have at $x_0, \ldots, x_{n+1}$

$$Q_n(x) - P_n(x) = f(x) - P_n(x) - (f(x) - Q_n(x))$$

has the same sign as $f(x) - P_n(x)$. Thus it has opposite sign at $n+2$ points, giving $n+1$ sign changes. But this is impossible since it has only n roots.

B.11 Jacobi Polynomials

Jacobi polynomials $P_n^{(\alpha,\beta)}(x)$ are orthogonal in the interval $[-1, 1]$ for the weight function

$$w(x) = (1-x)^\alpha (1+x)^\beta.$$

Thus the family of Jacobi polynomials includes both the Legendre polynomials

$$L_n(x) = P_n^{(0,0)}(x)$$

and the Chebyshev polynomials

$$T_n(x) = P_n^{\left(-\frac{1}{2},-\frac{1}{2}\right)}(x).$$

B.12 Fourier Series

Let $f(\theta)$ be periodic with period 2π and square integrable on $[0, 2\pi]$. Consider the approximation of $f(\theta)$

$$S_n(\theta) = \frac{1}{2}a_0 + \sum_{r=1}^{n}(a_r \cos r\theta + b_r \sin r\theta).$$

Let us minimize

$$||f - S_n||_2 = \left\{ \int_0^{2\pi} \left(f(\theta) - S_n(\theta)\right)^2 d\theta \right\}^{\frac{1}{2}}.$$

Let

$$J(a_0, a_1, \ldots, a_n, b_1, \ldots, b_n) = ||f - S_n||^2.$$

The trigonometric functions satisfy the orthogonality relations

$$\int_0^{2\pi} \cos(r\theta)\cos(s\theta)d\theta = \begin{cases} 0 & \text{if } r \neq s \\ \pi & \text{if } r = s \neq 0 \end{cases} \tag{B.1}$$

$$\int_0^{2\pi} \sin(r\theta)\sin(s\theta)d\theta = \begin{cases} 0 & \text{if } r \neq s \\ \pi & \text{if } r = s \neq 0 \end{cases} \tag{B.2}$$

$$\int_0^{2\pi} \sin(r\theta)\cos(s\theta)d\theta = 0. \tag{B.3}$$

At the minimum, allowing for these,

$$0 = \frac{\partial J}{\partial a_r} = -2\int_0^{2\pi} f(\theta)\cos(r\theta)d\theta + 2a_r \int_0^{2\pi} \cos^2(r\theta)d\theta$$

whence

$$a_r = \frac{1}{\pi}\int_0^{2\pi} f(\theta)\cos(r\theta)d\theta,$$

and similarly

$$b_r = \frac{1}{\pi}\int_0^{2\pi} f(\theta)\sin(r\theta)d\theta.$$

Also since $||f - S_n||^2 \geq 0$, we obtain Bessel's inequality

$$\frac{1}{2}a_0^2 + \sum_{r=1}^{n}\left(a_r^2 + b_r^2\right) \leq \frac{1}{\pi}\int_0^{2\pi} f^2(\theta)d\theta,$$

and since the right hand side is independent of n, the sum converges and

$$\lim_{r\to\infty} |a_r| = 0, \quad \lim_{r\to\infty} |b_r| = 0.$$

Also

$$\lim_{n\to\infty} ||f - S_n||^2 = 0.$$

If not, it would violate Weierstrass's theorem translated to trigonometric functions (Isaacson and Keller, p 230, p 198).

B.13 Error Estimate for Fourier Series

Let $f(\theta)$ have K continuous derivatives. Then the Fourier coefficients decay as

$$|a_r| = \mathcal{O}\left(\frac{1}{r^K}\right), \quad |b_r| = \mathcal{O}\left(\frac{1}{r^K}\right)$$

and

$$|f(\theta) - S_n(\theta)| = \mathcal{O}\left(\frac{1}{n^{K-1}}\right).$$

Integrating repeatedly by parts:

$$\begin{aligned} a_r &= \frac{1}{\pi}\int_0^{2\pi} f(\theta)\cos(r\theta)d\theta \\ &= -\frac{1}{r\pi}\int_0^{2\pi} f'(\theta)\sin(r\theta)d\theta \\ &= \frac{1}{r^2\pi}\int_0^{2\pi} f''(\theta)\cos(r\theta)d\theta \\ &\vdots \end{aligned}$$

Let $M = \sup|f^{(K)}(\theta)|$. Then since $|\cos(r\theta)| \le 1$ and $|\sin(r\theta)| \le 1$,

$$|a_r| \le \frac{2M}{r^K}$$

and similarly

$$|b_r| \le \frac{2M}{r^K}.$$

Also

$$\begin{aligned} |S_m(\theta) - S_n(\theta)| &\le \left|\sum_{r=n+1}^{m} (a_r \cos r\theta + b_r \sin r\theta)\right| \\ &\le \sum_{r=n+1}^{m} \frac{4M}{r^K} \end{aligned}$$

$$\leq 4M \int_n^\infty \frac{d\xi}{\xi^K}$$
$$= \frac{4M}{K-1} \frac{1}{n^{K-1}},$$

where the integral contains the sum of the rectangles shown in the sketch. Thus the partial sums $S_n(\theta)$ form a Cauchy sequence and converge uniformly to $f(\theta)$, and

$$|f(\theta) - S_n(\theta)| \leq \frac{4M}{K-1} \frac{1}{n^{K-1}}.$$

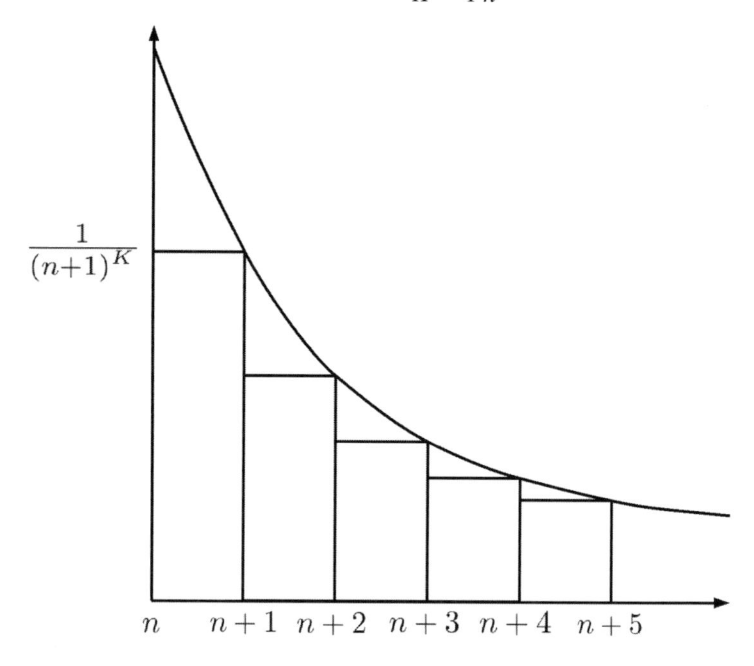

B.14 Orthogonality Relations for Discrete Fourier Sums

The trigonometric functions satisfy discrete orthogonality relations that correspond to the integral orthogonality relations (B.1), (B.2), and (B.3). These enable easy derivation of the trigonometric interpolation formulas in the next section. Consider the sum

$$S = \sum_{j=0}^{2n-1} e^{ir\theta_j} = \sum_{j=0}^{2n-1} \omega^{rj},$$

where

$$\theta_j = \frac{j\pi}{n}, \quad \omega = e^{i\frac{\pi}{n}}, \quad \omega^{2n} = 1.$$

Then

$$S = \frac{1 - \omega^{2nr}}{1 - \omega^r} = \begin{cases} 0 & \text{if } \omega^r \neq 1 \\ 2n & \text{if } \omega^r = 1 \end{cases}.$$

Therefore

$$\sum_{j=0}^{2n-1} e^{ir\theta_j} e^{-is\theta_j} = \begin{cases} 0 & \text{if } |r-s| \neq 0, 2n, 4n \\ 2n & \text{if } |r-s| = 0, 2n, 4n \end{cases}.$$

Taking the real part,

$$\sum_{j=0}^{2n-1} \cos(r-s)\theta_j = 0 \text{ if } |r-s| \neq 0, 2n, 4n$$

and

$$\sum_{j=0}^{2n-1} \cos(r+s)\theta_j = 0 \text{ if } |r+s| \neq 0, 2n, 4n.$$

Also

$$\sum_{j=0}^{2n-1} \cos r\theta_j \cos s\theta_j = \frac{1}{2} \sum_{j=0}^{2n-1} \left(\cos(r-s)\theta_j + \cos(r+s)\theta_j \right).$$

It is now easy to verify that

$$\sum_{j=0}^{2n-1} \cos r\theta_j \cos s\theta_j = \begin{cases} 0 & \text{if } r \neq s \\ n & \text{if } r = s \neq 0, n \\ 2n & \text{if } r = s = 0, n \end{cases}. \tag{B.4}$$

It can be similarly shown that

$$\sum_{j=0}^{2n-1} \sin r\theta_j \sin s\theta_j = \begin{cases} 0 & \text{if } r \neq s \\ n & \text{if } r = s \neq 0, n \\ 0 & \text{if } r = s = 0, n \end{cases} \tag{B.5}$$

and

$$\sum_{j=0}^{2n-1} \sin r\theta_j \cos s\theta_j = 0. \tag{B.6}$$

Note that $\sum_{j=0}^{2n-1} w^{rj}$ is the sum of $2n$ equally spaced unit vectors as sketched, except in the case when $\omega^r = 1$.

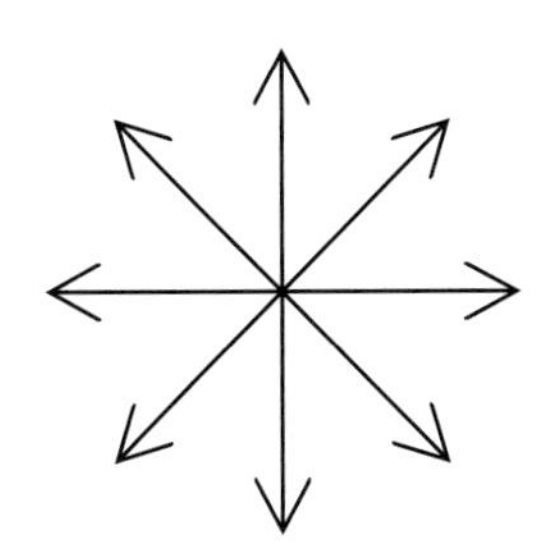

B.15 Trigonometric Interpolation

Instead of finding a least square fit, we can calculate the coefficients of the sum so that the sum exactly fits the function at equal intervals around the circle. Let A_r and B_r be chosen so that

$$U_n(\theta) = \frac{1}{2}A_0 + \sum_{r=1}^{n-1}(A_r \cos r\theta + B_r \sin r\theta) + \frac{1}{2}A_n \cos n\theta = f(\theta)$$

at the $2n + 1$ points $\theta_j = \frac{j\pi}{n}, j = 0, 2n$.

Using the orthogonality relations (B.4), (B.5), and (B.6) derived in the previous section, it follows on multiplying through by $\cos r\theta_j$ or $\sin r\theta_j$ and summing over θ_j that

$$A_r = \frac{1}{n}\sum_{j=0}^{2n-1} f(\theta_j)\cos r\theta_j$$

$$B_r = \frac{1}{n}\sum_{j=0}^{2n-1} f(\theta_j)\sin r\theta_j$$

for $r = 1, \ldots, n$. Note that writing

$$c_0 = \frac{1}{2}A_0$$

$$c_r = \frac{1}{2}(A_r - iB_r)$$

$$c_{-r} = \frac{1}{2}(A_r + iB_r),$$

the sum can be expressed as

$$U_n(\theta) = \sum_{r=-n}^{n-1} c_r e^{ir\theta} = \sum_{r=-n}^{n}{}' c_r e^{ir\theta},$$

where

$$c_r = \frac{1}{2n}\sum_{j=0}^{2n-1} f(\theta_j)e^{-ir\theta_j}$$

and $\sum'$ is defined as

$$\sum_{r=-n}^{n}{}' a_r = \sum_{r=-n}^{n} a_r - \frac{1}{2}\left(a_n + a_{-n}\right).$$

B.16 Error Estimate for Trigonometric Interpolation

In order to estimate the error of the interpolation, we compare the interpolation coefficients A_r and B_r with the Fourier coefficients a_r and b_r. The total error can then be estimated as the sum of the error in the truncated Fourier terms and the additional error introduced by replacing the Fourier coefficients by the interpolation coefficients. The interpolation coefficients can be expressed in terms of the Fourier coefficients by substituting the infinite Fourier series for $f(\theta)$ into the formula for the interpolation coefficients. For this purpose it is convenient to use the complex form. Thus

$$\begin{aligned} C_r &= \frac{1}{2n} \sum_{j=0}^{2n-1} f(\theta_j) e^{-ir\theta_j} \\ &= \frac{1}{2n} \sum_{j=0}^{2n-1} e^{-ir\theta_j} \sum_{k=-\infty}^{\infty} c_k e^{ik\theta_j} \\ &= \frac{1}{2n} \sum_{j=0}^{2n-1} \sum_{k=-\infty}^{\infty} c_k e^{-i(r-k)\theta_j}. \end{aligned}$$

Here

$$\sum e^{i(r-k)\theta_j} = \begin{cases} 2n & \text{if } r-k = 2qn \text{ where } q = 0, 1, \ldots \\ 0 & \text{if } r-k \neq 2qn. \end{cases}$$

Hence

$$C_r = c_r + \sum_{k=1}^{\infty} (c_{2kn+r} + c_{2kn-r})$$

and similarly

$$C_{-r} = c_{-r} + \sum_{k=1}^{\infty} (c_{2kn+r} + c_{2kn-r}).$$

Accordingly

$$A_r = C_r + C_{-r} = a_r + \sum_{k=1}^{\infty} (a_{2kn+r} + a_{2knr-r})$$

and similarly

$$B_r = i(C_r - C_{-r}) = b_r + \sum_{k=1}^{\infty} (b_{2kn+r} - b_{2kn-r}).$$

Thus the difference between the interpolation and Fourier coefficients can be seen to be an aliasing error in which all the higher harmonics are lumped into the base modes.

Now we can estimate $|A_r - a_r|$, $|B_r - b_r|$ using

$$|a_r| \leq \frac{2M}{r^K}, \quad |b_r| \leq \frac{2M}{r^K}.$$

We get

$$\begin{aligned}|A_r - a_r| &\leq \sum_{k=1}^{\infty} 2M \left| \frac{1}{(2kn - r)^K} + \frac{1}{(2kn + r)^K} \right| \\ &= \frac{2M}{(2n)^K} \sum_{k=1}^{\infty} \frac{1}{k^K} \left| \frac{1}{\left(1 - \frac{r}{2kn}\right)^K} + \frac{1}{\left(1 + \frac{r}{2kn}\right)^K} \right|.\end{aligned}$$

Since $r \leq n$, this is bounded by

$$\begin{aligned}\frac{2M}{(2n)^K} \sum_{k=1}^{\infty} \frac{1}{k^K}(2^k + 1) &\leq \frac{2(2^K + 1)M}{2^K n^K} \left(1 + \int_1^{\infty} \frac{d\xi}{\xi^K}\right) \\ &\leq \frac{2(2^K + 1)M}{2^K n^K} \left(1 + \frac{1}{K - 1}\right).\end{aligned}$$

Using a similar estimate for B_r, we get

$$|A_r - a_r| < \frac{5M}{n^K}, \quad |B_r - b_r| < \frac{5M}{n^K}$$

for $K \geq 2$. Finally we can estimate the error of the interpolation sum $u_n(\theta)$ as

$$\begin{aligned}|f(\theta) - u_n(\theta)| &\leq |f(\theta) - S_n(\theta)| + |S_n(\theta) - u_n(\theta)| \\ &< \frac{4M}{K - 1} \frac{1}{n^{K-1}} + \sum_{r=0}^{n} \{|A_r - a_r| + |B_r - b_r|\} \\ &< \frac{4M}{K - 1} \frac{1}{n^{K-1}} + \frac{10M(n + 1)}{n^K}.\end{aligned}$$

B.17 Fourier Cosine Series

For a function defined for $0 \leq \theta \leq \pi$, the Fourier cosine series is

$$S_c(\theta) = \sum_{r=0}^{\infty} a_r \cos r\theta,$$

where

$$\begin{aligned}a_0 &= \frac{1}{\pi} \int_0^{\pi} f(\theta) d\theta \\ a_r &= \frac{2}{\pi} \int_0^{\pi} f(\theta) \cos(r\theta) d\theta, r > 0.\end{aligned}$$

Hence since $\frac{d}{d\theta}\cos(r\theta) = 0$ at $\theta = 0, \pi$, the function will not be well approximated at $0, \pi$ unless $f'(\theta) = f'(\pi) = 0$. The cosine series in fact implies an even continuation at $\theta = 0, \pi$.

Now if all the derivatives of $f(\theta)$ of order $0, 1, \ldots, K-1$ are continuous, $f^{(p)}(0) = f^{(p)}(\pi) = 0$ for all odd $p < K$, and $f^{(K)}(\theta)$ is integrable, then integration by parts from the coefficients of $f^{(K)}(\theta)$ gives

$$|a_r| \leq \frac{M}{r^k},$$

with the consequence that the cosine series satisfies an error estimate similar to the error estimate for the full Fourier series derived in Section B.13.

B.18 Cosine Interpolation

Let $f(\theta)$ be approximated by

$$U_n(\theta) = \sum_0^n a_r \cos r\theta$$

and let

$$U_n(\theta_j) = f(\theta_j) \text{ for } j = 0, 1, \ldots n,$$

where

$$\theta_j = \frac{2j+1}{n+1}\frac{\pi}{2}.$$

Then

$$\sum_{j=0}^{n} \cos r\theta_j \cos s\theta_j = \begin{cases} 0 & \text{if } 0 \leq r \neq s \leq n \\ \frac{n+1}{2} & \text{if } 0 \leq r = s \leq n \\ n+1 & \text{if } 0 = r = s \end{cases}.$$

Now if we multiply the first equation by $\cos s\theta_j$ and sum, we find that

$$a_r = \frac{2}{n+1}\sum_{j=0}^{n} f(\theta_j)\cos r\theta_j \text{ for } 0 < r < n$$

$$a_0 = \frac{1}{n+1}\sum_{j=0}^{n} f(\theta_j).$$

B.19 Chebyshev Expansions

Let

$$g(x) = \sum_{r=0}^{\infty} a_r T_r(x)$$

approximate $f(x)$ in the interval $[-1, 1]$. Then $G(\theta) = g(\cos\theta)$ is the Fourier cosine series for $F(\theta) = f(\cos\theta)$ for $0 \leq \theta \leq \pi$, since

$$T_r(\cos\theta) = \cos r\theta$$

$$G(\theta) = g(\cos\theta) = \sum_{r=0}^{\infty} a_r \cos r\theta.$$

Thus

$$a_0 = \frac{1}{\pi}\int_0^{\pi} f(\cos\theta)d\theta = \frac{1}{\pi}\int_{-1}^{1} f(x)\frac{dx}{\sqrt{1-x^2}}$$

and

$$a_r = \frac{2}{\pi}\int_0^{\pi} f(\cos\theta)\cos r\theta d\theta = \frac{2}{\pi}\int_{-1}^{1} f(x)T_r(x)\frac{dx}{\sqrt{1-x^2}}.$$

Now from the theory of Fourier cosine series, if $f^{(p)}(x)$ is continuous for $|x| \leq 1$ and $p = 0, 1, \ldots, K-1$, and $f^{(K)}(x)$ is integrable,

$$|a_r| \leq \frac{M}{r^K}.$$

Since $|T_r(x)| \leq 1$ for $|x| \leq 1$, the remainder after n terms is $\mathcal{O}\left(\frac{1}{n^{K-1}}\right)$. Now

$$F'(\theta) = -f'(\cos\theta)\sin\theta$$

$$F''(\theta) = f''(\cos\theta)\sin^2\theta - f'(\cos\theta)\cos\theta$$

$$F'''(\theta) = -f'''(\cos\theta)\sin^3\theta + 3f''(\cos\theta)\sin\theta\cos\theta + f'(\cos\theta)\sin\theta,$$

and in general, if $F^{(p)}$ is bounded,

$$F^{(p)}(0) = F^{(p)}(\pi) = 0$$

when p is odd, since then $F^{(p)}$ contains the term $\sin\theta$. As a result, the favorable error estimate applies, provided that derivatives of f exist up to the required order.

B.20 Chebyshev Interpolation

We can transform cosine interpolation to interpolation with Chebyshev polynomials by setting

$$G_n(x) = \sum_{r=0}^{n} a_r T_r(x)$$

and choosing the coefficients so that

$$G_n(x_j) = f(x_j) \text{ for } j = 0, 1, \ldots n,$$

where

$$x_j = \cos\theta_j = \cos\left(\frac{2j+1}{n+1}\frac{\pi}{2}\right).$$

These are the zeros of $T_{n+1}(x)$. Now for any $k < n$, we can find the discrete least squares approximation. When $k = n$, because of the above equation, the error is zero, so the least squares approximation is the interpolation polynomial. Moreover, since $G_n(x)$ is a polynomial of degree n, this is the Lagrange interpolation polynomial at the zeros of $T_{n+1}(x)$.

B.21 Portraits of Fourier, Legendre, and Chebyshev

Figure B.1 Joseph Fourier (1768–1830).

Figure B.2 Adrien-Marie Legendre (1752–1833).

Figure B.3 Pafnuty Lvovich Chebyshev (1821–1894).

Appendix C Polynomial Interpolation, Differentiation, and Integration

C.1 Interpolation Polynomials

Suppose that we wish to approximate a smooth function $f(x)$ by a polynomial $P_n(x)$ of degree n over some interval. A natural approach is to make $P_n(x) = a_0 + a_1 x + \cdots + a_n x^n = f(x)$ at $n+1$ points $x_0, x_1, \ldots, x_n$. Then we have $n+1$ equations for the $n+1$ coefficients. Such a polynomial is called an interpolation polynomial. A solution exists because the determinant of the equations is the Vandemonde determinant

$$D = \begin{vmatrix} 1 & x_0 & \cdots & x_0^n \\ 1 & x_1 & \cdots & x_1^n \\ \vdots & \vdots & \ddots & \vdots \\ 1 & x_n & \cdots & x_n^n \end{vmatrix} = \prod_{i>j}(x_i - x_j) \neq 0.$$

This can be seen by noting that D vanishes if any x_i equals any x_j, but D is a polynomial of just the same order as the product. The interpolation polynomial is unique since if there were two such polynomials $P_n(x)$, $Q_n(x)$, then

$$P_n(x_i) - Q_n(x_i) = 0 \text{ for } i = 0, 1, \ldots, n.$$

But then $P_n(x) - Q_n(x)$ is an nth degree polynomial with $n+1$ roots, so it must be zero.

In practice it is easiest to construct $P_n(x)$ indirectly as a sum of specially chosen polynomials rather than solve for the a_j. To do this note that

$$\phi_{n,j}(x) = \frac{(x - x_0)(x - x_1)\ldots(x - x_{j-1})(x - x_{j+1})\ldots(x - x_n)}{(x_j - x_0)(x_j - x_1)\ldots(x_j - x_{j-1})(x_j - x_{j+1})\ldots(x_j - x_n)} \tag{C.1}$$

satisfies the relation

$$\phi_{n,j}(x_i) = \delta_{ij} = \begin{cases} 1 & \text{if} \quad i = j \\ 0 & \text{if} \quad i \neq j \end{cases}.$$

Then we can get

$$P_n(x) = \sum_{j=0}^{n} f(x_j)\phi_{n,j}(x), \tag{C.2}$$

which equals $f(x)$ at $x = x_0, x_1, \ldots, x_n$ and is the only such polynomial, as has just been shown.

The $\phi_{n,j}(x)$ are called Lagrange interpolation coefficients. We can write

$$\phi_{n,j}(x) = \frac{w_n(x)}{(x - x_j)w_n'(x_j)},$$

where

$$w_n(x) = (x - x_0)(x - x_1) \ldots (x - x_n).$$

C.2 Error in the Interpolating Polynomial

We shall show that if $f(x)$ has an $(n + 1)$th derivative and $P_n(x)$ is an interpolating polynomial, then the remainder is

$$f(x) - P_n(x) = \frac{(x - x_0)(x - x_1) \ldots (x - x_n)}{(n + 1)!} f^{(n+1)}(\xi),$$

where ξ is in the interval defined by $x_0, x_1, \ldots, x_n$.

Let $S_n(x)$ be defined by

$$f(x) - P_n(x) = w_n(x)S_n(x),$$

where

$$w_n(x) = (x - x_0)(x - x_1) \ldots (x - x_n).$$

Define

$$F(z) = f(z) - P_n(z) - w_n(z)S_n(x).$$

This is continuous in z and vanishes at $n + 2$ points $x_0, x_1, \ldots, x_n, x$. Thus by Rolle's theorem, $F'(z)$ vanishes at $n + 1$ points, $F''(z)$ vanishes at n points, ..., and finally $F^{(n+1)}(z)$ vanishes at one point ξ. But

$$\frac{d^{n+1}}{dz^{n+1}} P_n(z) = 0$$

$$\frac{d^{n+1}}{dz^{n+1}} w_n(z)S_n(x) = (n + 1)!\, S_n(x).$$

Thus

$$F^{(n+1)}(z) = f^{n+1}(z) - (n + 1)!\, S_n(x),$$

and setting $z = \xi$,

$$S_n(x) = \frac{1}{(n+1)!} f^{(n+1)}(\xi)$$

$$f(x) - P_n(x) = \frac{w_n(x)}{(n+1)!} f^{(n+1)}(\xi).$$

Since the error depends on $w_n(x)$, we naturally ask how the x_i should be distributed to minimize $w_n(x)$. Consider interpolation at $n+1$ points in the interval $[-1,1]$. Then, according to the minimax property of the Chebyshev polynomials proved in Section B.9, $||w_n(x)||_\infty$ will be minimized of we choose x_i as the zeros of $T_{n+1}(x)$, with the consequence that

$$w_n(x) = \frac{1}{2^n} T_{n+1}(x)$$

and

$$||w_n(x)|| = \frac{1}{2^n}.$$

C.3 Error of Derivative with Lagrange Interpolation

Suppose that

$$f(x) = P_n(x) + \frac{w_n}{(n+1)!} f^{(n+1)}(\xi).$$

Then

$$f'(x) = P_n'(x) + \frac{w_n'}{(n+1)!} f^{(n+1)}(\xi) + \frac{w_n}{(n+1)!} f^{(n+2)}(\xi) \frac{d\xi}{dx},$$

where

$$w_n' = \sum_{j=0}^{n} \prod_{k \neq j}^{n} (x - x_k).$$

Then at an interpolation point,

$$f'(x_j) = P_n'(x_j) + \prod_{k \neq j}^{n} (x_j - x_k) \frac{f^{(n+1)}}{(n+1)!}(\xi).$$

Accordingly, with equal intervals h,

$$f'(x_0) = P'(x_0) + \frac{h^n}{n+1} f^{(n+1)}(\xi)$$

since

$$(x_1 - x_0)(x_2 - x_0)\dots(x_n - x_0) = n!\,h^n.$$

With one interval,

$$f'(x_0) = \frac{f(x_1) - f(x_0)}{h} + \frac{h}{2} f''(\xi).$$

C.4 Error of the *k*th Derivative of the Interpolation Polynomial

Let $f^{n+1}(x)$ be continuous and let

$$R_n(x) = f(x) - P_n(x).$$

Then

$$R_n^{(k)}(x) = \prod_{j=0}^{n-k} (x - \xi_j) \frac{f^{(n+1)}(\xi)}{(n+1-k)!},$$

where the distinct points ξ_j are independent of x and lie in the intervals $x_j < \xi_j < x_{j+k}$ for $j = 0, 1, \dots, n-k$, and $\xi(x)$ is some point in the interval containing x and the ξ_j.

Proof $R_n = f - P_n$ has $n+1$ continuous derivatives and vanishes at $n+1$ points $x = x_j$, for $j = 0, 1, \dots, n$. Apply Rolle's theorem $k \leq n$ times. The zeros are then distributed as illustrated below.

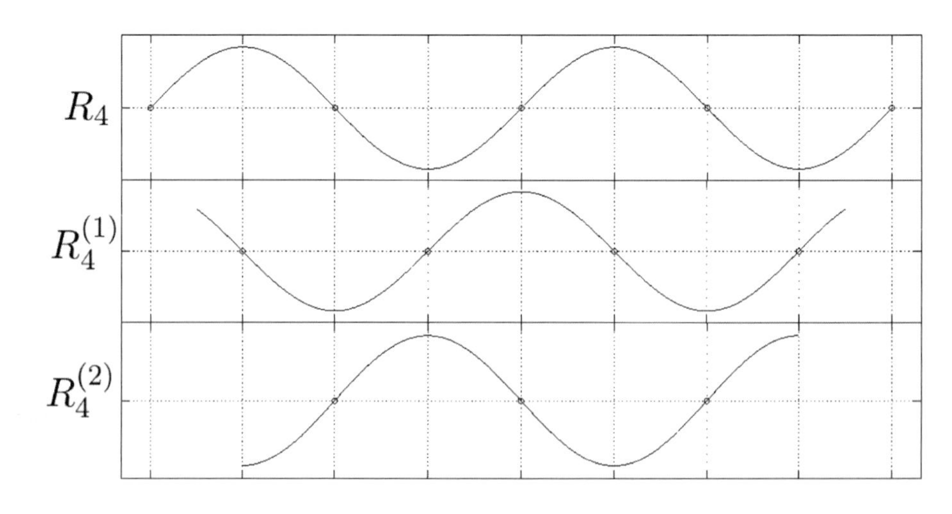

This leads to the following table for the distribution of the $n-k+1$ zeros ξ_j of $R_n^{(k)}$.

R	$R^{(1)}$	$R^{(2)}$	$\cdots$	$R^{(k)}$
x_0				
x_1	(x_0, x_1)			
x_2	(x_1, x_2)	(x_0, x_2)		
$\vdots$				
x_k	(x_{k-1}, x_k)	(x_{k-2}, x_k)	$\cdots$	(x_0, x_k)
$\vdots$				
x_n	(x_{n-1}, x_n)	(x_{n-2}, x_n)	$\cdots$	(x_{n-k}, x_n)

Define

$$F(z) = R_n^{(k)}(z) - \alpha \prod_{j=0}^{n-k} (z - \xi_j).$$

For any x distinct from the ξ_j, choose α such that

$$F(x) = 0.$$

Then $F(z)$ has $n-k+2$ zeros, and by Rolle's theorem $F^{(n-k+1)}(z)$ has one zero, η, say, in the interval containing x and the ξ_j.

Thus

$$\begin{aligned} 0 &= F^{(n-k+1)}(\eta) \\ &= R_n^{n+1}(\eta) - \alpha(n-k+1)! \\ &= f^{n+1}(\eta) - \alpha(n-k+1)! \end{aligned}$$

or

$$\alpha = \frac{f^{n+1}(\eta)}{(n-k+1)!}.$$

It follows that for a mesh width bounded by h

$$R_n^{(k)} \leq M h^{n-k+1},$$

where M is proportional to $\sup \left| f^{n+1}(x) \right|$ in the interval.

C.5 Newton's Form of the Interpolation Polynomial

The Lagrangian form of the interpolation polynomial has the disadvantage that the coefficients have to be recomputed if a new interpolation point is added. To avoid this, let

$$P_n(x) = \sum_{k=0}^{n} a_k w_{k-1}(x),$$

where

$$w_k(x) = (x - x_0)(x - x_1)\dots(x - x_k) \text{ and } w_{-1}(x) = 1.$$

We can determine the a_k so that

$$P_n(x_j) = f(x_j) \text{ for } j = 0, 1, \dots, n.$$

Suppose this has been done for P_{k-1}, and we now add x_k. Then

$$P_k(x) = P_{k-1}(x) + a_k w_{k-1}(x)$$

$$P_k(x_j) = P_{k-1}(x_j) = f(x_j) \text{ for } j = 0, 1, \dots, k-1$$

$$P_k(x_k) = P_{k-1}(x_k) + a_k w_{k-1}(x_k) = f(x_k),$$

where $a_k = \frac{f(x_k) - P_{k-1}(x_k)}{w_{k-1}(x_k)}$.

On the other hand, the coefficient of the highest order term in the Lagrange expression is

$$\sum_{j=0}^{n} \frac{f(x_j)}{w'_n(x_j)}$$

so by comparison

$$a_n = \sum_{j=0}^{n} \frac{f(x_j)}{w'_n(x_j)}.$$

Thus a_n is a linear combination of $f(x_j)$ for $j = 0, 1, \dots, n$. Multiply by

$$x_n - x_0 = x_n - x_j + x_j - x_0.$$

Then

$$\begin{aligned} a_n(x_n - x_0) &= -\sum_{j=0}^{n} \frac{f(x_j)}{w'_n(x_j)}(x_j - x_n) + \sum_{j=1}^{n} \frac{f(x_j)}{w'_n(x_j)}(x_j - x_0) \\ &= -\sum_{j=0}^{n} \frac{f(x_j)}{\prod_{k \neq j}(x_j - x_k)} + \sum_{j=1}^{n} \frac{f(x_j)}{\prod_{k \neq j}(x_j - x_k)}. \end{aligned}$$

These are just the expressions for a_{n-1} in the intervals x_0 to x_{n-1} and x_1 to x_n. Thus if we define

$$f[x_0, \dots, x_n] = a_n = \sum_{j=0}^{n} \frac{f(x_j)}{\prod_{k \neq j}(x_j - x_k)},$$

we find that

$$f[x_0, \dots, x_n](x_0 - x_n) = f[x_1, \dots, x_n] - f[x_0, \dots, x_{n-1}],$$

and starting from

$$f[x_0] = f(x_0)$$

we have

$$f[x_0, x_1] = \frac{f[x_1] - f[x_0]}{x_1 - x_0}$$

$$f[x_0, x_1, x_2] = \frac{f[x_1, x_2] - f[x_0, x_1]}{x_2 - x_0}.$$

$$\vdots$$

The square bracketed expressions are Newton's divided differences, and we now have Newton's form for the interpolation polynomial:

$$P_n(x) = f[x_0] + (x - x_0)f[x_0, x_1] + \cdots + (x - x_0)\dots(x - x_{n-1})f[x_0, \dots, x_n],$$

where each new term is independent of the previous terms.

To estimate the magnitude of the higher divided differences, we can use the result already obtained for the remainder.

THEOREM C.5.1 *Let $x_0, x_1, \dots, x_{k-1}$ be distinct points and let f be continuous in interval containing all these points. Then for some point ξ in this interval*

$$f[x_0, \dots, x_{k-1}, x] = \frac{f^{(k)}(\xi)}{k!}.$$

Proof From Newton's formula:

$$\begin{aligned} f(x) - P_{k-1}(x) &= (x - x_0)\dots(x - x_{k-1})f[x_0, \dots, x_{k-1}, x] \\ &= (x - x_0)\dots(x - x_{k-1})\frac{f^{(k)}(\xi)}{k!} \end{aligned}$$

by the remainder theorem. But x is distinct from $x_0, x_1, \dots, x_{k-1}$, so the theorem follows on dividing out the factors. □

C.6 Hermite Interpolation

We can generalize interpolation by matching derivatives as well as values at the interpolation points. Such a polynomial is called the osculating polynomial, and the procedure is called Hermite interpolation.

Let $H_{2n+1}(x)$ be a polynomial of degree $2n + 1$ such that

$$H_{2n+1}(x) = f(x_j) \text{ for } j = 0, 1, \dots, n$$

$$H'_{2n+1}(x) = f'(x_j) \text{ for } j = 0, 1, \dots, n.$$

The $2n + 2$ coefficients of $H_{2n+1}(x)$ can be found indirectly as in Lagrange interpolation.

Let $\psi_{n,j}(x)$ and $\gamma_{n,j}(x)$ be polynomials of degree $2n+1$ such that

$$\psi_{n,j}(x_i) = \delta_{ij}, \qquad \psi'_{n,j}(x_i) = 0, \quad \text{for } i = 0, 1, \ldots, n$$

$$\gamma_{n,j}(x_i) = 0, \qquad \gamma'_{n,j}(x_i) = \delta_{ij}, \quad \text{for } i = 0, 1, \ldots, n.$$

Then the Hermite interpolation polynomial is

$$H_{2n+1}(x) = \sum_{j=0}^{n} \left(f(x_j)\psi_{n,j}(x) + f'(x_j)\gamma_{n,j}(x) \right).$$

It can be directly verified by differentiation that the required polynomials are

$$\psi_{n,j}(x) = (1 - 2\phi'_{n,j}(x)(x - x_j))\phi^2_{n,j}(x)$$

and

$$\gamma_{n,j}(x) = (x - x_j)\phi^2_{n,j}(x),$$

where the $\phi_{n,j}(x)$ are the Lagrange interpolation coefficients defined by (C.1) and (C.2).

The error in Hermite interpolation is

$$f(x) - H_{2n+1}(x) = \omega_n^2(x)\frac{f^{2n+2}(\xi)}{(2n+2)!},$$

where

$$\omega_n(x) = (x - x_0)(x - x_1)\ldots(x - x_n).$$

This can be proved in the same way as the error estimate for Lagrange interpolation where now

$$F(\xi) = f(\xi) - H_{2n+1}(\xi) - \omega_n^2(\xi)S_n(x),$$

and on the first application of Rolle's theorem, $F'(\xi)$ has $2n+2$ distinct zeros.

C.7 Integration Formulas using Polynomial Interpolation

Convenient numerical differentiation and integration formulas can be obtained by differentiating or integrating the Lagrange interpolation polynomial $P_n(x)$ of degree n defined by (C.1) and (C.2). Suppose we approximate

$$\int_a^b f(x)dx$$

by

$$\int_a^b P_n(x)dx,$$

where typically the interpolation points lie within the interval of integration, possibly including the endpoints. Then

$$\int_a^b f(x)dx = \sum_{j=0}^{n} A_j f(x_j) + E(f),$$

where $E(f)$ denotes the error, and

$$A_j = \int_a^b \phi_{n,j}(x)dx.$$

The result is exact if $f(x)$ is a polynomial of degree $\leq n$, since then $P_n(x) = f(x)$. This is called precision of degree n.

C.8 Integration with a Weight Function

We may include a weight function $w(x) \geq 0$ in the integral. Then

$$\int_a^b f(x)w(x)dx = \sum_{j=n}^{n} A_j f(x_j) + E(f),$$

where now

$$A_j = \int_a^b \phi_{n,j}(x)w(x)dx.$$

Depending on $w(x)$, analytical expressions are not necessarily available for evaluating the coefficients A_j.

C.9 Newton Cotes Formulas

These are derived by approximating $f(x)$ by an interpolation formula using $n+1$ equally spaced points in $[a,b]$ including the end points. The first three such formulas are:

Trapezoidal rule

$$\int_a^b f(x)dx = \frac{h}{2}(f(a) + f(b)) + E(f), h = b - a,$$

Simpson's rule

$$\int_a^b f(x)dx = \frac{h}{3}f(a) + \frac{4h}{3}f(a+h) + \frac{h}{3}f(b) + E(f), h = \frac{b-a}{2},$$

and Simpson's $\frac{3}{8}$ formula

$$\int_a^b f(x)dx = \frac{3h}{8}f(a) + \frac{9h}{8}f(a+h) + \frac{9h}{8}f(a+2h) + \frac{3h}{8}f(b) + E(f), h = \frac{b-a}{3}.$$

C.10 Gauss Quadrature

When the interpolation points x are fixed, the integration formula has $n+1$ degrees of freedom corresponding to the coefficients A_j for $j = 0, \ldots, n$. Accordingly we can find values of A_j that yield exact values of the integral of all polynomials of degree n,

$$P_n(x) = a_0 + a_1 x + \cdots + a_n x^n,$$

since these also have $n+1$ degrees of freedom corresponding to the coefficients a_j for $j = 0, \ldots, n$. If we are also free to choose the integration points x_j, then we have $2n+2$ degrees of freedom corresponding to x_j and A_j, for $j = 0, \ldots, n$. Now it is possible to find values of x_j and A_j that enable exact integration of polynomials of degree $\leq 2n+1$. One could try to find the required values by solving $2n+2$ nonlinear equations

$$\sum_{j=0}^{n} A_j f(x_j) = \int_a^b f(x)dx,$$

where $f(x) = 1, x, \ldots, x^{2n+1}$ in turn. However, the required values can be found indirectly as follows.

Let $\phi_i(x)$ be orthogonal polynomials for the weight function $w(x)$, so that

$$\int_a^b \phi_j(x)\phi_k(x)w(x) = 0 \text{ for } j \neq k.$$

Choose x_j as the zeros of $\phi_{n+1}(x)$. Then integration using polynomial interpolation is exact for polynomials up to degree $2n+1$.

Proof If $f(x)$ is a polynomial of degree $\leq 2n+1$, it can be uniquely expressed as

$$f(x) = Q(x)\phi_{n+1}(x) + R(x),$$

where $Q(x)$ and $R(x)$ are polynomials of degree $\leq n$. Then

$$\int_a^b f(x)w(x)dx = \int_a^b Q(x)\phi_{n+1}(x)w(x)dx + \int_a^b R(x)w(x)dx = \int_a^b R(x)w(x)dx$$

since $\phi_{n+1}(x)$ is orthogonal to all polynomials of degree $< n + 1$.

Also the approximate integral is

$$\begin{aligned} I &= \sum_{j=0}^{n} A_j f(x_j) \\ &= \sum_{j=0}^{n} A_j Q(x_j)\phi_{n+1}(x_j) + \sum_{j=0}^{n} A_j R(x_j) \\ &= \sum_{j=0}^{n} A_j R(x_j) \end{aligned}$$

since $\phi_{n+1}(x_j)$ is zero for $j = 0, 1, \ldots, n$. But then

$$I = \int_a^b R(x)w(x)dx$$

exactly, since $R(x)$ is a polynomial of degree $\leq n$.

The degree of precision $2n+1$ is the maximum attainable. Consider the polynomial of $\phi_{n+1}^2(x)$ of degree $2n + 2$. Then

$$I = 0$$

since $\phi_{n+1}(x_j) = 0$ for $j = 0, 1, \ldots, n$. But

$$\int_a^b \phi_{n+1}^2(x)w(x)dx > 0$$

because $\phi_{n+1}^2(x) > 0$ except at the zeros $x_0, x_1, \ldots, x_n$.

When applied to an arbitrary smooth function $f(x)$, which can be expanded as

$$f(a) + (x - a)f'(a) + \cdots + \frac{(x - a)^{2n+2}}{(2n + 2)!} f^{(2n+2)}(\xi),$$

the error is proportional to a bound on $|f^{(2n+2)}(x)|$ because the preceding terms in the Taylor series are integrated exactly.

C.11 Formulas for Gauss Integration with Constant Weight Function

Table C.1. Formulas for Gauss integration with constant weight function.

Number of Points	x_j	A_j	Precision	Order
1	0	2	1	2
2	$\pm\frac{1}{\sqrt{3}}$	1	3	4
3	0	$\frac{8}{9}$	5	6
	$\pm\frac{\sqrt{15}}{5}$	$\frac{5}{9}$		
4	$\pm\frac{\sqrt{3-2\sqrt{\frac{6}{5}}}}{7}$	$\frac{18+\sqrt{30}}{36}$	7	8
	$\pm\frac{\sqrt{3+2\sqrt{\frac{6}{5}}}}{7}$	$\frac{18-\sqrt{30}}{36}$		
5	0	$\frac{128}{225}$	9	10
	$\pm\sqrt{5-2\sqrt{\frac{10}{7}}}$	$\frac{322+13\sqrt{70}}{900}$		
	$\pm\sqrt{5+2\sqrt{\frac{10}{7}}}$	$\frac{322-13\sqrt{70}}{900}$		

C.12 Error Bound for Gauss Integration

Let $H_{2n+1}(x)$ be the Hermite interpolation polynomial to $f(x)$ at the integration points. The Gauss integration formula is exact for $H_{2n+1}(x)$. Hence

$$\int_a^b H_{2n+1}(x)w(x)dx = \sum_{j=0}^{n} A_j f_j.$$

Thus

$$\int_a^b f(x)w(x)dx - \sum_{i=0}^{n} A_j f_j = \int_a^b (f(x) - H_{2n+1}(x))w(x)dx.$$

According to the error formula for Hermite interpolation,

$$f(x) - H_{2n+1}(x) = \tilde{P}_n^2(x)\frac{f^{(2n+2)(\xi)}}{(2n+2)!},$$

where

$$\tilde{P}_n(x) = (x - x_0)(x - x_1)\dots(x - x_n)$$

is the orthogonal polynomial normalized so that its leading coefficient is unity.

Then the error is

$$E = \frac{1}{(2n+2)!}\int_a^b w(x)\tilde{P}_n^2(x) f^{(2n+2)}(\xi)dx.$$

Since $w(x)\hat{P}_n^2(x) \geq 0$ the error lies in the range between

$$\min f^{(2n+2)}(\xi)A \text{ and } \max f^{(2n+2)}(\xi)A,$$

where

$$A = \frac{1}{(2n+2)!}\int_a^b w(x)\tilde{P}_n^2(x)dx,$$

and hence

$$E = Af^{(2n+2)}(\eta)$$

for some value of η in $[a, b]$.

C.13 Discrete Orthogonality of Orthogonal Polynomials

Let $\phi_j(x)$ be orthogonal polynomials satisfying

$$(\phi_j, \phi_k) = \int_a^b \phi_j(x)\phi_k(x)w(x)dx = 0 \text{ for } j \neq k.$$

Let x_i be the zeros of $\phi_{n+1}(x)$.

Then if $j \leq n, k \leq n$, $\phi_j(x)\phi_k(x)$ is a polynomial of degree $\leq 2n$. Accordingly (ϕ_j, ϕ_k) is evaluated exactly by Gauss integration. This implies that the ϕ_j satisfy the discrete orthogonality condition

$$\sum_{i=0}^{n} A_i\phi_j(x_i)\phi_k(x_i) = 0, \text{ given } j \neq k, j \leq n, k \leq n,$$

where A_i are the coefficients for Gauss integration at the zeros of ϕ_{n+1}.

C.14 Equivalence of Interpolation and Least Squares Approximation using Gauss Integration

Let $f^*(x)$ be the least squares fit to $f(x)$ using orthogonal polynomials $\phi_j(x)$ for the interval $[a, b]$ with weight function $w(x) \geq 0$. Then

$$f^*(x) = \sum_{j=0}^{n} c_j^*\phi_j(x),$$

where

$$c_j^* = \frac{(f, \phi_j)}{(\phi_j, \phi_j)} = \frac{\int_a^b f(x)\phi_j(x)w(x)dx}{\int_a^b \phi_j^2(x)w(x)dx}. \tag{C.3}$$

Let x_i be the zeros of $\phi_{n+1}(x)$, and let $P_n(x)$ be the interpolation polynomial to $f(x)$ satisfying

$$P_n(x_i) = f(x_i) \text{ for } i = 0, 1, \ldots, n.$$

$P_n(x)$ can be expanded as

$$P_n(x) = \sum_{k=0}^{n} \hat{c}_k \phi_k(x).$$

Then the $\hat{c}_k$ are determined by the conditions

$$\sum_{k=0}^{n} \hat{c}_k \phi_k(x_i) = f(x_i) \text{ for } i = 0, 1, \ldots, n. \tag{C.4}$$

Also the polynomials $\phi_j(x)$ satisfy the discrete orthogonality condition

$$\sum_{i=0}^{n} A_i \phi_j(x_i)\phi_k(x_i) = 0 \text{ given } j \neq k, j \leq n, k \leq n,$$

where A_i are the coefficients for Gauss integration.

Now multiplying (C.4) by $A_i\phi_j(x_i)$ and summing over i, it follows that

$$\hat{c}_j = \frac{\sum_{i=0}^{n} A_i f(x_i)\phi_j(x_i)}{\sum_{i=0}^{n} A_i \phi_j^2(x_i)}.$$

This is exactly the formula (C.3) for evaluating the least squares coefficient c_j^* by Gauss integration.

Alternative Proof The coefficients $\hat{c}_j$ are actually the least squares coefficients that minimize

$$J = ||f - \sum_{j=0}^{n} c_j \phi_j||^2$$

in the discrete semi-norm in which

$$\int \left(f(x) - \sum_{j=0}^{n} c_j \phi_j(x) \right)^2 w(x)dx$$

is evaluated by Gauss integration at the zeros x_i of $\phi_{n+1}(x)$ as

$$J = \sum_{i=0}^{n} A_i \left(f(x_i) - \sum_{j=0}^{n} c_j \phi_j(x_i) \right)^2 w(x_0).$$

But as a consequence of the interpolation condition (C.4), the coefficients $\hat{c}_j$ yield the value $J = 0$, thus minimizing J.

C.15 Gauss Lobato Integration

To integrate $f(x)$ over [−1,1], choose integration points at $-1, 1$ and the zeros of ϕ_{n-1}. These are the zeros of

$$(1 + x)(1 - x)\phi_{n-1}(x).$$

Then let

$$f(x) = Q(x)(1 - x^2)\phi_{n-1}(x) + R(x),$$

where if $f(x)$ is a polynomial of degree $\leq 2n - 1$, $Q(x)$ and $R(x)$ are polynomials of degree $\leq n - 2$. Now let the polynomials ϕ_j be orthogonal for the weight function

$$w(x) = 1 - x^2.$$

Then

$$I = \int_{-1}^{1} f(x)dx = \int_{-1}^{1} Q(x)\phi_{n-1}(x)w(x)dx + \int_{-1}^{1} R(x)dx = \int_{-1}^{1} R(x)dx$$

since ϕ_{n-1} is orthogonal to all polynomials of lower degree. Also the approximate integral is

$$\hat{I} = \sum_{j=0}^{n} A_j f(x_j),$$

where

$$A_j = \int_{-1}^{1} \phi_{n,j}(x)dx.$$

Then

$$\hat{I} = \sum_{j=0}^{n} A_j Q(x_j)(1 - x_j^2)\phi_{n-1}(x_j) + \sum_{j=0}^{n} A_j R(x_j) = \sum_{j=0}^{n} A_j R(x_j)$$

since $(1 - x_j^2)\phi_{n-1}(x_j) = 0$ for $j = 0, 1, \ldots, n$. But then

$$\hat{I} = I$$

since $r(x)$ is a polynomial of degree $< n$. Here ϕ_j are the Jacobi polynomials $P_j^{(1,1)}$, where $P_j^{(\alpha,\beta)}$ is orthogonal for the weight

$$(1 - x)^{\alpha}(1 + x)^{\beta}.$$

C.16 Portraits of Lagrange, Newton, and Gauss

Figure C.1 Joseph-Louis Lagrange (1736–1813).

Figure C.2 Sir Isaac Newton (1642–1726).

Figure C.3 Carl Friedrich Gauss (1777–1855).

Appendix D Potential Flow Methods

D.1 Boundary Integral Methods

The first major success in computational aerodynamics was the development of boundary integral methods for the solution of the subsonic linearized potential flow equation

$$\left(1 - M^2{}_{\infty}\right)\phi_{xx} + \phi_{yy} = 0. \tag{D.1}$$

This can be reduced to Laplace's equation by stretching the x coordinate by the factor $\sqrt{1 - M_{\infty}{}^2}$. Then, according to potential theory, the general solution can be represented in terms of a distribution of sources or doublets, or both sources and doublets, over the boundary surface. The boundary condition is that the velocity component normal to the surface is zero. Assuming, for example, a source distribution of strength $\sigma(Q)$ at the point Q of a surface S, this leads to the integral equation

$$2\pi\sigma_p - \int\int_S \sigma\left(Q\right) n_p.\nabla\left(1/r\right) = 0, \tag{D.2}$$

where P is the point of evaluation, and r is the distance from P to Q. A similar equation can be found for a doublet distribution, and it usually pays to use a combination. Equation (D.2) can be reduced to a set of algebraic equations by dividing the surface into quadrilateral panels, assuming a constant source strength on each panel, and satisfying the condition of zero normal velocity at the center of each panel. This leads to N equations for the source strengths on N panels.

The first such method was introduced by Hess and Smith in 1962 (Hess & Smith 1962). The method was extended to lifting flows, together with the inclusion of doublet distributions, by Rubbert and Saaris (1968). Subsequently higher order panel methods (as these methods are generally called in the aircraft industry) were introduced. A review has been given by Hunt (1980). An example of a calculation by a panel method is shown in Figure D.1. The results are displayed in terms of the pressure coefficient, defined as

$$c_p = \frac{p - p_{\infty}}{\frac{1}{2}\rho_{\infty}q_{\infty}^2}.$$

Figure D.2 illustrates the kind of geometric configuration that can be treated by panel methods.

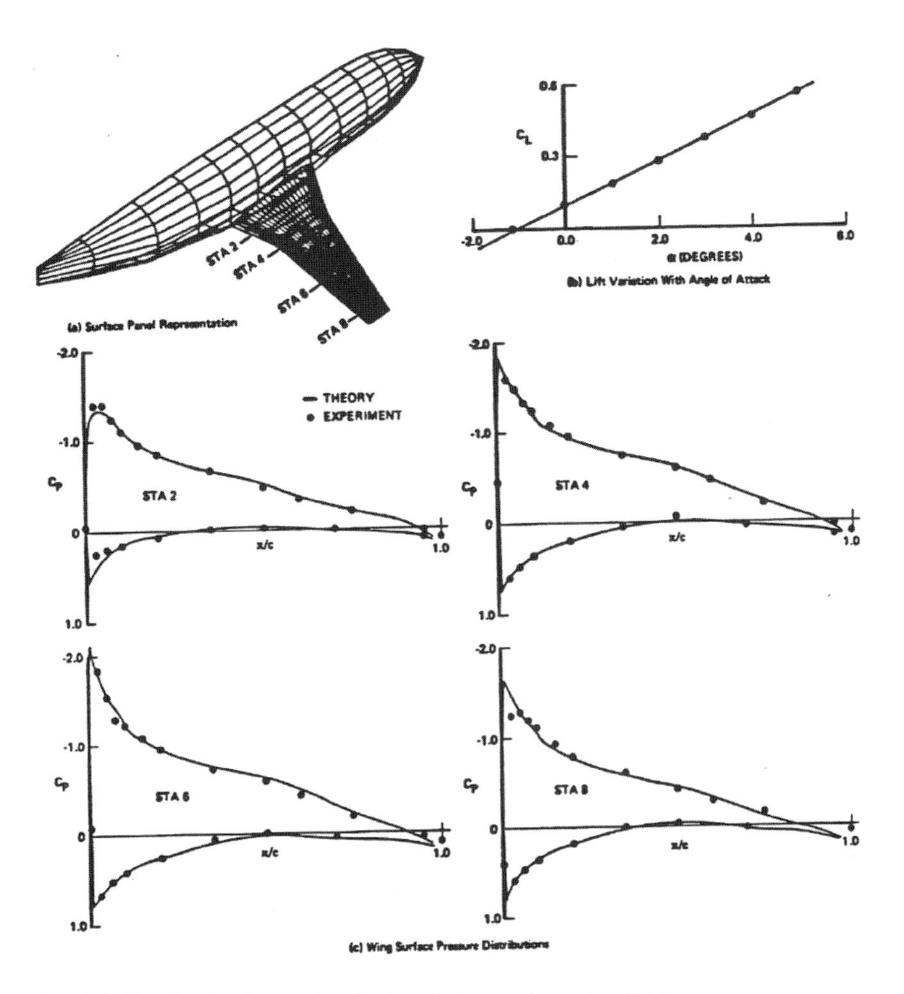

Figure D.1 Panel method applied to Boeing 747. Supplied by Paul Rubbert, the Boeing Company.

In comparison with field methods, which solve for the unknowns in the entire domain, panel methods have the advantage that the dimensionality is reduced. Consider a three-dimensional flow field on an $n \times n \times n$ grid. This would be reduced to the solution of the source or doublet strengths on $N = O(n^2)$ panels. Since, however, every panel influences every other panel, the resulting equations have a dense matrix. The complexity of calculating the $N \times N$ influence coefficients is then $O(n^4)$. Also $O(N^3) = O(n^6)$ operations are required for an exact solution. If one directly discretizes the equations for the three-dimensional domain, the number of unknowns is n^3, but the equations are sparse and can be solved with $O\ (n)$ iterations, or even with a number of iterations independent of n if a multigrid method is used.

Although the field methods appear to be potentially more efficient, the boundary integral method has the advantage that it is comparatively easy to divide a complex surface into panels, whereas the problem of dividing a three-dimensional domain

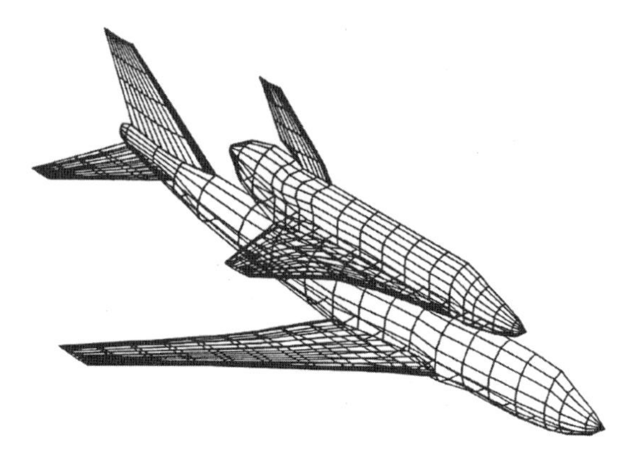

Figure D.2 Panel method applied to flow around Boeing 747 and space shuttle. Supplied by Allen Chen, the Boeing Company.

into hexahedral or tetrahedral cells remains a source of extreme difficulty. Moreover, the operation count for the solution can be reduced by iterative methods, while the complexity of calculating the influence coefficients can be reduced by agglomeration (Vassberg 1997). Panel methods have thus continued to be widely used both for the solution of flows at low Mach numbers for which compressibility effects are unimportant and also to calculate supersonic flows at high Mach numbers, for which the linearized equation (D.1) is again a good approximation.

D.2 Formulation of the Numerical Method for Transonic Potential Flow

The case of two-dimensional flow serves to illustrate the formulation of a numerical method for solving the transonic potential flow equation. With velocity components u, v and coordinates x, y, (2.6) takes the form

$$\frac{\partial}{\partial x}(\rho u) + \frac{\partial}{\partial y}(\rho v) = 0. \tag{D.3}$$

The desired solution should have the properties that ϕ is continuous and the velocity components are piecewise continuous, satisfying (D.3) at points where the flow is smooth, together with the jump condition

$$[\rho v] - \frac{dy}{dx}[\rho u] = 0 \tag{D.4}$$

across a shock wave, where [ ] denotes the jump, and $\frac{dy}{dx}$ is the slope of the discontinuity. That is to say, ϕ should be a weak solution of the conservation law (D.3), satisfying the condition,

$$\int\int \left(\rho u \psi_x + \rho v \psi_y\right) dx\, dy = 0, \tag{D.5}$$

for any smooth test function ψ that vanishes in the far field.

The general method to be described stems from the idea introduced by Murman and Cole (1971), and subsequently improved by Murman (1974), of using type-dependent differencing, with central difference formulas in the subsonic zone, where the governing equation is elliptic, and upwind difference formulas in the supersonic zone, where it is hyperbolic. The resulting directional bias in the numerical scheme corresponds to the upwind region of dependence of the flow in the supersonic zone. If we consider the transonic flow past a profile with fore and aft symmetry, such as an ellipse, the desired solution of the potential flow equation is not symmetric. Instead it exhibits a smooth acceleration over the front half of the profile, followed by a discontinuous compression through a shock wave. However, the solution of the potential flow equation (2.9) is invariant under a reversal of the velocity vector, $u_i = -\phi_{x_i}$. Corresponding to the solution with a compression shock, there is a reverse flow solution with an expansion shock, as illustrated in Figure D.3. In the absence of a directional bias in the numerical scheme, the fore and aft symmetry would be preserved in any solution that could be obtained, resulting in the appearance of improper discontinuities.

Since the quasi-linear form does not distinguish between conservation of mass and momentum, difference approximations to it will not necessarily yield solutions that satisfy the jump condition unless shock waves are detected and special difference formulas are used in their vicinity. If we treat the conservation law (D.3), on the other hand, and preserve the conservation form in the difference approximation, we can ensure that the proper jump condition is satisfied. Similarly, we can obtain proper solutions of the small-disturbance equation by treating it in conservation form.

The general method of constructing a difference approximation to a conservation law of the form

$$f_x + g_y = 0$$

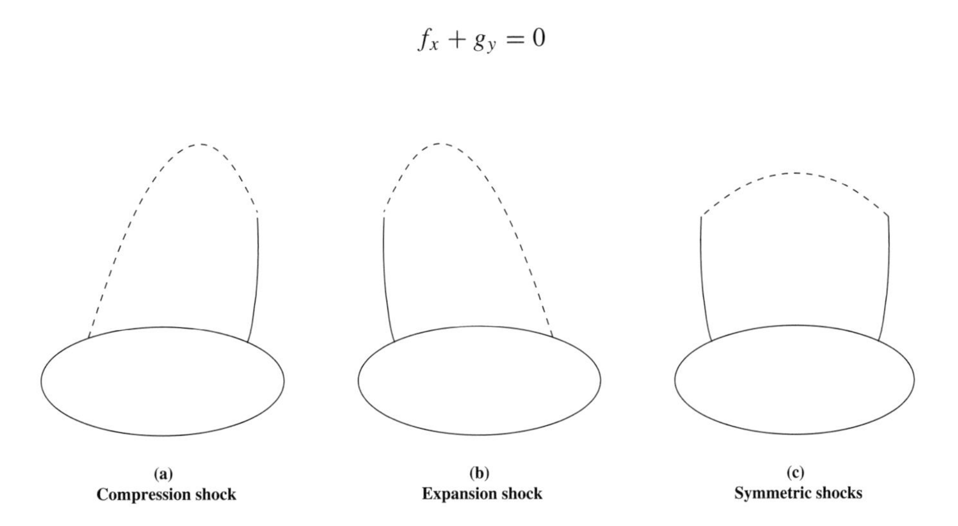

Figure D.3 Alternative solutions for an ellipse.

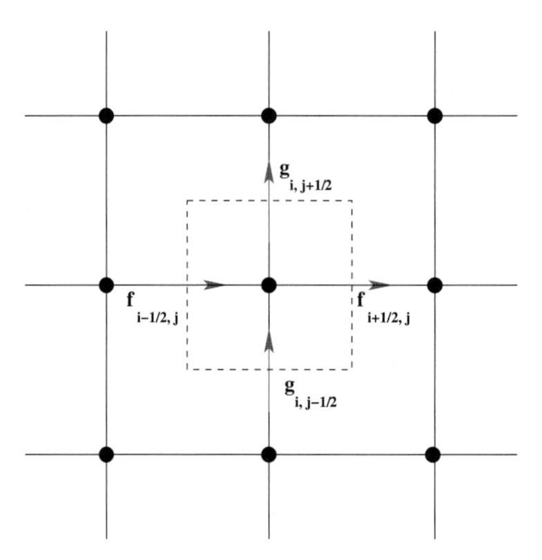

Figure D.4 Flux balance of difference scheme in conservation form.

is to preserve the flux balance in each cell, as illustrated in Figure D.4. This leads to a scheme of the form

$$\frac{F_{i+\frac{1}{2},j} - F_{i-\frac{1}{2},j}}{\Delta x} + \frac{G_{i,j+\frac{1}{2}} - G_{i,j-\frac{1}{2}}}{\Delta y} = 0, \tag{D.6}$$

where F and G should converge to f and g in the limit as the mesh width tends to zero. Suppose, for example, that (D.6) represents the conservation law (D.3). Then on multiplying by a test function ψ_{ij} and summing by parts, there results an approximation to the integral (D.5). Thus the condition for a proper weak solution is satisfied. Some latitude is allowed in the definitions of F and G, since it is only necessary that $F = f + O(\Delta x)$ and $G = g + O(\Delta x)$. In constructing a difference approximation, we can therefore introduce an artificial viscosity of the form

$$\frac{\partial P}{\partial x} + \frac{\partial Q}{\partial y},$$

provided that P and Q are of order Δx. Then the difference scheme is an approximation to the modified conservation law

$$\frac{\partial}{\partial x}(f + P) + \frac{\partial}{\partial y}(g + Q) = 0,$$

which reduces to the original conservation law in the limit as the mesh width tends to zero.

This formulation provides a guideline for constructing type-dependent difference schemes in conservation form. The dominant term in the discretization error introduced by the upwind differencing can be regarded as an artificial viscosity. We can, however, turn this idea around. Instead of using a switch in the difference scheme to introduce an artificial viscosity, we can explicitly add an artificial viscosity, which produces an upwind bias in the difference scheme at supersonic points. Suppose that we have a central difference approximation to the differential equation in conservation form. Then the conservation form will be preserved as long as the added viscosity is also in conservation form. The effect of the viscosity is simply to alter the conserved quantities by terms proportional to the mesh width Δx, which vanish in the limit as the mesh width approaches zero, with the result that the proper jump conditions must be satisfied. By including a switching function in the viscosity to make it vanish in the subsonic zone, we can continue to obtain the sharp representation of shock waves that results from switching the difference scheme.

There remains the problem of finding a convergent iterative scheme for solving the nonlinear difference equations that result from the discretization. Suppose that in the $(n+1)^{st}$ cycle, the residual R_{ij} at the point $i\Delta x, j\Delta y$ is evaluated by inserting the result $\phi_{ij}{}^{(n)}$ of the nth cycle in the difference approximation. Then the correction $C_{ij} = \phi_{ij}{}^{(n+1)} - \phi_{ij}{}^{(n)}$ is to be calculated by solving an equation of the form

$$NC + \sigma R = 0. \tag{D.7}$$

Here C and R denote vectors of the residuals and corrections over the entire mesh, N is a discrete linear operator, and σ is a scaling function. In a relaxation method, N is restricted to a lower-triangular or block-triangular form so that the elements of C can be determined sequentially. In the analysis of such a scheme, it is helpful to introduce a time-dependent analogy. The residual R is an approximation to $L\phi$, where L is the operator appearing in the differential equation. If we consider C as representing $\Delta t \phi_t$, where t is an artificial time coordinate, and $N\Delta t$ is an approximation to a differential operator D, then (D.7) is an approximation to

$$D\phi_t + \sigma L\phi = 0. \tag{D.8}$$

Thus we should choose N so that this is a convergent time-dependent process.

With this approach, the formulation of a relaxation method for solving a transonic flow is reduced to three main steps:

- Construct a central difference approximation to the differential equation.
- Add a numerical viscosity to produce the desired directional bias in the hyperbolic region.
- Add time-dependent terms to embed the steady-state equation in a convergent time-dependent process.

Methods constructed along these lines have proved extremely reliable. Their main shortcoming is a rather slow rate of convergence. In order to speed up the convergence, we can extend the class of permissible operators N.

D.3 Solution of the Transonic Small-Disturbance Equation

D.3.1 Murman Difference Scheme

The basic ideas can conveniently be illustrated by considering the solution of the transonic small-disturbance equation (Ashley & Landahl 1985)

$$A\phi_{xx} + \phi_{yy} = 0, \tag{D.9}$$

where

$$A = 1 - M_\infty^2 - (\gamma + 1)M_\infty^2 \phi_x.$$

The treatment of the small disturbance equation is simplified by the fact that the characteristics are locally symmetric about the x direction. Thus the desired directional bias can be introduced simply by switching to upwind differencing in the x direction at all supersonic points. To preserve the conservation form, some care must be exercised in the method of switching, as illustrated in Figure D.5.

Let p_{ij} be a central-difference approximation to the x derivatives at the point $i\Delta x, j\Delta y$:

$$\begin{aligned} p_{ij} &= (1 - M_\infty^2)\frac{\phi_{i+1,j} - \phi_{ij} - (\phi_{ij} - \phi_{i-1,j})}{\Delta x^2} \\ &\quad - (\gamma + 1)M_\infty^2 \frac{(\phi_{i+1,j} - \phi_{ij})^2 - (\phi_{ij} - \phi_{i-1,j})^2}{2\Delta x^3} \\ &= A_{ij}\frac{\phi_{i+1,j} - 2\phi_{ij} + \phi_{i-1,j}}{\Delta x^2}, \end{aligned} \tag{D.10}$$

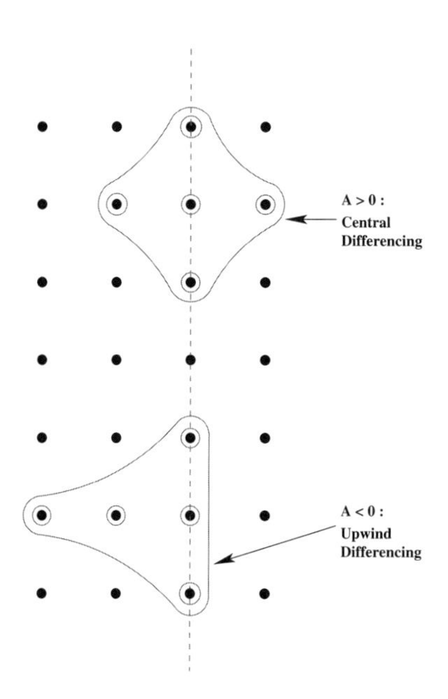

Figure D.5 Murman–Cole Difference Scheme: $A\phi_{xx} + \phi_{yy} = 0$.

where

$$A_{ij} = (1 - M_\infty^2) - (\gamma + 1)M_\infty^2 \frac{\phi_{i+1,j} - \phi_{i-1,j}}{2\Delta x} \tag{D.11}$$

approximates A. Also let q_{ij} be a central-difference approximation to ϕ_{yy}:

$$q_{ij} = \frac{\phi_{i,j+1} - 2\phi_{ij} + \phi_{i,j-1}}{\Delta y^2}. \tag{D.12}$$

Define a switching function μ with the value unity at supersonic points and zero at subsonic points:

$$\mu_{ij} = 0 \; if \; A_{ij} > 0; \quad \mu_{ij} = 1 \; if \; A_{ij} < 0. \tag{D.13}$$

Then the original scheme of Murman and Cole (1971) can be written as

$$p_{ij} + q_{ij} - \mu_{ij}(p_{ij} - p_{i-1,j}) = 0. \tag{D.14}$$

The added terms are an approximation to

$$-\mu \Delta x \frac{\partial}{\partial x}\left(A\phi_{xx}\right).$$

The contribution $-\Delta x A\phi_{xxx}$ in the supersonic region may be regarded as an artificial viscosity of order Δx that is added at all points of the supersonic zone. Since the coefficient $-A$ of $\phi_{xxx} = u_{xx}$ is positive in the supersonic zone, it can be seen that the artificial viscosity includes a term similar to the viscous terms in the Navier–Stokes equation.

Since μ is not constant, the artificial viscosity is not in conservation form, with the result that the difference scheme does not satisfy the conditions stated in the previous section for the discrete approximation to converge to a weak solution satisfying the proper jump conditions. To correct this, all that is required is to recast the artificial viscosity in a divergence form as

$$\Delta x \frac{\partial}{\partial x}\left(\mu A\phi_{xx}\right).$$

This leads to Murman's fully conservative scheme (Murman 1974)

$$p_{ij} + q_{ij} - \mu_{ij}p_{ij} + \mu_{i-1,j}p_{i-1,j} = 0. \tag{D.15}$$

At points where the flow enters and leaves the supersonic zone, μ_{ij} and $\mu_{i-1,j}$ have different values, leading to special parabolic and shock-point equations

$$q_{ij} = 0,$$

and

$$p_{ij} + p_{i-1,j} + q_{ij} = 0.$$

With the introduction of these special operators, it can be verified by directly summing the difference equations at all points of the flow field that the correct jump conditions are satisfied across an oblique shock wave.

D.3.2 Solution of the Difference Equations by Relaxation

The nonlinear difference equations (D.10, D.11, D.12, D.13, and D.14 or D.15) may be solved by a generalization of the line relaxation method for elliptic equations. At each point we calculate the coefficient A_{ij} and the residual R_{ij} by substituting the result ϕ_{ij} of the previous cycle in the difference equations. Then we set $\phi_{ij}^{(n+1)} = \phi_{ij}^{(n)} + C_{ij}$, where the correction C_{ij} is determined by solving the linear equations

$$\frac{C_{i,j+1} - 2C_{i,j} + C_{i,j-1}}{\Delta y^2} + (1 - \mu_{i,j})A_{i,j}\frac{-(2/\omega)C_{i,j} + C_{i-1,j}}{\Delta x^2} + \mu_{i-1,j}A_{i-1,j}\frac{C_{i,j} - 2C_{i-1,j} + C_{i-2,j}}{\Delta x^2} + R_{i,j} = 0 \quad \text{(D.16)}$$

on each successive vertical line. In these equations ω is the over-relaxation factor for subsonic points, with a value in the range 1 to 2. In a typical line relaxation scheme for an elliptic equation, provisional values $\tilde{\phi}_{ij}$ are determined on the line $x = i\Delta x$ by solving the difference equations with the latest available values $\phi_{i-1,j}^{(n+1)}$ and $\phi_{i+1,j}^{(n)}$ inserted at points on the adjacent lines. Then new values $\phi_{i,j}^{(n+1)}$ are determined by the formula

$$\phi_{ij}^{(n+1)} = \phi_{ij}^{(n)} + \omega(\tilde{\phi}_{ij} - \phi_{ij}^{(n)}).$$

By eliminating $\tilde{\phi}_{ij}$, we can write the difference equations in terms of $\phi_{ij}^{(n+1)}$ and $\phi_{ij}^{(n)}$. Then it can be seen that ϕ_{yy} would be represented by $(1/\omega)\delta_y{}^2\phi^{(n+1)} + (1 - (1/\omega))\delta_y{}^2\phi^{(n)}$ in such a process, where $\delta_y{}^2$ denotes the second central-difference operator. The appropriate procedure for treating the upwind difference formulas in the supersonic zone, however, is to march in the flow direction, so that the values $\phi_{ij}^{(n+1)}$ on each new column can be calculated from the values $\phi_{i-2,j}^{(n+1)}$ and $\phi_{i-1,j}^{(n+1)}$ already determined on the previous columns. This implies that ϕ_{yy} should be represented by $\delta_y{}^2\phi^{(n+1)}$ in the supersonic zone, leading to a discontinuity at the sonic line. The correction formula (D.16) is derived by modifying this process to remove this discontinuity. New values $\phi_{ij}^{(n+1)}$ are used instead of provisional values $\tilde{\phi}_{ij}$ to evaluate ϕ_{yy}, at both supersonic and subsonic points. At supersonic points, ϕ_{xx} is also evaluated using new values. At subsonic points, ϕ_{xx} is evaluated from $\phi_{i-1,j}^{(n+1)}, \phi_{i+1,j}^{(n)}$ and a linear combination of $\phi_{ij}^{(n+1)}$ and $\phi_{ij}^{(n)}$. In the subsonic zone the scheme acts like a line relaxation scheme, with a comparable rate of convergence. In the supersonic zone it is equivalent to a marching scheme, once the coefficients A_{ij} have been evaluated. Since the supersonic difference scheme is implicit, no limit is imposed on the step length Δx as A_{ij} approaches zero near the sonic line.

D.3.3 Nonunique Solutions of the Difference Equations for One-Dimensional Flow

Some of the properties of the Murman difference formulas are clarified by considering a uniform flow in a parallel channel. Then $\phi_{yy} = 0$, and with a suitable normalization of the potential, the equation reduces to

$$\frac{\partial}{\partial x}\left(\frac{{\phi_x}^2}{2}\right) = 0, \tag{D.17}$$

with ϕ and ϕ_x given at $x = 0$ and ϕ given at $x = L$. The supersonic zone corresponds to $\phi_x > 0$. Since ${\phi_x}^2$ is constant, ϕ_x simply reverses sign at a jump. Provided we enforce the entropy condition that ϕ_x decreases through a jump, there is a unique solution with a single jump whenever $\phi_x(0) > 0$ and $\phi(0) + L\phi_x(0) \geq \phi(L) \geq \phi(0) - L\phi_x(0)$.

Let $u_{i+1/2} = (\phi_{i+1} - \phi_i)/\Delta x$ and $u_i = (u_{i+1/2} + u_{i-1/2})/2$. Then the fully conservative difference equations can be written as

Elliptic:	${u^2}_{i+1/2} = {u^2}_{i-1/2}$	when $u_i \leq 0$	$u_{i-1} \leq 0,$	(a)
Hyperbolic:	${u^2}_{i-1/2} = {u^2}_{i-3/2}$	when $u_i > 0$	$u_{i-1} > 0,$	(b)
Shock Point:	${u^2}_{i+1/2} = {u^2}_{i-3/2}$	when $u_i \leq 0$	$u_{i-1} > 0,$	(c)
Parabolic:	$0 = 0$	when $u_i > 0$	$u_{i-1} < 0.$	(d)

These admit the correct solution, illustrated in Figure D.6 a, with a constant slope on the two sides of the shock. The shock-point operator allows a single link with an intermediate slope, corresponding to the shock lying in the middle of a mesh cell.

The nonconservative difference scheme omits the shock-point operator, with the result that it admits solutions of the type illustrated in Figure D.6b, with the shock too far forward and the downstream velocity too close to the sonic speed (zero with the present normalization). The direct switch in the difference scheme from (b) to (a)

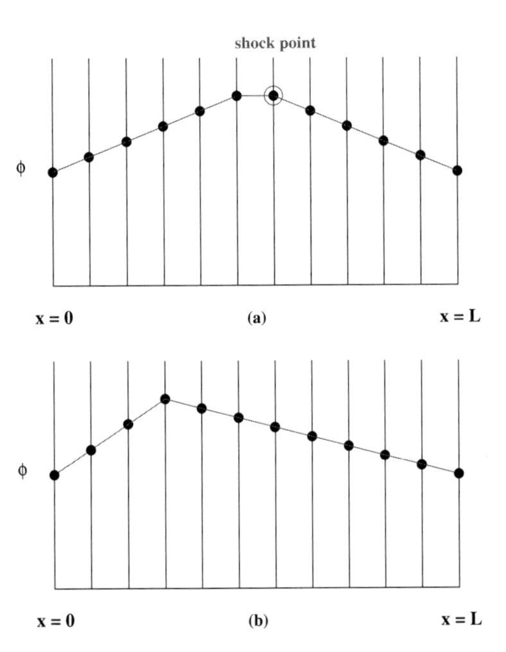

Figure D.6 One-dimensional flow in a channel. • value of ϕ at node i.

allows a break in the slope as long as the downstream slope is negative. The magnitude of the downstream slope cannot exceed the magnitude of the upstream slope, however, because then $u_{i-1} < 0$, and accordingly the elliptic operator would be used at the point $(i-1)\Delta x$. Thus the nonconservative scheme enforces the weakened shock condition,

$$\phi_i - \phi_{i-2} > \phi_i - \phi_{i+2} > 0,$$

which allows solutions ranging from the point at which the downstream velocity is barely subsonic up to the point at which the shock strength is correct. When the downstream velocity is too close to sonic speed, there is an increase in the mass flow. Thus the nonconservative scheme may introduce a source at the shock wave.

However, the fully conservative difference equations also admit various improper solutions. Figure D.7 a illustrates a saw-tooth solution with u^2 constant everywhere except in one cell ahead of a shock point. Figure D.7 b illustrates another improper solution in which the shock is too far forward. At the last interior point, there is then an expansion shock that is admitted by the parabolic operator. Since the difference equations have more than one root, we must depend on the iterative scheme to find the desired root. The scheme should ideally be designed so that the correct solution is stable under a small perturbation, and improper solutions are unstable. Using a scheme similar to (D.16), the instability of the saw-tooth solution has been confirmed in numerical experiments. The solutions with an expansion shock at the downstream boundary, on the other hand, are stable if the compression shock is too far forward by more than the width of a mesh cell. Thus there is a continuous range of stable improper solutions, while the correct solution is an isolated stable equilibrium point.

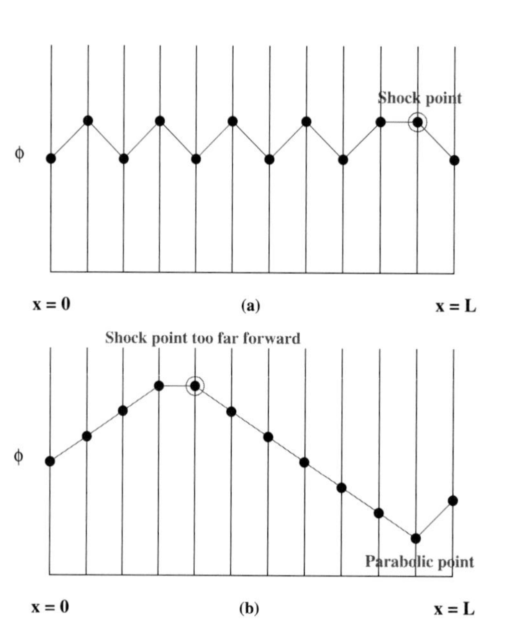

Figure D.7 One-dimensional flow in a channel: (a) Sawtooth solution (b) solution with downstream parabolic point.

D.4 Solution of the Exact Potential Flow Equation

D.4.1 Difference Schemes for the Exact Potential Flow Equation in Quasi-linear Form

It is less easy to construct difference approximations to the potential flow equation with a correct directional bias, because the upwind direction is not known in advance. Following Jameson (1974), the required rotation of the upwind differencing at any particular point can be accomplished by introducing an auxiliary Cartesian coordinate system that is locally aligned with the flow at that point. If s and n denote the local stream-wise and normal directions, then the transonic potential flow equation becomes

$$(c^2 - q^2)\phi_{ss} + c^2\phi_{nn} = 0. \tag{D.18}$$

Since u/q and v/q are the local direction cosines, ϕ_{ss} and ϕ_{nn} can be expressed in the original coordinate system as

$$\phi_{ss} = \frac{1}{q^2}\left(u^2\phi_{xx} + 2uv\phi_{xy} + v^2\phi_{yy}\right), \tag{D.19}$$

and

$$\phi_{nn} = \frac{1}{q^2}\left(v^2\phi_{xx} - 2uv\phi_{xy} + u^2\phi_{yy}\right). \tag{D.20}$$

Then, at subsonic points, central-difference formulas are used for both ϕ_{ss} and ϕ_{nn}. At supersonic points, central-difference formulas are used for ϕ_{nn}, but upwind difference formulas are used for the second derivatives contributing to ϕ_{ss}, as illustrated in Figure D.8.

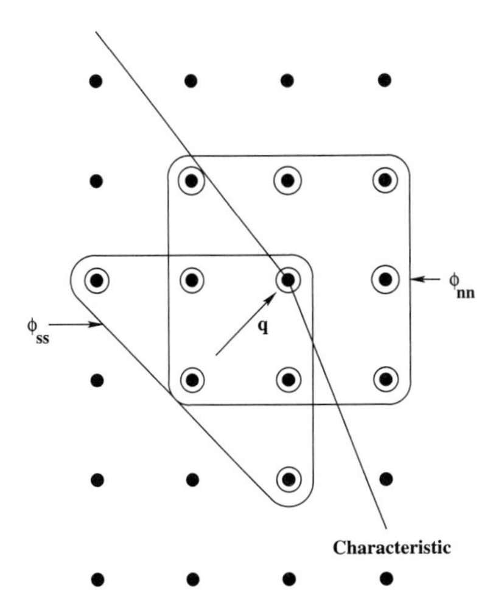

Figure D.8 Rotated difference scheme.

At a supersonic point at which $u > 0$ and $v > 0$, for example, ϕ_{ss} is constructed from the formulas

$$\phi_{xx} = \frac{\phi_{ij} - 2\phi_{i-1,j} + \phi_{i-2,j}}{\Delta x^2},$$

$$\phi_{xy} = \frac{\phi_{ij} - \phi_{i-1,j} - \phi_{i,j-1} + \phi_{i-1,j-1}}{\Delta x \Delta y},$$

$$\phi_{yy} = \frac{\phi_{ij} - 2\phi_{i,j-1} + \phi_{i,j-2}}{\Delta y^2}. \tag{D.21}$$

It can be seen that the rotated scheme reduces to a form similar to the scheme of Murman and Cole for the small-disturbance equation if either $u = 0$ or $v = 0$. The upwind difference formulas can be regarded as approximations to $\phi_{xx} - \Delta x \phi_{xxx}$, $\phi_{xy} - \frac{\Delta x}{2}\phi_{xxy} - \frac{\Delta y}{2}\phi_{xyy}$ and $\phi_{yy} - \Delta y \phi_{yyy}$. Thus at supersonic points the scheme introduces an effective artificial viscosity

$$\left(1 - \frac{c^2}{q^2}\right)\left[\Delta x(u^2 u_{xx} + uvv_{xx}) + \Delta y(uvu_{yy} + v^2 v_{yy})\right], \tag{D.22}$$

which is symmetric in x and y.

D.4.2 Difference Schemes for the Exact Potential Flow Equation in Conservation Form

In the construction of a discrete approximation to the conservation form of the potential flow equation, it is convenient to accomplish the switch to upwind differencing by the explicit addition of an artificial viscosity. Thus we solve an equation of the form

$$S_{ij} + T_{ij} = 0, \tag{D.23}$$

where T_{ij} is the artificial viscosity, which is constructed as an approximation to an expression in divergence form $\partial P/\partial x + \partial Q/\partial y$, where P and Q are appropriate quantities with a magnitude proportional to the mesh width. The central-difference approximation is constructed in the natural manner as

$$S_{ij} = \frac{(\rho u)_{i+1/2,j} - (\rho u)_{i-1/2,j}}{\Delta x} + \frac{(\rho v)_{i,j+1/2} - (\rho v)_{i,j-1/2}}{\Delta y}. \tag{D.24}$$

Consider first the case in which the flow in the supersonic zone is aligned with the x coordinate, so that it is sufficient to restrict the upwind differencing to the x derivatives. In a smooth region of the flow, the first term of S_{ij} is an approximation to

$$\frac{\partial}{\partial x}(\rho u) = \rho\left(1 - \frac{u^2}{c^2}\right)\phi_{xx} - \frac{\rho u v}{c^2}\phi_{xy}.$$

We wish to construct T_{ij} so that ϕ_{xx} is effectively represented by an upwind difference formula when $u > c$. Define the switching function

$$\mu = min\left[0, \rho\left(1 - \frac{u^2}{c^2}\right)\right]. \tag{D.25}$$

Then set

$$T_{ij} = \frac{P_{i+1/2,j} - P_{i-1/2,j}}{\Delta x}, \tag{D.26}$$

where

$$P_{i+1/2,j} = -\frac{\mu_{ij}}{\Delta x}\left[\phi_{i+1,j} - 2\phi_{ij} + \phi_{i-1,j} - \epsilon(\phi_{ij} - 2\phi_{i-1,j} + \phi_{i-2,j})\right]. \tag{D.27}$$

The added terms are an approximation to $\partial P/\partial x$, where

$$P = -\mu[(1-\epsilon)\Delta x \phi_{xx} + \epsilon \Delta x^2 \phi_{xxx}].$$

Thus, if $\epsilon = 0$, the scheme is first order accurate; but if $\epsilon = 1 - \lambda \Delta x$ and λ is a constant, the scheme is second order accurate. Also when $\epsilon = 0$, the viscosity cancels the term $\rho(1 - u^2/c^2)\phi_{xx}$ and replaces it by its value at the adjacent upwind point.

In this scheme the switch to upwind differencing is introduced smoothly because the coefficient $\mu \to 0$ as $u \to c$. If the first term in S_{ij} were simply replaced by the upwind difference formula

$$\frac{(\rho u)_{i-1/2,j} - (\rho u)_{i-3/2,j}}{\Delta x},$$

the switch would be less smooth because there would also be a sudden change in the representation of the term $(\rho uv/c^2)\phi_{xy}$, which does not necessarily vanish when $u = c$. A scheme of this type proved to be unstable in numerical tests.

The treatment of flows that are not well aligned with the coordinate system requires the use of a difference scheme in which the upwind bias conforms to the local flow direction. The desired bias can be obtained by modeling the added terms T_{ij} on the artificial viscosity of the rotated difference scheme for the quasi-linear form described in the previous section. Since (D.3) is equivalent to (D.18) multiplied by ρ/c^2, P and Q should be chosen so that $\partial P/\partial x + \partial Q/\partial y$ contains terms similar to (D.22) multiplied by ρ/c^2. The following scheme has proved successful. Let μ be a switching function that vanishes in the subsonic zone:

$$\mu = max\left[0, \left(1 - \frac{c^2}{q^2}\right)\right]. \tag{D.28}$$

Then P and Q are defined as approximations to

$$-\mu\left[(1-\epsilon)u\Delta x \rho_x + \epsilon u \Delta x^2 \rho_{xx}\right],$$

and

$$-\mu\left[(1-\epsilon)v\Delta y \rho_y + \epsilon v \Delta y^2 \rho_{yy}\right],$$

where the parameter ϵ controls the accuracy in the same way as in the simple scheme. If $\epsilon = 0$, the scheme is, first order accurate, and at a supersonic point where $u > 0$ and $v > 0$, P then approximates

$$-\Delta x\left(1-\frac{c^2}{q^2}\right)u\rho_x = \Delta x\frac{\rho}{c^2}\left(1-\frac{c^2}{q^2}\right)(u^2u_x + uvv_x).$$

When this formula and the corresponding formula for Q are inserted in $\partial P/\partial x + \partial Q/\partial y$, it can be verified that the terms containing the highest derivatives of ϕ are the same as those in (D.22) multiplied by ρ/c^2. In the construction of P and Q, the derivatives of P are represented by upwind difference formulas. Thus the formula for the viscosity finally becomes

$$T_{ij} = \frac{P_{i+1/2,j} - P_{i-1/2,j}}{\Delta x} + \frac{Q_{i,j+1/2} - Q_{i,j-1/2}}{\Delta y}, \tag{D.29}$$

where if $u_{i+1/2,j} > 0$, then

$$P_{i+1/2,j} = u_{i+1/2,j}\mu_{ij}[\rho_{i+1/2,j} - \rho_{i-1/2,j} - \epsilon(\rho_{i-1/2,j} - \rho_{i-3/2,j})],$$

and if $u_{i+1/2,j} < 0$, then

$$P_{i+1/2,j} = u_{i+1/2,j}\mu_{i+1,j}[\rho_{i+1/2,j} - \rho_{i+3/2,j} - \epsilon(\rho_{i+3/2,j} - \rho_{i+5/2,j})],$$

while $Q_{i,j+1/2}$ is defined by a similar formula.

D.4.3 Analysis of the Relaxation Method

Both the nonconservative rotated difference scheme and the difference schemes in conservation form lead to difference equations that are not amenable to solution by marching in the supersonic zone, and a rather careful analysis is needed to ensure the convergence of the iterative scheme. For this purpose it is convenient to introduce the time-dependent analogy proposed in Section D.2. Thus we regard the iterative scheme as an approximation to the artificial time–dependent equation (D.8). It was shown by Garabedian (1956) that this method can be used to estimate the optimum relaxation factor for an elliptic problem.

To illustrate the application of the method, consider the standard difference scheme for Laplace's equation. Typically, in a point over-relaxation scheme, a provisional value $\tilde{\phi}_{ij}$ is obtained by solving

$$\frac{\phi_{i-1,j}^{(n+1)} - 2\tilde{\phi}_{ij} + \phi_{i+1,j}^{(n)}}{\Delta x^2} + \frac{\phi_{i,j-1}^{(n+1)} - 2\tilde{\phi}_{ij} + \phi_{i,j+1}^{(n)}}{\Delta y^2} = 0.$$

Then the new value $\phi_{ij}^{(n+1)}$ is determined by the formula

$$\phi_{ij}^{(n+1)} = \phi_{ij}^{(n)} + \omega(\tilde{\phi}_{ij} - \phi_{ij}^{(n)}),$$

where ω is the over-relaxation factor. Eliminating $\tilde{\phi}_{ij}$, this is equivalent to calculating the correction $C_{ij} = \phi_{ij}^{(n+1)} - \phi_{ij}^{(n)}$ by solving

$$\tau_1 \left(C_{ij} - C_{i-1,j}\right) + \tau_2 \left(C_{ij} - C_{i,j-1}\right) + \tau_3 C_{i,j} = R_{ij}, \tag{D.30}$$

where R_{ij} is the residual, and

$$\tau_1 = \frac{1}{\Delta x^2},$$

$$\tau_2 = \frac{1}{\Delta y^2},$$

$$\tau_3 = \left(\frac{2}{\omega} - 1\right)\left(\frac{1}{\Delta x^2} + \frac{1}{\Delta y^2}\right).$$

Equation (D.30) is an approximation to the wave equation

$$\tau_1 \Delta t \Delta x \phi_{xt} + \tau_2 \Delta t \Delta y \phi_{yt} + \tau_3 \Delta t \phi_t = \phi_{xx} + \phi_{yy}.$$

This is damped if $\tau_3 > 0$, and to maximize the rate of convergence, the relaxation factor ω should be chosen to give an optimal amount of damping.

If we consider the potential flow equation (D.18) at a subsonic point, these considerations suggest that the scheme (D.30), where the residual R_{ij} is evaluated from the difference approximation described in Section D.4.1, will converge if

$$\tau_1 \geq \frac{c^2 - u^2}{\Delta x^2}, \quad \tau_2 \geq \frac{c^2 - v^2}{\Delta y^2}, \quad \tau_3 > 0.$$

Similarly, the scheme

$$\tau_1 \left(C_{ij} - C_{i-1,j}\right) + \tau_2 \left(C_{i,j+1} - 2C_{ij} + C_{i,j-1}\right) + \tau_3 C_{i,j} = R_{ij}, \tag{D.31}$$

which requires the simultaneous solution of the corrections on each vertical line, can be expected to converge if

$$\tau_1 \geq \frac{c^2 - u^2}{\Delta x^2}, \quad \tau_2 = \frac{c^2 - v^2}{\Delta y^2}, \quad \tau_3 > 0.$$

At supersonic points, schemes similar to (D.30) or (D.31) are not necessarily convergent (Jameson 1974). If we introduce a locally aligned Cartesian coordinate system and divide through by c^2, the general form of the equivalent time-dependent equation is

$$\left(M^2 - 1\right)\phi_{ss} - \phi_{nn} + 2\alpha\phi_{st} + 2\beta\phi_{nt} + \gamma\phi_t = 0, \tag{D.32}$$

where M is the local Mach number, and s and n are the stream-wise and normal directions. The coefficients α, β, and γ depend on the coefficients of the elements of C on the left-hand side of (D.30)and (D.31). The substitution

$$T = t - \frac{\alpha s}{M^2 - 1} + \beta n,$$

reduces this equation to the diagonal form

$$\left(M^2-1\right)\phi_{ss}-\phi_{nn}-\left(\frac{\alpha^2}{M^2-1}-\beta^2\right)\phi_{TT}+\gamma\phi_T=0.$$

Since the coefficients of ϕ_{nn} and ϕ_{ss} have opposite signs when $M>1$, T cannot be the time-like direction at a supersonic point. Instead, either s or n is time-like, depending on the sign of the coefficient of ϕ_{TT}. Since s is the time-like direction of the steady-state problem, it ought also to be the time-like direction of the unsteady problem. Thus, when $M>1$, the relaxation scheme should be designed so that α and β satisfy the compatibility condition

$$\alpha>\beta\sqrt{M^2-1}. \tag{D.33}$$

The characteristics of the unsteady equation (D.32) satisfy

$$\left(M^2-1\right)\left(t^2+2\beta nt\right)-2\alpha st-\left(\beta s-\alpha n\right)^2=0.$$

Thus the characteristic cone touches the $s-n$ plane. As long as condition (D.33) holds with $\alpha>0$ and $\beta>0$, it slants upstream in the reverse time direction, as illustrated in Figure D.9.

To ensure that the iterative scheme has the proper region of dependence, the flow field should be swept in a direction such that the updated region always includes the upwind line of tangency between the characteristic cone and the s–n plane.

A von Neumann analysis (Jameson 1974) indicates that the coefficient of ϕ_t should be zero at supersonic points, reflecting the fact that t is not a time-like direction. The mechanism of convergence in the supersonic zone can be inferred from Figure D.9. An equation of the form of (D.33) with constant coefficients reaches a steady state because with advancing time the cone of dependence ceases to intersect the initial time-plane. Instead it intersects a surface containing the Cauchy data of the steady-state problem. The rate of convergence is determined by the backward inclination of the most retarded characteristic,

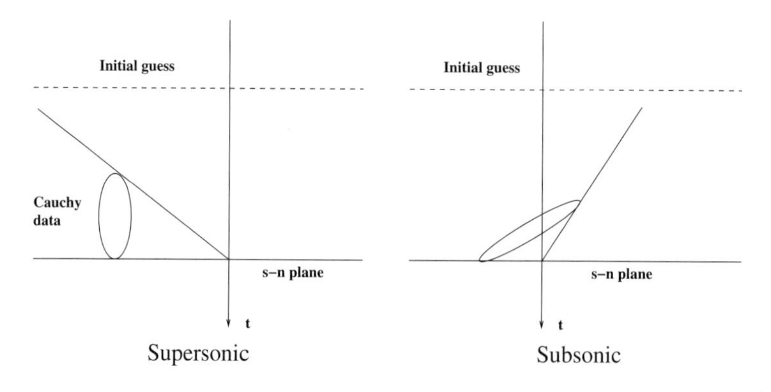

Figure D.9 Characteristic cone of equivalent time-dependent equation.

$$t = \frac{2\alpha s}{M^2 - 1}, \qquad n = -\frac{\beta}{\alpha} s,$$

and is maximized by using the smallest permissible coefficient α for the term in ϕ_{st}. In the subsonic zone, on the other hand, the cone of dependence contains the t axis, and it is important to introduce damping to remove the influence of the initial data.

D.5 Treatment of Complex Geometric Configurations

An effective approach to the treatment of two-dimensional flows over complex profiles is to map the exterior domain conformally onto the unit disk (Jameson 1974). Equation (D.3) is then written in polar coordinates as

$$\frac{\partial}{\partial \theta}\left(\frac{\rho}{r}\phi_\theta\right) + \frac{\partial}{\partial r}\left(r\rho\phi_r\right) = 0, \tag{D.34}$$

where the modulus h of the mapping function enters only in the calculation of the density from the velocity

$$\mathbf{q} = \frac{\nabla \phi}{h}. \tag{D.35}$$

The Kutta condition is enforced by adding circulation such that $\nabla\phi = 0$ at the trailing edge. This procedure is very accurate. Figure D.10 shows a numerical verification of Morawetz' theorem that a shock free transonic flow is an isolated point and that arbitrary small changes in boundary conditions will lead to the appearance of shock waves (Morawetz 1956).These calculations were performed by the author's program FLO6.

Applications to complex three-dimensional configurations require a more flexible method of discretization, such as that provided by the finite element method. Jameson and Caughey proposed a scheme using isoparametric bilinear or trilinear elements

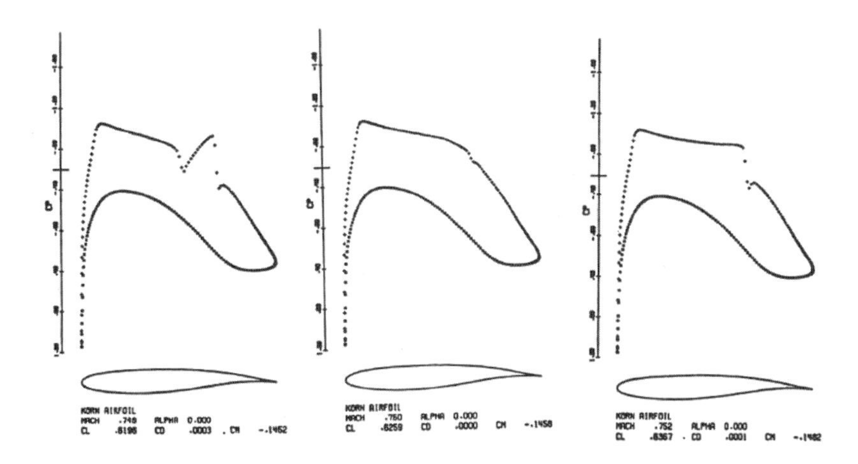

Figure D.10 Sensitivity of a shock free solution.

(Jameson & Caughey 1977, Jameson 1978). The discrete equations can most conveniently be derived from the Bateman variational principle. In the scheme of Jameson and Caughey, I is approximated as

$$I = \sum p_k V_k,$$

where p_k is the pressure at the center of the kth cell and V_k is its area (or volume), and the discrete equations are obtained by setting the derivative of I with respect to the nodal values of potential to zero. Artificial viscosity is added to give an upwind bias in the supersonic zone, and an iterative scheme is derived by embedding the steady state equation in an artificial time dependent equation. Several widely used programs (FLO27, FLO28, FLO30) have been developed using this scheme. Figure D.11 shows a result for a swept wing.

An alternative approach to the treatment of complex configurations was developed by Bristeau et al. (1980, 1981). Their method uses a least-squares formulation of the problem, together with an iterative scheme derived with the aid of optimal control

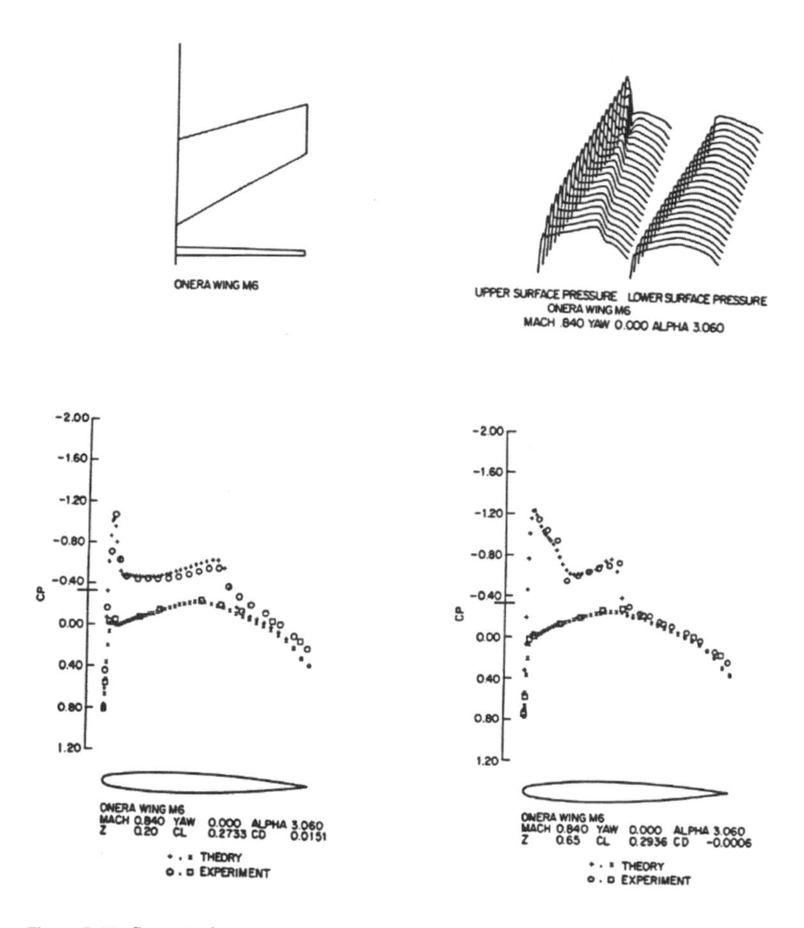

Figure D.11 Swept wing.

theory. The method could be used in conjunction with a subdivision into either quadrilaterals or triangles, but in practice triangulations were used.

The simplest conceivable least squares formulation calls for the minimization of the objective function

$$I = \int_S \psi^2 dS,$$

where ψ is the residual of (D.3), and S is the domain of the calculation. The resulting minimization problem could be solved by a steepest descent method in which the potential is repeatedly modified by a correction $\delta\phi$ proportional to $\frac{\partial I}{\partial \phi}$. Such a method would be very slow. In fact it simulates a time dependent equation of the form

$$\phi_t = -L^* L\phi,$$

where L is the differential operator in (D.3), and L^* is its adjoint. Much faster convergence can be obtained by the introduction of a more sophisticated objective function

$$I = \int_S \nabla\psi^2 dS,$$

where the auxiliary function ϕ is calculated from

$$\nabla^2\psi = \nabla.\left(\rho\nabla\phi\right).$$

Let g be the value of $\frac{\partial\phi}{\partial n}$ specified on the boundary C of the domain. Then this equation can be replaced by the corresponding variational form

$$\int_S \nabla\psi.\nabla v dS = \int_S \rho\nabla\ .\nabla v dS - \int_C g v dS,$$

which must be satisfied by ψ for all differentiable test functions v. This formulation, which is equivalent to the use of an H^{-1} norm in Sobolev space, reduces the calculation of the potential to the solution of an optimal control problem, with ϕ as the control function and ψ as the state function. It leads to an iterative scheme that calls for solutions of Poisson equations twice in each cycle. A further improvement can be realized by the use of a conjugate gradient method instead of a simple steepest descent method.

The least squares method in its basic form allows expansion shocks. In early formulations these were eliminated by penalty functions. Subsequently it was found best to use upwind biasing of the density. The method was extended at Avions Marcel Dassault to the treatment of extremely complex three-dimensional configurations including complete aircraft, using a subdivision of the domain into tetrahedra (Bristeau et al. 1985).

Appendix E Fundamental Stability Theory II

E.1 Conservative Schemes

The scheme is conservative if it conserves a global property of the solution such as the mass or energy. This implies that for some weights A_i, an explicit scheme satisfies the condition

$$\sum_i A_i v_i^{n+1} = \sum_i A_i v_i^n. \tag{E.1}$$

Suppose that the scheme is written as

$$A_j v_j^{n+1} = A_j v_j^n + \sum_{i \neq j} f_{ij}(v^n),$$

where $f_{ij}(v^n)$ is the flux from i to j. Then (E.1) is satisfied if

$$f_{ij}(v^n) = -f_{ji}(v^n) \tag{E.2}$$

since then

$$\begin{aligned}
\sum_j A_j v_j^{n+1} &= \sum_j A_j v_j^n + \sum_i \sum_j f_{ij}(v^n) \\
&= \sum_j A_j v_j^n + \frac{1}{2} \sum_i \sum_j \left(f_{ij}(v^n) + f_{ji}(v^n) \right) \\
&= \sum_j A_j v_j^n.
\end{aligned}$$

Consider an explicit scheme in the form

$$v_i^{n+1} = \sum_j a_{ij} v_j^n. \tag{E.3}$$

If this satisfies

$$\sum_i v_i^{n+1} = \sum_i v_i^n, \tag{E.4}$$

it follows that

$$\sum_i \sum_j a_{ij} v_j^n = \sum_j v_j^n.$$

If a_{ij} does not depend on v^n, then on setting $v_j^n = \delta_{kj}$, it follows that

$$\sum_i \sum_j a_{ij}\delta_{kj} = \sum_j \delta_{kj}$$

or

$$\sum_i a_{ik} = 1.$$

Suppose that the discrete values v_j^n correspond to the values of a sufficiently smooth function $u(x,t)$ at $x = j\Delta x$, $t = n\Delta t$. Then a Taylor series expansion yields

$$v + \Delta t v_t + \frac{\Delta t^2}{2} v_{tt} + \cdots = \sum_j a_{ij}\left(v + j\Delta x v_x + j^2 \frac{\Delta x^2}{2} v_{xx} + \cdots\right),$$

and (E.2) is consistent with a differential equation with no source term only if

$$\sum_j a_{ij} = 1. \tag{E.5}$$

Then (E.3) can be written as

$$v_i^{n+1} = v_i^n + \sum_{j \neq i} a_{ij}(v_j^n - v_i^n),$$

and (E.4) implies that

$$\sum_i \sum_j a_{ij}(v_j^n - v_i^n) = 0.$$

A sufficient condition for (E.1) and hence (E.4) to be satisfied is thus that A is symmetric,

$$a_{ji} = a_{ij}.$$

This is typically the case for a diffusion equation.

Semi-discrete schemes may similarly be considered conservative if they satisfy the condition that

$$\sum_i A_i \frac{dv_i}{dt} = 0 \tag{E.6}$$

for an appropriate set of weights A_i. If the scheme can be written in the form

$$A_j \frac{dv_j}{dt} = \sum_{i \neq j} f_{ij}(v),$$

where the fluxes $f_{ij}(v^n)$ satisfy condition (E.2), then (E.6) is conservative. If the semi-discrete scheme is written as

$$\frac{dv_i}{dt} = \sum_j a_{ij} v_j,$$

then

$$\sum_i \frac{dv_i}{dt} = 0$$

if and only if

$$\sum_i \sum_j a_{ij} v_j = 0.$$

If the coefficients a_{ij} do not depend on v, it follows on setting $v_j = \delta_{kj}$ that

$$\sum_i a_{ik} = 0.$$

Also (E.1) is consistent with a differential equation with no source term if

$$\sum_j a_{ij} = 0.$$

In this case the scheme is conservative if the symmetry condition (E.5) is satisfied.

E.2 Nonincreasing Maximum Norm

If

$$v_i^{n+1} = \sum_j a_{ij} v_j^n \tag{E.7}$$

or in matrix notation

$$v^{n+1} = Av^n,$$

then

$$\begin{aligned} \left|v_i^{n+1}\right| &\leq \sum_j \left|a_{ij}\right| \left|v_j^n\right| \\ &\leq \max_i \sum_j \left|a_{ij}\right| \, \left\|v^n\right\|_\infty . \end{aligned}$$

Also if

$$u_j = \operatorname{sgn}(a_{ij})$$

for the row for which this maximum is realized, then

$$v_i^{n+1} = \sum_j \left|a_{ij}\right|.$$

Thus

$$\left\|v^{n+1}\right\|_\infty \leq \left\|v^n\right\|_\infty$$

if and only if

$$\max_i \sum_j \left|a_{ij}\right| \leq 1 \tag{E.8}$$

or, in matrix notation

$$\|A\|_\infty \leq 1,$$

where $\|A\|_\infty$ is the induced norm

$$\sup_v \frac{\|Av\|_\infty}{\|v\|_\infty} = \max_i \sum_j \left|a_{ij}\right|.$$

Suppose that (E.7) is the evolution formula for a discrete scheme that approximates a differential equation, where the superscript $+$ denotes the updated value after advancing a time step Δt. Assuming that the discrete values v_j^n correspond to the values of a sufficiently smooth function $u(x,t)$ at $x = j\Delta x$, $t = n\Delta t$, a Taylor series expansion yields

$$v + \Delta t v_t + \frac{\Delta t^2}{2} v_{tt} + \cdots = \sum_j a_{ij} \left(v + j\Delta x v_x + j^2 \frac{\Delta x^2}{2} v_{xx} + \cdots \right).$$

Thus (E.7) represents a differential equation with no source term only if

$$\sum_j a_{ij} = 1. \tag{E.9}$$

Then (E.7) can be written as

$$v_j^{n+1} = \left(\sum_j a_{ij} \right) v_i^n + \sum_{j \neq i} a_{ij} \left(v_j^n - v_i^n \right) \tag{E.10}$$

$$= v_i^n + \Delta t \sum_{j \neq i} \hat{a}_{ij} \left(v_j^n - v_i^n \right), \tag{E.11}$$

where

$$a_{ij} = \Delta t \, \hat{a}_{ij},$$

and the sum is a discretization of the spatial derivatives.

If condition (E.9) is satisfied and $a_{ij} < 0$ for any j, then the sum of the absolute values $\left|a_{ij}\right|$ would exceed unity, violating condition (E.8). On the other hand, if

$$a_{ij} \geq 0, \tag{E.12}$$

then

$$\sum_j \left|a_{ij}\right| = \sum_j a_{ij} = 1.$$

Thus the solution of (E.7) has a nonincreasing maximum norm if conditions (E.9) and (E.12) are satisfied. It follows from (E.9) that

$$a_{ii} = 1 - \sum_{j \neq i} a_{ij} = 1 - \Delta t \sum_{j \neq i} \hat{a}_{ij}.$$

Thus (E.12) can only be satisfied if the time step satisfies the constraint

$$\Delta t \leq \frac{1}{\sum_{j \neq i} \hat{a}_{ij}}.$$

These results are valid both in the one-dimensional case and for initial boundary values problems on arbitrary multi-dimensional meshes.

In the one-dimensional case it is convenient to write the scheme in the form

$$v_i^{n+1} = v_i^n + \sum_{j=-\infty}^{\infty} c_{ij}\left(v_j^n - v_{j-1}^n\right),$$

where the evolution formula is expressed in terms of the differences $v_j^n - v_{j-1}^n$ over each mesh interval. Since (E.10) can be expanded as

$$\begin{aligned}
v_i^{n+1} = u_i &+ (a_{i,i+1} + a_{i,i+2} + a_{i,i+3} \ldots)\,(v_{i+1}^n - v_i^n) \\
&+ (a_{i,i+2} + a_{i,i+3} \ldots)\,(v_{i+2}^n - v_{i+1}^n) \\
&+ (a_{i,i+3} \ldots)\,(v_{i+3}^n - v_{i+2}^n) \\
&+ \cdots \\
&- (a_{i,i-1} + a_{i,i-2} + a_{i,i-3} \ldots)\,(v_i^n - v_{i-1}^n) \\
&- (a_{i,i-2} + a_{i,i-3} \ldots)\,(v_{i-1}^n - v_{i-2}^n) \\
&- (a_{i,i-3} \ldots)\,(v_{i-2}^n - v_{i-3}^n) \\
&- \ldots,
\end{aligned}$$

it follows that

$$c_{i,i+1} = \sum_{j=i+1}^{\infty} a_{ij}$$

$$c_{i,i+2} = \sum_{j=i+2}^{\infty} a_{ij}$$

$$\cdots$$

and

$$c_{i,i} = -\sum_{j=-\infty}^{i-1} a_{ij}$$

$$c_{i,i-1} = -\sum_{j=-\infty}^{i-2} a_{ij}$$

$$\cdots$$

Conversely, if c_{ij} is the first nonzero coefficient,

$$\begin{aligned} a_{i,j-1} &= -c_{ij} \\ a_{ij} &= c_{ij} - c_{i,j+1} \\ &\cdots \\ a_{i,i-1} &= c_{i,i-1} - c_{ii}, \end{aligned}$$

and if c_{ik} is the last nonzero coefficient,

$$\begin{aligned} a_{ik} &= c_{ik} \\ a_{i,k-1} &= c_{i,k-1} - c_{ik} \\ &\cdots \\ a_{i,i+1} &= c_{i,i+1} - c_{i,i+2} \end{aligned}$$

while

$$a_{ii} = 1 + c_{ii} - c_{i,i+1}.$$

Thus conditions (E.12) are satisfied if and only if

$$\begin{aligned} c_{i,i+1} \geq c_{i,i+2} \geq c_{i,i+3} \cdots \geq 0 \\ c_{ii} \leq c_{i,i-1} \leq c_{i,i-2} \cdots \leq 0 \end{aligned}$$

and

$$c_{i,i+1} - c_{ii} \leq 1.$$

E.3 Nondecreasing Maximum Norm

If

$$u_i = \sum_j a_{ij} v_j$$

and

$$|a_{ii}| - \sum_{j \neq i} |a_{ij}| \geq 1 \text{ for all } i,$$

then

$$\|u\|_\infty \geq \|v\|_\infty .$$

Proof let i be the index for which $|v_i|$ is the greatest, so $|v_i| = \|v\|_\infty$. Then

$$|u_i| = \left| a_{ii}\, v_i + \sum_{j \neq i} a_{ij}\, v_j \right|$$

$$\geq |a_{ii}|\,|v_i| - \sum_{j\neq i} \left|a_{ij}\right|\,\left|v_j\right|$$

$$\geq \left\{ |a_{ii}| - \sum_{j\neq i} \left|a_{ij}\right| \right\} \left|v_j\right|.$$

Thus

$$\|u\|_\infty \geq \min_i \left\{ |a_{ii}| - \sum_{j\neq i} \left|a_{ij}\right| \right\} \|v\|_\infty .$$

E.4 Implicit Schemes with Nonincreasing Maximum Norm

Consider the general scheme

$$\sum_i b_{ij} v_j^{n+1} = \sum_j a_{ij} v_j^n. \tag{E.13}$$

Assuming that the discrete values v_j^n correspond to a sufficiently smooth function $u(x,t)$ at $x = j\Delta x$, $t = n\Delta t$, a Taylor series expansion yields

$$\sum_j b_{ij}(v + \Delta t\ v_t + j\Delta x\ v_x + \cdots) = \sum_j a_{ij}(v + j\Delta x\ v_x + \cdots).$$

Consistency with a differential equation with no source term implies that

$$\sum_j b_{ij} = \sum_j a_{ij},$$

and without loss of generality (E.13) can be scaled so that

$$\sum_j b_{ij} = \sum_j a_{ij} = 1. \tag{E.14}$$

Let

$$u = Av.$$

Then according to the result of Section E.3

$$\|u\|_\infty \geq \left\|v^{n+1}\right\|_\infty \tag{E.15}$$

if

$$b_{ii} \geq 1 + \sum_{j\neq i} \left|b_{ij}\right|. \tag{E.16}$$

But according to (E.14)

$$b_{ii} = 1 - \sum_{j\neq i} b_{ij},$$

and if any $b_{ij} > 0, j \neq i$, this would be violated. On the other hand, if

$$b_{ij} \leq 0 \text{ when } j \neq i, \tag{E.17}$$

then (E.16) is satisfied and (E.15) holds. Now, if in addition

$$a_{ij} \geq 0, \tag{E.18}$$

then

$$\|u\|_\infty \leq \|v^n\|_\infty .$$

Thus the implicit scheme (E.13) is stable in the maximum norm

$$\|v^{n+1}\|_\infty \leq \|v^n\|_\infty$$

if conditions (E.14), (E.17), and (E.18) are satisfied.

E.5 Semi-discrete Schemes with Nonincreasing Maximum Norm

Consider the general semi-discrete scheme

$$\frac{dv_i}{dt} = \sum_j a_{ij} v_j. \tag{E.19}$$

Assuming that the discrete values $v_j(t)$ correspond to a sufficiently smooth function $v(x,t)$ at $x = j\Delta x$, a Taylor series expansion yields

$$\frac{dv_i}{dt} = \sum_j a_{ij} \left(v + j\Delta x v_x + j^2 \frac{\Delta x^2}{2} v_{xx} + \cdots \right),$$

and (E.19) is consistent with a differential equation with no source term only if

$$\sum_j a_{ij} = 0.$$

Then (E.19) can be written as

$$\frac{dv_i}{dt} = \left(\sum_j a_{ij} \right) v_i + \sum_{j \neq i} a_{ij}(v_j - v_i)$$

or

$$\frac{dv_i}{dt} = \sum_{j \neq i} a_{ij}(v_j - v_i).$$

Suppose that

$$a_{ij} \geq 0 \, , \, j \neq i. \tag{E.20}$$

Then if v_i is a maximum,

$$v_j - v_i \leq 0$$

and

$$\frac{dv_i}{dt} \leq 0. \tag{E.21}$$

Similarly if v_i is a minimum,

$$\frac{dv_i}{dt} \geq 0. \tag{E.22}$$

Now $\|v\|_\infty$ can increase only if the maximum increases or the minimum decreases. But

$$\left|\frac{dv_i}{dt}\right| \leq \|A\|_\infty \|v\|_\infty;$$

so if $|v_i| \leq \|v\|_\infty$ there is a time interval $\epsilon > 0$ during which it cannot become the maximum or minimum, while if $|v_i| = \|v\|_\infty$ it follows from (E.21) and (E.22) that

$$\left|\frac{dv_i}{dt}\right| \leq 0.$$

Thus (E.20) is sufficient to ensure that $\|v\|_\infty$ does not increase. If (E.20) is not satisfied, then by choosing $v_i = 1$; $v_j = 1$ if $a_{ij} \geq 0$, $v_j = 0$ if $a_{ij} < 0$, one obtains

$$\frac{dv_i}{dt} > 0$$

and $\|v\|_\infty$ will increase.

E.6 Nonincreasing Euclidean Norm

Suppose that

$$v_i^{n+1} = \sum_j a_{ij} v_j^n \tag{E.23}$$

or, in matrix notation

$$v^{n+1} = Av^n,$$

where

$$\sum_j a_{ij} = 1, \tag{E.24}$$

so that the scheme is consistent with a differential equation with no source term.

Then

$$\|v^{n+1}\|_2^2 = {v^{n+1}}^T v^{n+1} = {v^n}^T A^T A v^n.$$

Since $A^T A$ is symmetric and real, it can be reduced to a diagonal form M containing the eigenvalues μ_i of $A^T A$ as its elements by a unitary transformation

$$U^T A^T A U = M\,,\ U^T U = I.$$

Set

$$w = U^T v.$$

Then

$$w^T w = v^T U U^T v = v^T v = \|v\|_2^2.$$

Also

$$v^T A^T A v = w^T M w = \sum_i \mu_i w_i^2$$

and

$$\|v^{n+1}\|_2^2 \leq \max_i |\mu_i| w^T w.$$

Thus

$$\|v^{n+1}\|_2 \leq \|v^n\|_2$$

if

$$\max_i |\mu_i| \leq 1.$$

Suppose that

$$\sum_i a_{ij} = 1 \tag{E.25}$$

so that the scheme is conservative, and

$$a_{ij} \geq 0. \tag{E.26}$$

Set

$$C = A^T A.$$

Then

$$\sum_j c_{ij} = \sum_j \sum_k a_{ki} a_{kj} = \sum_k a_{ki} = 1 \tag{E.27}$$

and

$$c_{ij} = \sum_k a_{ki} a_{kj} \geq 0. \tag{E.28}$$

By Gershgorein's theorem the eigenvalues of C lie in the union of circles with centers

$$x_i = c_{ii},\ y_i = 0$$

and with radii

$$r_i = \sum_{j \neq i} |c_{ij}|.$$

It follows from (E.27) and (E.28) that

$$r_i = 1 - c_{ii}.$$

Then every eigenvalue μ_i lies in the unit disk, and condition (E.24) is satisfied. Thus the solution of (E.23) has a nonincreasing Euclidean norm if conditions (E.24), (E.25), and (E.26) are satisfied.

E.7 Implicit Schemes with Nonincreasing Euclidean Norm

Consider the general scheme

$$\sum_j b_{ij} v_j^{n+1} = \sum_j a_{ij} v_j^n, \tag{E.29}$$

which satisfies the condition

$$\sum_j b_{ij} = \sum_j a_{ij} = 1 \tag{E.30}$$

if it is consistent with a differential equation with no source term. Then (E.29) could be written as

$$v_i^{n+1} + \sum_{j \neq i} b_{ij}(v_j^{n+1} - v_i^{n+1}) = v_i + \sum_{j \neq i} a_{ij}(v_j^n - v_i^n)$$

or in matrix form as

$$\left(I + \frac{\Delta t}{2}\hat{B}\right) v^{n+1} = \left(I + \frac{\Delta t}{2}\hat{A}\right) v^n,$$

where

$$\frac{\Delta t}{2}\hat{A} = A - I\,, \quad \frac{\Delta t}{2}\hat{B} = B - I.$$

Suppose that $\hat{B} = -\hat{A}$. Then the scheme reduces to the trapezoidal rule

$$v^{n+1} = v^n + \frac{\Delta t}{2}\hat{A}(v^{n+1} + v^n).$$

Multiplication by $(v^{n+1} + v^n)^T$ yields

$$\|v^{n+1}\|_2^2 - \|v^n\|_2^2 = (v^{n+1} + v^n)^T (A - I)(v^{n+1} - v^n). \tag{E.31}$$

Suppose that A is symmetric

$$a_{ij} = a_{ji} \tag{E.32}$$

so that the scheme is conservative, and that

$$a_{ij} \geq 0. \tag{E.33}$$

It follows from (E.29) that

$$\sum_{j \neq i} |a_{ij}| = \sum_{j \neq i} a_{ij} = 1 - a_{ii}.$$

Since $A - I$ is symmetric, it has real eigenvalues. By Gershgorin's theorem the eigenvalues lie in the union of the circles with centers at

$$x_i = -(1 - a_{ii}), \; y_i = 0$$

and with radii

$$r_i = \sum_{j \neq i} |a_{ij}| = 1 - a_{ii}.$$

Thus the eigenvalues are nonpositive, and the quadratic form

$$(v^{n+1} + v^n)^T (A - I)(v^{n+1} - v^n) \leq 0.$$

It follows that the solution of (E.31) has a nonincreasing Euclidean norm if conditions (E.30), (E.32), and (E.33) are satisfied.

E.8 Semi-discrete Schemes with Nonincreasing L_2 Norm

Consider the general semi-discrete scheme

$$\frac{dv_i}{dt} = \sum_j a_{ij} v_j, \tag{E.34}$$

where

$$\sum_j a_{ij} = 0 \tag{E.35}$$

so that (E.35) is equivalent to

$$\frac{dv_i}{dt} = \sum_j a_{ij}(v_j - v_i).$$

Now the scheme will be conservative if

$$a_{ji} = a_{ij} \tag{E.36}$$

since then

$$\sum_i \frac{dv_i}{dt} = \sum_i \sum_j a_{ij}(v_j - v_i) = \frac{1}{2} \sum_i \sum_j \{a_{ij}(v_j - v_i) - a_{ji}(v_i - v_j)\} = 0.$$

Let

$$J = \frac{1}{2} \|v\|_2^2 = \frac{1}{2} \sum_i v_i^2.$$

Then

$$\frac{dJ}{dt} = \sum_i v_i \frac{dv_i}{dt} = \sum_i \sum_j a_{ij} v_i v_j.$$

Suppose that

$$a_{ij} \geq 0\,,\, j \neq i. \tag{E.37}$$

According to (E.35)

$$a_{ii} = -\sum_{j \neq i} a_{ij}.$$

Hence it follows from (E.37) that

$$\sum_{j \neq i} |a_{ij}| = |a_{ii}|.$$

Now by Gershgorin's theorem, the eigenvalues of the symmetric matrix A with coefficients a_{ij} lie in the union of circles with centers at

$$x = a_{ii} < 0\,,\; y = 0$$

and with radii

$$r_i = \sum_{j \neq i} |a_{ij}| = |a_{ii}|.$$

Thus the eigenvalues λ_i of A are nonpositive. Since A is real and symmetric, it can be reduced to a diagonal form Λ containing the eigenvalues of A as its elements by a unitary transformation

$$U^T A U = \Lambda\,,\; U^T U = I.$$

Then the form

$$v^T A v = w^T U^T A U w = \sum_i \lambda_i w_i,$$

where

$$w = U^T v\,,\; v = Uw,$$

and since $\lambda_i \leq 0$, it follows that

$$\frac{dJ}{dt} \leq 0.$$

Thus the solution of the semi-discrete scheme (E.34) has a nonincreasing Euclidean norm if it satisfies conditions (E.35), (E.36), and (E.37).

The first two conditions correspond to a conservative scheme that is consistent with a differential equation with no source term. This result holds for the initial boundary value problem on an arbitrary multi-dimensional mesh.

E.9 Nonincreasing L_1 Norm

If

$$v_i^{n+1} = \sum_j a_{ij} v_j^n \tag{E.38}$$

or in matrix notation

$$v^{n+1} = Av^n, \tag{E.39}$$

then

$$\|v^{n+1}\|_1 = \sum_i |v_i^{n+1}| = \sum_i s_i v_i^{n+1},$$

where s_i is the sign of v_i

$$s_i = \begin{cases} 1 & if \quad v_i^n > 0 \\ 0 & if \quad v_i^n = 0 \\ -1 & if \quad v_i^n < 0 \end{cases}$$

Thus

$$\|v^{n+1}\|_1 = \sum_i s_i \sum_j a_{ij} v_j^n = \sum_j v_j^n w_j,$$

where

$$w_j = \sum_i a_{ij} s_i$$

and

$$|w_j| \le \sum_i |a_{ij}|.$$

Then

$$\|v^{n+1}\|_1 \le \max_j |w_j| \sum_j |v_j^n| \le \|v^n\|_1$$

if

$$\max_j \sum_i |a_{ij}| \le 1 \tag{E.40}$$

or, in matrix notation

$$\|A\|_1 \le 1,$$

where $\|A\|_1$ is the induced norm

$$\sup_v \frac{\|Av\|_1}{\|v\|_1} = \max_j \sum_i |a_{ij}|.$$

Suppose that (E.38) represents a differential equation with no source term, with the consequence that

$$\sum_j a_{ij} = 1, \tag{E.41}$$

and (E.38) can be written as

$$v_i^{n+1} = \left(\sum_j a_{ij}\right) v_i^n + \sum_{j \neq i} a_{ij}(v_j^n - v_i^n) = v_i^n + \Delta t \sum_{j \neq i} \tilde{a}_{ij}(v_j^n - v_i^n),$$

where

$$a_{ij} = \Delta t\, \tilde{a}_{ij}.$$

Suppose also that the scheme is conservative

$$\sum_i v_i^{n+1} = \sum_i v_i^n \tag{E.42}$$

and that it is positive

$$a_{ij} \geq 0. \tag{E.43}$$

Suppose that $v_i^n \geq 0$ for all i. Then it follows from (E.38) that $v_i^{n+1} \geq 0$ for all i. Now (E.42) implies that

$$\|v^{n+1}\|_1 = \sum_i |v_i^{n+1}| = \sum_i v_i^{n+1} = \sum_i v_i^n = \sum_i |v_i^n| = \|v^n\|_1.$$

If there are both positive and negative values v_i^n, conditions (E.41), (E.42), and (E.43) are not sufficient to ensure that the L_1 norm does not increase, as can be seen from the following example:

$$\begin{bmatrix} 0 & 1 & 0 & 0 \\ 0 & 1 & 0 & 0 \\ 0 & 0 & 1 & 0 \\ 0 & 0 & 1 & 0 \end{bmatrix} \begin{bmatrix} 0 \\ 1 \\ -1 \\ 0 \end{bmatrix} = \begin{bmatrix} 1 \\ 1 \\ -1 \\ -1 \end{bmatrix},$$

in which $\|v\|_1$ is doubled. In the case that the scheme satisfies the condition

$$\sum_i a_{ij} = 1, \tag{E.44}$$

which also assumes that it is a conservative scheme that satisfies (E.42), then if $a_{ij} < 0$ for any i, the sum of the absolute values $|a_{ij}|$ would exceed unity, violating condition (E.9). On the other hand, (E.43) and (E.44) now imply that

$$\sum_i a_{ij} = \sum_i |a_{ij}| = 1.$$

Thus the solution of (E.40) has a nonincreasing L_1 norm if conditions (E.43) and (E.44) are satisfied.

It also follows from (E.41) that (E.40) can be written as

$$v_i^{n+1} = \left(\sum_j a_{ij}\right) v_i^n + \sum_{j \neq i} a_{ij}(v_j^n - v_i^n) = v_i^n + \Delta t \sum_{j \neq i} \tilde{a}_{ij}(v_j^n - v_i^n),$$

where

$$a_{ij} = \Delta t\, \tilde{a}_{ij}.$$

Then (E.44) implies that

$$a_{jj} = 1 - \sum_{i \neq j} a_{ij} = 1 - \Delta t \sum_{i \neq j} \tilde{a}_{ij}.$$

Thus (E.43) can only be satisfied if the time step satisfies the constraint

$$\Delta t \leq \frac{1}{\sum_{i \neq j} \tilde{a}_{ij}}.$$

These results are valid both in the one-dimensional case and for initial boundary value problems on arbitrary multi-dimensional meshes, provided that the boundary conditions also satisfy the hypotheses.

E.10 Nondecreasing L_1 Norm

Suppose that

$$u_i = \sum a_{ij} v_j$$

and

$$|a_{jj}| - \sum_{i \neq j} |a_{ij}| \geq 1 \text{ for all } j. \tag{E.45}$$

Then

$$\|u\|_1 \geq \|v\|_1.$$

Proof Let s_i be the sign of u_i. Then

$$\|u\|_1 = \sum_i s_i u_i = \sum_i \sum_j s_i a_{ij} v_j = \sum_j w_j v_j,$$

where

$$w_j = \sum_i a_{ij} s_i$$

and

$$|w_j| \geq |a_{jj}| - \sum_{i \neq j} |a_{ij}|,$$

and if (E.45) holds then

$$\min_j |w_j| \geq 1.$$

It follows that

$$\|u\|_1 \geq \min_j |w_j| \sum_j |v_j| \geq \|v\|_1.$$

E.11 Implicit Schemes with Nonincreasing L_1 Norm

Consider the general implicit scheme

$$\sum_j b_{ij} v_j^{n+1} = \sum_j a_{ij} v_j^n,$$

which satisfies the condition

$$\sum_j b_{ij} = \sum_j a_{ij} = 1$$

if it is consistent with a differential equation with no source term. If the scheme also satisfies the conditions

$$\sum_i b_{ij} = \sum_i a_{ij} = 1, \tag{E.46}$$

it is conservative,

$$\sum_j v_j^{n+1} = \sum_j v_j^n.$$

Let

$$u = Av^n.$$

Then according to the result of Section E.10

$$\|u\|_1 \geq \|v^{n+1}\|_1$$

if

$$|b_{jj}| \geq 1 + \sum_{i \neq j} |b_{ij}|. \tag{E.47}$$

If (E.46) holds,

$$b_{jj} = 1 - \sum_{i \neq j} b_{ij}.$$

Then (E.47) is satisfied only if

$$-\sum_{i \neq j} b_{ij} \geq \sum_{i \neq j} |b_{ij}|,$$

and this is violated if any $b_{ij} > 0, j \neq i$.

On the other hand it is satisfied if

$$b_{ij} \leq 0\,,\, j \neq i. \tag{E.48}$$

Also, if

$$a_{ij} \geq 0\,,\, j \neq i \tag{E.49}$$

then

$$\|u\|_1 \leq \|v^n\|_1.$$

Thus the implicit scheme is stable in the L_1 norm

$$\|v^{n+1}\|_1 \leq \|v^n\|_1$$

if conditions (E.46), (E.48), and (E.49) are satisfied. Condition (E.46) is not necessary, in general, however, for (E.11) to hold.

E.12 Semi-discrete Schemes with Nonincreasing L_1 Norm

Consider the general semi-discrete scheme

$$\frac{dv_i}{dt} = \sum_j a_{ij} v_j,$$

where

$$\sum_j a_{ij} = 0 \tag{E.50}$$

so that (E.12) is equivalent to

$$\frac{dv_i}{dt} = \sum_{j \neq i} a_{ij}(v_j - v_i).$$

Suppose also that

$$a_{ij} = a_{ji} \tag{E.51}$$

so that the scheme is conservative.

The L_1 norm can be expressed as

$$\|v\|_1 = \sum_i |v_1| = \sum_i s_i v_i,$$

where s_i is the sign of v_i

$$s_i = \begin{cases} 1 & \text{if } v_i > 0 \\ 0 & \text{if } v_i = 0 \\ -1 & \text{if } v_i < 0 \end{cases}$$

Now if $v_i \neq 0$,

$$\frac{d}{dt}|v_i| = s_i \frac{dv_i}{dt} + v_i \frac{ds_i}{dt} = s_i \frac{dv_i}{dt}.$$

If $v_i(t)$ changes sign at $t = t_1$, then there is a jump in $\frac{d}{dt}|v_i|$, which is correctly accounted for by the jump in s_i, so (E.50) is still valid. Thus

$$\frac{d}{dt}\|v\|_1 = \sum_i \sum_j s_i a_{ij} v_j = \sum_j w_j v_j,$$

where

$$w_j = \sum_i a_{ij} s_i.$$

Then

$$\frac{d}{dt}\|v\|_1 \leq 0$$

if w_j has the opposite sign to v_j.

This will be the case if

$$-a_{jj} \geq \sum_{i \neq j} |a_{ij}|.$$

But it follows from (E.50) and (E.51) that

$$-a_{jj} = \sum_{i \neq j} a_{ij},$$

and together these imply that

$$\sum_{i \neq j} a_{ij} \geq \sum_{i \neq j} |a_{ij}|,$$

which is violated if $a_{ij} < 0$ for any $j \neq i$ and satisfied if

$$a_{ij} \geq 0\,, j \neq i. \tag{E.52}$$

Thus the solution of the semi-discrete scheme (E.12) has a nonincreasing L_1 norm if conditions (E.50), (E.51), and (E.52) are satisfied. The same conditions also ensure that the solution has nonincreasing L_2 and L_∞ norms.

Appendix F Turbulence Models

F.1 Algebraic Models

Cebeci–Smith Model

The eddy viscosity is

$$\nu_t = \begin{cases} \nu_{t_i}, & y \le y_m, \\ \nu_{t_o}, & y > y_m, \end{cases}$$

where y_m is the smallest value of y for which $\nu_{T_i} = \nu_{T_o}$.

Inner layer:

$$\nu_{t_i} = l_{\text{mix}}^2 \left[\left(\frac{\partial U}{\partial y} \right)^2 + \left(\frac{\partial V}{\partial x} \right)^2 \right]^{1/2},$$

$$l_{\text{mix}} = \kappa y \left[1 - e^{-y^+/A^+} \right].$$

Outer layer:

$$\nu_{t_o} = \alpha U_e \delta_v^* F_{\text{Kleb}}(y;\delta),$$

$$F_{\text{Kleb}}(y;\delta) = \left[1 + 5.5 \left(\frac{y}{\delta} \right)^6 \right]^{-1}.$$

Closure coefficients:

$$\kappa = 0.40, \quad \alpha = 0.0168, \quad A^+ = 26 \left[1 + y \frac{dp/dx}{\rho u_\tau^2} \right]^{-1/2},$$

$$\delta_v^* = \int_0^\delta (1 - U/U_e) dy.$$

Baldwin–Lomax model

Inner layer:

$$\nu_{t_i} = l_{\text{mix}}^2 |\omega|,$$

$$l_{\text{mix}} = \kappa y \left[1 - e^{-y^+/A_o^+} \right].$$

Outer layer:

$$\nu_{t_o} = \alpha C_{cp} F_{\text{wake}} F_{\text{Kleb}}(y; y_{\max}/C_{\text{Kleb}}),$$

$$F_{\text{wake}} = \min[y_{\max} F_{\max}; C_{\text{wk}} y_{\max} U_{\text{dif}}^2 / F_{\max}],$$

$$F_{\max} = \frac{1}{\kappa}\left[\max_y(l_{\text{mix}}|\omega|)\right],$$

where $y_{\max}$ is the value of y at which $l_{\text{mix}}|\omega|$ achieves its maximum value. Closure coefficients:

$$\left.\begin{array}{lll} \kappa = 0.40, & \alpha = 0.0168, & A_o^+ = 26 \\ C_{cp} = 1.6, & C_{\text{Kleb}} = 0.3, & C_{\text{wk}} = 1 \end{array}\right\}$$

$$F_{\text{Kleb}}(y;\delta) = \left[1 + 5.5\left(\frac{y}{\delta}\right)^6\right]^{-1},$$

with δ replaced by $y_{\max}/C_{\text{Kleb}}$.

$$\omega = \left[\left(\frac{\partial V}{\partial x} - \frac{\partial U}{\partial y}\right)^2 + \left(\frac{\partial W}{\partial y} - \frac{\partial V}{\partial z}\right)^2 + \left(\frac{\partial U}{\partial z} - \frac{\partial W}{\partial x}\right)^2\right]^{1/2}.$$

F.2 One Equation Models

Spalart–Allmaras

The model PDEs are

$$\frac{\partial \bar{\rho}}{\partial t} + \frac{\partial}{\partial x_i}(\bar{\rho}\tilde{u}_i) = 0,$$

$$\frac{\partial}{\partial t}(\bar{\rho}\tilde{u}_i) + \frac{\partial}{\partial x_j}(\bar{\rho}\tilde{u}_j\tilde{u}_i) = -\frac{\partial \bar{p}}{\partial x_i} + \frac{\partial}{\partial x_j}\left(2(\bar{\mu} + \mu_t)\tilde{S}_{ji}\right),$$

$$\frac{\partial}{\partial t}\left[\bar{\rho}\left(\tilde{e} + \frac{1}{2}\tilde{u}_i\tilde{u}_i\right)\right] + \frac{\partial}{\partial x_j}\left[\bar{\rho}\tilde{u}_j\left(\tilde{h} + \frac{1}{2}\tilde{u}_i\tilde{u}_i\right)\right] = \frac{\partial}{\partial x_j}\left(2(\bar{\mu} + \mu_t)\tilde{S}_{ji}\tilde{u}_i\right) + \frac{\partial}{\partial x_j}\left[\left(\frac{\bar{\mu}}{Pr} + \frac{\mu_t}{Pr_t}\right)\frac{\partial \tilde{h}}{\partial x_j}\right],$$

$$\frac{\partial}{\partial t}(\bar{\rho}\nu_{\text{sa}}) + \frac{\partial}{\partial x_j}(\bar{\rho}\tilde{u}_j\nu_{\text{sa}}) = c_{b1} S_{\text{sa}}\bar{\rho}\nu_{\text{sa}} - c_{w1} f_w \bar{\rho}\left(\frac{\nu_{\text{sa}}}{d}\right)^2 + \frac{1}{\sigma}\frac{\partial}{\partial x_k}\left[(\bar{\mu} + \bar{\rho}\nu_{\text{sa}})\frac{\partial \nu_{\text{sa}}}{\partial x_k}\right] + \frac{c_{b2}}{\sigma}\bar{\rho}\frac{\partial \nu_{\text{sa}}}{\partial x_k}\frac{\partial \nu_{\text{sa}}}{\partial x_k},$$

where d is the distance to the nearest no-slip wall, and

$$\bar{\mu} = \mu_0 \left(\frac{\tilde{T}}{T_0}\right)^{3/2} \frac{T_0 + S}{\tilde{T} + S}, \quad \tilde{S}_{ij} = \tilde{s}_{ij} - \frac{1}{3}\tilde{s}_{kk}\delta_{ij}, \quad \tilde{s}_{ij} = \frac{1}{2}\left(\frac{\partial \tilde{u}_i}{\partial x_j} + \frac{\partial \tilde{u}_j}{\partial x_i}\right),$$

$$\bar{p} = \bar{\rho} R \tilde{T}, \quad \tilde{e} = c_v \tilde{T}, \quad \tilde{h} = c_p \tilde{T} = \tilde{e} + \frac{\bar{p}}{\bar{\rho}},$$

$$\mu_t = \bar{\rho}\nu_t = \bar{\rho}\nu_{\text{sa}} f_{v1}, \quad S_{\text{sa}} = \Omega + \frac{\nu_{\text{sa}}}{\kappa^2 d^2} f_{v2}, \quad \Omega = \sqrt{2\tilde{\Omega}_{ij}\tilde{\Omega}_{ij}}, \quad \tilde{\Omega}_{ij} = \frac{1}{2}\left(\frac{\partial \tilde{u}_i}{\partial x_j} - \frac{\partial \tilde{u}_j}{\partial x_i}\right),$$

$$f_{v2} = 1 - \frac{\chi}{1 + \chi f_{v1}}, \quad f_{v1} = \frac{\chi^3}{\chi^3 + c_{v1}^3}, \quad \chi = \frac{\nu_{\text{sa}}}{\tilde{\nu}},$$

$$f_w = g\left(\frac{1 + c_{w3}^6}{g^6 + c_{w3}^6}\right)^{1/6}, \quad g = r + c_{w2}\left(r^6 - r\right), \quad r = \frac{\nu_{\text{sa}}}{S_{\text{sa}}\kappa^2 d^2}.$$

The constants c_v and c_p are fluid properties. The constants μ_0, T, and S are the calibration parameters appearing in Sutherland's law, and c_{b1}, c_{b2}, c_{v1}, σ, c_{w1}, c_{w2}, c_{w3}, and κ are the SA model calibration parameters.

F.3 Two Equation Models

Standard $k - \epsilon$ model

The standard $k - \epsilon$ model is as follows: Kinematic eddy viscosity:

$$\nu_T = \frac{C_\mu k^2}{\epsilon}.$$

Turbulent kinetic energy:

$$\frac{\partial k}{\partial t} + u_j \frac{\partial k}{\partial x_j} = \tau_{ij}\frac{\partial u_i}{\partial x_j} - \epsilon + \frac{\partial}{\partial x_j}\left[\left(\nu + \frac{\nu_T}{\sigma_k}\right)\frac{\partial k}{\partial x_j}\right].$$

Dissipation rate:

$$\frac{\partial \epsilon}{\partial t} + u_j \frac{\partial \epsilon}{\partial x_j} = C_{\epsilon 1}\frac{\epsilon}{k}\tau_{ij}\frac{\partial u_i}{\partial x_j} - C_{\epsilon 2}\frac{\epsilon^2}{k} + \frac{\partial}{\partial x_j}\left[\left(\nu + \frac{\nu_T}{\sigma_k}\right)\frac{\partial \epsilon}{\partial x_j}\right].$$

Closure coefficients and auxiliary relations:

$$C_{\epsilon 1} = 1.44, \quad C_{\epsilon 2} = 1.92, \quad C_\mu = 0.09, \quad \sigma_k = 1.0, \quad \sigma_\epsilon = 1.3,$$

$$\omega = \frac{\epsilon}{C_\mu k}, \quad l = \frac{C_\mu k^{3/2}}{\epsilon}.$$

RNG $k-\epsilon$ model

Modified closure coefficients:

$$C_{\epsilon 2} \equiv \tilde{C}_{\epsilon 2} + \frac{C_\mu \lambda^3 (1 - \lambda/\lambda_0)}{1 + \beta \lambda^3}, \quad \lambda \equiv \frac{k}{\epsilon}\sqrt{2 S_{ij} S_{ji}}.$$

$$C_{\epsilon 1} = 1.42, \quad \tilde{C}_{\epsilon 2} = 1.68, \quad C_\mu = 0.085, \quad \sigma_k = 0.72, \quad \sigma_\epsilon = 0.72,$$

$$\beta = 0.012, \quad \lambda_0 = 4.38.$$

$k-\omega$ model (Wilcox)

The model PDEs are

$$\frac{\partial \bar{\rho}}{\partial t} + \frac{\partial}{\partial x_i}(\bar{\rho}\tilde{u}_i) = 0,$$

$$\frac{\partial}{\partial t}(\bar{\rho}\tilde{u}_i) + \frac{\partial}{\partial x_j}(\bar{\rho}\tilde{u}_j\tilde{u}_i) = -\frac{\partial \bar{p}}{\partial x_i} + \frac{\partial}{\partial x_j}\left(2(\bar{\mu} + \mu_t)\tilde{S}_{ji} - \frac{2}{3}\bar{\rho}k\delta_{ji}\right),$$

$$\frac{\partial}{\partial t}(\bar{\rho}\tilde{E}) + \frac{\partial}{\partial x_j}(\bar{\rho}\tilde{u}_j\tilde{H}) = \frac{\partial}{\partial x_j}\left(2(\bar{\mu} + \mu_t)\tilde{S}_{ji}\tilde{u}_i - \frac{2}{3}\bar{\rho}k\delta_{ji}\tilde{u}_i\right)$$
$$+ \frac{\partial}{\partial x_j}\left[\left(\frac{\bar{\mu}}{Pr} + \frac{\mu_t}{Pr_t}\right)\frac{\partial \tilde{h}}{\partial x_j}\right] + \frac{\partial}{\partial x_j}\left[\left(\bar{\mu} + \sigma^* \frac{\bar{\rho}k}{\omega}\right)\frac{\partial k}{\partial x_j}\right],$$

$$\frac{\partial}{\partial t}(\bar{\rho}k) + \frac{\partial}{\partial x_j}(\bar{\rho}\tilde{u}_j k) = \left(2\mu_t \tilde{S}_{ij} - \frac{2}{3}\bar{\rho}k\delta_{ji}\right)\frac{\partial \tilde{u}_i}{\partial x_j} - \beta^* \bar{\rho}k\omega + \frac{\partial}{\partial x_j}\left[\left(\bar{\mu} + \sigma^* \frac{\bar{\rho}k}{\omega}\right)\frac{\partial k}{\partial x_j}\right],$$

$$\frac{\partial}{\partial t}(\bar{\rho}\omega) + \frac{\partial}{\partial x_j}(\bar{\rho}\tilde{u}_j\omega) = \alpha\frac{\omega}{k}\left(2\mu_t \tilde{S}_{ij} - \frac{2}{3}\bar{\rho}k\delta_{ji}\right)\frac{\partial \tilde{u}_i}{\partial x_j} - \beta\bar{\rho}\omega^2$$
$$+ \sigma_d \frac{\bar{\rho}}{\omega}\frac{\partial k}{\partial x_j}\frac{\partial \omega}{\partial x_j} + \frac{\partial}{\partial x_j}\left[\left(\bar{\mu} + \sigma\frac{\bar{\rho}k}{\omega}\right)\frac{\partial \omega}{\partial x_j}\right],$$

where

$$\bar{\mu} = \mu_0 \left(\frac{\tilde{T}}{T_0}\right)^{3/2} \frac{T_0 + S}{\tilde{T} + S}, \quad \tilde{S}_{ij} = \tilde{s}_{ij} - \frac{1}{3}\tilde{s}_{kk}\delta_{ij}, \quad \tilde{s}_{ij} = \frac{1}{2}\left(\frac{\partial \tilde{u}_i}{\partial x_j} + \frac{\partial \tilde{u}_j}{\partial x_i}\right),$$

$$\bar{p} = \bar{\rho}R\tilde{T}, \quad \tilde{E} = \tilde{e} + \frac{1}{2}\tilde{u}_i\tilde{u}_i + k, \quad \tilde{e} = c_v\tilde{T},$$

$$\tilde{H} = \tilde{h} + \frac{1}{2}\tilde{u}_i\tilde{u}_i + k, \quad \tilde{h} = c_p\tilde{T} = \tilde{e} + \frac{\bar{p}}{\bar{\rho}},$$

$$\mu_t = \frac{\bar{\rho}k}{\hat{\omega}}, \quad \hat{\omega} = \max\left(\omega, C_{lim}\sqrt{\frac{2}{\beta^*}\tilde{S}_{ij}\tilde{S}_{ij}}\right),$$

$$\beta = \beta_0 f_\beta, \quad f_\beta = \frac{1 + 85\chi_\omega}{1 + 100\chi_\omega},$$

$$\chi_\omega = \left| \frac{\Omega_{ij}\Omega_{jk}\hat{S}_{ki}}{(\beta^*\omega)^3} \right|, \quad \hat{S}_{ki} = \tilde{s}_{ki} - \frac{1}{2}\frac{\partial \tilde{u}_m}{\partial x_m}\delta_{ki}, \quad \sigma_d = \begin{cases} 0, & \dfrac{\partial k}{\partial x_j}\dfrac{\partial \omega}{\partial x_j} \le 0 \\ \sigma_{d0}, & \dfrac{\partial k}{\partial x_j}\dfrac{\partial \omega}{\partial x_j} > 0 \end{cases}.$$

The constants c_v and c_p are fluid properties. The constants μ_0, T, and S are the calibration parameters appearing in Sutherland's law, and β^*, σ^*, α, β_0, σ, and σ_{d0} are the $k - \omega$ (Wilcox) calibration parameters.

$k - \omega$ model (Menter/SST)

The model PDEs are

$$\frac{\partial \bar{\rho}}{\partial t} + \frac{\partial}{\partial x_i}(\bar{\rho}\tilde{u}_i) = 0,$$

$$\frac{\partial}{\partial t}(\bar{\rho}\tilde{u}_i) + \frac{\partial}{\partial x_j}(\bar{\rho}\tilde{u}_j\tilde{u}_i) = -\frac{\partial \bar{p}}{\partial x_i} + \frac{\partial}{\partial x_j}\left(2(\bar{\mu} + \mu_t)\tilde{S}_{ji} - \frac{2}{3}\bar{\rho}k\delta_{ji}\right),$$

$$\begin{aligned} \frac{\partial}{\partial t}(\bar{\rho}\tilde{E}) + \frac{\partial}{\partial x_j}(\bar{\rho}\tilde{u}_j\tilde{H}) &= \frac{\partial}{\partial x_j}\left(2(\bar{\mu} + \mu_t)\tilde{S}_{ji}\tilde{u}_i - \frac{2}{3}\bar{\rho}k\delta_{ji}\tilde{u}_i\right) \\ &+ \frac{\partial}{\partial x_j}\left[\left(\frac{\bar{\mu}}{Pr} + \frac{\mu_t}{Pr_t}\right)\frac{\partial \tilde{h}}{\partial x_j}\right] + \frac{\partial}{\partial x_j}\left[\left(\bar{\mu} + \sigma\frac{\bar{\rho}k}{\omega}\right)\frac{\partial k}{\partial x_j}\right], \end{aligned}$$

$$\frac{\partial}{\partial t}(\bar{\rho}k) + \frac{\partial}{\partial x_j}(\bar{\rho}\tilde{u}_j k) = \left(2\mu_t\tilde{S}_{ij} - \frac{2}{3}\bar{\rho}k\delta_{ji}\right)\frac{\partial \tilde{u}_i}{\partial x_j} - \beta^*\bar{\rho}k\omega + \frac{\partial}{\partial x_j}\left[\left(\bar{\mu} + \sigma\frac{\bar{\rho}k}{\omega}\right)\frac{\partial k}{\partial x_j}\right],$$

$$\begin{aligned} \frac{\partial}{\partial t}(\bar{\rho}\omega) + \frac{\partial}{\partial x_j}(\bar{\rho}\tilde{u}_j\omega) &= \frac{\alpha}{\nu_t}\left(2\mu_t\tilde{S}_{ij} - \frac{2}{3}\bar{\rho}k\delta_{ji}\right)\frac{\partial \tilde{u}_i}{\partial x_j} - \beta\bar{\rho}\omega^2 \\ &+ \frac{\partial}{\partial x_j}\left[(\bar{\mu} + \sigma_\omega\mu_t)\frac{\partial \omega}{\partial x_j}\right] + 2(1 - F_1)\bar{\rho}\sigma_{\omega 2}\frac{1}{\omega}\frac{\partial k}{\partial x_j}\frac{\partial \omega}{\partial x_j}, \end{aligned}$$

where

$$\bar{\mu} = \mu_0\left(\frac{\tilde{T}}{T_0}\right)^{3/2}\frac{T_0 + S}{\tilde{T} + S}, \quad \tilde{S}_{ij} = \tilde{s}_{ij} - \frac{1}{3}\tilde{s}_{kk}\delta_{ij}, \quad \tilde{s}_{ij} = \frac{1}{2}\left(\frac{\partial \tilde{u}_i}{\partial x_j} + \frac{\partial \tilde{u}_j}{\partial x_i}\right),$$

$$\begin{aligned} \bar{p} &= \bar{\rho}R\tilde{T}, \quad \tilde{E} = \tilde{e} + \frac{1}{2}\tilde{u}_i\tilde{u}_i + k, \quad \tilde{e} = c_v\tilde{T}, \\ \tilde{H} &= \tilde{h} + \frac{1}{2}\tilde{u}_i\tilde{u}_i + k, \quad \tilde{h} = c_p\tilde{T} = \tilde{e} + \frac{\bar{p}}{\bar{\rho}}, \end{aligned}$$

$$\begin{aligned} \mu_t &= \frac{a_1\bar{\rho}k}{\max(a_1\omega, \Omega F_2)}, \quad \Omega = \sqrt{2\tilde{\Omega}_{ij}\tilde{\Omega}_{ij}}, \\ F_2 &= \tanh(\arg_2^2), \quad \arg_2 = \max\left(\frac{2\sqrt{k}}{0.09\omega d}, \frac{500\tilde{\nu}}{d^2\omega}\right), \end{aligned}$$

$$\sigma_k = F_1\sigma_{k1} + (1 - F_1)\alpha_2, \quad \sigma_\omega = F_1\sigma_{\omega 1} + (1 - F_1)\sigma_{\omega_2}, \quad \beta = F_1\beta_1 + (1 - F_1)\beta_2,$$

$$\alpha = F_1\alpha_1 + (1 - F_1)\alpha_2, \quad \alpha_1 = \frac{\beta_1}{\beta^*} - \sigma_{\omega 1}\frac{\kappa^2}{\sqrt{\beta^*}}, \quad \alpha_2 = \frac{\beta_2}{\beta^*} - \sigma_{\omega 2}\frac{\kappa^2}{\sqrt{\beta^*}},$$

$$F_1 = \tanh(\mathrm{arg}_1^4), \quad \mathrm{arg}_1 = \min\left(\mathrm{arg}_2, \frac{4\bar{\rho}\sigma_{\omega 2}k}{\mathrm{CD}_{k\omega}d^2}\right),$$

$$\mathrm{CD}_{k\omega} = \max\left(2\bar{\rho}\sigma_{\omega 2}\frac{1}{\omega}\frac{\partial k}{\partial x_j}\frac{\partial \omega}{\partial x_j}, 0\right),$$

and d is the distance to the nearest wall. The constants c_v and c_p are fluid properties. The constants μ_0, T, and S are the calibration parameters appearing in Sutherland's law, and σ_{k1}, σ_{k2}, $\sigma_{\omega 1}$, $\sigma_{\omega 2}$, β_1, β_2, β^*, σ^*, κ, and a_1 are the SST calibration parameters.

References

S. Abarbanel & D. Gottlieb (1981). "Optimal time splitting for two- and three-dimensional navier-stokes equations with mixed derivatives." *Journal of Computational Physics* **41**(1): 1–33.

Ira H. Abbott & A. E. Von Doenhoff (1959). *Theory of Wing Sections, Including a Summary of Airfoil Data*. Dover Publications.

R. Abgrall (2006). "Residual distribution schemes: Current status and future trends." *Computers and Fluids* **35**: 641–669.

M. J. Aftosmis, J. E. Melton, & M. J. Berger (1995). "Adaptation and surface modeling for Cartesian mesh methods." AIAA paper 95-1725-CP, AIAA 12th Computational Fluid Dynamics Conlerence, San Diego, CA.

E. Ahlstrom, R. Gregg, J. Vassberg, & A. Jameson (2000). "G-Force: The design of an unlimited class Reno racer." AIAA paper 4341, AIAA 18th Applied Aerodynamics Conference, Denver, CO.

B. Ahrabi, D. J. Mavriplis, & M. Brazell (2019). "Accelerating Newton method continuation for CFD problems." AIAA paper 2019-0100, AIAA SciTech Meeting, San Diego, CA.

S. Allmaras (1993). "Analysis of a local matrix preconditioner for the 2-D Navier-Stokes equations." AIAA paper 93-3330, AIAA 11th Computational Fluid Dynamics Conference, Orlando, FL.

S. Allmaras (1995). "Analysis of semi-implicit preconditioners for multigrid solution of the 2-D Navier-Stokes equations." AIAA paper 95-1651, AIAA 12th Computational Fluid Dynamics Conference, San Diego, CA.

S. Allmaras (1997). "Algebraic smoothing analysis of multigrid methods for the 2-D compressible Navier-Stokes equations." AIAA paper 97-1954, AIAA 13th Computational Fluid Dynamics Conference, Snowmass, CO.

J. J. Alonso & A. Jameson (1994). "Fully-implicit time-marching aeroelastic solutions." AIAA paper 94-0056, AIAA 32nd Aerospace Sciences Meeting, Reno, NV.

J. J. Alonso, L. Martinelli, & A. Jameson (1995). "Multigrid unsteady Navier-Stokes calculations with aeroelastic applications." AIAA paper 95-0048, AIAA 33rd Aerospace Sciences Meeting, Reno, NV.

W. K. Anderson & V. Venkatakrishnan (1997). "Aerodynamic design and optimization on unstructured grids with a continuous adjoint formulation." AIAA paper 97-0643, AIAA 35th Aerospace Sciences Meeting, Reno, NV.

W. K. Anderson, J. L. Thomas, & D. L. Whitfield (1986). "Multigrid acceleration of the flux split euler equations." AIAA paper 86-0274, AIAA 24th Aerospace Sciences Meeting, Reno, NV.

J. H. Argyris (1954a). "Energy theorems and structural analysis: A generalized discourse with applications on energy principles of structural analysis including the effects of temperature and non-Linear stress-strain relations." *Aircraft Engineering and Aerospace Technology* **26**(11):383–394.

J. H. Argyris (1954b). "Flexure-torsion failure of panels: A study of instability and failure of stiffened panels under compression when buckling in long wavelengths." *Aircraft Engineering and Aerospace Technology* **26**(6):174–184.

J. H. Argyris (1954c). "The open tube: A study of thin-walled structures such as interspar wing cut-outs and open-section stringers." *Aircraft Engineering and Aerospace Technology* **26**(4):102–112.

J. H. Argyris (1960). *Energy Theorems and Structural Analysis: A Generalised Discourse with Applications on Energy Principles of Structural Analysis Including the Effects of Temperature and Nonlinear Stress-Strain Relations.* Butterworth.

P. Arminjon & A. Dervieux (1989). "Construction of TVD-like artificial viscosities on 2-dimensional arbitrary FEM grids." INRIA Report 1111.

D. N. Arnold (1982). "An interior penalty finite element method with discontinuous elements." *SIAM Journal on Numerical Analysis* **19**(4):742–760.

H. Ashley & M. Landahl (1985). *Aerodynamics of Wings and Bodies.* Dover Publications.

T. J. Baker (1986). "Mesh generation by a sequence of transformations." *Applied Numerical Mathematics* **2**:515–528.

N. S. Bakhvalov (1966). "On the convergence of a relaxation method with natural constraints on the elliptic operator." *USSR Computational Mathematics and Mathematical Physics* **6**(5):101–135.

A. Balan, G. May, & J. Schöberl (2012). "A stable high-order spectral difference method for hyperbolic conservation laws on triangular elements." *Journal of Computational Physics* **231**(5):2359–2375.

J. Bardina, J. H. Ferziger, & W. C. Reynolds (1980). "Improved subgrid scale models for large-eddy simulation." AIAA paper 80-1357, AIAA 13th Fluid & Plasma Dynamics Conference, Snowmass, CO.

J. Bardina & C. K. Lombard (1987). "Three dimensional hypersonic flow simulations with the CSCM implicit upwind Navier-Stokes method." AIAA paper 87-1114, AIAA 8th Computational Fluid Dynamics Conference, Honolulu, HI.

T. J. Barth (1994). "Aspects of unstructured grids and finite volume solvers for the Euler and Navier Stokes equations." In *von Karman Institute for Fluid Dynamics Lecture Series Notes 1994-05*, Brussels.

T. J. Barth & D. C. Jespersen (1989). "The design and application of upwind schemes on unstructured meshes." AIAA paper 89-0366, AIAA 27th Aerospace Sciences Meeting, Reno, NV.

F. Bassi & S. Rebay (1997). "A high-order accurate discontinuous finite element method for the numerical solution of the compressible Navier–Stokes equations." *Journal of Computational Physics* **131**(2):267–279.

F. Bassi & S. Rebay (2000). "A high order discontinuous Galerkin method for compressible turbulent flows." In Bernardo Cockburn, George E. Karniadakis, & Chi-Wang Shu (eds.), *Discontinuous Galerkin Methods*, pp. 77–88. Springer.

F. Bauer, P. Garabedian, & D. Korn (1972). "A theory of supercritical wing sections, with computer programs and examples." *Lecture Notes in Economics and Mathematical Systems 66* **1**:81–83.

A. Bayliss & E. Turkel (1982). "Far field boundary conditions for compressible flows." *Journal of Computational Physics* **48**(2):182–199.

O. Baysal & M. E. Eleshaky (1992). "Aerodynamic design optimization using sensitivity analysis and computational fluid dynamics." *AIAA Journal* **30**(3):718–725.

R. W. Beam & R. F. Warming (1976). "An implicit finite difference algorithm for hyperbolic systems in conservation form." *Journal of Computational Physics* **23**:87–110.

A. Belov, L. Martinelli, & A. Jameson (1995). "A new implicit algorithm with multigrid for unsteady incompressible flow calculations." AIAA paper 95-0049, AIAA 33rd Aerospace Sciences Meeting, Reno, NV.

J. A. Benek, P. G. Buning, & J. L. Steger (1985). "A 3-D chimera grid embedding technique." AIAA paper 85-1523, AIAA 7th Computational Fluid Dynamics Conference, Cincinnati, OH.

J. A. Benek, T. L. Donegan, & N. E. Suhs (1987). "Extended chimera grid embedding scheme with applications to viscous flows." AIAA paper 87-1126, AIAA 8th Computational Fluid Dynamics Conference, Honolulu, HI.

M. Berger & R. J. LeVeque (1989). "An adaptive Cartesian mesh algorithm for the Euler equations in arbitrary geometries." AIAA paper 89-1930.

H. Bijl, M. H. Carpenter, V. N. Vatsa, & C. A. Kennedy (2002). "Implicit time integration schemes for the unsteady compressible Navier–Stokes equations: laminar flow." *Journal of Computational Physics* **179**(1):313–329.

N. Bons & J. Martins (2020). Aerostructural Wing Design Exploration with Multidisciplinary Design Optimization. Aerospace 7(118).

P. D. Boom & D. W. Zingg (2013). "High-order implicit time integration for unsteady compressible fluid flow simulation." AIAA paper 2013-2831, AIAA 21st Computational Fluid Dynamics Conference, San Diego, CA.

J. P. Boris & D. L. Book (1973). "Flux-corrected transport. I. SHASTA, A fluid transport algorithm that works." *Journal of Computational Physics* **11**(1):38–69.

J. Boussinesq (1877). *Essai sur la théorie des eaux courantes*. Mémoires présentés par divers savants à l'Académie des Sciences de l'Institut National de France, Tome XXIII. No. 1.

A. Brandt (1977). "Multi-level adaptive solutions to boundary value problems." *Mathematics of Computation* **31**:333–390.

M.O. Bristeau, O. Pironneau, R. Glowinski, J. Periaux, P. Perrier, & G. Poirier (1980). "Application of optimal control and finite element methods to the calculation of transonic flows and incompressible viscous flows." In *Proceedings of the IMA Conference on Numerical Methods in Applied Fluid Dynamics*, pp. 203–312. (A 82-15826 04-34) Academic Press.

M. O. Bristeau, R. Glowinski, J. Periaux, P. Perrier, G. Poirier, & O. Pironneau (1981). "Transonic flow simulations by finite elements and least square methods." In *Proceedings of the 3rd International Conference on Finite Elements in Flow Problems, Banff, Alberta*, pp. 11–29.

M. O. Bristeau, O. Pironneau, R. Glowinski, J. Periaux, P. Perrier, & G. Poirier (1985). "On the numerical solution of nonlinear problems in fluid dynamics by least squares and finite element methods (II). Application to transonic flow simulations." In J. St. Doltsinis (ed.), *Proceedings of the 3rd International Conference on Finite Element Methods in Nonlinear Mechanics, FENOMECH 84, Stuttgart, 1984*, pp. 363–394, North Holland.

A. E. Bryson & Y-C. Ho (1975). *Applied optimal control: Optimization, estimation and control*. Hemisphere.

J. C. Butcher (1964a). "Integration processes based on Radau quadrature formulas." *Mathematics of Computation* **18**(86):233–244.

J. C. Butcher (1964b). "Implicit Runge-Kutta processes." *Mathematics of Computation* **18**(85):50–64.

J. C. Butcher (1987). *The Numerical Analysis of Ordinary Differential Equations: Runge-Kutta and General Linear Methods*. Wiley-Interscience.

J. C. Butcher (2003). *Numerical Methods for Ordinary Differential Equations*. J. Wiley Ltd.

C. Canuto, M.Y. Hussaini, A. M. Quarteroni, & T. A. Zang Jr. (1988). *Spectral Methods in Fluid Dynamics*. Springer-Verlag.

C. Canuto, M. Y. Hussaini, A. M. Quarteroni, & T. A. Zang Jr. (2007a). *Spectral Methods, Evolution to Complex Geometries and Applications to Fluid Dynamic*. Springer-Verlag.

C. Canuto, M. Y. Hussaini, A. M. Quarteroni, & T. A. Zang Jr. (2007b). *Spectral Methods, Fundamentals in Single Domains*. Springer-Verlag.

P. Castonguay, P. E. Vincent, & A. Jameson (2012). "A new class of high-order energy stable flux reconstruction schemes for triangular elements." *Journal of Scientific Computing* **51**(1):224–256.

P. Castonguay, D. M. Williams, P. E. Vincent, & A. Jameson (2013). "Energy stable flux reconstruction schemes for advection–diffusion problems." *Computer Methods in Applied Mechanics and Engineering* **267**:400–417.

C. Castro, C. Lozano, F. Palacios, & E. Zuazua (2007). "Systematic continuous adjoint approach to viscous aerodynamic design on unstructured grids." *AIAA Journal* **45**(9):2125–2139.

D. A. Caughey (1987). "A diagonal implicit multigrid algorithm for the Euler equations." AIAA paper 87-453, AIAA 25th Aerospace Sciences Meeting, Reno, NV.

S. R. Chakravarthy (1984). "Relaxation methods for unfactored implicit upwind schemes." AIAA paper 84-0165, AIAA 22nd Aerospace Sciences Meeting, Reno, NV.

P. Chandrashekar (2013). "Kinetic energy preserving and entropy stable finite volume schemes for compressible Euler and Navier-Stokes equations." *Communications in Computational Physics* **14**(5):1252–1286.

R. W. Clough (1960). "The finite element method in plane stress analysis." In *Proceedings of the Second ASCE Conference on Electronic Computation, Pittsburgh, PA*.

R. W. Clough (2004). "Early history of the finite element method from the view point of a pioneer." *International Journal for Numerical Methods in Engineering* **60**(1):283–287.

B. Cockburn, S. Hou, & C-W. Shu (1990). "The Runge-Kutta local projection discontinuous Galerkin finite element method for conservation laws. IV. The multidimensional case." *Mathematics of Computation* **54**(190):545–581.

B. Cockburn, S-Y. Lin, & C-W. Shu (1989). "TVB Runge-Kutta local projection discontinuous Galerkin finite element method for conservation laws III: one-dimensional systems." *Journal of Computational Physics* **84**(1):90–113.

B. Cockburn & C-W. Shu (1989). "TVB Runge-Kutta local projection discontinuous Galerkin finite element method for conservation laws. II. General framework." *Mathematics of Computation* **52**(186):411–435.

B. Cockburn & C-W. Shu (1998a). "The local discontinuous Galerkin method for time-dependent convection-diffusion systems." *SIAM Journal on Numerical Analysis* **35**(6):2440–2463.

B. Cockburn & C-W. Shu (1998b). "The Runge-Kutta discontinuous Galerkin method for conservation laws V: multidimensional systems." *Journal of Computational Physics* **141**(2):199–224.

B. Cockburn & C-W. Shu (2001). "Runge-Kutta discontinuous Galerkin methods for convection-dominated problems." *Journal of Scientific Computing* **16**(3):173–261.

L. Collatz (1966). *The Numerical Treatment of Differential Equations*, Vol. 1. Springer, 3rd edn.

G. J. Cooper & A. Sayfy (1983). "Additive Runge-Kutta methods for stiff ordinary differential equations." *Mathematics of Computation* **40**(161):207–218.

G. J. Cooper & A. Sayfy (1980). "Additive methods for the numerical solution of ordinary differential equations." *Mathematics of Computation* **35**(152):1159–1172.

R. Courant (1943). "Variational methods for the solution of problems of equilibrium and vibrations." *Bulletin of the American Mathematical Society* **49**(1):1–23.

M. G. Crandall & A. Majda (1980). "Monotone difference approximations for scalar conservation laws." *Mathematics of Computation* **34**(149):1–21.

P. I. Crumpton & M. B. Giles (1995). "Implicit time accurate solutions on unstructured dynamic grids." AIAA paper 95-1671, AIAA 12th Computational Fluid Dynamics Conference, San Diego, CA.

G. Dahlquist (1956). "Convergence and stability in the numerical integration of ordinary differential equations." *Mathematica Scandinavia* **4**(1):33–53.

G. Dahlquist (1963). "A special stability problem for linear multistep methods." *BIT* **3**:27–43.

P. J. Davis (1975). *Interpolation and Approximation*. Dover Publications.

S. S. Davis (1982). "NACA 64A010 (NASA Ames model) oscillatory pitching," AGARD report 702, AGARD, January 1982.

B. Delaunay (1934). "Sur la Sphere vide." *Bulletin of the Academy of Sciences of the USSR VII: Class Scil, Mat. Nat.* pp. 793–800.

M. Desai & K. Ito (1994). "Optimal controls of Navier-Stokes equations." *SIAM Journal on Control and Optimization* **32**(5):1428–1446.

S. M. Deshpande, N. Balakrishnan, & S. V. Raghurama Rao (1994). "PVU and wave-particle splitting schemes for Euler equations of gas dynamics." *Sadhana* **19**(6):1027–1054.

J. R. Dormand & P. J. Prince (1980). "A family of embedded Runge-Kutta formulae." *Journal of Computational and Applied Mathematics* **6**(1):19–26.

P. A. Durbin & B. A. Pettersson Reif (2011). *Statistical Theory and Modeling for Turbulent Flows*. Wiley.

T. D. Economon, F. Palacios, S. R. Copeland, T. W. Lukaczyk, & J. J. Alonso (2016). "SU2: An open-source suite for multiphysics simulation and design." *AIAA Journal* **54**(3):828–846.

P. R. Eiseman (1979). "A multi-surface method of coordinate generation." *Journal of Computational Physics* **33**:118–150.

J. Elliot & J. Peraire (1996). "Practical 3D aerodynamic design and optimization using unstructured meshes." AIAA paper 96-4710, 6th AIAA/NASA/USAF Multidisciplinary and Optimization Symposium, Seattle, WA.

B. Engquist & A. Majda (1977). "Absorbing boundary conditions for numerical simulation of waves." *Proceedings of the National Academy of Sciences* **74**(5):1765–1766.

B. Engquist & S. Osher (1981). "One-sided difference approximations for nonlinear conservation laws." *Mathematics of Computation* **36**(154):321–351.

L. E. Eriksson (1982). "Generation of boundary-conforming grids around wing-body configurations using transfinite interpolation." *AIAA Journal* **20**:1313–1320.

A. J. Favre (1965). "Review on space-time correlations in turbulent fluids." *Journal of Applied Mechanics* **32**(2):241–257.

R. P. Fedorenko (1964). "The speed of convergence of one iterative process." *USSR Computational Mathematics and Mathematical Physics* **4**:227–235.

E. Fehlberg (1969). "Low-order classical Runge-Kutta formulas with stepsize control and their application to some heat transfer problems." Technical Report NASA-TR-R-315, NASA Technical Report, NASA Marshall Space Flight Center; Huntsville, AL.

T. C. Fisher & M. H. Carpenter (2013). "High-order entropy stable finite difference schemes for nonlinear conservation laws: Finite domains." *Journal of Computational Physics* **252**: 518–557.

D. V. Gaitonde & M. R. Visbal (1998). "High-order schemes for Navier-Stokes equations: Algorithm and implementation into FDL3DI." Technical Report AFRL-VA-WP-TR-1998-3060, Air Force Research Laboratory, Wright-Patterson AFB.

P. R. Garabedian (1956). "Estimation of the relaxation factor for small mesh size." *Mathematical Tables and Other Aids to Computation* pp. 183–185.

M. Germano (1986a). "Differential filters for the large eddy numerical simulation of turbulent flows." *Physics of Fluids* **29**(6):1755–1757.

M. Germano (1986b). "Differential filters of elliptic type." *Physics of Fluids* **29**(6):1757–1758.

M. Germano, U. Piomelli, P. Moin, & W. H. Cabot (1991). "A dynamic subgrid-scale eddy viscosity model." *Physics of Fluids A: Fluid Dynamics (1989-1993)* **3**(7):1760–1765.

M. Giles, M. Drela, & W. T. Thompkins (1985). "Newton solution of direct and inverse transonic Euler equations." AIAA paper 85-1530, Cincinnati.

H. Glauert (1926). *The Elements of Aerofoil and Airscrew Theory.* Cambridge University Press.

S. K. Godunov (1959). "A difference method for the numerical calculation of discontinuous solutions of hydrodynamic equations." *Matematicheskii Sbornik* **47**:271–306. Translated as JPRS 7225 by U.S. Dept. of Commerce, 1960.

D. Gottlieb & S. A. Orszag (1977). *Numerical Analysis of Spectral Methods.* Society for Industrial and Applied Mathematics.

S. Gottlieb, D. I. Ketcheson, & C-W. Shu (2009). "High order strong stability preserving time discretizations." *Journal of Scientific Computing* **38**(3):251–289.

S. Gottlieb, D. I. Ketcheson, & C-W. Shu (2011). *Strong Stability Preserving Runge-Kutta and Multistep Time Discretizations.* World Scientific.

S. Gottlieb & C-W. Shu (1998). "Total variation diminishing Runge-Kutta schemes." *Mathematics of Computation of the American Mathematical Society* **67**(221):73–85.

S. Gottlieb, C-W. Shu, & E. Tadmor (2001). "Strong stability-preserving high-order time discretization methods." *SIAM Review* **43**(1):89–112.

A. R. Gourlay & A. R. Mitchell (1966). "A stable implicit difference scheme for hyperbolic systems in two space variables." *Numerische Mathematik* **8**:367–375.

J-L. Guermond, R. Pasquetti, & B. Popov (2011). "Entropy viscosity method for nonlinear conservation laws." *Journal of Computational Physics* **230**(11):4248–4267.

B. Gustaffson, H-O. Kreiss, & J. Oliger (2013). *Time-Dependent Problems and Difference Methods.* John Wiley & Sons, Ltd.

B. Gustafsson (1975). "The convergence rate for difference approximations to mixed initial boundary value problems." *Mathematics of Computation* **29**(130):396–406.

B. Gustafsson, H-O. Kreiss, & A. Sundström (1972). "Stability theory of difference approximations for mixed initial boundary value problems. II." *Mathematics of Computation* **26**(119):649–686.

W. Haase, E. Chaput, E. Elsholz, M. A. Leschziner, & U. R. Mueller (1997). *ECARP: European Computational Aerodynamics Research Project: Validation of CFD Codes and Assessment of Turbulence Models*, Vol. 58. Friedr Vieweg & Sohn Verlagsgesellschaft.

W. Hackbusch (1978). "On the multi-grid method applied to difference equations." *Computing* **20**:291–306.

M. G. Hall (1985). "Cell vertex multigrid schemes for solution of the Euler equations." In K. W. Morton & M. J. Baines (eds.), *IMA Conference on Numerical Methods for Fluid Dynamics, University Reading*, pp. 303–345. Oxford University Press.

A. Harten (1983). "High resolution schemes for hyperbolic conservation laws." *Journal of Computational Physics* **49**:357–393.

A. Harten, B. Engquist, S. Osher, & S. R. Chakravarthy (1987). "Uniformly high order accurate essentially non-oscillatory schemes, III." *Journal of Computational Physics* **71**(2):231–303.

A. Harten, J. M. Hyman, P. D. Lax, & B. Keyfitz (1976). "On finite-difference approximations and entropy conditions for shocks." *Communications on Pure and Applied Mathematics* **29**(3):297–322.

A. Harten, P. D. Lax, & B. Van Leer (1983). "On upstream differencing and Godunov-type schemes for hyperbolic conservation laws." *SIAM Review* **25**:35–61.

W. D. Hayes (1947). *Linearized supersonic flow*. Ph.D. thesis, California Institute of Technology.

P. W. Hemker & S. P. Spekreijse (1984). "Multigrid solution of the steady Euler equations." In *Proceedings of the Oberwolfach Meeting on Multigrid Methods.*

P. Henrici (1962). *Discrete Variable Methods in Ordinary Differential Equations*, Vol. 1. Wiley.

J. L. Hess & A. M. O. Smith (1962). "Calculation of the non-lifting potential flow about arbitrary three dimensional bodies." *Douglas Aircraft Report, ES* **40622**.

J. S. Hesthaven, S. Gottlieb, & D. Gottlieb (2007). *Spectral Methods for Time-Dependent Problems*, vol. 21. Cambridge University Press.

J. S. Hesthaven & T. Warburton (2008). *Nodal Discontinuous Galerkin Methods: Algorithms, Analysis, and Applications*, Vol. 54. Springer.

J. E. Hicken & D. W. Zingg (2008). "Parallel Newton-Krylov solver for the Euler equations discretized using simultaneous approximation terms." *AIAA Journal* **46**(11):2773–2786.

R. M. Hicks & P. A. Henne (1979). "Wing design by numerical optimization." AIAA paper 79-0080, AIAA 17th Aerospace Sciences Meeting, New Orleans, LA.

R. M. Hicks, E. M. Murman, & G. N. Vanderplaats (1974). "An assessment of airfoil design by numerical optimization." NASA TM X-3092, NASA Ames Research Center, Moffett Field, CA.

J. C. Huan & V. Modi (1994). "Optimum design for drag minimizing bodies in incompressible flow." *Inverse Problems in Engineering* **1**:1–25.

T. J. R. Hughes, L. P. Franca, & M. Mallet (1986a). "A new finite element formulation for computational fluid dynamics: I. Symmetric forms of the compressible Euler and Navier-Stokes equations and the second law of thermodynamics." *Computer Methods in Applied Mechanics and Engineering* **54**(2):223–234.

T. J. R. Hughes, M. Mallet, & M. Akira (1986b). "A new finite element formulation for computational fluid dynamics: II. Beyond SUPG." *Computer Methods in Applied Mechanics and Engineering* **54**(3):341–355.

B. Hunt (1980). "The mathematical basis and numerical principles of the boundary integral method for incompressible potential flow over 3-D aerodynamic configurations." *Numerical Methods in Applied Fluid Dynamics* pp. 39–135.

H. T. Huynh (2007). "A flux reconstruction approach to high-order schemes including discontinuous Galerkin methods." AIAA paper 2007-4079, AIAA 18th Computational Fluid Dynamics Conference, Miami, FL.

H. T. Huynh (2009). "A reconstruction approach to high-order schemes including discontinuous Galerkin for diffusion." AIAA paper 2009-403, AIAA 47th Aerospace Sciences Meeting, Orlando, FL.

A. Iserles (1982). "Order stars and a saturation theorem for first-order hyperbolics." *IMA Journal of Numerical Analysis* **2**(1):49–61.

A. Jameson (1974). "Iterative solution of transonic flows over airfoils and wings, including flows at mach 1." *Communications on Pure and Applied Mathematics* **27**:283–309.

A. Jameson (1978). "Remarks on the calculation of transonic potential flow by a finite volume method." In *Proceedings of the Conference on Computational Methods in Fluid Dynamics, Institute of Mathematics and Applications.*

A. Jameson (1979). "Acceleration of it's transonic potential flow calculations on arbitrary meshes by the multiple grid method." AIAA paper 79-1458, AIAA 4th Computational Fluid Dynamics Conference, Williamsburg, VA.

A. Jameson (1983). "Solution of the Euler equations by a multigrid method." *Applied Mathematics and Computation* **13**:327–356.

A. Jameson (1984). "A non-oscillatory shock capturing scheme using flux limited dissipation." Lectures in Applied Mathematics, B. E. Engquist, S. Osher, and R. C. J. Sommerville, (eds.), A.M.S., Part 1, 22:345–370, 1985.

A. Jameson (1986a). "Multigrid algorithms for compressible flow calculations." In W. Hackbusch & U. Trottenberg (eds.), *Lecture Notes in Mathematics*, vol. 1228, pp. 166–201. Springer-Verlag.

A. Jameson (1986b). "A vertex based multigrid algorithm for three-dimensional compressible flow calculations." In T. E. Tezduar & T. J. R. Hughes (eds.), *Numerical Methods for Compressible Flow - Finite Difference, Element And Volume Techniques*. ASME Publication AMD 78.

A. Jameson (1988). "Aerodynamic design via control theory." *Journal of Scientific Computing* **3**:233–260.

A. Jameson (1989). "Computational aerodynamics for aircraft design." *Science* **245**(4916): 361–371.

A. Jameson (1990). "Automatic design of transonic airfoils to reduce the shock induced pressure drag." In *Proceedings of the 31st Israel Annual Conference on Aviation and Aeronautics, Tel Aviv*, pp. 5–17. Citeseer.

A. Jameson (1991). "Time dependent calculations using multigrid, with applications to unsteady flows past airfoils and wings." AIAA paper 91-1596, AIAA 10th Computational Fluid Dynamics Conference, Honolulu, HI.

A. Jameson (1993). "Artificial diffusion, upwind biasing, limiters and their effect on accuracy and multigrid convergence in transonic and hypersonic flows." AIAA paper 93-3359, AIAA 11th Computational Fluid Dynamics Conference, Orlando, FL.

A. Jameson (1994). "Optimum aerodynamic design via boundary control." AGARD FDP/Von Karman Institute Lecture Notes on Optimum Design Methods in Aerodynamics. AGARD Report 803, pages 3-1 to 3-33, 1994.

A. Jameson (1995a). "Analysis and design of numerical schemes for gas dynamics 1, artificial diffusion, upwind biasing, limiters and their effect on multigrid convergence." *International Journal of Computational Fluid Dynamics* **4**:171–218.

A. Jameson (1995b). "Analysis and design of numerical schemes for gas dynamics 2, artificial diffusion and discrete shock structure." *International Journal of Computational Fluid Dynamics* **5**:1–38.

A. Jameson (1995c). "Optimum aerodynamic design using the control theory." *Computational Fluid Dynamics Review* pp. 495–528.

A. Jameson (1997). "Reengineering the design process through computation." AIAA paper 97-0641, AIAA 35th Aerospace Sciences Meeting and Exibit, Reno, NV.

A. Jameson (2008). "The construction of discretely conservative finite volume schemes that also globally conserve energy or entropy." *Journal of Scientific Computing* **34**(2):152–187.

A. Jameson (2010). "A proof of the stability of the spectral difference method for all orders of accuracy." *Journal of Scientific Computing* **45**(1-3):348–358.

A. Jameson (2017a). "Evaluation of fully implicit Runge–Kutta schemes for unsteady flow calculations." *Journal of Scientific Computing* **73**(2):819–852.

A. Jameson (2017b). "Origins and further development of the Jameson–Schmidt–Turkel scheme." *AIAA Journal* **55**(5):1487–1510.

A. Jameson & J. J. Alonso (1996). "Automatic aerodynamic optimization on distributed memory architectures." AIAA paper 96-0409, AIAA 34th Aerospace Sciences Meeting and Exhibit, Reno, NV.

A. Jameson & T. J. Baker (1983). "Solution of the Euler equations for complex configurations." *Proceedings of the AIAA 6th Computational Fluid Dynamics Conference*, Denver, CO. pp. 293–302.

A. Jameson, T. J. Baker, & N. P. Weatherill (1986). "Calculation of inviscid transonic flow over a complete aircraft." AIAA paper 86-0103, AIAA 24th Aerospace Sciences Meeting, Reno, NV.

A. Jameson & D. A. Caughey (1977). "A finite volume method for transonic potential flow calculations." In *Proceedings of the AIAA 3rd Computational Fluid Dynamics Conference*, pp. 35–54, Albuquerque, NM.

A. Jameson & D. A. Caughey (2001). "How many steps are required to solve the Euler equations of steady, compressible flow: In search of a fast solution algorithm." AIAA paper 2001-2673, AIAA 15th Computational Fluid Dynamics Conference, Anaheim, CA.

A. Jameson & S. Kim (2003). "Reduction of the adjoint gradient formulas for aerodynamic shape optimization." *AIAA Journal* **41**(11):2114–2129.

A. Jameson & P. D. Lax (1986). "Conditions for the construction of multipoint total variation diminishing schemes." *Applied Numerical Mathematics* **2**:335–345.

A. Jameson, L. Martinelli, & N. A. Pierce (1998). "Optimum aerodynamic design using the Navier–Stokes equations." *Theoretical and Computational Fluid Dynamics* **10**(1-4): 213–237.

A. Jameson & D. J. Mavriplis (1987). "Multigrid solution of the Euler equations on unstructured and adaptive grids." In S. McCormick (ed.), *Multigrid Methods, Theory, Applications and Supercomputing. Lecture Notes in Pure and Applied Mathematics*, vol. 110, pp. 413–430.

A. Jameson, N. Pierce, & L. Martinelli (1997). "Optimum aerodynamic design using the Navier-Stokes equations." AIAA paper 97-0101.

A. Jameson, W. Schmidt, & E. Turkel (1981). "Numerical solution of the Euler equations by finite volume methods using Runge–Kutta time stepping schemes." AIAA paper 1981–1259, 14th AIAA Fluid and Plasma Dynamics Conference, Palo Alto, CA.

A. Jameson & E. Turkel (1981). "Implicit schemes and LU decompositions." *Mathematics of Computation* **37**(156):385–397.

M. Jayaram & A. Jameson (1988). "Multigrid solution of the Navier-Stokes equations for flow over wings." AIAA paper 88-0705, AIAA 26th Aerospace Sciences Meeting, Reno, NV.

G-S. Jiang & C-W. Shu (1996). "Efficient implementation of weighted ENO schemes." *Journal of Computational Physics* **126**(1):202–228.

G. Jothiprasad, D. J. Mavriplis, & D. A. Caughey (2003). "Higher-order time integration schemes for the unsteady Navier–Stokes equations on unstructured meshes." *Journal of Computational Physics* **191**(2):542–566.

A. Katz & A. Jameson (2009). "Multicloud: Multigrid convergence with a meshless operator." *Journal of Computational Physics* **228**(14):5237–5250.

D. I. Ketcheson, C. B. Macdonald, & S. Gottlieb (2009). "Optimal implicit strong stability preserving Runge–Kutta methods." *Applied Numerical Mathematics* **59**(2):373–392.

I. P. E. Kinmark (1984). "One step integration methods with large stability limits for hyperbolic partial differential equations." In R. Vichnevetsky & R. S. Stepleman (eds.), *Advances in Computer Methods for Partial Differential Equations*, vol. V, pp. 345–349, IMACS.

D. A. Knoll & D. E. Keyes (2004). "Jacobian-free Newton-Krylov methods: A survey of approaches and applications." *Journal of Computational Physics* **193**(2):357–397.

A. Kolmogorov (1941a). "The local structure of turbulence in incompressible viscous fluid for very large Reynolds' numbers." In *Doklady Akademiia Nauk SSSR*, vol. 30, pp. 301–305.

A. Kolmogorov (1941b). "The logarithmically normal law of distribution of dimensions of particles when broken into small parts." In *Doklady Akademiia Nauk SSSR*, vol. 30, pp. 301–305.

D. A. Kopriva & J. H. Kolias (1996). "A conservative staggered-grid Chebyshev multidomain method for compressible flows." *Journal of Computational Physics* **125**(1):244–261.

H-O. Kreiss (1968). "Stability theory of difference approximations for mixed initial boundary value problems. I." *Mathematics of Computation* **22**(119):703–714.

H-O. Kreiss & J. Oliger (1973). *Methods for the approximate solution of time dependent problems*. Global Atmospheric Research Programme (Publication Series No. 10).

N. Kroll & J. K. Fassbender (2006). *MEGAFLOW - Numerical Flow Simulation for Aircraft Design: Results of the second phase of the German CFD initiative MEGAFLOW, presented during its closing symposium at DLR, Braunschweig, Germany, December 10 and 11, 2002*. Notes on Numerical Fluid Mechanics and Multidisciplinary Design. Springer.

Y. Kuya, K. Totani, & S. Kawai (2018). "Kinetic energy and entropy preserving schemes for compressible flows by split convective forms." *Journal of Computational Physics* **375**: 823–853.

M. H. Lallemand & A. Dervieux (1987). "A multigrid finite-element method for solving the two-dimensional Euler equations." In S. F. McCormick (ed.), *Proceedings of the Third Copper Mountain Conference on Multigrid Methods, Lecture Notes in Pure and Applied Mathematics*, pp. 337–363, Copper Mountain.

H. Lamb & R. Caflisch (1993). *Hydrodynamics*. Cambridge Mathematical Library. Cambridge University Press.

J. D. Lambert (1991). *Numerical Methods for Ordinary Differential Systems: The Initial Value Problem*. John Wiley & Sons.

A. M. Landsberg, J. P. Boris, W. Sandberg, & T. R. Young (1993). "Naval ship superstructure design: Complex three-dimensional flows using an efficient, parallel method." *High Performance Computing 1993: Grand Challenges in Computer Simulation*.

B. E. Launder, G. J. Reece, & W. Rodi (1975). "Progress in the development of a Reynolds-stress turbulence closure." *Journal of Fluid Mechanics* **68**(03):537–566.

P. D. Lax & R. D. Richtmyer (1956). "Survey of the stability of linear finite difference equations." *Communications on Pure and Applied Mathematics* **9**(2):267–293.

P. D. Lax & B. Wendroff (1960). "Systems of conservation laws." *Communications on Pure and Applied Mathematics* **13**:217–237.

S. K. Lele (1992). "Compact finite difference schemes with spectral-like resolution." *Journal of Computational Physics* **103**(1):16–42.

A. Leonard (1974). "Energy cascade in large-eddy simulations of turbulent fluid flows." In *Turbulent Diffusion in Environmental Pollution*, vol. 1, pp. 237–248.

M. A. Leschziner (2003). *Turbulence Modeling for Aeronautical Flows*. VKI Lecture Series, Von Karman Institute for Fluid Dynamics.

H. W. Liepmann & A. Roshko (1957). *Elements of Gas Dynamics*. John Wiley & Sons.

M. J. Lighthill (1945). "A new method of two dimensional aerodynamic design." Rand M 1111, Aeronautical Research Council.

D. K. Lilly (1968). "Models of cloud-topped mixed layers under a strong inversion." *Quarterly Journal of the Royal Meteorological Society* **94**(401):292–309.

J. L. Lions (1971). *Optimal Control of Systems Governed by Partial Differential Equations*. Springer-Verlag. Translated by S. K. Mitter.

M-S. Liou (1996). "A Sequel to AUSM: AUSM+." *Journal of Computational Physics* **129**(2):364–382.

M-S. Liou (2006). "A sequel to AUSM, Part II: AUSM-up for all speeds." *Journal of Computational Physics* **214**(1):137–170.

M-S. Liou (2011). "Open problems in numerical fluxes: Proposed resolutions." AIAA paper 3055, AIAA 20th Computational Fluid Dynamics Conference, Honolulu, HI.

M-S. Liou (2012). "Unresolved problems by shock capturing: Taming the overheating problem." *7th International Conference on Computational Fluid Dynamics* pp. 2012–2203.

M-S. Liou & C. J. Steffen (1993). "A New flux splitting scheme." *Journal of Computational Physics* **107**:23–39.

F. Liu & A. Jameson (1992). "Multigrid Navier-Stokes calculations for three-dimensional cascades." AIAA paper 92-0190, AIAA 30th Aerospace Sciences Meeting, Reno, NV.

X-D. Liu, S. Osher, & T. Chan (1994). "Weighted essentially non-oscillatory schemes." *Journal of Computational Physics* **115**(1):200–212.

Y. Liu, M. Vinokur, & Z. J. Wang (2006). "Spectral difference method for unstructured grids I: basic formulation." *Journal of Computational Physics* **216**(2):780–801.

R. Löhner (2008). *Applied Computational Fluid Dynamics Techniques: An Introduction Based on Finite Element Methods*. Wiley.

R. Löhner, K. Morgan, J. Peraire, & M. Vahdati (1987). "Finite element flux-corrected transport (FEM-FCT) for the Euler and Navier–Stokes equations." *International Journal for Numerical Methods in Fluids* **7**(10):1093–1109.

R. Löhner & P. Parikh (1988). "Generation of three-dimensional grids by the advancing front method." *International Journal for Numerical Methods in Fluids* **8**(10):1135–1149.

C. Lozano (2012). "Adjoint viscous sensitivity derivatives with a reduced gradient formulation." *AIAA Journal* **50**(1):203–214.

C. Lozano (2019). "Watch your adjoints! Lack of mesh convergence in inviscid adjoint solutions." *AIAA Journal* **57**(9):3991–4006.

C. C. Lytton (1987). "Solution of the Euler equations for transonic flow over a lifting aerofoil–the Bernoulli formulation (Roe/Lytton method)." *Journal of Computational Physics* **73**(2):395–431.

R. W. MacCormack (1969). "The effect of viscosity in hyper-velocity impact cratering." AIAA paper 69-354.

R. W. MacCormack (1985). "Current status of numerical solutions of the Navier–Stokes Equations." AIAA paper 85-0032, *AIAA 23rd Aerospace Sciences Meeting, Reno*, NV.

R. W. MacCormack (1997). "A new implicit algorithm for fluid flow." AIAA paper 97-2100, AIAA 13th Computational Fluid Dynamics Conference, Snowmass, CO.

R. W. MacCormack & A. J. Paullay (1972). "Computational efficiency achieved by time splitting of finite difference operators." AIAA paper 72-154, AIAA 33rd Aerospace Sciences Meeting.

C. A. Mader, G. K. W. Kenway, A. Yildirim, & J. R. R. A. Martins (2020). "ADflow: An open-source computational fluid dynamics solver for aerodynamic and multidisciplinary optimization." *Journal of Aerospace Information Systems* pp. 1–20.

L. Martinelli (1987). *Calculations of Viscous Flows with a Multigrid Method*. Ph.D. thesis. Princeton University, Princeton, N.J.

L. Martinelli & A. Jameson (1988). "Validation of a multigrid method for the Reynolds-averaged equations." AIAA paper 88-0414, AIAA 26th Aerospace Sciences Meeting, Reno, NV.

L. Martinelli, A. Jameson, & E. Malfa (1992). "Numerical simulation of three-dimensional vortex flows over delta wing configurations." In M. Napolitano & F. Solbetta (eds.), *Proceedings of the 13th International Confrence on Numerical Methods in Fluid Dynamics*, pp. 534–538, Rome. Springer Verlag, 1993.

K. Mattsson (2003). "Boundary procedures for summation-by-parts operators." *Journal of Scientific Computing* **18**(1):133–153.

D. J. Mavriplis (2019). Progress in CFD Discretizations, Algorithms and Solvers for Aerodynamic Flows. AIAA paper 2019-2944, AIAA Aviation Forum, Dallas, TX.

D. J. Mavriplis & A. Jameson (1990). "Multigrid solution of the Navier–Stokes equations on triangular meshes." *AIAA Journal* **28**(8):1415–1425.

D. J. Mavriplis & M. Long (2014). "NSU3D results for the Fourth AIAA Drag Prediction Workshop." *Journal of Aircraft* **51**(4):1161–1171.

D. J. Mavriplis & L. Martinelli (1991). "Multigrid solution of compressible turbulent flow on unstructured meshes using a two-equation model." AIAA paper 91-0237, AIAA 29th Aerospace Sciences Meeting, Reno, NV.

D. J. Mavriplis & V. Venkatakrishnan (1996). "A 3D agglomeration multigrid solver for the Reynolds-averaged Navier–Stokes equations on unstructured meshes." *International Journal of Numerical Methods in Fluids* **23**:1–18.

G. May & A. Jameson (2005). "Unstructured algorithms for inviscid and viscous flows embedded in a unified solver architecture, Flo3xx." AIAA paper 2005-0318, AIAA 43rd Aerospace Sciences Meeting & Exhibit, Reno, NV.

M. S. McMullen (2003). *The application of non-linear frequency domain methods to the Euler and Navier–Stokes equations*. Ph.D. thesis, Stanford University.

M. S. McMullen, A. Jameson, & J. J. Alonso (2002). "Application of a non-linear frequency domain solver to the Euler and Navier–Stokes equations." AIAA paper 2002-0120, AIAA 40th Aerospace Sciences Meeting and Exhibit, Reno, NV.

J. A. Meijerink & H. A. van der Vorst (1977). "An iterative solution method for linear systems of which the coefficient matrix is a symmetric M-matrix." *Mathematics of Computation* **31**(137):148–162.

N. D. Melson, M. D. Sanetrik, & H. L. Atkins (1993). "Time-accurate Navier–Stokes calculations with multigrid acceleration." In *Proceedings of the Sixth Copper Mountain Conference on Multigrid Methods*, Copper Mountain.

J. E. Melton, S. A. Pandya, & J. L. Steger (1993). "3D Euler flow solutions using unstructured cartesian and prismatic grids." AIAA paper 93-0331, Reno, NV.

C. B. Millikan (1938). "A critical discussion of turbulent flows in channels and circular tubes." In *Proc. 5th International Congress of Applied Mechanics*, vol. 386.

C. S. Morawetz (1956). "On the non-existence of continuous transonic flows past profiles I." *Communications on Pure and Applied Mathematics* **9**(1):45–68.

W. A. Mulder (1989). "A new multigrid approach to convection problems." *Journal of Computational Physics* **83**:303–323.

W. A. Mulder (1992). "A high-resolution Euler solver based on multigrid, semi-coarsening, and defect correction." *Journal of Computational Physics* **100**:91–104.

E. M. Murman (1974). "Analysis of embedded shock waves calculated by relaxation methods." *AIAA Journal* **12**:626–633.

E. M. Murman & J. D. Cole (1971). "Calculation of plane steady transonic flows." *AIAA Journal* **9**:114–121.

R. H. Ni (1982). "A multiple grid scheme for solving the Euler equations." *AIAA Journal* **20**:1565–1571.

R. A. Nicolaides (1975). "On multiple grid and related techniques for solving discrete elliptic systems." *Journal of Computational Physics* **19**(4):418–431.

J. Oliger & A. Sundstrom (1978). "Theoretical and practical aspects of some initial boundary value problems in fluid dynamics." *SIAM Journal on Applied Mathematics* **35**(3):419–446.

S. A. Orszag & M. Israeli (1974). "Numerical simulation of viscous incompressible flows." *Annual Review of Fluid Mechanics* **6**(1):281–318.

S. Osher (1984). "Riemann solvers, the entropy condition, and difference approximations." *SIAM Journal on Numerical Analysis* **121**:217–235.

K. Ou, C. Liang, & A. Jameson (2010). "A high-order spectral difference method for the Navier–Stokes equations on unstructured moving deformable grids." AIAA paper 2010-541, 48th AIAA Aerospace Sciences Meeting, Orlando, FL.

K. Ou, C. Liang, S. Premasuthan, & A. Jameson (2009). "High-order spectral difference simulation of laminar compressible flow over two counter-rotating cylinders." AIAA paper 2009-3956, AIAA 27th Applied Aerodynamics Conference, San Antonio, TX.

V. Parthasarathy, Y. Kallinderis, & K. Nakajima (1995). "A hybrid adaptation method and directional viscous multigrid with prismatic-tetrahedral meshes." AIAA paper 95-0670, AIAA 33rd Aerospace Sciences Meeting, Reno, NV.

S. V. Patankar & D. B. Spalding (1972). "A calculation procedure for heat, mass and momentum transfer in three-dimensional parabolic flows." *International Journal of Heat and Mass Transfer* **15**(10):1787–1806.

J. Peraire, J. Peiro, L. Formaggia, K. Morgan, & O. C. Zienkiewicz (1988). "Finite element Euler computations in three dimensions." *International Journal for Numerical Methods in Engineering* **26**(10):2135–2159.

J. Peraire, J. Peirö, & K. Morgan (1992). "A 3D finite-element multigrid solver for the Euler equations." AIAA paper 92-0449, AIAA 30th Aerospace Sciences Conference, Reno, NV.

J. Peraire & P-O. Persson (2008). "The Compact Discontinuous Galerkin (CDG) method for elliptic problems." *SIAM Journal on Scientific Computing* **30**(4):1806–1824.

P-O. Persson, D. J Willis, & J. Peraire (2010). "The numerical simulation of flapping wings at low Reynolds numbers." AIAA paper 2010-724, AIAA 48th Aerospace Sciences Meeting, Orlando, FL.

N. A. Pierce & M. B. Giles (1997). "Preconditioning compressible flow calculations on stretched meshes." *Journal of Computational Physics* **136**:425–445.

N. A. Pierce, M. B. Giles, A. Jameson, & L. Martinelli (1997). "Accelerating three-dimensional Navier–Stokes calculations." AIAA paper 97-1953, AIAA 13th Computational Fluid Dynamics Conference, Snowmass, CO.

O. Pironneau (1984). *Optimal Shape Design for Elliptic Systems.* Springer-Verlag.

S. B. Pope (2000). *Turbulent Flows.* Cambridge University Press.

M. J. D. Powell (1981). *Approximation Theory and Methods.* Cambridge University Press.

L.. Prandtl (1904). "Über flüssigkeitsbewegung bei sehr kleiner reibung (On fluid motion with very small friction)." In *Proceedings of 3rd International Mathematics Congress, Heidelberg.*

L. Prandtl (1925). "Bericht über untersuchungen zur ausgebildeten turbulenz." *Zeitschrift für Angewandte Mathematik und Mechanik* **5**(2):136–139.

L. Prandtl & O. G. Tietjens (1957). *Applied Hydro and Aeromechanics*, vol. 1. Dover Publications.

S. Premasuthan, C. Liang, A. Jameson, & Z. J. Wang (2009). "p-multigrid spectral difference method for viscous compressible flow using 2D quadrilateral meshes." AIAA paper 2009-950, AIAA 47th Aerospace Sciences Meeting including The New Horizons Forum and Aerospace Exposition, Orlando, FL.

A. Pueyo & D. W. Zingg (1998). "Efficient Newton-Krylov solver for aerodynamic computations." *AIAA Journal* **36**(11):1991–1997.

T. H. Pulliam (2011). "High order accurate finite-difference methods: As seen in OVERFLOW." AIAA paper 2011-3851, AIAA 20th Computational Fluid Dynamics Conference, Honolulu, HI.

T. H. Pulliam & J. L. Steger (1985). "Recent improvements in efficiency, accuracy and convergence for implicit approximate factorization algorithms." AIAA paper 85-0360, AIAA 23rd Aerospace Sciences Meeting, Reno, NV.

J. J. Quirk (1994). "A contribution to the great Riemann solver debate." *International Journal for Numerical Methods in Fluids* **18**:555–574.

R. Radespiel, C. Rossow, & R. C. Swanson (1989). "An efficient cell-vertex multigrid scheme for the three-dimensional Navier–Stokes equations." In *Proceedings of the AIAA 9th Computational Fluid Dynamics Conference*, pp. 249–260, Buffalo, NY. AIAA paper 89-1953-CP.

D. Ray & P. Chandrashekar (2013). "Entropy stable schems for compressible Euler equations." *International Journal of Numerical Analysis and Modeling, Series B* **4**(4):335–352.

W. H. Reed & T. R. Hill (1973). "Triangular mesh methods for the neutron transport equation." *Los Alamos Report LA-UR-73-479.*

J. Reuther, J. J. Alonso, J. C. Vassberg, A. Jameson, & L. Martinelli (1997). "An efficient multiblock method for aerodynamic analysis and design on distributed memory systems." AIAA paper 97-1893, AIAA 13th Computational Fluid Dynamics Conference.

J. J. Reuther, A. Jameson, J. J. Alonso, M. J. Rimllnger, & D. Saunders (1999). "Constrained multipoint aerodynamic shape optimization using an adjoint formulation and parallel computers, part 2." *Journal of Aircraft* **36**(1):61–74.

L. F. Richardson (1911). "The approximate arithmetical solution by finite differences of physical problems involving differential equations, with an application to the stresses in a masonry dam." *Philosophical Transactions of the Royal Society of London. Series A, Containing Papers of a Mathematical or Physical Character* **210**:307–357.

L. F. Richardson (1922). *Weather Prediction by Numerical Process*. Cambridge University.

R. D. Richtmyer & K. W. Morton (1994). *Difference Methods for Initial-Value Problems*. Krieger Publishing Co., 2nd edn.

H. Rieger & A. Jameson (1988). "Solution of steady three-dimensional compressible Euler and Navier–Stokes equations by and implicit LU scheme." AIAA paper 88-0619, AIAA 26th Aerospace Sciences Meeting, Reno, NV.

A. Rizzi & H. Viviand (1981). "Numerical methods for the computation of inviscid transonic flows with shock waves." In *Proceedings of the GAMM workshop, Stockholm*.

T. W. Roberts (1990). "The behavior of flux difference splitting schemes near slowly moving shock waves." *Journal of Computational Physics* **90**(1):141–160.

P. L. Roe (1981). "Approximate Riemann solvers, parameter vectors, and difference schemes." *Journal of Computational Physics* **43**(2):357–372.

C-C. Rossow (2007). "Efficient computation of compressible and incompressible flows." *Journal of Computational Physics* **220**(2):879–899.

P. E. Rubbert (1998). "The Boeing airplanes that have benefited from Antony Jameson's CFD technology." In D. A. Caughey & M. M. Hafez (eds.), *Frontiers of Computational Fluid Dynamics 1998*. World Scientific Publishing Company Incorporated.

P. E. Rubbert & G. R. Saaris (1968). "A general three-dimensional potential flow method applied to V/STOL aerodynamics." SAE paper 680304.

V. Vasil'evich Rusanov (1961). "Calculation of interaction of non-steady shock waves with obstacles." *USSR Computational Mathematics and Mathematical Physics* pp. 267–279.

S. J. Ruuth & R. J. Spiteri (2002). "Two barriers on strong-stability-preserving time discretization methods." *Journal of Scientific Computing* **17**(1–4):211–220.

Y. Saad & M. H. Schultz (1986). "GMRES: A generalized minimal residual algorithm for solving nonsymmetric linear systems." *SIAM Journal on Scientific and Statistical Computing* **7**(3):856–869.

P. Sagaut (2002). *Large Eddy Simulation for Incompressible Flows*. Springer.

S. S. Samant, J. E. Bussoletti, F. T. Johnson, R. H. Burkhart, B. L. Everson, R. G. Melvin, D. P. Young, L. L. Erickson, & M. D. Madson (1987). "TRANAIR: A computer code for transonic analyses of arbitrary configurations." AIAA paper 87-0034.

K. Sawada & S. Takanashi (1987). "A numerical investigation on wing/nacelle interferences of USB configuration." AIAA paper 87-0455, AIAA 25th Aerospace Sciences Meeting, Reno, NV.

H. Schlichting (1933). "Laminare strahlausbreitung." *ZAMM-Journal of Applied Mathematics and Mechanics/Zeitschrift für Angewandte Mathematik und Mechanik* **13**(4):260–263.

H. Schlichting & K. Gersten (1999). *Boundary Layer Theory*. Springer.

A. Sheshadri & A. Jameson (2016). "On the stability of the flux reconstruction schemes on quadrilateral elements for the linear advection equation." *Journal of Scientific Computing* **67**(2):769–790.

C-W. Shu & S. Osher (1988). "Efficient implementation of essentially non-oscillatory shock-capturing schemes." *Journal of Computational Physics* **77**:439–471.

J. Smagorinsky (1963). "General circulation experiments with the primitive equations: I. The basic experiment." *Monthly Weather Review* **91**(3):99–164.

R. E. Smith (1983). "Three-dimensional algebraic mesh generation." In *Proceedings of the AIAA 6th Computational Fluid Dynamics Conference*, Danvers, MA. AIAA paper 83-1904.

W. A. Smith & J. M. Weiss (1995). "Preconditioning applied to variable and constant density flows." *AIAA Journal* **33**(11):2050–2057.

G. A. Sod (1978). "A survey of several finite difference methods for systems of nonlinear hyperbolic conservation laws." *Journal of Computational Physics* **27**(1):1–31.

R. L. Sorenson (1986). "Elliptic generation of compressible three-dimensional grids about realistic aircraft." In J. Hauser & C. Taylor (eds.), *International Conference on Numerical Grid Generation in Computational Fluid Dynamics*, Landshut.

R. L. Sorenson (1988). "Three-dimensional elliptic grid generation for an F-16." In J. L. Steger & J. F. Thompson (eds.), *Three-Dimensional Grid Generation for Complex Configurations: Recent Progress*. AGARDograph.

J. C. South & A. Brandt (1976). "Application of a multi-level grid method to transonic flow calculations." In T. C. Adamson & M. F. Platzer (eds.), *Proc. of Workshop on Transonic Flow Problems in Turbomachinery*, pp. 180–206. Hemisphere, 1977.

R. V. Southwell (1946). *Relaxation Methods in Theoretical Physics*. Clarendon Press.

S. P. Spekreijse (1987). "Multigrid solution of monotone second-order discretizations of hyperbolic conservation laws." *Mathematics of Computation* **49**:135–155.

R. J. Spiteri & S. J. Ruuth (2002). "A new class of optimal high-order strong-stability-preserving time discretization methods." *SIAM Journal on Numerical Analysis* **40**(2):469–491.

J. L. Steger & D. S. Chaussee (1980). "Generation of body-fitted coordinates using hyperbolic partial differential equations." *SIAM Journal on Scientific & Statistical Computing* **1**: 431–437.

J. L. Steger & R. F. Warming (1981). "Flux vector splitting of the inviscid gas dynamic equations with applications to finite difference methods." *Journal of Computational Physics* **40**:263–293.

P. K. Subbareddy & G. V. Candler (2009). "A fully discrete, kinetic energy consistent finite-volume scheme for compressible flows." *Journal of Computational Physics* **228**(5): 1347–1364.

R. C. Swanson, E. Turkel, & C-C. Rossow (2007). "Convergence acceleration of Runge–Kutta schemes for solving the Navier–Stokes equations." *Journal of Computational Physics* **224**(1):365–388.

S. Ta'asan, G. Kuruvila, & M. D. Salas (1992). "Aerodynamic design and optimization in one shot." AIAA paper 92-005, AIAA 30th Aerospace Sciences Meeting, Reno, NV.

S. Tatsumi, L. Martinelli, & A. Jameson (1995). "A new high resolution scheme for compressible viscous flows with shocks." AIAA paper 95-0466, 33rd Aerospace Sciences Meeting, Reno, NV.

T. Theodorsen (1931). "Theory of wing sections of arbitary shape." *NACA TR-411* .

J. F. Thompson, F. C. Thames, & C. W. Mastin (1974). "Automatic numerical generation of body-fitted curvilinear coordinate system for field containing any number of arbitrary two-dimensional bodies." *Journal of Computational Physics* **15**:299–319.

J. F. Thompson, Z. U. A. Warsi, & C. W. Mastin (1982). "Boundary-fitted coordinate systems for numerical solution of partial differential equations: A review." *Journal of Computational Physics* **47**:1–108.

E. F. Toro (2009). *Riemann Solvers and Numerical Methods for Fluid Dynamics: A Practical Introduction*. Springer.

E. F. Toro, M. Spruce, & W. Speares (1994). "Restoration of the contact surface in the HLL-Riemann solver." *Shock Waves* **4**(1):25–34.

E. Turkel (1987). "Preconditioned methods for solving the incompressible and low speed equations." *Journal of Computational Physics* **72**:277–298.

M. J. Turner, R. W. Clough, H. C. Martin, & L. J. Topp (1956). "Stiffness and deflection analysis of complex structures." *Journal of the Aeronautical Sciences* **23**(6):805–823.

E. R. Van Driest (2012). "On turbulent flow near a wall." *Journal of the Aeronautical Sciences (Institute of the Aeronautical Sciences)* **23**(11).

M. Van Dyke (1964). *Perturbation Methods in Fluid Mechanics*. Academic Press.

B. Van Leer (1973). "Towards the ultimate conservative difference scheme I. The quest of monotonicity." In *Proceedings of the Third International Conference on Numerical Methods in Fluid Mechanics*, pp. 163–168. Springer.

B. Van Leer (1974). "Towards the ultimate conservative difference scheme. II. Monotonicity and conservation combined in a second order scheme." *Journal of Computational Physics* **14**:361–370.

B. Van Leer (1977a). "Towards the ultimate conservative difference scheme III. Upstream-centered finite-difference schemes for ideal compressible flow." *Journal of Computational Physics* **23**(3):263–275.

B. Van Leer (1977b). "Towards the ultimate conservative difference scheme. IV. A new approach to numerical convection." *Journal of Computational Physics* **23**(3):276–299.

B. Van Leer (1979). "Towards the ultimate conservative difference scheme. V. A second-order sequel to Godunov's method." *Journal of Computational Physics* **32**(1):101–136.

B. Van Leer, W. T. Lee, & P. L. Roe (1991). "Characteristic time stepping or local preconditioning of the Euler equations." AIAA paper 91-1552, AIAA 10th Computational Fluid Dynamics Conference, Honolulu, HI.

R. S. Varga (2000). "Alternating-direction implicit iterative methods." In *Matrix Iterative Analysis*, pp. 235–274. Springer.

J. C. Vassberg (1997). "A fast surface-panel method capable of solving million-element problems." AIAA paper 1997-0168, AIAA 35th Aerospace Sciences Meeting, Reno, NV.

J. C. Vassberg & A. Jameson (2010). "In pursuit of grid convergence for two-dimensional Euler solutions." *Journal of Aircraft* **47**(4):1152–1166.

V. Venkatakrishnan (1988). "Newton solution of inviscid and viscous problems." AIAA paper 88-0413, AIAA 26th Aerospace Sciences Meeting, Reno, NV.

V. Venkatakrishnan (1996). "A perspective on unstructured grid flow solvers." *AIAA Journal* **34**:533–547.

V. Venkatakrishnan & D. J. Mavriplis (1996). "Implicit method for the computation of unsteady flows on unstructured grids." *Journal of Computational Physics* **127**:380–397.

J. P. Veuillot & H. Viviand (1979). "Pseudo-unsteady method for the computation of transonic potential flows." *AIAA Journal* **17**:691–692.

G Vijayasundaram (1986). "Transonic flow simulations using an upstream centered scheme of Godunov in finite elements." *Journal of Computational Physics* **63**(2):416–433.

P. E. Vincent, P. Castonguay, & A. Jameson (2011). "A new class of high-order energy stable flux reconstruction schemes." *Journal of Scientific Computing* **47**(1):50–72.

P. E. Vincent & A. Jameson (2011). "Facilitating the adoption of unstructured high-order methods amongst a wider community of fluid dynamicists." *Mathematical Modelling of Natural Phenomena* **6**(3):97–140.

T. Von Kármán (1930). "Mechanische änlichkeit und turbulenz." *Nachrichten von der Gesellschaft der Wissenschaften zu Göttingen, Mathematisch-Physikalische Klasse* **1930**: 58–76.

G. Voronoi (1908). "Nouvelles applications des parametres continus a la theorie des formes quadratiques. Deuxieme memoire: Recherches sur les parallelloedres primitifs." *Journal für die reine und angewandte Mathematik* **134**:198–287.

E. L. Wachspress (1963). "Extended application of alternating direction implicit iteration model problem theory." *Journal of the Society for Industrial & Applied Mathematics* **11**(4): 994–1016.

Z. J. Wang (2002). "Spectral (finite) volume method for conservation laws on unstructured grids. basic formulation: Basic formulation." *Journal of Computational Physics* **178**(1):210–251.

Z. J. Wang (2007). "High-order methods for the Euler and Navier–Stokes equations on unstructured grids." *Progress in Aerospace Sciences* **43**(1):1–41.

Z. J. Wang & H. Gao (2009). "A unifying lifting collocation penalty formulation including the discontinuous Galerkin, spectral volume/difference methods for conservation laws on mixed grids." *Journal of Computational Physics* **228**(21):8161–8186.

R. F. Warming, R. M. Beam, & B. J. Hyett (1975). "Diagonalization and simultaneous symmetrization of the gas-dynamic matrices." *Mathematics of Computation* **29**(132):1037–1045.

N. P. Weatherill & C. A. Forsey (1985). "Grid generation and flow calculations for aircraft geometries." *Journal of Aircraft* **22**:855–860.

R. T. Whitcomb (1956). "A study of the zero lift drag-rise characteristics of wing-body combinations near the speed of sound." NACA Report TR 1273.

R. T. Whitcomb (1974). "Review of NASA supercritical airfoils." In *9th ICAS Congress*, no. 74-10, Haifa.

R. T. Whitcomb (1976). "A design approach and selected wind-tunnel results at high subsonic speeds for wing-tip mounted winglets." NASA Technical Note D-8260.

D. C. Wilcox (2006). *Turbulence Modeling for CFD*, vol. 3. DCW industries La Cañada, CA.

D. M. Williams, P. Castonguay, P. E. Vincent, & A. Jameson (2013). "Energy stable flux reconstruction schemes for advection-diffusion problems on triangles." *Journal of Computational Physics* **250**:53–76.

D. M. Williams & A. Jameson (2014). "Energy stable flux reconstruction schemes for advection–diffusion problems on tetrahedra." *Journal of Scientific Computing* **59**:721–759.

F. D. Witherden & A. Jameson (2018). "On the spectrum of the Steger-Warming flux-vector splitting scheme." *International Journal for Numerical Methods in Fluids* **87**(12):601–606.

F. D. Witherden, A. Jameson, & D. W. Zingg (2017). "Chapter 11 - The design of steady state schemes for computational aerodynamics." In R. Abgrall & C-W. Shu (eds.), *Handbook of Numerical Methods for Hyperbolic Problems*, vol. 18 of *Handbook of Numerical Analysis*, pp. 303–349. Elsevier.

H. C. Yee (1985). "On symmetric and upwind TVD schemes." In *Proceedings of the 6th GAMM Conference on Numerical Methods in Fluid Mechanics*, Göttingen.

S. Yoon & A. Jameson (1987). "Lower-upper symmetric-Gauss–Seidel method for the Euler and Navier–Stokes equations." AIAA paper 87-0600, AIAA 25th Aerospace Sciences Meeting, Reno, NV.

D. M. Young (2003). *Iterative Solution of Large Linear Systems*. Dover Publications.

O. C. Zienkiewicz (1995). "Origins, milestones and directions of the finite element method–A personal view." *Archives of Computational Methods in Engineering* **2**(1):1–48.

O. C. Zienkiewicz (2004). "The birth of the finite element method and of computational mechanics." *International Journal for Numerical Methods in Engineering* **60**(1):3–10.

Index